THE ROUTLEDGE HANDBOOK OF ETHICS IN FORENSIC LINGUISTICS

The Routledge Handbook of Ethics in Forensic Linguistics is the first comprehensive reference work to explore the ethical dimensions of forensic language analysis across a range of applied and academic contexts.

With the use of linguistic and phonetic evidence now commonplace in legal investigations and court proceedings, questions of consent, bias, responsibility, and professional integrity have become increasingly urgent. This volume brings together over 30 original chapters by leading scholars and practitioners in forensic linguistics and speech science to critically examine these issues. Chapters span a wide range of research and practice settings, from expert testimony and investigative consultancy to academic publishing, teaching, and public engagement. This handbook addresses ethical questions across diverse linguistic and legal systems and offers both conceptual frameworks and practical guidance for navigating the ethical challenges of forensic language work.

This authoritative resource is essential reading for researchers, practitioners, and students in applied linguistics, sociolinguistics, phonetics, criminology, psychology, and legal studies.

I. M. Nick holds a PhD (English Linguistics); an MA (German Linguistics); a MSc (Forensic Psychology); and the German "Habilitation". Former Chair of the Linguistic Society of America's Committee on Ethnic Diversity in Linguistics, she's President of the Germanic Society for Forensic Linguistics. Her forensic linguistic research investigates criminal aliases, the Holocaust, threat communications, and suicide.

Kirsty E. Blewitt is Lecturer in Criminology and Sociology at Northumbria University (UK). Her multimodal linguistic research focuses on language use in courtroom interactions, the police utilization of language, and survivor narratives in domestic violence and abuse cases.

ROUTLEDGE HANDBOOKS IN APPLIED LINGUISTICS

Routledge Handbooks in Applied Linguistics provide comprehensive overviews of the key topics in applied linguistics. All entries for the handbooks are specially commissioned and written by leading scholars in the field. Clear, accessible and carefully edited *Routledge Handbooks in Applied Linguistics* are the ideal resource for both advanced undergraduates and postgraduate students.

THE ROUTLEDGE HANDBOOK OF TEACHING ENGLISH AS AN INTERNATIONAL LANGUAGE
Edited by Ali Fuad Selvi and Nicola Galloway

THE ROUTLEDGE HANDBOOK OF LANGUAGE TEACHER ACTION RESEARCH
Edited by Anne Burns and Kenan Dikilitaş

THE ROUTLEDGE HANDBOOK OF LANGUAGE AND RACE
Edited by Sinfree Makoni, Bassey E. Antia, and Stephanie Rudwick

THE ROUTLEDGE HANDBOOK OF TRANSLANGUAGING IN THE GLOBAL SOUTH
Edited by Kleber Aparecido da Silva and Leketi Makalelato

THE ROUTLEDGE HANDBOOK OF LANGUAGE AND SOCIAL MEDIA AROUND THE WORLD
Edited by Caroline Tagg, Korina Giaxoglou and Kristin Vold Lexander

THE ROUTLEDGE HANDBOOK OF ETHICS IN FORENSIC LINGUISTICS
Edited by I. M. Nick and Kirsty E. Blewitt

For a full list of titles in this series, please visit: www.routledge.com/Routledge-Handbooks-in-Applied-Linguistics/book-series/RHAL

THE ROUTLEDGE HANDBOOK OF ETHICS IN FORENSIC LINGUISTICS

Edited by I. M. Nick and Kirsty E. Blewitt

Routledge
Taylor & Francis Group
LONDON AND NEW YORK

Designed cover image: Getty Images

First published 2026
by Routledge
4 Park Square, Milton Park, Abingdon, Oxon OX14 4RN

and by Routledge
605 Third Avenue, New York, NY 10158

Routledge is an imprint of the Taylor & Francis Group, an informa business

British Library Cataloguing-in-Publication Data
A catalogue record for this book is available from the British Library

ISBN: 9781032488424 (hbk)
ISBN: 9781032488431 (pbk)
ISBN: 9781003391074 (ebk)

DOI: 10.4324/9781003391074

Typeset in Times New Roman
by Deanta Global Publishing Services, Chennai, India

My deepest gratitude goes towards Günter, Dennis, Frank, and Larry…the Northern Stars in my life.

Iman Nick

For Sarah, my love and lighthouse. You are the reason.

Kirsty Blewitt

CONTENTS

List of Illustrations *xi*
Notes on Editors and Contributors *xii*
Abbreviations and Acronyms *xviii*
Acknowledgements *xxi*

1 Introduction 1
I. M. Nick

SECTION I
Providing Expertise as a Consultant **15**

2 The Integrity of Forensic Analysis 17
Robin T. Bowen

3 Ethics and Professional Conduct Oversight in Forensic Linguistic Expert Evidence 27
Isabel Picornell

4 The Expert's Role in the Justice System is to Provide Impartial and Transparent Evidence, Not to Perform Social Advocacy 43
Richard Rhodes and Tina Cambier-Langeveld

5 Ethical Issues Encountered in Forensic Linguistic Practice: Case Examples Inspired by Real Life Cases 53
Tina Cambier-Langeveld and Richard Rhodes

6 Maintaining Professional Distancing in Social Justice Cases Involving Non-Native Speakers (NNSs) 63
Margaret van Naerssen

7 An Intelligence Scaffold to Aid Ethical Considerations in Forensic Linguistic Casework 76
Ria Perkins, James Rosie, and Sharon Glaas

8 Psychological Biasing in Forensic Linguistic Research: Is Such Information Obtained Expertly and Ethically? 88
Ray Bull and Iris Blandón-Gitlin

9 Countering Pseudolinguistic Claims: The Ethical Challenges Facing Forensic Linguists 105
Tanya Karoli Christensen and Sofia Navarro Beck

10 Disinformative Linguistics: Using Linguistic Science for Unethical Purposes 117
Timothy Habick

11 Ethics and the Linguistic Practice of Dutch Police Interrogations: The Right to Silence 130
Martha Komter

SECTION II
Serving as an Expert Witness **143**

12 Linguists and Lawyers: Ethical Challenges in an Adversary System 145
Robin Conley Riner and John M. Conley

13 The Expert Witness and the Ethics of Advocacy 159
Michael O'Laughlin

14 Challenging Legal Assumptions: Ethics, Due Process, and Forensic Linguistic Research 173
Mel Greenlee

15 The Forensic Linguist's Ethical Duty to Explain Language Use to Triers of Fact 191
Dakota Wing

16 The Golden Rule and the Golden Thread: The Ethics of Communicating the Criminal Standard of Proof 206
Chris Heffer

17 Ethical Uncertainties and Inconsistencies in Diminished Responsibility Rulings: The Case for Fictionalism 225
Felicity Deamer and Sam Wilkinson

18 Tightening the Rein on Expert Evidence: Ethical Reasons for Judicial Gatekeeping in the Canadian Context and Implications for Forensic Linguistics 239
Jennifer Glougie

19 Ethical Expert Practice: Engaging with Everyday Linguistic Harms 254
Kelly Elizabeth Wright

20 A Critical Evaluation of Adjustments and Accommodations in the Legal System in England and Wales for Sex Abuse Crimes Involving Adult Witnesses with an Intellectual Disability 266
Michelle Aldridge and Tina Pereira

21 The Problematic Nature of L1 and Translation in US Criminal Courts: Access, Ethics, and Errors 279
Mel Greenlee

SECTION III
Researching, Teaching, Publishing, and Providing for the Next Generation 297

22 "Too Close to Home": A Dialogue on the Ethics of Forensic Linguists' Proximity to Their Data 299
Grace Sullivan Buker

23 Navigating Legality, Linguistic Rights, and Social Justice in Linguistic Research with Deaf Refugees 312
Nina Sivunen and Johanna Ennser-Kananen

24 Ethical Research Practices with Highly Stigmatized Populations: Engaging with Individuals Convicted of Sexual Offenses 328
Stina Lindegren

25 Ethical Considerations when Working with Political Extremist Threat Communications 343
I. M. Nick and Ulrike Preiss

26 Ethical Considerations in Corpus-Based Legal Interpretation Analysis 363
Brett Hashimoto and Derek Christian Haderlie

27 Get Your Hands Off My Idiolect! A Spectrum Model of Misuse of Idiolectal Information as a Heuristic Tool 377
Peter S. Harrison and Dominic Watt

28 Analyzing the Language of Non-Human Animals: Why Should Forensic Linguists Care? 398
Meike M. de Boer

29 "So This Is How You Can Manipulate Witnesses!": Ethical Considerations in the Teaching and Dissemination of Research in Forensic Linguistics 409
Karoline Marko

30 Teaching Sensitive Topics in Forensic Linguistics: Ethical Implications for Students and Lecturers 425
Samuel Larner

31 Welcome Forensic Linguists!: Ethical Forensic Linguistics Pedagogy and Classroom Practice 437
Christine Jacknick and M Peregrine Balmat

32 Ethics and Publishing in Forensic Linguistics: Policing Credibility 450
Sarah Frances Gordon and Bernardo Turnbull Plaza

33 Diversity in Forensic Linguistics: A Call to Fulfill the Promise 462
I. M. Nick

34 Concluding Thoughts 483
I. M. Nick

Case Law Index *495*
Subject-Author Index *498*

LIST OF ILLUSTRATIONS

Figures

7.1	The Intelligence Cycle	80
26.1	Framework of Responsible Application of Interpretative Methods	371
27.1	Spectrum of Idiolect Misuse (Without Vocal Element)	389
27.2	Spectrum of Idiolect Misuse (Vocal Element Alone)	390
27.3	Spectrum of Idiolect Misuse (Verbal and Vocal Element)	392
33.1	Demographic Composition of the US STEM Workforce in 2011 and 2021 by Percentage	469

Tables

14.1	A Comparison of Miranda and Faretta Waiver Features	174
26.1	Categories of Criticisms of CLIA	365
31.1	Our Pedagogical Approach	440
33.1	Gender and Ethnoracial Identity of US University Graduates with a Research Doctorate in Language and Linguistics by Percentage in 2012 and 2022	470

NOTES ON EDITORS AND CONTRIBUTORS

Michelle Aldridge is the Deputy Head of School for Cardiff University's School of English, Communication, and Philosophy. An Academic Fellow of Cardiff University's Crime and Security Research Institute, her research explores vulnerable people in the justice system. She is on the Editorial Board of the *International Journal of Applied Linguistics*.

M Peregrine Balmat is Associate Professor at the Borough of Manhattan Community College, City University of New York (CUNY). Using discourse analysis and ethnography, her research examines how practitioners in social institutions such as police departments put policies into practice. She is co-editor and contributor to *Analysing Social Work Communication: Discourse in Practice*.

Iris Blandón-Gitlin is Psychology Professor at California State University, Fullerton. Her research focuses on social-cognitive influences on memories, suggestibility, confessions, deception detecting, police lineup techniques, and jurors' perceptions of police interrogation tactics. She has served as an expert consultant in approximately 450 criminal cases and trains legal and investigative experts.

Kirsty E. Blewitt is Lecturer in Criminology and Sociology at Northumbria University (UK). Her multimodal linguistic research focuses on language use in courtroom interactions, the police utilization of language, and survivor narratives in domestic violence and abuse cases.

Robin T. Bowen is Assistant Professor in the Forensic and Investigative Science Department, West Virginia University. She is the primary developer of Forensic Educational Alliance; former Director for Forensic and Investigative Science Outreach; and past advisory member under the National Science and Technology Council Subcommittee on Forensic Science.

Ray Bull is Criminal Investigation Professor, Derby University. He has written expert reports in ca. 180 cases. Elected Honorary Fellow of the British Psychological Society (2010), he was President of the European Association of Psychology and Law (2014–2017). In 2018, he joined the Steering Committee writing the UN's investigative interviewing guidelines.

Tina Cambier-Langeveld is the senior linguist for language analysis in asylum procedures for the Dutch Immigration and Naturalisation Service. Since 1999, she has been an expert consultant at the Netherlands Forensic Institute and teaches forensic speech science at Leiden University. She is honorary chair of the International Association for Forensic Phonetics and Acoustics.

John M. Conley is a William Rand Kenan Jr. Law Professor (University of North Carolina, Chapel Hill). He is the former Editor-in-Chief of *Duke Law Journal* and co-edits the Law and Society book series (Chicago UP). In 2019, the third edition of his book, *Just Words: Law, Language, and Power*, was released.

Robin Conley Riner is Anthropology Professor at Marshall University. She investigates language, jurors' experiences, and decision-making in death-penalty cases. Her book, *Confronting the Death Penalty* (OUP, 2016), explores jurors' use of language and empathy for defendants. Her latest co-authored book is *Case Studies in Language and Social Justice* (Routledge, 2019).

Meike M. de Boer is a forensic linguist and owner of the consultancy, Lingua Forensica. Her research investigates language use in the darknet and child sexual abuse. She initiated the first forensic linguistic society within the Dutch Criminological Society [Nederlandse Vereniging voor Criminologie] and serves on several national animal rights organizations.

Felicity Deamer is Senior Lecturer in Aston University's Institute for Forensic Linguistics (UK). Her work examines the intersection between the law, psychiatry, and language. In her work, she has investigated linguistic data in expert psychiatric reports submitted as evidence during murder trials, police interviews, 999 calls, and anonymous online fora users.

Johanna Ennser-Kananen is Research Fellow of the Academy of Finland, University of Jyväskylä. Her research explores socio-linguistically sustaining pedagogies for migrant teachers and anti-oppressive language education for migrant learners. She is co-editor of the *Routledge Handbook of Educational Linguistics* and has published extensively (e.g., *The International Journal of Language Studies*).

Sharon Glaas, PhD, is a Civil Servant and an Honorary Senior Research Fellow at the University of Birmingham, UK. Her research and teaching has focused on persuasive and evaluative language in political and media discourse, as well as the application of corpus linguistics and computational methods to language analysis.

Jennifer Glougie, PhD, University of British Columbia. Her dissertation examined the linguistic expression of evidentiality in police interviews. For over a decade, she was a lawyer in private practice. She is Associate Chair of British Columbia's Labour Relations Board and President of the International Association for Forensic and Legal Linguistics.

Sarah Frances Gordon is Executive Editor of *Psicología Iberoamericana*. She holds a doctorate in Psychology from the University of Cape Town, where she specialized in violence against women. After a postdoctoral fellowship at the University of Stellenbosch, she relocated to Mexico, where she teaches trauma and emergency psychology (Universidad Iberoamericana).

Mel Greenlee holds a PhD in Linguistics and a JD (University of California, Berkeley). Before her 20-year legal defense career (California Appellate Project), she conducted research at CUNY's Centro de Estudios Puertorriqueños, UCLA, and Stanford. Her investigations explore interpreter error, prosecutors' arguments, and the role of language in determining defendant competence for self-representation.

Timothy Habick, PhD, in Linguistics (University of Illinois, Urbana), specializes in evaluating the language used in high-stakes communications, especially for the purpose of creating fair assessments. Through his firm, Reasoning, Inc., he serves as an expert consultant on cases involving, for example, defamation, authorship, copyrights, trademarks, and professional licensure surveys.

Derek Christian Haderlie is Assistant Professor of Philosophy at Brigham Young University. His primary research includes metaethics, epistemology, and the philosophy of religion. His work centers on issues surrounding metaphysical explanation as it relates to metaethical issues. In 2020, he obtained his PhD in Philosophy from the University of Texas at Austin.

Peter S. Harrison (University of York's Law School) has served as a litigator in the English High Court and Court of Appeal; the Trial and Appeal Divisions of the Federal Court of Canada; and the European Court of Justice. His publications investigate ownership questions and the individual voice, AI-driven voice-cloning, and synthesis.

Brett Hashimoto is Assistant Professor of Linguistics (Brigham Young University). His research centers on legal linguistics, language assessment, and pragmatics. He obtained his PhD in Applied Linguistics (Northern Arizona University). From 2019 to 2020, he was a Corpus Linguistic Research Fellow at Brigham Young University's J. Reuben Clark Law School.

Chris Heffer is Reader in Forensic Linguistics at the Centre for Language and Communication Research (Cardiff University). He is an Advisory Board member of the Aston Institute of Forensic Linguistics and is on the Editorial Board of the OUP Language and Law series. He is co-editor of *Legal-Lay Communication* (OUP, 2013).

Christine Jacknick is Professor of Academic Literary and Linguistics at the Borough of Manhattan Community College, City University of New York (CUNY). She obtained her PhD in Applied Linguistics from Teachers College, Columbia University. Her research involves utilizing conversation analysis to examine discursive practices, agency, and power in institutional settings.

Tanya Karoli Christensen, PhD, is Associate Professor at the Department for Nordic Studies and Linguistics, University of Copenhagen. In her forensic linguistic research, she applies methods from semantics, pragmatics, sociolinguistics, and corpus linguistics to analyze linguistic data in legal cases. She is an expert forensic linguistic consultant for police and private clients.

Martha Komter is a retired research fellow at the Netherlands Institute for the Study of Crime and Law Enforcement. Her research examines communication in institutional settings such as courtroom proceedings and police interrogations. Her publications include *Dilemmas in the Courtroom: A Study of Trials of Violent Crime in the Netherlands*.

Samuel Larner, PhD, is Senior Lecturer in Forensic Linguistics (Manchester Metropolitan University). He is Co-head of the Centre for Applied Pragmatics and Forensic Linguistics. His key publications have centered on forensic authorship analysis, using corpus-based approaches to the identification of formulaic sequences as a potential linguistic cue to deception.

Stina Lindegren is a postdoctoral researcher at the Department of Social Work (Uppsala University, Sweden). Her research focuses on perpetration-centered approaches to sexual abuse prevention. She has over 13 years of clinical experience, including probation supervision, assessments, and treatment of men convicted of sex offenses and intimate partner violence.

Karoline Marko, PhD, is a postdoctoral researcher and coordinator of the Forensic Linguistics Certificate Program at the University of Graz (Austria) where she has taught Forensic Linguistics to law students since 2022. Her main research interests lie in authorship analysis, internet linguistics, sociolinguistics, and the language of conspiracy theorists.

Sofia Navarro Beck holds a BA and an MA in Linguistics from Aarhus University (Denmark). Her research includes conversation analysis and computer-mediated communication. She is a PhD student at Aarhus University. Her dissertation investigates the linguistic strategies of sexual offenders during the online grooming of minors.

I. M. Nick holds a PhD (English Linguistics), an MA (German Linguistics), an MSc (Forensic Psychology), and the German "Habilitation". Former Chair of the Linguistic Society of America's Committee on Ethnic Diversity in Linguistics, she is President of the Germanic Society for Forensic Linguistics. She investigates criminal aliases, suicide messages, and threat communications.

Michael O'Laughlin is Director of Interpreter Training (Boston University). After working many decades as a court interpreter in California and Massachusetts, he consulted on over 300 cases involving language proficiency and origin assessments. His testimony was instrumental in freeing Angel Echavarria who'd been wrongfully imprisoned 21 years for a murder.

Tina Pereira, PhD, is accredited as a Registered Intermediary in the justice system in England and Wales, facilitating communication between vulnerable witnesses, defendants, suspects, and respondents in criminal and family courts. She has worked in Crown, Magistrate, and Youth courts with witnesses and victims, defendants, and respondents.

Ria Perkins, PhD, is a Civil Servant for the Ministry of Defence, and is an Honorary Research Fellow at the Aston Institute for Forensic Linguistics in Birmingham, UK. Her casework speciality is authorship profiling, and her research interests include the language of persuasion and power, and Other Language Influence Detection.

Isabel Picornell is Visiting Research Fellow (Aston Institute for Forensic Linguistics, UK). A certified Fraud Examiner and forensic linguistic consultant, her research focuses on authorship in deceptive contexts. She co-edited *Methodologies and Challenges in Forensic Linguistic Casework* (2022) and was President of the International Association for Forensic Linguistics (2021–2023).

Ulrike Preiss holds a BA in English and German Linguistics; and an MA in Linguistics from the Universities of Cologne and Hamburg. In 2022, she completed her PhD at Heidelberg University in Germany. Her doctoral dissertation involved a discriminative analysis of German right-wing and left-wing extremist threat communications.

Richard Rhodes is senior forensic consultant at the Forensic Speech and Acoustics Laboratory, J. P. French Associates. He is an expert witness and consultant in criminal court proceedings on speaker comparison, transcription, and other types of forensic phonetic casework. He is also Associate Lecturer for Forensic Speech Sciences (University of York, UK).

James Rosie is a former government social scientist and anthropologist. He has specialized in introducing and applying a variety of analytical approaches and methodologies to acute, high impact areas and operational decision-making in both domestic and international contexts. His research specialities include strategic communication, critical thinking, and applied ethical frameworks.

Nina Sivunen is a PhD researcher in the Centre of Applied Language Studies at the University of Jyväskylä, Finland. Her doctoral research focuses on deaf asylum seekers in Finland. She is particularly interested in signed narratives, deaf experiences, ethics, sign language practices, linguistic human rights, and language ideologies.

Grace Sullivan Buker received her doctorate in Linguistics at Georgetown University. Her dissertation examines discrimination against African American Language speakers in the US courtroom. Her forensic linguistic research combines interactional sociolinguistics and discourse analysis. She currently teaches linguistics at Northeastern University where she encourages students to seek justice through research.

Bernardo Turnbull Plaza is Editor-in-Chief of *Psicología Iberoamericana*. He holds a BA in Social Psychology (Universidad Autónoma Metropolitana Iztapalapa, Mexico), an MA in Clinical Psychology (Chapman University, US), and a PhD in Social Sciences (University of Sussex, UK). A Psychology professor at Universidad Iberoamericana, he advocates for accessible, impactful publishing.

Margaret van Naerssen is an Expert in Second Language Development and English Language Proficiency Assessment. Since 1997, her forensic linguistics consultancy primarily has involved non-native English speakers. Working with the defense and prosecution in criminal and civil cases, she has testified in five federal courts, hearings, depositions, and bench trials.

Dominic Watt is Senior Lecturer in Forensic Speech Science (University of York, UK). He has an MA in Linguistics (Edinburgh) and a PhD in Phonetics (Newcastle). A forensic phonetic consultant (J. P. French Associates), he provides expertise for both the prosecution and defense, on speaker comparison, disputed content analysis, and earwitness testimony.

Dakota Wing is a PhD Candidate in Applied Linguistics (York University, Canada) and holds an MA in Forensic Linguistics (Hofstra University, US). His research includes police interviews, school shooters' discourse, and extremist propaganda. A regular criminal, civil, and investigative case consultant, he has worked with the Forensic Linguistics Capital Case Innocence Project.

Sam Wilkinson is Lecturer in the Department of Social and Political Sciences, Philosophy, and Anthropology (Exeter University, UK). He holds a PhD (University of Edinburgh) and an MA in Philosophy and Cognitive Science (Jean Nicod Institute in Paris). His research explores trauma and psychosis, as they relate to perception and cognition.

Kelly Elizabeth Wright, PhD, is an experimental sociolinguist, lexicographer, and Assistant Professor in the Department of Language Science (University of Wisconsin–Madison, USA). A scholar-activist, she is committed to promoting linguistic justice and interdisciplinarity both inside and outside of the academy. Through metalinguistic interviews and sociophonetic experimentation, her research has demonstrated discrimination in journalism, the housing market, and academia.

ABBREVIATIONS AND ACRONYMS

AAAL	American Association for Applied Linguistics
AAC	Augmentative and Alternative Communication
AAL	African American Languages
AAVE	African American Vernacular English
ABE	Achieving Best Evidence
ACS	American College of Surgeons
AFSP	Association of Forensic Science Providers (UK)
AI	Artificial Intelligence
AIFL	Aston Institute for Forensic Linguistics
AMRC	Academy of Royal Medical Colleges
ASL	American Sign Language
ASR	Automated Speech Recognition
BAAL	British Association of Applied Linguistics
BMCC	Borough of Manhattan Community College
CLIA	Corpus-based Legal Interpretation analysis
COCA	Corpus of Contemporary American English
COPE	Committee on Public Ethics
CORG	Communication of Rights Group
CPR	Civil Procedure Rules (UK)
CPR	Criminal Procedure Rules (England and Wales)
CPS	Crown Prosecution Service
CRP	Culturally Responsive/Relevant Pedagogy
CSP	Culturally Sustaining Pedagogy
CUNY	City University of New York
DEI	Diversity, Equity, Inclusion (US)
DET	Detective
DNA	Deoxyribonucleic acid
DOJ	United States Department of Justice
EAD	Expert Advisers Database
EAL	English as an Additional Language
EASO	European Asylum Support Office

ENFSI	European Network of Forensic Science Institutes
EWHC	England and Wales High Court
FBI	Federal Bureau of Investigation (US)
FinSL	Finnish Sign Language
FL	Forensic Linguistics
FPA	Foreign Policy Analysis
FSA	Forensic Science Activity
FSR	Forensic Science Regulator (UK)
GMC	General Medical Council
GSFL	Germanic Society for Forensic Linguistics
HE	Higher Education
HEI	Higher Education Institutions (UK)
HEP	Hierarchy of Expert Performance.
IAFLL	International Association for Forensic and Legal Linguistics
IAFPA	International Association for Forensic Phonetics and Acoustics
ILAC	International Laboratory Accreditation Cooperation
IRC	International Rescue Committee
IS	International Sign
ISO	International Organization for Standardization
KIV	Knowing, Intelligent, and Voluntary test
KLFI	Komunitas Linguistik Forensik Indonesia
L1	First language learned
L2	Second or additional language learned
LEP	Limited English Proficient speaker
LSA	Linguistic Society of America
LSU	Linear Sequential Unmasking
LWOP	Life without the possibility of parole
MIGRI	Finnish Immigration Services
MPT	Medical Practitioner Tribunal
MUSE	Mainstream US English
NAJIT	National Association of Judiciary Interpreters and Translators
NAS	National Academy of Science
NCA	National Crime Agency
NIST	National Institute of Standards and Technology
NNS	Non-native speakers
NRC	National Research Council (US)
NYPD	New York City Police Department
OPFPU	Ontario Pediatric Forensic Pathology Unit
OSAC	Organization of Scientific Area Committee for Forensic Science
PAFLL	Philippine Association for Forensic and Legal Linguistics
PCAST	President's Council of Advisors on Science and Technology (US)
Pro Per or Pro Se Defendant	in propria persona defendant or self-represented defendant
RI	Registered Intermediary

SAC	Scientific Area Committees
SUS	Suspect
TESOL	Teachers of English to Speakers of Other Languages
TIP	Trauma-informed Pedagogy
TOEFL	Test of English as a Foreign Language
UMA	Ulkomaalaisasioiden asiankäsittelyjärjestelmä—Electronic case management system for non-permanent residents, Finland
UNHCR	United Nations Office of the High Commissioner for Refugees
	United States
WFD	World Federation of the Deaf
WID	Witnesses with Intellectual Disability
WOS	Web of Science
YJCEA	The Youth Justice and Criminal Evidence Act

ACKNOWLEDGEMENTS

We would like to give our heartfelt thanks to all of the authors who have contributed to this volume. Your pioneering work, dedication, and perseverance are an inspiration to us all. Without you, this project would not have been possible. We would also like to thank our family for their constancy, support, and guidance. It is because of you that we are who we are today.

1
INTRODUCTION

I. M. Nick

Introduction

In 2012, during the tenth Biennial Conference of the International Association for Forensic Linguistics (IAFL), Ronald Butters gave a closing address before stepping down from the IAFL Presidency. During his remarks, he highlighted the many accomplishments the IAFL had achieved with its sister society, the International Association for Forensic Phonetics (IAFP). For example, both the IAFL and IAFP were about to celebrate their 20th anniversary; and their scholarly publication, the *Journal of Language, Speech, and the Law*, was thriving. The societies, like the field, were coming into their own. At the same time, there were many tasks which stubbornly remained on the perennial to-do list. Chief amongst them was addressing the many challenging ethical questions confronting the field and its growing number of practitioners. As Butters stressed in his address, devoting sustained critical attention to ethics was vital to ensuring the field's longevity and credibility—both essential given the increasing importance of forensic linguistics to helping to resolve matters before the court.

The rising demand for linguists to provide expert testimony for courtroom deliberations was a part of a general boom witnessed across the forensic sciences (Saks et al. 2003). For better or worse, there was a widespread belief in the general public that forensics could provide the proverbial "smoking gun", even in the most contentious of cases. Acutely aware of this effect, it became increasingly commonplace for legal counsel to invest considerable time and finances to locate and hire scientists whose testimony could support their arguments and impress jurors, judges, and magistrates. It was during this time that many linguists who would later be celebrated as founders in the field were regularly called in to provide their expert opinion on cases. Over time, the accumulating body of casework became the basis for a series of ground-breaking academic publications (e.g., Baugh 2018; Solan, Ainsworth, Shuy 2015; Shuy 2012; Tiersma and Solan 2012; Shuy 2011; Solan, Tiersma, and Gales 2010; Solan 2010a/b; Tiersma 2010; Tiersma and Solan 2010; Eades 2008; Gibbons and Turell 2008; Cao 2007; Coulthard and Johnson 2007; Shuy 2007; Tiersma 2006; Cotterill 2003; Gibbons 2003; Hollien 2002; McMenamin 2002; Rose 2002; Solan 2002; Tiersma 1999; Solan 1995; Solan 1993; Eades 1992; Baldwin and French 1990; O'Barr 1983).

DOI: 10.4324/9781003391074-1

On January 9, 2009, during the 83rd annual meeting of the Linguistic Society of America, a special symposium was held. Organized by Ronald Butters and Edward Finegan, in cooperation with Roger Shuy, the event was specifically geared toward providing assistance to "linguists who may be taking on their first consulting assignment" and would address the "numerous pitfalls both practical and ethical" (LSA Conference Meeting Handbook 2009: 89). The two-and-a-half-hour forum featured the following panel of speakers and papers: (1) Janet Ainsworth (Seattle University School of Law), "The Consumer's Perspective: Ethical, Technical, and Practical Considerations Lawyers Face in Using Linguistic Experts"; (2) Ronald R. Butters (Duke University), "The Forensic Linguist's Professional Credentials"; (3) Edward Finegan (University of Southern California), "An Expert Linguist's Truth: Always Whole and Nothing But?"; (4) Geoffrey Nunberg (University of California Berkeley), "The Relationship between Case-Drive Linguistic Research and Scholarly Publication"; and (5) Gail Stygall (University of Washington), "Guiding Principles: Forensic Linguistics and Codes of Ethics in Other Fields and Professions". That symposium subsequently became the basis for a Special Issue of the *Journal of Speech, Language, and the Law*. In addition, the Symposium served as a powerful springboard for the development of a formal "Code of Ethics for Forensic Linguistic Consultancy" by the Linguistic Society of America (LSA).

A year after the Symposium, in 2010, the *Routledge Handbook on Forensic Linguistics was* released. Edited by Malcolm Coulthard and Alison Johnson (now May), this 673-page reference featured chapters from some of the rising stars of the international forensic linguistics scene. Even now, well over a decade after its initial release, this reference remains one of the most seminal in the field today. Taken all together, this scholarship has been central to attracting and educating generations of emerging linguistic scholars. Still, despite the many impressive scholarly achievements in researching and applying linguistics for legal purposes, much of the initial headway made in ethical issues seemed to fall by the wayside. Even in the prodigious *Routledge Handbook*, as Butters lamented during his final remarks, the word "ethics" only appeared on less than a handful of pages. The same was true of many of the other widely used introductory works on forensic linguistics that were published about this period (e.g., Olsson 2013, 2004; Gibbons 2003; McMenamin 2002). As Butters was quick to defend, however, this should not be seen as evidence that the forensic linguistic community never considered or talked about "ethical issues or debate standards and practices"; however, it must be conceded, Butters continued, that the subject of ethics was still "generally far from central" (353). Ironically perhaps, one of the reasons for this perpetual lag may have been the speed with which the discipline was developing. Indeed, in many ways, the rising prominence and popularity of the discipline was proving to be a double-edged sword to the discipline.

On the one hand, the increased attention had succeeded in giving linguistics in general a positive public make-over. No longer perceived as being an exclusive intellectual niche where only lovers of philosophy, logic, and mathematics meet, linguistics was more and more being recognized as a dynamic discipline with real-world applications to address real-world problems. Within the United States and elsewhere, this change in image was pivotal in helping linguistics join the expanding diversity of forensic disciplines that could provide potentially useful paid consultants. On the other hand, the rising demand for and reliance upon individuals willing to provide expert analysis of language evidence brought into sharp relief many long-standing issues within the field. For example, there was still no general consensus over what made a linguist an "expert"...or not. For that matter, there was no clear definition of what a "forensic linguist" was or did.

In the early years, the term "forensic linguist" was routinely used to describe any linguist who applied their specialized expertise in service of the court. Consequently, before applying their skills to analyze forensic cases, many expert forensic linguists like John Baugh, Susan Berk-

Seligson, Diana Eades, Peter French, and Roger W. Shuy first obtained their PhD in linguistics or phonetics and phonology with specializations in sociolinguistics, dialectology, and discourse analysis. By the same token, scholars like Peter Tiersma, Kate Haworth, and Lawrence Solan, whose work examines the relationship between language and the law, obtained a PhD in linguistics as well as a law degree. In some ways, this educational trajectory is quite analogous to what one would see in medicine, where the first step is to earn a degree in general medicine. Then, after obtaining an expert understanding of the features and operation of the body, some doctors decide to go on to specialize in a particular area of medicine such as gynecology, oncology, nephrology, or podiatry. While this two-step process has its advantages, it also was grown out of necessity, given the lack of programs offering a degree in forensic linguistics.

Since then, the situation has changed of course. Thanks in no small measure to the commitment and brilliance of the field's pioneers and the sustained international networking efforts of the International Association for Forensic and Legal Linguistics IAFL(L) and the International Association for Forensic Phonetics and Acoustics (IAFPA), there are now academic programs in which students can take courses, complete modules, obtain certificates, or earn a university degree in forensic linguistics. These differing educational pathways, however, have further complicated resolving lingering ethical questions surrounding qualifications, certification, and licensure.

For unlike other disciplines such as medicine and psychology—two fields which have long and prestigious histories of providing scientific expertise to help resolve pressing legal questions—forensic linguistics is not regulated by overarching governing boards. There is no central institution which awards licenses or issues certifications to affirm that a linguist in question has successfully met a transparent set of professional qualifications; nor is there any organizational power within the field that has the power to discipline members who have been found to have violated the professional regulations and ethical standards of the discipline.

Without these professional guardrails, there is very little the field can do to thwart blatantly unqualified individuals from calling themselves a "forensic linguist" and offering to provide "expert testimony". By the same token, there is almost nothing to prevent or sanction academics who had, in fact, obtained a degree in linguistics from an accredited university from committing any number of unethical actions. These potential infractions include but are not limited to offering testimony on subjects clearly outside of their area of expertise; employing questionable or inappropriate methods to generate, handle, or analyze the language evidence at hand; making exaggerated or unsubstantiated assertions of certainty, often without acknowledging rates of error or methodological limitations; agreeing to provide expert testimony favorable to the client in exchange for exorbitant amounts of money or other perks such as unfettered access to sensitive data, or prominence in the media.

Without the power to censure, suspend, or revoke professional licensure, there is little that the field of forensic linguistics can do to protect itself or, even more importantly, the community-at-large from such professional acts of malpractice, aside from appealing to people's goodwill and internal sense of right and wrong. This does not mean to imply that, without these external controls, the field is riddled with opportunists, charlatans, and general ne'er-do-wells. Indeed, as case after case has demonstrated, the sophisticated analyses linguists are uniquely trained to provide can prove pivotal in helping to successfully convict the guilty and, just as importantly, exonerate the innocent. In the final analysis, it is precisely because of the high stakes involved that it is imperative to more systematically address some of the ethical questions that had been nagging the field and its members.

Historically, aside from the chronic lack of time and resources from which most academics suffer, there have been a number of factors that have complicated taking on these and other vexing ethical issues. The simple fact that forensic linguistics is applied in different countries, under different cultural norms, by different legal systems, with different regulations concerning the use of expert consultants, who have all been trained in different education systems, certainly has not made the development of a common set of ethical codes easy. Added to those hurdles are the unique characteristics of and circumstances surrounding every case, which make different forensic demands on the linguistic consultant. Sometimes, "consultancy" consists of providing general advisement. Other times, consultancy involves providing an official written report of findings obtained and conclusions drawn after a labor-intensive, in-depth analysis of the language evidence. And in still other instances, consultants will be asked to present and/or defend their professional assessment and opinions in a court of law. These varying assignments and degrees of case involvement significantly complicate the task of creating a universally applicable set of guidelines.

Further adding to this complexity is tremendous variation in the specific legal-linguistic questions forensic linguists are asked to examine as consultants. Quite often, depending on the specifics of the case, a specific kind of linguistic expertise will be required. The following list features just a small sampling of the highly specialized forensic services which linguists have provided:

- assessing taunting letters sent to police by a person(s) claiming responsibility for a series of grisly sexual murders and isolating linguistic features that correlate with the author's probable age, gender, education, native language, and regional dialect;
- deciphering the last words of a dying victim whose voice had been captured on a recording;
- analyzing the socio-onomastic features of a wordmark to determine if regulations concerning propriety language had been violated;
- testifying on cases of alleged discrimination and explaining to juries how prejudices may be unconsciously triggered by the names of applicants for housing, employment, or education;
- comparing and contrasting translations of international laws across different languages in order to identify potentially significant linguistic deviations which may have implications for implementation and interpretation;
- analyzing the commercial names appearing in a trademark for possible violations of regulations against disparagement;
- investigating the writings of a decedent in order to identify linguistic features that have been found to correlate strongly with suicide ideation; and
- comparing the original testimony of witnesses against the versions provided by courtroom interpreters to identify potential discrepancies in meaning and content.

Despite all being subsumed under the "forensic linguistics", each of these cases requires a different set of specialized skills and training that goes beyond the traditional boundary of "general linguistics". This is an important point. Depending on the specific question being asked and the type of language evidence involved, a specific area of specialized linguistic expertise may be necessitated. The fact that an individual has obtained a degree in linguistics does not automatically qualify them to offer an expert opinion on any and every subject that falls within the field. Just as it would be more desirable to seek the expert opinion of a toxicologist than a neighborhood general practitioner to consult for a case of suspected fatal poisoning, the same is true where linguistic consultancy is concerned. Ideally then, ethical guidelines for forensic linguistics would address the necessity of only accepting casework within their specific sphere of knowledge, expertise, and skill set. Just as importantly, it would also stipulate that in order to retain the title of "expert", linguists must remain abreast of the theoretical, methodological, and technological advances relevant for their specific area of competence.

To address this and other best practices, in the mid 2000s, both the IAFL and IAFPA developed Codes of Practice. Not surprisingly, a review of these guidelines reveals a great deal of overlap between them. Each, for example, contains a list of best practices regarding professional integrity, competence, confidentiality, accuracy, reliability of methods, and analysis. Another similarity of the Codes is that each is specifically designed for members "who carry out forensic casework" (IAFP 2020). While consultancy is an important aspect of many forensic linguists' work, however, it is not the only one.

Today, requests for the expert services of forensic linguists come from a variety of sources. The courtroom is just one of them. As the field has continued to evolve, the demand for expertise of forensic linguists has expanded to encompass a wide variety of professional arenas and clientele, both public and private. Accordingly, the understanding of what specifically a forensic linguist may do has dramatically evolved as well. Although there continues to be much formal debate over where the field begins and ends (Wright and Picornell 2024), it can generally be agreed that the work of forensic linguists is by no means restricted to consultancy, be it inside or outside of the courtroom. Instead, it also involves many other professional spheres such as researching, publishing, teaching, and mentoring the next generation of scholars. And within each of these spheres there are pressing ethical questions which also demand attention. It is the recognition of this need that became part of the impetus for this book.

Genesis of the Handbook

The original idea for this book came from many years of in-depth private conversations, in various constellations, between the editors, Janet Ainsworth and Lawrence Solan. These discussions eventually led to the organization of a special panel which took place in 2016 at the annual meeting of the Linguistic Society of America. At that time, I. M. Nick was serving as the LSA's Chair of the Committee for Ethnic Diversity in Linguistics (CEDL). In this position, she convened a special panel on Hate Speech. Guest speakers for the event included the forensic linguist and language policy specialist William Eggington (Brigham Young University); the human rights lawyer and current Director of the Dangerous Speech Project Susan Benesch (American University); and Joseph Tuman (San Francisco State University), Professor of Political and Legal Communications and expert on First Amendment protections of Freedom of Expression.

Before a packed audience of LSA attendees, the panelists discussed the legal challenges of addressing linguistic forms of expression that caused injury and offense to individuals and communities—many of whom were already amongst the most vulnerable in our society. In the discussion that ensued, many questions revolving around the ethical responsibilities of the linguist were raised. Was there a space where social activism could be performed without violating historical norms which demanded a professional's unwavering allegiance to objectivity? Was remaining impartial in the face of blatant injustice, in fact, a form of complicity and therefore in direct violation of the ethical mandate codified by the LSA itself? For example, in paragraph 2 of its official Ethics Statement, the LSA states:

> It is the responsibility of linguists, individually and collectively, to ensure that we uphold the ideals and principles of equality and anti-discrimination in our professional interactions—in

> our language and in our behavior, on a personal level, in our interactions with others, in our institutional roles, and in our academic commitments.
>
> *(https://www.lsadc.org/lsa_ethics_statement)*

The discussion was spirited and the insights offered were deeply thought-provoking; and just as importantly, the sharing of experiences, concerns, questions, and suggestions was respectful and kind. That positive experience led to several other riveting talks and the gradual germination of an idea: a handbook devoted to forensic linguistics and ethics.

The Aim of the Handbook

The vision was to create a multi-authored reference which would address many of the difficult ethical challenges which forensic linguists face. We were all agreed, however, that the ultimate purpose of such a work would not be to call for, or worse yet generate, a rigid set of regulations. Rather, the goal was to create a reference that would stimulate more open, honest, respectful dialogue about ethics and forensic linguistics.

As soon became apparent, however, for many people the term "ethics" serves as a powerful trigger word, sparking a range of strong emotions—from mild feelings of discomfort and defensiveness to full-on indignation and anger, with accusations of self-righteousness and sanctimony. As Dror and Charloton (2006) observe, discussions of ethics are always a "sensitive issue that often meets with defensive responses" (604). Part of the reason for this kind of backlash may be the (un)conscious associations which many make between ethics and other, even more triggering concepts such as "morality", "virtuousness", "propriety", and "decency" as well as "inadequacy", "incompetence", "negligence", "fraud", "failure", and "error". The strength of such associations, combined with the increasingly acrimonious state of public discourse, made the idea of putting together a book on ethics a daunting task to say the least. However, as we all came to agree, it was precisely because so many people had become so afraid to talk about these kinds of issues that there was an even greater need for such a reference. As Cinaglia, Rabie-Ahmed, and De Costa (2024) observe in their review of discussions surrounding ethics and Applied Linguistics, "despite a growing number of ethically-oriented methodological publications, there has been less focus on ethical considerations specific to different sub-fields" (2). In the resulting absence of a forum for transparent, honest, and caring discussion of ethics within forensic linguistics—far beyond the private conversations and insular discussions that sporadically take place between colleagues—many have been left to fend for themselves, without the guidance they needed or support they deserved.

For the purposes of this publication, a rather broad, pragmatic, working definition of "ethics" offered by Cinaglia and Rabie-Ahmed (2024) is used. According to them, ethics in Applied Linguistics "can be understood as a set of virtues guiding human behavior to ensure that an individual's actions respect and positively impact both the self and the greater society" (310). The term "ethics" is not used here as a prescriptive model of conduct, but rather it refers to the attempt to maximize the potential positive effects of applying forensic linguistics, while minimizing the possible ill-effects, both immediate and long-ranging. As readers will discover, although all of the chapters in this work are in agreement with this general goal, the pathways which the contributors take to reach that end differ, sometimes dramatically—so much so in fact that the conclusions reached and the suggestions made by one author may be in direct contradiction to another. This lack of agreement is a direct reflection of the complexity presented by ethical dilemmas. Instead of attempting to mask that complexity, this reference seeks to bring it out into the open. Readers may find themselves wholeheartedly agreeing with one position and vehemently disagreeing with another. In either case, it is hoped that the chapters presented here will spark more transparent and deliberate engagement with forensic linguistics, with all of its potential and limitations.

Many of the names in this reference will be familiar to readers. A concerted effort was made to include scholars who have had long and illustrious careers within forensic linguistics. In fact, many of these trailblazers contributed to the original *Routledge Handbook on Forensic Linguistics* (e.g., Michelle Aldridge, Janet Ainsworth, Mel Greenlee, Chris Heffer, and Lawrence Solan). Many of the scholars who generously assisted in this Handbook, either by contributing a chapter or serving as reviewers, play major leadership roles in the IAFLL and IAFPA. Examples include the current presidents, Jennifer Glougie and Tina Cambier-Langeveld of the two representative sister societies, as well as past and present members of their governing boards (e.g., Malcolm Coulthard, Felicity Deamer, Isabel Picornell, Richard Rhodes, Dominic Watt). In addition, this reference also includes co-authored contributions from many emerging scholars who are just beginning to make their own mark. Finally, given the inherently interdisciplinary nature of the research and application of forensic linguistics, this reference also contains multiple contributions from scholars from closely allied fields (e.g., criminology, sociology, criminal justice, law enforcement, government and politics, psychology, translation and interpreting, etc.). As a result, this Handbook has been designed to satisfy a wide range of readers who no doubt face similar, if not identical, ethical challenges in their own work.

By the same token, a concerted effort was made to include contributions from a wide demographic range of authors. The decision to welcome such a diversity of authors was by no means window dressing. Instead it was motivated by the shared understanding that decisions over what is or is not ethical are highly sensitive to not only the experiences, positioning, norms, and roles of the individuals involved, but also the surrounding social, political, legal, historical, and linguistic contexts. As Jenks and Lee (2025) caution: "Ethical procedures are inherently subjective, context-dependent, and must accordingly be discussed in relation to the subjects from which the ethics aims to govern" (297). It is sincerely hoped that the diversity of perspectives presented here will therefore have enhanced relevance to the growing global contexts in which forensic linguistics is applied.

The Organization

The Handbook has been arranged in three sections. Each one is devoted to a different aspect of forensic linguistic work.

Section I: Providing Expertise as a Consultant

Section I is devoted to examining ethical issues that may arise when providing services as a consultant. The section begins with a contribution by Bowen who points out that, within the forensic sciences, professional ethics is rarely black and white, but perpetually gray. Nevertheless, as Bowen goes on to explain, the fact that ethics is rarely straightforward heightens the necessity of adhering to the professional norms of best practice. Only then can the professional work to mitigate mistakes and errors which can have devastating repercussions. However, as Picornell explains in her chapter, the endorsement of such precautions may not be enough. The institution of regulatory oversight may be required to help ensure the ethical conduct of forensic linguistic experts. Until such time that a system of professional oversight is instituted, forensic linguists are ethically bound to operate with the highest degree of integrity possible.

According to Rhodes and Cambier-Langeveld, this goal can only be accomplished when forensic linguist consultants provide completely transparent analyses devoid of all partiality. The attempt to combine or supersede that responsibility by playing—no matter how well-intentioned—the role of social advocate is neither ethically acceptable nor defensible. As the co-authors go on to explain in a second contribution to this Handbook, their conclusion is not based upon theory alone, but is born from years of casework in which they routinely faced difficult ethical dilemmas. As Cambier-Langeveld and Rhodes argue, maintaining the principles of impartiality and objectivity were key to resolving these conflicts. This position is echoed by van Naerssen. In her chapter, she details how professional distancing was essential to providing effective, valid, and reliable data that proved of critical value in fairly adjudicating social justice cases involving non-native speakers. Although in retrospect, the pathways taken by these experts may appear clear, in the moment, successfully navigating these ethical challenges can be extremely difficult.

In view of that fact, Perkins, Rosie, and Glaas offer readers an intelligence scaffold designed to aid in navigating what can feel like a veritable minefield of ethical questions. Importantly, as Bull and Blandón-Gitlin rightly point out, the difficulty in deciding what is the most ethical action to take comes not only from the circumstances of a case, but also the decision-makers themselves. As human beings, forensic linguists are invariably subject to a host of psychological biases which can affect their memory and perception, judgment and behavior. Bull and Blandón-Gitlin offer readers scientifically founded suggestions for mitigating the potential unwanted effects upon forensic linguistic analysis. Just as forensic linguists have an ethical responsibility to challenge faulty assumptions and unsubstantiated premises in their own work, as Karoli Christensen and Beck Navarro argue in the next chapter, they also have an ethical responsibility to challenge the misconceptions about language and language-users within the society-at-large. Operating from this premise, the authors call attention to the proliferation of pseudoscientific "linguistic" claims and the subsequent adoption of dubious law enforcement practices. In view of the potentially devastating consequences, Karoli Christensen and Navarro Beck argue that forensic linguists do their part to counter the spread and impact of "junk linguistic science". Unfortunately, as Habick discusses in the following chapter, the need to guard against linguistic misinformation is not limited to calling attention to unethical practices outside of the field. The intentional dissemination of ideas and concepts that are lacking in factual basis is also a problem within the field. Although infrequent, as Habick warns, instances of ethical misconduct may expose the individual professional to legal jeopardy and pose a serious threat to the overall credibility of the discipline. Individually and together, these chapters powerfully underscore, the field simply does not have the luxury of assuming that things are proceeding as they should. It is only when the potential for error and abuse is acknowledged that effective precautionary measures can be taken and the real potential benefit of the field's societal contribution can came to fruition. On this positive note, Section I ends with a chapter that discusses the very real positive impact which forensic linguistics can make. Building on years of research comparing official reports and audio recordings of police interrogations with suspects, Komter calls attention to the disturbing discrepancies and makes concrete policy recommendations to help see that citizens' rights are respected and that criminal justice systems obtain the accurate and reliable intelligence needed to protect the society's general welfare.

Section II: Serving as an Expert Witness

This section begins with a contribution by Conley Riner and Conley who examine the specific ethical challenges facing linguists and lawyers in adversary court systems, such as found in the United States. Staying within the United States, O'Laughlin's chapter critically interrogates the ethics of advocacy for the US American Expert Witness. As O'Laughlin explains, the scientific principle of neutrality stands in opposition to the expectation that an expert witness will work alongside the attorney who hired them. In the following chapter, Greenlee details how differing professional objectives and norms can give rise to clashes between linguists and lawyers. Greenlee then goes on to make concrete suggestions to help linguists avoid these clashes and provide accurate, fact-based testimony for evidentiary purposes. According to Greenlee, the key to successful collaboration lies in explaining how language operates within a given social context. A similar conclusion is reached by Wing. Using concrete examples from US case law, Wing argues that it is not only expeditious, but also ethically imperative that forensic linguists inform non-expert triers of fact how language actually does and does not work.

The collective insights provided in these chapters are then effectively complemented by Heffer. In his chapter, he addresses the contentious legal construct "beyond a reasonable doubt" in British courtroom deliberations. To be sure, much has been written about the problems of effectively communicating to juries what is meant by this legal construct; and the literature is filled with possible remedies such as altering the phrasing or introducing formula (Keane and McKeown 2019; Solan 1999; Kagchiro and Stanton 1985; Tribe 1971). However, Heffer's chapter takes an entirely different and novel approach. Starting from the Golden Rule of doing unto others as we would have them do unto us, Heffer critically considers what judicial authorities may ethically owe juries when they give them formal instructions. In the following chapter, Deamer and Wilkinson contend with another potentially disputatious legal construct "diminished responsibility". Using case examples from authentic UK court proceedings, the co-authors illustrate how forensic linguistic analysis can help to uncover significant disparities in judicial interpretations and rulings. The following chapter shifts focus from the courtrooms of the British Isles and returns to the North American continent. As Glougie explains in her chapter, in the Canadian legal system, expert evidence is only considered admissible if and only if the adjudicator believes admitting evidence will do more good than harm. As Glougie demonstrates, this protective threshold has direct ethical implications for forensic linguists who present expert evidence in and for Canadian court proceedings.

A similar principle of benevolence is at the heart of the chapter contributed by Wright who contends that experts in forensic linguistics are ethically bound to take active steps to avoid causing injury. According to Wright, the historical reluctance or refusal to recognize this responsibility to actively employ a proper standard of care contributes, every day, to significant and long-standing harms suffered by many of the most vulnerable members of society. Precisely this point is taken up by Aldridge and Pereira. In their co-authored contribution, the two critically examine the legal system of England and Wales and describe how adult witnesses with an intellectual disability have been mistreated in cases involving sex abuse. To increase the protections provided to these vulnerable witnesses, this chapter suggests the introduction of several protective special measures. Increasing protections to marginalized peoples is also a subject of the last chapter in Section II. Here, Greenlee calls much-needed attention to some of the chronic problems concerning court translations in US criminal proceedings. This chapter concludes Section II.

Section III: Researching, Teaching, Publishing, and Providing for the Next Generation

Section III moves beyond the consultancy and focuses on other areas of forensic linguistic work: namely, researching, teaching, and publishing. The section begins with Sullivan Buker's exploration of the challenges experts face when analyzing emotionally confronting forensic material. Using interview material gathered from leading members of the forensic linguistic community,

Sullivan Buker challenges the notion that stifling empathy and compassion is a prerequisite for having a successful, productive career. Instead, what is needed are coping strategies to process confronting material. Practicing mindfulness places the forensic expert in a much better position to conduct their analyses reliably, accurately, responsibly, and therefore ethically.

Remaining aware of one's professional and personal limits is particularly important when faced with real-life ethical dilemmas far beyond that which generic codes of conduct typically address. The gulf between theory and praxis is vividly illustrated by Sivunen and Ennser-Kananen's contribution. Based on their work with deaf refugees, the two share how excruciatingly difficult it can be to find oneself caught between rigid institutional policies and naked human suffering. This chapter is followed by an equally riveting contribution by Lindegren who details the ethical challenges that can arise when conducting research with individuals convicted of sex crimes. Importantly, as Lindegren highlights, the ethical dilemmas encountered are not only a function of the confronting details of the offenses or the biographies of persons convicted of committing them. Many of the ethical issues come from the society's negative prejudices and stereotypes about anyone connected to these crimes, including the forensic researcher. In the following chapter, Nick and Preiss focus on another form of violent crime: political extremism. Using their separate yet complementary research experiences, the co-authors discuss the methodological and ethical issues involved in procuring, securing, analyzing, and reporting highly sensitive criminological data, while protecting research participants and colleagues, loved ones, and oneself from potential harm. As Sullivan Buker, Sivunen and Ennser-Kananen, Lindegren, and Nick and Preiss all stress, taking precautions to protect one's physical, mental, and emotional health is not simply a matter of good common sense, if one is to have a long, productive, and fulfilling career in forensic linguistics. It is also an ethical prerequisite for making a substantive scientific and societal contribution.

In contrast to these previous chapters in Section III, Hashimoto and Haderlie place their focus on methodology. More specifically, they examine some of the pressing ethical issues involved in using linguistic corpora to interpret legal texts. To assist the growing number of forensic linguists interested in adopting this approach, the co-authors offer an innovative framework for assessing the ethical soundness of their methodology. In the next chapter, Harrison and Watt tackle another technological advance that promises to revolutionize much for forensic linguistic work: namely, artificial intelligence (AI). At the same time that forensic scientists are exploring the applications of AI, so too are bad actors. Within forensic speech sciences, particularly troubling has been the emergence of increasingly realistic AI-generated "voices" to an increasing variety of crimes—from small-scale scams designed to extort money from targeted individuals to large-scale crime syndicates that generate deep-fakes of politicians' voices in an attempt to manipulate elections, the potential dangers are real and growing. In their chapter, Harrison and Watt examine what makes an individual's voice distinctive and present an original scale scientists can use to determine the relative degree of misuse. As the co-authors stress, this is an area of research which demands increased attention. In the next chapter, de Boer presents yet another subject-area deserving attention from the forensic linguistic community: the analysis of non-human animal voices. For some readers, this subject may seem an odd fit for a Handbook on Forensic Linguistics. However, in point of fact, the proposition is not without precedent. From forensic botany and forensic entomology, the knowledge gained from the examination of non-human life-forms has long played an important role in crime scene analysis. Moreover, the voices of non-human animals have already been used to provide compelling evidence in court proceedings. Against this background, de Boer argues that experts trained in theories, technologies, and methodologies of forensic speech science are particularly well-suited to provide voice analyses of non-human species.

The ability of forensic linguistics to react to new trends is perhaps one of the reasons why this field continues to attract so much student interest. Although not all of those who take a forensic linguistic course will go on to pursue a career in the field, the knowledge and insights students gain can still contribute to their understanding of and appreciation for the complex relationship between language and the law. However, it would be naïve to think that everyone who completes a forensic linguistic course will put their newly won knowledge to good use. There is a certain percentage, however small, who may use the information gained to cause themselves and/or others harm. This potential for misuse and abuse is the subject of Marko's chapter. It asks what ethical responsibility the forensic linguistic educator has when disseminating information in the classroom and offers concrete suggestions for minimizing the risk involved in sharing sensitive information to audiences which may contain persons with malicious intent. Of course, the potential risks of teaching forensic subjects are not limited to so-called "bad actors". Forensic linguistics by its very nature often involves highly confronting, potentially traumatizing, material. In this regard, forensic linguistics is by no means unusual, however. Students may be confronted with difficult subject matter in a variety of disciplines, including the allied fields of criminal justice, psychology, criminology, and law. In contrast to these fields where much has been published on how to sensitively, responsibly, and ethically teach students (e.g., Harding 2022; Dalton 2020; Klein 2017; Winstone and Kinchin 2017; Carello and Butler 2015; Scriver and Kennedy 2015; Zurbriggen 2011; Dalton 2010; Heath 2005; Rogers and Webb 1991; Michalowski 1977), this issue has received much less formal attention in forensic linguistics. In Larner's chapter, this oversight is addressed. More specifically, Larner examines the ethical conundrum of teaching a subject that is devoted to reducing societal harms at the same time that the instruction itself may cause persons harm. As both Larner and Marko stress, in order to teach effectively and ethically, it is essential to avoid viewing their students as a homogeneous mass. No matter how large class sizes may become, each course is made up of individuals with their own histories, needs, agendas, resources, vulnerabilities, and talents.

Jacknick and Balmat agree: by not only recognizing, but openly embracing the natural diversity of the classroom environment, teaching forensic linguistics can become a positive and transformative experience for students and instructors alike. To illustrate this potential, the co-authors share their experiences in teaching forensic linguistics from a trauma-informed pedagogical perspective that is both culturally responsive and culturally sustaining. While Marko, Larner, and Jacknick and Balmat all draw attention to the ethical necessity of recognizing and respecting the different perspectives and gifts students can contribute to the field, Gordon and Turnbull Plaza draw attention to the ethical imperative of extending the same welcome to our international colleagues. However, as the co-authors sagely stress, to facilitate an environment in which ideas and information are freely and equitably exchanged, far more is required than simply unlocking the door and waiting for marginalized communities to enter.

What is needed is sustained action to dismantle the long-standing institutional and institutionalized barriers to multinational, multicultural, and multilingual participation. To highlight the destructive effects of the continuing impasses in forensic linguistics, Gordon and Turnbull Plaza focus on one of the most glaring quadrants of discriminatory policies in higher education: publishing. Importantly, the obstacles to access faced by many of the world's scientific experts do irreparable damage to not only individual careers , but also to the disciplines these professionals represent. As Gordon and Gutierrez Fierros (2022) warn, exclusionary practices in scientific publishing contribute to "incomplete scientific understanding and [hinder] international collaborations on global challenges" (84).

Moreover, as Nick then goes on to detail, the negative repercussions of inequity are not only evident in publishing. They permeate every aspect of the discipline. Luckily, there is a simple and effective remedy to this problem: diversity. The promotion of diversity has long been recognized as essential to ensuring the health and longevity of the forensic sciences across the board. As Wagstaff and Laporte (2018) explain in a report published by the scholarly journal of the US National Institute of Justice: "to strengthen the forensic sciences, it is imperative that we train and employ more forensic scientists and foster an environment in which uniqueness and difference are embraced and valued" (1). As Nick reasons, what is true for the Forensic Sciences in general is also true for the discipline of forensic linguistics. Using a combination of governmental reports and scientific findings, Nick argues that forensic linguistics must make good on the years of institutional promises to promote diversity, equity, and inclusion. Honoring this commitment is imperative for helping to ensure not only the survival of the field into the next century, but also social justice for the individuals and people the field is ethically bound to serve. The Handbook then ends with a series of overarching concluding remarks.

Concluding Observations

Again, the purpose of this collection is not to raise an accusatory voice of condemnation. Doing so would not only be entirely inappropriate, but truly unnecessary. One of the reasons why forensic linguistics has continued to grow as much as it has over the past few decades, is the honor, integrity, intelligence, and ethics of many of its most prominent members. Their collective and individual commitment to personal and professional excellence stands as powerful inspiration for how valuable the knowledge, experience, and skill of a trained forensic linguistic expert can be in safe-guarding our communities. It is within that spirit of duty that the original idea of this book was conceived and each chapter was contributed. It is the sincere hope that this work will stimulate productive, open, respectful, and sensitive discussions. This Handbook is then an invitation for those in the field to do what we do best. Discuss, debate, reason, disagree, compromise, listen, re-consider, learn, teach, develop, grow, innovate, question, and then start the entire wondrous process of respectful communication all over again.

References

Baldwin, John, and Peter French. 1990. *Forensic Phonetics*. London: Pinter.

Baugh, John. 2018. *Linguistics in Pursuit of Justice*. Cambridge: Cambridge University Press.

Cao, Deborah. 2007. *Translating Law*. New York: Multilingual Matters.

Carello, Janice, and Lisa Butler. 2015. "Practicing What We Teach: Trauma-informed Educational Practice." *Journal of Teaching in Social Work* 35, no. 3: 262–278.

Cinaglia, Carlo, and Amr Rabie-Ahmed. 2024. "Ethical Dilemmas of Graduate Students Negotiating New Roles and Responsibilities: The Importance of Mindful Engagement, Self-care, and Reflexivity." In *Ethical Issues in Applied Linguistics*, 310–327. Edited by Carlo Cinaglia, Amr Rabie-Ahmed, and Peter De Costa. John Benjamins.

Cinaglia, Carlo, Amr Rabie-Ahmed, and Peter De Costa, eds. 2024. *Ethical Issues in Applied Linguistics Scholarship*. Amsterdam: John Benjamins.

Cotterill, Janet. 2003. *Language and Power in Court: A Linguistic Analysis of the O.J. Simpson Trial*. London: Palgrave MacMillan.

Coulthard, Malcolm, and Alison Johnson. 2007. *An Introduction to Forensic Linguistics: Language in Evidence*. New York: Routledge.

Dalton, Derek. 2020. "Trigger Warnings in Criminology Teaching Contexts: Some Reflections Based on Ten Years of Teaching a Sensitive Topic." In *Teaching and Learning Criminology*, 87–108. Edited by D. Palmer. New York: Palgrave Macmillan.

Dalton, Derek. 2010. "Crime, Law, and Trauma: A Personal Reflection of the Challenges and Rewards of Teaching Sensitive Topics to Criminology Students." *Enhancing Learning in the Social Sciences* 2, no. 3: 1–18.
Dror, Itiel, and David Charlton. 2006. "Why Experts Make Errors." *Journal of Forensic Identification* 56, no. 4: 600–616.
Eades, Diana. 2008. *Courtroom Talk and Neocolonial Control.* New York: Mouton de Gruyter.
Eades, Diana. 1992. *Aboriginal English and the Law: Communication with Aboriginal English-Speaking Clients: A Handbook for Legal Practitioners.* Brisbane: Queensland Law Society.
Gibbons, John, ed. 2013. *Language and the Law.* New York: Routledge.
Gibbons, John. 2003. *Forensic Linguistics: An Introduction to Language in the Justice System.* Oxford: Blackwell Publishing.
Gibbons, John, and M. Therese Turell, eds. 2008. *Dimensions of Forensic Linguistics.* Amsterdam: John Benjamins.
Gordon, Sarah, and Emiliano Gutierrez Fierros. 2022, September. "Combatting Exclusionary Language Practices in Science Publishing: A DEI Concern." *Science Editor* 45, no. 3: 84–85.
Harding, Natacha. 2022. "Reasonably Uncomfortable: Teaching Sensitive Material Sensitively." In *Teaching Criminology and Criminal Justice*, 199–122. Edited by Suzanne Young and Katie Strudwick. New York: Palgrave Macmillan.
Heath, Mary. 2005. "Encounters with the Volcano: Strategies for Emotional Management in Teaching the Law of Rape." *The Law Teacher* 39, no. 2: 129–149.
Hollien, Harry Francis. 2002. *Forensic Voice Identification.* New York: Academic Press.
International Association for Forensic Phonetics and Acoustics. 2020. ."Code of Practice." Accessed March 1, 2025. Retrieved from https://www.iafpa.net/about/code-of-practice/.
Jenks, Christopher, and Won Lee. 2025. "Managing Publication Expectations and Collaborations: On the Ethics of Co-authoring in Applied Linguistics." In *Ethical Issues in Applied Linguistics Scholarship*, 296–309. Edited by Carlo Cinaglia, Amr Rabie-Ahmed, and Peter De Costa. Amsterdam: John Benjamins.
Kagehiro, Dorothy, and W. Clark Stanton. 1985. "Legal vs. Quantified Definitions of Standards of Proof." *Law and Human Behavior* 9: 159–178.
Keane, Adrian, and Paul McKeown. 2019. "Time to Abandon 'Beyond Reasonable Doubt' and 'Sure': The Case for a New Direction on the Criminal Standard and Why and How it should be Used." *Criminal Law Review* 6: 505–527.
Klein, Jennifer. 2017. "Learning About the Labelled: Teaching A Course on Sexual Offenders While Accounting for Students Who May Be Abuse Survivors." *Journal of Criminal Justice Education* 29, no. 1: 79–95.
Linguistic Society of America. 2016, January 7–10. *Meeting Handbook of the Linguistic Society of America.* Washington, DC. Accessed March 1, 2025. Retrieved from https://www.lsadc.org/files/LSA%202016%20Annual%20Meeting%20Handbook.pdf.
Linguistic Society of America. 2009, January 8-11. *Meeting Handbook of the Linguistic Society of America.* San Francisco, California. Accessed March 1, 2025. Retrieved from https://www.lsadc.org/files/2009_Annual_Meeting_Handbook.pdf
McMenamin, Gerald. 2002. *Forensic Linguistics: Advances in Forensic Stylistics.* New York: CRC Press.
Michalowski, Ray. 1977. "A Gentle Pedagogy: Teaching Critical Criminology in the South." *Crime and Social Justice* 7: 69–73.
O'Barr, William. 1983. *Linguistics Evidence: Language, Power, and Strategy in the Courtroom.* New York: Academic Press.
Olsson, John. 2013. *Wordcrime: Solving Crime Through Forensic Linguistics.* London: Continuum.
Olsson, John. 2004. *Forensic Linguistics: An Introduction to Language, Crime, and Law.* London: Continuum.
Rogers, Dwight, and Jaci Webb. 1991. "The Ethic of Caring in Teacher Education." *Journal of Teacher Education* 42, no. 3: 173–181.
Rose, Phil. 2002. *Forensic Speaker Identification.* London: Taylor and Francis.
Saks, M., D. Risinger, Robert Rosenthal, and William Thompson. 2003. "Context Effects in Forensic Science: A Review and Application of the Science to Crime Laboratory Practice in the United States." *Science and Justice* 43, no. 2: 77–90.
Scriver, Stacey, and Kieran Kennedy. 2015. "Delivering Education About Sexual Violence: Reflections on the Experience of Teaching a Sensitive Topic in the Social and Health Sciences." *Irish Educational* Studies 35, no. 2: 195–211.

Shuy, Roger W. 2012. *The Language of Sexual Misconduct Cases*. Oxford: Oxford University Press.
Shuy, Roger W. 2011. *The Language of Perjury Cases*. Oxford: Oxford University Press.
Shuy, Roger W. 2007. *Creating Language Crimes: How Law Enforcement Uses and (Misuses)Language*. Oxford: Oxford University Press.
Solan, Lawrence. 2010a. *The Language of Statutes: Laws and the Interpretation*. Chicago: Chicago University Press.
Solan, Lawrence. 2010b. "Blame, Praise, and the Structure of Legal Rules." *Brooklyn Law Review* 75: 517–544.
Solan, Lawrence. 2002. "The Linguist on the Witness Stand: Forensic Linguistics in American Courts." *Language* 78, no. 2: 221–239.
Solan, Lawrence. 1999. "Can the Legal System Use Experts on Meaning?" *Tennessee Law Review* 66: 1167–1199.
Solan, Lawrence. 1995. "Judicial Decisions and Linguistic Analysis: Is There a Linguist in the Court?" *Washington University Law Quarterly* 73: 1069–1080.
Solan, Lawrence. 1993. *The Language of Judges*. Chicago: Chicago University Press.
Solan, Lawrence, Janet Ainsworth, and Roger W. Shuy. 2015. *Speaking of Language and Law*. Oxford: Oxford University Press.
Solan, Lawrence, Peter Tiersma, and Tammy Gales. 2010. *Speaking of Crime: The Language of Criminal Justice*. Chicago: University of Chicago Press.
Tiersma, Peter. 2010. *Parchment, Paper, Pixels: Law and the Technologies of Communication*. Chicago: Chicago University Press.
Tiersma, Peter. 2006. "Some Myths About Legal Language." *Journal of Law, Law, Culture and Humanities* 2: 9–50.
Tiersma, Peter. 1999. *Legal Language*. Chicago: Chicago University Press.
Tiersma, Peter, and Lawrence Solan, eds. 2012. *The Oxford Handbook of Language and Law*. Oxford: Oxford University Press.
Tiersma, Peter, and Lawrence Solan. 2010. *Speaking of Crime: The Language of Criminal Justice*. Chicago: Chicago University Press.
Tribe, Lawrence H. 1971. "Trial by Mathematics: Precision and Ritual in the Legal Process." *The Harvard Law Review* 84, no. 6: 1329–1393.
Wagstaff, Iris, and Gerald Laporte. 2018. "The Importance of Diversity and Inclusion in the Forensic Sciences." *National Institute of Justice Journal* 279: 1–11.
Winstone, Naomi, and Ian Kinchin. 2017. "Teaching Sensitive Issues: Psychological Literacy as an Antidote to Pedagogic Fragilty." *Psychology Teaching Review* 23, no. 1: 15–29.
Wright David, and Isabel Picornell. 2024. "Semiotic Perspectives on Forensic and Legal Linguistics: Unifying Approaches in the Language of the Legal Process and Language in Evidence." *International Journal for the Semiotics of Law/ Reveue international de Sémiotique juridue* 37: 293–304.
Zurbriggen, E. 2011. "Preventing Secondary Traumatization in the Undergraduate Classroom: Lessons from Theory and Clinical Practice." *Psychology Trauma: Theory, Research, Practice, and* Policy 3, no. 3: 223–228.

SECTION I

Providing Expertise as a Consultant

2

THE INTEGRITY OF FORENSIC ANALYSIS

Robin T. Bowen

Introduction

A central goal of forensic science is to sustain reliability and the key to achieving this aim is standardization. One of the first formal concerns over the consistency within forensic science disciplines arose when the National Research Council of the United States National Academy of Science (NAS) published a report in 2009 titled, "Strengthening Forensic Science in the United States: A Path Forward". The chief concerns identified the variability between factors such as methods, error rates, and general acceptance across forensic science disciplines. Forensic science facilities demonstrate vast differences in "capacity, oversight, staffing, certification, and accreditation across federal and state jurisdictions" (NAS 2009: 14). Considerations for the law in regard to admissibility and reliability of forensic evidence in trials are the extent to which the forensic discipline is based upon reliable scientific methods, allows for accurate analysis, and relies on human interpretation that could be subject to error, bias, or a lack of reliable procedures or standards. The recommendations specified by the report identified the need for greater scientific objectivity in forensic science disciplines, specifically through research, to establish the limits and measures of performance, to address the sources of variability and potential bias, and to develop rigorous protocols (Venter 2020; NAS 2009: 6).

In 2016, the President's Council of Advisors on Science and Technology (PCAST) released a report, "Forensic Science in Criminal Courts: Ensuring Scientific Validity of Feature-Comparison Methods" which examined the status of forensic science. The committee addressed the lack of studies employed to assess error rates and suggested that the lack of empirical testing indicated a weakness in the scientific culture of forensic science. The conclusion of the PCAST report identified two important needs for validity and reliability in forensic science: clarity regarding standards of forensic methods and evaluation of current methods. Though the forensic community acknowledged some limitations in research, overall, the report met dissatisfaction for its omissions and unsupported statements regarding scientific validity (FBI 2015). In addition, while professional groups identified the value of empirical testing, they also acknowledged the importance of experience and judgment in many forensic disciplines (OSAC Friction Ridge Subcommittee).

The focus on standardization with an emphasis on minimum requirements and protocols that support results of forensic analysis that are reliable, reproducible, and minimize human

DOI: 10.4324/9781003391074-3

factors conveyed in the NAS report prompted the creation of the Organization of Scientific Area Committees for Forensic Science (OSAC). In collaboration with the National Institute of Standards and Technology (NIST) and the US Department of Justice (DOJ), OSAC formed in 2014 to "establish standards and best practices within and between [forensic science] disciplines" that address terminology, methods, and training practices (NIST 2024). The standardization that occurs through the work of OSAC varies by discipline as each has its own training and research needs, methods, reporting processes, and potential for bias (Johnson 2018). The OSAC is comprised of seven Scientific Area Committees (SACs) that include Biology, Chemistry: Seized Drugs and Toxicology, Chemistry: Trace Evidence, Digital/Multimedia, Medicine, Physics/Pattern Interpretation, and Scene Examination that provide direction to the discipline-specific subcommittees. The discipline-specific subcommittees identify, draft, and develop standards for review.

While forensic linguistics is not yet a discipline specifically identified by the OSAC, it has connections to both the Digital/Multimedia and Physics/Pattern Interpretation committees. The Speaker Recognition Subcommittee, which is part of the Digital/Multimedia SAC, focuses on standards and guidelines related to the practice of speaker recognition, voice data collection, measurement, transmission, and retrieval. It includes recommendations for the physical preparation of the location, selection of appropriate recording hardware and audio formats, and possible methods for interviewers to elicit the desired type and amount of speech from subjects. It does not cover the methods used to analyze the resulting recordings and does not deal with details related to the handling, transmission, storage, or preservation of the collected data. It is another source for information on expectations for forensic linguistics. Research and development needs for this subcommittee include Effect of Specific Data Variations on Forensic Speaker Recognition Results, Standard Forensic Speaker Recognition Test Data, and Synthetic Speech Detection (NIST 2023). The Friction Ridge Subcommittee, which is part of the Physics/Pattern Interpretation SAC, focuses on standards and guidelines related to examination of friction ridge detail from hands and feet. While this forensic discipline may seem like a significant departure from linguistic evidence, if considered in the context of "patterns" rather than the location of the pattern evidence, the principles are quite similar. The Friction Ridge Subcommittee's focal areas include improving the quality and consistency of examination practices, normalizing differing decisions/conclusions for subjective interpretations, and regulating the process for verification. Questions regarding verification may include how many steps are needed, what happens when there is a conflicting conclusion, when should the process be blind, when is an outside agency part of the review, etc. (OSAC 2016). According to official assessments by these and other oversight organizations, forensic disciplines may encounter unique ethical challenges while also navigating the standard dilemmas faced in the forensic profession.

Forensic scientists represent several scientific areas made up of subdisciplines that comprise the field including Biology, Digital/Multimedia, Chemistry, Pattern Interpretation, Linguistics, and others. In addition to the variety of scientific areas, disciplines, jurisdictions, and types of forensic professions, there is also a wide range of education, training, and experiences that lead to one's proficiency as an expert witness. Most professionals learn skills both formally, through undergraduate and graduate education and formal training, and informally, through mentorship and professional associations (Bowen 2024). Applications of science include theory building, inferences based on existing knowledge to form hypotheses that predict new outcomes, and use of results for human purposes. Forensic science utilizes those skills to provide information for the adjudication of criminal activity, as either a testifying expert or a consulting expert. Forensic scientists are not legal advocates, although they work closely with criminal justice systems. Their role within these systems is to remain impartial and report the findings based on science with clear, informative accounts. Factors that challenge impartiality include individual interpretations, pressures, biases, conflicting professional cultures, differences in standards, and gray areas. Experts must point out the nature of scientific findings, including the potential for error, and acknowledge contradictory evidence when it exists (Bowen 2024).

Forensic Linguistics and Ethics

Although a comparatively new discipline within forensic sciences, forensic linguistics has quickly developed into a thought-provoking area of science with research being conducted in an expanding number of subdisciplines (e.g., pragmatics, author/speaker identification, forensic stylistics, discourse analysis, and forensic interpretation and translation) (Alhumsi 2019; Tkacukova 2019; Correa 2013) as well as forensic speech sciences. When an emergent scientific field as diverse and dynamic as forensic linguistics begins to define itself, it may require clarification as to what is or is not part of the discipline, including ethical behaviors for conduct. Of course, "ethical behavior" is a vague concept because there are no specific conduct rules for forensic science disciplines within the legal system. When people are uninformed about the standards for appropriate behaviors in professional situations, there is an increased chance for unethical behavior to occur. The scientific community presents many opportunities for excellence, which unfortunately contributes to the frequency of conflicts of effort as people try to build their credentials, affiliations, and stay current in the field. In addition, forensic science is a profession where personal perspectives and professional opinions are not easily distinguished, which can further complicate the idea of following broad guidelines (Bowen 2024). In forensic linguistics, there are professional organizations that provide guidelines through codes of practice, codes of conduct, and discipline specific ideals:

> **The International Association for Applied Linguistics** (IAAL) (IAAL) follows the American Association for Applied Linguistics (AAAL) Ethical Guidelines, which includes information regarding research, teaching, and service and are noted as a frame of reference for members.
>
> **The International Association for Forensic and Legal Linguistics** (IAFLL) offers its members a 'Code of Practice' that includes ideals for professional integrity, objectivity, competency, confidentiality, and conflicts of interest.
>
> **The International Association for Forensic Phonetics and Acoustics** (IAFPA) provides a 'Code of Practice' that contains elements of professional competence and integrity, including experience, limitations, and conflicts of interest, and professional practice, including information about analyses, interpretation, and providing professional opinions.

A primary contribution of these and other scholarly societies is to provide helpful guidelines and basic professional principles.

Standardization within the discipline is important for objectivity; however, practitioners should also be prepared to address the subjective elements of analysis within their discipline. As in other forensic disciplines, forensic linguistics is concerned with answering questions in search of the truth. Practitioners who offer their expertise must rely on their knowledge, experience, and skill to provide the best evidence possible to the triers of fact. When courts pose questions, it is important that judges, lawyers, and experts work together in the common search for justice. Although forensic linguists may serve as expert witnesses contracted by either the defense or prosecution, it is imperative that they present the linguistic evidence factually and without prejudice for a particular "side". Courts around the world accept linguistic expert testimony, though the way courts ensure reliability of the evidence varies. In the United States, courts focus on the need for valid and reliable methods, while in the United Kingdom courts experts are required to certify their analyses were conducted in a neutral, impartial manner. Expert testimony requires that practitioners present information to the jury in a way that is distinct from legal advocates, though that may not appear clearly to jurors due to the brutality of the adversary process. The adversary system is cynical regarding scientific uncertainty, can be aggressive on behalf of clients, and at times, may view the expert witness as an extension of the legal advocate who has requested their service (Solan 2020). The court affords forensic linguistics experts the opportunity to provide detailed information in a non-narrative context. Linguistics experts should explain, rather than articulate, accounts and provide succinct responses to open-ended questions to educate the triers of fact. To accomplish this goal, experts receive the highest proportion of extended turns speaking. With this professional privilege comes ethical responsibility.

Admissibility of expert testimony varies but is commonly based on the ability and duty of experts' knowledge to assist the fact finder, the relevance of the information provided, the proper qualifications of the expert, and the commitment to assist the fact finders, which overrides any obligation to people who instruct and/or pay them (Fadden and Solan 2015). When forensic linguists serve as experts, it is imperative that they use the opportunity to apply precision and clarity to language evidence that may have been vague without their expertise. The forensic linguist's role is to provide objective and reliable information to the decision-makers. They must be aware of the standard procedures, norms, and regulations of legal proceedings, including admissibility and grounds for appeals. Necessary topics include techniques for expert testimony, ethics of litigation, and responsibilities of experts with a specific focus on resistance to advocacy and appropriate presentation of information (O'Barr 2014). Regarding this last skill, experts must also be proficient in accurately communicating findings in the most effective way possible, without unintentionally amending results or emphasizing certainty (Fadden and Solan 2015). Standardization of the methodologies of some subfields in forensic linguistics may allow an agreement to form about the function to the legal community, which would create a "best practices" approach for experts to maintain consistency in the courtroom (Solan 2020).

Linguists serve as expert witnesses to assist the jury in understanding evidence. One complex aspect of expert testimony is achieving a balance between authority and comprehensibility in that experts need to teach the jury without oversimplifying, which could result in alienation and possible lack of respect, while they remain within the strict boundaries of the specific questions asked of them (Coulthard, May, and Rui Sousa-Silva 2020; Solan 2020; Fadden and Solan 2015). Experts may also be more likely to give desired answers in response to leading questions, though impartiality is paramount. When utilizing discursive forms of evidence, it is important to avoid overstatement of results as such evidence is unregulated, unpredictable, and inadequately supported, making it potentially less credible than other forms of forensic evidence (O'Barr 2014). Experts must remain unbiased, though they also must present themselves as more credible than other experts by style rather than substance, which is an inherent pressure for experts (Fadden and Solan 2015). Learning how to give brief yet informative answers is not merely a skill, it is an ethical imperative to help guard against legal advocates manipulating the information provided and skewing the later decision-making process. The forensic linguist, like all forensic scientists, has a duty to describe the evidence *as it is*, whereas the lawyer has a duty to describe the evidence in the *most favorable light* for their client (Lucas 2007). Experts are constrained by the questions asked, which dictate the quantity and sometime quality of information they may provide when testifying.

One of the early landmark cases was United States v. Granderson (1994), discussed by Kaplan et al., who published an academic opinion regarding cases where the interpretation of phrases should have been inadmissible based on standards within linguistics. The phrases were ambiguous but presented as simultaneously having two different meanings regarding sentencing and imprisonment. Mr. Granderson pleaded guilty to a charge of destroying mail, which carried a maximum sentence of six months imprisonment. The judge instead sentenced him to five years of probation. Later, he violated his probation when caught in possession of cocaine, where the law instructed the Court to "revoke the sentence of probation and sentence the defendant to not less than one third the original sentence". The Court believed "original sentence" referred to the *probation* and imposing "not less than one third" would reduce the penalty as it had more than 20 months left. The decision was that the correct interpretation required the court to sentence him to *20 months in jail* even though that sentence was more than three times greater than the original maximum (Kaplan 1995). The Supreme Court Justices acknowledged the opinion and noted that based on linguistics, the interpretation changed the final meaning. Ultimately, though, it is the purview of the court to determine the meaning of the information provided. A forensic linguist serving as an expert witness may only provide specialist information to the court and assist in the legal decision-making process. The opportunity to serve in this capacity is greatly reliant on the forensic linguist's ability to provide neutral, fact-based evidence and not assume the role of adjudicator. Courts may disallow forensic linguists' testimony where judges feel the expert analysis may impede their decision-making power. This may be especially true when judges have the perception that the linguist is attempting to do the job of the judge and/or jury by providing *meaning* or opinions on ultimate issues, which suggest a verdict (Fadden and Sloan 2015; Kaplan et al. 1995).

Expert witness opinions are not included in the legal system without concern. Some issues faced by language experts include the trouble expressing opinions as semantically encoded opinions rather than in terms of probability (Coulthard, May, and Rui Sousa-Silva 2020). Another issue is that modern judges, juries, and lawyers may distrust experts' objectivity because historically experts were part of the criminal justice system due to their connection to prosecutors (Bowen 2024). The next issue concerns general skepticism about experts' usefulness, the limits imposed on their efforts, and their ethical principles. Limitations of linguistic testimony must be considered; such evidence is not enough to convict or exonerate on its own, it is not infallible evidence as it is subject to human interpretation, and experimentation or analysis may be difficult to demonstrate in the courtroom, so assumptions may be necessary (Correa 2013). In regard to value, jurors may grant expert testimony greater weight than other evidence. Finally, there are widely varying qualifications for courts to deem experts as such. When both sides present experts, jurors cannot differentiate the levels of expertise or objectivity and will not know if they properly weighed the experts' contributions. Worse still, when experts encounter opposing experts who provide unscientific testimony, they have an ethical responsibility to counter it, despite possible confusion to the jury, potential harm to the case, or the appearance of advocacy for a side rather than the truth (Bowen 2024; Fadden and Solan 2015). Consulting forensic linguistic experts must acquire the skill to deal with these "unusual interactional rules" both professionally and ethically (Coulthard, May, and Rui Sousa-Silva 2020). For example, the adversary process does not allow for the exchange of information between experts when their analyses of the facts conflict. Such discussion is a normal and important part of science as it ultimately helps to overcome uncertainty; however, it does not align with proper court procedure. In the courtroom, a lawyer's questioning will highlight the weaknesses of the opposing experts' opinions, while simultaneously presenting the strongest case on their clients' behalf (Bowen 2024; Fadden and Solan 2015) while their expert opinion seemingly contradicts that of the opposing expert. Forensic linguistic experts may be unaware of such discursive antagonism that takes place, so to avoid issues the forensic linguist expert must master the ability to testify accurately and with great discretion.

The fact that forensic linguists are testifying to their expertise in the courtroom is not only a reflection of lawyers becoming more aware of the potential uses for such analyses but also that the court systems widely accept linguistic analyses as scientifically reliable (Coulthard, May, and Rui Sousa-Silva 2020; Solan 2020). It is a positive indication that practitioners in this field have largely demonstrated the value of their expertise as forensic scientists, while skillfully negotiating the legal and ethical challenges of the criminal justice system. The contributions of forensic linguists may begin at the scene of a crime and go through many investigative stages until a verdict is reached in the courtroom. Throughout the progression of evidence gathering, processing, and analysis, there are many opportunities for ethical issues to occur. A common problem for practitioners in forensic disciplines is the expectation that they will function according to the mandates of their consumers, which is often law enforcement or prosecutors (Bowen 2024). While one would hope that forensic scientists, officers, investigators, lawyers, judges, and jurors are intrinsically ethical by nature, the reality is that these individuals have varied morality, differing goals and professional standards, and contrasting interests in the evidence of a case. Institutional discourse, practices, and expectations may generate unbalanced relationships that can lead to inequality and injustice between professionals and their respective work cultures. Maintaining the necessary professional balance requires understanding that differences in purpose exist. It is imperative that the forensic linguist become cognizant of the unique, sometimes conflicting, aspects of professional cultures that work closely together in forensic settings. Professional awareness of the influence created by intersecting professions is an important component of professional ethics. After examining the nuances of professional relationships and challenges for language scientists, one can now explore factors that increase the chance for professional misconduct.

Ethical Misconduct

All decisions forensic scientists make have broader implications on people's lives, so professional ethical considerations are essential. When exploring the pressures that lead to unethical behavior and the types of misconduct that occur, it is important to note that many examples of misconduct do not occur because someone is inherently unethical, more so because they did not choose the best course of action for the situation (Bowen 2024). Many factors can increase the likelihood of unethical behavior in forensic science disciplines. The primary ethical concerns in forensic linguistics are like other disciplines and include variations in training and examiner competency, subjective evidence that is reliant on the quality of the sample and examiner interpretation, overstated results, and examiners who do not properly acknowledge error rates or limitations of the science. The general unfamiliarity of lawyers and judges to science is another ethical challenge, which may create difficulties during the presentation of evidence to jurors. It is important that practitioners are aware of factors that contribute to misconduct to prevent them before they arise, address problems with enhanced preparation, and impede misconduct before it significantly affects the professional integrity of the forensic science discipline and profession.

Types of misconduct include ignoring, misrepresenting, or misusing scientific evidence. Misuse of expert testimony may also occur, for example, when lawyers seek biased experts, suppress test results, or overstate expert findings (Bowen 2024). Forensic science experts face many direct and indirect professional pressures. In examinations of misconduct amongst forensic scientists, cases occurred when experts crossed the boundaries of scientific neutrality and exhibited advocacy. Signs of advocacy may include exclusively working with one side in the adversary system; making statements of absolute certainty; not describing evaluation procedures, data, or results; and making unwarranted personal attacks on opposing witnesses. The negative repercussions of ethical transgressions extend far beyond the individual forensic expert. Cases of unethical behavior create negative publicity and affect the credibility and reliability of the entire profession (Bowen 2024). In the aftermath of such misconduct, judges and juries, for example, have become generally suspicious of experts' intentions. Although the association may not be intentional, individual acts of misconduct can create the impression that the field itself is susceptible to bias.

Evidence of bias in comparison-based disciplines commonly manifests itself in one of four ways: (1) conclusions guided by individual expectations rather than factual evidence; (2) new information is dismissed as not being strong enough to change original conclusions; (3) insignificant details are found to affect judgments by triggering examiners' emotions, stereotypes, or feelings of accountability; or (4) evidence is presented in a manner determined more by personal conviction than rational logic (PCAST 2016; NAS 2009). To counteract the influence of bias, it is recommended that comparison disciplines within forensic science examine evidence in a random order of potential matches to guard against examiners distinguishing between known and questioned evidence through sequence alone (Koen and Bowers 2023; Kassin, Dror, and Kukucka 2013). For the same reason, it is recommended to utilize blind comparisons to offset the effects of cognitive bias. In addition, when investigating evidence where patterns are reliably perceptible, forensic examiners must keep in mind the possibility of erroneous or coincidental matches. While not all these precautions may be possible to implement regularly in forensic linguistic analyses, the integration of at least some of the methodological strategies can be of value (Bowen 2024). Objective analysis is the basis for maximally reliable forensic linguistic evidence.

Historically, forensic science avoided discussions of bias based on the notion that practitioners were inherently ethical and that basic individual checks were enough to assure unbiased work. Experience and science have shown otherwise. The truth is bias may unintentionally or intentionally affect the performance or judgment of forensic scientists at every stage of their work. For example, without proper quality controls, the forensic scientist may selectively attend to pieces of evidence that tend to confirm their initial hypotheses and reject those that do not (i.e., confirmation bias). They may also make decisions that are based on their (un)conscious beliefs and untested assumptions—filtering information through one's own experiences (Casad 2023). Such *cognitive bias* may "affect the reliability and validity of one's observations and conclusions" (OSAC Lexicon 2019). People tend to be doubtful of information that is inconsistent with their beliefs, so the impact of bias is not limited to forensic experts (Bowen 2024). However, given their role in high-stakes decision-making, bias in an examiner's perceptions and conclusions can have particularly dramatic repercussions. Acknowledging the potential influence of cognitive and confirmation biases is therefore paramount to limiting their effects. Although bias can affect all forensic scientific endeavors, it may be especially problematic in disciplines where examiners heavily rely on comparisons for their analyses.

Case Example

The focus on ethics in forensic science is essential, particularly as concerns arise about the analysis and testimony that led to past convictions. In 2012, discussions began about the Justice Department's awareness of flawed forensic hair matches that resulted in convictions of potentially innocent suspects, dating back to the 1970s; however, they had not thoroughly reviewed those cases (Hodge 2020). A review of the Federal Bureau of Investigation (FBI) cases that included morphological examinations of hair evidence began in response to claims of inappropriate use of hair evidence, a lack of reliability and validity of methods used, and a high chance of false positive results (Wilkinson 2020). The FBI conducted the review in collaboration with the US

Department of Justice, the Innocence Project, and the National Association of Criminal Defense Lawyers (Bowen 2024; Bonventre 2021; FBI 2015).

A report issued in 2015 regarding the FBI's microscopic hair analysis testimony showed that between 1985 and 1999, nearly 3,000 cases from 41 states utilized microscopic hair analysis. Independent reviews to evaluate protocols, practices, and procedures revealed that 90% of microscopic hair analysis cases included errors based on the language used to characterize the strength of the evidence rather than sourcing errors (Morgan 2023). Examples included erroneous statements, definitive associations between questioned and known samples, inadequate examiner training and educational requirements, lack of standardized procedures and high-quality research in both theory and validity of methods, and poor proficiency testing (Bowen 2024; Wilkinson 2020).

Microscopic examinations allow examiners to evaluate the morphological characteristics of hair and obtain basic information about the sample to begin the process of comparison (Hodge 2020). Although people acknowledged flaws in microscopic hair examinations, it was sometimes used over the more reliable DNA testing methods because at the time, DNA was more expensive, time consuming, and potentially not accessible for examiners (FBI 2015). Today, we know that hair analysis is a valid technique, but there are no uniform standards, which allow examiners to state that hair "matches", so they cannot assert such a conclusion with any degree of scientific certainty (Bowen 2024; Hodge 2020). Forensic disciplines have reliable procedures so now it is imperative that examiners understand and can properly communicate the strength of the evidence, analysis, and results, as well as the possible error rates, limitations, and standards used.

How might this case of misconduct in hair analysis benefit the forensic linguist? In addition to the basic ethical concerns in forensic science, there are many elements to consider in cases like this, no matter the discipline:

- Examiners: concerns include questions regarding training and competency for evidence and expert testimony, cognitive and confirmation bias, use of reliable and valid methods, prosecutor advocacy, understanding the limits of evidence, and exposure to professional pressures within their agency or from prosecutors. Was this a case of correct analysis but the results were oversold to jurors?
- Justice Department: concerns include abuse of authority, lack of oversight, lack of investigation of potential problems, discouraging proper methods and/or reporting, lack of support for examiners, and potential bias/advocacy. Who requires a higher level of responsibility for such outcomes, the examiners or managers?
- Prosecutors: concerns include pressuring experts, purposefully selecting biased experts, utilizing hair evidence to guarantee a conviction over stronger types of evidence, and knowledge that experts were overstating weight of conclusions. To what degree should they have a responsibility for the actions of the experts?

As forensic science creates standards and best practices that address terminology, methods, and training practices, it is emerging as a stronger profession. Addressing complex issues such as bias and error rates will allow experts to present the most valid and reliable evidence in the courtroom.

Conclusion

The field of forensic linguistics has developed considerably over the last 30 years and continues to diversify. While some practitioners see their role as chiefly descriptive, others see their purpose in applying their expertise to help remedy societal issues. Whatever stance one takes, moving forward, the entire forensic linguistics community must invest in continuing to advance professional ethical standards. Growing the field will require increasing formal education opportunities for university degrees, encouraging the development of professional scholarly organizations to support continuing education, and developing formal codes of conduct to prevent or address common ethical issues unique to the discipline. These developments are paramount as interest in, and the need for, the field will doubtlessly continue to increase given the rapid advancements in digital communication (Coulthard, May, and Rui Sousa-Silva 2020). To meet these challenges, forensic linguists may consider adapting standards from related forensic science disciplines dealing with pattern and/or speaker recognition. Collaborations with other disciplines can not only help to advance the various branches of forensic science; by working together, they can also help to keep the ethical bar elevated throughout the criminal justice process.

References

Alhumsi, Mohammad. 2019. "Key Aspects in Relation to Forensic Linguistics." *International Journal of Linguistics, Literature and Translation* 2, no. 5: 83–86.

The American Association for Applied Linguistics (AAAL). 2023. https://www.aaal.org/ethics-guidelines.

Bonventre, Catherine L. 2021. "Wrongful Convictions and Forensic Science." *Wiley Interdisciplinary Reviews: Forensic Science* 3, no. 4: e1406.

Bowen, R. T. 2024. *Ethics and the Practice of Forensic Science*. Boca Raton: CRC Press.

Casad, B. J., and J. E. Luebering. 2023. s. v. "Confirmation Bias." *Encyclopedia Britannica*, March 31, 2023. https://www.britannica. com/science/confirmation-bias.

Correa, Maite. 2013. "Forensic linguistics: An Overview of the Intersection and Interaction of Language and Law." *Kalbų Studijos* 23: 5–13.

Coulthard, Malcolm, Allison May, and Rui Sousa-Silva, eds. 2020. *The Routledge Handbook of Forensic Linguistics*. London: Routledge.

Executive Office of the President's Council of Advisors on Science and Technology (PCAST). 2016, September. *Report to the President: Forensic Science in Criminal Courts: Ensuring Scientific Validity of Feature-comparison Methods*. PCAST.

Fadden, Lorna, and Lawrence M. Solan. 2015 "Expert Witness Communication." In *Communication in Investigative and Legal Contexts: Integrated Approaches from Forensic Psychology, Linguistics and Law Enforcement*, 209–228. Edited by Gavin Oxburgh, Trond Myklebust, Tim Grant, and Rebecca Milne. John Wiley & Sons.

FBI. 2015 "Testimony on Microscopic Hair Analysis Contained Errors in at Least 90 Percent of Cases in Ongoing Review."

Hodge, Samuel D. Jr., and Amelia Holjencin. 2020. "A Post-Mortem Review of Forensic Hair Analysis–A Technique Whose Current Use in Criminal Investigations Is Hanging on by a Hair." *Saint Louis University Law Journal* 64, no. 2: 219–240.

The International Association for Forensic and Legal Linguistics (IAFLL). 2023. https://www.iafl.org/.

Johnson, Steve. 2018. "Where Are We with OSAC: An IAI Perspective." [PowerPoint Slides]. Presented at The International Association for Identification 2018 Educational Conference, San Antonio, Texas.

Kaplan, Jeffrey P., Georgia M. Green, Clark D. Cunningham, and Judith N. Levi. 1995 "Bringing Linguistics into Judicial Decision Making: Semantic Analysis Submitted to the US Supreme Court." *International Journal of Speech Language and the Law* 2, no. 1: 81–98.

Kassin, S. M., I. E. Dror, and J. Kukucka. 2013. "The Forensic Confirmation Bias: Problems, Perspectives, and Proposed Solutions." *Journal of Applied Research in Memory and Cognitio* 2, no. 1: 42–52.

Koen, Wendy J., and C. Michael Bowers, eds. 2023. *Forensic Science Reform: Protecting the Innocent*. Boston: Academic Press.

Lucas, Doug M. 2007, July 12. "Forensic Science and Ethics—An Essential Association." Presented at NERFI, MSP DNA Course.

Morgan, John. 2023. "Wrongful Convictions and Claims of False or Misleading Forensic Evidence." *Journal of Forensic Sciences* 68, no. 3: 908–961.

National Academy of Sciences (NAS), National Research Council. 2009. *Strengthening Forensic Science in the United States: A Path Forward.* National Academies Press.

National Institute of Science and Technology (NIST). 2024, April 13. "About Us." https://www.nist.gov/organization-scientific-area-committees-forensic-science/about-us.

National Institute of Science and Technology (NIST). 2023, June 13. "OSAC Subcommittees." https://www.nist.gov/organization-scientific-area-committees-forensic-science/digitalmultimedia-scientific-area-committee.

O'Barr, William M. 2014. *Linguistic Evidence: Language, Power, and Strategy in the Courtroom*. New York: Elsevier.

OSAC Friction Ridge Subcommittee. 2016, December 14.

OSAC Lexicon. 2019. s. v. "Cognitive bias." Retrieved May 13, 2023. https://lexicon.forensicosac.org.

Solan, Lawrence M. 2020. "The Forensic Linguist: The Expert Linguist Meets the Adversarial System." In *The Routledge Handbook of Forensic Linguistics*, 349–363. Edited by Malcolm Coulthard Alison May and Rui Sousa- Silva. London: Routledge.

Song, Ci. 1981. *The Washing Away of Wrongs: Forensic Medicine in Thirteenth-century China*. Ann Arbor: University of Michigan Press.

Tkacukova, Tatiana. 2019. "Forensic Linguistics and Language and the Law." In *An Introduction to Applied Linguistics*, 190–204. Edited by David Coulson, Norbert Schmitt, and Jonathon Clenton. London: Routledge.

Venter, Casper Henderik. 2020. "A Critical Review of the Current State of Forensic Science Knowledge and Its Integration in Legal Systems." PhD Dissertation. University of South Africa.

Wilkinson, Laura, and Claire Gwinnett. 2020. "An International Survey into the Analysis and Interpretation of Microscopic Hair Evidence by Forensic Hair Examiners." *Forensic Science International* 308: 110158.

3

ETHICS AND PROFESSIONAL CONDUCT OVERSIGHT IN FORENSIC LINGUISTIC EXPERT EVIDENCE

Isabel Picornell

Introduction: A Brief History of Forensic Linguistics and the Appearance of "Experts"

The origins of forensic linguistics lay in the need to apply linguistic knowledge and expertise to analyze linguistic aspects in investigative casework. The earliest identifiable case of "expert" critical textual analysis dates to 1440 with the work of the Italian philologist Lorenzo Valla. Valla produced the most famous exposé of a literary fraud, refuting the 4th-century CE authorship of *The Donations of Constantine*, arguing that the linguistic evidence showed the text to be an 8th-century forgery (Valla 1440). Similarly, in the 17th-century, the French classical scholar David Blondel identified forgeries in a 9th-century CE collection of papal texts, the *Decretals of Pseudo-Isidore*. By tracking down the sources on which the forgeries were based, Blondel (1628) demonstrated that letters ascribed to the early 4th-century popes contained phrases and legal terminology which did not come into use until the 6th-century and later.

Modern forensic linguistic casework is often said to have originated in the late 1960s, although earlier judicial reports refer to provision of linguistic analysis and evidence. One such example comes from the United States (*Detroit Co v. Gagliardi* 1934) in which the court allowed into evidence the comparison of certain spellings and grammatical constructions to corroborate conclusions on handwriting (Shim 2022). In another early case, in Australia (*Stuart v. R* 1959), the court permitted evidence on language variation to be used to argue that the confession of an Indigenous Australian had been fabricated (Eades, Fraser and Heydon 2023).

The term "forensic linguistics" was first introduced a decade later, in Jan Svartvik's seminal paper, "The Evans Statements: A Case for Forensic Linguistics" (Svartvik 1968). The article was based on the Swedish linguist's 1966 expert report to the Royal Court of Justice Public Inquiry into the 1950 conviction of Timothy Evans for the murder of his daughter, Geraldine. In his report, Svartvik pointed to stylistic inconsistencies between Evans' various statements and argued that material sections relating to Evans' role in the murder of his daughter had been more likely authored by the interviewing police officer. The Inquiry Judge subsequently concluded that "it is more probable than not that Evans did not kill Geraldine" (Brabin 1966: 6). Unfortunately,

DOI: 10.4324/9781003391074-4

that judgment and the subsequent pardon came too late for Evans, who had already been hanged in 1950 for Geraldine's murder (Carradice 2012).

In his Preface to "The Evans Statements", Svartvik stressed the need for more research in forensic linguistic analysis, commenting that his analysis highlighted the inadequacy of knowledge on how language was used in various contexts. He also linked the term "forensic linguistics" with benefits for social justice, writing:

> Firstly, [this case] has provided the linguist with one of those rare opportunities of making a contribution that might be directly useful to society. Such a statement does not imply that utility should be the goal of linguistic research, only that, once in a while, it is good for the linguist to know that he can be useful, and that applied linguistics need not be identical with language teaching or machine translation.
>
> *(Svartvik 1968: 4)*

Over the next 25 years, early pioneers in forensic linguistic analysis from Germany, the United States, the United Kingdom, and Australia (e.g., Hannes Kniffka, Roger Shuy, Malcolm Coulthard, Peter French, and Diana Eades) expanded the discipline further, applying a broader range of linguistic knowledge to legal casework. This expansion fostered the development of professional associations: in 1991, the International Association for Forensic Phonetics (IAFP) was founded, followed in 1992 by the International Association of Forensic Linguists (IAFL), formed to encourage and facilitate the sharing of research.[1] Since then, forensic linguistics has developed into an expertise, with universities offering modules as well as graduate and postgraduate degrees in forensic linguistics and forensic speech sciences; and national organizations such as the Germanic Society for Forensic Linguistics (GSFL), the Komunitas Linguistik Forensik Indonesia (KLFI), and the Philippine Association for Forensic and Legal Linguistics (PAFLL) emerging to promote language-specific research.

Linguists, translators, lawyers, and even judges have become active in the discipline of forensic linguistics and have transformed practice through their efforts such as publishing critiques in legal journals on judicial language and deficient translations; working with law enforcement to improve interviewing skills; promoting changes in communication to increase laypersons' understanding of the specialized language used in criminal justice systems; and providing expert linguistic evidence in court (e.g., Berk-Seligson 2017; Coulthard, Johnson, and Wright 2017; Heydon 2005; Solan 2005; Tiersma and Solan 2002; Solan 1993). The increasing use of forensic linguistic analysis as expert evidence is a particularly important indication of the general acceptance of the specialist knowledge and training required to practice in this field and its increasing acceptance as a forensic science. These achievements demonstrate just how far the field has come and continues to develop.

This chapter discusses the role of forensic linguistic experts and expert witnesses and explores some of the pressing ethical issues which accompany this expanding role and the responsibility forensic linguists carry in their service to the criminal justice system. Although emphasis is placed on procedures in Anglosphere jurisdictions (England and Wales, United States, Canada, and Australia), the discussion provided here is germane internationally.

In this chapter, the term "expert" is used to refer to an individual with "expertise": a high degree of skill or knowledge, by training or experience, in a certain subject. "Expert witnesses" assist the court by providing evidence and opinion on matters within their expertise. In this respect, their role differs from that of "lay witnesses" or "witnesses of fact". While justice systems vary in the amount of guidance they offer on the role and obligations of expert witnesses, it is argued here that the ethical obligations of experts—whether they are in the role of consultants for clients or witnesses for the courts—are the same.

Ethical Frameworks for Forensic Linguistic Experts

Guidelines and Regulations from the Law

To combat partisan and unethical expert evidence, the nations of England, Wales, and to a lesser extent, Scotland, have set out in considerable detail the obligations of expert witnesses in the rules of civil and criminal procedure. The Criminal Procedure Rules 2020 (CrimPR 19.2) state, for example, that experts must give "opinion which is (i) objective and unbiased (ii) within the expert's area or areas of expertise". The Procedure Rules and subsequent case law reinforce the need for experts to not only avoid but also be seen to avoid the appearance of conflicts of interest, partisanship, or bias. Further, both criminal (CrimPR 19.2) and civil rules (CPR 35.3(2)) of procedure highlight that the expert's duty to the court overrides any obligation to the person from whom the expert receives instructions or by whom they are paid. To reinforce this point, in 2011 the Supreme Court of England and Wales overturned 400 years of practice and abolished immunity against prosecution for experts who provide negligent expert evidence in litigation (*Jones v. Kaney* 2011).

In contrast, rules in the United States are often perceived to be less prescriptive. Rule 702 of the Federal Rules of Evidence defines an expert as "A witness who is qualified as an expert by knowledge, skill, experience, training, or education". However, the rules do not formally define where the expert's duty lies. Neither do they include a specific requirement for experts to be independent. This is a significant omission as certain practices acceptable within the lawyer-expert relationship in the United States, such as permitting lawyers to have direct input into expert reports, may make it more difficult for experts to maintain their independence.

In Canada, case law is more focused on the scrutiny of expert evidence than the duties of the expert. However, the evolution of case law has created the requirement that experts be "properly qualified" (*R v. Mohan* 1994), by experience as well as formal training (*R. v. N.O.* 2009), and that their evidence be limited to areas within their expertise (*Consulate Ventures Inc. v. Amico Contracting & Engineering* 2010). Canadian case law has yet to clearly address the issue of independence and impartiality to prevent experts acting as advocates for the parties that retain them; although a 2015 ruling of the Nova Scotia Court of Appeal held that "whether the expert's opinion would not change regardless of which party retained him or her" was a determining factor in the admissibility of expert evidence (*White Burgess Langille Inman v. Abbott and Haliburton Co* 2015: para 32).

Guidelines and Regulations from the Discipline

Currently, the forensic linguistic profession does not require any special certification for individuals to practice as a forensic linguist. Beyond the degree level attained, there is no requirement for continuing professional education to maintain a level of expertise or to promote their services as an "expert", unless the individual is employed in an academic position and the university or college requires it.

Although professional associations dedicated to promoting forensic linguistic research and establishing scientific quality standards of analysis do exist, institutional oversight laying out experts' obligations to their profession and to their role as forensic linguistic "experts" has yet to be codified. Forensic linguists are not alone in this regulatory vacuum; however, there are increasing institutional efforts to define "experts" and oversee how they conduct themselves in the judicial process through professional standards of conduct. One such example is by the Academy of Royal Medical Colleges, a voluntary professional body, which has issued guidance to healthcare clinical professionals who provide expert opinion or act as expert witnesses in court and tribunals. The Academy states that while, as a voluntary body, they are not in a position to enforce compliance with the guidance, they make clear that any member "who fails to meet the standards set out in the guidance is not considered to be meeting the expectations or demonstrating the values of their profession" (AMRC 2019: 3). Such codes of ethics and conduct lay out the moral principles and procedures that professional associations expect their members to adhere to as part of their professional culture. These codes are becoming increasingly part of the fabric of professions and, in some cases, permit the profession's association to take enforcement action against members who break the codes when behaving unethically in judicial contexts.

Little information is available on how often this happens, but judgments from court cases where experts have sued their organizations following disciplinary actions against them for unethical conduct suggest that sanctions by professional associations are relatively rare. However, where they have occurred, courts have affirmed the organizations' authority to punish its members for violation of their professional codes. In the United States, the 7th Circuit Court of Appeals held that an association could discipline its members for improper expert witness evidence (*Austin v. American Association of Neurological Surgeons* 2001). In 2019, the UK High Court issued a contempt ruling against a doctor for being dishonest and reckless in his expert evidence (*Liverpool Victoria Insurance v. Zafar* 2019). The Medical Practitioner Tribunal (MPT) against the doctor that followed found that the High Court's ruling demonstrated that he had violated the Guidance on expert witnesses issued by the General Medical Council (GMC 2013) and suspended his registration for one year. The suspension was referred to the Divisional Court for review, who upheld the MPT's decision (*GMC v. Zafar* 2020). Similarly, in 2023, the GMC ordered the erasure of a psychologist from the medical register for misleading the Crown Court by not having the required expertise to undertake the assessment, for copying sections from another expert's report into her own, and for lying in her evidence (MPTS 2023).

Why should professional associations formally codify the ethical behavior of its members who act as expert witnesses? After all, forensic science has a history of preference for self-regulation, and governments appear to have been reluctant to change this. In the United States, one of the final acts of the Department of Justice under the Trump administration was to reject a call for regulation by the President's Council of Advisors on Science and Technology which criticized the flawed use of unvalidated forensic science to secure convictions in court (PCAST 2016). In the United Kingdom, cost-cutting has resulted in police forces under pressure to find ever cheaper forensic science services producing faster results from a dwindling number of forensic labs, resulting in the flaws and errors that a lack of training, skills, and capacity produces (Samuels 2022). Although the Forensic Science Regulator (FSR) has sought to maintain and improve standards, the process remains voluntary; the provider's non-compliance with standards can be taken into consideration by courts in the amount of weight that may be given to the evidence, but it does not give rise to criminal or civil liability (CPS 2023). Furthermore, guidance issued by the FSR for expert witnesses applies to how results are reported and not to the ethical behavior of the experts themselves (FSR 2020).

Expert witness guidance already exists to varying degrees. Some jurisdictions, such as the United Kingdom (CrimPR19 and CPR35), Australia (Expert Evidence Practice Notes), and Canada (Federal Court Rules 52.2), already provide guidance that expert witnesses agree to be bound by, although this is not always mandatory. Still, profession-specific codes of ethics can make a positive difference.

Self-regulation requires a greater level of awareness, control, and resistance to pressure that not all individuals may possess, and one should not underestimate the pressures on experts to compromise their integrity and independence. A 2019 UK survey of expert witnesses found that "44% of respondents said they had come across experts who professed expertise in an area in which they were either not qualified or did not warrant expertise"; 25% reported "they had been asked or felt pressurized to change their report in a way that damages their impartiality by an instructing party"; 41% "indicated that they had come across an expert they considered to be a 'hired gun'" (Solon 2019).

Forensic linguists who intend to enter the expert witness arena would first become aware of the pressures they may be faced with through their profession's code of ethics. Those who find themselves being pressured to take their instructing party's position, who are asked to amend their expert reports, who find themselves being coached as to what to say, would be able to fall back on their profession's codes for guidance. This guidance would ideally state that compromising their independence and objectivity in this manner is unethical and, in some jurisdictions, such as in the United Kingdom, would also expose them to criminal prosecution. The guidelines could also provide concrete recommendations for experts who find themselves being pressured by clients to behave unethically, such as by suggesting they consider removing themselves from the case. The same code of ethics would prevent the expert from changing sides by pointing out that this would create a conflict of interest with the potential to compromise the expert's objectivity.

The code would also instill a sense of responsibility and pride among practitioners about what it means to be a forensic linguistic professional. Conversely, the lack of a code potentially sends a message that the association is prepared to accept that their members may offer their services as consultants and experts without adequate training and expertise (Koehn 2006). Having a code of ethics governing the conduct of members could also assist lawyers and judges in assessing the professionalism and expertise of a forensic linguistic "expert", as signing up to a set of standards of behavior and competence could give credibility to expert witnesses as it demonstrated their commitment to maintaining their continuing professional development (GMR Consulting 2021).

Ultimately, a code of ethics and professional conduct written for forensic linguists is important because it would give the field moral legitimacy. A code of ethics is a public statement of accountability, of a willingness by the profession to be held to approved standards of ethical and professional behavior. In addition, by setting out what constitutes acceptable qualifications, behavior, and analysis, a professional code would help to define the intellectual rigor of the field (Schiffman 1999), reassure clients and the court that they can expect certain standards of professional and ethical behavior from member-practitioners, and give them recourse if these standards are violated. However, until such a time that a set of codified standards for the entire discipline is enacted, there are general ethical norms that are to be expected of professionals in this field. In the following segments, attention is given to five.

Specific Ethical Principles Required for Forensic Linguists Serving as Experts

The Ethics of Competence: Expertise

The ethical considerations of being a forensic linguistic expert begin at the time they receive the initial inquiry. The expert must then decide whether it is ethical to take on a case as a consultant in an investigation or as an expert witness for court. Most casework does not actually end up in court, but the potential always exists for an expert's investigative report to evolve into an expert witness report. Consequently, experts should be mindful of the ethical judgments they will be making, consciously and subconsciously, from that first approach onwards.

One of the primary responsibilities of experts when confronted with the request to take on a case is determining whether the scope of their expertise qualifies them as an "expert witness" for the case in question. Forensic linguistics covers a wide area of analytical domains, from forensic phonetics to discourse analysis, authorship analysis to threat assessment, and beyond. No forensic linguist can be an expert in the whole field. Consequently, deciding whether they have the expertise in the particular field of inquiry required for the case is one of the first ethical challenges facing experts asked to share their expertise.

It holds to reason that experts who accept instructions should have the technical competency and knowledge to perform the analyses required. In many jurisdictions, persons with specialist knowledge are given considerable latitude to determine whether they have the prerequisite level or area of competence. Courts have generally resisted requiring formal qualifications to claim "expert" status, creating difficulties for those in need of one. In England and Wales, the 2020 Criminal Procedure Rules require experts to describe in their reports the qualifications that make them an expert on the topic being considered so that the court may "decide whether the expert's opinion is sufficiently reliable" (Part 19.4(h)). The "sufficiently reliable" opinion is particularly important in areas of expert evidence such as forensic linguistics because judgments regularly have to be made on inferences and interpretation of the data.

No individual should hold themselves out to be an expert in matters in which they have limited knowledge or experience, and certainly not if their peers with similar knowledge and experience do not (Schiffman 1999). Furthermore, an expert can only be "expert" in a narrow field, for example, experience in one or more related subdomains of linguistics. For the forensic linguist to take instruction without having the full expertise to do so is unethical. It is reckless and irresponsible when they hold themselves out as having the knowledge on which others rely to administer justice. It is also a betrayal of the loyalty forensic linguists owe to the science of linguistics (Ainsworth 2022).

Just as important is being aware of the limits of one's expertise. For example, a forensic linguist testifying on an authorship analysis case is not an expert on how cognitive decline may affect a person's authorship style.[2] Guidance in England and Wales from the 2020 Criminal Procedure Rules requires experts to draw attention to questions which would require them to go outside their area of expertise when giving evidence in court (Part 19.2(3)(b)). An "I don't know. That is outside my area of expertise" is an acceptable and ethically responsible response.

Finally, it is imperative that one remains wary of "conflicts of commitments" that may arise. Current commitments or future demands can severely limit an expert's capacity to take on additional projects. Academics have multifaceted professional duties—from lecturing, researching, and publishing, to supervising and serving on various committees and boards. Professionals who fail to realize their limits may be tempted to take on casework for which they do not actually have the time. As a result, the extra pressure of consultancy may lead to favoring efficiency over thoroughness, which can lead to reduced critical analysis and compromise the quality of their expert services (Ask, Granhag, and Rebelius 2011).

The Ethics of Communication: Comprehensibility

To maintain loyalty to linguistics and to justice, forensic linguistic experts have the ethical obligation to make themselves—their analysis and interpretation of the results— understandable to their instructing party and to the triers of fact. As opposed to a specialist audience of their peers, experts in court are communicating with a generalist audience who have little, if any, training in linguistics. This creates the potential for miscarriages of justice, either arising from deliberate ambiguity to disguise invalid or partisan forensic analysis or from failure of judges and juries to understand the evidence (Garrett and Neufeld 2009).

Research tells us that lawyers, judges, and juries often struggle to understand technical scientific evidence, lack the expertise to assess evidence reliability (Edmond 2020; Wojcikiewicz 2013; NRC 2009), and frequently have difficulty understanding conclusions in expert reports (Keijser and Elffers 2012). Keijser and Elffers' (2012) study found that judges, lawyers, and even forensic professionals rated their comprehension of verbal and visual presentations of likelihood ratio expressions higher than could be justified by results. A 1997 survey of the Australian judiciary (Lonergan 2000), in which over 50% of the country's 478 judges responded, found that 70% of judges reported hearing expert evidence they did not understand and 20% also found it difficult to evaluate differing expert opinions introduced by the parties involved. A similar survey of 400 US judges in 1998 found that over 90% of them did not understand the scientific concepts of falsifiability and error rates and did not seem to recognize (or acknowledge) their lack of understanding of the *Daubert* criteria and how to apply them in admissibility decision-making (Gatowski et al. 2001). A 2009 survey of Polish judges found that a great many of them did not have sufficient knowledge to assess scientific evidence (Wojcikiewicz 2013). Another Australian study (Howes et al. 2013) into the readability of forensic science conclusions in expert reports concluded that the language used was too difficult for an intended generalist audience to read and understand.

This failure to effectively communicate can be considered unethical as the point of such reports is to assist the triers of fact with comprehensible information. In a context which involves persuading the court of the credibility of an expert, jargon-laden and difficult-to-comprehend language is counterproductive. Research conducted with mock juries demonstrates that juries attribute more trustworthiness (and hence, credibility) to the expert giving testimony in clear straightforward language. When juries find complex scientific language too difficult to process, they default to other heuristic cues, such as academic credentials and fees, to judge the persuasiveness of the expert's evidence (Cooper and Neuhouse 2000; Cooper, Bennett, and Sukel 1996). Other research (Oppenheimer 2006) also suggests that language complexity and intelligence are perceived as being negatively associated: those who use straightforward language in written texts are often judged to be more intelligent than those who use complex language. Studies involving written texts across different genres and involving different types of judgments and word manipulation, robustly demonstrated that "needless complexity led to negative evaluations" (Oppenheimer 2006: 151), including the perception of being a "hired gun" (Cooper and Neuhaus 2000).

Experts should also take on the role of educators, both in their written reports and in their oral testimony if called to speak. Solan (1998) argues that the proper role for a forensic linguist in court should be that of a "semantic tour guide". In this role, the expert acts, not as a party advocate, but as a science-educator: someone who conveys the knowledge of the field to help the triers of fact understand and assess the relative strengths and weaknesses of the evidence.

It is also imperative to stress that experts are not in court to usurp the role of the judge and jury by telling them what their conclusions should be. While it is permissible in the United States for experts to opine on the *ultimate issue* (the legal issue at stake) at the invitation of the judge, expert opinion on ultimate issues in UK courts is not allowed, and doing so may lead to their evidence being rejected. In a New Zealand case, a psychologist's expert evidence of the reliability of a confession was excluded because the appeal judges felt the psychologist had overstepped his role as expert witness:

> It would be open to Professor Gudjonsson to give evidence of his opinion as to why, by reason of his psychological assessment of the appellant, Pora might be disposed to make an unreliable confession but, in the Board's view, it is not open to him to assert that the confession is in fact unreliable.
>
> (Pora v. R (New Zealand) *[2015] para 24)*

The role of the expert is to answer questions on their report to assist the triers of fact to understand their evidence and make an informed decision. For the most part, a simple "yes" and "no" answer may suffice. However, the expert also needs to be aware of the risks posed when lawyers reduce the complexity of linguistics to "yes" or "no" questions. In such cases, a "yes but qualified" or "a qualified no" response should suffice, signaling the need to clarify; and only if invited, providing structured and reasoned understanding of how language works under the case circumstances. It is not for the expert to argue the case or provide answers to questions that have not been asked.

The Ethics of Independence: Neutrality

However much an expert takes on an instruction intending to adopt the role of the objective educator, the pressure to provide partisan evidence is immense. An expert's organizational environment—cultural pressures, efficiency goals, the pressure to publish, and other sources of stress and anxiety—have an effect on their behavior and judgment (e.g., Chin, Lutsky, and Dror 2019; Murrie and Boccaccini 2015; Murrie et al. 2013; Sanders 2009; Sanders 2007; Krafka et al. 2002). Situations such as having a long-standing or financial association with a party in a legal dispute can create "adversarial allegiance", which influences judgments. A profession's ethical code would provide guidance on avoiding the corrupting influences of such allegiances.

Forensic expert opinion tends to be partisan toward the side that pays them (Krafka et al. 2002) or with whom they have a continuing relationship (e.g., regularly working for the prosecution or defense), particularly when analysis is less structured and requires more subjective interpretation (Murrie and Boccaccini 2015; Murrie et al. 2013). Contingency fees, where payment is dependent on the outcome, are of particular concern. Not only do such fees increase the potential to produce partisan results (Murrie et al. 2013) but they also increase the likelihood that an expert's opinion may be viewed as biased (Cooper and Neuhaus 2000). The UK courts agree. In *R (on the application of Factortame) v. Secretary of State for Transport* (2002), the court ruled that:

> To give evidence on a contingency fee basis gives an expert, who would otherwise be independent, a significant financial interest in the outcome of the case. As a general proposition, such an interest is highly undesirable.
>
> *(para 73)*

This judgment was reinforced in *Gardiner & Theobald LLP v. Jackson* (2018) where the Court indicated that it was difficult to conceive how working on a contingency fee basis was compatible with the expert's obligation to act objectively and independently at any stage of the expert's engagement. The common law rule in most US jurisdictions is that it is improper for a lawyer to engage an expert witness on a contingency fee (Wertlieb 2016). This prohibition is enshrined as a disciplinary rule in the Code of Professional Responsibility of the American Bar Association (DR 7-109 C) and judges in *Person v. Ass'n of Bar of City of New York* (1977) upheld the right of the New York State Bar Association to enforce the rule, stating: "We are not convinced, however, that there is no danger of the inducement of false expert testimony by such contingency arrangements" (para 538). Experts would also be advised to maintain a healthy skepticism about the justness of the cause of the party retaining them (Finegan 1993). Holding deeply personal views or being heavily invested in a "noble cause" may create "conflicts of conscience", resulting in unconscious bias that could motivate experts to lean toward a particular interpretation of the information.

Experts tend to think of bias as influencing the interpretation of analysis, but the problem is much broader. Bias can influence the selection of data: how much data is enough or whether the right sort of data is collected. Bias can also determine the selection of a methodological approach. Simply having pride in or comfort with one's methodology can create reliance on expectations based on past experiences, reducing the thoroughness of analysis, and lead to tunnel vision (Dror 2020).

Research also demonstrates that forensic scientists are blind to their own biases; they see it in others but do not recognize it in themselves (Pronin, Lin, and Ross 2002) or, if they do, persist in an illusion of control (Dror 2020) that the bias may be countered by pure force of will (Kukucka et al. 2017; Pronin and Kugler 2007). Many believe bias to be an ethical problem of unethical people, not realizing that cognitive biases actually impact honest and dedicated analysts. Training and experience intended to improve expertise and facilitate recognition and decision-making can make experts more susceptible to biases by creating more a priori assumptions and expectations (Dror 2020) or create overconfidence in their findings (Dror et al. 2018). Training based on other casework or bringing experience from a previous case to new casework can create expectations that impact the interpretation of new incoming data (de Lange, Heilbron, and Kok 2018). The training and education that create "experts" may increase efficiency in certain aspects of expertise such as pattern recognition and seamlessly integrating a number of analytical steps, but, paradoxically, it also introduces errors arising from selective information bias (discarding information as irrelevant) and tunnel vision (focusing on a narrow field of features) that arise because such expert analysis is cognitively automated to be efficient (see studies review in Dror, Pascual-Leone, and Ramachandran 2011).

Irrespective of the discipline, analysts are most vulnerable to bias when they are required to interpret information to reach an opinion (Haber and Haber 2013). This is called "evidence elasticity" (Ask, Rebelius, and Granhag 2008). Forensic linguists may be particularly vulnerable because conclusions are rarely clear-cut and tend to be subjective, and it is often the case that opinions evolve as new information is considered.

In forensic linguistic cases, the nature of the material is such that it is difficult to protect experts from contextual information which might influence their decision-making. Contextual information may create expectations and influence experts' inclination in a particular direction. The sequence in which information is provided also creates the potential for bias. Similar to semantic priming, the "anchoring bias" (Yversky and Kahneman 1974)—the tendency to rely on the first bits of evidence provided regardless of their accuracy—can influence judgments on subsequent information that is encountered. For example, in a case involving linguistic profiling, the name of a suspect may bias the expert toward a particular language family.

While it is impossible to achieve objectivity, this should not detract from the goal of working toward it. Personal awareness of bias is important if experts are to combat it, and codes of ethics can play a role in drawing awareness to and reducing bias by requiring education on bias fallacies and mitigation strategies, and reminding experts to consider all the information, alternative hypotheses, and a range of conclusions instead of a single target conclusion.

The Ethics of Preparedness: Integrity

While cognitive biases do arise subconsciously and affect even the most honest and professional person, this is different to the biased conclusions proffered by experts who lack scientific integrity and deliberately engage in misconduct (Young and Goodman-Delahunty 2021). Academics, experts, and judges have long argued that the cost of unethical expert evidence is substantial (e.g., EWI 2021; Hayes 2018; Sanders 2009). Biased and inexpert expert witnesses are unethical because they hold themselves up as having the expertise and knowledge that triers of fact rely on to reach the truth; consequently, such experts are detrimental to achieving the right outcome in court because they reduce the ability of the triers of fact to arrive at correct conclusions. Surveys of US lawyers and judges in 1991 and 1998 found their most common complaints to be about experts becoming advocates for the side that hired them (Krafka et al. 2002). Similarly, a 1999 Australian survey of judges on expert evidence found that 85% said they "had encountered partisanship in expert witnesses", 40% of whom said "this was a significant problem for the quality of fact finding in court" (Freckelton, Reddy, and Selby 1999).

The problem persists today. Lord Hodge, Deputy President of the England and Wales Supreme Court, made this observation on the state of expert witness evidence: "There are plenty of not very good experts around. Some soi-disant [self-styled] experts are worse than not very good, they do great harm" (EWI 2021). Mr. Justice Williams, a High Court Judge, criticized experts for lack of preparation, poor presentation, and lack of accessible language, for failing to abide by the expert guidance codes, and for straying outside their expertise (EWI 2023).

One reason for this problem persisting could be the continuing lack of professional codes of ethics. Such codes would help enhance the quality and transparency of expert witnesses as well as protect them: a profession's code of ethics would act as a shield by providing a certain amount of protection for their members against pressures to alter their perspectives to suit their instructing party; at the same time, it would serve as a sword by offering guidance procedures when these norms are violated (Roberts and Stockdale 2018). This dual function is especially important when practices differ between jurisdictions.

For example, in certain jurisdictions, such as the United States and Canada, it is acceptable practice for lawyers to coach witnesses and expert witnesses and to have input into the format and content of an expert's reports, so that the expert's evidence is framed in a way most favorable for their client (Fadden and Solan 2015; Ainsworth 2010). In the United Kingdom, these practices are explicitly forbidden in guidance and case law and would result in the expert's evidence being excluded and public censure by the judge in court or/and in their written judgment. Coaching introduces the potential for contamination of witness evidence (*R v. Momodou & Limani* 2005). It is one matter for lawyers to request that the expert clarify sections of a report without changing its substance; it is another to enhance or suppress evidence in favor of a particular interpretation. The judge in *Liverpool Victoria Insurance Co. Ltd v. Zafar* (2019), in which the expert made a false statement in his report at his instructing solicitor's suggestion, decided this was not only unethical but also amounted to misconduct and sentenced him to prison suspended for two years.

The Ethics of Truth: Accuracy

One of the issues with the adversarial justice system is that it is based on the theory that the truth is most likely to emerge through a battle of partisan adversaries. This relies heavily on the power of persuasion, which does not always equate to the search for Truth. Jurisdictions' rules of evidence determine what information is admitted in court, and trial outcomes do not always rest on the facts of the case. There is also the issue of who has the more convincing expert witness and the better lawyer who can shape the admissible facts into a truthful, albeit not the whole truth, narrative.

Lawyers attempt to keep experts on a tight rein with narrow questioning focus, and experts are torn about answering the questions asked of them without misleading judges and juries (Gutheil et al. 2003). Experts raise concerns about being expected to present their evidence in a manner that is favorable to clients' interests but is scientifically inaccurate or misleading. Television's representation of the analytical process and evidential weight of forensic evidence results in juries requiring unreasonable levels of certainty. Consequently, lawyers may attempt to find experts who claim to provide such definite evidence, whether or not they can actually do so.

The courts already rely on a number of legal measures to mitigate against biased and partisan experts, one of these being the gatekeeping role of the judge in deciding on the admissibility of experts and evidence depending on their value to the trial. However, this presupposes that judges have the requisite scientific knowledge to understand the evidence provided or acquire the expertise relatively quickly. Judges are also meant to facilitate effective cross-examination to test the reliability of the expert and expose weak expert evidence, thereby assisting juries to reach informed decisions. However, this also relies on the presumption that all parties understand the evidence. Cross-examination exposes juries to technical arguments for which they may lack the skill and insight to evaluate (Young and Goodman-Delahunty 2021), and generalist judges do not have the knowledge to untangle arguments on complicated linguistic issues from opposing parties (Domitrovich 2016). Faced with such obstacles, what can be done to encourage expert witnesses to provide accurate testimony?

In England and Wales, expert witnesses are already required in civil and criminal cases to swear to a "Statement of Truth", which is included in their reports. By incorporating this statement, experts affirm that they have a genuine belief in the truth of what they have said in their expert report; and they understand the consequences of being found to have made a dishonest statement. Expert reports in England and Wales also require the inclusion of a "Statement of Compliance", which is used to confirm that experts are aware of their duty to the court and have complied with it. To assert compliance, experts are expected to have read the relevant Practice Directions and may even be examined on that knowledge in court. If an expert has signed a Statement of Compliance and is later found to be unfamiliar with the Directions, then the court can rule that the expert lied in signing the Statement of Truth. Any false statement supported by a Statement of Truth is such a severe offense that jail sentence of up to two years can be expected (*Liverpool Victoria Insurance Co Ltd v. Zafar* 2019). Incompetent experts who fail in their duty to the court and waste court time can also expect to have costs awarded against them as in *Thimmaya v. Lancashire NHS Trust* (2020). The judge ruled:

> But Mr Jamil owed important, and significant, duties to the Court. He failed comprehensively in those duties from November 2017 onwards. As a result, a public body has incurred significant unnecessary costs. Whilst it would not be right to use him as an example to send a message to experts, it is right that experts should all understand the importance of their duties to the court and the potential consequences if they fail in them.
>
> *(para 19)*

The court determined that the consultant's conduct as an expert had been "improper, unreasonable, or negligent", and he was ordered to pay the defendant's costs amounting to over £88,800.

Conclusion

The purpose of this chapter has been to highlight the complex ethical issues involved with being an "expert" generally and an expert witness specifically. Differences in culture, rules of evidence, and general procedures across the world make it impossible to legislate on the behavior of experts universally (Hayes 2018). Consequently, there are strong potential benefits for a profession-specific code of ethics for forensic linguists to guide them through the minefields that exist. As was discussed here, being an expert includes undertaking the tasks of being "expert". This in turn requires not only having the knowledge and expertise to perform the required analysis, it also necessitates being competent in one's obligation to the court and learning to grapple with the intricacies of law and court proceedings without relying on lawyers to advise them.

The legal and ethical pitfalls, duties, and obligations of being an expert are not learned in the normal course of obtaining forensic linguistic expertise. Neither are they set out within the courts' codes of expert witness conduct but are judged by the standards of the discipline of linguistics. This means that, currently, forensic linguists who wish to work in this capacity are often left to learn as they go. This is unacceptable for both the individual and the field as a whole.

Every time a forensic linguistic expert provides an investigative report or gives expert evidence in court, they represent the knowledge that the field has to offer on a specific aspect of forensic linguistics. If a forensic linguist is seen to have taken a case without having the requisite expertise, or is less than independent or, even worse, is a "hired gun", then the entire profession suffers. Even alleged unethical behavior tarnishes the reputation of forensic linguistics as a forensic science and causes it to be regarded as "flawed". This in turn may create suspicion about its value as potential assistance to the court. And the damage does not end there. Unethical behavior produces unreliable evidence which, if accepted, may mislead those who rely on it and create the potential for less accurate outcomes in court and miscarriages of justice.

The purpose of forensic linguistic research is to arrive at the knowledge that helps provide a social benefit and enhances the delivery of justice. Toward that end, it is imperative that forensic linguists become and remain aware of the pitfalls and dangers before they arise. Only with this information and awareness can an expert steadfastly exhibit ethical standards of behavior. This is where an overarching profession-specific code of ethics would be invaluable. In the meantime, it is important that professional organizations in the discipline take an ethical stance, adopt and publicize a code of ethics that they stand by, and demonstrate a willingness to uphold it. While such codes will not deter individuals' intent on deliberate misconduct, they will go a long way toward mitigating unintentional honest mistakes.

Notes

1 IAFP has since changed its name and abbreviation to the International Association for Forensic Phonetics and Acoustics (IAFPA) and IAFL to the International Association for Forensic and Legal Linguistics (IAFLL).

2 From my own personal experience in cross-examination.

Case Law

Austin v American Association of Neurological Surgeons 253 F.3d 967 (7th Cir. 2001)
Consulate Ventures Inc. v. *Amico Contracting & Engineering (1992) Inc.*, 2010 ONSC 2181
Daubert v Merrell Dow Pharmaceuticals, Inc. 509 U.S. 579 (1993)
Detroit Fire & Marine Ins. Co v. Gagliardi, 32 P.2d 832 (Colo. 1934)
General Medical Council and Professional Standards Authority for Health and Social Care v. Asef Azar [2020] EWHC 846 (Admin)
Jones v Kaney [2011] UKSC 13
Kannankeril v. Terminix Int'l, Inc., 128 F.3d 802, 809 (3d Cir. 1997)
Kumho Tire Co. v. Carmichael, 119 S.Ct. 1167 (1999)
Liverpool Victoria Insurance Co. Ltd v. Zafar [2019] EWCA Civ. 392

Person v. Ass'n of Bar of City of New York, 554 F.2d 534 (2d Cir. 1977)
Pora v R (New Zealand) [2015] UKPC 9
R. v. J.-L.J., 2000 SCC 51 (CanLII) [2000] 2 SCR 600
R. v. Mohan, 1994 CanLII 80 (SCC) [1994] 2 SCR 9
R v. Momodou & Limani [2005] EWCA Crim 177
R. v. N.O., 2009 ABCA 75 (CanLII)
Thimmaya v Lancashire NHS Foundation Trust [2020] 1 WLUK 437
Tyus v. Urban Search Management, 102 F.3d 256 (7th Cir. 1996)
Stuart v R [1959] HCA 27; (1959) 101 CLR 1 (19 June 1959)
White Burgess Langille Inman v. Abbott and Haliburton Co., 2015 SCC 23 (CanLII) [2015] 2 SCR 182

References

Academy of Medical Royal Colleges. 2019, May. *Acting as an Expert or Professional Witness: Guidance for Healthcare Professionals*. https://ficm.ac.uk/sites/ficm/files/documents/2021-10/expert_witness_guidance_final.pdf.

Ainsworth, Janet. 2022. "Serving Science and Serving Justice: Ethical Issues Faced by Forensic Linguists in Their Role as Expert Witnesses." In *Language as Evidence: Doing Forensic Linguistics*, 35–53. Edited by Victoria Guillén-Nietoand Dieter Stein. Cham: Springer International Publishing.

Ainsworth, Janet. 2010. "A Lawyer's Perspective: Ethical, Technical, and Practical Considerations in the Use of Linguistic Expert Witnesses." *The International Journal of Speech Language and Law* 16, no. 2: 279–291.

Ask, Karl, Pär Anders Granhag, and Anna Rebelius. 2011. "Investigators under Influence: How Social Norms Activate Goal-directed Processing of Criminal Evidence." *Applied Cognitive Psychology* 25, no. 4: 548–553.

Ask, Karl, Anna Rebelius, and Pär Anders Granhag. 2008. "The 'Elasticity' of Criminal Evidence: A Moderator of Investigator Bias." *Applied Cognitive Psychology: The Official Journal of the Society for Applied Research in Memory and Cognition* 22, no. 9: 1245–1259.

Berk-Seligson, Susan. 2017. *The Bilingual Courtroom: Court Interpreters in the Judicial Process*. Chicago: University of Chicago Press.

Blondel, David. 1628. *Pseudo-Isidorus et Turrianus vapulantes*. éditeur non identifié. https://books.google.co.uk/books?hl=en&lr=&id=vdZXRjrfaTwC&oi=fnd&pg=PA77&dq=Pseudo-Isidorus+et+Turrianus+vapulantes&ots=vWCJSFYJo9&sig=T54KUjVqJQxFAWSLNlaXVx-3hkk&redir_esc=y#v=onepage&q=Pseudo-Isidorus%20et%20Turrianus%20vapulantes&f=false. Commentary available at https://ccel.org/s/schaff/encyc/encyc09/htm/iv.v.lxxiv.htm.

Carradice, Phil. 2012, January 16. "The Execution of Timothy Evans." *BBC*. Retrieved from https://www.bbc.co.uk/blogs/wales/entries/cdc56160-91eb-366d-ae0e-5d9c5a676fe2.

Chin, Jason M., Michael Lutsky, and Itiel E. Dror. 2019. "The Biases of Experts: An Empirical Analysis of Expert Witness Challenges." *Manitoba Law Journal* 42: 21.

Cooper, Joel, Elizabeth A. Bennett, and Holly L. Sukel. 1996. "Complex Scientific Testimony: How do Jurors Make Decisions?" *Law and Human Behavior* 20: 379–394.

Cooper, Joel, and Isaac M. Neuhaus. 2000. "The 'Hired Gun' Effect: Assessing the Effect of Pay, Frequency of Testifying, and Credentials on the Perception of Expert Testimony." *Law and Human Behavior* 24, no. 2: 149–171.

Coulthard, Malcolm, Alison Johnson, and David Wright. 2017. *An Introduction to Forensic Linguistics: Language in Evidence*. London: Routledge.

Crown Prosecution Service. 2023. *Forensic Science Regulator Act 2021 and the Forensic Science Regulator's Code of Practice*. Retrieved from https://www.cps.gov.uk/legal-guidance/forensic-science-regulator-act-2021-and-forensic-science-regulators-code-practice.

De Lange, Floris P., Micha Heilbron, and Peter Kok. 2018. "How do Expectations Shape Perception?" *Trends in Cognitive Sciences* 22, no. 9: 764–779.

Domitrovich, Stephanie. 2016. "Fulfilling Daubert's Gatekeeping Mandate through Court-appointed Experts." *Journal of Criminal Law & Criminology* 106, no. 1: 35–48.

Dror, Itiel E. 2020. "Cognitive and Human Factors in Expert Decision Making: Six Fallacies and the Eight Sources of Bias." *Analytical Chemistry* 92, no. 12: 7998–8004.

Dror, Itiel E., Jeff Kukucka, Saul M. Kassin, and Patricia A. Zapf. 2018. "When Expert Decision Making Goes Wrong: Consensus, Bias, the Role of Experts, and Accuracy." *Journal of Applied Research in Memory and Cognition* 7, no. 1: 162–163.

Dror, Itiel E., Alvaro Pascual-Leone, and Vilayanur Ramachandran. 2011. "The Paradox of Human Expertise: Why Experts Get it Wrong." In *The Paradoxical Brain*, 177–188. Edited by Narinder Kapur. Cambridge: Cambridge University Press.

Eades, Diana, Helen Fraser, and Georgina Heydon. 2023. *Forensic Linguistics in Australia: Origins, Progress and Prospects*. Cambridge: Cambridge University Press.

Edmond, Gary. 2020. "Forensic Science and the Myth of Adversarial Testing." *Current Issues in Criminal Justice* 32, no. 2: 146–179.

Expert Witness Institute. 2023. "Sir Michael Davies Lecture Keynote Address by Mr Justice Williams." Retrieved from https://www.ewi.org.uk/News/sir-michael-davies-lecture-2023-the-criticism-of-expert-witnesses-in-the-courts#:~:text=%E2%80%9CCriticism%20in%20a%20judgment%20can,judge%20is%20not%20necessarily%20criticism.

Expert Witness Institute. 2021. "Conference Keynote Address by Lord Hodge." Retrieved from https://www.ewi.org.uk/what-the-court-expects-of-a-competent-expert-witness.

Fadden, Lorna, and Lawrence M. Solan. 2015. "Expert Witness Communication." In *Communication in Investigative and Legal Contexts*, 209–228. Edited by Gavin Oxburgh, Tor Age Myklebust, and Rebecca Milne. New York: Wiley Blackwell.

Federal Court of Australia Expert Evidence Practice Note (GPN-EXPT). 2016. Retrieved from https://www.fedcourt.gov.au/law-and-practice/practice-documents/practice-notes/gpn-expt.

Federal Court Rules SOR/98-106 (Canada) Rule 52.2. Retrieved from https://laws-lois.justice.gc.ca/eng/regulations/sor-98-106/page-1.html.

Finegan, Edward. 1993. "Ethical Considerations for Expert Witnesses in Forensic Linguistics." *Issues in Applied Linguistics* 4, no. 2: 179–187.

Forensic Science Regulator. 2020. *Forensic Science Regulator Guidance: Expert Report Guidance FSR-G-200,* Issue 4. Retrieved from https://assets.publishing.service.gov.uk/media/5ea6fa0586650c0314ac74f0/Statement_and_Report_Guidance_Appendix_Issue_4.pdf.

Freckelton, Ian R., Prasuna Reddy, and Hugh Selby. 1999. *Australian Judicial Perspectives on Expert Evidence: An Empirical Study*. Australian Institute of Judicial Administration Incorporated.

Garrett, Brandon L., and Peter J. Neufeld. 2009. "Invalid Forensic Science Testimony and Wrongful Convictions." *Virginia Law Review*, 95, no. 1: 1–97.

Gatowski, Sophia I., et al. 2001. "Asking the Gatekeepers: A National Survey of Judges on Judging Expert Evidence in a Post-Daubert World." *Law and Human Behavior* 25 no. 5: 433–458.

General Medical Council. 2013, April. *Acting as a Witness in Legal Proceedings*. Updated 2020. Retrieved from https://www.gmc-uk.org/-/media/documents/gmc-guidance-for-doctors---acting-as-a-witness-in-legal-proceedings_pdf-58832681.pdf.

GMR Consulting. 2021, June. "EWI President Discusses the Critical Role of the Expert Witness." Retrieved from https://www.gmrconsulting.com/articles-news/45-employment/1464-ewi-president-discusses-the-critical-role-of-the-expert-witness.

Gutheil, Thomas G., et al. 2003. "'The Whole Truth' Versus 'the Admissible Truth': An Ethics Dilemma for Expert Witnesses." *Journal of the American Academy of Psychiatry and the Law* 31: 422–4227.

Haber, Ralph Norman, and Lyn Haber. 2013. "The Culture of Science: Bias and Forensic Evidence." *Journal of Applied Research in Memory and Cognition* 2, no. 1: 65–67.

Hayes, Alexander. 2018. "Buying Justice: The Cost of Expert Witnesses in Jurisdictions Across the World." *Idaho Critical Legal Studies Journal* 12: 1–24.

Heydon, Georgina. 2005. *The Language of Police Interviewing*. Hampshire: Palgrave Macmillan.

Howes, Loene M., et al. 2013. "Forensic Scientists' Conclusions: How Readable Are They for Non-scientist Report Users?" *Forensic Science International* 231, no. 1–3: 102–112.

De Keijser, Jan, and Henk Elffers. 2012. "Understanding of Forensic Expert Reports by Judges, Defense Lawyers and Forensic Professionals." *Psychology, Crime and Law* 18, no. 2: 191–207.

Koehn, Daryl. 2006. *The Ground of Professional Ethics*. New York: Routledge.

Krafka, Carol, et al. 2002. "Judge and Attorney Experiences, Practices, and Concerns Regarding Expert Testimony in Federal Civil Trials." *Psychology, Public Policy, and Law* 8, no. 3: 309–332.

Kukucka, Jeff, Saul M. Kassin, Patricia A. Zapf, and Itiel E. Dror. 2017. "Cognitive Bias and Blindness: A Global Survey of Forensic Science Examiners." *Journal of Applied Research in Memory and Cognition* 6, no. 4: 452–459.

Lonergan, Wayne. 2000. "Australian Judicial Perspective on Expert Evidence." *Law Society Journal* 38, no. 7: 54–57.

Medical Practitioners Tribunal Service. 2023, December. "Record of Determination – Dr Seshni Moodliar." Retrieved from https://www.mpts-uk.org/-/media/mpts-rod-files/dr-seshni-moodliar-15-dec-23.pdf.

Murrie, Daniel C., and Marcus T. Boccaccini. 2015. "Adversarial Allegiance Among Expert Witnesses." *Annual Review of Law and Social Science* 11, no. 1: 37–55.

Murrie, Daniel C., Marcus T. Boccaccini, Lucy A. Guarnera, and Katrina A. Rufino. 2013. "Are Forensic Experts Biased by the Side that Retained Them?" *Psychological Science* 24, no. 10: 1889–1897. https://doi.org/10.1177/0956797613481812

National Research Council of the National Academies. 2009. *Strengthening Forensic Science in the United States: A Path Forward*. Washington, DC: The National Academic Press.

Oppenheimer, Daniel M. 2006. "Consequences of Erudite Vernacular Utilized Irrespective of Necessity: Problems with Using Long Words Needlessly." *Applied Cognitive Psychology* 20, no. 2: 139–156.

Part 1 – https://www.thetext.co.uk/Evans%20Statements%20Part%201.pdf

Part II – https://www.thetext.co.uk/Evans%20Statements%20Part%202.pdf

President's Council of Advisors on Science and Technology. 2016. *Report to the President: Forensic Science in Criminal Courts: Ensuring Scientific Validity of Feature-Comparison Methods*. Washington, DC: Executive Office of the President of the United States. https://obamawhitehouse.archives.gov/sites/default/files/microsites/ostp/PCAST/pcast_forensic_science_report_final.pdf

Pronin, Emily, and Matthew B. Kugler. 2007. "Valuing Thoughts, Ignoring Behavior: The Introspection Illusion as a Source of the Bias Blind Spot." *Journal of Experimental Social Psychology* 43, no. 4: 565–578.

Pronin, Emily, Daniel Y. Lin, and Lee Ross. 2002. "The Bias Blind Spot: Perceptions of Bias in Self Versus Others." *Personality and Social Psychology Bulletin* 28, no. 3: 369–381.

Roberts, Paul, and Michael Stockdale. 2018. "Introduction: Forensic Science, Evidential Reliability and Institutional Reform." In *Forensic Science Evidence and Expert Witness Testimony: Reliability through Reform*, 1–26. Edited by Paul Roberts and Michael Stockdal. Cheltenham, UK: Edward Elgar Publishing.

Samuels Alec. 2022. "Forensic Science Regulator: The Forensic Science Regulator Act 2021." *Medico-Legal Journal* 90, no. 2: 86–89.

Sanders, Joseph. 2009. "Science, Law, and the Expert Witness." *Law and Contemporary Problems* 72, no. 1: 63–90.

Sanders, Joseph. 2007. "Expert Witness Ethics." *Fordham Law Review* 76, no. 3: 1539–1584.

Shiffman, Melvin A. 1999. *Ethics in Forensic Science and Medicine: Guidelines for the Forensic Expert and the Attorney*. New York: Charles C Thomas Publisher.

Shim, Abigail. 2022. "Forensic Linguistics: Science or Fiction?" *Hastings Law Journal* 74, no. 1: 207–234.

Solan, Lawrence M. 2005. "Vagueness and Ambiguity in Legal Interpretation." In *Vagueness in Normative Texts*, 73–96. Edited by Vijay K. Bhatia. Bern: Peter Lang.

Solan, Lawrence M. 1998. "Linguistic Experts as Semantic Tour Guides." *Forensic Linguistics* 5: 87–106.

Solan, Lawrence M. 1993. *The Language of Judges*. Chicago: University of Chicago Press.

Solon, Mark. 2019. "The Times Bond Solon Expert Witness Survey 2019." Retrieved from https://www.bondsolon.com/media/169523/expert-witness-survey-report-2019.pdf

Svartvik, Jan. 1968. *The Evans Statements: A Case for Forensic Linguistics*. Gothenburg: University of Gothenburg Press.

The Case of Timothy John Evans Home Office Report by Hon Mr Justice Brabin. PP/WWS/H/3/3. Wellcome Collection. 1966. Retrieved from https://wellcomecollection.org/works/tx9cv44

The Civil Procedure Rules Practice Directions, Part 35 – Experts and Assessors. Retrieved from https://www.justice.gov.uk/courts/procedure-rules/civil/rules/part35/pd_part35

Tiersma, Peter, and Lawrence M. Solan. 2002. "The Linguist on the Witness Stand: Forensic Linguistics in American Courts." *Language* 78, no. 2: 221–239.

The Criminal Procedure Rules 2020 No. 759 (L. 19) Part 19. Retrieved from https://www.legislation.gov.uk/uksi/2020/759/part/19

Valla, Lorenzo. 1440. *The treatise of Lorenzo Valla on the Donation of Constantine*. Project Gutenberg. Retrieved from https://www.gutenberg.org/files/70092/70092-h/70092-h.htm

Young, Gerald, and Goodman-Delahunty, Jane. 2021. "Revisiting *Daubert*: Judicial Gatekeeping and Expert Ethics in Court." *Psychological Injury and Law* 14: 304–315.

Tversky, Amos, and Daniel Kahneman. 1974. "Judgment under Uncertainty: Heuristics and Biases." *Science* 185: 1124–1131. Retrieved from https://www.science.org/doi/pdf/10.1126/science.185.4157.1124

Wertlieb, Neil J. 2016. "Ethics Issues in the Use of Expert Witnesses." *The Professional Lawyer* 24, no. 3: 35–46.

Wojcikiewicz, Jozef. 2013. "Judges' Attitude Towards Scientific Evidence." *Revija za kriminalistiko in kriminologijo* 64: 249–255.

Yversky, Amos, and Daniel Kahneman. 1974. "Judgment Under Uncertainty: Heuristics and Biases". *Science*, 185, 1124-1131. https://www.science.org/doi/pdf/10.1126/science.185.4157.1124

4

THE EXPERT'S ROLE IN THE JUSTICE SYSTEM IS TO PROVIDE IMPARTIAL AND TRANSPARENT EVIDENCE, NOT TO PERFORM SOCIAL ADVOCACY

Richard Rhodes and Tina Cambier-Langeveld

Our Position

As practicing forensic phoneticians, we provide expert evidence in criminal cases on a regular basis. Usually, our task is to evaluate the strength of evidence provided by the analysis of speech and audio recordings; this might be in relation to the comparison of voices in different recordings, assessing and transcribing what different speakers are saying, or investigating the authenticity of a recording, for example. It is from this experience and point of view that we will address the question of whether forensic practitioners can act as social advocates or campaigners as part of their professional practice.

In this chapter, we argue that forensics may inspire social advocacy, but not vice versa. When giving expert evidence in court, all external factors—that is, anything which is not directly within the forensic evidence being examined—should be regarded as a potential source of bias. In this chapter, we observe that cognitive biases are of a different nature to personal, social, or political biases. Various strategies to tackle cognitive bias have been proposed in the literature, but this chapter addresses the latter, more personal, ethical factors.

In our conception, the primary role of the forensic expert is to: (a) determine what questions the linguistic evidence can or cannot assist in answering; (b) indicate what weight may be attached to the evidence in question; and (c) try to ensure that the evidence is understood properly by the various end-users in different justice systems. We strongly argue that it is explicitly *not* within the expert's role to judge, or unduly influence, the ultimate question of guilt, how suspects or victims may be treated fairly, or indeed any aspect of the wider justice system, particularly if this is guided by their own moral compass.

Producing expert forensic evidence that is targeted toward helping what the expert perceives to be the "right side" is highly likely to be unethical. The expert must perform their role in the context of the justice system as a whole. To think otherwise is to lose perspective on how social—and certainly legal or criminal—justice works. A fair and functioning justice system is the mechanism

DOI: 10.4324/9781003391074-5

by which most societies achieve justice; the expert's role within that system is to provide objective evidence in a transparent manner. Typically, this objective expertise will be available to all parties in a case. If the expert strays from that role in order to achieve what they perceive to be a "just" outcome, they risk that very purpose.

Forensic Expert Evidence versus Social Advocacy

Definitions

This chapter requires definitions of some key terms, such as "forensic" and "social advocate". Giving an explicit definition of these terms helps us clarify to what extent they are similar, overlapping, or distinct from each other.

Common definitions of "forensics" or "forensic science" refer to "the application of scientific principles and techniques to matters of criminal justice especially as relating to the collection, examination, and analysis of physical evidence" (Merriam-Webster.com). Forensic evidence is used by decision-making authorities in criminal and other types of law; and during criminal investigation in particular, it is governed by the legal standards of admissible evidence and criminal procedure (see further under the Guidance sections below). Forensic experts may collect, analyze, and generate evidence during the course of an investigation or in a prosecution/trial process. They may also testify in court about their findings.

Forensic experts may also testify in court on issues in a case without having directly examined or tested any evidence. The goal of such expert testimony should still be to assist decision-making authorities in making fair and just decisions. For example, an expert linguist can ensure that folk-linguistic beliefs are replaced by a more scientific perspective on language. Another example is that forensic expert testimony may be necessary to reveal the shortcomings of evidence presented by others to ensure that undue faith is not placed in evidence of questionable strength.

"Social advocacy" "includes action aimed at the realization of a just society, which respects and is protective of human rights, is inclusive of a plurality of interests, and is responsive to the most marginalized members of a society" (Cohen, Vega, and Watson 2001); it can be defined as "the act of supporting a particular cause that is aimed at improving the well-being of a specific group of individuals or targeting a societal problem" (https://www.lawentrance.com/article/social-advocate.html). Often, this involves parties speaking on behalf of—"advocating for"—disadvantaged individuals, groups, or cultures; this might also involve engaging in campaigns, direct action, or activism in order to promote more awareness, or to affect change in policy or society. These causes or ideals often have aspects of social justice at their heart: for example, equality, inclusion, empowerment, human and civil rights. They can also be directed at specific issues, for example, the environment and the climate crises, treatment of specific groups or cultures, or structural barriers to fair treatment of people. Social advocacy is typically motivated by a personal conviction that change in these areas is needed.

Distinctions

Forensic analysis of linguistic/phonetic evidence and giving expert testimony belong together in that they apply to individual legal cases, and they are governed by rules and principles set out for forensic expert evidence. They also go together in practice: testifying in court in relation to an analysis of evidence will often go hand in hand with testifying about the limits of a particular method or the general state of the art in the field, for example. If the purpose of the forensic expert's work is to assist a decision-maker or trier of fact to make a justified decision, this purpose can only be achieved through a balanced and scientifically valid evaluation of the strength of the evidence in that case. The forensic expert should refrain from expressing an opinion or conclusion on ultimate issues, i.e., the binary decisions that have to be made concerning the fundamental question of guilt or innocence; they should also avoid other issues that are for the court to decide, including the assumption that a person or item is the source of the evidential trace, the credibility of a witness, or the acceptability of what is presented as expert evidence, for example. These issues are specifically and formally for the legal system and the trier(s) of fact to decide.

Supporting social ideals to initiate social change is a very different activity to the forensic activities described above. Social advocacy addresses wider social issues rather than single-issue instances. It is also very different in that it is motivated by personal ideals and beliefs. We do not imply that such work is any less valid or less needed. There are many ways that linguists can positively influence the operation of justice: by challenging and investigating issues of language and power in courtrooms and police interviews, accent/language discrimination, effective and fair translation and interpreting, and many more.

Linguistics is a broad field, and by applying their knowledge and expertise, linguists may positively impact the justice system in many ways. However, we argue that this application of linguistics to the issues in our society should be viewed as a separate activity to forensic casework. In our opinion, the two should be considered as having very different aims, rules, and spheres of influence. We acknowledge that there may be individual cases which raise more general issues for society, but forensic evidence in these cases should still be considered impartially and not on the basis of wider social issues.

We find it difficult to say anything about the role and impact of forensics on social advocacy, as the latter is not within our area of expertise. We do have a clear opinion on the impact vice versa: there should be none. If the linguist is carrying out forensic casework, their conclusions and advice should be based on the evidence alone, and not any pre-existing assumptions or biases about the issues that may be present in the case. It is essential that the expert's professional ethics are detached from, and unaffected by, their personal ethics. We will present principles of sound forensic evidence as laid down in formal guidance documents for forensic practitioners in support of this position.

Principles and Guidance on Sound Forensic Evidence

Four Key Principles

The role of the expert in the forensic context is to help a court make better-informed decisions, by assessing evidence which requires specialist knowledge and experience which is outside that of the trier of fact (judge or jury). The expert's primary duty is to this court (or other decision-making body), and not the party who commissions the work, or any other group. In Rhodes and Cambier-Langeveld (in press), we presented four key principles which are the foundation of how forensic practitioners should operate:

Impartiality means that decisions are made fairly and on the basis of objective criteria rather than on the basis of bias, i.e. an a priori inclination towards the position of the prosecution or the defense (or any other group or cause).

Transparency guarantees that evidence from a forensic expert is accessible to the multiple parties to a criminal case, and those parties can see how and why the expert has come to their conclusions.

Validity is the extent to which a conclusion or measurement is well-founded and reflects the ground truth of what actually happened.

Effectiveness requires that the forensic evidence is produced and communicated in such a way that it has an appropriate and useful impact on the case.

The two principles that apply most centrally to this chapter are impartiality and transparency. The primary duty of forensic experts is that they act impartially; this is often the first tenet in codes of conduct or guidance for practitioners, and for good reason. They must also present their work transparently, so that decision-makers can assess their findings on the basis of "the whole truth". If a forensic expert were to act in a criminal case in furtherance of a social cause, they would not be acting **impartially**; they would be acting based on a conflict of interest. Furthermore, they would also violate the principle of **transparency** as it would appear that their findings were based on the evidence, when in fact they were based on, or influenced by, other matters. In this situation, the evidence also runs the risk of being **invalid**, reflecting as it would the pre-existing social belief, rather than the truth of what happened. All of these factors would lead to expert evidence which does not serve its central purpose **effectively**.

Guidance: Legal

The key principle of impartiality, sometimes also referred to as balance, objectivity, or neutrality, is embodied in sources of guidance for experts across many disciplines. There are many international and UK sources which outline this point, and which question the reliability of forensic or expert evidence that is based on inferences or conclusions not based on fair consideration of the examination results. We have limited these to sources written in English as UK authorities provide a plethora of guidance, one author operates in the United Kingdom, and also so that no translation is necessary; we observe that other countries have similar rules for forensic evidence and forensic experts to follow.

The Criminal Procedure Rules in England and Wales state that:

> 19.2.(1) An expert must help the court to achieve the overriding objective,
> (a) by giving opinion which is—
> (i) objective and unbiased, and
> (ii) within the expert's area or areas of expertise

And in the Practice Directions which apply these rules, courts are advised:

> in considering reliability, and especially the reliability of expert scientific opinion, the court should be astute to identify potential flaws in such opinion which detract from its reliability, such as:
>
> (a) being based on a hypothesis which has not been subjected to sufficient scrutiny (including, where appropriate, experimental or other testing), or which has failed to stand up to scrutiny;
> (b) being based on an unjustifiable assumption;
> (c) being based on flawed data;
> (e) relying on an inference or conclusion which has not been properly reached.

Further, experts are required to declare that their work is free of potential conflicts of interest and that they have observed their primary duty to the court, in the following defined format:

1. I understand that my duty is to help the court to achieve the overriding objective by giving independent assistance by way of objective, unbiased opinion on matters within my expertise, both in preparing reports and giving oral evidence....
3. I know of no conflict of interest of any kind, other than any which I have disclosed in my report.
4. I do not consider that any interest which I have disclosed affects my suitability as an expert witness on any issues on which I have given evidence.

Guidance: Regulatory

Impartiality, and acting without conflicts, is a central part of the UK Forensic Science Regulator's (2025) Codes of Practice. This is to be a core tenet of how the organization operates, and each practitioner is required to individually comply:

34.1.2 As a practitioner undertaking a Forensic Science Activity you shall:

a. Recognise your overriding duty is to the court and to the administration of justice.
b. Act with honesty, integrity, objectivity and impartiality.
c. Comply with the legal obligations imposed on practitioners (and specifically expert witnesses) in England and Wales.
d. Declare, at the earliest opportunity, any personal, business, financial and/or other interest that could be perceived as a potential conflict of interest.

The legal obligations on experts also include impartiality; these are described in the Regulators guidance on Legal Obligations (2020b) [FSR-I-400, based on England and Wales case law and Criminal Procedure Rules]:

5.5.1 The expert's duties are owed to the court and override any obligation to the person who instructs or pays the expert. The expert is to maintain professional objectivity and impartiality at all times.[1]

5.5.6 The requirement for independence and objectivity may make it inappropriate for an expert to act as an adviser to a party and a witness in the same case.[2]

An impartial mindset is also a fundamental protection against potential cognitive bias effects, and so it is reflected in the Regulator's guidance (2020a) [FSR-G-217] on the issue:

2.1.1 A key requirement of the Forensic Science Regulator's Codes of Practice and Conduct for Forensic Science Providers and Practitioners (the Codes) is that they [...] act with honesty, integrity, objectivity and impartiality [...]

2.1.2 However, many fields of forensic science include subjective assessment and comparison stages that are potentially susceptible to subconscious personal bias (cognitive contamination), which in turn could undermine the objectivity and impartiality of the forensic process.

It is also cited as a foundation stone of the process experts should follow in evaluating evidence. In the Association of Forensic Science Providers, there is the Standards for the formulation of evaluative forensic science expert opinion (2009) which states:

> 3 The expert will base [their] opinion upon the four principles: balance, logic, robustness and transparency. An expert will not usually give [their] opinion on issues that do not require expert knowledge.

The regulatory guidance is not just restricted to the UK systems. The international standard ISO/IEC 17025:2017 applied to forensic testing laboratories has impartiality at its heart:

> 4.1 Impartiality:
> 4.1.1 Laboratory activities shall be undertaken impartially and structured and managed so as to safeguard impartiality
> 4.1.2–4.1.5 [The guidance continues to describe how laboratory management shall be committed to impartiality and its maintenance, shall identify risks to impartiality, and detail how they minimize or eliminate these risks.]

Guidance: Specific Forensic Linguistic/Phonetic Standards

Professional organizations for forensic linguistic practitioners naturally follow the same principles in their Codes of Practice.

The International Association for Forensic Phonetics and Acoustics (IAFPA 2020)—Code of Practice:

> 2.1 Members must act in all circumstances with integrity, fairness, and impartiality.
> 2.6. Members should not take on casework where there is a conflict of interest.

Linguistic Society of America (LSA)—Code of Ethics for Linguists in Forensic Linguistics Consulting; and the International Association for Forensic and Legal Linguistics (IAFLL n.d.) Code of Practice:

> I.A. Integrity: Mindful of their obligations to furnish the justice system with valid, reliable, accurate, and unbiased linguistic information and analyses, consultants must recognize that their duty is to provide evidence that will assist the trier of fact in achieving a just result—a duty that overrides any obligation owed to the retaining lawyers, litigants, or other clients who have engaged them.

None of the guidelines quoted above contain the terms "social justice", "social advocacy", "human rights", or "equality". We firmly believe this is appropriate because these are not parts of the legal role of forensic experts.

A notable exception is the Germanic Society for Forensic Linguistics (GSFL 2017), whose Articles of Association, specifying the organization's regulations and purpose, mention "the ideals of social justice" as its main pursuit and its commitment to "adding to individual, national, communal, and societal betterment". They make no reference to the analysis of evidence and giving expert testimony, and no mention of the terms "impartiality", "neutral", "balance", "unbiased". GSFL thus firmly places itself in the realm of social advocacy and not in the field of providing forensic linguistic expertise in court. While all of the above is perfectly valid and transparent in a sense (at least for those taking the trouble to study the relevant documents), we do wonder whether GSFL is fully aware of how their goals are fundamentally different from other organizations with "forensic" in their name. One could argue that GSFL should rename itself to reflect that it is concerned with social and legal language issues and not with providing or analyzing forensic evidence.

Interestingly, the International Association for Forensic Linguistics (IAFL) recently renamed itself and is now the International Association for Forensic **and Legal** Linguistics (IAFLL). The purpose and aims as published on IAFLL's website show that this organization now indeed intends to envelop both forensics and social advocacy in specifically legal settings, although this is not reflected yet in their Code of Practice. We sincerely hope that this chapter and this book will serve as an acute reminder that we need to clearly delineate the essential differences between the role of a linguist as a forensic expert versus a social advocate, and to understand how these differences affect the duties of the linguist in each role.

Social Advocacy: An Issue in Forensic Linguistics in Particular?

Formal forensic guidance applies to many different fields of expertise. The open question regarding social advocacy made us wonder whether this question is equally likely to arise in other areas. We suspect that there are inherent properties of language that make forensic linguistics particularly sensitive to the issues addressed in this book.

Language can not only form a forensic trace that can be linked to its source or shed light on what happened but is also pivotal in communication and human interaction. The specific legal setting adds to the complexity and importance of proper communication. It also increases the chances of language being misused or misinterpreted, given that the setting involves severe power imbalances between individuals with very different social backgrounds. Most communication in the legal process takes place between and across lay-people and actors with different ranges of legal experience, increasing the risk of miscommunication. The ways in which a linguist may serve justice through investigating issues of language are thus extraordinarily broad. Some of the issues that a linguist may address come very close to ultimate issues of credibility or guilt. Few other fields have such a degree of overlap between evidential traces, factors affecting social justice, and the determination of ultimate issues. The question whether a linguist could act as "social advocate" and "forensic expert" at the same time could thus very well be a question that is more likely to arise in the field of linguistics than in other fields.

Moreover, evaluation of linguistic evidence generally allows for more subjective interpretation than other types of forensic evidence, due to the variability inherent in language use. Few other disciplines deal with traces that can vary within one source to the same extent as linguistic samples can, depending on topic, emotion, communicative setting, health, etc. The largely subjective nature of the conclusions given by a forensic linguist, as well as the perceptual mechanisms involved, allow room for all sorts of biases to affect the conclusion. Instead of giving linguists leeway to let ideals of social justice influence their forensic casework, however, we would argue that mitigating against biasing effects is an even more pressing issue in forensic linguistics than it is in other fields.

Biases

It is essential that expert evidence remains as free of biases as is practicably possible. We note that cognitive biases are of a very different nature to personal, social, and political biases (Dror 2009).

As a topic which has been very much in the zeitgeist, various strategies to tackle cognitive biases have been proposed in the literature.[3] These biases are generally considered to be unconscious and/or structural; they include, for example, confirmation bias or "tunnel vision", contextual information biases, role effects, expectancy effects, ordering or anchoring effects, and reconstructive effects.

Myriad, detailed countermeasures have been suggested within the literature. In general, they focus on removing or mitigating the sources of bias or the risks of bias by using processes which protect against them. For example, if irrelevant case information contains facts about a suspect's criminal record or other types of inculpatory evidence, it should not be viewed by the expert. If the order in which samples are analyzed could affect the outcome, this process should be carefully managed to avoid this. If there is a danger of approaching the evidence from one viewpoint, or only assimilating information which supports that viewpoint, these risks to impartiality can be mitigated by considering the likelihood of finding the evidence given two competing hypotheses which represent the view of both prosecution and defense.

While much has been written on cognitive biases and how to avoid them, the attention paid to more personal types of bias has been much more limited, the discussion much quieter. Motivational and role effects are mentioned, but they tend to refer to the organizations which forensic practitioners are part of, or where experts see their role within the prosecution or defense "team". Political/social biases may be more conscious, yet they can also be purposely hidden, often behind biased methodology; these biases are therefore harder to pinpoint and address (see Hoskin, Cambier-Langeveld, and Foulkes 2019 for an example from language analyses in asylum cases). Accusations at this level are also quickly perceived as ad hominem attacks rather than efforts to maintain balance.

This chapter might promote a further discussion about what safeguards we should be taking to mitigate personal or ethical biases that might affect forensic casework. Should we, as experts and those managing organizations providing forensic services, screen potential analysts' personal beliefs and their ability to work and think independently of their belief or ethical system? Forensic training currently includes a strong focus on impartiality and bias management, but should it also include awareness training on the kinds of ethical issues raised in this chapter and the wider volume? We consider that training for all forensic experts, whether established or newly qualified, should recognize that the expert's duty is to consider, analyze, and present the evidence neutrally.

Conclusion

This chapter was written to address the "question of whether dispassionate science is always in the best interests of social justice". It can be healthy to occasionally question the basic premises underlying what we do professionally and personally. However, as forensic practitioners, we firmly maintain and support the doctrine that provision of forensic evidence should be clearly distinct from social advocacy.

This is not to say that there is no place for addressing legal and social language issues in a more general sense. There is a vital place for campaigning, critical method assessment and method development, as well as discussions of social wrongs and issues that revolve around language use. There are clearly issues in society which require examination, re-balancing, or campaigning from linguists; our field cannot operate in a social vacuum. However, if a linguist decides to act as social advocate to address a certain legal linguistic issue, it is questionable whether they can at the same time act as an impartial and neutral expert in casework in which that issue is at play. Such a linguist would have either the very real impact of, or the appearance of having, an a priori opinion working against their independence; this linguist would have difficulty in showing that their opinion on the general issue cannot have interfered with the weighing of evidence in a particular case.

There is no question in our minds that forensic evidence must be examined and evaluated without bias, without emotions concerning the case, and without a preferred outcome in mind. This is the "number 1 rule" in official guidance on offering forensic expertise, as well as in codes of practice put together by the field of forensic linguistics itself. Ultimately this rule serves to clearly delineate the roles, tasks, and responsibilities of each contributor to the processes of justice. The forensic linguist, just like any other expert, should stay away from ultimate issues that are for the trier of fact to decide upon. The expert can help the trier of fact with making these decisions justly by offering expert advice on the meaning and weight of the evidence available. To guarantee an objective evaluation of the strength of the evidence that lies before the expert, the expert should do everything in their power to prevent having a predetermined goal related to an ultimate issue in mind during the analysis. Experts are given a privileged position in that they can give evidence of opinion; the expert is simply not at liberty to misuse this privilege to influence the judicial system toward their personal beliefs.

Returning to our core question, we consider that dispassionate forensic science is always in the best interests of justice systems. While there is a place for linguists in the achievement of social justice, this place lies not in the forensic or legal process, or the courtroom, where the role of the expert is rightly to assist the justice system by impartially evaluating the strength of the evidence in the case at hand.

Notes

1 In *St Helens Council v. M & F* [2018] EWHC Fam 1, the court summarized the position of the expert as follows: "39. The role of expert witness is a privileged one, allowing the witness to give an opinion on questions outside the court's own expertise. Such opinions are likely to be influential on the court's thinking. Where they are soundly based, they serve the interests of justice. Where they are not, the opposite is the case. Judges and juries can be misled, parties may come to believe that their cases are stronger or weaker than they truly are, and subsequent assessments may be confounded".

2 See *Anglo Group plc, Winther Brown & Co Ltd v. Winter Brown & Co Ltd, BML (Office Computers) Ltd, Anglo Group plc, BML (Office Computers) Ltd* [2000] EWHC Technology 127.

3 For a summary, see Rhodes and Cambier-Langeveld (in press); for detailed overviews, see Kassin, Dror, and Kukucka 2013, and UK FSR Cognitive Bias Effects Relevant to Forensic Science Examinations, FSR-G-217. For bias mitigation strategies for forensic speech analysis specifically, see Rhodes, Earnshaw and Nuttall, 2024.

References

Association of Forensic Science Providers [UK]. 2009. "Standards for the Formulation of Evaluative Forensic Science Expert Opinion." *Science & Justice* 4, no. 3: 161–164.

Cohen, D., R. de la Vega, and G. Watson. 2001. *Advocacy for Social Justice*. Bloomfield, CT: Kumarian Press.

Criminal Procedure Rules and Practice Directions [England and Wales]. 2020. "Part 19 Expert Evidence." https://www.legislation.gov.uk/uksi/2020/759/part/19/made.

Dror, I. E. 2009. "How Can Francis Bacon Help Forensic Science? The Four Idols of Human Biases." *Jurimetrics* 50: 93–110.

Forensic Science Regulator. 2025, March. *Forensic Science Activities: Statutory Code of Practice*, version 2. Retrieved from https://www.gov.uk/government/publications/statutory-code-of-practice-for-forensic-science-activities.

Forensic Science Regulator. 2020a. "Cognitive Bias Effects Relevant to Forensic Science Examinations FSR-G-217, Issue 2." Retrieved from https://www.gov.uk/government/publications/cognitive-bias-effects-relevant-to-forensic-science-examinations.

Forensic Science Regulator. 2020b. "Legal Obligations FSR-I-400, Issue 8." Retrieved from https://www.gov.uk/government/collections/fsr-legal-guidance.

GSFL. 2017. "The Articles of Association for the Establishment of the Germanic Society for Forensic Linguistics as a Non-profit Registered Society." Retrieved from https://germanicsocietyforensiclinguistics.org/policy-and-governance/.

Hoskin, J., T. Cambier-Langeveld, and P. Foulkes. 2019. "Improving Objectivity, Balance and Forensic Fitness in LAAP: A Response to Matras." *The International Journal of Speech, Language and the Law* 26, no. 2: 257–277.

IAFPA. 2020. "Code of Practice." Retrieved from https://www.iafpa.net/the-association/code-of-practice/.

IAFLL. n.d. Retrieved from https://www.iafl.org/about-iafl/.

ISO/IEC 17025. 2017. *General Requirements for the Competence of Testing and Calibration Laboratories.* Retrieved from https://www.iso.org/ISO-IEC-17025-testing-and-calibration-laboratories.html.

Kassin, S. M., I. E. Dror, and J. Kukucka. 2013. "The Forensic Confirmation Bias: Problems, Perspectives, and Proposed Solutions." *Journal of Applied Research in Memory and Cognition* 2, no. 1: 42–52.

Merriam-Webster.com Medical Dictionary, s.v. "Forensic Science." Accessed April 26, 2024. Retrieved from https://www.merriam-webster.com/medical/forensic%20science

Rhodes, R., Earnshaw, K., and Nuttall, B. 2024. Managing Bias Risks: Information Management Strategies for Forensic Speech and Audio Analysis. *The International Journal of Speech, Language and the Law*, *31*(2), 328-349.

Rhodes, R., and T. Cambier-Langeveld. In press. "Guidance for Practitioners." In *Oxford Handbook of Forensic Phonetics*. Edited by F. Nolan, K. McDougall, and T. Hudson. Oxford: Oxford University Press.

5

ETHICAL ISSUES ENCOUNTERED IN FORENSIC LINGUISTIC PRACTICE

Case Examples Inspired by Real Life Cases

Tina Cambier-Langeveld and Richard Rhodes

Introduction

The role of a forensic expert is to assist the relevant justice system with matters that would be considered outside of the skill and knowledge of the triers of fact. In our view, the primary role of the forensic expert is to: (a) determine what questions the linguistic evidence can or cannot assist in answering; (b) indicate what weight may be attached to the evidence in question; and (c) try to ensure that the evidence is understood properly by the various end-users in different justice systems.

Providing any type of forensic evidence or testimony to the trier of fact is governed by clear and explicit rules. We note the existence of the following documents in the English language as general guidance on the provision of forensic evidence:

- National legal admissibility requirements for expert evidence such as the Criminal Procedure Rules for England and Wales (2020), the "Frye standard" set by the US District of Columbia Court of Appeals (1923), and the "Daubert criteria" produced by the United States Supreme Court (1993)
- Standards set by the Association of Forensic Science Providers (2009)
- Codes of practice for forensic practitioners set by national bodies such as the UK Forensic Science Regulator (2020, 2025) and the US National Research Council (2009)
- Accreditation standards set by the International Organization for Standardization (ISO), in particular ISO/IEC 17025 (2017), and by the International Laboratory Accreditation Cooperation (ILAC), in particular ILAC G19:06/2022 (2022)

While many of the above-mentioned guidelines have been designed for use by a wide range of experts working within the forensic sciences, specific guidance and standards for the field of linguistic and phonetic analyses also exist:

- Documents from the European Network of Forensic Science Institutes (ENFSI) such as the 'Best practice manual for the methodology of forensic speaker comparison' (2022)

DOI: 10.4324/9781003391074-6

- Codes of practice for members of professional organizations such as the International Association for Forensic Phonetics and Acoustics (2020), the International Association for Forensic and Legal Linguistics (no specified date of publication), and the Linguistic Society of America (2011)

Taken together, these documents provide professional guidance for forensic linguists who offer their expertise to practitioners within the law enforcement and the criminal justice system. However, the guidance is of a general nature and does not prevent ethical dilemmas from still arising in actual practice. In this chapter we aim to illustrate some of the ethical challenges practitioners may face, and it will illustrate how even the most well-intentioned experts may find themselves taking less than perfect courses of action. Such situations arise even when experts involved in forensic casework are aware of the relevant standards, their legal obligations, and the necessity of maintaining a neutral stance.

For example, in real life it is not always easy to recognize when the instructing party is influencing the outcome of an analysis by the way in which they present the case to the expert or by the way in which the question is formulated. Furthermore, it is not easy to acknowledge that case information known to the expert may compromise their objectivity, despite their professionalism in general. And therein lies an important point. In the experience of the current authors, forensic evidence that does not abide by the key principle of impartiality is often not produced out of malicious intent. Rather it tends to stem either from ignorance on the potential effects of cognitive bias on human judgments, or from the human desire to "do good" by contributing to what the expert believes is the just outcome of the case as a whole. Yet elsewhere in this book (see Chapter 4), and in line with general guidance, we argue that biases toward what is perceived to be "the just outcome" should not be allowed to influence expert evidence.

This chapter presents several hypothetical cases which illustrate the kinds of ethical issues that the authors have encountered in their work as forensic speech scientists. Both authors have each provided expert evidence in hundreds of cases and have also witnessed the work of colleagues who have provided expert testimony. While the examples presented in this chapter are inspired by real casework, circumstances from several cases are synthesized and particular details have been changed.[1] For each case we will put forward a supposed case and outline what one might (be tempted to) do as an expert. We then discuss the central issue in the case and suggest what we think is a more appropriate course of action. In some cases, although the case is imagined, the inappropriate response of practitioners to these issues is sadly less than hypothetical.

These examples are meant to serve as food for thought. Given that ethical or moral questions are involved, it is inevitable that there are gray areas. We (the authors) acknowledge that our view on the most appropriate action may not be universally accepted or shared by all. Moreover, there is very rarely only one right way to do things (and there are certainly myriad wrong ways). However, our perspective is informed by both relevant guidance and our varied experience working as forensic experts.

Hypothetical Case Examples

Case 1: A Highly Publicized Kidnapping Case

The case: A teenage girl from a high-class family was abducted. Her parents received a number of phone calls demanding ransom money. All except the first phone call were recorded. Despite the fact that the parents paid a substantial amount of ransom money, a few weeks later the body of their daughter was found. She had been raped and killed. Her body was then left behind in a small wooded area a few miles from the victim's house. DNA traces of the offender were found but did not give a match within the DNA database of convicted criminals.

The investigation initially led nowhere until months later, when all residents living within a few miles of the victim's home were asked to voluntarily submit a DNA sample. The ethical questions raised by this drastic measure sparked widespread media coverage. Yet the unusual measure proved effective: through the large-scale DNA testing initiative, it was possible to identify two innocent individuals who were genetically strongly related to the perpetrator. A suspect was then arrested and a full DNA match between the suspect and the DNA recovered from the crime scene was established. Yet much to everyone's dismay, the suspect refused to give any statements and continued to deny involvement.

In search of additional evidence and in order to establish whether the suspect was also the ransomer (implying that he must have received the large sum of money), an expert was called in to perform a speaker comparison between the ransom calls as disputed materials and newly recorded telephone conversations with the suspect as reference material.

The analysis: The expert, well aware of the situation as sketched above, set out to analyze the materials. In both sets of material, there were accent features consistent with an origin from a particular geographical region. The speech was found to be fluent and indicative of someone who was well-educated. Apart from these broad similarities, the disputed sample sounded markedly different: the voice was much lower and deeper, the vowel formants were lower overall, the speech rate was slower, the rhythm was more staccato. Yet all of these differences could be due to voice disguise in the disputed samples, for example, by purposely lowering the larynx. The expert thus concluded that there was nothing that would preclude the samples from coming from the same speaker. The expert further stated that it was very well possible that the disputed samples had been produced by the suspect.

The ethical issue in this case: To what extent might the conclusion have been influenced by the a priori expectation that the suspect is indeed the ransomer, or by the implicit (conscious or even subconscious) desire to help the prosecution with this case?

Our comments: In high-profile cases, there is an increased chance that experts are hired who are willing to testify in court to achieve what is perceived as social justice. Knowledge about the nature of the offense and the other evidence in the case can be detrimental to the impartiality and objectivity of the expert (Saks et al. 2003). The only practical solution is to apply some form of context information management, such that the request for services and the supply of materials are received by someone who will not be serving as the forensic analyst (Forensic Science Regulator 2020a; Mattijssen et al. 2016; Kassin, Dror, and Kukucka 2013). This precaution helps to ensure that case information is withheld from the expert(s) performing the analysis. In this example it might also be worthwhile to first present the expert with an edited version of the disputed speech in which any words/utterances that would link it to the actual case had been stripped away. The expert could then be asked to determine whether there are signs of disguise (and if so, what kind). This would reveal whether there are evidence-internal indications of a certain type of disguise, instead of an assumption or post-hoc rationalization of disguise that is prompted by differences found with the reference material.

Implementing a form of context information management may be somewhat of a challenge for lone consultants. In a high-profile and widely publicized case as presented above, with an unavoidable bias against the suspect due to known other evidence, we would advise a lone practitioner to ask the client for permission to set up a case-specific collaboration with another expert. This would allow precautions like the ones we suggest to be taken, which is absolutely critical in this case.

Case 2: Review of a Voice Recognition Procedure

The case: The telephone conversations of a group of young adults were wire-tapped by law enforcement who believed they were involved in criminal activities. The group appeared to share a number of mobile phones among them, so that the telephone numbers could not be linked to a single user. It was only during these telephone conversations that it became clear what each of the individuals' roles were in a series of smaller and larger criminal activities. Two community police officers, familiar with the young individuals, attributed speech in the recorded telephone conversations to the various individuals. The officers claimed to have been able to identify the speakers on the basis of their nicknames and voice recognition. The individuals were charged, and the case went to court. The defense, eager to call the reliability of the officers' non-expert voice recognitions into question, called upon an expert to review the voice recognition process in this case.

The analysis: The expert set out to establish whether the process followed by the officers was reliable, and in particular, whether it met the criteria for expert forensic analysis. Given the fact that the police officers lacked the required expertise and training and had not performed any linguistic, acoustic, or auditory feature-based analyses, the expert concluded that the voice recognitions did not meet the criteria for expert speaker comparison, were therefore without scientific basis, and must accordingly be deemed scientifically invalid and unreliable.

The ethical issue in this case: Did the expert provide expert evidence/testimony in this specific case or did the expert de facto argue against reliance on voice recognition by police officers in general?

Our comments: The conclusion that the voice recognitions did not meet the criteria for expert speaker comparison was in fact known from the outset, since the process did not involve experts to begin with. The expert appears to have interpreted and addressed the question as a general issue rather than as a request to offer advice on how to evaluate the case-specific evidence. However, the expert's conclusion was presented as if it were based on an evaluation of the voice recognition process in this case. Consequently, it may not be transparent that the argument applies to any voice recognitions by the police.

We can think of at least two alternative courses of action that might have been more helpful in this situation. The first is to provide the court with guidance on how to evaluate the reliability of the evidence that lies before them. The expert could have pinpointed the type of information needed to determine whether the recognitions came about independently from prior information or not, which is crucial to determining the value of the officers' evaluations. These include how the community police officers had been instructed, how the materials they assessed had been presented to them, how much information they were given, how often the recognitions were based on the use of nicknames, and whether independent double-checks were done. Furthermore, it would be helpful to explain in some detail that recognition of familiar voices (as in the officers' assessments) is not the same thing as expert analytical comparison of unknown voices: these processes take place in different parts of the brain (Kreiman and Sidtis 2011).

A second alternative course of action would have been to provide additional evidence in the form of re-analysis of a subset of the data. The expert could have asked the defense to provide specific examples where the attribution of speech to a certain individual was disputed, as well as some undisputed reference material. The expert could then perform an expert auditory-acoustic analysis. Such an exercise would serve first to pinpoint exactly where the police voice recognition might have gone wrong according to the defense, then to provide an independent expert opinion on the matter, and finally to indicate whether the police voice recognitions are in line with the results of expert analysis in a subset of the data.

Case 3: Screaming and Possession of a Firearm

THE CASE: A street brawl between two men ends fatally. One of them must have carried an illegal handgun. A security camera recorded the sound of the fight, but no images. The defendant claimed that the deceased had taken out a gun at some point during the fight, but after some struggling the defendant managed to grab it and shoot. On being presented with the audio sample, he claimed that it was he who screamed for help and at some point yelled, "he has a gun!". The prosecution maintained that it was the deceased who was screaming for help. An expert is called in to address two questions: (1) is it possible to perform speaker comparison to determine who was screaming for help? and (2) is it possible to confirm the utterance, "he has a gun!"?

THE ANALYSIS: The expert decided to have the audio transcribed by three listeners who had no knowledge of the alleged content. The result is that only one of them comes up with "(...) a gun", so it is not possible to confirm the alleged utterance. The utterance is produced in falsetto and the rest consists of panicked, high-pitched screaming and words/sounds like "help" and "aah". These materials are clearly not suitable for a speaker comparison, and so the expert reports accordingly.

The expert also noticed that the screaming stops very abruptly, mid-word, immediately after the gunshot. Not just before, not a few seconds after. The expert pondered whether this fact might be considered more in line with the scenario in which the screaming comes from a person who was shot, than with the scenario in which the screaming comes from a person who had to struggle for the gun and managed to shoot. Yet commenting on this observation falls outside the remit of the phonetician, nor was it part of the assignment. The expert reminds himself that he should stay within his area of expertise and decides not to report this particular observation.

THE ETHICAL ISSUE IN THIS CASE: While limiting oneself to the given explicit assignment is a defendable course of action, does it not fall within the expert's responsibility to ensure that no relevant evidence contained within the materials presented to the expert is missed by the trier of fact?

OUR COMMENTS: We note that in some jurisdictions it is common for the trier of fact to rely almost entirely on written statements and reports, and it is highly uncommon that the underlying audio is listened to, plainly due to lack of appropriate equipment. In these jurisdictions, the expert might be well-advised to check if the available written report(s) would give the reader any indication of how abruptly the screaming stops. If this is not the case, the expert could choose to briefly make note of this observation in the report and suggest that in this case the trier of fact may want to listen to the recording to draw their own conclusions.

Generalizing from this example, we note that the duty to help the court in reaching justifiable conclusions based on the available evidence can sometimes conflict with the demand to stay within one's area of expertise. While our opinion on this may be open to dispute, we feel that the expert has a duty not only to perform their tasks but also to indicate important evidence from the speech or language sample that runs the risk of being missed by the trier of fact.

Case 4: Provision of a Counter Report in an Asylum Case

THE CASE: A linguist specializing in an African language and its dialects is approached by a lawyer acting for an asylum seeker. The expert is asked to review a language analysis report and to provide a counter report. Such language analysis reports are used by immigration services to investigate whether the applicant's linguistic repertoire can confirm the claimed region of origin. The initial report concluded that the asylum seeker does not speak the language variety of the Southern part of the claimed country of origin, although the person claimed to have come from that area. According to the lawyer, the asylum seeker is traumatized and has not been able to testify consistently to the authorities about the details of his travel route and other experiences. The instructing lawyer informs the linguist that the government lets native speakers do "language analyses" to make their negative decisions look stronger for the court and that these native speakers have a rather simplistic approach to language: for example, they rarely if ever consider secondary influences on a person's language. In this particular case, according to the lawyer, the asylum seeker's mother came from the Northern part of the country, although he was born and raised in the South. In addition, due to the civil war in his home country, the asylum seeker claimed he spent most of his life indoors, listening to a radio station that broadcasted in the national standard which is based on a Northern language variety.

The analysis: The expert examined the initial report. Indeed, the report can hardly be considered a scientific report. It lists a few contrastive examples of pronunciation, word choice, and grammar, but does not contain an exposé on the dialects in question, nor does it provide any linguistic discussion of the relevant features. When listening to the speech sample, the expert was able to identify quite a few well-known Southern features along with many standard features. The expert thus wrote a counter report which concluded that the applicant's language variety was in line with the region of origin claimed; and that the standard features identified were to be expected given the applicant's background. The expert's counter report also included a sharp critique of the initial report.

The ethical issues in this case: Is it possible that the conclusion reached in the counter report regarding the speaker's origin is influenced by the expert's feeling that the asylum authorities are too restrictive and unjustly withholding protection from those who need it? What should the expert have known about the way in which language analyses in the asylum procedure are performed before offering critique? How accurate and reliable was the information provided by the instructing party in this case?

Our comments: There is always the possibility that strong emotions or opinions may (un)consciously affect a person's judgment. In this case, there is a risk that the expert's critical attitude toward the government's policies or treatment of asylum seekers in general may have affected the expert's evaluation of the evidence. In this case, the linguist should at the very least have been more critical of many of the instructing lawyer's claims. For instance, the lawyer's claim that governmental language analysis reports are performed by native speakers to strengthen their negative decisions suggests that the lawyer may be inclined to present a somewhat biased picture of the situation, perhaps in an effort to instill a counter-bias in the linguist. Second, the linguist might have been more wary of the defense lawyer's tendency to accept claims of secondary influences as facts. The assertions of the Northern origin of the mother and the predilection toward listening to a certain radio channel could be true. However, they could also have been made up to account for the initial language analysis results. We note that an expert should thus be careful to accept claims of secondary influences at face value. The expert should further be aware that a defense lawyer may be expected to argue vigorously for a client. Thus it may be wise to limit both the extent of interaction with the lawyer and the nature of the information provided, to shield the expert from biasing influences.

The entire notion of providing a "counter report" may actually be viewed as undesirable, since it can easily lead to the misinterpretation that it is the expert's task to "counter" the initial report. A more neutral term would be the provision of a "second opinion". It is also questionable whether a critique of the initial report is appropriate or necessary, especially when the expert knows little about the criteria imposed on these reports by external factors. One needs to be aware that forensic reports are not scientific papers, and their content is often restricted. In asylum cases like this, institutional demands and/or a general policy to keep information away from future applicants as much as possible may well be the reason that reports do not contain an exposé of the linguistic features underlying the analysis (Cambier-Langeveld 2012). Again, providing an independent second opinion on the question at hand would be more helpful to the trier of fact. Should a review be required, this should be preceded by an independent analysis.

Case 5: A Request Not to Write a Report

THE CASE: A linguist is approached by a defense lawyer who has just received a report with an unfavorable conclusion for their client. The linguist is asked to perform a preliminary examination. After preliminary independent analysis and a review of the first report, the expert informs the defense that he agrees with the findings as well as the conclusions of the first report. The defense then says they will not be needing anything further, so the linguist does not need to write a report.

A few weeks later, the linguist finds out that the defense still chooses to argue against the conclusions of the first report. Moreover, they found themselves another expert who was apparently more inclined to provide a conclusion that was favorable to their client. Somewhat frustrated, the linguist wonders whether he should contact the prosecution office or the court to let them know the outcome of his preliminary examination.

THE ETHICAL ISSUE IN THIS CASE: In this case, the linguist is running into what some perceive as an unbalanced justice system: while in most jurisdictions the prosecution should weigh the evidence in all fairness and is moreover obliged to share their information with the defense, in many jurisdictions the defense has no such disclosure obligations. How does the responsibility of the expert to help and inform the court balance out against the right of the defense in some jurisdictions not to submit information that is unfavorable for their client?

OUR COMMENTS: We note that, however unfair it may seem, the forensic practitioner should not attempt to go against the workings of the justice system. It is often within the role of the defense to only submit evidence that is favorable for their client. Whether this means that they should be allowed to go "expert-shopping" is not for linguists to decide upon.

To be clear, the issue at hand is different from the one presented in hypothetical case 3, where we suggested that it is the expert's duty to raise the trier of fact's attention to evidence within the audio materials if this evidence runs the risk of being overlooked. In case 3, the reason for this evidence possibly being overlooked is the typical reliance on written representations of the audio sample, while it is the audio that constitutes the actual evidence. We argued that pointing to the original audio evidence to ensure inclusion of non-linguistic clues in the evaluation of the evidence would fall within the duty of the expert. On the other hand, informing the court that the defense has withheld information—presuming that this is within their rights within the relevant jurisdiction—would constitute damaging the rights of the defense, and thereby the justice system that is in place. We argue that this is clearly not within the expert's remit. We could imagine as an acceptable course of action that the expert sets out to invoke a legal or even social debate on the question whether "expert-shopping" is a form of professional misconduct or not, but we think this should be done without reference to a specific case.

Case 6: Audio Evidence in a Rape Trial

THE CASE: A forensic phonetician was instructed by the defense team in a rape and sexual assault case to transcribe a series of WhatsApp voice note recordings between the complainant and the defendant. The expert was given a letter of instructions which also described how the defense lawyers cannot fully understand most of the speech, but they think that the messages are flirtatious and characterize the complainant as inviting sexual attention, either undermining the complainant's character or discrediting their account of the events that took place. The letter requested the audio to be transcribed so that the defense can pursue a narrative that the complainant invited, and therefore consented to, the sexual activity.

THE ETHICAL ISSUE IN THIS CASE: An expert might not personally agree with how this type of defense is used in rape cases; they may feel that, in general, it is one of the reasons that effective prosecution rates for sexual offenses are so low, and it is damaging to victims of crime. This might lead the expert toward an interpretation that was not neutral: they might develop a preference for marking certain speech from the complainant as "unintelligible", rather than presenting a potential transcription of words that may be viewed as possibly flirtatious, for example. Although most sources talk about motivation bias effects as favoring the instructing party, experts should also be aware that anti-client biases can exist as well.

OUR COMMENTS: This hypothetical case demonstrates how management of information can protect the forensic expert against different influences induced by extraneous case information or material that is not directly relevant to their examinations, thereby protecting their impartiality. If the instructions had been filtered by someone other than the expert, and if any material not directly relevant had been removed, this effect would not have been induced. Lone practitioners can provide instructions to potential clients prior to any assignment, as to the type of information that is required to develop an effective case strategy, versus extraneous case information that should not be presented to the expert. If transcriptions already exist, these should also be kept away from the expert at least during their first pass (i.e., when making their own independent transcription).

Discussion: General Observations from the Hypothetical Cases

Some general lessons may be learnt from the case examples presented above. Principally, the expert should be protected from biasing information concerning the case at hand. Police agencies may usually consult an expert to confirm their pre-existing version of events. However, the expert should not allow their analysis and conclusions to be led by such expectations. Indirect communication of these expectations, or even pressure toward them, is one reason why direct communication between the analyst and the instructing police team is best avoided. This can be achieved by having the person(s) responsible for the data analysis be different than the person who has contact with the contracting party. Lone practitioners are advised to be as explicit as possible about the information that is required for an assignment, as well as to emphasize that other types of extraneous information (irrelevant for the analysis) should *not* be provided.

We also note that a forensic expert should keep a critical eye on the instructions and information received from the instructing party. It is crucial that the expert makes a sharp contrast between facts and unverified claims, even if the instructing party does not. The formulation of the question to be addressed by the analysis is also critical and can steer toward an a priori desired conclusion. There are principled ways of deciding whether the question posed is appropriate for the forensic expert to address: the evidence assessment should not have a predetermined outcome but instead should have the potential to support either side, it should have truth finding to contribute to fair decision-making at its heart, and the expert should have the expertise to analyze the evidence in light of the question to be addressed while the trier of fact does not.

Reviews of existing expert reports and conclusions are normally carried out on behalf of the defense in respect of prosecution reports. In this situation, one may wonder if it is helpful to offer critical reviews without giving an actual second opinion on the evidence itself, or without being able to offer a solution which is both practicable and more effective. Methods are not infallible, nor do they have to be; forensic practitioners tend to be well aware of the limitations of their field, yet they choose to offer the court guidance to the best of their ability for the benefit of justice. It is easy to criticize from the sideline, but an expert considering this would first need to think about the criteria that apply. An academic without forensic training and case experience may not be in the best position to fully understand and appreciate the practical challenges involved in policing, trials, and forensic analysis.

Finally, we observe that highly publicized cases increase the risk of biased expert opinion. We can think of several reasons why this may be so. First, the pressure to "do well" is higher, while insulation against knowledge about the case (such as the nature of other evidence in the case) is harder. Second, high-profile cases tend to contain explosive and/or emotional subject matter, which can induce stronger conclusions in expert decisions (Dror, Charlton, and Peron 2006). Third, high-profile cases appear to attract experts who are not necessarily experienced forensic experts, but become involved with motivational biases. These motivational biases might have to do with the publicly desired outcome of the case or with the chance to promote oneself by playing a newsworthy part in a high-profile case. Obviously, we believe that a neutral stance is important in every case including high-profile ones.

Conclusion

The case examples discussed in this chapter illustrate how the context in which the forensic expert works has the potential to influence the expert's approach. Various countermeasures against cognitive bias have been proposed in forensic literature. These are mainly practical and methodological in nature and generally involve forms of context information management aimed at shielding the expert from biasing information (e.g., Mattijssen et al. 2016; Dror et al. 2015). However, where speech evidence is involved, case information cannot always be withheld from the expert, since it is often in the speech signal itself.

Consequently, within forensic speech sciences, the development of additional methodological measures against bias is essential (see Rhodes, Earnshaw, and Nuttall, 2024). At the same time, there is an overarching need for a more open discussion about how to strike a balance between being a socially engaged human being by nature, presumably with a passion for one's profession, and maintaining a professional and neutral (dispassionate) attitude in the interest of forensic science. This discussion is critical because general guidance (in the form of a manual or a code of practice) does not always provide the best course of action in a specific situation. The case examples presented in this chapter show a narrow glimpse of the wide variety of ethical issues forensic linguists may encounter in real life.

Opportunities like the present book to discuss non-methodological, sensitive, and often personal issues are rare. Awareness of the existence of such issues is important to newcomers to the field and established scholars alike. The quality of forensic linguistic evidence lies not just in the linguistic and phonetic expertise of the expert but also in the expert's ability to deal with the mental and ethical challenges inherent to working in a forensic and real-life context.

Note

1 In fact, considering whether to publish our experience from real cases presented an ethical issue in itself. The guidance the authors tend to follow is that public comment on cases is best avoided, especially as cases can always go to appeals. Preventing recognition of actual cases was also important to avoid indirect identification of other experts involved. However, there are clear benefits in sharing best practice and raising issues for discussion. As a field of forensic linguistics/phonetics, we should always be looking for opportunities to learn from our collective experience. The method we have chosen here represents our solution for addressing all of these practical and ethical concerns.

Case Law

Daubert v. Merrell Dow Pharmaceuticals, 509 U.S. 579. 1993.
Frye v. United States, 293 Fed. 1013 (D.C. Cir.) 1923.

References

Association of Forensic Science Providers [UK]. 2009. "Standards for the Formulation of Evaluative Forensic Science Expert Opinion." *Science & Justice* 49, no. 3: 161–164.

Cambier-Langeveld, T. 2012. Clarification of the Issues in Language Analysis: a Rejoinder to Fraser and Verrips. *The International Journal of Speech, Language and the Law 19*(1), 95–108.

Criminal Procedure Rules and Practice Directions [England & Wales]. 2020. "Part 19 Expert Evidence." Retrieved from https://www.legislation.gov.uk/uksi/2020/759/part/19/made

Dror, I. E, D. Charlton, and A. Peron. 2006. "Contextual Information Renders Experts Vulnerable to Making Erroneous Identifications." *Forensic Science International* 156: 174–178.

Dror, I. E., et al. 2015. "Letter to the Editor- Context Management Toolbox: A Linear Sequential Unmasking (LSU) Approach for Minimizing Cognitive Bias in Forensic Decision Making." *Journal of Forensic Sciences* 60, no. 4: 1111–1112.

ENFSI. 2022. *Best Practice Manual for the Methodology of Forensic Speaker Comparison ENFSI-FSA-BPM-003*. Retrieved from https://enfsi.eu/about-enfsi/structure/working-groups/documents-page/documents/best-practice-manuals/

Forensic Science Regulator. 2025, October . *Forensic Science Activities: Statutory Code of Practice*, version 2. Retrieved from https://www.gov.uk/government/publications/statutory-code-of-practice-for-forensic-science-activities

Forensic Science Regulator. 2020a. "Cognitive Bias Effects Relevant to Forensic Science Examinations FSR-G-217, Issue 2." Retrieved from https://www.gov.uk/government/publications/cognitive-bias-effects-relevant-to-forensic-science-examinations

Forensic Science Regulator. 2020b. "Legal Obligations FSR-I-400, Issue 8." Retrieved from https://www.gov.uk/government/collections/fsr-legal-guidance

IAFLL. n.d. *Website*. Retrieved from https://www.iafl.org/about-iafl/

IAFPA. 2020. *Code of Practice*. Retrieved from https://www.iafpa.net/the-association/code-of-practice/

ILAC. 2022. *Modules in a Forensic Science Process* ILAC-G19:06/2022. Retrieved from https://ilac.org/publications-and-resources/ilac-guidance-series/

ISO/IEC 17025. 2017. *General Requirements for the Competence of Testing and Calibration Laboratories.* Retrieved from https://www.iso.org/ISO-IEC-17025-testing-and-calibration-laboratories.html

Kassin, S. M., I. E. Dror, and J. Kukucka. 2013. "The Forensic Confirmation Bias: Problems, Perspectives, and Proposed Solutions." *Journal of Applied Research in Memory and Cognition* 2, no. 1: 42–52.

Kreiman, J., and D. Sidtis. 2011. *Foundations of Voice Studies: An Interdisciplinary Approach to Voice Production and Perception*. Chichester: Wiley-Blackwell.

Linguistic Society of America. 2011. "Code of Ethics for Linguists in Forensic Linguistics Consulting." https://lsaethics.wordpress.com/category/forensic-linguistics-ethics-statement/

Mattijssen, E. J., et al. 2016. "Implementing Context Information Management in Forensic Casework: Minimizing Contextual Bias in Firearms Examination." *Science and Justice* 56: 113–122.

National Research Council. 2009. *Strengthening Forensic Science in the United States: A Path Forward.* Washington, DC: The National Academies Press. Retrieved from http://www.nap.edu/catalog/12589.html

Rhodes, R., K. Earnshaw, and B. Nuttall. 2024. Managing Bias Risks: Information Management Strategies for Forensic Speech and Audio Analysis. *The International Journal of Speech, Language and the Law, 31*(2), 328-349.

Saks, M. J., D. Risinger, R. Rosenthal, and W. Thompson. 2003. "Context Effects in Forensic Science: A Review and Application of the Science of Science to Crime Laboratory Practice in the United States." *Science & Justice* 43: 77–90.

6

MAINTAINING PROFESSIONAL DISTANCING IN SOCIAL JUSTICE CASES INVOLVING NON-NATIVE SPEAKERS (NNSS)

Margaret van Naerssen

Introduction

In this chapter, I address selected ethical issues I have encountered over 20 years of work as an applied linguist on Non-Native Speaker (NNS) cases in the US legal system.[1] In this area, ethical issues can develop out of tensions involving court and professional standards that require scientific reasoning and objectivity, as well as linguistic examinations of the individual communication needs of NNSs of English.

Providing these professional services can be said to inherently involve social justice from the political ideological vantage point that everyone deserves equal economic, political, and social rights and opportunities. This fundamental right has been widely championed/advocated/endorsed by a variety of organizations devoted to equality from international agencies such as the United Nations General Assembly (1966) to private charitable organizations like the San Diego Foundation (2024). One main principle all these groups share is that social justice involves equal access. In a legal context, when a person does not have equal access to legal resources, social injustice may result. Likewise, access to justice can be impacted by how well a NNS can or cannot effectively communicate about a topic in the language of the court. Thus, in NNS cases fair language access can affect social justice.

Scientists working on NNS cases may be asked to address social crises. When that happens, they need to "follow the science" even in the face of evolving government policies and distrust by society. Applied linguists, as scientists, might also face distrust or resistance when presenting their findings as experts in courts. To work effectively, applied linguists should try to identify the misbeliefs or folk beliefs about NNSs and their language(s) held by participants in the judiciary and law enforcement who may not hear, or want to hear, linguistic explanations.

One reason for such resistance is that language-related issues can feel deeply personal rather than scientific. While it is important that applied linguists who consult on NNS cases recognize this fact, they must remember that it is their professional duty to dedicate their skills and experience in linguistics to scientifically address issues in the real world. As I usually work on cases involving single individuals, each case tests my ability to address real-world problems. In each

DOI: 10.4324/9781003391074-7

NNS case, my task is straightforward: I need to examine the language evidence in the legal process in order to ascertain the degree of language access to the legal system. Only then is it possible to answer the question of whether access to social justice was available.

The linguistic expert's role is only to assist the court in understanding the nature of the language evidence, not to give an opinion on guilt or innocence. To maintain professional credibility, the expert needs to remain objective and use appropriate *scientific* reasoning to maintain the necessary professional distance and not become too involved in a case.

The goal of this chapter is to support linguists working on NNS cases to maintain the professional position of distancing, so that they can objectively and scientifically examine the language evidence to determine whether a NNS has had fair language access to justice, thus, access to social justice.

Despite my long-term professional adherence to this ethical code, I was once challenged about my fairness. In essence, the person said: "So, you've worked for so many years with Non-Native Speakers of English. Then, you must be biased in your work to favor them!!" I was shocked at the overtly stated assumption about a potential ethical violation on my part. I responded by briefly listing two advantages I have due to my familiarity with NNS communications. The familiarity helps me consider what is relevant to the legal question. It can also help me spot possible inconsistencies in the NNS's language in case the NNS might be trying to manipulate the system. The challenge stopped there. I had maintained professional objectivity. That experience taught me a lot about my future work in the field.

Previewing the chapter organization, I begin by addressing two general legal contexts involving the need for professional distancing along with examples of ethical issues for linguists to consider. The chapter then provides supportive resources, i.e., references to two professional codes of ethics and a set of guidelines. A closing section addresses the role of folk beliefs about NNS language.

Distancing Contexts and Ethics in NNS Forensic Cases

The main body of this chapter addresses two general legal contexts involving the ethical need for maintaining professional distancing. These contexts occur when (1) becoming engaged on a case; and (2) conducting language assessment of NNSs. While not all inclusive, these issues are representative of contexts colleagues and I have encountered in NNS cases. Rather than discuss the issues *abstractly*, a variety of concrete examples illustrate the legal contexts and professional ethical issues. As the discussion will demonstrate, in each of these situations, masking one's emotions can become a distancing tool to avoid both contaminating expert reports and behaving inappropriately in court. This is an essential point to remember. No matter how great the emotional pull of the social justice issues involved, the job of an expert is to objectively assist the fact-finders in the court (e.g., the judge and/or jury) to understand the language evidence so that they can make informed decisions. As a linguistics expert consultant and/or witness, you are not allowed to provide an opinion on guilt or innocence. To help ensure that you maintain the necessary professional distance in the two contexts mentioned above, each one of them is discussed separately below.

Becoming engaged on a case

This section begins by examining five selected ethical topics around a linguistics expert's decision to become engaged with a NNS legal case: (1) conducting a bias self-check; (2) checking your qualifications; (3) developing linguistic questions; (4) separating your expert linguistic opinion from an attorney's agenda; and (5) gathering information about the NNS and the case without learning too much. Each of these precautions is examined individually below.

Conducting a Bias Self-Check

A bias self-check can help you decide whether to take a NNS case. Ask yourself what your own personal biases are and whether you can overcome them. It's natural to have personal biases, including subconscious ones about the defendant and the nature of the alleged activities.

Below is one common claim about language and NNSs (van Naerssen 2013). First read it to conduct your own bias self-check. Then read a possible counter response.

> An adult living and working in the US for 10 years should be able to understand and speak English. If such a person claims not to understand English at all or only very little, the person is lying.

Based on first language (L1) and second language (L2) development research, length of residence (LOR) is not the sole predictor of language proficiency. It is especially important to consider the opportunities for regular, meaningful L2 interaction. Other factors to consider are the use of L1 and L2 in the NNS's place of residence and work; the NNS's dependence on L1 or L2 among family and friends; and the NNS's age of first L2 exposure. NNS adults may also be self-conscious about their limited English communication skills and find it easier to simply self-report having limited English proficiency without having the negative intention of lying.

Lightbown and Spada (2013) provide an accessible basic reference for teachers and language practitioners: *How Languages Are Learned.* Their discussion explores popular (mis)beliefs about L2 speakers' language. This reference can be a useful starting point for linguists to do their own self-checks in legal contexts. Self-checks can help linguists spot biases held by others during NNS cases. It can also help linguists maintain professional distancing in NNS cases. Below is a bias self-check I did.

Although already experienced with NNS cases, early in my expert work, I was asked to work on an alleged rape case involving a NNS male. This was my first case involving possible rape. I disclosed this fact to the female military attorney. I also informed her that, of course, I would not know ahead of time what the evidence would show. As a female linguist I hesitated about taking the case. I asked myself how objective could I be? I began to gather general information about the context of the case. I learned that in a US military context, fraternization between military personnel (e.g., dating, sexual relations, extra-marital affairs, etc.) could lead to court-martial. I also learned that the language evidence in this case would involve direct/indirect contradictory details of the intimacies involved and the US military's code for defining "rape". I realized that the only way I'd know if I could maintain professional distance would be to take the case. If, at some point, I truly felt I could not remain objective enough, of course, ethically, I'd owe it to the attorney and defendant to withdraw. In the end, I stayed on the case.

It is also possible to perform bias self-checks later in a case, after you have agreed to serve as a linguistics consultant. For example, I alerted the hiring attorney's team to general concerns I had about the content and style in my report. Several concerns I had were whether I had gone beyond the scope of my assignment, what might not be allowed in court, and whether there were parts of the report that weren't clear. Attorneys are allowed to generally address such questions but cannot tell an expert what to say. Through my carefully worded questions, their indirect responses helped me make the necessary adjustments to maintain the necessary professional distance.

Checking Your Qualifications

Applied Linguistics has a wide range of subdisciplines potentially relevant to NNS communications including language development/acquisition (Lightbown and Spada 2013), sociolinguistics, phonetics and phonology, and language assessment. An applied linguist might be an expert in a number of these areas. However, that expertise might not include the formal assessment of the oral interactive language proficiency of NNSs.

In the, USA, no official court certification is required to serve as an expert witness in linguistics. Instead, the attorneys who hire linguistics experts present them in court. If you are hired to serve in this capacity, counsel from the other side might challenge your credentials. This voir dire is your job interview, with the judge as the gatekeeper. Attorneys and judges are not always sure what linguists can opine on. This explains why, as a linguistics expert, you will soon become accustomed to the usual question: "So—how many languages do you speak?"

During voir dire, briefly explain the areas of linguistics you work in. If you're working with the defense, also consider noting, for balance, any work experience you've had with the prosecution or law enforcement. You do not want to be seen as a "stooge" or "hired gun" for the defense. Also, if asked in court how much you are being paid for your testimony, Shuy has personally advised colleagues to reply, "I'm not getting paid for my opinion. I'm getting paid for my expertise and time". If you have already been an expert witness, a judge or attorney may assume you are broadly qualified for all language matters. Or the judge, for whom language is a daily tool, may decide no expert on language is even needed. Shuy (2006) reminds us to show the judge what more a linguist can offer.

For our discussion here however, there is a more important question at hand. Beyond presenting yourself as a linguist, ask yourself if you have the narrower specialization(s) relevant to the specific issues in the case. The question you need to ask yourself is not whether you are an expert in linguistics but "Am I the right expert?" (van Naerssen 2023).

If, overall, you are qualified but have limited expertise on one relevant issue, it is advisable to let your hiring attorney know you intend to consult a relevant expert to strengthen a point in your report. This is a strategy which I have found helpful in the past. In one case, recognizing my limitations, I consulted two colleagues—a chemistry professor and a corpus linguistics expert—to help me prepare a demonstration involving a statistical argument. Although they could not be informed of any details of the case, their scientific expertise was an invaluable resource in my report.

The case involved a complex sting operation to break up a criminal operation. Local law enforcement had identified a "foreign" owner of a convenience store who they believed might be involved in stealing bottles of a well-known brand of liquid detergent to sell to a complicit dealer who would then re-sell the "liquid gold" to buy drugs. The goal of the sting operation was to break up this lucrative interstate criminal network. There was one major legal challenge, however. Under federal regulations, in a sting operation conducted by law enforcement, the "target" must understand that the proposed activity is "illegal". A future prosecutor needs to prove that the target knowingly agreed to carry out the criminal activity. So, for a successful sting, it was essential to determine if the target understood enough English to know what was happening. This was where my NNS expertise came in.

The evidence I examined involved cell phone A–V recordings made by a local law enforcement officer (LEO) who visited the suspect's store to set up the sting. The recordings included vague exchanges about "taking" bottles off trucks, interspersed with "you know what I mean?" During the exchanges, the suspect did not make any direct statement that would clearly indicate an interest in buying stolen bottles. In hopes of gathering the necessary evidence, the LEO decided to say the word "steal" himself during his interaction with the suspect.

In the initial store visit, the LEO very quietly muttered the word "steal" in a rushed interaction. Although there was no recorded response from the NNS, in a later visit he eventually did pay for the bottles of detergent before the LEO referred again to "steal". The NNS was then placed under arrest. The question facing the court was whether the defendant had actually understood and agreed to buy stolen goods; or had he not understood what the LEO was offering? My job on this case was to evaluate the NNS's English language skills.

I assessed his proficiency to be at a very low level. This finding was consistent with the NNS's very minimal responses in the police cell phone recordings. Nevertheless, the question remained: How likely was it that the defendant had understood that the LEO's use of the word "steal" referred to participation in an illegal activity? To help answer that question, I consulted a colleague. This top expert in corpus linguistics did a database search of the Corpus of Contemporary American English (COCA) for forms (lemmas) of "steal" and "take". According to the analysis, the frequency of "take" was significantly higher than "steal". I then consulted a chemistry professor colleague to help me project how many grains of rice would be needed to represent the frequency of "steal" lemmas relative to "take" lemmas. In my proposed courtroom demonstration, the grains of rice representing forms of "take" would be noisily poured into a glass bowl. Then, a single red-colored grain of rice would be dropped into the bowl. This grain would illustrate the likelihood of the NNS's exposure to the word "steal". The purpose of this demonstration was to help the court understand the statistical information in my report.

As part of the legal discovery process, my report was given to the prosecution. Several days later, as I was preparing for my long trip to the courthouse, I was notified that the charge of illegal activity had been dropped. "Your report caused the other side to cave!" While I would have liked to have given the demonstration in court, what was most important was the outcome for the NNS defendant who had been arrested on the basis of a poorly conducted sting operation. I attribute this social justice victory, in large part, to the decision to call in additional expertise.

This is not the only instance where it can be useful to consult with other experts. Sometimes, even when you are completely qualified for taking on a case, it is advisable to consult other linguists to increase the overall reliability of your assessments. Reputable testing companies, for example, minimally require two raters for formal national and international placement purposes. Hiring attorneys may not always appreciate the importance of such confirmation, seeing it as an added unnecessary expense. When that happens, it is important to explain the necessity of your request.

For example, in a sensitive Federal Bureau of Investigation (FBI) case, the attorney did not want to pay for a second examiner to provide an oral English proficiency assessment in a distant East Coast city where the defendant lived. As the case involved alleged conspiracy in a terrorist plot, I felt it was professionally indispensable to have a second rater. I knew that the court might view my single assessment as lacking the necessary professional objectivity. For that reason, I persisted with my request for an additional rater. The attorney finally agreed to hiring a second examiner.

To facilitate the additional assessment, I then contacted the organization that had developed the test I would administer. I asked that they identify their strongest examiner in the defendant's region and send that name directly to the defense attorney. This step ensured that I would not know the second assessor's name. I also requested that this second assessor's report be sent directly to the attorney to avoid any question of assessment independence. Once the other examiner's language assessment had been completed, I traveled to the defense attorney's office. When a federal marshal brought the defendant to the attorney's office, I administered the same test that had been previously used by the other examiner and gave my results to the attorney. I then submitted my oral proficiency report to the attorney. The two sets of scores from the same independently administered test matched almost perfectly, with only a .01 difference in one category.

Developing Linguistic Questions

The linguistics questions an expert develops set the boundaries for the expert's work and for maintaining appropriate distance. To establish a mutual understanding of the focus, I take the following steps for developing the linguistic questions for the commissioned linguistic analyses. After the attorney tells me the legal issue/question, I propose linguistic questions (van Naerssen, 2009). We operationalize these questions together making them mutually comprehensible and professionally acceptable; and then we identify the language evidence available. During this exchange, I also take the opportunity to inform the attorney about what a linguist can do and hopefully to clear up possible misbeliefs about NNSs. We also determine whether or not certain probability phrases are acceptable in that court system.

Hypothetical wording can be appropriate since the expert cannot get into the mind of a suspect and was not at the scene of the incident. Hypothetical wording also automatically distances the expert from the suspect. Taking all of these steps into account, a legal-linguistic question for a NNS case might look like the following:

> In the communications in the police interview of (date), how likely is it that a person with Mr. X's assessed English proficiency could have accurately understood and made his "voice" understood by the police, without an interpreter?

Separating Your Expert Linguistic Opinion from an Attorney's Agenda

Will your hiring attorney respect your professional opinion or will you be pressed to shape your opinion to fit the agenda of the case? Will you follow your professional instincts and ethics? While an expert might want the work, the expert should keep alert and not be inappropriately drawn too far into the case. This is easier to say than do, especially after becoming familiar with the evidence involved. Ideally, attorneys are supposed to wall off experts so they are not part of the legal team.

However, in practice, this does not always happen. This lack of distance can occur for several reasons. Sometimes attorneys who simply don't know what a linguist can and cannot do might ask you to opine on matters outside your area of expertise. This can happen, for example, with lawyers who are relatively inexperienced or who have never worked with a linguist before. When this happens, it is imperative that you clearly articulate the limits of your expertise. Boundary setting may also become necessary when dealing with experienced attorneys who decide to have a linguist as a part of the legal team, with the professional status of a consultant instead of a witness.

If you sense an attorney might be pushing overtly for only one outcome, as the expert it is important to directly state how linguistic analyses can and cannot inform the case. This boundary needs to be clarified both at the beginning of the assignment and repeatedly throughout the case, as needed. To maintain the necessary professional distance, it is important to keep in mind one's professional ethics and to listen to one's instincts. A concrete example from my casework illustrates how I put this advice into practice.

In the first example, an attorney asked how much I would charge for doing an initial "look-see" to determine whether to take the case. However, the lawyer did not tell me what kind of evidence was involved nor how much there was. Furthermore, as is often the case, the attorney also didn't know what linguists do. To provide an estimate, I offered to skim just a few pages of the transcribed phone calls he wanted examined and get back to him.

After reviewing several pages, I knew something was strange about this small sampling of interactions. I knew I could not provide analyses that the attorney seemed to want for his client. I also wanted to alert him to the limits of further linguistic analyses. I wrote a brief informal report of these points to accompany the materials I returned. Later I learned that the case was highly political. Also, in an informal chat with a law enforcement contact, I discovered that it was likely that both persons on the calls were probably faking their identities. I was relieved that I had followed my professional instincts.

Gathering Information about the NNS and the Case without Learning Too Much

In NNS cases, it's very likely that the evidence will involve interactive communication. When examining this type of evidence, it is important for the linguistics expert to set up protective walls to avoid becoming too involved with the data. While it is natural to want to learn details about a case from the attorney, you should be mindful not to learn so much that your objectivity is compromised. Establish a protective wall by reminding the attorney not to tell you too much about the case. Another protective wall comes from resisting the temptation to do information searches about the alleged crime. Remember your job as the linguistics expert is to work with the evidence available and to consult with the attorney about other general information from case records that might be needed (e.g., a defendant's educational background, known literacy ability, immigration status, length of residence (LOR) in the new country, and so on.

Attorneys might not always know what information is needed to appropriately assess a NNS's English. When this happens, it is important that the linguistics expert communicate what is required. For example, tell the attorney what instruments are needed or, if necessary, what communication tasks will need to be developed. The expert may also need to probe for the availability of any other language samples being used as evidence, if relevant. These might include texts of official waiver scripts or permissions, recordings, and transcripts. The goal in a NNS case is to obtain only as much information as is necessary to provide a solid linguistic assessment.

Conducting Language Assessment of NNSs: Selected Ethical Alerts

The scope of this chapter does not allow for extensive discussion of the history and development of the field of language assessment. Nor is this intended as training for language assessment experts. However, since NNS cases frequently involve issues surrounding the NNS's abilities in interactive communication in legal contexts, selected ethical concerns are discussed here.

As mentioned earlier, in NNS cases, the linguistics expert might be asked to report to legal practitioners (judges and attorneys) who may have their own agendas and bring their own misbeliefs about language assessment. They might assume, for example, that the best (most commonly used) formal language tests must be reliable. What they may not realize, however, is that a high reliability rating may be rendered useless if the tests are not valid for the NNS communications in question, for example, in a road-side traffic stop.

To help ensure the validity of a language proficiency assessment, a linguistics expert needs significant training and experience in second/foreign language proficiency assessment. This goes to the earlier question under checking your professional qualifications, namely "Am I the right expert?" If you find you are lacking the specific expertise necessary, you should inform the attorney and consult outside experts who can fill in this gap. As explained earlier, this is a strategy that I have used many times in past NNS cases.

Where NNS language assessments are concerned, there are several follow-up questions that are critical. For example, before you agree to take on a NNS case, ask yourself if you are qualified to select and administer appropriate assessments. Are you qualified to objectively review language assessment reports by counter-experts? Are you able to gather additional language samples and develop elicitation tasks that are grounded in principles in my field? This is an important point as it would be scientifically inadequate and therefore ethically unacceptable to just use ad hoc conversations that may be provided by a lawyer.[2]

The answers to these questions do not only depend on your professional qualifications, however. They also relate to your personality. Can you be assertive with an attorney about the need to do the language assessment BEFORE looking at the evidence? Can you fend off attorneys pressing for "just a brief phone sample" to save time and money? Are you able to consider both sides of an argument if challenged in court? Can you be flexible about language assessment conditions while maintaining the underlying principles in assessment supporting objectivity?

Language assessment experts need to think "outside the box" of large-scale assessments under monitored conditions. As a linguistics expert, I have been to asked to conduct assessments in a wide range of contexts—from the corner alcove of a public library and in the hallway outside the attorney's office, to conference video calls in hotel rooms, in a defendant's kitchen, and in a variety of incarceration settings (e.g., jails, federal detention centers, and a maximum-security prison). Each had its unique conditions and challenges. However, no matter the place nor the space, applying professional standards and maintaining professional distancing remain essential.

Roger Shuy (2006), a long-respected sociolinguist and early international leader in applied linguistics in legal settings since the late 1990s, provides an important insight on language assessments in legal contexts. In a personal communication we had some time ago, discussing NNS cases, he made the following point: While the field is quite broad, working on "live cases" with persons with limited English proficiency is probably one of the few instances in which an expert might actually be in direct communication with a suspect or defendant. This contact can occur, for example, during the language assessment process or in the gathering of additional language samples. This proximity offers the linguistics expert a unique advantage where language assessment is concerned, but it also comes with an important ethical consideration. It is important to keep an open mind during the evaluation process.

A linguist testifying in court needs to be prepared for challenging questions that may contain negative misbeliefs about a NNS's language, e.g., "Have you considered the possibility that the NNS is 'faking' not being able to understand English for a legal advantage?" A legal advantage might be gained for the defendant, for example, when a NNS is given the Miranda warning by a police officer before a fact-finding interview. *If* it is shown that the NNS did not understand the language of the caution, *then* evidence gathered during the interview might be thrown out. To stop this from happening, the prosecution might argue that the NNS is only pretending not to have understood the officer's legal warning.

Although the term "faking" implies popular negative beliefs about the truthfulness of NNSs, an important linguistic question should be addressed: "Was the NNS intentionally pretending a lower than truthful language proficiency for a legal advantage?" While linguists cannot get into the mind of another person to assess "true intention", they can provide objective linguistic evidence, e.g., language assessment data or additional samples of the NNS's English to possibly reveal linguistic (in)consistencies in language use in unplanned communications. Also to be considered are contextually relevant sociolinguistic factors like power relationships and the communicative setting,. Thus, the linguist must demonstrate professional distancing and objectively show that both sides of a claim have been considered (van Naerssen 2015, 2014, 2013).

Legal Standards, Selected Professional Codes of Ethics, and Guidelines Relevant to NNSs

Selected standards, professional codes of ethics, and guidelines can help the linguistics expert retain objectivity and maintain professional distancing. These standards and codes can also provide criteria for linguists considering social justice concerns. In this section I will examine those which are particularly pertinent to NNS cases.

Court Standards and Rules of Evidence

Courts operate under national and local laws with standards (rules and procedures) that experts must follow. These legal standards also inform experts about what is expected in their reports and in court testimonies. In US courts, the trial judge relies on these standards to assess whether an expert witness's scientific testimony is based on scientifically valid reasoning which can properly be applied to the facts at issue.

The Daubert standard is the test currently used in US federal courts and in some state courts. In federal courts, it replaced the earlier Frye standard which still applies in some states (National Civil Justice Institute 2023).[3] As a linguistics expert, you need to adhere to the rules of evidence. This means in your testimony and report-writing, it is essential to not only assert that linguistics is a scientific discipline but also demonstrate that you are qualified as an expert. You also would explain your specific area of linguistics, and show that your testimony is applicable to the facts at issue.

Selected Professional Codes of Ethics

International Association for Forensic and Legal Linguistics (IAFLL): Code of Practice

The IAFLL's "Code of Practice" notes that there are three guiding principles of ethical conduct for members "who engage in forensic linguistics research and legal consulting and testimony". These are Integrity; Objectivity and Professional Competence; and Confidentiality and Conflict of Interest. Along with these basic principles, the IAFLL has a full Statement of Professional Ethics for Forensic Linguistic Consulting. According to the IAFLL website (iafll.org) this Code "is intended to serve as a guide to its members and is made public so that others may become aware of recommended standards for professional and ethical practice in forensic linguistics".

Some of my colleagues and I now use a signed witness statement at the end of our expert reports as a Statement of Professional Ethics. Once, an attorney in a large law firm observed that he hadn't seen such a statement in other expert reports. My response was that since many in the legal field do not know much about our field, we feel it is important to provide such a statement, to legitimize our field in the eyes of the law. This practice shows we support professional ethical standards such as maintaining professional distancing.

National Association of Judiciary Interpreters and Translators (NAJIT): Code of Ethics and Professional Responsibilities

Linguistics experts in legal cases involving NNSs frequently work with judiciary translators or interpreters. It is therefore appropriate for them to become familiar with all eight Canons of the NAJIT Code that shape the professional work culture of fellow language practitioners, judiciary interpreters, and court translators. However, of special interest for this chapter is Canon 2 which is devoted to "Impartiality and Conflicts of Interest". This Canon specifically addresses how interpreters and translators are expected to maintain their objectivity and impartiality, thus enacting professional distancing.

According to Canon 2, court interpreters and translators are to remain impartial and neutral in proceedings where they serve, and they must avoid all unnecessary contact with the involved parties. They are also to abstain from commenting on matters in proceedings in which they serve. They are expected to immediately disclose to the Court as well as all interested parties when they become aware of any real or potential conflict of interest. These regulations are excellent guidelines for applied linguists who serve as expert witnesses for the Court or offer their linguistic expertise to attorneys.

Communication of Rights Group Guidelines (Independent, International)

The Communication of Rights Group (CoRG) is an independent, international group made up of 21 linguists, psychologists, lawyers, and interpreters. It was originally convened by the internationally recognized applied linguists Diana Eades and Aneta Pavlenko. CoRG completed a set of best practices for communicating rights to NNSs called *Guidelines for Communicating Rights to Non-Native Speakers of English in Australia, England and Wales, and the USA* .[4] The *Guidelines* were based on CoRG's collective experience and research on how NNSs comprehend or do not comprehend their rights in English language legal contexts.

Since the *Guidelines* were first issued, they have been endorsed and/or made available by multiple professional organizations. For instance, the American Association of Applied Linguistics as well as the IAFLL host a copy of this resource on their respective websites. The *Guidelines* are also broadly disseminated through professional conferences, publications, and court reports. The document is routinely used by individual professionals not listed as signatories. When reviewing evidence in NNS cases in which police cautions become an issue, the *Guidelines* are now also being used by some experts as an analysis framework for maintaining objectivity.

For linguists who take on NNS cases, there are different parts of the *Guidelines* that offer relevant advice. In Part A, for example, it is recommended that a standardized Plain English (Clear English) version be used in both oral and written forms of communication. This recommendation is relevant not only to suspects who are first language speakers of English. It is also pertinent to non-native speakers of English applied linguists who have experience with NNSs; who can show how the use of Plain English Language can help ensure that police cautions are clearly communicated; and who can show that failure to use Plain Language can hinder the effective communication of those rights.

In Part B of the *Guidelines*, the key concern is *how* suspects are informed of their rights. For example, suspects are usually informed much too quickly for meaningful comprehension, even by native speakers. While linguistics experts cannot ethically claim to know exactly how much comprehension is happening in the minds of NNSs (or of anyone), the *Guidelines* recommend key supporting conditions. These can also be referenced for analyzing the relevance or validity of language evidence and for challenging misbeliefs.

Selected (Mis)beliefs and Folk Beliefs about Non-Native Speakers

Popular folk (mis)beliefs can frequently intrude into NNS legal cases. Sometimes, they might even appear to carry more weight in the legal system than the linguist's use of scientific reasoning. The expert's findings might seem distant from and not match what a non-linguist believes about NNSs and their language (Greenlee 2023; Niedzielski and Preston 2003). Folk beliefs may block the judge and jury from taking in their testimony and reports. Greenlee (2023) refers to the dangers of erroneous sociolinguistic notions in legal contexts involving NNSs. As Green (2021) reminds us, informal, personal evaluations of NNS language differ from formal language assessment. The failure to distinguish the two can have serious consequences where social justice is concerned.

> From a scientific perspective, folk beliefs about language are, at best, innocent misunderstandings of language, or at worst, the bases of prejudice, leading to the continuation, reformulation, rationalization, justification, and even the development of a variety of social injustices.
>
> *(Niedzielski and Preston 2003: 1)*

There is a wide range of misconceptions about language and language users. Where NNSs are concerned, misbeliefs about a need for interpreters are common (Eades, Fraser, and Heydon 2023; English 2020; Pavlenko, Hepford, and Jarvis 2019). In my own NNS casework, I recall a case in particular where the faulty assumptions of the judge and a police officer resulted in a mismatch with my findings. Based on the English oral proficiency assessments that had been done in my evaluation and other communication factors, my report supported the defendant's likely need for an interpreter.

Police communications had initially been conducted at dusk during a roadside traffic stop. The suspect was placed in the backseat of the patrol car. The police officer spoke to the suspect from the front seat of the patrol car. Later that evening, a formal police interview of the Spanish first language speaker was done at the police station. No interpreting support was provided for the suspect as the police officer reportedly assumed that the suspect understood all communications. The suspect was formally charged with resisting arrest.

I served in court as an expert linguistics witness during the trial. As one part of my continuing professional development for working with NNS adults, I had felt it was important to learn more about the process of becoming a US citizen. I took two workshops offered by US immigration officials at an international convention for Teachers of English to Speakers of Other Languages (TESOL). I also read research on US citizenship related to the English proficiency of NNSs and on the US Oath of Allegiance. Thus, I knew there is no formal English language proficiency test that an applicant must pass to become a US citizen.

During the trial, the prosecution read aloud the US Oath of Allegiance[5] and noted how complex the language is. The prosecuting attorney then asked me to agree to the following assertion: since the defendant is a US citizen, he had to have understood the lengthy and complex Oath, and therefore he would not have needed an interpreter in the police interactions. I countered the attorney's inaccurate assumption. I referred to my assessment of the defendant's low level of English language proficiency. Based on my observations of citizenship ceremonies, I also pointed out that the Oath is read aloud by the administrator, clause by clause, and these clauses are then slowly repeated by the citizenship applicant(s). As I then explained, this is neither a natural nor a meaningful interactive communication context.

The judge asked the arresting police officer if the defendant had been given interpreting support. "No, he didn't need any", the officer replied. The judge then asked how he knew the defendant had understood. "I just knew", the officer explained. The judge accepted that answer, and the defendant's statements during the police investigation were admitted into evidence. Perhaps the judge was blinded by his own folk beliefs about the NNS's English.

Discussion and Conclusion

As stated at the beginning, a goal of this chapter is to show how linguists can provide opportunities for social justice through helping to ensure fair language access for NNSs, while upholding their ethical obligation to maintain their objectivity through professional distancing. As I was considering a way to close this chapter, I spotted a blog by Hilda Zavala that was posted on the website of the aforementioned NAJIT organization. Zavala's reflections really "spoke" to the linguist in me and perhaps they will to you as well. In this blog, Zavala (2021) reflects on finding a balance between wanting to "help" Limited English Speakers (LEPs), NNSs, yet remaining professionally detached. She speaks from years of experience as a state certified/ approved Spanish court interpreter, a past NAJIT board member, and an active member of the professional interpreting and translating community. Outside of court, she has found multiple outlets for helping these marginalized communities. Within the court, she stresses that the key to promoting social justice is remaining professionally detached. Only then can we provide the scientifically sound linguistic services to help the course of justice.

Notes

1 Interested readers are encouraged to consult the work of McNamara and Roever (2006) which draws attention to socio-ethical concerns. The language assessment community has continued research in this area of concern.
2 Assessment tasks should not replicate actual case communications. If formal language test tasks are not relevant to a NNS's background, alternate scenarios could be simulated, with similar linguistic functions.
3 For further information on Daubert see https://www.law.cornell.edu/wex/daubert_standard. Since there is still some variation among the US states referencing Frye in some form, applied linguists who are serving as an expert court witness may wish to confirm, with their hiring attorney, the status of Daubert and Frye in the legal jurisdiction involved. Another possible resource is the National Civil Justice Institute (NCIJ.org).
4 At the time the *Guidelines* was being developed, it was not practical time-wise to include other countries, but the document might be adapted, as appropriate, to other legal systems. A full and final copy of the 2016 document can be found in Eades and Pavlenko (2016). As of the writing of this chapter, the *Guidelines* has not been revised. Also, in References, see Communication of Rights Group, 2015.
5 The text is as follows: "I hereby declare, on oath, that I absolutely and entirely renounce and abjure all allegiance and fidelity to any foreign prince, potentate, state, or sovereignty, of whom or which I have heretofore been a subject or citizen; that I will support and defend the Constitution and laws of the United States of America against all enemies, foreign and domestic; that I will bear true faith and allegiance to the same; that I will bear arms on behalf of the United States when required by the law; that I will perform noncombatant service in the Armed Forces of the United States when required by the law; that I will perform work of national importance under civilian direction when required by the law; and that I take this obligation freely, without any mental reservation or purpose of evasion; so help me God" (US Department of Homeland Security 1952).

References

Communication of Rights Group (CoRG), 2015. Independent, international group of 21 linguists, psychologists, lawyers, and interpreters. which developed "Guidelines for Communicating Rights to Non-Native Speakers of English In Australia, England and Wales, and the US.". The Guidelines are posted on the websites of several professional organizations and can be retrieved, e.g., from, https://www.aaal.org and https://www.aaal.org/about-applied-linguistics.help_outline.

Eades, Diana, Helen Fraser, and Georgina Heydon. 2023. *Forensic Linguistics in Australia: Origins, Progress and Prospects.* Elements in Forensic Linguistics. Cambridge: Cambridge University Press.

Eades, Diana, and Aneta Pavlenko. 2016. "Translating Research into Policy: New Guidelines for Communicating Right to Non-native Speakers." *Language and Law/ Linguagem e Direito* 3, no. 2: 45–64.

English, Fiona. 2020. "Non-native Speakers in Detention: Assessing the English Language Proficiency of Non-native Speakers in Detention: An Expert Account." In *The Routledge Handbook of Forensic Linguistics*, 465–484. Edited by Malcolm Coulthard, Alison May, and Rui Sousa-Silva. New York: Routledge.
Green, Anthony. 2021. *Exploring Language Assessment and Testing*. New York: Routledge.
Greenlee, Mel. 2023, March 18. "Language Mythology in the Criminal Court." Paper at American Association for Applied Linguistics, Portland, Oregon.
International Association of Forensic and Legal Linguistics (IAFL). n.d. "Code of Practice." Retrieved from https://iafll.org/wp-content/uploads/2023/07/IAFL_Code_of_Practice_1-1.pdf.
Lightbown, Patsy, and Nina Spada. 2013. *How Languages Are Learned.* 4th ed. Oxford: Oxford University Press.
McNamara, Tim, and Carsten Roever. 2006. *Language Testing: The Social Dimension.* Oxford: Blackwell.
National Association of Judiciary Interpreters and Translators (NAJIT). n.d. *Code of Ethics and Professional Responsibilities*. Retrieved from http://www.najit.org/ethics.html.
National Civil Justice Institute. 2023, July 11. "State-by State Compendium Standards of Evidence." Accessed August 7, 2024. Retrieved from https://ncji.org/wp-content/uploads/2024/01/Evidence-Standards-by-State-7.12.23.pdf.
Niedzielski, Nancy, and Dennis Preston. 2003. *Folk Linguistics.* Berlin: Mouton de Gruyter.
Pavlenko, Aneta, Elizabeth Hepford, and Scott Jarvis. 2019. "An Illusion of Understanding: How Native and Non-native Speakers of English Understand (and Misunderstand) their Miranda Rights." *International Journal oof Speech, Language, and the Law* 26, no. 2: 181–207.
San Diego Foundation. N.d. "Community Impact Statement." Accessed July 10, 2024. Retrieved from https://www/sdfoundation.org.
Shuy, Roger. 2006. *Linguistics in the Courtroom: A Practical Guide*. Oxford: Oxford University Press.
United Nations General Assembly. 1966. *The International Covenant on Civil and Political Rights*. Retrieved from https://treaties.un.org/Pages/Home.aspx?clang=en.
US Department of Homeland Security. 1952. "The 1952 Immigration and Nationality Act." Retrieved from uscis.gov.
van Naerssen, Margaret. 2023, March 18. "On Being a Linguist." Paper at American Association for Applied Linguistics, Portland, Oregon.
van Naerssen, Margaret. 2015. "Language Proficiency and Its Relationship to Language Evidence." In *Cultural Issues in Criminal Defense*, 31–100. Edited by Ama Appiah. Huntington, NY: Juris Publishing.
van Naerssen, Margaret. 2014. "The Interface of Language Assessment and Forensic Contexts." In *The Companion to Language Assessment*, 1534–1552. Edited by Anthony Kunnan. Hoboken, NJ: Wiley.
van Naerssen, Margaret. 2013. "Is Litigant Faking Low English Proficiency? Ask a Forensic Linguist!" Plenary at 2013 Pennsylvania State Trial Judges Conference.
van Naerssen, Margaret. 2009. "Going from Language Proficiency to Linguistic Evidence in Court Cases." In *Language Testing Matters: Investigating the Wider Social and Educational Impact of Assessment*, 36–58. Edited by Lynda Taylor and Cyril Weir. Cambridge, UK: Cambridge University Press.
Zavala, Hilda. 2021. "To Be or Not to Be, That Is the Question…" Blog, NAJIT. National Association for Judiciary Interpreters and Translators. Accessed August 9, 2024. Retrieved from https://najit.org/to-be-or-not-to-be-that-is-the-question/.

7

AN INTELLIGENCE SCAFFOLD TO AID ETHICAL CONSIDERATIONS IN FORENSIC LINGUISTIC CASEWORK

Ria Perkins, James Rosie, and Sharon Glaas

Introduction

Using a case study, this chapter takes the discussion of ethics grounded in theory and looks at how ethical considerations may manifest. Rather than a critique of either ethical theory or established practice as expert witnesses, the aim of the chapter is to highlight some of the ethical challenges and dilemmas that a practicing forensic linguist might encounter when undertaking casework. First, a discussion is offered on the theoretical aspects of ethics and how this relates to forensic linguistic practice. Many of the processes of forensic linguistic work may appear straightforward; however, as this chapter demonstrates, there are numerous points at which ethical dilemmas can occur and where ethical decisions are required. The second part of the chapter draws on the concept of the intelligence cycle and offers a novel guiding scaffold to aid in the identification of ethical issues and facilitate ethical thinking throughout casework. To demonstrate the utility of the guiding scaffold, the chapter then works through an exemplar case. This real-life case illustrates the ethical dimensions of decisions made throughout forensic linguistic consultation.

The purpose of this chapter is to provide a forensic linguist who undertakes casework with an initial understanding of ethics and how ethical principles may apply. However, it is beyond the current scope to provide specific guidance or training for any ethical situation one may encounter or what particular challenge may represent "ethical dilemmas", where there is no clear "correct" ethical route forward. This quandary may result from a tension between differing ethical frameworks, for example, following ethics rules (deontology) may not lead to an ethical, or good, outcome (consequentialism). Alternatively, a tension could arise within a framework when, for example, two ethical rules are in conflict, or when a single decision may lead to two equally undesirable outcomes.

Further complicating the issue for the forensic linguist who serves as a consultant for a police investigation is the fact that ethics for law enforcement often focuses on the procedural correctness of investigation practices, which includes the collection of evidence within a wider legal framework. The activities of a forensic linguist may fall within this wider framework of law enforcement activity. However, as will be demonstrated in this chapter, the ethical considerations of the forensic

 DOI: 10.4324/9781003391074-8

linguist may at times differ from those law enforcement professionals conducting an investigation. Of course, there are clear legal guidelines that regulate how law enforcement officers, and investigations in general, are to be conducted. Any forensic linguist who serves as an expert witness and provides evidence for the court is bound by such governmental regulations. That said, a greater majority of the forensic linguist's time will usually be spent operating within academia and will be subject to the ethical frameworks of their institution (e.g., a department, a university or college, a research think-tank, etc.). Beyond those institutional guidelines, the lack of an accreditation, certification, or licensing system for forensic linguistics means that its practitioners are often forced to seek out for themselves the ethical and legal frameworks within which they are working. As this chapter will show, it is imperative that forensic linguists do so with careful consideration.

Defining Ethics

Before moving on to discuss theories of ethics, it is important to note that there is considerable debate around how one defines and distinguishes between "ethics" and "morals". Typically, "morality" is defined as the wider norms and values of society that will in turn determine what that society sees as right and wrong (Gert and Gert 2020), although sub-cultures or groups may have moral norms that diverge to some degree from the society in which they are situated. "Ethics", though, tend to focus more on an individual's acts or behaviors. Nevertheless, there still exist wider concepts of what is considered moral. This is an important consideration because public acceptance, which is likely to be based on societal perceptions of morality, may be a key consideration of what is deemed ethical in a given context. Indeed, according to the UK Statistics Authority (2022), it is one of six ethical principles used for conducting qualitative research.[1] Further discussion delineating ethics and morals is beyond the scope of this chapter. However, for ease of understanding, this chapter treats the construct of "ethics" as interchangeable with "morals" in the sense of determining either how "right" or "correct" a given course of action may be. Within the field of moral philosophy, there are several approaches to the concept of ethics, for example Kant focuses on universal moral laws (Kant, trans. Gregor 1998), Scanlon (1998) speaks about our obligations to offers, and Aristotle describes ethical behaviors (2014). Thinkers have suggested, developed, and used these approaches since classical times to articulate and explain how to recognize whether or not an action is ethically "right", for example Kant's "Categorical Imperative": "Live your life as though your every act were to become a universal law" (Kant, ibid.). Below, we outline the three principal schools of thought, to provide a grounding for our approach and a common understanding for the rest of this chapter.

Consequentialism and Outcomes

A consequentialist view, as the term implies, adopts the view that the "rightness" of an action is determined by the consequences or outcomes of that action. In other words, the degree of "good" or "utility" an action yields is a measure of its rightness (Sinnott-Armstrong 2022).

One of the most straightforward challenges to this approach is that it can rapidly move to absurd or extreme situations, where terrible acts (such as torture) are permissible on the basis that the ends justify the means. Furthermore, this argument also forces decision-makers to rely upon either hindsight or their ability to foresee the future to determine the ethics of their actions. The more long term or complex the decision, the harder it becomes for the decision-maker to reliably anticipate the consequences of their decisions and the rightness of their actions. A separate but equally complex challenge is in identifying who has the authority to determine which consequences make a course of action "right" or "ethical". For example, is it public good, individual privacy, or the minimization of harm which should be prioritized?

Deontology and Rules

The second of the two major schools of thought takes the view that the essence of morality lies in a series of duties, or rules, which require, allow, or prohibit certain actions. For example, in the Judeo-Christian tradition, if an individual obeys the Ten Commandants, they are "good", and if they transgress, they are "bad" (Alexander and Moore 2021); or in the Confucian tradition, Xunzi asserts that rule and ritual is necessary to overcome the evil of human nature (Xunzi, in Knoblock 1994).

Certainly, when adopting a rules-based approach, ethical challenges may be presented by multiple sets of rules which all require observance. For example, everyone would first follow the laws of the land and, if working in an official capacity, wider policy guidance and strictures in place. Ethical dilemmas emerge when these rules demand differing or even opposing courses of action.[2]

One advantage of a deontological approach is its apparent simplicity. However, it is not always clear from where a set of rules originates, or indeed who regulates or enforces them, or what actions one should take in novel situations where the rules are either unclear or as yet non-existent. There is also the challenge that simple adherence to a set of rules may deprive individuals of the opportunity to develop their own critical and moral thinking. This issue may have particular relevance to discussions around the development and adoption of ethical guidelines for forensic linguists.

Virtues and Behaviors

Both consequentialist and deontological approaches place the moral value outside of the individual, within outcomes and rules respectively. However, thinkers within the Western classical tradition, as far back as Aristotle, have suggested that moral value instead lies in the behaviors and character, or characteristics, of the individual (Hursthouse and Pettigrove 2022). Although leading to moral and ethical relativism, rather than absolute or objective morality, this idea does lend itself to consideration of the different behaviors required and expected from different professional groups. The UK Police Code of Ethics, for example, states that "[e]thical *behavior* comes from the values, beliefs, attitudes and knowledge that guide the judgments of each individual" and its policing principles "underpin and strengthen the existing procedures and regulations for ensuring standards of professional *behavior* for both police officers and police staff [emphasis added]" (College of Policing 2014: iv). It is also necessary to recognize the importance Aristotle places on balance. For him, an excess of a characteristic is just as bad as its lack. For example, while a deficiency of courage could be considered cowardice, an excess may lead to foolhardiness which is just as bad. Following this Aristotelian logic, a "moral saint" would not necessarily benefit either the individual or the wider society any more than a person bent on moral transgression (Wolf 1982). While the arguments against moral deficiency are perhaps evident, the importance of balance may be much less so and will therefore be taken up further in the following section.

Ethics in Forensic Linguistics

Consequentialist, deontic, and virtue approaches to ethics all have something to offer when considering courses of action in a given context. Any specific ethical dilemma may lend itself to one of these approaches. By the same token, an ethical issue may be viewed from any of the three perspectives. Conclusions drawn from each perspective may be in harmony or they may highlight tensions that may require resolution, particularly if different actors adopt different approaches. As such, it is the application of these theoretical principles that will be the main concern for the discussion of ethics in Forensic Linguistics. This discussion is particularly important given that there are no formal, internationally recognized and enforceable regulations for ethical practice in casework. While the field of Forensic Linguistics currently lacks an ethical framework for casework, that is not to say that it operates in a moral vacuum. There are perhaps two spheres of practice from which many forensic linguists may draw: academia and law enforcement.

Many Forensic Linguists are both academics and forensic practitioners. Whilst they usually undertake casework separately from their academic careers, they are used to working within university ethical research frameworks. All research has ethical implications (Baldwin et al. 2014). Furthermore, there may be an expectation that casework undertaken should be conducted in keeping with university practices. Where case work involves contributing to an investigation, law enforcement officers will be working within ethical and legal guidelines enshrined in law and these practices will also have to be considered by the linguist. Nevertheless, we suggest that specific guidelines related to casework are needed and that these would work together with the ethics of academia and law enforcement.

Guidance for Ethical Working Practices

The Intelligence Cycle

The intelligence cycle is a simplified, idealized model used to demonstrate the main stages in how intelligence investigations are conducted. As noted by Phythian (2013), various versions of the cycle have been proposed. However, the five main sections are usually the following: (1) establishment of requirements and *direction* for the investigation; (2) *collection* of data; (3) *processing* of that data; (4) *analysis*; and (5) *dissemination* of conclusions. As shown in Figure 7.1, each stage within an intelligence cycle is subject to continuous review by the analyst.

It is understood that this framework does not represent an entirely accurate model of how intelligence/investigative activities are conducted. Critics commonly point out that the model oversimplifies the complexity of investigations and the reality of practice (Hulnick 2006). Despite these limitations, the framework provides an accessible understanding of some of the key stages of investigation, if at a relatively high level of abstraction. This conceptualization, in turn, allows us to consider the different stages as which ethical decisions come into play for the forensic linguist.

A Preliminary Guiding Scaffold

The five individual stages of the intelligence cycle map well onto the stages of forensic linguistic casework. The cycle can therefore be readily adapted to create a guiding scaffold of the stages of forensic linguistic casework: (1) identification of need for forensic linguistic input; (2) collection of specific data (and relevant baseline data); (3) cleaning and preparation of data; (4) analysis and assessment; and (5) communication of findings. Where the intelligence cycle is very much cyclical, these stages could be considered more as a linear progression. However, as with the intelligence cycle, it is entirely possible for a forensic linguist to move both forward and backward, repeating or possibly even jumping over certain stages. The framework is intended to draw out different types of activities, rather than provide an accurate map of any given case.

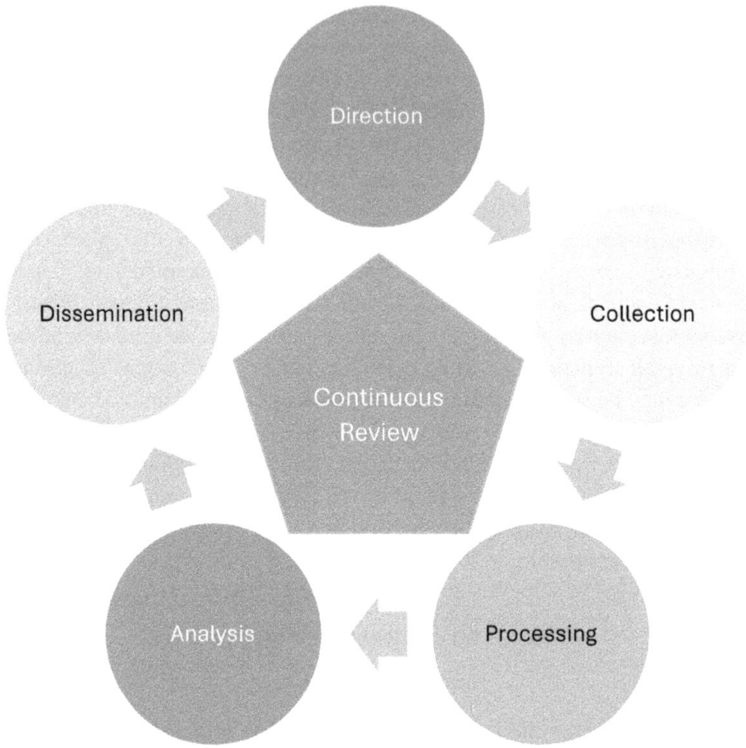

Figure 7.1 The Intelligence Cycle

Guiding Scaffold in Action

Case Study Introduction

The specific case explored in this chapter is taken from the casework of one of the authors. While seconded to a regional police unit as part of a research project, the forensic linguist was approached by investigators to help on a collection of ongoing cases. The work was undertaken pro bono. The case under discussion involved a somewhat anonymous user posting on a dark-web pedophile forum, seemingly detailing regular horrific attacks on a child in their care. The content of the messages being posted by the user (both text and images) caused considerable concern for law enforcement.[3]

The investigators on the case had worked closely with forensic linguists before, both through research, training, and casework, and were therefore unusually attuned to potential linguistic clues. They were also aware of what linguistic analysis could and could not do. In addition, they were familiar with different linguists' specializations and, in particular, with the previous research and work by the forensic linguist in question. This meant that the question as framed to the linguist was appropriate, and the data was appropriate (both in terms of being suitable for linguistic analysis and demonstrating that it likely contained analyzable features). They were also confident that this was the right analyst to approach, and the analyst would re-direct them elsewhere if necessary. In the case under discussion, the investigators had noticed features in both the content of the online messages and the language used, which indicated to them that the author might be from continental

Europe and that English was not the individual's first language. More specifically, they suspected that the author might be a native speaker of some other Germanic language like German or Dutch.

A further unusual feature of this case is that linguistic involvement was being sought to assist police prioritization and direction—rather than to significantly further the investigation (though that was of course welcome if possible). The content and language clues had indicated to the investigators that the author was likely German or Dutch, and not located in the United Kingdom. The forensic linguistic analysis and input was therefore being sought to feed in to an upcoming decision meeting, specifically to aid the decision of which authority(ies) should be notified (there was also a lack of funding at this stage of the case). The case priorities were therefore different to many other cases; speed, and simplicity of findings, were the key foci (rather than a high level of certainty, and a report detailing the fine nuances of the analyses that could be cross-examined).

The following sections discuss the ethical considerations of this forensic linguistic casework with relation to each stage of the guiding scaffold. The discussion covers general points that are to be considered and relevant aspects of the exemplar case. The school of ethics that this most links to will be outlined in *italics* to demonstrate how these different lenses map over to the different stages.

The Stages of the Guiding Scaffold

Identification of Forensic Linguistic Need

Deontologically, there are very few rules that dictate when and where forensic linguists can consult in cases. This gap is largely filled with *virtues and behaviors*-based approaches. In the United Kingdom, an expert who might be able to assist in investigations can be registered on the Expert Advisers Database (EAD) of the National Crime Agency (NCA). Further, forensic linguists will likely explicitly adhere to the ethical guidelines of bodies such as the International Association of Forensic and Legal Linguists (IAFLL) or British Association of Applied Linguistics (BAAL). If the case is likely to proceed to court, then the consulting forensic linguist is legally bound (*deontology*) by the rules governing expert witnesses in their country. In England and Wales, these are the *C*ivil and *C*riminal *P*ractice *R*ules, respectively, CPR 35 and CrimPR19. However, in the United Kingdom, there is no single body that is responsible for exerting professional or technical oversight for the field of forensic linguistics. It is upon the individual practitioner to be aware of these rules and the implications for their work (*virtues and behaviors*).

In practice, the first step is identifying if a forensic linguist is needed and if forensic linguistic analysis would add to the case. There are several questions that must be answered at this first step:

- Is there a linguistic question (or what is the linguistic question)?
- Is there sufficient relevant data to address the question?
- What forensic linguistic expertise is needed to address the question?

Investigators with significant experience at working with linguists (such as in the case described above) may be adept at developing an approximate answer to these questions. However, often an in-depth working knowledge of linguistics is needed to fully formulate the answers to these. This is where a forensic linguist can be exceedingly useful in the role similar to that of a case manager. This service links to *virtues and behaviors* as it is up to the individual forensic linguist(s) to work with the relevant parties to ensure that these first-stage questions are adequately addressed. It is critical that forensic linguists be honest and self-reflective about whether they are best placed to help, and to recognize potential biases.

With relation to this case, many forensic linguists might have been uncomfortable in taking a case with foci that favor operational practices rather than academic and methodological detail. While this reticence is exceedingly understandable, there is the ethical question of the potential *consequences* of *not* taking such a case. By virtue of the linguist having been seconded to the police, they had gained an understanding of operational practices and priorities (as well as developed two-way trust). That meant that they understood (*virtues and behaviors*) that if the linguistic analysis could be undertaken in the time available, the decision of which authorities to pass the case to (and hence which authorities would investigate it further) could be based on more rigorous and scientifically grounded information (*consequences*). Were a forensic linguist not to be involved, it might hamper the potential successful resolution of the investigation, which could result in horrific real-world harm to the victim(s). However, the potential negative *consequences* of not taking the case also have to be balanced against the risk of emotional harm to the analyst (*consequences*). Being able to recognize which cases might be too much is a very individual question and requires a lot of self-honesty that is not easy to develop (*virtues and behaviors*). In this case, the analyst was already working with the material in question and confident of their support network.

Collection of Case and Baseline Data

In this case the available data had already been collected as part of the initial investigation. Given operational priorities, there wouldn't be an option to collect more. Therefore, determining whether the existing data would be sufficient for a basic linguistic analysis was one of the early steps in determining whether the analyst would be able to contribute to the case. This stage was therefore subsumed in (or arguably preceded) the forensic linguistic direction stage.

A further complication of the material was its content and the form it took. Working with data relating to sensitive subjects (in this case pedophilic content) comes with special handling instructions (*deontology*). In this case, the data contained information that some forensic linguists would not be comfortable with (*virtues and behaviors*). Secondary or vicarious trauma is a very genuine risk that forensic linguists are by no means immune to (Elstein and Kredens 2023; Giménez, Elstein, and Queralt 2020).

On this occasion, the forensic linguist was not only already familiar with data of this type and from this source but also had an active support network. The circle was composed of supportive investigators working in the same location; a therapist who specifically supported people working with that data; and linguistic colleagues who were familiar with the impact of working with such data. It is important to note that different people will find different content and aspects difficult, and further, this might change and vary across time and contexts. It is important for an analyst to regularly check in with themselves, and to be open and honest with themselves and those around them. Further, creating a community culture that supports a positive pro-active approach to mental health and dealing with difficult content is in our view vital (*virtues and behaviors*).

Another issue involved in working with sensitive data is that the handling instructions (*deontology*) might dictate where it can be analyzed. These restrictions might be geographical (e.g., the data can't be taken out of a certain building) or technological (e.g., the data can only be stored or used on certain devices). These regulations can have unintended consequences and raise practical difficulties (*consequentialism*): Is the forensic linguist able to work during the hours allowed, at that specific location? How long will it take to get the forensic linguist access to the data? Does the forensic linguist have access to the software needed to undertake the analysis? In this exemplar case, because the forensic linguist had already been working with the investigative unit on research, these logistical issues were not a problem.

It is important to note that for many people simply having their data collected may be considered a suspicion and an invasion of privacy (*consequences*). Consequently, anyone who collects data must respect individuals' right to privacy (*behavior*) and weigh this against the need to collect the data. Additionally in the age of artificial intelligence and forms of generative software, data is being used in ways that were unimaginable a few years ago. As a result, it would be prudent to at least be aware of the possibility of new future harms and consequences which may arise from the collected data.

Cleaning and Preparation

Within this context, the term "processing" refers to the preparation of data prior to analysis. As a broad rule of thumb, the more data is collected, the more processing is required before analysis can be conducted. When discussing the cleaning and preparation of linguistic data, it might be natural to focus on how this is cleaned ready for linguistic analysis. However, with sensitive data, other steps might be needed so that it can be shared with the forensic linguist. This process might involve removing images, videos, or other particularly sensitive information. This is very much a *deontological* issue, with legal data handling aspects that must be considered, and then adhered to. Further, data handlers need to trust that a forensic linguist has the necessary *virtues* to report any breaches, though there would also be legal implications and sanctions (*consequences*) if they violated that trust.

Cleaning data so that forensic linguists can put it on their own work devices (rather than those owned by law enforcement) is time-consuming for investigators. Therefore, there needs to be a balance and discussion to understand what further analysis this might enable (for example, the analyst might be able to spend more time working on the material, and they might be able to support their analysis with certain software). This awareness of the wider context and costs, and balancing the implications of this, requires understanding and open discussion from the analyst (*virtues and behaviors*). In this exemplar case, as the linguist was already working with the investigators on similar data, the process was already established. Further the linguist was already aware of the parameters and variations in handling this data (*rules*) such as the fact that the handling instructions changed if the data included images and videos.

Different people will have different priorities when cleaning and processing data. A data analyst will typically want the material presented differently than a forensic linguist will. Further, a law enforcement agent will likely focus on the classification or anonymity of the material, rather than its readiness for analysis (*deontology*). Dealing with these differences is best achieved in an on-going discussion with all parties involved. Another consideration is the combination of data, which might elevate the classification or risk. For example, through a combination of different, anonymized datasets, it may become possible to de-anonymize and identify individuals and their protected characteristics. Awareness and monitoring (*virtues and behaviors*) of this risk, and potential *consequences* (which could also contravene the laws and data handling *rules*) combines all three schools of ethics. The forensic linguist must therefore be cognizant of the potential for tension between competing ethical issues.

Analysis and Assessment

With relation to analysis and assessment, there are two key interrelated aspects that are normally considered: (1) minimizing bias; and (2) choosing the most appropriate analytical approaches. The question of determining the most appropriate analytical approaches will vary on a case-by-case basis and is predicated on finding the right forensic linguist to undertake the case in the first place. The right analyst should be able to determine the right approaches,- and identify if they need to bring in extra expertise for support (*virtues and behaviors*). This assistance might come from a professional with specific language expertise; different linguistic expertise (such as computational linguistics approaches or phonetic expertise); or non-linguistic expertise (such as statistics or data analytics). Consideration should also be given to licensing and the roles in which products are licensed (to an organization or individual) for.

Avoiding bias in linguistic analysis is a broad area that is beyond the scope of this chapter but nonetheless is exceedingly relevant for ethical practices. We suggest that the first step in minimizing bias is recognizing that bias exists (*virtues and behaviors*) and where and how it might occur within the specific case (and within the analyst themselves). This can then lead to mitigating the bias through methodological steps.

One approach to mitigate bias, as outlined in Grant and Grieve (2022), is through multiple linguists working together. In that situation (as in our own experience) the main linguist involved with the first three stages of the case might be a different one to the linguist that undertakes the analysis and assessment stage (and potentially the communication of findings). In situations with multiple analysts, it would be necessary that there was a shared ethical approach and understanding, though this would need to be predicated on trust and experience, as the nature of the arrangement is often to prevent one linguist being aware of all the wider information, and hence reducing potential bias.

In the exemplar case study detailed above, there was only one forensic linguist to undertake the analysis. The time constraints meant that there was not the option to bring in professionals with complementary expertise; nor was there time available to broaden the analyses by, for example, integrating wider computational approaches. The time restrictions were compounded by the operational priorities and the fact that the case was pro bono.

Communication of Findings

The case under discussion was unusual where the communication of the findings was concerned, in that what was most useful to the investigators was not a standard full report. In fact, the time and expense such a report would have required, would have made forensic linguistic involvement untenable (*consequentialism*). Based on the mutual trust that had already been established (*virtues and behaviors*), the forensic linguist was confident that the investigators would respect the bounds of what could be said based on the analysis by not, for example, assuming higher levels of certainty than were possible. Here the ethics at play were not just those of the linguist, but also the linguist's perceptions of and experience with the investigators' ethics. It should be noted that limitations and cautions were still provided by the forensic linguist, despite the brevity of the concluding remarks and the fact they were aimed at guiding an investigation. Appropriate discussion of limitations is vital in a more conventional report (*virtues and behaviors*) because a lack of understanding can result in severe negative *consequences*, but these limitations also need to be communicated when guiding investigations too.

A case does not necessarily finish with the communication of findings. In this case, the forensic linguist reported the findings to investigators, who fed them in to the decision conference. In the end, the case was passed to German law enforcement. Over a year later, one of the former investigators informally informed the forensic linguist that the perpetrator had been found and prosecuted, and that their profile matched the one that had been conveyed. It is very common *not* to find out what has happened after the forensic linguistic findings have been delivered. This lack of definite conclusion can have a negative psychological impact (*consequence*) on analysts, which should be weighed into considerations before accepting a case (stage 1).

Further ethical behavior from the analyst with relation to the case has to continue beyond the natural conclusion of the case. As already stated, most expert forensic linguists are also academics, and this comes with a pressure to publish. This is not just to further their own career, but also to help shape the wider field. Sharing case findings can also contribute to the wider field. If forensic linguists did not publish information about their cases, the field would not have been established, nor would it attract people to study, research, practice, and fund work in the area. Further, on-going discussion of issues faced in practice is necessary to help shape research and further the discipline. Not publishing findings would hamper progress within forensic linguistics and prevent its ability to assist investigations and judicial processes (*consequentialism*). However, the benefits of publishing need to be balanced against a wider law enforcement perspective. It is possible that to discuss their work an academic might want to publish details that would not be in breach of any contractual or data handling rules (deontology and rules), but that might still have negative consequences for the investigators or wider law enforcement agencies (for example, through accidentally exposing investigative practices that might be of interest to criminals seeking to prevent detection). It is therefore incumbent upon forensic linguists to consider these perspectives and seek out the relevant permissions (*virtues and behaviors*).

Discussion and Conclusion

In further discussion and dissemination of information around this case it should also be remembered that success in a case does not necessarily indicate a good methodology, nor a solid ethical practice. It should further be highlighted that the decisions made at each stage of this case were unique to the specific context. A different forensic linguist might make different decisions, without affecting the ethical nature of those decisions. Even the same forensic linguist might make different decisions given a different context. For example, if presented with the same case at a different time, the forensic linguist might decline the case if they determined that they did not currently have the resilience to deal with it (*virtues and behaviors*). Alternatively other aspects of their workload or casework should take priority as they might produce more good (*consequences*). They might determine potential detrimental personal impact to be too great (*consequences*) or that it might impact their neutrality (*virtues, consequences, and deontology*). Such a decision would be incredibly ethical and should be lauded.

The intelligence cycle, and the modified application suggested here, can provide a guiding scaffold to help structure and inform a linguist's ethical thinking and approaches. Using an exemplar case, this chapter highlights some important considerations for each stage, which are relevant to forensic linguistic practitioners and those that work with them (e.g., police investigators). This is not all-exhaustive and the case selected is, by design, one that is more investigatory than evidentiary, as it demonstrates different considerations to cases normally discussed.

This chapter has examined some of the potential ethical issues which a forensic linguist may encounter at different stages when conducting casework. It is hoped that this chapter has illustrated that engaging in casework is not a single activity, but rather a series of linked activities, with a single aim, each of which is likely to present differing ethical considerations. Equally, it is not the intent to present the three ethical framings here as "correct", either singly or in combination, and within each framing there are different ways to approach and apply their principles. Instead, they are presented as tools or approaches to help guide a forensic linguist when they encounter a possible ethical issue.

The intent of this chapter was not to provide a checklist of how to approach each stage, nor provide a "map" of where ethical considerations should apply. Rather, it was to stimulate thinking as to the complexity of ethical practice and the importance of careful consideration at every stage of the process. Ethical considerations remain a question throughout the lifecycle of a case (and beyond); this includes across the stages of: identification of forensic linguistic need, collection of data, cleaning and preparation of case data, analysis and assessment, and communicating findings. Therefore the intelligence cycle, and the modified process suggested here, should not be regarded as a solution. Rather, they represent a starting point for ethical awareness and one possible means of providing a guiding scaffold to help inform a linguist's ethical thinking and approaches, as well as help them identify when such thinking is required.

Notes

1 The UK Statistical Authority's six principles are: Public Good; Confidentiality/Data Security; Methods and Quality; Legal Compliance; Public Views and Engagement; and Transparency.

2 It is recognized that it is possible for laws to be immoral or unjust; however, such a discussion is beyond the scope of this chapter.

3 Due to the sensitive nature of the case, it is not possible to discuss specifics in detail, nor to give any of the data—indeed for numerous reasons, as will be discussed below, it would be unethical to do so. For reasons of space, we will limit our focus to the ethical considerations that arose. For a wider discussion on the methodologies applied in the case, please see Perkins (2022).

References

Alexander, Larry, and Michael Moore. 2021. "Deontological Ethics." In *The Stanford Encyclopedia of Philosophy*. Edited by Edward N. Zalta. Stanford University. Accessed July 25, 2023. Retrieved from https://plato.stanford.edu/archives/win2021/entries/ethics-deontological.

Aristotle. 2014. *Nicomachean Ethics*. Translated by Roger Crisp. Cambridge, UK: Cambridge University Press.

Baldwin, J. R., et al. 2014. *Intercultural Communication for Everyday Life*. Chichester: Wiley Blackwell.

College of Policing. 2014. *Code of Ethics: A Code of Practice for the Principles and Standards of Professional behavior for the Policing Profession of England and Wales*. College of Policing. Accessed July 25, 2023. Retrieved from https://assets.college.police.uk/s3fs-public/2021-02/code_of_ethics.pdf.

Elstein, Solly, and Krzysztof Kredens. 2023. "Occupational Stress in Forensic Linguistic Practice." *Journal of Applied Linguistics and Professional Practice* 17, no. 1: 50–72.

Gert, Bernard, and Joshua Gert. 2020. "The Definition of Morality." In *The Stanford Encyclopedia of Philosophy*. Edited by E. N. Zalta and U. Nodelman. Metaphysics Research Lab, Stanford University. Accessed July 25, 2023. Retrieved from https://plato.stanford.edu/entries/morality-definition/.

Giménez, Roser, Solly Elstein, and Sheila Queralt. 2020. "The Pandemic and the Forensic Linguistics Caseworker's Wellbeing: Effects and Recommendations." *International Journal of Speech, Language & the Law* 27, no. 2: 1–22.

Grant, Tim, and Jack Grieve. 2022. "The Starbuck Case: Methods for Addressing Confirmation Bias in Forensic Authorship Analysis." In *Methodologies and Challenges in Forensic Linguistic Casework*, 13–28. Edited by Isabel Picornell, Ria Perkins, and Malcolm Coulthard. Hoboken, NJ: Wiley Blackwell.

Hulnick, Arthur S. 2006. "What's Wrong with the Intelligence Cycle." *Intelligence and National Security* 21, no 6: 959–979.

Hursthouse, Rosalind, and Glen Pettigrove. 2022. "Virtue Ethics." In *The Stanford Encyclopedia of Philosophy*. Edited by Edward N. Zalta and Uri Nodelman. Stanford University. Accessed July 25, 2023. Retrieved from https://plato.stanford.edu/archives/win2022/entries/ethics-virtue.

Kant, Immanuel. 1998. *Groundwork of the Metaphysics of Morals*. Translated by Mary J. Gregor. Cambridge, UK: Cambridge University Press.

Perkins, Ria. 2022. "Other Language Influence Detection: Profiling the Native Language of a Dark web Pedophile." In *Methodologies and Challenges in Forensic Linguistic Casework*, 63–76. Edited by Isabel Picornell, Ria Perkins, and Malcolm Coulthard. Hoboken, NJ: Wiley Blackwell.

Phythian, Mark. 2013. "Introduction: Beyond the Intelligence Cycle?" In *Understanding the Intelligence Cycle*, 1–8. Edited by Mark Phythian. Abingdon, UK: Routledge.

Scanlon, Thomas M. 1998. *What We Owe to Each Other*. Harvard, MA: Harvard University Press.

Sinnott-Armstrong, Walter. 2022. "Consequentialism." In *The Stanford Encyclopedia of Philosophy*. Edited by Edward N. Zalta and Uri Nodelman. Stanford University. Accessed July 25, 2023. Retrieved from https://plato.stanford.edu/archives/win2022/entries/consequentialism/.

UK Statistics Authority. 2022. *Ethical Considerations Associated with Qualitative Research Methods*. Accessed July 25, 2023. Retrieved from https://uksa.statisticsauthority.gov.uk/publication/ethical-considerations-associated-with-qualitative-research-methods/pages/2/.

Wolf, Susan. 1982. "Moral Saints." *The Journal of Philosophy* 79, no. 8: 419–439.

Xunzi, Xun Kuang. 1994. *Xunzi: A Translation and Study of the Complete Works*. 3: books 17–32. Translated by John Knoblock. Stanford: Stanford University Press.

8

PSYCHOLOGICAL BIASING IN FORENSIC LINGUISTIC RESEARCH

Is Such Information Obtained Expertly and Ethically?

Ray Bull and Iris Blandón-Gitlin

Introduction

In their professional roles as consultants (e.g., to police, lawyers) within the criminal justice system or as evidence-gatherers for courts, tribunals, etc., forensic linguists (FL) should be providing reliable and verifiable information that ought to be unbiased by personal opinion or expectation and thus free from bias (Coulthard and Johnson 2007). However, the materials that forensic linguists are typically provided with may already have been unduly influenced by a number of biases and limitations, some of which could be unethical (French and Harrison 2007). This chapter presents a psychological overview of the ways in which such limitations and biases may affect the activities of FL experts. This chapter will first provide an overview of some of the ways memory and perception can profoundly influence personal recall and thereby affect subsequent forensic linguistic analysis. The chapter will then review how interviewing and questioning in various settings of the criminal justice system can influence what is said or not said by people, including vulnerable persons. It ends with a section on possible biases which could apply to forensic linguists' own activities.

Perception: A Selective and Interpretative Process

Perception is a mental process through which we make sense of our external sensory experiences. Albright (2015) defines it as "a process by which the fundamental ambiguity of sensation is resolved through the use of contextual cues, which enable identification of the causes of sensation and attributions of meaning" (37). Perception enables us to understand the vast and complex array of sensations we receive through our sensory systems. However, given the sheer volume and complexity of sensory input, we cannot possibly interpret all of our sensations. Thus, perception is, by necessity, selective and interpretative. It is also constrained by what is available to the perceptual systems and each individual's capacity to process information. In this intricate process, as the philosopher William James (1890) eloquently put it: "Whilst part of what we perceive comes through our senses from the object before us, another part (and it may be the larger part) always comes (in Lazarus's phrase) out of our own head" (72). This insight highlights the subjectivity of perception,

 DOI: 10.4324/9781003391074-9

reminding us that what we perceive is not just a passive reception of the external world but also an active construction by the mind, filled with ample personal interpretation.

Perhaps the most striking phenomenon in selective perception research is "inattentional blindness". This occurs when a surprisingly salient event which takes place right in front of the perceiver is not encoded. In Simons and Chabris's (1999) most famous study on inattentional blindness, participants were asked to watch a video of people passing a basketball and count the passes. During the video, a person in a gorilla suit walked through the scene. Despite how unusual that event was, the majority of the participants did not notice the gorilla at all. A gorilla entering a basketball game can be missed because it is not expected in most real-world contexts. However, this perceptual error can also occur in expected and consequential events. For example, in a training study of routine police vehicle stops, officers had to interact with a driver sitting inside a vehicle. In some conditions, a gun was placed on the dashboard, and the critical question was whether officers would notice the weapon. Results showed that a non-trivial percentage of officers failed to see the gun on the dashboard (Simons and Schlosser 2017). These real-world studies, as well as extensive laboratory research (Hutchinson et al. 2022), illustrate just how much selective attention can limit what we perceive and, consequently, our memories (and accounts) of these complex events.

According to researchers who investigate inattentional blindness, this phenomenon is due to a failure to capture information in clear view: it is not the result of a faulty memory (Ward and Scholl 2015). Other studies in perception have demonstrated that precisely the opposite may occur as well; namely, we may add or "perceive" information that is actually not present in a complex event, but is consistent with our expectations. In a classic study by Brewer and Treyens (1981), the researchers explored people's perception and interpretation of information. In their experiment, the researchers took participants to a university office and left them there without any specific instructions. Afterward, the research participants were taken to another room and asked to recall as many items as possible from the office. The results showed that many participants were likely to remember objects that fit with their schema of an office (e.g., pencils) and also to "misremember" schema-consistent objects that had not been present in the office. For example, in one recall test, a third of the participants claimed to have seen books in the office they had just visited, when in fact the books had deliberately been removed from the office before their arrival. Participants also forgot or misremembered items that were present but did not fit the office schema, and were less consistent with their expectations (e.g., wine bottle, jar of creamer, and football-player doll).

Building on the impact of schema-based expectations on perception and memory, Hellman and Memon (2016) demonstrated that attributional biases regarding a perpetrator's motives to commit a crime can significantly alter eyewitness recollections. In their study, participants watched a film sequence created from the movie *Monster*, where a female protagonist killed four men for no clear reason. Observers then read a fictional review framing the protagonist's actions as either dispositional (e.g., motivated by hate) or situational (e.g., acting in self-defense). Finally, in a recognition test, eyewitnesses were asked to identify items on a list purportedly derived from the film. The list included three types of information: (a) true statements (e.g., The perpetrator got changed after the first murder); (b) false statements with a dispositional bias (e.g., The perpetrator had a skull tattoo on her arm); and (c) false statements with a situational bias (e.g., One of the men threatened the perpetrator with a knife). Results in two experiments showed that false recognition was consistent with the attributional bias from the post-event fictitious review. That is, whereas those in the dispositional film review group were more likely to falsely recall dispositional-based items, those in the situational group were more likely to falsely recall situational-based items. Moreover, both groups did not falsely recall neutral items, strongly suggesting that observers were prone to accepting false information that matched their attribution and able to dismiss what did not. Hellman and Memon (2016) noted that attributions are connected to our mental frameworks, which can shape our expectations, guide our attention, and form our memories. Therefore, when eyewitnesses encounter misleading information that aligns with their expectations about why a crime was committed, it can distort their memory of the event.

Similarly, Holst and Pezdek (1992) examined expectation-based mental frameworks called "scripts" in a forensic-relevant context. They assessed people's perception for three types of robberies (a convenience store robbery, a bank robbery, and a mugging) to determine consistency of these scripts. The researchers found that despite the demographic diversity of the participants, there was a high degree of agreement regarding what they considered to be a typical sequence of events during robberies. These scripts were then part of a hypothetical robbery case presented to mock jurors. They listened to an audio recording of the mock trial where an eyewitness to the robbery testified. The main results showed that jurors were likely to assume certain script-consistent details (e.g., cased the store/bank, used force) were given by the eyewitness when in fact they had not been. As this research on perception shows, unconscious cognitive processes can negatively affect the accuracy of recall and produce testimonial evidence that is flawed at best, or completely erroneous at worse (for more on the effects of events' scripts on recall and judgments see García-Bajos et al. 2012; Lenton and Bryan 2005; García-Bajos and Migueles 2003).

As noted, perception is not only a selective but also an interpretative process in which people use contextual cues to make sense of sensory information, especially when that information is ambiguous. Much like schemas or scripts, context provides a framework for people to understand what they sense via their sensory systems. Albright (2015) provides a humorous example to illustrate this point. Consider the phrase "Pas de lieu Rhone que nous". Within the linguistic context of French, it may appear meaningless. However, if perceived in an English-speaking context, it could be interpreted as "Paddle your own canoe". Once the phrase is interpreted as English, the words, its sounds, and the imagery associated with it can become a memory that may be quite different from the actual sensory stimulus. Albright's example demonstrates how the same sensory data can be interpreted and recalled differently depending on the contextual cues available to the respective observers. In forensic contexts, there may be multiple examples of this effect. For instance, police reports might inadvertently emphasize or omit case details if the interviewee was the victim versus the suspect. Victims' perception of the crime could be influenced by the emotional and situational context of the incident, potentially affecting the accuracy of their recall. For forensic interpreters, context is essential; they must navigate not only linguistic differences but also cultural and situational nuances to ensure accurate communication during crime investigation interviews. The risk here is that any misinterpretation or emphasis could alter the perceived narrative of the crime, affecting the investigation and the work of forensic linguists.

In their investigation, Dror and colleagues (2022) demonstrated just how powerful context can influence perception, outweighing even the evidence from the stimulus itself. For their study, they examined how contextual information affects manner of death determinations made by forensic professionals, such as medical examiners and forensic pathologists. The professionals were shown various autopsy photographs that depicted a single gunshot wound. The forensic experts were then asked to deduce the most probable cause of death (i.e., homicide, suicide, accident, natural, or undetermined). To manipulate context, the researchers presented a narrative of the case suggestive of either homicide or suicide. For example, in the homicide narrative the critical details relayed that law enforcement agents confronted the decedent with gunfire, and the decedent seemed to have raised a gun toward the police when police opened fire. The decedent was never out of sight, and law enforcement found the decedent with a shot to the right side of the head. In the suicidal narrative, all details were the same, except that the decedent sheltered in place after the gunfire exchange and was out of sight for some time. Later the decedent was found dead with the gunshot wound to the right side of the head. The results of the study revealed that the contextual information had a significant impact on the cause of death the experts reported. Notably, the decisions of a substantial number of these professionals were influenced by the suggestive contextual narrative even more than by the autopsy information provided. In fact, participants were found to frequently alter their initial conclusions they had originally drawn from the autopsy evidence to align with the contextually suggested cause of death subsequently presented in the narrative.

Together, these studies suggest that the interpretation and subsequent recall of information can significantly vary with experiences and contexts the individual might have encountered before or during the perceptual process. Consequently, it is imperative for forensic linguists not only to consider the linguistic evidence in isolation. It is also important that they take into account the context in which that forensic evidence was procured, as the clarity, accuracy, and completeness of the linguistic evidence can be critically affected by the characteristics surrounding the encoding of the initial experience or situation being communicated.

Memory: A Product of Dynamic Construction

Simons and Chabris (2011) surveyed a large segment of the general population representative of the US population to gauge their understanding of memory functionality. When presented with the following assertion, "Human memory works like a video camera, accurately recording the events we see and hear so that we can review and inspect them later", 63% of the participants agreed. Similarly, when presented with the statement that "People generally notice when something unexpected enters their field of view, even when they're paying attention to something else", a significant proportion (77.6%) of the respondents agreed. These findings were later replicated with other large-scale studies (e.g., Simons and Chabris 2012). Such beliefs about memory are in direct contrast to the scientific findings. People's perception and memory do not function like a perfect video recorder, capturing, storing, and replaying experiences exactly as they occurred. Instead, memory is a highly constructive process in which information is actively created and stored (as noted in the previous section) as well as dynamically retrieved. This process of memory encoding and retrieval is also influenced by a whole host of psychological factors (Tulving and Craik 2000) that are beyond the scope of this chapter. In the next segment, however, we focus on significant factors that are of high relevance for forensic linguists.

Memory Storage: Effects of Post-Event Factors

Decay

The term "memory decay" refers to the process by which information stored in the memory deteriorates over time when it is not used or retrieved. Research by Ebbinghaus (1913) demonstrated that, over time, in the absence of rehearsal, memory degrades in a "forgetting curve" (Murre and Dros 2015). This seminal work demonstrated that a significant amount of information could be lost even within an hour of learning and stabilized over time in a loosely created L-shaped curve. Although contemporary research has not yet determined the exact mechanisms of memory decay, there is much evidence to suggest that forgetting is due to the deterioration of neural connections; or the "proactive interference" of responsible neurological pathways by similar information also stored in the memory; or the "retroactive interference" of the neuronal connections by newly acquired information (Davis and Zhong 2017). An important implication of this research for forensic work is that decay may lead to bias and distortion if a perceiver is not interviewed immediately after an incident.

In the domain of forensic linguistics, this research on memory degradation suggests that linguistic evidence may be significantly compromised if initially obtained after various delays.

It is worth noting that memories of emotionally charged and personally significant events tend to be more resistant than "regular" memories to forgetting. This resistance is primarily because these important events may be more frequently retrieved as the perceiver thinks about the occurrences in private or discusses with others. In fact, research on traumatic memories has found that these memories often remain vivid and persistent over time (Berntsen and Thomsen 2005). However, such memories are not completely resistant to forgetting (Pezdek and Taylor 2002). For example, it has been found that while the central details of traumatic events are often well-remembered, peripheral details may be less well-remembered. This differential recall is thought to occur because the individual's attention is narrowly focused on the central aspects of the trauma: as a result, less attention is paid and memory is allocated for peripheral details (Engelhard, et al. 2019). Experimental evidence provides compelling information about the effects of post-event factors on memory.

Christianson and Hubinette (1993), for example, conducted a study with witnesses of actual bank robberies. Between 4 to 15 months after the robbery, the researchers interviewed the witnesses about their emotional reactions at the time of the crime. The memories of the study participants were also probed for criminological details (e.g., the robbers' words, appearance, and behavior). These details recalled were compared with the initial police reports. The results showed a high level of accuracy where the details of the robbery were concerned. However, details about the circumstances surrounding the crime such as the day of the week it occurred, and perpetrator's appearance and language, were far less accurate. In the experiment, it was also found that bank tellers who had been direct victims of a robbery had a better memory overall than other bank employees and customers who had been bystanders. Nevertheless, the study provided compelling evidence that critical aspects of memories formed after emotional real-life events can be recalled accurately even after a long interval. Similar findings were reported by Odinot et al. (2009) who investigated memories of a supermarket robbery. Three months after the event, a comparison of the remembered details with the actual video recording of the robbery found the study participants' recall to be 84% correct.

Importantly, it is not emotion alone which seems to be responsible for victims and witnesses experiencing greater resistance to memory decay after these traumatic events. Real-life event studies have also demonstrated that the initial police interview may also play a pivotal role in shaping the longevity and accuracy of these memories (Bull and Milne 2020). Witnesses sometimes provide their first account of the crime immediately following the event. This immediacy can potentially reinforce the memory traces of the details witnessed (Bull and Milne 2020). For crime witnesses who are not interviewed immediately after the event, their recall may show some degradation, despite the fact that their heightened emotional state during the crime may initially have facilitated their encoding and storage of details. This means that details that are not strongly encoded or reinforced through subsequent deliberation or discussion (without interference) might never find their way into the linguistic evidence they provide during interviews conducted some time after the offense. Their recall may be marked by significant omissions and/or inaccuracies. This imperfect recall underscores the inherent fallibility of memory and must be carefully considered by forensic linguists during their analyses.

Post-Event Information

The process of introducing information after an event can significantly alter an individual's memory and lead to misremembering or creating entirely new false memories. This phenomenon, known as the "Misinformation Effect", has been widely studied (Loftus and Klemfuss 2023; Loftus 2005). One of the pioneers in this area of research is cognitive psychologist Elizabeth Loftus whose work has demonstrated how simple suggestions can alter memories and their accounts (Loftus 2005). For example, in a groundbreaking investigation, Loftus and Palmer (1974) showed participants a film of a car accident. They then asked them to estimate the speed of the vehicles. The wording of the question drastically impacted the participants' responses. When asked to rate how fast the cars were traveling when they "smashed into" each other, the respondents gave higher speed estimates than those who were asked to rate the speed when the cars "contacted" each other. Furthermore, those participants who were presented with the "smashed into" phrasing were also more likely to report having seen broken glass, even though none was present in the film. Similar studies replicating this "misinformation effect" have shown how post-event information can distort recollections of specific details, an entire event, and even the fundamental essence of an event (Cullen et al. 2020; Loftus 2005; Roediger et al. 1996). Furthermore, Loftus and Bernstein (2005) have found that people can easily develop richly detailed false memories of entirely fabricated events, pointing to the incredible malleability of human memory. The implication of this research for forensic linguists is that it is important to remember that the memory-based evidence they are provided with may well not be an entirely accurate account. It could even be completely, albeit unintentionally, false.

The interplay of emotion and repetition in creating memory distortions following misinformation exposure was carefully examined by Nahleen and colleagues (2021). In their experiment, participants were initially shown a film depicting a traumatic event (a car accident due to a distracted teen and the demise of the passengers). The next day they were given misleading details about the film. These misleading details were presented by other "eyewitnesses" who gave descriptions of the same traumatic film with either emotionally charged or neutral content which was either repeated or mentioned once. Subsequently, the participants' memory of the original film was tested. The findings showed that misinformation increased the memory distortion for the trauma event when compared to a control condition. More specifically, across all of the experimental conditions, the authors reported that any degree of misinformation exposure led to a 36–59% memory error rate. Interestingly, emotionally charged misinformation led to more memory distortion (e.g., non-presented scenes of cars colliding) than neutral information.

Importantly, there is much evidence that misinformation is likely to continue to influence memory, even if corrected. Ecker and colleagues (2011), for example, examined how the strength of misinformation encoding and the subsequent correction or retraction influenced the persistence of misinformation in memory. Their results showed that the stronger the initial misinformation was encoded, the greater the continued influence of the misinformation, even after it had been corrected. This resistance to correction persisted even when participants acknowledged the correction and could accurately recall it. This study highlights the resilience of misinformation in memory and points to potentially enduring errors in the language evidence that forensic linguists may be given to analyze.

Memory Retrieval: Influences on Accuracy

Retrieval-Induced Forgetting and Retrieval Failure

The process of recollection fundamentally entails the reconstruction of past experiences, which is achieved through the retrieval of a blend of old and new information. This reconstruction often involves people refining or editing memories to improve their coherence. It also regularly necessitates forgetting to streamline and manage the information. Thus, as odd as it may sound, forgetting is an essential optimizing tool for remembering. By the same token, the act of remembering can paradoxically lead to forgetting (Anderson, et al. 1994). Such "retrieval-induced forgetting" occurs when, during the process of memory retrieval about a specific event, we unintentionally overlook or forget other pertinent information. These memory phenomena can affect the comprehensiveness and accuracy of recollections. For example, recalling the name of one individual met at a conference could inhibit the ability to remember the names of other attendees. In a forensic context this could occur when a witness who distinctly recalls the tattoo on the arm of a robbery suspect may experience inhibition in their ability to recall other details or suspects from the scene, such as the clothing or facial features of an accomplice. This effect often intensifies as the temporal gap between the event and its recollection increases (Murayama et al. 2014). It is not the passage of time alone which can lead to difficulties in memory retrieval. As we accumulate new experiences and perspectives over time, these fresh insights can obstruct our access to older, once-familiar memories. From this perspective, new knowledge can effectively "block" recollection of information obtained under different circumstances, resulting in what psychologists call "retrieval failure". These mechanisms highlight the complexities of obtaining accurate information from human memory. The potential effect of these cognitive processes is crucial for forensic linguists to consider when questions about the accuracy and reliability of linguistic evidence arise. While retrieval-induced forgetting and retrieval failure are everyday phenomena that affect the quality of memory retrieval, there are other memory impediments that are specific to crime investigation. These will be discussed in the following section.

Co-Witness Influence

This term refers to a situation where one person's account influences another's memory. When witnesses discuss an event with each other, it can have both positive and negative effects on their memories of that event. On one hand, discussing the event can strengthen the memory trace, making the memory more resilient to forgetting. On the other hand, co-witness discussions can introduce erroneous information that add in partially or completely inaccurate detail to the witnesses' memories of the event, especially if the internal memory representation is weak (Horry et al. 2012). The importance of this potential memory contamination is not to be underestimated. Witnesses often do discuss with others in the aftermath of an event. This assertion is not simply based on anecdotal observation but has also been demonstrated in scientific research. For example, a survey of students who had witnessed a serious event revealed that when there was a co-witness present, most students discussed the event with the co-witness (Paterson and Kemp 2006). In another survey, with 60 crime witnesses participating in a police identification procedure, 88% reported having witnessed the event with one or multiple co-witnesses (Skagerberg and Wright 2008). Of these, 58% revealed that they had discussed the crime and suspect details with at least one co-witness. A product of this shared communication is "memory conformity" (Gabbert et al. 2003). Also called "social contagion of memory" (Roediger et al. 2001), this phenomenon involves individuals' recollections of an event becoming more similar to each other after they discuss the event. This process can happen consciously or subconsciously and can lead to significant distortions in individual recall.

The strength of this effect was demonstrated by Gabbert and colleagues (2003). In their study, they had research volunteers watch a video that depicted someone stealing money from a wallet. What the study participants did not know is that the researchers had created two almost identical versions of the film, but one of the two did not actually show the act of stealing. Participants were either individually assessed (control group) or placed in pairs (co-witness group). The paired participants were informed that they had observed the same video, but in fact they had each watched a different version of the film. The participants in the control group were asked to recall the crime individually, while the co-witness group was first allowed to discuss what they had seen in pairs. After the discussions, all of the participants were asked to complete a questionnaire designed to evaluate their memory of the event. The findings revealed that 71.6% of participants from the co-witness group reported at least one detail that they did not actually see in the video. Furthermore, 60% asserted that the target in the film was guilty, even though they had not actually witnessed the act of theft.

Horry and colleagues (2012) took their research one step further. In their study, they investigated how social influence can impact memory reports, even when the memory is confidently held. To find an answer, they conducted a series of experiments in which participants were first introduced to a highly credible or less credible partner who, unbeknownst to them, was a part of the research team. This so-called "confederate" was consistently found to impact the judgment of the study participants. For example, it was found that the participants often conformed to the highly credible confederate's false identifications of stimuli. Even when they had initially expressed high confidence about their own recall, the participants were still ready to conform to others' suggestions and discount their own internal cues to accuracy. As will be described in the next section, these experimental results are strongly mirrored by the research that has been done on police interviewing.

Suggestive Interviewing

There is extensive psychological research demonstrating the effects of suggestive questioning on memory errors. In this section we focus on the extreme end of this suggestibility: when individuals are induced to provide rich accounts of false memories. In 2015, Shaw and Porter examined how successful certain suggestive interview techniques such as visualization, false evidence, and social influence, were in making participants believe they had committed a crime that never happened. Over multiple interviews, the researchers presented participants with narratives about events in their teenage years. The task for the participants was to recall details about those target events. One narrative was of a real event that had happened to the participant and was based on information provided by the participant's parents. The other narrative was a researcher-fabricated account of a crime the participant had allegedly committed. By the end of the last interview, a significant proportion (70%) of participants had come to develop a false memory of the criminal act and could describe their non-existent criminal encounter. Participants recalled intricate details such as who had been with them at the time, what clothing they had worn, or how they had felt—even though the event had never occurred. Incredibly, the details of these false memories were just as vivid, rich, and emotionally intense as the memories of the real events. In light of such research, some modern professional law enforcement agencies have implemented training procedures designed some to eradicate the use of highly suggestive witness questioning techniques. Nevertheless, these findings emphasize the power of suggestion during the process of memory retrieval for forensic evidence gathering.

The Effects of Criminal Investigators' Interviewing on What Interviewees Say

Psychological research has found that criminal investigators in law enforcement are not free from biases (and that they sometimes behave unethically). For example, Ask and Granhag (2007) found that experienced police interviewers evaluated witness testimony that confirmed their investigative hypothesis as being more reliable and credible. As the researchers explain, "evidence that is inconsistent with a desired belief is subject to more cognitive analysis and scrutiny than is consistent evidence" (565). There is a large body of psychological research that corroborates this assertion.

For example, in a 2020 study, Komter focused on how the written reports of police interviews with suspects transformed what was actually said in the interviews. In many countries around the world, audio(-video) recordings or transcripts of police-suspect interviews are not, as yet, made. Instead, a written account of the interviews is made either during or after a suspect interview takes place. Sometimes these "statements" are made by defense lawyers who document their conversations with clients. Typically, however, this linguistic evidence is made by the police. Such is the case in Denmark. When comparing Dutch police audio recordings of suspect interviews with the related "official" police written accounts which had been partly prepared during the interviews, Komter made several important discoveries: (1) much of what the interviewers had actually said had not been included in the written accounts; (2) open-ended questions tended to interfere with "the interrogators' attempts to elicit 'recordable' answers" (227); (3) the officers who were responsible for preparing the official written statements tended to "select those items from the talk that are likely to contribute to making a case" (233). Based on these findings, Komter concluded that the written statements produced by the Dutch police were a "compromise" between reality and what the police thought was necessary (226). As Komter (2012) explained, "meanings are removed from the suspect's statement" (731).

Similar conclusions have been drawn by other researchers who have studied written statements prepared by law enforcement (e.g., Ferraz, de Almeida, and Drew 2020; de Keijser, Malsch, Kranendonk, and de Gruijter 2012). One of the major reasons why law enforcement officials may (un) consciously shape their written record of suspect-interviews is their awareness of procedural requirements as well as the prosecution's needs and the court's expectations (Milne, Roberts, Westera. and Kebbell 2011). The shaping of linguistic evidence may take several forms. Examples include omitting details the officer perceives as irrelevant, highlighting other details which are considered important for "proving a point", and presenting the statement in a simple chronological order, although the suspect's narrative may have been non-sequential (Mason and Rock 2020; Haworth 2018). However, the potential influence of investigator bias on linguistic evidence does not end there. The very questions police and other law enforcement officials ask (or do not ask) interviewees to answer may also be influenced by their hypotheses. In addition, the way in which criminal investigators interview witnesses and interrogate suspects can influence the linguistic evidence they gather.

At present, a large number of law enforcement agencies around the world use confrontational approaches that discourage free narration when interrogating suspects. In their 2012 examination of Canadian police interviews with suspects, Snook et al. discovered, for example, that very few of the questions asked by the police were open-ended and that closed questions constituted 70%. A free narrative was requested in only a seventh of the interviews. The continuing pervasiveness of such approaches is unfortunate when one considers their relative ineffectiveness. To date, there is a large body of research demonstrating that a dominant, confession-seeking approach in which the interrogator does the majority of the talking and uses closed or focused question types will often result in less information being provided by interviewees than may be obtained when officers adopt a non-confrontational, information-gathering approach which poses open-ended questions and encourages the suspect to talk unimpeded (May et al. 2021; Bull 2019a; Miller, Redlich, and Kelly 2018; Bull 2013). These findings are in line with other studies into law enforcement interview techniques. Consistently, it has been found that criminal investigators can gather more relevant information much more effectively by adopting interviewing approaches that focus on encouraging a dialogue between the investigator and the suspect (Bull and Milne 2020; Bull and Rachlew 2019; Walsh and Bull 2010). Such is the case with the "PEACE", a method of investigative interviewing that involves five stages: (1) *p*lanning and preparation; (2) *e*ngaging and explaining; (3) gaining a full *a*ccount of events; (4) obtaining *c*losure by summarizing suspect statements; and (5) *e*valuating the results of the interview process. A key element of the "PEACE" method is building and maintaining rapport with interviewees. This feature has been found to be associated with suspects providing a greater amount of information (Walsh and Bull 2012). Interviewers' demonstrations of empathy have also been positively correlated with the procurement of suspect information (Baker-Eck, Bull and Walsh 2021). Similarly, Leahy-Harland and Bull's (2017) study of interviewer strategies with suspects found that information-gathering was facilitated by interviewers showing rapport/empathy, presenting evidence, and requesting the suspect's attention. However, just the opposite was achieved when investigators explicitly asked suspects to account for themselves and tell the truth, emphasized the seriousness of the offense, or stressed the situational futility that the suspects found themselves in.

In investigative interviews with suspected terrorists, Surmon-Böhr et al. (2020) also found that the style of interviewing substantially affected what the suspects said in that interviewers' "commitment to creating an accepting, empathic atmosphere was associated with increased engagement" (1019). Similarly, Goodman et al. (2014) examined types of strategies used to try to facilitate disclosure from an international sample of "high-value detainees" and found that the disclosing of meaningful information was more complete in response to non-coercive strategies such as building rapport and showing respect. How an investigative interview is conducted can not only influence what an interviewee decides to say. If conducted without skill, an interview can also influence what an interviewee subsequently recalls. In a comprehensive analysis of a police interrogation in the United States, Hill (2003) revealed how apparently incriminating details in a suspect's account occurred in response to several (unethical) leading questions. For example, "The most significant detail, restraint by tying, is provided by the police, while the presence of bruise marks on the ears, is elicited by the police after multiple leading questions" (41). Of course, law enforcement officers are not the only professionals in the criminal justice system who have been found to routinely employ leading and potentially biasing questions.

The Effects of Lawyers' Questioning on What Witnesses Say in Court

One of the most common linguistic strategies employed by courtroom lawyers, especially when cross-examining, is to pose leading questions. The prevalence of this technique is long-standing. In their examination of the advice given to defense lawyers for conducting cross-examinations, Bull and Griffiths (2019) found the following two recommendations: (1) "Asking only leading questions is perhaps the oldest rule…it is a good one…allows the cross-examiner to be forceful" (readingschools.org) and (2) "Use leading questions which legitimately call only for a 'yes' or 'no' answer" (see https://www.yumpu.com/en/document/view/11449308/a-checklist-of-winning-cross-examination-conceptsand-techniques). Similarly, in his best-selling book on advocacy entitled *The Devil's Advocate: A Short Polemic on How to Be Seriously Good in Court*, Morley (2009) stated that "Rightly or wrongly, adversarial advocacy is not really an enquiry into the truth. Perhaps the adversarial system should be about finding out what really happened. But it isn't" (12) and "Always ask leading questions", "Never ask an open question" (157). Such advice has been shown to have a powerful influence on the types of questions posed in court and, by extension, the quality of linguistic evidence gathered. For example, Kebbell, Deprez, and Wagstaff (2003) examined lawyers' questioning used in UK rape trials. Their research determined that the lawyers' questions allowed little opportunity for witnesses to provide complete accounts of the alleged events. Lively et al. (2023) analyzed the questions put to witnesses by Canadian judges in 22 criminal cases. They found that most of the questions were clarifications (37%) followed by facilitators (i.e., verbal gestures that encourage the flow of conversation such as "Um-hmm", "Yes", "Okay") (17%) and closed yes/no (10%). Fewer than 1% of questions were open-ended, even though the longest answers were provided in response to these open-ended questions.

Questions that severely restrict or discourage elaboration by the witness are not the only source of potential bias of linguistic evidence in the courtroom. Andrews and Lamb (2017) analyzed transcripts of 36 child abuse criminal proceedings involving 56 alleged victims conducted between 2009 and 2014. They determined that defense lawyers asked linguistically complex questions significantly more often than did prosecutors, especially "when prompting children with option-posing and suggestive prompts" (189). Some lawyers' utterances to the children contained up to eight questions, ten sentences, and over 150 words (some words containing up to 15 letters). These complex questions resulted in more unresponsiveness, expressions of uncertainty, and self-contradictions among these child witnesses. These results are particularly important when one considers the fact that high levels of detail, consistency, plausibility, and coherence, along with low levels of contradictions, are very important for determining whether an account is considered valid, accurate, and/or reliable (Bull 2019b). In view of their findings, Andrews and Lamb (2017) concluded that the questions the lawyers asked the children sometimes appeared to "exploit their developmental limitations" and were in violation of questioning guidelines designed to "elicit truthful testimony" (191). Of course, it is important to stress that the use of leading, confrontational, complex, questioning in the courtroom may diminish the accuracy, reliability, and richness of linguistic evidence provided by most witnesses, regardless of their age. However, as will be shown in the following section, this potential for introducing bias may be particularly great for the most vulnerable members of our society.

Questioning and Interviewing the "Vulnerable" in the Criminal Justice System

It is well known that a meaningful proportion of underage victims, witnesses, and suspects have limited cognitive (e.g., memory) and communication skills. These restrictions have direct implications for social interactions that take place across a variety of environments, including forensic settings. For example, Johnstone et al. (2016) determined that around 70% of adolescent defendants had listening-comprehension skills below the norm for ten-year-olds. Similarly, it has been reported that of children in the UK youth justice system, over half had significant speech, language, or communication difficulties, and a third of young people in police custody have a mental health disorder (Barnado's 2023). Even minors who do not have cognitive or communicative restrictions are still at a decided disadvantage when they enter the adult world of the criminal justice system. For that reason, around the world, children in forensic settings are commonly automatically deemed to be "vulnerable'" and obtaining linguistic evidence from them, therefore, requires special skills (Melinder, Magnusson and Gilstrap 2021; Bull 2010). As research has shown, when interviewed by a trained professional, children can provide credible, accurate, and relevant linguistic evidence of great forensic importance. Pipe et al. (2013), for example, compared the investigative interviews of suspected child sexual abuse victims conducted either before or after a large sample of investigators (in the US state of Utah) had been trained to use an interviewing protocol based on relevant research. They reported the following findings: (1) deciding to bring charges was more likely for post-training interviews (53%) than for pre-training interviews (42%); and (2) guilty verdicts more often occurred in post-training cases. A study in Italy by Roma et al. (2011) compared the contents of children's "testimony" in sexual abuse cases in which courts rendered either a guilty or not guilty verdict. The types of child discourse content that significantly differentiated between the types of verdict were quantity of details (including unusual and superfluous details), description of the interaction, reproduction of conversation, and spontaneous corrections/additions. In Norway, Myklebust and Bjorklund (2009) found that in court/legal proceedings regarding alleged child abuse, a factor that discriminated between cases that resulted in convictions and cases that did not was the length of children's answers to open questions in police interviews. That being the case, adult linguistic behaviors, which may tend to restrict the length of children's responses in the criminal justice system, can be highly problematic.

Evidence of this assertion can be seen in a 1993 study conducted by Flin et al. in Scotland. Their observational investigation involved 89 children who testified in 40 criminal trials in which the principal charges were assault, murder/attempted murder, or sex-related. Three-quarters of the children were bystander witnesses and 25% were alleged victims. In 12% of examinations-in-chief, vocabulary was used that some of the children seemed not to understand. During the cross-examinations, that percentage jumped to 40% although the children rarely explicitly said they did not understand. Clearly, lapses in a witness's level of comprehension, be they disclosed or undisclosed, can significantly bias the linguistic evidence they provide. Scientists have shown that this problem is not limited to underage witnesses. Kebbell, Hatton, and Johnson (2004), for example, compared UK court transcripts for rape, sexual assault, and assault trials involving witnesses with intellectual disabilities with a sample of similar cases involving witnesses from the general population. The accounts of witnesses with intellectual disabilities were shorter and more likely to agree with leading questions than were those of the witnesses from the general population.[1]

In their analyses of UK police interviews, Farrugia and Gabbert (2020) compared outcomes gained with suspects who had mental health disorders and those without vulnerability. The researchers reported that the vulnerable suspects more frequently sought clarification of the questions being put to them, especially for open questions. However, when the criminal investigators were examined, overall there were no differences regarding the rates of interviewers asking various question types between the two groups of suspects.[2] Such a finding implies that the mental health disorders of the interviewees had little effect on the interviewing strategy used by the investigators. What was different was the ways in which the vulnerable and non-vulnerable interviewees responded to the interviewing. Given the prevalence of vulnerable people in our societies, the implications of these findings are quite significant. According to a 2019 report published by the UK national newspaper *The Guardian*, of the "around 1 million" police interviews with adult suspects in the preceding year, approximately 17% or 17,000 involved vulnerable persons. In a Dutch study by Geijsen et al. (2018), a sample of suspects in police custody were psychologically screened for intellectual disabilities, mental disorders, abnormal mental states, and substance abuse. It was found that 69% of the suspects could be deemed vulnerable.

Given the size and special needs of vulnerable people in the criminal justice system, in 2001, the government in England and Wales commissioned Professor Ray Bull (and others) to write the relevant guidance document for obtaining the best evidence possible. After reviewing the relevant body of literature, the following conclusion was reached: "It cannot be over-emphasised that responses to leading questions referring to central facts of the case that have not already been described...earlier by the person...are likely to be of very limited evidential value in criminal proceedings" (77). This guidance contrasts sharply with the advice frequently given to lawyers and the practices of some law enforcement agencies. Failure to follow such evidence-based guidelines in the criminal justice system may result in the continued production of linguistic evidence that is less complete and more affected by biases.

Possible Bias in Forensic Linguists

Forensic linguists should remain entirely neutral to serve the processes of justice. However, as this chapter has shown, much of the linguistic evidence that they are provided with for their analysis may already have been influenced by biases (and sometimes by unethical behavior).

In addition to taking these precautions, forensic linguists also need to guard against introducing bias into their own work. This is by no means an easy task, given that all people are biased to some extent (Watkins and Harvey 2020). The first step to minimizing the influence of bias is to admit that one is susceptible. Unfortunately, far too often, people are unaware of their own blindspots and operate under the assumption that they themselves are not biased, but everyone else is. Forensic experts are no exception, despite the fact that many may erroneously believe that "they can consciously control their impartiality and remain objective" (Areh 2020: 101607).[3]

As research shows, the decision-making of even highly trained, greatly experienced experts may also be affected by (un)conscious biases (Huang and Bull 2020; Wilcock, Bull, and Milne 2008). For example, Kersholt and Eikelboom (2007) found professional crime analysts were unduly influenced by an investigation team's prior interpretation. The question is, what can forensic linguists do to avoid the problem of introducing bias into their work? Several useful, evidence-based suggestions can be found in the ground-breaking work of Itiel Dror (e.g., Dror 2018; Dror and Murrie 2018; Dror 2016). Based on his investigations into how to optimize the performance of forensic scientists, Dror developed a "Hierarchy of Expert Performance" or HEP. According to the HEP model, Dror (2016) identified three factors that are relevant to expert performance. The first HEP factor is *reliability*. It relates to an expert's performance not being influenced by irrelevant matters such as the accent of a suspect being interviewed by a police officer (Dixon, Mahoney, and Cocks 2002). There may not always be consensus about whether and/or which factors are relevant to forensic linguistic analysis. However, there should always be transparency (e.g., in writing) as to what factors were considered and how they contributed to the analysis. Even when these factors might be understood or applied differently by various individuals or within different jurisdictions, providing complete transparency with regard to the decisions made and the procedures followed is essential for guarding against the hidden effects of biases.

The second HEP factor is *consistency*. To achieve this element of expert performance, it can be useful for forensic linguists to compare their analysis with either separate analyses they performed sometime later or preferably with the contemporaneous and independent analyses performed by other forensic linguists.

The third HEP factor importantly differentiates the observations from the *conclusions*, that is (i) how the data/information is gathered and (ii) how this is evaluated. In the forensic science domain, it is usually reasonably straightforward to distinguish observations from conclusions. For instance, a fingerprint examiner will first carefully observe the minutia characteristics in the friction ridge of the fingerprints (observation) and then draw a conclusion as to whether they "match" those on another print. In forensic linguistics, it should be made very clear how the observations lead to the conclusion. (For more on this, see Huang and Bull 2020).

Conclusion

Although it may not be possible to nullify or avoid the effects of all bias, striving to achieve the three HEP factors described above can assist forensic linguists in keeping their work as free from bias as possible. With that ultimate goal in mind, in this chapter, our intention was to help make forensic linguists more aware of potential biases embedded within the material they examine, as well as increase sensitivity to their own possible biases. As the research presented in this chapter shows, biases can significantly impact how information about a crime event is encoded, stored, recalled, conveyed, recorded, interpreted, and analyzed. By becoming and remaining aware of these potential biases, forensic linguists can navigate their tasks with a greater degree of reliability and accuracy, thus contributing to more effective and fair outcomes in legal processes.

Notes

1 For an overview of language impairments among young offenders, see Anderson, Hawes, and Snow (2016) and for an overview of how and why suspects' developmental immaturity renders them vulnerable to questioning, see Cleary (2017).
2 The only exception was so-called "echo questions" in which the interviewers echoed what the suspect had just communicated and were used more often with suspects who had mental health disorders.
3 See also MacLean et al. (2019).

References

Albright, T. D. 2015. "Perceiving". *Daedalus* 144, no. 1: 22–41.

Anderson, M. C., R. A. Bjork, and E. L. Bjork. 1994. "Remembering Can Cause Forgetting: Retrieval Dynamics in Long-term Memory." *Journal of Experimental Psychology: Learning, Memory, and Cognition* 20: 1063–1087.

Anderson, S., D. Hawes, and P. Snow. 2016. "Language Impairments Among Youth Offenders: A Systematic Review." *Children and Youth Services Review* 65: 195–203.

Andrews, S., and M. Lamb. 2017. "The Structural Linguistic Complexity of Lawyers' Questions and Children's Responses in Scottish Criminal Trials." *Child Abuse & Neglect* 65: 182–193.

Areh, I. 2020. "Forensic Assessment May Be Based on Common Sense Assumptions Rather Than Science." *International Journal of Law and Psychiatry* 71: 101607.

Ask, K., and P. Granhag. 2007. "Motivational Bias in Criminal Investigators' Judgments of Witness Reliability." *Journal of Applied Social Psychology* 37, no. 3: 561–591.

Baker-Eck, B., R. Bull, and D. Walsh. 2021. "Investigative Empathy: Five Types of Cognitive Empathy in a Field Study of Investigative Interviews with Suspects of Sexual Offences." *Investigative Interviewing: Research and Practice* 11*:* 28-39.

Barnardo's. 2023. Barnardo's Annual Report 2023. chrome-extension://efaidnbmnnnibpcajpcglclefindmkaj/https://www.barnardos.ie/wp-content/uploads/2024/07/Barnardos-Annual-Report-2023-Final.pdf.

Berntsen, and D.K. Thomsen 2005. "Personal Memories for Remote Historical Events: Accuracy and Clarity of Flashbulb Memories Related to World War II." *Journal of Experimental Psychology: General* 134: 242–257.

Brewer, W. F., and J. C. Treyens. 1981. "Role of Schemata in Memory for Places." *Cognitive Psychology* 13, no. 2: 207–230.

Bull, R. 2019a. "Roar or PEACE: Is It a Tall Story?" In *Routledge International Handbook of Legal and Investigative Psychology.* Edited by R. Bull and I. Blandon-Gitlin. London: Routledge.

Bull, R. 2019b. "Credible Accounts: What They Are and How to Obtain Them." *Linguistics and the Human Sciences* 12: 166–176.

Bull, R. 2013. "What Is 'Believed' or Actually 'Known' About Characteristics That May Contribute to Being a Good/Effective Interviewer?" *Investigative Interviewing: Research and Practice* 5: 128–143.

Bull, R. 2010. "The Investigative Interviewing of Children and Other Vulnerable Witnesses: Psychological Research and Working/Professional Practice." *Legal and Criminological Psychology* 15: 5–23.

Bull, R., and A. Griffiths. 2019. "Advocating the 'PEACE' Method: Will It Make People Cross?" In *Access to Justice for Vulnerable Peopl.* Edited by P. Cooper and L. Hunting. London: Wildy.

Bull, R., and R. Milne. 2020. "Recommendations for Collecting Event Memory Evidence." In *Memory and Sexual Misconduct: Psychological Research for Criminal Justice*. Edited by J. Pozzulo, E. Pica, and C. Sheahan. New York: Routledge.

Bull, R., and A. Rachlew. 2019. "Investigative Interviewing: From England to Norway and Beyond." In *Interrogation and Torture: Research on Efficacy, and Its Integration with Morality and Legality*. Edited by S. Barela, M. Fallon, G. Gaggioli, and J. Ohlin. Oxford: Oxford University Press.

Christianson, S.-Å., and B. Hübinette. 1993. "Hands Up! A Study of Witnesses' Emotional Reactions and Memories Associated with Bank Robberies.". *Applied Cognitive Psychology* 7: 365–379.

Cleary, H. 2017. "Applying the Lessons of Developmental Psychology to the Study of Juvenile Interrogations: New Directions for Research, Policy, and Practice." *Psychology, Public Policy and Law* 23: 118–130.

Coulthard, M., and A. Johnson. 2007. *An Introduction to Forensic Linguistics: Language in Evidence*. New York: Routledge.

Cullen, H. J., H. Paterson, and C. van Golde. 2020. "Stopping Crime? The Effect of Crime Re-enactments on Eyewitness Memory." *Psychiatry, Psychology, and Law: An Interdisciplinary Journal of the Australian and New Zealand Association of Psychiatry, Psychology and Law* 28, no. 2: 286–309.

Davis, R. L., and Y. Zhong. 2017. "The Biology of Forgetting: A Perspective." *Neuron* 95, no. 3: 490–503.

de Keijser, J., M. Malsch, R. Kranendonk, R., and M. de Gruijter. 2012. "Written Records of Police Interrogation: Differential Registration as Determinant of Statement Credibility and Interrogation Quality." *Psychology, Crime & Law* 18, no. 7: 613–629.

Dixon, J. A., B. Mahoney, and R. Cocks. 2002. "Accents of Guilt? Effects of Regional Accent, Race, and Crime type on Attributions of Guilt." *Journal of Language and Social Psychology* 21, no. 2: 162–168.

Dror, I. E. 2016. "A Hierarchy of Expert Performance." *Journal of Applied Research in Memory and Cognition* 5, no. 2: 121–127.

Dror, I. E. 2018. "Biases in Forensic Experts." *Science* 360, no. 6386: 243.

Dror, I. E., and D. Murrie. 2018. "A Hierarchy of Expert Performance Applied to Forensic Psychological Assessments." *Psychology, Public Policy, and Law* 24, no. 1: 11–23.

Dror, I. E., D. Wolf, G. Phillips, S. Gao, Y. Yang, and S. Drake. 2022. "Contextual Information in Medicolegal Death Investigation Decision-making: Manner of Death Determination for Cases of a Single Gunshot Wound." *Forensic Science International: Synergy* 5: 00285.

Ebbinghaus, H. 1913. (1885). *Memory: A Contribution to Experimental Psychology*. New York: Teachers College, Columbia University.

Ecker, U. K., S. Lewandowsky, B. Swire, and D. Chang. 2011. "Correcting False Information in Memory: Manipulating the Strength of Misinformation Encoding and its Retraction.". *Psychonomic bulletin & review* 18, no 7: 570–578.

Engelhard, I. M., R. McNally, and K. van Schie. 2019. "Retrieving and Modifying Traumatic Memories: Recent Research Relevant to Three Controversies." *Current Directions in Psychological Science* 28, no. 1: 91–96.

Farrugia, L., and F. Gabbert. 2020. "Vulnerable Suspects in Police Interviews: Exploring Current Practice in England and Wales." *Journal of Investigative Psychology and Offender Profiling* 17, no. 1: 17–30.

Ferraz de Almeida, F., and P. Drew. 2020. "The Fabric of Law-in-action: 'Formulating' The Suspect's Account During Police Interviews in England." *International Journal of Speech, Language & the Law* 27, no. 1: 35–58.

Flin, R., R. Bull, J. Boon, and A. Knox. 1993. "Child Witnesses in Scottish Criminal Trials." *International Review of Victimology* 2: 309–329.

French, P., and P. Harrison. 2007. "Positions Statement Concerning Use of Impressionistic Likelihood Terms in Forensic Speaker Comparison Cases, with a Foreword by Peter French and Philip Harrison." *International Journal of Speech Language and The Law* 14: 137–144.

Gabbert, F., A. Memon, and K. Allan. 2003. "Memory Conformity: Can Eyewitnesses Influence Each Other's Memories for an Event?" *Applied Cognitive Psychology* 17, no. 5: 533–543.

García-Bajos, E., and M. Migueles. 2003. "False Memories for Script Actions in a Mugging Account." *European Journal of Cognitive Psychology* 15, no. 2: 195–208.

García-Bajos, E., M. Migueles, and A. Aizpurua. 2012. "Bias of Script-driven Processing on Eyewitness Memory in Young and Older Adults." *Applied Cognitive Psychology* 26, no. 5: 737–745.

Geijsen, K., C. de Ruiter, and N. Kop. 2018. "Identifying Psychological Vulnerabilities: Studies on Police Suspects' Mental Health Issues and Police Officers' Views." *Cogent Psychology* 5 no. 1: 1462133.

Goodman-Delahunty, J., N. Martschuk, and M. Dhami. 2014. "Interviewing High Value Detainees: Securing Cooperation and Disclosures." *Applied Cognitive Psychology* 28, no. 6: 883–897.

Haworth, K. 2018. "Tapes, Transcripts and Trials: The Routine Contamination of Police Interview Evidence." *The International Journal of Evidence & Proof* 22, no. 4: 428–450.

Hellmann, D. F., and A. Memon. 2016. "Attribution of Crime Motives Biases Eyewitnesses' Memory and Sentencing Decisions." *Psychology, Crime & Law* 22, no 10: 957–976.

Hill, M. D. 2003. "Identifying the Source of Critical Details in Confessions." *International Journal of Speech, Language and the Law* 10, no. 1: 23–61.

Holst, V. F., and K. Pezdek, K. 1992. "Scripts for Typical Crimes and Their Effects on Memory for Eyewitness Testimony." *Applied Cognitive Psychology* 6: 573–587.

Horry, R., M. Palmer, M. Sexton, and N. Brewer. 2012. "Memory Conformity for Confidently Recognized Items: The Power of Social Influence on Memory Reports." *Journal of Experimental Social Psychology* 48, no. 3: 783–786.

Huang, C-Y., and R. Bull. 2020. "Applying Hierarchy of Expert Performance (HEP) to Investigative Interview Evaluation: Strengths, Challenges and Future Directions." *Psychiatry, Psychology and Law* 28: 255–273.

Hutchinson, D. T., K. Pammer, K. Bandara, and B. N. Jack. 2022. "A Tale of Two Theories: A Meta-analysis of the Attention Set and Load Theories of Inattentional Blindness." *Psychological Bulletin* 148, no. 5/6: 370–396.

James, W. 1890. *The Principles of Psychology*. Vol. 2. Henry Holt and Company.

Johnstone, K., K. Prentice, H. Whitehead, L. Taylor, R. Watts , and T. Tranah. 2016. "Assessing Effective Participation in Vulnerable Juvenile Defendants." *Journal of Forensic Psychiatry and Psychology* 27: 802–818.

Kebbell, M., S. Deprez, and G. Wagstaff. 2003. "The Direct and Cross-examination of Complainants and Defendants in Rape Trials: A Quantitative Analysis of Question Type." *Psychology, Crime and Law* 9, no. 1: 49–59.

Kebbell, M., C. Hatton, and S. Johnson. 2004. "Witnesses with Intellectual Disabilities in Court: What Questions are Asked and What Influence do They Have?" *Legal and Criminological Psychology* 9, no. 1: 23–35.

Kerstholt, J. H., and A. Eikelboom. 2007. "Effects of Prior Interpretation on Situation Assessment in Crime Analysis." *Journal of Behavioral Decision Making* 20, no. 5: 455–465.

Komter, M. 2020. "Practical Narratives in the Criminal Law Process: The Suspect's Statement." *Conflicting Narratives of Crime and Punishment*, 219–236. Edited by M. Althoff, B. Dollinger, and H. Schmidt. Cham: Palgrave Macmillan.

Komter, M. 2012. "The Career of a Suspect's Statement: Talk, Text, Context." *Discourse Studies* 14, no. 6: 731–752.

Leahy-Harland, S., and R. Bull. 2017. "Police Strategies and Suspect Responses in Real-life Serious Crime Interviews." *Journal of Police and Criminal Psychology* 32: 138–151.

Lenton, A. P., and A. Bryan. 2005. "An Affair to Remember: The Role of Sexual Scripts in Perceptions of Sexual Intent." *Personal Relationships* 12: 483–498.

Lively, C. J., L. Fallon, L., B. Snook, and W. Fahmy. 2023. "Objection, Your Honour: Examining the Questioning Practices of Canadian Judges." *Psychology, Crime & Law* 29: 677–695.

Loftus, E. F., and D. M. Bernstein. 2005. "Rich False Memories: The Royal Road to Success." In *Experimental Cognitive Psychology and Its Applications,* 101–113. Edited by A. F. Healy. American Psychological Association.

Loftus, E. F., and J. Klemfuss. 2023. "Misinformation – Past, Present, and Future." *Psychology, Crime & Law* 30, no. 4: 312–318.

Loftus, E. F., and J. Palmer. 1974. "Reconstruction of Automobile Destruction: An Example of the Interaction Between Language and Memory." *Journal of Verbal Learning & Verbal Behavior* 13, no. 5: 585–589.

MacLean, N., T. Neal, R. Morgan, and D. Murrie. 2019. "Forensic Clinicians' Understanding of Bias." *Psychology, Public Policy, and Law* 25, no. 4: 323–330.

Mason, M., and F. Rock, eds. 2020. *The Discourse of Police Interviews*. Chicago: University of Chicago Press.

May, L., E. Gewehr, J. Zimmermann, Y. Raible, and R. Volbert. 2021. "How Guilty and Innocent Suspects Perceive the Police and Themselves: Suspect Interviews in Germany." *Legal and Criminological Psychology* 26, no. 1: 42–61.

Melinder, A., M. Magnusson, and L. Gilstrap. 2021. "What Is a Child-appropriate Interview? Interaction Between Child Witnesses and Police Officers. *International Journal on Child Maltreatment: Research, Policy, and Practice* 3, no. 4: 369–392.

Miller, J. C., A. Redlich, and C. Kelly. 2018. "Accusatorial and Information-gathering Interview and Interrogation Methods: A Multi-country Comparison." *Psychology, Crime & Law* 24, no. 9: 935–956.

Milne, B., K. Roberts, N. Westera, and M. Kebbell. 2011. "Interviewing Witnesses: Do Investigative and Evidential Requirements Concur?" *The British Journal of Forensic Practice* 13: 103–113.

Morley, I. 2009. *The Devil's Advocate: A Short Polemic on How to Be Seriously Good in Court.* London: Sweet and Maxwell.
Murre, J. M., and J. Dros. 2015. "Replication and Analysis of Ebbinghaus' Forgetting Curve." *PloS One* 10, no. 7: e0120644.
Murayama, K., T. Miyatsu, D. Buchli, and B. Storm. 2014. "Forgetting as a Consequence of Retrieval: A Meta-analytic Review of Retrieval-induced Forgetting." *Psychological Bulletin* 140, no. 5: 1383–1409.
Myklebust, T., and R. Bjorklund. 2009. "The Child Verbal Competence Effect in Court: A Comparative Study of Field Investigative Interviews of Children in Child Sexual Abuse Cases." *Journal of Investigative Psychology and Offender Profiling* 6: 117–128.
Nahleen, S., R. Nixon, and M. Takarangi. 2021. "The Role of Belief in Memory Amplification for Trauma Events." *Journal of Behavior Therapy and Experimental Psychiatry* 72: 101652.
Odinot, G., G. Wolters, and P. van Koppen. 2009. "Eyewitness Memory of a Supermarket Robbery: A Case Study of Accuracy and Confidence After 3 Months." *Law and Human Behavior* 33, no. 6: 506–514.
Paterson, H. M., and R. Kemp. 2006. "Co-witnesses Talk: A Survey of Eyewitness Discussion." *Psychology, Crime & Law* 12, no. 2: 181–191.
Pezdek, K., and J. Taylor. 2002. "Memory for Traumatic Events in Children and Adults." In *Memory and Suggestibility in the Forensic Interview,* 165–183. Edited by In M. L. Eisen, J. A. Quas, and G. S. Goodman. New York: Lawrence Erlbaum Associates Publishers.
Pipe, M-E., Y. Orbach, M. Lamb, C. Abbott, and H. Stewart. 2013. "Do Case Outcomes Change When Investigative Interviewing Practices Change?" *Psychology, Public Policy, and Law* 19: 179–190.
Roediger, H. L. III, J. D. Jacoby, and K. B. McDermott. 1996. "Misinformation Effects in Recall: Creating False Memories Through Repeated Retrieval." *Journal of Memory and Language* 35, no. 2: 300–318.
Roediger, H. L. III, M. Meade, and E. Bergman. 2001. "Social Contagion of Memory." *Psychonomic Bulletin & Review* 8, no. 2: 365–371.
Roma, P., et al. 2011. "Validity of Criteria-Based Content Analysis (CBCA) at Trial in Free-narrative Interviews." *Child Abuse & Neglect* 35: 613–620.
Shaw, J., and S. Porter. 2015. "Constructing Rich False Memories of Committing Crime." *Psychological Science* 26, no. 3 : 291–301.
Simons, D. J., and C. Chabris. 2012. "Common (Mis)beliefs About Memory: A Replication and Comparison of Telephone and Mechanical Turk Survey Methods." *PloS One* 7, no. 12: e51876.
Simons, D. J., and C. F. Chabris. 1999. "Gorillas in Our Midst: Sustained Blindess for Dynamic Events.". *Perception* 28, no 9: 1059–1074.
Simons, D. J., and C. Chabris. 2011. "What People Believe About How Memory Works: A Representative Survey of the U.S. population." *PloS One* 6, no. 8: e22757.
Simons, D. J., and M. Schlosser. 2017. "Inattentional Blindness for a Gun during a Simulated Police Vehicle Stop." *Cognitive Research: Principles and Implications* 2, no. 1: 37.
Skagerberg, E. M., and D. Wright. 2008. "The Prevalence of Co-witnesses and Co-witness Discussions in Real Eyewitnesses." *Psychology, Crime & Law* 14, no. 6: 513–521.
Snook, B., K. Luther, H. Quinlan, H., and R. Milne. 2012. "Let'em Talk! A Field Study of Police Questioning Practices of Suspects and Accused Persons." *Criminal Justice and Behavior* 39, no. 10: 1328–1339.
Surmon-Böhr, F., L. Alison, P. Christiansen, and E. Alison. 2020. "The Right to Silence and the Permission to Talk: Motivational Interviewing and High-value Detainees." *American Psychologist* 75: 1011–1021.
Tulving, E., and F. Craik , eds. 2000. *The Oxford Handbook of Memory.* Oxford: Oxford University Press.
Walsh, D., and R. Bull. 2012. "Examining Rapport in Investigative Interviews with Suspects: Does Its Building and Maintenance Work?" *Journal of Police and Criminal Psychology* 27: 73–84.
Walsh, D., and R. Bull. 2010. "What Really Is Effective in Interviews with Suspects? A Study Comparing Interview Skills Against Interview Outcomes." *Legal and Criminological Psychology* 15: 305–321.
Ward, E. J., and B. Scholl. 2015. "Inattentional Blindness Reflects Limitations on Perception, Not Memory: Evidence from Repeated Failures of Awareness." *Psychonomic Bulletin & Review* 22, no. 3: 722–727.
Watkins, T., and L. Harvey. 2020. "I am Not Biased. It Is Everyone Else's Problem." *Spinal Cord* 58: 389–390.
Wilcock, R., R. Bull, and R. Milne. 2008. *Witness Identification in Criminal Cases.* Oxford: Oxford University Press.

9

COUNTERING PSEUDOLINGUISTIC CLAIMS

The Ethical Challenges Facing Forensic Linguists

Tanya Karoli Christensen and Sofia Navarro Beck

Introduction

Ethics is concerned with morality and norms for behavior, or, in basic terms, what we ought to do or not do. Given that forensic linguistics deals both with the letter of the law and with the prosecution of transgressions of the law, ethical issues form a steady undercurrent to everything forensic linguists do as professionals and academics. However, there are also ethical challenges they face that are related to the actions of others outside of the profession—persons who (ab)use linguistics to make unsubstantiated claims about the intelligence that can be extracted from language data. This chapter reviews some of the pseudolinguistic schemes that are marketed to law enforcement, judicial systems, and the general public to convince them that simple, easy-to-use tools can reveal who is lying and who is truthful, who is guilty and who is innocent. We deconstruct three typical claims from operatives in this marketplace regarding language and linguistics to demonstrate how they are "epistemically irresponsible" (Heffer 2020: 178). This chapter argues that the marketing of such pseudolinguistic methods in the realm of legal prosecutions poses a significant ethical challenge to forensic linguists who not only seek to debunk false claims but also offer informative and understandable, evidence-based counter-narratives that are rooted solidly in linguistic science.

This chapter cannot, and does not attempt to, represent the full breadth and extent of pseudolinguistic offers on the market. It would be impossible to account for all those being promoted outside of the public sphere. The same is true of the many previously debunked methods that are being peddled under new names (or acronyms, as the fashion would have it). One of the pragmatic challenges that faces the forensic linguist is to remain vigilant for new developments that may cause harm. The difficulty in keeping abreast of the latest pseudoscientific developments may be greater in countries with few forensic linguists to share the burden of discovering and countering misleading claims about language in the legal realm. It is therefore important for forensic linguists to work as an international research community and share information about pseudolinguistic methods. By the same token, forensic linguists must warn practitioners and the public alike about the potential risks to the dispensation of justice if such legitimate linguistic methods are improperly used or illegitimate methods are used at all.

DOI: 10.4324/9781003391074-10

Pseudolinguistic Methods and Language Evidence

Language is used for legal (and illegal) purposes in myriad ways, and forensic linguistics is continually branching out to cover an increasing number of them. For some time now, linguistic expertise has been applied to improve investigative and judicial procedures and to offer expert testimony in civil and criminal cases. However, there are other areas where language and the law clearly intersect but the science of linguistics has (as yet) had little to no influence on current practices. This state of affairs leaves such areas open to linguistically untrained actors who may or may not genuinely believe that they understand how language works. For instance, in the area of policing, interrogations are still being conducted with coercive techniques for verbal "deception detection" and "credibility analysis". What many of these law enforcement agencies may not realize is that none of these methods is based on a real understanding of linguistic theory. Although many champions of these methods frequently use linguistic terminology, theory, or results, the fact is, they often do so in a faulty and misleading manner. As will be shown in the following section, this misuse of linguistics can have significant negative consequences.

Police Interviews

Internationally, there are many competing approaches to police interviews on offer today. To distinguish them, a dividing line is often drawn between two main groups: (1) "accusatorial" (e.g., Meissner et al. 2012) or "confession-based" (e.g., Davis and Leo 2013) methods; and (2) "information-gathering" or "cognitive" methods (Memon and Gawrylowicz 2019). To the first group belongs the often-discussed "Reid Technique" (Inbau and Reid 2011 and later; Jane and Buckley 1999) and to the other belong the English-Welsh *PEACE* model (where the abbreviation stands for *Planning and Preparation, Engage and Explain, Account Clarification and Challenge, Closure, Evaluation*) (Clarke and Milne 2001; Cherryman and Bull 2000) and the Norwegian *KREATIV* model (*Kommunikasjon "Communication", Rule of law, Ethics and empathy, Active consciousness, Trust through openness, and Information, Verified through science*) (Fahsing and Jakobsen 2017). Studies in the United Kingdom and elsewhere have demonstrated that intelligence gathered from cognitive interviewing is richer and more reliable than intelligence gathered from confession-based methods (e.g., Meissner et al. 2012; Memon, Meissner, and Joanne Fraser 2010).

The Reid Technique, named after its primary inventor John E. Reid, is widely used throughout the globe. According to the guidelines outlined in the "Nine Steps of Interrogation", a prerequisite for using this method is that "the investigator is reasonably certain of the suspect's guilt" (Inbau and Reid 2011: 5). Such a presumption of guilt need not rely on independent investigatory evidence but can be generated by the initial non-accusatory "Behavior Analysis Interview". As has been shown repeatedly, the indicators of truth and falsity suggested for that pre-interrogation interview are scientifically unsubstantiated and operate upon the reductionist notion that only "guilty suspects" lie (Vrij et al. 2022; Heydon 2019; Hirsch 2014). Unsurprisingly, there are plenty of examples of false confessions that have resulted from the deployment of the Reid Technique against innocent people (see, e.g., Hirsch 2014; Gudjonsson and Pearse 2011; Moore and Fitsimmons 2011; Ofshe and Leo 1997).

The corporation behind the Reid Technique, Reid & Associates, Inc., is very aware of the decades of criticisms that have been lodged against their methods, as evidenced by the long list of responses on their "Critics Corner". Nevertheless, in an examination of the adjustments made to the Reid Technique over the years, Hirsch (2014) finds that its tactics still carry grave risks of "breaking" innocent people and making them confess to crimes they did not commit. In the following section, examples are given to demonstrate some of the primary problems with the Reid Technique and how these issues are directly related to fundamental misunderstandings and misrepresentations about language and language users.

Abuse of Language within the Reid Technique

A simple look at the training programs offered by Reid & Associates (https://reid.com n.d.) gives cause for concern. Within "The Reid Nine Steps of Interrogation" participants will learn what is summarized as "Handling Denials", "Overcoming Objections", "Challenging the Suspect's Values and Traits", and "Committing the Suspect to the Crime". Given that these interrogation tactics are dispersed throughout the nine steps, they leave precious little room for suspects to change the interrogators' decision that they are guilty.

The concern that the Reid Technique can be used for unethical coercion is heightened when the actual content of the courses is examined. The excerpt below is taken directly from the Reid Technique's "Step Three: Handling Denials":

> A guilty suspect will oftentimes introduce his denials by asking permission to speak with such phrases as, "May I say something?" or "If you just let me say one thing…" Were the investigator to allow the suspect to make his full statement, each of these introductory remarks would be followed by the denial, "I didn't do it." Therefore, whenever a suspect asks to speak during this phase of the interrogation, the investigator should respond by saying, "Andy, just hold on for a minute. Let me explain to you exactly what we know."
>
> *(Reid & Associates n.d.)*

In this excerpt, it is not only taken for granted that a polite request to speak will inevitably be followed by a denial of guilt but also assumed that such a combination of speech acts will only occur with guilty suspects. There is no consideration of the variation exhibited in politeness formulae (e.g., Schneider and Placencia 2017) or their interaction with the asymmetrical and stressful situation of being subjected to a high-stakes police interrogation (e.g., Archer 2017), nor of the possible influence of cross-cultural misunderstandings (e.g., Sifianou and Blitvich 2017). It is simply assumed that, aside from an interviewee's guilt, there are no other reasons for someone in police custody to preface their questions with variants of "May I say something?" Later in the explanation of Step Three, this excerpt contrasts the guilty suspect's behavior with that of the innocent:

> The innocent suspect, however, will generally not ask permission to speak when he wants to make a statement. Instead, he will, without any display of etiquette, promptly and unequivocally state that he had absolutely nothing to do with the theft. Innocent denials will strengthen in time, and the suspect will begin to assert control over the interrogation. In the event of such a development, the investigator must begin to modify the intensity of his position and begin a probing search for some other information such as the suspect's suspicions about who may have committed the crime. This will help explain to the suspect why he was questioned. It is very rare for an innocent suspect to move past this denial state. He remains steadfast in the assertion of his innocence. The sincerity with which he presents his case is a clear indication of his truthfulness.
>
> *(Reid & Associates n.d.)*

Here, it is unequivocally asserted that innocent people will maintain their innocence over time, despite repeated accusations by law enforcement. This is consistently countered by experiments in social psychology (e.g., Kassim 2015; Kassim and Kiechel 1996).

Criticisms of the Reid Technique are not limited to their questionable interpretations of interviewee language and behavior. It has been frequently reported that law enforcement employing this method have subjected suspects to hours-long interrogations in which officers lie to suspects about available evidence against them. Reid & Associates deny that such tactics are promoted in their programs (Buckley 2022) but there are multiple reports that counter those claims. An excellent example comes from a case that took place in New York in 2014 (https://www.ei-ie.org/en/item/19574:denmark-and-usa-malthe-thomsen-dismissed-of-all-charges). A young Danish man, Malthe Thomsen, was accused by a co-worker of sexual misconduct against children at the pre-school where he was interning. Mr. Thomsen was put through a seven-hour interrogation during which the officers made repeated reference to video evidence allegedly proving his guilt. After six hours he gave a written confession, which he later retracted, claiming coercion. Charges were later dropped as the video showed no evidence of sexual misconduct and the co-worker was determined to have acted in revenge after she was fired from the pre-school.

In short, confession-based methods of police interrogation carry a high risk of false confessions based on misrepresentations of what suspect language looks like.

Deception Detection

So-called "deception detection" is a highly appealing concept within law enforcement since evaluating the validity of witness statements is crucial for successful investigations. Disproving the faulty claims of witnesses risks wasting valuable resources that could have been invested in more promising avenues of investigation. At the same time, failing to uncover untruth carries a risk of convicting innocent people and letting the real perpetrators go free. To minimize these risks, any tool that offers a simple and supposedly foolproof method of detecting untrue statements will understandably be very attractive to investigators.

SCAN

One of the oft-discussed methods of deception detection is the misleadingly named *Scientific Content Analysis*, or *SCAN*. It was developed in the 1980s by Avinoam Sapir, a polygraph examiner. Sapir designed the tool based on the premise "that the structure and contents of a subject's statement reveal whether there's an attempt at deception" (LSI Laboratory for Scientific Interrogation, Inc.). Like many other approaches to veracity assessment, SCAN is partially based on the "Undeutsch Hypothesis". Named for its inventor, Udo Undeutsch, this hypothesis proposes that false statements differ in content and quality from truthful statements because they are not based on actual memories (Vrij 2015). Based on this premise, SCAN works by having a subject write down their account of events in a statement, which the SCAN-trained officer then analyzes. This analysis is done by matching the statement against a set of linguistic criteria that indicate either truthfulness or deception.

However, one of the main criticisms of SCAN that is repeatedly reported in the scientific literature is its lack of a defined set of linguistic criteria against which a statement should be judged. For example, Bogaard et al. (2013b: 2) report that when one of the authors attended the SCAN course, "no checklist of criteria is offered, nor does the SCAN manual contain such a list". In other words, the tool is not standardized, meaning that the results of a SCAN analysis will vary according to the person conducting the analysis. To find out what criteria are most frequently used by SCAN-trained officers, Bogaard et al. (2013b) analyzed 82 sexual abuse case statements from the Netherlands where SCAN had been applied by a police officer. They found that there are largely 12 criteria that drive SCAN analyses, 7 of which indicate truthfulness (denial of allegations, social introductions, structure of the statement, emotions, objective and subjective time, first person singular pronouns, and past tense verbs), while the remaining 5 indicate deception (change in language, spontaneous corrections, lack of conviction or memory, out of sequence, and extraneous and missing information) (Bogaard et al. 2016: 4, 2013b). Furthermore, they found that although the police officers examining these statements had received SCAN training, their inter-rater reliability was extremely low, suggesting "that the SCAN method is insufficiently developed as a forensic tool" (Bogaard et al. 2013b: 15). In fact, the available literature predominantly argues against its use given its lack of empirical support and its susceptibility to contextual bias, i.e., that surrounding circumstances may influence the analysis and interpretation of the statement (Goormans, Mergaerts, and Vandeviver 2022; Heydon 2019; Bogaard et al. 2016; Vrij et al. 2015; Bogaard et al. 2013a; Bogaard et al 2013b; for a comprehensive review see Vrij et al. 2022).

911-Call Analysis

Another questionable method of deception detection is "911-call analysis". Surfacing in the early 2000s, this method was developed by former US police officer Tracy Harpster. According to Harpster, 911-call analysis provides a reliable way for investigators to determine whether the caller is telling the truth (Harpster and Adams 2017; Harpster, Adams, and Jarvis 2009; Harpster 2006). This method offers 8 variables, all based on the Undeutsch hypothesis, that are associated with innocence, e.g., "plea for help" and "self-corrections", for example, explicitly saying "I need help" during the call, and 12 variables that are associated with guilt, e.g., offering "extraneous information" and exhibiting "inappropriate politeness", such as saying *thank you* to the 911-call operator (Harpster, Adams, and Jarvis 2009). Based on these variables, Harpster devised a scale called "Considering Offender Probability in Statements" or COPS. The purpose of this scale is to assist police in evaluating 911 calls. To date, Harpster has declined to disclose the data on which COPS is based, referencing a concern that researchers would "steal" his research to "profit" from it (Murphy 2022). As with both SCAN and the Reid Technique, research examining 911-call analysis is scarce. The few reports that have been published appear to stem from authors affiliated with the tool in question and may therefore be biased in favor of it (e.g., Elms and Keatley 2023; O'Donnell et al. 2023; Cromer et al. 2018).

All in all, existing research into deception detection methods such as SCAN and 911-call analysis shows that they are difficult to deploy consistently and that they produce conflicting and misleading results. Even more troubling, however, is the fact that there is no scientific backing for the claim that any of these methods can reliably detect whether someone is being deceptive based on linguistic features. Instead, it seems that the reported effectiveness of such tools is more a function of confirmation bias than any underlying correlation between linguistic features and deception. The continuing popularity of these tools in law enforcement would seem to be driven by a widespread lack of basic knowledge about how linguistic interactions work (Heydon 2019; Picornell 2013). By sharing their expertise and experience with law enforcement, forensic linguists can help decrease not only the continued use of dubious language analysis methods but also the potential for avoidable miscarriages of justice.

An Analysis of Pseudolinguistic Claims

Before we begin our data analysis, we must address an important issue regarding the sources for our data. While we believe it is important to identify and critique the use of pseudolinguistic methods, we also recognize the potential harm that could be caused by singling out individuals advocating such methods. Moreover, the market for deception detection and credibility analysis is constantly changing, making it quite difficult as researchers to determine exactly who is the primary proponent of a given claim. For these reasons, we have chosen not to reveal the specific sources for our data. The main focus of this study is to provide a description of the tendencies within a field that is marred by pseudoscience. Consequently, we limit ourselves to paraphrasing some of the recurring claims we have encountered throughout a year-long survey of several private companies that offer credibility assessments or deception detection services, based on tools such as the Reid Technique or SCAN, and promote their services to law enforcement, prosecution, and other authorities. Our focus here is on deconstructing the claims to show why they are problematic from a theoretical and practical viewpoint. By avoiding the naming of specific individuals or companies, we hope to encourage a broader dialogue around the need for rigorous and ethical standards in forensic linguistic practices. We believe that this approach will ultimately serve the interests of the broader forensic linguistic community. The five claims are presented below.

Claim One: The Invisibility of Function Words

Function words such as prepositions, pronouns, and adverbs are "invisible" to the brain and have little meaning because they are the "glue" that holds meaningful words together. Function words are uttered and perceived unconsciously because they are located in a different place in the brain. They are interesting when evaluating credibility because they reveal useful information about our relationships and feelings toward ourselves and others.

Note: This supposition appears to draw on Pennebaker (2013).

Claim Two: The Fixed Meaning of First Person Pronouns

The pronoun *we* can never designate a negative relationship, because *we* is a word that indicates cooperation and community. The personal pronoun *I* is a word which grants responsibility as well as revealing that the person's focus is on themselves. So, if someone says: "My wife and I went shopping" it indicates emotional attachment. But if they say: "I went shopping with my wife" it indicates distance.

Note: This supposition appears to stem from SCAN training (Heydon 2019).

Claim Three: The Suspicious Variation of Terminology

Referring to individuals inconsistently throughout a written statement, e.g., referring to your alleged rapist interchangeably as *someone*, *the guy*, *the boy*, and *the man*, is problematic and can indicate deception.

Note: This supposition also appears to stem from SCAN training (Goormans, Mergaerts, and Vandeviver 2022).

Claim Four: The Suspicious Use of Redundant Words

If someone is confident that what they say is true, they will have no need to convince their recipient of their confidence. They would just deliver the information as is, with no "superfluous" words, that is, words that can be deleted from a sentence without making it grammatically incorrect. Superfluous words can indicate deception in, e.g., emergency calls.

Note: This supposition appears to draw partly upon studies of linguistic modality that make the point that "there is no epistemically stronger statement than a categorical [non-modalized] assertion" (Lyons 1977: 809).

Claim Five: The Unambiguous Protestation of Innocence

People mostly speak directly and unambiguously about the topic at hand, and they will say explicitly if they are innocent. In other words, if someone is falsely accused of a crime, it should not be difficult for them to deny the accusation in a clear and distinct way.

Note: This supposition appears to stem from the Reid Technique (see above).

Epistemically Irresponsible Discourse

Although the above five claims seem to be empirically driven, in the communication of them, the proponents of the claims employ (at least) two subtypes of what Heffer (2020: 182) terms epistemically irresponsible discourse pathologies, namely those of *distortion* (i.e., linguistically misrepresenting the evidence), *withholding* (i.e., omitting, evading, and even blocking counter-evidence), and *dogma* (i.e., closed-mindedness that disregards counter-evidence). Let us illustrate this by deconstructing each claim.

Claim One. Aside from the obvious contradiction that function words "have little meaning" while still somehow revealing important information about our relationships and feelings, we can identify several other problematic statements. The first sentence is partially taken (without reference) from Chung and Pennebaker (2007: 345) who write:

> We can begin to detect linguistic style by paying attention to "junk words"—those words that do not convey much in the way of content. These junk words, usually referred to as function words or particles, serve as the cement that hold content words together.

It is important to note that Chung and Pennebaker (2007) make the above statement in the specific context of analyzing linguistic style. However, the proponents of Claim One misuse this assertion and apply it to evaluating credibility.

The claim that function words are "invisible" to the brain belongs to what Heffer (2020: 182) impishly describes as a type of claim that is "evidentially worthless because its meaning is unclarifiable". The final statements that function words are "uttered and perceived unconsciously" and that they are "located in a different place in the brain" are cases of distortion by misstatement. We know from research in psycholinguistics that cognitive capacities such as speech processing are "effortless, reliable, fast, unconscious and require no instruction" (Kemmerer 2015: 111), so it is arguably irrelevant to claim that the brain perceives function words unconsciously, because *all* speech processing happens at an unconscious level. And while research has shown that different areas in the brain are indeed activated in the perception of different word classes (Diaz and McCarthy 2009), it is misleading to present this as an indication that words are "located" in separate "places in the brain" and that function words are thereby collectively good indicators of deception. Overall, Claim One is an example of how psycholinguistic research can be misrepresented for the purpose of making unsubstantiated promises about deception detection.

As has been noted previously by other research (e.g., Heydon 2019), research into linguistic markers of deception tends to be conducted by psychologists rather than linguists and is usually part of "a broader study of deceptive behaviour" in which "the explanations for the use of linguistic cues indicating deception or truthfulness are explained in relation to theories of cognition or behaviour" (Heydon 2019: 62). However, oftentimes individuals marketing specific deception detection tools will ascribe psychological motives or unconscious reasons to the linguistic choices individual speakers make, without contextualizing them within a broader cognitive or psychological theory. Let us examine this in further detail by deconstructing Claim Two.

Claim Two. In our survey, we find this claim phrased in ways that indicate the original source to be *The Secret Life of Pronouns: What Our Words Say About Us* (2013), a popular science book written by social psychologist James W. Pennebaker to demonstrate how the text-analysis tool LIWC (Linguistic Inquiry and Word Count) can be used. LIWC is a computational tool designed to examine large quantities of texts and works on the assumption that our language reflects our internal psychological states, emotions, ways of thinking, and social concerns (Pennebaker 2013). In Pennebaker's version, Claim Two was made in the context of determining the emotional state of an author over time, by computationally analyzing their written statements. But we find that it is frequently applied to analyses of spoken language in the context of deception detection, thus becoming distorted partly due to a lack of distinction between written and spoken language.

Claim Three. This claim appears to be a variation of a claim cited from SCAN training material, where an object is first referenced using a full noun phrase and then taken up again in the form of the pronoun *it* (Heydon 2019: 63). The underlying notion is that the writer inadvertently reveals an inconsistency in their narrative by shifting either between different noun phrases or a noun phrase and a pronoun. However, as Heydon (2019: 63) points out, these analyses often suffer from a lack of basic understanding of language, i.e., the use of anaphora is not a symptom of deception but rather of perfectly normal language use.

Claims Four and Five. These claims are problematic from a purely theoretical viewpoint, in part because they assume a link between psychological states and linguistic behavior using contrived examples. The assumption behind Claims Four and Five is that there is a clear distinction between "innocent" and "guilty" linguistic behavior, an assumption we recognize from both 911-call analysis and the Reid Technique, and for which there is no empirical evidence.

For example, the statements that "most people speak directly and unambiguously about the topic at hand" and that truth-tellers "would deliver the information as it is, with no extra add-ons" conflict with decades of linguistic research in fields such as conversation analysis (CA) about how the practices of spoken language "fit together in highly intricate ways" (Sidnell 2010: 1). Speakers design their turns specifically for their interlocutor, and even more specifically they design it to align to the project undertaken in the previous turn (Schegloff 2007). In other words, "the response (…) that promotes the accomplishment of the activity underway is, typically, 'preferred'" (Sidnell 2010: 77). *Preference* here does not refer to the psychological preferences of the speakers in a conversation, but to the structural preferences that organize much of talk-in-interaction. Dispreferred responses are often delayed by inter-turn gaps, turn initial delays, insertions of whole sequences like accounts, palliatives, tokens of appreciation, apologies, etc., that are oriented toward mitigating the valence of a dispreferred turn. These structural preferences are normative, meaning that they are dictated by the social norms we cultivate and care about. However, in the context of credibility assessment they are often interpreted as being indicative of deception, perpetuating the notion that "truth" and "lies" differ on a linguistic level. In spite of this, the assumption that a speaker will "focus on the objective (…) instead of using extraneous, distancing, or evasive measures" (Harpster, Adams, and Jarvis 2009: 78) is, unfortunately, very common.

Evaluation of the Claims

If we look at the claims collectively and in the context of the promotion of deception detection or veracity assessment methods, we will argue that the proponents of the claims are being epistemically *negligent*, because they have "a duty of epistemic care" but fail to "investigate in accordance with that duty" and fail to hedge their commitment to the claims (Heffer 2020: 202). In his definition of epistemic irresponsible discourses Heffer (2020) draws on philosopher Harry Frankfurt's notion of *bullshitting* which is a type of insincere speech which "deceives the hearer about the speaker's knowledge of what is said (the hearer falsely assumes that the speaker knows what she is talking about)" (Heffer 2020: 5). The present authors know that at least one of the proponents of Claims One–Five has been made aware that there is no scientific backing for their claims, meaning that they warrant the truth of what they say but conceal "the fact that [they have] no grounds for that warrant" (Heffer 2020: 5).

A Note on Pennebaker (2013)

One of the main issues that arises when dealing with these types of pseudolinguistic claims is that it can be difficult to determine their origins. In connection with this study, one of the authors familiarized herself with Pennebaker's *The Secret Life of Pronouns* (2013) and found several claims about linguistic lie detection. We feel compelled to address this issue briefly here, because it appears from our survey of the deception detection marketplace that Pennebaker (2013) is a likely academic source used by many proponents of deception detection, credibility, and validity assessment tools. This is valuable information for forensic linguists who seek to examine their claims further.

In the sixth chapter of the book, titled "Lying Words", Pennebaker (2013) makes several assertions about linguistic deception detection that we can recognize from tools like SCAN and the Reid Technique (such as *use of pronouns*, *passive constructions*, *avoiding answering a question*, *cognitive words*, etc.). For example, in a sub-section titled "The language of lying, liars, and scoundrels: in search of the linguistic lie detector" Pennebaker (2013: 140) states that it is theoretically possible to develop a linguistic lie detector tool and later adds: "not surprisingly, the language of lies is generally seen in the function words we use" (Pennebaker 2013: 169). This is a concerning statement given that the psychological experiments he uses to support that claim have low accuracy rates (51–73%), as referenced by himself.

Despite the large body of scientific evidence that refutes the assertions made about linguistic deception detection in *The Secret Life of Pronouns* (2013), tools like SCAN and the Reid Technique continue to be used. Although Pennebaker (2013) does not make these claims in the context of promoting specific lie detection tools, he has (perhaps unwittingly) laid a non-linguistic foundation for those interested in developing and using such tools.

Conclusion: Ethical Considerations When Addressing Pseudolinguistics

When confronted with pseudoscience, we face the simultaneous challenge of defending the validity of our own scientifically grounded approach, while addressing and countering misleading claims. To counter misinformation and maintain public trust, it is crucial to engage in effective science communication (Boudry et al. 2015). As argued above, forensic linguists have an ethical responsibility to identify and publicly illuminate the rift between evidence-based forensic linguistics and pseudolinguistic practices within law enforcement. As many other scientific fields are discovering, the task of educating decision-makers (to say nothing of the general public) is often easier said than done.[1] As the medical scientists Capuano and Killu (2020: 246) note, "there are many cases in which providing education about pseudoscientific practice is insufficient to dissuade a caregiver from pursuing these treatments". This same kind of resistance can be found in the area of deception detection. As Heydon (2019: 63) explains, forensic linguistic research has produced a

> substantial and authoritative corpus of research literature dating back several decades that challenges the validity of lie detection, linguistic or otherwise, [which] seems to have had a depressingly small impact on the use of such methods in law enforcement and the justice system more broadly.

Although effective science communication is always a challenge, it is particularly so when researchers are faced with countering popular pseudoscientific theories, methods, and tools. The best weapon for fighting these faulty beliefs and practices remains education. Part of the ethical responsibility of the forensic linguist is to explain the true complexities of language use, the necessities of applying rigorous, transparent scientific methods, and the importance of insisting upon evidence-based findings as the basis for real-world policies. Another part is to address and preferably also correct faulty analyses promoted to sell tools and training that rely on language evidence. We urge that such corrections are offered in a respectful and constructive manner; no one is convinced by a competing account if they are shamed into doing it. In essence, when addressing pseudolinguistic claims in the public sphere, the forensic linguist will have to balance advocacy and objectivity. Admittedly, accomplishing these goals is no easy task, particularly when one considers that pseudoscientific claims about language-based deception detection are often also held by individuals and organizations in the criminal justice system. However, this fact does not alter the ethical imperative that forensic linguists be at the forefront of countering pseudoscientific claims about language and language users, since a failure to do so can have devastating consequences in terms of wrongfully convicting the innocent and allowing the guilty to go free.

Note

1 Many branches of science are familiar with pseudoscientists and their tendency "to reject results that are widely accepted by specialists working within the pertinent area of science" (Cleland and Brindell 2013: 183). There are a number of public controversies over scientific claims "in which scientists attempt to justify hypotheses and theories" (Cleland and Brindell 2013: 183), but fail to settle the discussion publicly after the initial damage has been done. A well-known example from the world of medicine is the debate about whether vaccines can cause autism in children. Although this claim has long since been debunked, it is still a matter of public debate (Taylor, Swerdfeger, and Eslick 2014).

References

Archer, Dawn. 2017. "(Im) politeness in Legal Settings." In *The Palgrave Handbook of Linguistic (Im) Politeness*, 713–737. Edited by Jonathan Culpepper, Michael Haugh, and Dánel Z. Kádár. New York: Palgrave Macmillan.

Bogaard, Glynis, Ewout H. Meijer, Aldert Vrij, Nick J. Broers, and Harald Merckelbach. 2013a. "Contextual Bias in Verbal Credibility Assessment: Criteria-Based Content Analysis, Reality Monitoring and Scientific Content Analysis." *Applied Cognitive Psychology* 28: 79–90.

Bogaard, Glynis, Ewout H. Meijer, Aldert Vrij, Nick J. Broers, and Harald Merckelbach. 2013b. "SCAN is Largely Driven by 12 Criteria: Results from Sexual Abuse Statements." *Psychology, Crime and Law* 20, no. 5: 430–449.

Bogaard, Glynis, Ewout H. Meijer, Aldert Vrij, and Harald Merckelbach. 2016. "Scientific Content Analysis (SCAN) Cannot Distinguish Between Truthful and Fabricated Accounts of a Negative Event." *Frontiers in Psychology* 7, article 243: 1–7.

Boudry, Maarten, Stefaan Blancke, and Massimo Pigliucci. 2015. "What Makes Weird Beliefs Thrive? The Epidemiology of Pseudoscience." *Philosophical Psychology* 28, no. 8: 1177–1198.

Buckley, Joseph. 2022. "Clarifying Misrepresentations About Law Enforcement Interrogation Techniques." Accessed June 22, 2023. Retrieved from https://reid.com/pdfs/Clarifying-Misrepresntations-of-Law-Enforcement-Interrgoation-Techniques-Feb-2022_2022-03-18-140724_nabu.pdf

Capuano, Angela M., and Kim Killu. 2020. "Understanding and Addressing Pseudoscientific Practices in the Treatment of Neurodevelopmental Disorders: Considerations for Applied Behavior Analysis Practitioners." *Behavioral Interventions* 36, no. 1: 242–260.

Cherryman, Julie, and Ray Bull. 2000. "Reflections on Investigative Interviewing." In *Core issues in Policing*, 194–210. Edited by Frank Leishman, Barry Loveday, and Stephen P. Savage. New York: Pearson Education.

Chung, Cindy, and James Pennebaker. 2007. "The Psychological Function of Function Words." In *Social Communication*, 343–359. Edited by Klaus Fiedler. New York: Psychology Press.

Clarke, Colin, and Rebecca Milne. 2001. *A National Evaluation of the PEACE Investigative Interviewing Course*. London: Home Office.

Cleland, Carol E., and Sheralee Brindell. 2013. "Science and the Messy, Uncontrollable World of Nature." In *Philosophy of Pseudoscience: Reconsidering the Demarcation Problem*, 183–202. Edited by Massimo Pigliucci, et al. Chicago: University of Chicago Press.

Cromer Jon D., JoAnne Brewster, Kethera Fogler, and Michael Stoloff. 2018. "911 Calls in Homicide Cases: What Does the Verbal Behavior of the Caller Reveal?" *Journal of Police and Criminal Psychology* 34, no. 1: 156–164.

Davis, Deborah, and Richard A. Leo. 2013. "The Problem of Interrogation-induced False Confession: Sources of Failure in Prevention and Detection." In *Handbook of Forensic Sociology and Psychology*, 47–75. Edited by Stephen John Morewitz and Mark L. Goldstein. New York: Springer.

Diaz, Michele T., and Gregory McCarthy. 2009. "A Comparison of Brain Activity Evoked by Single Content and Function Words: An fMRI Investigation of Implicit Word Processing." *Brain Research* 1282: 38–49.

Elms, Serena, and David Keatley. 2023. "The Linguistic Disposition of Emergency Callers: Using Proximity Coefficients to Identify the Sentiments of Guilty and Innocent Emergency Callers." *Journal of Police and Criminal Psychology*, 40: 312–322.

Fahsing, Ivar A., and Kristina Kepinska Jakobsen. 2017. "Investigative Interviewing of Suspects in Scandinavia." In *International Developments and Practices in Investigative Interviewing and Interrogation, volume 2: Suspects*, 212–224. Edited by David Walsh, Gavin Oxburgh, Allison Redlich, and Trond Myklebust. New York: Routledge.

Goormans, Isabo, Lore Mergaerts, and Christophe Vandeviver. 2022. "SCANning for Truth. Scholars' and Practitioners' Perceptions on the Use(fulness) of Scientific Content Analysis in Detecting Deception during Police Interviews." *Psychology, Crime & Law*. 30, no. 9: 971–993.

Gudjonsson, Gisli H., and John Pearse. 2011. "Suspect Interviews and False Confessions." *Current Directions in Psychological Science* 20, no. 1: 33–37.

Harpster, Tracy. 2006. *The Nature of 911 Homicide Calls: Using 911 Homicide Calls to Identify Indicators of Innocence and Guilt.* Bowling Green: Bowling Green State University.

Harpster, Tracy, and Susan H. Adams. 2017. *Analyzing 911 Homicide Calls: Practical Aspects and Applications*. Boca Raton: CRC Press.

Harpster, Tracy, Susan H. Adams, and John P. Jarvis. 2009. "Analyzing 911 Homicide Calls for Indicators of Guilt and Innocence—An Exploratory Analysis." *Homicide Studies* 13: 69–93.

Heffer, Chris. 2020. *Insincerity, Irresponsibility, and the Judgment of Untruthfulness*. Oxford: Oxford University Press.

Heydon, Georgina. 2019. "Lie Detection and Linguistics." In *Researching Forensic Linguistics: Approaches and Applications*, 60–77. Edited by Georgina Heydon. London: Routledge.

Hirsch, Alan. 2014. "Going to the Source: The New Reid Method and False Confessions." *Ohio State Journal of Criminal Law* 11, no. 2: 803–826.

Inbau, Fred, and John E. Reid. 2011. *Criminal Interrogation and Confessions*. Baltimore: The Williams & Wilkins Company.

Jane, Brian C., and Joseph P. Buckley. 1999. *The Investigator Anthology: A Compilation of Articles and Essays about The Reid Technique of Interviewing and Interrogation*. Chicago: John E. Reid & Associates.

Kassin, Saul M. 2015. "The Social Psychology of False Confessions." *Social Issues and Policy Review* 9: 25–51.

Kassin, Saul M., and Katherine L. Kiechel. 1996. "The Social Psychology of False Confessions: Compliance, Internalization, and Confabulation." *Psychological Science* 7, no. 3: 125–128.

Kemmerer, David. 2015. *Cognitive Neuroscience of Language*. New York: Psychology Press.

LSI Laboratory for Scientific Interrogation, Inc. Retrieved from https://www.lsiscan.com/id37.htm.

Lyons, John. 1977. *Semantics*. Vols. 1 and 2. Cambridge: Cambridge University Press.

Meissner, Christian A., Allison D. Redlich, Sujeeta Chatt, and Susan Brandon. 2012. "Interview and Interrogation Methods and their Effects on True and False Confessions." *Campbell Systematic Reviews* 8, no. 2: 1–53.

Memon, Amina, and Julie Gawrylowicz. 2019. "The Cognitive Interview." In *The Handbook of Communication Skills*, 511–530. Edited by Owen Hargie. New York: Routledge.

Memon, Amina, Christian A. Meissner, and Joanne Fraser. 2010. "The Cognitive Interview: A Meta-analytic Review and Study space Analysis of the Past 25 Years." *Psychology, Public Policy, and Law* 16, no. 4: 340–372.

Moore, Timothy E., and C. Lindsay Fitzsimmons. 2011. "Justice Imperiled: False Confessions and the Reid Technique." *Criminal Law Quarterly* 57: 509–542.

Murphy, Brett. 2022. "They Called 911 for Help. Police and Prosecutors Used a New Junk Science to Decide They Were Liars: Tracing the Fallacy of 911 Call Analysis through the Justice System, from Quantico to the Courtroom." *ProPublica*. Retrieved from https://www.propublica.org/article/911-call-analysis-fbi-police-courts/amp

O'Donnell, Daniel E., et al. 2023. "911 Calls in Mysterious Disappearances of Children: Indicators of Veracity and Deception." *Applied Cognitive Psychology* 37, no. 3: 578–589.

Ofshe Richard J., and Richard A. Leo. 1997. "The Decision To Confess Falsely: Rational Choice and Irrational Action." *Denver Law Review* 74: 979–983.

Pennebaker, James W. 2013. *The Secret Life of Pronouns: What Our Words Say About Us*. Bloomsbury Press.

Picornell, Isabel. 2013. *Cues to Deception in a Textual Narrative Context: Lying in Written Witness Statements*. Birmingham: Aston University.

Reid and Associates. n.d. Accessed May 15, 2023. Retrieved from http://archive.reid.com/educational_info/critictechnique.html

Schegloff, Emanuel A. 2007. *Sequence Organization in Interaction: A Primer in Conversation Analysis*. Cambridge: Cambridge University Press.

Schneider, Klaus P., and Maria Elena Placencia. 2017. "(Im) politeness and Regional Variation." In *The Palgrave Handbook of Linguistic (Im)politeness*, 539–570. Edited by Jonathan Culpepper, Michael Haugh, and Dánel Z. Kádár. New York: Palgrave Macmillan.

Sidnell, Jack. 2010. *Conversation Analysis: An Introduction*. New York: Wiley-Blackwell.

Sifianou, Maria, and Garcés-Conejos Blitvich. 2017. "(Im) politeness and Cultural Variation." In *The Palgrave Handbook of Linguistic (Im)politeness*, 571–599. Edited by Jonathan Culpepper, Michael Haugh, and Dánel Z. Kádár. New York: Palgrave Macmillan.

Taylor, Luke E., Amy L. Swerdfeger, and Guy D. Eslick. 2014. "Vaccines are not Associated with Autism: An Evidence-based Meta-analysis of Case-control and Cohort Studies." *Vaccine* 32, no. 29: 3623–3629.

Vrij, Aldert. 2015. "Verbal Lie Detection Tools: Statement Validity Analysis, Reality Monitoring and Scientific Content Analysis." In *Detecting Deception: Current Challenges and Cognitive Approaches*, 3–35. Edited by Pär Anders Granhag, Aldert Vrij, and Bruno Verschuere. New York: John Wiley & Sons.

Vrij, Aldert, Paul Taylor, and Isabel Picornell. 2015. "Verbal Lie Detection." In *Communication in Investigative and Legal Contexts: Integrated Approaches from Forensic Psychology, Linguistics and Law Enforcement*, 259–286. Edited by Gavin Oxburgh, Trond Myklebust, Tim Grant, and Rebecca Milne. New York: Wiley-Blackwell.

Vrij, Aldert, et al. 2022. "Verbal Lie Detection: Its Past, Present and Future." *Brain Sciences* 12, no. 1644: 1–20.

10
DISINFORMATIVE LINGUISTICS
Using Linguistic Science for Unethical Purposes

Timothy Habick

Introduction

A good deal of forensic linguistic work involves analyzing documents to determine the interpretation, if any, that is best supported by linguistic evidence. In forensic practice, of course, litigants already know which interpretation they desire; with any luck a forensic linguist operating competently and ethically will demonstrate that the desired interpretation is legitimate. Legitimacy can be ascribed only to rule-governed interpretations, however. A linguist's task is to identify the relevant rules and apply them without succumbing to pressure from clients. A legitimate interpretation is distinguished from opinion solely by the operation of the applicable rules of communication. Deception of any type is illegitimate, of course, because it depends on violations of those rules.

Language is the most complex of human behaviors, so it is naturally governed by vast numbers of rules, patterns, norms, and conventions, which operate simultaneously at many levels of analysis. These rules are not arbitrary: everyone unconsciously knows and follows them under baseline conditions. Despite the boggling variety and complexity of the rules, linguists can describe their operations and predict their outcomes with reasonable accuracy. Not all communicative phenomena can be explained by rules, of course, but are best understood as patterns of variable features with measurable likelihoods of occurrence for a given speaker in a given social setting and a particular level of formality (Labov 2006). The frequency of occurrence of a feature is another way of describing a rule governing that feature, so these patterns of variation are called frequency rules.

When speakers of a shared dialect converse, each party instantly evaluates the overall communicative adequacy of the other parties' statements at each conversational step. The typical response to another party's apparent rule violation is a correction or a request for clarification. People might be unable to explain the applicable rules, but they know them quite well, even without formal linguistic training. Similarly, healthy people competently perform the sequence of muscular movements, both voluntary and involuntary, needed to swallow food, yet cannot explain how they do that without choking. Just as otolaryngologists can explain the mechanics of swallowing, linguists are trained to identify the applicable rules of communicative sufficiency and can discuss them within coherent theoretical frameworks.

An utterance with exactly one legitimate interpretation is a successful communication to the extent that the meaning understood is reasonably close to the meaning intended. Communicative

DOI: 10.4324/9781003391074-11

success occurs when speaker and listener follow identical rules. The rules include semantic, morphological, and syntactic rules governing how to fit concepts, words, and sentences together as well as pragmatic rules governing how to make each communicative contribution responsive to the physical, social, emotional, intellectual, and historical environment.

Successful communication, according to Grice (1989: 26–58), implies compliance with certain foundational principles or *maxims*, which essentially stipulate that natural human conversations follow unspoken guidelines related to the quality, quantity, relation, and manner of information exchanged. The baseline expectation for exchanges of information is that the participants will be reflexively truthful, informative to the right degree, relevant to current concerns, and clear to the intended audience in the actual circumstances. Successful communications occur only when speakers and listeners respect applicable communicative rules.

According to Grice, when someone clearly violates a maxim—intentionally or not—other people become confused, suspicious, or amused. It is impossible for speakers—or listeners—to communicate cooperatively while intentionally violating the maxims, unless the violations are for comedic, rhetorical, poetic, or other artistic purposes. Cooperative communication is a two-way street, relying on rule-governed encoding of messages but also on rule-governed *de*coding of them. Under normal circumstances, competent users of a given dialect communicate in compliance with applicable grammatical rules. People misspeak occasionally and make grammatical errors that they quickly correct. But only people with artistic aspirations, language disorders, or questionable motives flout the maxims. As Ortoni and Gupta (2019: 149) note, a deceptive speaker "neither wants nor expects the hearer to recognize that one or more…maxims have been violated". In other words, a deceptive speaker "hopes that the hearer will erroneously assume that the speaker is being conversationally cooperative". That goal entails the use of strategies for deception.

Unlike ordinary communications, which are typically based on enormous amounts of unspoken context, high-stakes documents must explain many essential aspects of context that normally remain unstated. By comparison, ordinary communications can be successful without such manicured clarity because the context permits follow-up interactions among interlocutors and because listeners, as Grice points out, often know what the next most appropriate contribution to a conversation should be. High-stakes documents are designed to communicate effectively to the intended audience, especially when further clarification is impossible.

Interpretations of high-stakes texts compliant with logical, linguistic, and pragmatic principles are by definition legitimate, meaning compliant with applicable rules. Some sentences, in fact, can have more than one legitimate interpretation. Fanciful, wishful, and cynical interpretations are illegitimate in that they rely on violations of applicable rules or they deny (linguistic) reality in some way. Promoting an illegitimate interpretation is not necessarily unethical; it is unethical, and disinformative, to *knowingly* promote an illegitimate interpretation.

Uncommon interpretations are different from illegitimate ones, however. In uncommon but valid interpretations, linguistic structures lead directly by rule to the intended meaning; in invalid interpretations, some relevant rule must be broken to reach that meaning.

The term *legitimate* means compliant with applicable rules. That does not mean, however, that interpretations violating an out-of-context rule are necessarily illegitimate. Pragmatic rules, for example, outrank logical rules in certain communicative contexts. That is demonstrated by the fact that the sentence *I could care less* is regularly intended (and understood) to mean its denial, *I couldn't care less*, even by listeners who recognize a rule violation at the level of logical analysis.

High-stakes communications include formalized linguistic acts that bind speakers, writers, and signatories to the truthfulness of their assertions. These communicative acts are considered *performatives* because they not only describe an event but actually create it by the utterance of certain words, such as those for vows of marriage and citizenship (Austin 1975). When we communicate performatively, we attest to certain truths and principles, we engage ourselves to behave in a manner consistent with our promises, and we agree to be held to our literal word. "Literal" means by the letter, exactly as written, without unstated reservations. When reciting performatively, we use formally approved wordings designed to be specific about the rights and privileges accorded and the duties entailed.

Performatives are often fossilized in a standardized, even archaic form of language, not in casual vernaculars. In order to be maximally accessible to their audience, however, many high-stakes communications use a standardized version of a contemporary dialect and avoid archaisms. The important performative effects of contracts, attestations, and declarations—especially those made before a court of law—require a form of language that guarantees communicative success, at least within specific contexts and the intended readerships.

All forensic cases require experts to present a rule-governed, best-supported, or most likely interpretation of data. In forensic linguistic cases, the data can be sentences with meanings in dispute. In other cases, communications are analyzed to identify their most likely author. In each case, the linguist uses forensic tools and the theoretical tradition of linguistic science to craft appropriate argumentation.

Many contracts and high-stakes examinations are immunized from serious forensic challenges because they were crafted according to linguistic principles to achieve an unnatural level of clarity, pertinence, and efficiency, allowing exactly one legitimate interpretation in context. Such preventive linguistic interventions eliminate misunderstandings before they arise.

Opportunities to use forensic linguistic methods to resolve real-world communicative disputes entail weighty professional duties, often including the requirement to sign a performative declaration like this one upon submitting a report for a trademark case:

> The undersigned being warned that willful false statements and the like are punishable by fine or imprisonment, or both, under 18 U.S.C. 1001, and that such willful false statements... may jeopardize the validity of the application or document or any registration resulting therefrom, declares that all statements made of his own knowledge are true and that all statements made on information and belief are believed to be true.
>
> *(28 U.S. Code §1746—Unsworn declarations under penalty of perjury)*

A linguist's decision to accept a forensic case implies fiduciary duties to the client's legitimate best interests, a legal duty to make good faith, truthful assertions before the court, a professional duty of loyalty to the integrity of the science of linguistics, and even a personal duty to the linguist's own needs, goals, values, ethics, and sense of justice. In this chapter, ethical behavior for forensic linguists is defined as involving activities in which a linguist's professional and personal best interests are aligned with those of society in general. A linguist's professional best interests in turn are ideally aligned with the goals of linguistic science, and the goals of linguistic science are aligned with the highest standards of scientific inquiry.

The interests of particular clients, however, are not necessarily aligned with those of a just society, let alone linguistic science. In the real world, some clients are actually incorrect, misinformed, cynical, criminal, or immoral. Nevertheless, the ethical imperative of the forensic linguist remains unchanged. When available data cannot support a client's desired outcome, the ethical and professional response from a forensic scientist is to straightforwardly inform the client without twisting or suppressing data to produce apparent support for the desired outcome. The overarching ethical principle appears simple: argue for cases when the positions are supported by linguistics, and decline cases when the positions cannot be supported. As Ainsworth (2022) mentions, however, it is often impossible for researchers to determine which route is appropriate until after they have become fully engaged with the legal team and have examined the relevant linguistic information. Once that step is reached, there are two ethical choices for the linguist: withdraw from the case gracefully or explain (preferably orally) why the client's proposed position is untenable. As this chapter explains, however, some practitioners select neither of those ethical paths. Instead, they choose a third, unethical option: disinformative linguistics.

Ethical Reasoning

The purpose of expert witness reports is to present evidence from which conclusions of some consequence can be drawn. As high-stakes documents presented to a court of law, these reports must meet high standards of good-faith argumentation.

The term *logical reasoning* would seem redundant if all reasoning complied with fact-based logic. It does not. The term *reasoning* means only the process of adducing explanations for a set of phenomena. Some explanations have fatal flaws that make no logical sense upon close examination, though they might satisfy casual observers. Some flawed methods of argumentation have been categorized since antiquity as logical fallacies: the typical ways that reasoning can go astray, by logical error or by intentional design (Pirie 2006).

Reasoning can go astray even when the reasoner tries to argue in good faith. People are vulnerable to accept flawed reasoning that redounds to the benefit of positions they already espouse, which sometimes results in the paradoxical phenomenon of self-deception (Beier 2019). Lawyers, of course, are trained not only to recognize deceptive argumentation but also to keep their own arguments on an impeccably logical track. Despite this training, however, even the most ethical and careful of lawyers might overlook deceptive maneuvers in technical reports when the maneuvers rely on technical knowledge or manipulations of technical terms.

The goal of linguistic science is to examine communicative phenomena and draw conclusions about the rules that govern them and the patterns that explain them. Published linguistic studies consist of data collection, data analyses, and comments concerning theoretical implications; they contain sequences of reasoning, but rarely a scent of fallacy, deception, or polemics. Discourse structures of persuasion are uncommon in linguistic studies because the purpose is to explain, not to convince, let alone deceptively. Litigants commission expert reports solely to convince the court of the correctness of their position; forensic linguists can ethically help them to discover—not create—the linguistic facts that serve their purpose.

Fact-based logical reasoning is concerned with the criteria for determining whether an argument's conclusion has been demonstrated, supported, weakened, or undermined. Logical analysis of argumentation exposes erroneous patterns of reasoning that could be used to achieve unfair benefit. Authors of misleading texts often use confusing and obtuse argumentation to convince others that their position is correct, reliable, unbiased, and irrefutable. A key to this subterfuge is the skillful employment of logical fallacies. As the next segment demonstrates, intentional distortions of language are not only encountered in informal communications, they also occur in formal high-stakes documents presented to courts of law. Forensic linguistic reports are no exception. Forensic linguists might be tempted to resort to disinformative tactics when they find themselves contracted to argue for a linguistically untenable position. That would be an unethical choice if embraced knowingly. However, some clients might deliberately mislead a consulting linguist. Forensic experts can be financially coddled by clients who design a comfortable bubble of disinformation to skew their perception. A forensic linguist's most ethical path is to avoid such conceptual entrapments, which entails an unbiased evaluation of each client's narratives and motivations. Each case is a test of character, for both client and linguist.

The Path of Disinformation

Disinformative linguistics is an attempt to support an interpretive position despite the lack of sufficient appropriate linguistic evidence. In disinformative practice, the apparent absence of pertinent evidence is ignored, and the reader's attention is sidetracked by dishonest techniques and misleading argumentation. Disinformative linguistics differs from misinformed linguistics by the linguist's intention. A misinformed linguist argues in good faith but makes honest mistakes of analysis. It is possible, for example, for a linguist to perform an authorship identification study using analytical tools in a technically correct way. However, if pertinent sociolinguistic facts are misunderstood or overlooked, the subsequent analysis could be flawed from the outset. When that occurs, the resulting report will be inaccurate, misleading, or misguided, but not by cynical design. Such cases involve ill-informed, but not disinformative, practice.

Unethical reports, however, are not merely factually or methodologically deficient. They are structured to inveigle the reader with deceptive techniques like untruthful implicatures (Meibauer 2019) as well as logical fallacies. Reliance on fallacious arguments violates an expert's professional obligation to tell the truth. Such practices are not only unethical, they could constitute professional malpractice, let alone malfeasance. It is imperative to call attention to disingenuous practices that can cause perceptual and conceptual distortions in the minds of judges and jurors. Despite the preeminence of their decision-making duties, triers of fact might not detect deceptive tactics in complex argumentation about esoteric topics. Judges observe the principle of cooperative communication and assume that experts do too.

Readers legitimately expect high-stakes documents to communicate without syntactic, semantic, or pragmatic garden-pathing (Patson et al. 2009). When pertinent ambiguities occur, cooperative readers indulgently expect them to be unintentional—a matter more of incompetence than disinformative trickery (Turner and Horn 2018). In one case, a high-stakes professional licensure renewal survey contained misleading structures that (unintentionally) entrapped a good-faith respondent in apparent lies. The relevant authority withdrew its accusation after I exposed the survey's serious communicative flaws. Not all high-stakes documents are constructed as if the respondents' professional lives depended on their clarity. Such failings are typically negligent oversights, not malfeasance.

Technical errors occur and editorial mistakes slip in, but pervasive disinformative maneuvers suggest intentionality. As Shuy (2010: 36) mentions, the recycling of topics (or in this case fallacies) substantiates that the writer intended to communicate in just that way. Fallacious argumentation presented by experts is designed to take unjust advantage and qualifies as scientific dishonesty (Nylenna et al. 1999). Linguists presenting truthful, pertinent evidence within well-formed arguments do not require fallacies.

Forensic Linguistic Support for a Malefactor

I never imagined that a client for one of my defamation cases would eventually be charged, arrested, convicted, and sentenced to many years in prison. The judge said in court that the defendant's corrupt activities created very substantial financial losses to the public. I realized too late that this client, a lawyer himself, had brought his lawsuit in a cynical attempt to identify and harass employees who wrote negative employer reviews on an online forum warning job seekers not to work for his law firm. The attributes asserted about the plaintiff were clearly negative, and the linguistic facts were ostensibly damning: if everything the anonymous critics had asserted about him were *false*, then this would have been an excellent defamation case. However, practically everything the plaintiffs had asserted about my client turned out to be true. As a linguist operating in good faith, I evaluated the linguistic facts at issue in light of the legal criteria for defamation. I relied on the information provided by the lawyers and assumed that they had fact-checked the situation. That was the last time I made that mistake.

In defamation cases, as in most cases, it is helpful to know who, if anyone, is speaking a good-faith version of the truth, and who, if anyone, is lying. It is advisable to avoid working for liars once they are identified. Unfortunately, that is sometimes impossible, rendering the job of forensic linguists ethically precarious. By contrast, lawyers are required to provide the best possible case for their clients, even habitual liars. It is challenging, often perhaps impossible, to provide diligent, appropriate professional services to individuals who are actively involved in disinformative activities. For forensic linguistic experts, that difficulty is greatly magnified because disinformation is the antithesis of language science.

In retrospect, I should have recognized the possibility that some firms might neglect their own due diligence to determine the truth about potential clients. I *did* ask my client-lawyers if there were any possibility that the allegations against the plaintiff were true. They dismissed the defendants as disgruntled former employees. Acting under the assumption that my client was truthful, I proceeded to examine the evidence. The defendants publicly insulted the plaintiff with vibrant images and metaphors. They characterized the plaintiff as a "true bottom feeder" and a "predator". They characterized his law firm as a "gulag", "an insane asylum", "a sweat shop", "a laughing stock", "a hell hole", and "a sinking ship". They also made unadorned assertions that the plaintiff had broken the law and was "unethical", "irrational", "crazy", and a "con". My report presented analyses showing that many of the comments were communicated through syntactic structures typically used for assertions of fact and semantic structures typically used to communicate attributes injurious to a person's reputation.

Shortly before I completed my report, another lawyer asked me to write a rebuttal for the same case. I responded that I had a conflict. It would be hard, I mused, for the rebuttal linguist to impeach the clear linguistic evidence of defamation pursuant to the relevant law, but I looked forward to the rebuttal.

The rebuttal arrived several days before I received news of a momentous event: the plaintiff had fled the country with assets stolen from his law firm's clients. The plaintiff's actions had already proved that the rebuttal was correct in that there were no grounds for defamation. The defendants had told the truth when they called him a con man. The plaintiff promptly abandoned his ill-starred defamation suit, eventually becoming a criminal defendant himself.

Though my report and its rebuttal immediately became legally moot and irrelevant, I treat rebuttals as my peer-reviewed report cards. I am always eager to learn from them. The rebuttal for this case, however, seemed superficial, consisting of 11 pages, compared to my report's 86 pages of detailed analyses, most of which the rebuttal ignored or distorted. In fact, the rebuttal employed disinformative techniques to mask its lack of pertinent evidence. My analysis led to the conclusion that the defendants' statements would be clearly defamatory under the assumption that the statements were false. The rebuttal argued that the statements did not qualify as defamatory because they were pragmatically appropriate expressions of free speech: the defendants' assertions were mere opinions, thus not provably false, and were expressed in a forum created for the dissemination of opinions. Many of the defendants' comments, however, were clearly fact-assertive with reputation-destroying attributes. Employing the fallacy of suppressed evidence, the rebuttal simply ignored those assertions of fact, focusing on the metaphors and rhetorical flourishes.

The rebuttal then presented a quotation that conflated the relevant state's legal definition of defamation with nonauthoritative commentary about metaphorical language. The actual code does not make the assertions that the rebuttal implied. Readers who took the rebuttal at its literal word without checking the original text of each reference could come away with the mistaken notion that materials containing metaphors, hyperbole, or other rhetorical devices can never qualify as defamatory. That maneuver employed the fallacy of presenting irrelevant material as if it were useful to an argument, plus a type of selective or distorted quotation that McClone and Baryshevtsev (2019) identify as lying. That might have been dismissed as a careless mistake, but the rebuttal continued with an argument that used the fallacy of equivocation: illegitimately changing the meaning of a key term within an argument.

My report made this (Gricean) point:

> Successful communications are defined as those following the cooperative principle. This refers to a tacit communicative contract specifying that speakers and writers endeavor to be clear, truthful, relevant, and appropriately informative and that listeners and readers have a legitimate right to assume that speakers and writers are so endeavoring.

The rebuttal responded:

> The speech event self-defines the contributed reviews as containing both possible criticism and praise. Complaints and other negative opinions are an integral part of the stated purpose of this website. Contrary to [the] discussion [above], therefore, the cooperative principle does not require the writers to be cooperative with the persons or companies about which they write their reviews.

That is not just a *non sequitur* and crafty distortion of Grice's model; it looks like a wholly false representation concocted to create confusion. Grice never defined (and I never redefined) *cooperative communication* to mean being charitable or kind to people who are the *topic* of a conversation. (It is cooperative to speak truthfully about them, yes.)

Grice was concerned with the reasonable and legitimate expectations that *participants* in a conversation have about an interlocutor's appropriate contributions at each point in a conversation and the inferences drawn when someone violates a maxim. The rebuttal shifted the discussion to a convenient misinterpretation of Grice's position, thereby sowing confusion around professional terminology nowhere in contention—all without proof. It is difficult to conceive that a professional linguist truly believes that distorted version of cooperative communication.

I had cited Grice's framework primarily to explain that the defendants' defamatory comments were particularly *successful* through their clear, unambiguous wording. The rebuttal's comment above was not only incorrect, it addressed a tangential point as if it were the main point. That disinformative ploy introduced the inflation of conflict fallacy; that is, suggesting, especially through equivocation, mislabeling, and misattribution, that experts in a field are in such disagreement about basic knowledge and terminology that the entire field cannot be trusted to provide useful evidence.

The rebuttal correctly noted that the defendants' negative comments were situationally appropriate. The question was not whether the setting was right, however; it was perfect. The question likewise was not whether the comments were kind or vicious. The pertinent questions were

(1) whether any comments qualified as assertions (some did) (2) whether those assertions were reputation-injurious and widely published (they clearly were), and (3) whether those assertions were false (no, they were true).

The first two questions define the linguist's proper domain. Linguists are not engaged to investigate the truth of defamatory statements, only whether they satisfy certain *linguistic* criteria. The proper role of the rebuttal was to show somehow that my *linguistic* evidence failed to meet the criteria.

The rebuttal linguist and I apparently agreed that a linguist's proper contribution was *not* to investigate the factual truth of the allegations, just their communicative structures.

The rebuttal argued on the side of justice for blameless defendants; ironically, it relied on disinformative confusions, to the detriment of the field of forensic linguistics. My support for the now-incarcerated plaintiff was hardly the field's finest hour either, yet I tried to provide scientifically sound, logical analyses.

What would *I* have done if I actually could have rebutted my own report? That is impossible to say, but an ethical path would have been to acknowledge that the data clearly satisfy the linguistic criteria for defamation. The next step would be to advise the legal team that the plaintiff wins the linguistic argument with no contest, *but only if we assume the defendants are lying*. The defendants *seemed* to be telling the truth. It might not be a linguist's role to verify that truth, but not every defamation case needs to be a make-work project for forensic linguists. There was never any need for a linguistic rebuttal for that particular case, just further investigation into the corruption the defendants allegedly witnessed at the plaintiff's law firm. That truth, if demonstrated, would have served the defendants better than intentionally befuddled linguistic argumentation.

Honest differences of opinion are common in forensic science. Often a case is so complex that two linguists could argue for conflicting positions without misrepresenting science, committing methodological errors, or acting unethically. There are cases so complex that entirely different conclusions could be reached, depending on the analytical method. In such cases, two opposing sides could understandably and ethically argue their points. However, if experts rely on sophistry rather than science to present their argumentation, they risk sabotaging their careers and damaging the entire field. The role of forensic linguistics is to present relevant linguistic facts, not to win arguments at any cost.

Good-Faith Service for Bad-Faith Actors

In the case presented above, I argued in good faith, following linguistic evidence exactly where it led. But was that ethical? Unbiased, competent linguistic analysis is a prerequisite for the ethical practice of forensic linguistics. I tried to meet that criterion, but was it enough? Are linguists ethically obligated to dig for the truth of a situation, and does omitting that step render further work unethical? How long are we ethically obligated to dig? Can the digging occur during the linguistic analysis or should it be completed first? No one, linguists included, will ever grasp the full truth of certain situations. Sometimes the best a linguist can do is the requested linguistic analysis, letting the facts of the situation fall where they may in a court of law. The importance of these questions extends beyond abstract debates. The way one answers these questions has real-world consequences.

Malefactors hide behind cynical linguistic representations. Linguists are in the best possible position to detect and identify discrepancies at all levels of the grammar between linguistic representations and consensus reality. To perform effectively, linguists must assess the pragmatic reality, which often includes the litigants' conversations and narratives. Those narratives, if elicited in communicatively cooperative environments, can provide contexts that inform an interpretation of the linguistic facts available for the case. That would certainly help, in some cases, to define the idiolects of relevant individuals—the specific rules, tendencies, preferences, and idiosyncrasies governing their use of language.

It seems counterintuitive to isolate forensic linguists from the individuals whose communicative patterns are at issue, but that is what occurs in many defamation cases. Linguists are instructed to consider limited sets of documents without consulting the individual's other known communications, let alone engaging with alleged victims. If an alleged perpetrator has a history of disseminating falsehoods, the person's contested communications could exhibit traces of the same patterns. Such corroborating evidence can be helpful in evaluating a case. With that in mind, Poynter (1997) provides this advice: "Do your best to get all the facts, whether they support your case or not. You want to see the entire file on the case" (58). Poynter admits there are discovery implications to providing certain documents to expert witnesses: "One reason your client-attorney might hold back information is because the other side might find it in your file. Your files are open, the attorney's are not" (58).

It is sometimes challenging for linguists to obtain the information they need, but ethics demands that they try—with due caution. Expert witnesses might damage their scientific independence by having direct contact with litigants. Sometimes that occurs from the outset, as when a litigant directly contacts the expert to request services. A common question at deposition is whether the witness has ever met the litigants. Whether the answer is yes or no, it should be followed by "I based my conclusions solely on linguistic evidence".

Good-Faith Service for Unpopular Entities

Ethical considerations do not necessarily preclude linguists from helping entities that tend to cause people distress while performing their lawful work. Debt collection agencies, for example, are typically detested by those reluctant to pay their debts, yet they can operate legally in the United States within a set of codified boundaries. Collection agencies help to reduce losses for companies with "uncollectible" accounts receivable by buying the right to pursue those debts. Expert witness work for businesses operating in compliance with relevant laws does not entail a violation of ethical responsibilities.

With that rationale, I agreed to take a debt collector as a client. The case concerned the standard wording that collection agencies must include in letters to debtors with long-term delinquency. *Time-barred debt* is a financial term of art for debt so old that collection companies are no longer permitted to sue the debtor to collect. (In the United States, statutes of limitation vary from an indulgent three years in Alaska to ten years in Rhode Island.) A debtor with time-barred debt is not relieved of the debt, however. While time-barred debts remain unpaid, collection companies may continue to request payment and make unfavorable credit reports, but cannot sue the debtor. An old debt loses its time-barred status, however, if the debtor admits responsibility for the debt by word or deed. Such an "act of recognition" of the debt restarts the clock in that regard, and the agency once again may sue the debtor for the entire unpaid balance.

The issue in this case was not the reasonableness of a law designed to protect the rights of both creditors and debtors and to prevent abuse by either party. The point of contention in the litigation was that the plaintiff's debt apparently lost time-barred status because the debtor had done something qualifying as an act of recognition. Here is the relevant paragraph from the actual time-barred debt letter:

> The law limits how long you can be sued on debt. Because of the age of your debt [the agency] will not sue you for it. If you do not pay the debt, [the agency] may report or continue to report it to the credit reporting agencies as unpaid. This is a communication from a debt collector and is an attempt to collect a debt. Any information obtained will be used for that purpose.
>
> *(Consumer Financial Protection Board 2020: 8)*

US debt collection agencies are required to use those sentences. The only difference is that some states require *cannot sue* rather than *will not sue*. In the relevant state for this case, the law specifies the phrase *will not sue*. The plaintiff misunderstood (or paid insufficient attention to) the somewhat counterintuitive message in the time-barred debt notification. It remained unclear just why, but the plaintiff responded in some way (like paying a token amount of the debt) that triggered reclassification of the debt to not-time-barred status.

That act of recognition was not in the plaintiff's best interest, but it was not an error of the collection agency, which can lawfully encourage debtors to pay their debts regardless of the age of those debts. The agency issued a standard, legally compliant notification using the precise wording that the US Consumer Financial Protection Board (CFPB) recommended as user-friendly and accessible. That recommended wording was based on a careful, formal survey and statistical analysis.

The plaintiff's linguist presented the following argument:

> [The *will not* word choice] allows for an interpretation that although the company states that they will not sue, they may change their mind in the future. This usage of *will not* implies that the current inclination of the company is to not sue the consumer. An inclination, however, can easily change.

It is a pragmatic error to apply a meaning appropriate only in casual communications to a high-stakes context requiring formal communication. In the formal register of contracts, a statement that a party *will not* perform an action indicates the party's commitment to a future state of affairs in which that action is precluded from occurrence. There is no ambiguity in that statement. It is written to be interpreted literally, as stated, and there is no warrant to apply a casual interpretation of the modal verb *will* indicating a nonbinding intention.

A professional debt collection letter is a genre of high-stakes document presenting relevant statements of financial and legal facts, not inclinations, opinions, or expressions of whimsy. International and domestic finance depends on parties being held to their literal word for their written commitments. In this context, the *will* modal auxiliary verb is regularly, and legitimately, interpreted as a commitment, not as an inclination that a writer might not feel like fulfilling at some future time. This is a description of what happens in such communications, not what should happen.

After my rebuttal was submitted, the plaintiff's complaint was withdrawn immediately. In this case, the plaintiff's linguist seemed inadequately knowledgeable about the formal register of high-stakes documents. Arguing for mistaken notions is not necessarily unethical; it is unethical to argue for positions that one knows to be incorrect.

Some people fall into debt through no fault of their own; their indebtedness could seem unfair. This case revolved solely around cooperative communication and the legitimate interpretation of high-stakes texts. The origin and fairness of the plaintiff's indebtedness were legally—and ethically—irrelevant.

Linguistic Certainty

The forensic linguist's challenge is often to determine the meaning of sentences in context (Borg 2004). Linguists can rank-order proposed interpretations according to their congruence with communicative rules of the relevant dialect. The rules are based on the totality of linguistic research, plus corpus investigations (Coulthard 2013). Communicative principles identify legitimate pragmatic expectations (Grice 1975, 1989). A self-evident yet instructive guiding principle that promotes responsible language policy is as follows: "*Any interpretation that satisfies applicable linguistic, logical, and pragmatic rules is treated as valid, even if it differs from the writer's intent or the most common reading.*" (Habick 2025). That foundational principle of linguistic interpretation is true *a priori* because it is basically tautological. This is an important point in the discussion of linguistic certainty in high-stakes documents.

Linguists can design sentences to have exactly one legitimate interpretation in context. There often *can* be certainty in linguistic judgments, especially when contextual and dialectal parameters are tightly controlled and when readers recognize their duty to interpret high-stakes documents cooperatively, not cynically. Interpretations requiring at least one pertinent rule violation are ranked impossible (invalid)—at least for a specified dialect and register. Interpretations compliant with relevant rules are classified as irrefutably valid, or uncommon yet still valid. Some unlikely interpretations can be entirely legitimate (rule-governed) logically, linguistically, and pragmatically (Atlas 2005).

Linguists can promote justice by demonstrating that an uncommon interpretation is valid and thus acceptable as a correct response. A coherent, responsible language policy for the construction of high-stakes documents (Habick 2025) favors linguistic structures that narrow the distance between intended and interpreted meanings as well as structures that efficiently communicate those meanings (Levshina 2023). These considerations are indispensable aspects of ethical product design.

The principles enshrined in professional codes of ethics imply the types of causes for which linguists *should* advocate (Hume 1739). The real-world circumstances of each case, however, require evaluation before the side of social justice can be identified. Yes, forensic linguists should use linguistic data to illustrate truthful patterns in a way that facilitates social justice. But it is sometimes impossible to determine on which side, if any, social justice will sit. Moreover, justice might not always reside on the same side as dispassionate linguistic evidence (linguistic truth). Attitudes toward a client's cause, let alone pressure to increase billable hours, cannot be permitted to muddle linguistic judgment.

Linguists, of all professionals, are uniquely positioned to foil disinformation; they certainly have no business creating it. Linguistics is the science of the construction and transfer of meaning. Techniques intended to thwart cooperative transfers are thus quintessentially inappropriate tools for forensic linguistic reports. Their use to gain unfair advantage is professionally self-disqualifying. Disinformative techniques—even for socially irreproachable causes—pave, at best, an unjust road toward justice. Meticulous adherence to scientific principles in the collection and evaluation of evidence is a forensic linguist's surest course.

Questioning the Goal-Free Evaluation of Unidentified Aerial Phenomena

As Solan (2023: 350) points out, lawyers must be careful to avoid lying in court yet "are not required to be sincere in their attacks" against expert witnesses, even those who submitted reports with sound, logically well-structured arguments. Indeed, many lawyers seem unafraid to dabble in false innuendo, as in ad hominem attacks to impugn the credibility of an expert.

A lawyer for the opposition in one of my cases, for example, asked during a Frye hearing if I knew that Michael Scriven not only writes about logic but also about UFOs. My report had cited Scriven's widely respected book *Reasoning* (1976). I truthfully answered *No*; I was unaware that Scriven sometimes used argumentation about flying saucers to illustrate flawed reasoning. I probably should have objected to the irrelevant innuendo, but the lawyer's rhetorical question caught me off-guard. The thinly veiled purpose of the lawyer's question was to make a link, however far-fetched, between my report and another author's (falsely implied) fascination with flying saucers. Revealingly, the opposition lawyers avoided direct discussion of all linguistic evidence in my report and offered no linguistic argumentation favoring their client's side of the case. Given that deficit, the lawyer deflected by discussing UFOs.

The adversarial system might be beneficial in keeping experts on their toes to avoid the reality and appearance of controversial associations and activities. Unscrupulous lawyers can and will allude to them—and reframe them as needed—to impugn the expert's character and credibility. Lawyers serve as officers of the court, however; they *do* have a primary ethical obligation to uphold the law. That means arguing fairly and on the point of the case at hand, not shifting the discussion to distracting irrelevancies. Lawyerly disinformative tactics, even if technically legal, undermine public trust in the system of justice as a whole.

Regardless of the lamentable rhetoric wielded with impunity by at least some lawyers, expert witnesses have an ethical duty to painstaking fairness in representing and evaluating their opponents' evidence, as well as their own. Forensic linguists, even those sporting law degrees, cannot ethically forward cynical interpretations and disinformative arguments to a judge. Yes, linguists practicing as expert witnesses must hold themselves to a higher standard—the standard fitting for all true experts—namely the duty to communicate competently and cooperatively in their professional work, and particularly in representations presented to a court of law.

Conclusion

It is imperative, both practically and ethically, for forensic linguists to be conversant in the cunning techniques of disinformative linguistics—not to deploy them, of course, but to recognize them in vivo. On the practical side, this awareness is helpful in composing rebuttals: the very presence of disinformative tricks in a report is an adverse credibility indicator. Disinformation suggests a position with an insufficient amount of legitimate linguistic support. On the ethical side, such awareness reminds practitioners of their duty to determine as early as possible for each engagement whether they can make informed, legitimate, and valid arguments for the client. If they cannot, then they should refrain from contorting reality and kicking sand in the eyes of the opposing side in unseemly efforts to bolster unsupportable linguistic claims.

References

Ainsworth, Janet. 2022. "Serving Science and Serving Justice: Ethical Issues Faced by Forensic Linguists in their Role as Expert Witnesses." *Language as Evidence: Doing Forensic Linguistics* 35–53. Edited by Victoria Guillén-Nieto and Dieter Stein. Cham: Palgrave Macmillan.

Atlas, Jay. 2005. *Logic, Meaning and Conversation*. Oxford: Oxford University Press.

Austin, John L. 1975. *How to Do Things with Words*. Oxford: Oxford University Press.

Beier, Kathi. 2019. "Lying and Self-deception." *The Oxford Handbook of Lying*, 203–213. Edited by Jörg Meibauer. Oxford: Oxford University Press.

Borg, Emma. 2004. *Minimal Semantics*. Oxford: Oxford University Press.

Consumer Financial Protection Board (CFPB). 2020. *Disclosure of Time-Barred Debt and Revival: Findings from the CFPB's Quantitative Disclosure Testing*. Washington, DC: CFPB.

Coulthard, Malcom. 2013. "On the Use of Corpora in the Analysis of Forensic Texts." *International Journal of Speech Language and the Law* 1, no. 1: 27–43.
Grice, Paul. 1989. *Studies in the Way of Words*. Cambridge: Harvard University Press.
Grice, Paul. 1975. "Logic and conversation." In *Syntax and Semantics 3: Speech Acts*, 26–40. Edited by Peter Cole and Jerry Morgan. New York: Academic Press.
Habick, Timothy. 2025. *AICPA Language Policy: Principles of Linguistic Integrity in High Stakes Assessment.* Ewing, NJ: Association of International Certified Public Accountants.
Hume, David. 1739. *A Treatise of Human Nature*. London: John Noon.
Labov, William. 2006. *The Social Stratification of English in New York City.* Cambridge: Cambridge University Press
Levshina, Natalia. 2023. *Communicative Efficiency*. Cambridge: Cambridge University Press.
Meibauer, Jörg. 2019a. "Lying, Implicating, and Presupposing." In *The Oxford Handbook of Lying,* 193–202. Edited by Jörg Meibauer. Oxford: Oxford University Press.
Meibauer, Jörg, ed. 2019b. *The Oxford Handbook of Lying.* Oxford: Oxford University Press.
McClone, Matthew, and Max Baryshevtsev. 2019. "Lying and Quotation." In *The Oxford Handbook of Lying*, 315–325. Edited by Jörg Meibauer. Oxford: Oxford University Press.
Nylenna, Magne, Daniel Andersen, Gisela Dahlquist, Matti Sarvas, and Asbjørn Aakvaag. 1999. "Handling of Scientific Dishonesty in the Nordic Countries." *The Lancet* 354: 57–61.
Ortoni, Andrew, and Swati Gupta. 2019. "Lying and Deception." In *The Oxford Handbook of Lying*, 149–169. Edited by Jörg Meibauer. Oxford: Oxford University Press.
Patson, N. D., E. S. Darowski, N. Moon, and F. Ferreira. 2009. "Lingering Misinterpretations in Garden-path Sentences: Evidence from a Paraphrasing Task." *Journal of Experimental Psychology: Learning, Memory, and Cognition* 35, no. 1: 280–285.
Pirie, Madsen. 2006. *How to Win Every Argument. The Use and Abuse of Logic*. London: Continuum International.
Poynter, Dan. 1997. *Expert Witness Handbook: Tips and Techniques for the Litigation Consultant*. Santa Barbara: Para.
Scriven, Michael. 1976. *Reasoning*. New York: McGraw-Hill.
Searle, J. R. 1969. *Speech Acts*. Cambridge: Cambridge University Press.
Shuy, Roger W. 2010. *The Language of Defamation Cases*. Oxford: Oxford University Press.
Solan, Lawrence. 2023. "The Forensic Linguist: The Expert Linguist Meets the Adversarial System." In *The Routledge Handbook of Forensic Linguistics*, 349–363. Edited by Malcolm Coulthard, Allison May, and Rui Sousa-Silva. New York: Routledge.
Turner, Ken Peter, and Laurence Horn. 2018. *Pragmatics, Truth and Underspecification: Toward an Atlas of Meaning*. Leiden: Koninklijke Brill NV.

11

ETHICS AND THE LINGUISTIC PRACTICE OF DUTCH POLICE INTERROGATIONS

The Right to Silence

Martha Komter

Introduction

A fundamental ethical ground rule of the criminal law process is the protection of citizens from the power of the state and its functionaries from performing actions that, in other contexts, would be considered serious crimes such as depriving citizens of their liberty, their money, or even sometimes their life. In view of such potential harms, the state's monopoly of power of the state necessitates the implementation of laws to protect citizens by possible abuses of governmental power. In other words, the relationship of citizens with the powerful state is regulated by laws that not only authorize the state to intrude deeply into peoples' lives but also aim to undo the effects of these intrusions. The suspect's right to silence is one of the measures assumed to perform the latter task.

Police interrogators are therefore confronted with two legal contradictions. On the one hand, there is the state-sanction permission to temporarily deprive suspects of their freedom; and on the other, there is the obligation to protect suspects from harm. This is not the only contradiction that police interrogators are confronted with. Within the Netherlands, the mixture of adversarial and inquisitorial elements in the criminal law process leads to suspects becoming both a source of information and a defending party. The treatment of suspects as source of information is deeply affected by the knowledge of their position as defending party. Suspects are invited to cooperate with the police as source of information about "the facts". At the same time, however, they are also expected to deny or avoid mentioning the more harmful elements of those facts in order to escape or mitigate punishment.

Another contradiction facing Dutch law enforcement is the conflict between the "presumption of innocence" and the "reasonable assumption of guilt" which, according to Section 27.1 of the Dutch Code of Criminal Procedure, is the legal prerequisite for prosecuting a person as a suspect in the Netherlands. For Dutch police officers, the desire to solve a criminal case may interfere with their treatment of suspects as "innocent-until-further-notice". A legal solution to the simultaneous and contradictory premises of their guilt and innocence is the suspects' right to silence.

However, in and of itself, the right to silence constitutes a blatant violation of normal conversational practice by disregarding the force of interaction and ignoring the principle of adjacency

 DOI: 10.4324/9781003391074-12

pairing as a basic building block of conversation. Explained briefly, questions and answers belong together. So, when an answer remains absent, this is "noticeable" (Sacks 1984; Schegloff and Sacks 1973) and a source of negative inferences (Heydon 2007). Moreover, questions and answers also work as actions, depending on the context in which they are uttered (Drew 2013). Within the context of a police interrogation, the actions performed can be, and often are, at odds with one another. As the police officers' questions are produced in order to elicit the information necessary to build a case, they can be understood as accusatory; and as the appropriate response to an accusation is a defense, not answering questions can be understood as having no defense or agreeing with the accusation, despite suspects' guaranteed right to silence.

It has been observed that contradictory tasks and interests are the preconditions of dilemmas (Komter 1998; Rawlins 1992; Billig et al. 1988). As dilemmas would not be dilemmas if they could be solved, it is not a question of the choice between solving crime or protecting the suspect. As both these tasks have to be performed simultaneously, the participants must find a way of doing both or of appearing to do both (Komter 1998).

The materials for this chapter consist of 34 audio-recorded and transcribed police interrogations, and the police reports made of those interrogations. These materials were collected over two periods of time. Before the turn of the century, 20 police interrogations were audio-recorded and the reports made of these interrogations were collected. Between 2007 and 2009, additional materials were collected,[1] resulting in 14 audio-recorded police interrogations and their accompanying reports.

To examine these materials, a Conversation Analysis (CA) approach was used which involved the investigation of turn-taking and the sequential organization of talk (Sacks 1984; Sacks et al. 1974). According to CA, each turn in talk displays the speaker's understanding of the previous turn and projects a range of activities available to the next speaker. It is not the analyst's interpretations or intuitions that count, but the interpretation of the participants themselves, as shown in the sequential organization of their talk, which is then an important analytic resource for the analyst. The sequential analysis of written texts can be grounded in findings from CA studies of talk, because readers can understand interactional moves and narrative coherence in a text similarly to the way participants do in face-to-face interaction (Clayman 1990).

The specific aim of this chapter is to examine how the participants in Dutch police interrogations manage ethical dilemmas resulting from their legal contradictory requirements. Toward that end, this chapter will examine the text of the law regarding the right to silence, discuss examples of how police officers and suspects put the right to silence into practice, and analyze the ways in which police officers construct the text of suspects' statements from the talk in the interrogation.

Coercion and Voluntariness

In the Netherlands, the suspect's right to silence is codified in Section 29 of the Code of Criminal Procedure. It consists of three sub-sections. Section 29.1 comprises the main principle. According to this law,

> in all instances where someone is being questioned as a suspect, the questioning judge or police officer refrains from everything that can be understood as obtaining a statement of which it cannot be said that it is made voluntarily. The suspect is not obliged to answer questions.

This section may be referred to as either the "right to silence" or the "prohibition against pressure". These two different names relate to the two opposing actors the law is designed to regulate. On the one side, there are the suspects and on the other, the agents of the state such as judges and police officers who have been imbued with the power to question them. The questioners have an *obligation* (to refrain from applying pressure) and the suspects have the *right* (to refuse to cooperate with their own conviction and incrimination). The idea behind this doubled mandate is that, if there is no pressure, then the suspect can talk freely. When suspects do not give the information requested, their questioners may be tempted to put them under pressure. To protect against this abuse of power, Section 29.1 permits suspects to invoke their right to silence. Thus, the right to silence can be understood as a kind of "safety valve" to reduce the pressure when police officers' questioning is in danger of crossing the line between fact-gathering and coercion.

Ideally, during suspect interrogations there should be no coercion at all. However, if it occurs, Section 29.1 offers the suspect a way out. This ambiguity is revealed in the rather awkward wording of Section 29.3: "refrains from everything that can be understood as obtaining a statement of which it cannot be said that it is obtained voluntarily". Note that the words "pressure" and "coercion" are absent. This may be because, in practice, some degree of pressure is considered inevitable and acceptable (Stevens and Verhoeven 2010). The purpose of Section 29 is to maintain a precarious balance between disregarding and recognizing the opposing forces of coercion and voluntariness.

A precondition for the implementation of the right to silence is the notification to suspects that they have this right. This caution is prescribed both verbally (Section 29.2), and in writing (29.3). The law makes it sound as if compliance is simple. For example, Section 29.2 states: "Before the interrogation the suspect is notified that he is not obliged to answer questions". Dutch police officers routinely put Section 29.2 into practice by repeating the text of the law, almost verbatim: "you are not obliged to answer questions". There are two reasons why this strategy may be problematic. As mentioned before, it is difficult for suspects to claim their right to silence because giving "no answer" may be understood as having "no defense". Consequently, the delivery of the caution grants suspects a right that may backfire against them when they claim it. Second, the aim of the police interview is to gather information from a person suspected to be a source of first-hand knowledge of facts about a criminal case. It is therefore impractical, to say the least, that police officers must tell suspects before the interrogation even begins that they are not obliged to respond.

The next excerpt shows the problems of a police interrogator (P) with delivering a caution while hoping to retrieve information. The author's English translation and the original Dutch text are provided below:

(1a) Author's English translation
1 P: You're not obliged to answer questions.
2: The Code of Criminal Procedure requires that I tell you that.
3: In other words you don't need to make a statement.
4: In view of the facts and circumstances that the police has come across
5: it is better to do it. In your own interest.[2]

(1b) Dutch original
1: P: Je bent niet verplicht te antwoorden
2: Het wetboek van strafvordering verplicht mij je dat te vertellen
3: met andere woorden je hoeft geen verklaring af te leggen.
4: Gezien de feiten en omstandigheden die de politie is tegen gekomen
5: is het beter het wel te doen. In je eigen belang.

In the first instance, P distances himself from his task by suggesting that he is only doing his job (line 2). He addresses the suspect as someone who is sensible enough to realize that there is no use denying what happened as the police have enough evidence for a conviction (line 4). This suggests that it will be in the suspect's own interest to tell everything rather than being exposed as a liar. In other words, the suspect might as well confess.

P manages the dilemma of coercion and voluntariness by doing service to both. He offers the suspect voluntariness (line 1) while pressuring him to talk (lines 4 and 5). He advises the suspect both not to answer the questions (lines 1 and 3) and to answer them (line 5). This contradictory message reflects the ambiguity of the officer's and the suspect's positions. The police officer has to provide information about the right to remain silent and encourage cooperation; and the suspect is granted a right that he is dissuaded to invoke.

Silence and Truth

In the Dutch criminal law system, witnesses are required by law to swear before questioning that they will tell the truth. Suspects are not required to take that oath. In the following example, a police officer deployed this legal difference strategically by suggesting that a witness is more likely to tell the truth than a suspect:

(2a) Author's English translation
1 P: You must look at it like this too so uh an uh a statement,
2 the witness he is under oath, he must tell the truth,
3 As a suspect you need not tell the truth
4 S: Yes I tell the truth because I have uh nothing to hide.
5 I've told you everything and that is the truth.

(2b) Dutch original
1 P: Je moet het ook zo zien eh een eh een verklaring,
2 de getuige hij staat onder ede, hij moet de waarheid spreken,
3 als verdachte hoef je niet de waarheid te spreken
4 S: Ja ik spreek de waarheid want ik heb niks te verbergen.
5 Ik heb u alles verteld en dat is de waarheid.

Of course police officers know from experience that witnesses do not always tell the truth. When it is a suspect's word against a witness's, witnesses are treated as the more credible, as they are assumed to have less to gain from lying. "Telling the truth" may be understood as equivalent to having "nothing to hide" (line 4). However, the suspect (S) is also advised that he need not tell the truth (line 3).

The right not to answer questions or to refuse to tell the truth is sometimes interpreted by police officers as "the right to lie". This conceptual equivalency is apparent in the excerpt below:

(3a) Author's English translation
1. P: You're allowed to lie.
2. S: I'm allowed to lie.
3. P: You don't need to make a statement
4. but <u>he</u> (1) can only tell the truth
5. S: Yes, but only he doesn't do that.

6. P: Well.
7. S: That was just the problem.

(3b) Dutch original
1. P: Jij mag liegen.
2. S: Ik mag liegen
3. P: Jij hoeft geen verklaring af te leggen
   Maar hij (1) kan alleen de waarheid spreken.
4. S: Ja, maar dat doet ie alleen niet.
5. P: Nou.
6. S: Dat was juist het probleem,

The dilemma of silence and truth is not solved by the police officer when he tells the suspect that he is allowed to lie (line 1). On the contrary, he suggests that he does not believe the suspect by emphasizing that the witness, who gave a different version of the events, "can only tell the truth" (line 4). The dilemma is also not solved by S, who cannot convince the police officer of his honesty by asserting that the witness is lying (line 5) or simply announcing that he is telling the truth (excerpt 2, line 4).

Although dilemmas of coercion and voluntariness and silence and truth have been discussed here in separate sections, they are closely interconnected and pervade interactions during police interrogations. It is because the law recognizes the risks of putting suspects under pressure to confess that they are granted the right to silence. And it is because the right to silence can be taken as authorization to lie that the invocation of this right may be considered a hindrance to truth-finding. In the next section, I shall juxtapose the instructions of the law with how police officers put them into practice.[3]

Talk and Text

Section 29.3 of the Dutch Criminal Code requires police officers to provide documentation of their investigation. A police report is a mandated part of a case file and serves as an official piece of evidence. As such, these reports may be used by law enforcement during the course of the investigation. Later on in the criminal law process, during court proceedings, the case file functions as direct source of evidence and thus as proxy for the appearance of witnesses. As Section 29.3 stipulates:

> the statements of the suspect, especially those that contain a confession of guilt, are written down in the report of the interrogation as much as possible in his own words. The notification stated in the second section is incorporated in the report.

While Section 29.1 prohibits the use of coercion to obtain a confession, the focus of Section 29.3 is on regulating the reporting of the activities of the interrogators.

Given the importance of police reports for the criminal justice system in the Netherlands, there has been mounting pressure to increase their transparency. Two phrases stand out: "especially those that contain a confession of guilt"; and "as much as possible in his own words". The vagueness of "especially" raises the questions of how, apart from a confession, the other items of the suspect's statement should be reported. The phrase "as much as possible" triggers questions about what would prevent police officers from writing down the suspect's "own words". The final sentence shows the importance of written evidence in the Dutch criminal law process.

If the caution is not given, or if it is given but not written down, the suspect's reported statement becomes invalid.

The requirement of reporting the suspect's statement "in his own words" is clearly meant to respect the suspect's version of the events. However, it disregards the impact of the interrogators' input and the effects of interaction upon the report writing. Therefore, in the aftermath of a series of miscarriages of justice in the beginning of this century (Brants 2012), a proposal for a new version of the Code of Criminal Procedure was drawn up. In this version, Section 29.3 was re-written. The previous phrase "as much as possible in his own words" was changed into: "as completely as possible and as much as possible in the question-answer format".[4] The singling out the suspects' confessions, the desirability of completeness, and the preference for the Q-A style in the police report all suggest growing recognition of the fact that suspects' confessions are best understood in the context of the police officers' questions. This contextualization, in turn, helps to provide greater transparency not only where police report writing is concerned but also where instances of police coercion may possibly have occurred.

Despite these potential benefits, in practice, the revision of Section 29.3 has become partially superfluous given procedural improvements in reporting police cautions. In some police stations, for example, the reporting officer is now required to fill in a prompt during the initial form-filling stage of the interrogation. Afterward, the following sentence automatically appears on the screen: "After I/we informed the suspect that he was not obliged to answer questions about what I/we wishes to hear, he stated:" (van Charldorp 2011). The texts in the police reports that are meant to represent the delivery of the caution often differ from the talk on which they are based. This difference is illustrated by the following exchange between an arresting officer and a person suspected of breaking weapons laws:

(4a) Author's English translation
1. P: Right you're not obliged to answer questions,
2. we're going to question you about the law of weapons and ammunition,
3. ((5 seconds of writing))
4. you were arrested during a traffic control,
5. among other things, a firearm has been found in your car,
6. S: It isn't my car.

(4b) Dutch original
1. P: Goed je bent niet tot antwoorden verplicht,
2. we gaan je horen (1) over de wet wapens en munitie,
3. ((typt, 5 seconden))
4. je bent tijdens een controle aangehouden,
5. der is onder andere een vuurwapen in je auto gevonden,
6. V: 't is mijn auto niet.

After the delivery of the caution, P informs the suspect of the upcoming topic (line 2). He then turns to the papers on his desk in front of him and starts writing. He does not give the suspect an opportunity to talk, as his attention is directed to the papers and his typing (line 3). S does not respond until line 6, when he contradicts P's prior utterance (lines 5–6). With the denial of his ownership of the car, S implicitly denies the charge suggested in line 2, his ownership of a weapon.

In contrast to the interaction above, as shown below, the report subsequently issued by the police officer contains many differences.

(5a) Author's English translation
1 After we informed the suspect that he was not obliged to answer questions
2 and about what we wished to question him he stated:
3 "I understand that I am not obliged to answer questions
4 And about what you wish to question me.

(5b) Dutch original
1 Nadat wij de verdachte hadden medegedeeld dat hij niet tot antwoorden verplicht was
2 En waarover wij hem wensten te horen verklaarde hij:
3 Ik begrijp dat ik niet tot antwoorden verplicht ben
4 en waarover u mij wenst te horen.

The delivery of the caution follows the text of the law, but the report of it does not represent the talk in the interrogation. In the interrogation, P immediately continues after the caution with information about the upcoming topic (excerpt 4, line 2).

A salient difference between my transcript and that of the police is that the suspect is here introduced as speaker and as responding to the caution. Apparently, the five-second pause (excerpt 4, line 3) was understood as a token of the suspect's understanding and acceptance of the caution (excerpt 5, lines 3–4). Almost all my police reports contain a response of the suspect to the caution. In almost none of the interrogations does this correspond with the talk.

These examples have shown the mechanical nature of both the delivery and the report of the caution. Both the talk and the report of the talk are produced routinely as part of the job. The talk and the report can be understood as specimens of "good work" and "proper procedures" (Zimmerman 1969). The insertions in the report of the suspect's expressions of acceptance and understanding can be read as their active adoption of a willingness to talk and, by implication, as an absence of pressure. As a result, everything that the suspect is reported to have said subsequently can be understood as having been produced voluntarily (Komter 2019; Heydon 2007).

Invoking the Right to Silence

This section discusses how a suspect invokes his right to silence in a longer fragment of talk. To do this, excerpts from a criminal case file are presented. In this case, the suspect is accused of having stolen a mobile phone in a metro station. The excerpt begins after the police interrogator confronted the suspect with the existence of CCTV images that showed the suspect turning right after leaving the metro station. The suspect maintains that he turned left. This difference is important as the alleged victim turned right after leaving the station. As shown below, the suspect's claim that he went in the opposite direction is important to support his claim to innocence.

(6a) Author's English translation
1 You ask me where I went to first after the exit doors.
2 I went to the left.
3 you tell me that I went to the right after I walked out of the door on the right.
4 That is on camera.

(6b) Dutch original
1 U vraagt mij waar ik na de deuren van de uitgang eerst naartoe ben gegaan.
2 Ik ben naar links gegaan.
3 U vertelt mij dat ik naar rechts ben gegaan nadat ik de rechterdeur uit ben gelopen.
4 Dat staat op de camera.

Lines 1 and 3 show a kind of reporting called the "recontextualised monologue style" (Komter 2019; 2006). In this writing style, the actions of the interrogator are transformed into assertions of those actions "you ask me" (line1), "you tell me" (line 3). This style looks like a monologue but the interaction can to some extent be reconstructed. After this exchange, there is further discussion about whether the suspect turned left or right, after which the police officer continues:

(7a) Author's English translation
1 P: So those camera images are lying
2 S: I first went to the left
3 P: So I'm asking you,
4 S: Sir, I sa- uh right to silence okay?
5 P: Why?
6 (2.0)
7 why right to silence?
8 S: right to silence sir I can- I am
9 P: because it is getting too difficult for you?
10 S: [I'm obliged to be silent [so I'm silent.
11 [you have the right to be silent.
12 S: Okay I have the right,
13 P: yes,
14 S: I'm silent.
15 P: You don't want to talk any more.
16 S: No.

(7b) Dutch original
1 P: dus die camerabeelden liegen
2 V: ik ging eerst naar links.
3 P: ik vraag dus aan jou,
4 V: meneer ik sa- eh zwijgrecht oké?
5 P: waarom?
6 (2.0)
7 waarom zwijgrecht?
8 V: zwijgrecht meneer ik mag- ik ben-,
9 P: omdat het te moeilijk voor je wordt?
10 S: ik ben verplicht om te zwijgen[dus zwijg ik.
11 P: [je hebt het recht om te zwijgen.
12 V: oké ik heb het recht,
13 P: ja,
14 V: ik zwijg.
15 P: je wilt niet meer praten.
16 V: nee.

As camera images are understood as "telling the truth", P's ironic comment (line 1) suggests that it must be the suspect who is lying. However, the suspect persists that he went to the left (line 2). When P appears to pursue the matter (line 3), S invokes his right to silence (line 4). P does not appear to accept this, and instead asks for an explanation (line 5). After two seconds, when S does not immediately reply, P repeats his request (line 7). Then S makes another attempt to invoke his right to silence (line 8). He looks for the right word (lines 8, 10), but chooses the wrong one. P corrects him in overlap (line 11) and S invokes his right to silence once more, this time correctly (lines 12, 14). P appears to accept S's claim to his right to silence by expressing how S will put his position into practice (line 15). His acceptance is confirmed. All uncertainties now seem to have disappeared. P summarizes, shows his understanding, and S agrees (lines 15, 16). However, in what follows P takes a step back:

(8a) Author's English translation
17 P: it is surely getting too difficult for you.
18 S: yes it is getting too difficult for me.
19 P: Oh
20 S: It is getting too difficult for me.
21 (2)
22 P: Why then.
23 [Why is it getting too difficult?
24 S: [right to silence.
25 P: (2)
26 P: Then I'm going to ask you a different question is that allowed?

(8b) Dutch original
17 P: het wordt je te moeilijk zeker.
18 V: ja het wordt me te moeilijk.
19 P: Oh.
20 V: Het wordt me te moeilijk.
21 (2)
22 P: waarom dan.
23 [waarom wordt het je te moeilijk?
24 V: [zwijgrecht.
25 (2)
26 P: dan ga ik je andere vraag stellen, mag dat wel?

The interaction has become repetitive. P has rejected S's invocation of his right to silence by giving an interpretation of S's motives. However, this time S agrees with this interpretation (lines 18, 20). When P persists in asking for clarification (lines 22, 23), S invokes his right to silence once more (line 24). P finally accepts this statement and proposes a new line of questioning (line 26).

The excerpts (7–8) show the tensions and dilemmas discussed earlier. When S is treated as a liar after being confronted with the "solid evidence" of camera images, he takes recourse to his right to silence and thereby attempts to defuse the effects of pressure from the interrogator. However, that strategy does not really work because the police interrogator does not immediately accept the suspect's claims until after a long discussion. Thus, the suspect's invocation of his right to silence appears to increase rather than decrease the pressure.

After this episode, the two interrogators discuss what to ask next. They decide to ask S about what he told a fellow officer in a previous interrogation the day before. In the segment below, the officer's official report of their entire interrogation with the suspect is given. This transcribed audio recording of the interrogation shown in excerpt (7) as well as the subsequent pivot in questioning are reported as follows:

(9a) Author's English translation
5 I tell you, I went to the left from the metro station.
6 I further invoke my right to silence.
7 Question: Yesterday you signed a waiver for a female colleague,
8 can you remember what you told her?

(9b) Dutch original
5 Ik zeg u, ik ging naar links vanaf het metrostation.
6 Ik beroep mij verder op mijn zwijgrecht.
7 Vraag: Gisteren heb je een afstandsverklaring getekend bij een vrouwelijke collega,
8 kan je je herinneren wat je haar hebt verteld?

To put it mildly, excerpt (9) is a meagre rendition of the talk transcribed in excerpt (7 and 8). There is much that has been altered and abbreviated. Line 5, for example, corresponds with line 2 of the transcribed audio recording (excerpt 7); and line 6 represents S's repeated attempts to invoke his right to silence (lines 4, 8, 10, 12, 14).

There are several possible explanations for the underreporting of interrogators' conduct. The Q-A style of the police report does not accommodate "third position actions". The assumption is that police interrogations consist of a chain of questions and answers, where the police officers are in a first position, in charge of asking the questions and thereby deciding on the topics. Suspects, who are in a second position, are expected to answer questions, after which control is returned to the interrogator, who typically continues with a next question. In settings such as police interrogations, where the suspects' answers are typically understood as defensive and unreliable, police officers frequently make use of the opportunity to fill the slot for their next turn after the suspect's answer by inserting a third position action such as a comment or evaluation of that answer (Komter 2019).

In the Q-A style police reports in my materials, only about 16% of the questions are reported (de Boer 2014). This can be related to the following circumstances. First, as the typing takes place during the interrogations, there is not enough time to report everything, and a selection is made on the basis of "typability" (resulting from time constraints due to the differential pace of typing and talking) and "recordability" (perceived relevance for the case) (Komter 2019). Second, interrogators may be reluctant to put their more adversarial actions on display.

Consider excerpt (8) once more. It shows S's contributions that he went to the left and that he invoked his right to silence (lines 5 and 6). And it also shows that P changed the topic (line 7). The sequential placement of this change of topic, however, suggests that S's invocation of his right to silence was accepted immediately. Thus, because of the absence of third position actions in the Q-A format, the reader can be misguided by this reporting style. P's inquisitive and antagonistic actions tend to remain unreported.

Conclusion and Discussion

There is—inevitably—a gap between ethics and their implementation (e.g., Mondada 2014; Speer and Stokoe 2014; Kubanyiova 2008; Guillemin and Gillam 2004). Ethics rest on general principles and ideas which are put into practice in a variety of settings, and in interaction with others (Egido 2020; International Center for Academic Integrity 2019; De Costa 2015). The general ethical principles underlying suspects' right to silence are to respect their autonomy and dignity and to protect them from possible abuses of the power of the state.

By comparing the text of the law, the corresponding police reports, and my transcripts of the relevant audio-recorded interrogation, the aim of this chapter was to highlight the manner in which the participants in police interrogations put the right to silence into practice. As was shown, the right to silence is fundamentally dilemmatic. Police officers are confronted with the dilemma of how to get suspects to tell the truth without putting them under excessive pressure, while (unquestioningly respecting) suspects' invocation of their right to silence. At the same time, suspects face the problem of attempting to invoke their right to silence without increasing the pressure placed upon them or creating the impression that they have something to hide.

There are two moments when the suspects' right to silence is put into practice: (1) at the beginning of the interrogations when police officers deliver the caution; and (2) during the course of the police questioning, when suspects claim their right. In both instances, there are marked differences between the talk in the police reports and in my transcripts.

Police officers deliver this caution in a routine and mechanical manner at the beginning of their interrogations. In the subsequent police report, the delivery of this caution is also reported in a standard, routine, and mechanical manner. Consequently, the police reports of these cautions can be best understood as typical examples of "procedural correctness" and "good work" rather than as correct representations of the talk. Indeed, there are significant differences between the police reports and my transcripts of these episodes. For example, suspects, I found, usually listened passively to the caution, although in the official transcripts, they were reported as expressing their acceptance and understanding of the caution. Along with implying an absence of pressure from police officials, these descriptions overstate suspects' cooperation, willingness to talk, and the voluntariness of their statement for the rest of the interrogation.

I also found that when suspects invoke their right to silence, this often leads to discussions before the police officers eventually accept and report the suspects' invocation. Earlier, I proposed the metaphor of a "safety valve" to describe the suspects' exercising their right to silence as a kind of tool for reducing pressure. However, my materials show that invoking the right to silence does not automatically reduce pressure. On the contrary, when suspects invoke their right to silence, this appears to elicit comments or evaluations by interrogators. Instead of neutralizing the pressure, claiming their rights seems to generate more pressure on suspects. Indeed, the vagueness of the law and the contradictory tasks and interests of the participants may induce police officers to put suspects under pressure and to misrepresent the talk in the interrogations, which may ultimately lead to flawed police reports and wrongful convictions (Malsch et al. 2015; Brants 2012).

Recent changes in Dutch law on the right to silence show increasing appreciation for transparency in police reports and a growing recognition of the significance of police-suspect interactions. Previously, when the monologue style of interrogation was common, the law expressed a preference for reporting "the suspect's own words". Then it was felt that the actions of police interrogators should play a greater role in the reports, and that the suspects' statement should be reported "as much as possible in the Q-A style" and "as completely as possible".

The inclusion of the police officers' questions in official reports has certainly contributed to a greater transparency. However, the Q-A style has its limitations. The focus on the Q-A format may discourage reporting actions other than questions and answers. This understates the antagonism of the talk. Moreover, the more complete and lengthy a report is, the more unmanageable and time-consuming it can become to read it. The new version of the Code of Criminal Procedure also expresses the desirability of audio and video recordings of police interrogations, especially in cases involving more serious offenses. However, the introduction of mandatory audio and video recordings of police interrogations has not contributed much to the transparency of police interrogations, as time constraints and work overload discourage judges and prosecutors from consulting the recordings. For this reason, the new version of the law proposes replacing written reports with audiovisual recordings and short written summaries. Judges have expressed their acceptance of interrogation summaries, as long as they are a "factual representation" of the talk (Franken 2010). Nevertheless, the proposed changes would ultimately reduce the transparency of the criminal process and increase the risk of distorting the information conveyed (Malsch 2022).

From an ethical point of view, it is understandable that the law grants suspects the right to silence in order to protect them from excessive police coercion; and it is desirable that it requires police officers to adequately report the talk in the interrogations. However, as this chapter has shown, what actually takes place during the interaction of suspects and police interrogators may transgress the tenor of the law.

The motivation of this chapter is to understand the complicated processes involved, and not to criticize. Thus, for professionals in the field, it is hoped that this chapter will help foster an awareness of the factors that make police interrogations vulnerable to criticism. Towards that end, this chapter can offer insights into the working circumstances that may affect the conduct of police interrogators. For forensic linguists, this chapter demonstrates the value of working with recordings of authentic police interrogations. In doing so, forensic linguists can help make a substantive contribution to understanding the criminal justice system. Ultimately, the research conducted can be used to provide better protective guardrails for the criminal justice system. By revealing the real conditions under which confessions are elicited, forensic linguistic research provides an ethical service to the society by helping to protect suspects against coercion; encourage law enforcement to gather truthful information in a proper manner; and ensure juries and judges that any decisions rendered are lawful and just.

Notes

1 Collected by Tessa van Charldorp (2011).

2 The transcription departs from the standard Jeffersonian transcript notation for reasons of readability, and because the excerpts are translated from Dutch. Those features of the standard transcript notation have been preserved that are compatible with the translation.The written texts of the police reports are presented in bold typescript.

P	Police interrogator
S	Suspect
Full stop.	Falling intonation
Comma,	Slightly rising intonation
Question mark?	Rising intonation
Underlining	Emphasis
Square brackets [ ]	Overlap
(3)	Pause of three seconds etc.
Dash–	Sudden break off
Double brackets (( ))	Transcriber note

3 This is not to say that I shall compare the text of the police reports with the talk in the interrogation. I can only compare the police reports with my own transcripts (Komter 2022).

4 This will enter into force on April 1, 2029.

References

Billig, Michael, et al. 1988. *Ideological Dilemmas: A Social Psychology of Everyday Thinking.* New York: Sage.

de Boer, M. 2014. "Ik zal het in die bewoordingen op papier zetten. De verschillen tussen politie verhoren en processen-verbaal" [I Shall Put it on Paper in These Words: The Differences Between Police Interrogations and Police Reports]. *NSCR Report.*

Brants, Chrisje. 2012. "Wrongful Convictions and Inquisitorial Process: The Case of the Netherlands." *University of Cincinnati Law Review* 80, no. 4: 1069–1114.

van Charldorp, Tessa. 2011. "From Police Interrogation to Police Record." PhD dissertation, VU University, Amsterdam.

Clayman, Steven. 1990. "From Talk to Text: Newspaper Accounts of Reporter-Source Interactions." *Media, Culture & Society* 12: 79–103.

De Costa, P. 2015. "Ethics and Applied Linguistics Research." In *Research Methods in Applied Linguistics: A Practical Resource*, 245–258. Edited by B. Paltridge and A. Phakiti. London: Bloomsbury.

Drew, P. 2013. "Conversation Analysis and Social Action." *Journal of Foreign Languages* 37, no. 3: 2–11.

Egido, A. A. 2020. "Concepts of Ethics in Applied Linguistics Research." *Horizontes de Linguistics Aplicada* 19, no. 2: 118–136.

Franken, A. A. 2010. "Regels voor het strafdossier" [Rules for the Criminal Case File]. *Delikt en Delinkwent* 40, no. 4: 403–418.

Guillemin, M., and L. Gillam. 2004. "Ethics, Reflexivity, and 'Ethically Important Moments' in Research." *Qualitative Inquiry* 10, no. 2: 261–280.

Heydon, Georgina. 2007. "When Silence Means Acceptance: Understanding the Right to Silence as a Linguistic Phenomenon." *Alternative Law Journal* 32, no. 3: 147–151.

International Center for Academic integrity. 2019. Retrieved from 2019_ICAI-Fundamental-Values_R12.pdf (academicintegrity.org).

Komter, Martha. 2022. "Institutional and Academic Transcripts of Police Interrogations." *Frontiers in Communication*. Retrieved from https://www.frontiersin.org/articles/10.3389/fcomm.2022.797145/full.

Komter, Martha. 2019. *The Suspect's Statement. Talk and Text in the Criminal Process.* Cambridge: Cambridge University Press.

Komter, Martha. 2006. "From Talk to Text. The Interactional Construction of a Police Record." *Research on Language and Social Interaction* 39, no. 3: 301–328.

Komter, Martha. 1998. *Dilemmas in the Courtroom: A Study of Trials of Violent Crime in the Netherlands.* New Jersey: Lawrence Erlbaum.

Kubanyiova, M. 2008. "Rethinking Research Ethics in Contemporary Applied Linguistics: The Tension Between Macroethical and Microethical Perspectives in Situated Research." *The Modern Language Journal* 92, no. 4: 503–518.

Malsch, Marijke, et al. 2015. *Kijken, Luisteren, Lezen: De invloed van beeld, geluid en schrift op het oordeel over verdachtenverhoren.* [Look, Listen, Read: The Influence of Image, Sound and Writing on Assessments of Suspect Interrogations: Politie en Wetenschap]. Amsterdam: NSCR Publication.

Malsch, Marijke. 2022. "Het verhullende proces-verbaal en ander ondeugdelijk gedrag van de politie." [The concealing police report and other improper conduct of the police] *Nederlands Juristenblad* 8: 556–561.

Mondada, L. 2014. "Anonymization as a Participant's Concern and a Participant's Practice." *Human Studies* 37, no. 2: 179–209.

Rawlins, William. 1992. *Friendship Matters: Communication, Dialectics, and the Life Course.* New York: Aldine de Gruyter.

Sacks, Harvey. 1984. "Notes on Methodology." In *Structures of Social Action: Studies in Conversation Analysis,* 2–27. Edited by John Heritage, and J. Maxwell Atkinson. Cambridge: Cambridge University Press.

Sacks, Harvey, et al. 1974. "A Simplest Systematics for the Organization of Turn-taking for Conversation." *Language* 50: no. 4; 696–735.

Schegloff, Emanuel, and Harvey Sacks.1973. "Opening up Closings." *Semiotica* 8, no. 4: 289–327.

Speer, S., and E. Stokoe. 2014. "Ethics in Action: Consent-gaining Interactions and Implications for Research Practice." *British Journal of Social Psychology* 20, no. 2: 54–73.

Stevens, Lonneke, and Willem-Jan Verhoeven. 2010. *Raadsman bij het politieverhoor* [Legal Advisers during Police Interrogations]. *Den Haag: Boom Juridische uitgevers.*

Zimmerman, D. 1969. "Record-keeping and the Intake-process in a Public Welfare Agency." In *On record: Files and Dossiers in American Life*, 319–354. Edited by Stanton Wheeler. New York: Russell Sage.

SECTION II

Serving as an Expert Witness

12

LINGUISTS AND LAWYERS

Ethical Challenges in an Adversary System

Robin Conley Riner and John M. Conley

Introduction

In this chapter, we compare the sometimes-competing ethical systems of lawyers and forensic linguists who work as experts. To do so, we provide specific examples of how professional lawyers and linguists work through their respective ethical obligations. We bring both academic and practical perspectives to this topic. One of us (Conley) was a trial lawyer earlier in his career, teaches legal ethics in law school, and remains active in law practice as a consultant. The other (Riner) is a legal and linguistic anthropologist who has served as an expert witness and advisor to lawyers in several cases, including those involving the death penalty.

Legal anthropology uses the ethnographic method—intensive, embedded, qualitative study of a society or one of its components through "participant observation"—to investigate diverse cultures' systems for defining and dealing with deviance and resolving disputes. Its roots go back to the 1800s. In fact, some of the earliest recognizable works of anthropology dealt with law (e.g., Maine 1861; Malinowski 1985 [orig. 1926]). Those early works often examined whether small-scale societies could be said to have law or argued for value-laden evolutionary theories with Western legal systems at the pinnacle. More recent approaches have asked fundamental questions about the nature of law and legal systems across a wide range of cultures. This work has helped to reveal a complex web of formal and informal systems working simultaneously and often in conflict, with attendant opportunities for resistance and change (e.g., Merry 2006). Throughout its history, much of legal anthropology has focused on law's language—a trend that may have accelerated in recent years (Conley, O'Barr, and Riner 2019; Richland 2008).

Linguistic anthropology, which has much in common with sociolinguistics, studies language and communication in cultural context (Duranti, George, and Riner 2023). Building on an early focus on documenting and preserving endangered languages, the contemporary field uses detailed ethnographic evidence to understand language as a cultural resource and communication, including but not limited to language, as a cultural practice. When applied to legal contexts, linguistic anthropology can show how the language that disputants, decision-makers, and others use can reflect, manipulate, and sometimes challenge their respective cultural understandings (Conley, O'Barr, and Riner 2019). This application has resulted in a convergence: as legal anthropologists

DOI: 10.4324/9781003391074-14

have increasingly focused on language, linguistic anthropologists have produced more research on law.

In the adversary system used in the United States, and in many other countries with legal systems derived from England's, the fundamental ethical duty of a lawyer in any case, be it civil or criminal, is clear, at least in theory: to advocate zealously for the client within the bounds of the law. The forensic linguist's ethical position is different. When serving as an expert, the forensic linguist is called on to give an objective, evidence-based opinion about an issue relevant to the case. Questions that a forensic linguist might address include, for example, whether a defendant's recorded speech could be interpreted as a confession; whether two co-defendants were really agreeing to kill someone; or whether a relevant speech community understands a contested trademark as a generic term (Solan and Tiersma 2005). The linguist's opinion is likely to advance the client's position (or the lawyer would not offer it), but an expert is not expected to function as a zealous advocate.

These differences in the respective duties of the lawyer and the linguistic expert can create recurring ethical dilemmas for forensic linguists. This chapter specifically discusses four of these routine ethical challenges: (1) How should linguists navigate the different ethical demands of advocacy and research in an adversarial system?; (2) What are the consequences for forensic linguists if they reach conclusions contrary to the client's interests?; (3) Can expert linguists ethically gather data simultaneously for both advocacy and research?; and (4) What rights, if any, do expert linguists have in publicizing the research they conduct on a case?

We explore these questions from our varied perspectives as researcher, expert, teacher, and lawyer. We draw on the formal and advisory codes of legal ethics of professional linguistics organizations (such as they are), as well as our own and others' writing on the subject. And, importantly, we integrate our personal experiences in the field. The chapter focuses heavily on US litigation, for several reasons. The United States has much more litigation than any other country (Ramseyer and Rasmusen 2010) and, consequently, the law relating to expert witnesses is especially well developed in the United States. In addition, our respective experiences have all been with the US legal system. Despite the fact that the specific legal points we make are specific to the United States, the broader conclusions are generalizable to the so-called "common law countries".

The Legal Duties of Lawyers and Experts

In this section, we analyze the legal and professional duties that apply to lawyers and the experts they engage, including forensic linguists. By "legal" duties, we mean those obligations directly imposed by statutes and rules, the violation of which can lead to such sanctions as financial penalties, default judgments in favor of the other side, and fines or even imprisonment for contempt of court. "Professional" duties are those imposed on lawyers by state rules of professional conduct. These usually follow the American Bar Association's Model Rules. Sanctions can be imposed by state bars and can include suspension or revocation of the lawyer's license to practice. The authority to take these actions is delegated to state bars by their respective state legislatures. Many of the legal duties and the consequences of violating them apply to expert witnesses as well as lawyers. By contrast, there are no professional obligations—in the sense that we have just defined them—that apply anywhere to forensic linguists. Finally, we use the term "ethics" to refer to moral norms that have no legal authority. We note preliminarily that in 2009, both the forensic linguist and onetime public defender Janet Ainsworth as well as former president of the International Association for Forensic and Legal Linguistics Edward Finegan commented on some of the same issues we discuss in this section in a special issue of *The International Journal of Speech, Language and the Law* that was devoted to ethical questions. Differences between our conclusions and theirs may be largely attributable to amendments to the US Federal Rules of Civil Procedure and Federal Rules of Evidence which have been followed by some states since 2009.

In this chapter, we follow the definition of an "expert witness" contained in Rule 702 of the Federal Rules of Evidence and most state evidence codes: "[a] witness who is qualified as an expert by knowledge, skill, experience, training, or education". Whereas most witnesses can testify only to facts, an expert is permitted to offer an opinion on a subject on which "the expert's scientific, technical, or other specialized knowledge will help the trier of fact to understand the evidence or to determine a fact in issue". This is usually interpreted to mean a subject that will be beyond the knowledge of an average juror. The opinion must be "based on sufficient facts or data" that have been reliably applied to the facts of the case. The opponent may challenge the admissibility of the expert's testimony in a preliminary *Daubert* hearing (established by the US Supreme Court in its 1993 *Daubert v. Merrell Dow Pharmaceuticals, Inc.* decision) where, out of the hearing of the jury, the proposed expert will be examined and cross-examined, and the judge will rule on the admissibility of the testimony under the Rule 702 standard.

The fundamental professional duty of a lawyer representing a client before any court or tribunal is that of "zealous advocacy" of the client's position, within the bounds of the law. In fact, it is so basic that it is one of the first things mentioned in the American Bar Association's Model Rules of Professional Conduct, in the Preamble: "As advocate, a lawyer zealously asserts the client's position under the rules of the adversary system" (American Bar Association 2018: Preamble [2]). This duty is not absolute, however, as the bounds of the law can be significant. Lawyers cannot, for example, sponsor testimony that they know to be perjurious, though they can rarely disclose that knowledge (American Bar Association 2018: Rule 3.3). Moreover, although all criminal defendants, and some civil litigants, have a legal right to representation, individual lawyers can refuse to accept clients. Lawyers might decide, for example, that they will not defend persons who are accused of child abuse. But once a lawyer has taken on a case, the pursuit of the client's cause must be unwavering (American Bar Association 2018: Rule 1.3, comment 1). A criminal defense lawyer, for example, cannot decide that the client is guilty and fail to cross-examine prosecution witnesses vigorously. Nor may a lawyer's personal opinion of the client or the client's cause be grounds for diminishing the energy of the advocacy. In addition, once in, it is difficult for a lawyer to get out of a case without the client's consent, especially in criminal cases. The Rules of Professional Conduct list the limited circumstances in which unilateral withdrawal might be required (usually where there is a conflict of interest) or permissible, such as when "the client persists in a course of action that the lawyer reasonably believes is criminal or fraudulent" or "the client insists upon taking action that the lawyer considers repugnant or with which the lawyer has a fundamental disagreement" (American Bar Association 2018: Rule 1.16(2, 4)). Note that the latter provision refers to "taking action" going forward, not to an already established position of the client.

When seeking to engage expert assistance in a case, a lawyer-advocate does so with the expectation that the expert will ultimately support their client's position. With that hope in mind, a lawyer will actively search for an expert with a track record of testifying or writing in ways that may advance the client's position. However, a lawyer may not make the expert's compensation dependent on the conclusions the expert reaches. Nor may a lawyer tell an expert what conclusion to reach. That would not only be unethical but also bad strategy, as any such communication may be discoverable by the other side (see below) once that same expert testifies.

In most instances, an expert who delves into the case and reaches an unhelpful conclusion will simply be thanked, paid for the time taken to provide a draft report, told *not* to write up an official final report (so as to forestall the creation of a potentially discoverable—and unhelpful—document), and is thereafter ignored. It is not uncommon for a dismissed expert to be found, usually accidentally, by the other side and asked to work on the case. When this occurs, experts typically decline. As Ainsworth (2009) puts it from an ethical perspective, "it is surely not permissible to defect to the other side" (218). There is, however, no generally applicable legal rule that imposes a duty of confidentiality on an expert or bars an expert from switching sides. Moreover, Ainsworth's ethical position is debatable: if the expert has taken a principled position, what is unethical about making it available to the other side? In individual cases, the expert may have a contract with the original lawyer—written, oral, or implied—that prohibits the expert from disclosing or using information learned about the case or the expert's own work product. From a practical perspective, since lawyers who usually work on either the plaintiff/prosecution or defense side have regular communication networks, a turncoat expert may have difficulty getting future work.

Even so, to guard against this happening, some well-financed lawyers may "freeze" experts. This practice involves finding and hiring multiple experts and then requiring them all to sign confidentiality agreements that effectively prevent them from working with the other side. This practice is particularly problematic when there are relatively few experts qualified to testify on a highly specialized topic. Though this may be ethically dubious, if the lawyer is reasonably careful, it will be difficult to prove that this practice violates any legal or professional conduct rule.

A lawyer's just-mentioned duties in discovery (Federal Rules of Civil Procedure 26(a)(2), (b)(4)) are also highly relevant to the lawyer-expert relationship and the expert's litigation experience. Like lawyers, forensic linguists and other experts can choose the cases they want to take on. Unlike lawyers, experts can also—in theory—withdraw from a case whenever they want, for any reason or none at all. Nonetheless, in a US federal civil case, and in most state cases, the lawyer is required to disclose, well before the trial, the identity of expert witnesses whom the lawyer "may use at trial" (Federal Rules of Civil Procedure 26(a)(2)(b)). The "may use at trial" language is critical, as the disclosure requirement does not cover experts who are initially contacted but then discarded. In all circumstances relevant here, once identified as a witness who may testify at trial, the expert will be required to provide a comprehensive written report that will be available to the other side and the court. The report must include all of the expert's opinions, together with bases and reasons for the conclusions drawn; all facts and data considered by the expert; all relevant exhibits; and the expert's qualifications and case experience within the last four years. The lawyer, in collaboration with the expert, has an ongoing duty to supplement the report upon learning that "in some material respect the [report] is incomplete or incorrect" (Federal Rules of Civil Procedure 26(e)). Significantly, draft reports do not have to be disclosed.

The opposing party has the right to examine under oath before trial (i.e., "depose") "any person who has been identified as an expert whose opinions may be presented at trial" (Federal Rules of Civil Procedure 26(b)(4)(a)). As written, this right to depose a disclosed expert seems absolute: that is, once triggered, it is apparently not abrogated even if the expert has changed his or her mind about participating or has done further research and reached a different conclusion. Given the broad range of the already-required disclosures, just about everything that the expert has done will be fair game for questioning at the deposition. Even lawyer-expert communications can be inquired into if they relate to expert compensation, facts, data, or assumptions that the lawyer conveyed to the expert and that the expert took into account (Federal Rules of Civil Procedure 26(b)(4)(C)). If the opponent finds the deposition testimony useful, the opponent may use the transcript at trial or conceivably compel a subsequently reluctant expert to testify in person. The filing of a report thus means that the expert is in for the duration of the case, if the opposing party insists on it.

There are similar reciprocal disclosure requirements in federal criminal cases (Federal Rules of Criminal Procedure 16(a)(1)(G), (b)(1)(C), (c)), although there is no comparable right to take depositions. Prosecutors and defense lawyers must make disclosures about expert testimony that they "intend to use" at trial. The specified content of these disclosures is largely identical to what is required in civil cases. The principal difference—a highly significant one—is that the disclosure is written by the lawyer, with no report from the expert required. Nonetheless, as in civil cases, this can impose practical limits on experts' options to change their minds or extricate themselves from the case.

To illustrate these rules, one of us (Conley) was asked on two separate occasions many years ago—once in a civil case, once in a criminal case—to perform a linguistic analysis that the lawyers hoped would support their client's position. Conley thought the issues in each case were significant and shared the lawyers' opinion that his research might advance the client's cause. But in each case, it did not, and he so advised the retaining lawyers—orally. Because his participation had not yet been disclosed, he simply remained silent thereafter. Because he never reached the point of becoming an expected witness and filing a required report, the lawyers had no obligation to disclose his unhelpful findings. However, if, for example, he had reached an initially supportive conclusion, his work had been disclosed, and then further research had upset his initial findings, he probably could not have simply walked away. The new findings would have had to be disclosed in a supplemental filing, undoubtedly attracting the attention of the other side, and he might have been captured by the adversary system.

A further interesting question that Conley did not confront is whether he might have changed sides. In the criminal case, he was retained by the defense, but suppose that it had been the prosecution. Suppose further that his research had undercut the prosecution's case and supported the defendant's innocence. Could he have volunteered his research to the defense lawyers, or perhaps submitted it directly to the court? As discussed above, there seems to be no clear legal or ethical barrier to doing so. In fact, an expert in that position might feel a strong ethical duty to prevent the conviction of an apparently innocent person. (The prosecution might also have a duty to disclose such exculpatory evidence to the defense, but that is a complicated legal question well beyond the scope of this chapter.)

This case highlights the fact that, although linguists and lawyers may work closely together, they are subject to entirely different sets of ethical norms and regulations. Linguists doing pure research are free to determine at any time that their project is not worthy of further work or publication and abandon it. That freedom may not be available in the forensic context. This is not the only role difference between lawyers and linguists. Whereas lawyers are expected to be zealous advocates for their clients in the courtroom, linguists working as experts are expected to be neutral purveyors of knowledge. Outside the courtroom, however, linguists may choose to be zealous advocates about issues related to their research. For example, they may advocate for legislation to protect a language whose speakers they are studying, or they may use their research findings as evidence when advocating for a desired political outcome, such as criminal justice reform. Such activism may, however, have an impact on how a linguist is perceived and evaluated as a legal expert. Lawyers who seek to hire a forensic linguist for their case must therefore consider not only the expert's expertise but also their reputation both inside and outside of the courtroom. In the next section, we examine the professional and ethical standards that may apply to forensic linguists.

The Professional and Ethical Duties of Forensic Experts

The following sections outline and critically discuss existing ethical guidelines for forensic linguists. Before doing so, it is important to mention that the field of linguistics does not require any licensure or special training to serve in the legal system. There is no professional or institutional body that defines or qualifies a linguist to serve as a legal expert. Linguists who serve as experts or consultants for cases typically do so because their specific area of expertise lines up with a language issue that could benefit a party in a case, such as proficiency in a particular dialect or technical acumen in analyzing acoustic patterns that can be used to help identify speakers.

Given the lack of professional licensure or qualifications for linguistics experts, there are scant ethical guidelines for those serving in such a capacity. In a 2006 book, Roger Shuy shone light on the lack of ethical guidance for consulting forensic linguists and some troubling practices that were resulting from that lacuna, including problematic relationships with attorneys, compensation issues, and practicing "experts" lacking sufficient expertise to be reliable. Three years later, Shuy and others convened a symposium entitled "Ethical Issues in Forensic Linguistics Consulting" at the annual conference of the Linguistic Society of America. The symposium culminated in a special issue of *The International Journal of Speech, Language and the Law* that was devoted to ethics in forensic linguistics. The symposium and special issue addressed such fundamental questions as what makes someone an expert and whether an expert's opinion should convey the whole truth of one's findings, including what the analysis could not reveal (Shuy 2009).

Soon thereafter, several academic societies developed ethical codes for linguists serving as legal experts: e.g., the Linguistic Society of America (LSA) (2011); the International Association for Forensic Linguists, now the International Association for Forensic and Legal Linguistics (2018); and the International Association for Forensic Phonetics and Acoustics (2020). Most recently, in a new edition of *Speaking of Crime*, Solan and colleagues (Solan, Tiersma, and Gates 2024) have reviewed newly developed computational techniques for forensic voice analyses and the best practices that some of the linguists doing this work have proposed. In the most comprehensive of these proposals, Morrison and Thompson (2017) offer 13 recommendations that are both methodological and ethical—e.g., that linguistic experts should document their decisions and make their notes available to all parties. As will be described in the following sub-sections, the ethical codes provided by these professional organizations, as well as writings about ethics in forensic linguistics, tend to concentrate on a few primary issues: qualifications and expertise, impartiality and objectivity, and confidentiality and conflicts of interest (Stygall 2009).

Qualifications and Expertise

Forensic linguistics literature emphasizes that linguists should be qualified to give expert testimony on the specific subject at issue in a case. This is only reasonable: One would not expect an expert in American literature to be qualified to provide expert testimony on forensic acoustics. Judges and attorneys are typically not aware of these fine-grained nuances in linguistic expertise. Discerning one's suitability for a particular case is thereby an ethical obligation that lies with the linguist themself, as "[i]ncompetent testimony is unethical" (Butters 2009: 23). Forensic linguists should therefore immediately disclose when their specialized expertise does not align with the language issue for which an attorney is seeking an opinion.

Fairness and Impartiality

Forensic linguistics guidelines assert that those who serve as courtroom experts should conduct "fair and impartial" reviews of all the materials provided to them (Stygall 2009: 262). "Fairness" as a standard is not defined explicitly in existing codes on forensic ethics or in the forensic linguistics literature; it is often used in forensic expert codes of practice as an undefined companion to impartiality (e.g., ACS 2004). In most legal contexts, fairness is considered to be assured by impartiality. A party to litigation, for instance, is thought to have had a fair trial if the judge makes an impartial decision. In the case of expert testimony from a linguist, an impartial review of data is considered to be both a required element and an indication of fairness.

While the idea of "impartiality" as a realistically achievable scientific and legal standard has been widely critiqued (e.g., Geyh 2013; Lacey 1999), maintaining its spirit is important to the practice of forensic linguistics. The standard of impartiality relied on by forensic linguists requires that experts review a body of data without bias toward either side of a case. A helpful test, Shuy (2006) suggests, is to ask yourself if your analysis would be the same if you were doing it for the other side in a case. It can also help to request pleadings and memoranda from both sides of a case as well as the reports of all other experts in order to provide an impartial opinion and to "identify the evidence that might support or refute each party's perspective" (Shuman and Greenberg 2003: 222). It is important that forensic experts follow their own academic fields' research standards when conducting analyses and offering opinions. Linguistic experts should follow the methodological standards upheld within their particular specialty. It is crucial to note that these standards are not identical across linguistics subfields. Linguistic anthropology and sociolinguistics, for instance, often rely on qualitative analysis of ethnographic data such as observations and interviews, whereas corpus linguistics follows more closely the research standards of statistics and other quantitative methods.

Even before the forensic linguist begins the process of analysis, the goal of impartiality may be compromised by the way in which they receive the data from the retaining attorney. For instance, attorneys might only share information that is relevant to or supports their perspective on a case, rather than sharing all information that could be relevant to a linguistic analysis. As revealed in this scenario, a linguist's ethical and scholarly obligations can clash with the obligations and practices of attorneys. As Finegan writes,

> as a scientist, a linguist may be committed to seeking the truth, but as an expert she may be so limited in access to information and perspectives that her ordinary scientific commitment to seek and report the truth may get co-opted by the retaining attorneys working toward a partisan outcome.
>
> *(Finegan 2009: 269)*

Experts must realize, Finegan continues, that their understanding of a case is filtered through the story told by one side's attorneys. He suggests that linguists try to imagine additional stories and contexts for the facts they are reviewing; this ultimately aids in achieving an impartial review of the facts provided. In the end, an expert must remember that their primary obligation is to the trier of fact, not to the attorneys who retained them, nor even to their client. All analysis and resulting reports should thus be completed independent of any legal argument or potential outcome of a case.

There are other ways in which academic codes of ethics may contradict how analysis operates in a legal context. For example, the Linguistics Society of America's Ethics Statement (2019) asserts that all participants in research must consent to their participation and that data storage and sharing should maintain participants' confidentiality. In the case of a trial, however, an expert may be (and in many cases probably will be) analyzing the language of someone who did not consent to such analysis—for example, someone who was wiretapped or recorded without their knowledge or consent. Linguists acting as experts must use their own judgment to navigate these contradictions.

Objectivity

Another frequently mentioned ethical guideline for forensic linguists is the maintenance of objectivity. Shuy (2006), for instance, states that forensic linguists should "analyze the data in the case as objectively as possible" (2006: 125). Ainsworth (2009) similarly asserts that experts have an obligation to be "objective" in their analysis, regardless of any requests or desires from attorneys. The LSA's *Code of Ethics in Forensic Linguists Consulting* (2011) foregrounds the duty of an expert to provide "objective scientific evidence that will assist the court in arriving at its conclusions" (2011: I.A.). In a section entitled "Objectivity and Professional Competency", the LSA concretizes this advice by recommending that forensic linguistic experts report all methods used in their analysis and provide, when possible, a measurable level of certainty of their findings (ibid.: II).

In analyzing these guidelines, it is important to stress that the standard of "objectivity" is not the same as "impartiality". This point is often overlooked. While impartiality involves forming an opinion by incorporating multiple viewpoints beyond one's own, objectivity (a potentially unreachable ideal) relies on "facts" that are not influenced by interpretation (Arendt 2006 [1961]). Given this distinction, it is possible for a linguistics expert to provide an impartial review and analysis of the evidence provided, while not meeting the standard of objectivity as it is typically understood by courts. More specifically, many of the methods that forensic linguists use to reach their findings may fail to reach the standard of objectivity that judges and lawyers expect from "science". Qualitative methods which might inform a linguist's expertise on a particular language variety or context of language use do not yield wholly objective findings, as they rely on the researcher's own, unique experience. Even among quantitative methods, the lack of methodological standardization raises significant ethical questions about objectivity (Solan 2012). There are not even definitive standards for what constitutes adequate linguistic data for providing forensic evidence, other than the LSA guideline to "take due account of…the technical and professional methods available at the time" (2011: II.B.). Thus, while a forensic linguistics expert may unproblematically attest to having reviewed the materials provided in a case impartially, a universal standard of objectivity may be unrealistic.

Confidentiality

The issue of confidentiality is highlighted in all the codes listed above. Again and again, experts are advised not to disclose confidential information that is acquired through their work on a case "unless there is a legal obligation to do so" (LSA 2011). Confidentiality standards become potentially more complex when linguistics experts develop the research they conducted for a case into a publication or presentation. The LSA code (2011) expressly encourages experts to share their work with other linguists in the interest of scientific advancement. Perhaps tellingly, this suggestion is not included in the code of the International Association of Forensic Linguists (IAFL), which otherwise is largely duplicative of the LSA code. This omission may reflect the extreme difficulty experts may face in striking a balance between their ethical and legal obligations toward both research and the law. When disseminating research findings obtained from earlier casework, for instance, both the LSA and IAFL codes require forensic linguists to obtain confidentiality waivers from any party who is a subject of the research or, when that is not possible, to present their results in such a way that obscures the identities of the parties.

As experts, linguists are subject to the complex legal duties of confidentiality and disclosure discussed earlier in this chapter. As researchers, they are bound by project-specific researcher-participant confidentiality agreements, which are enforced by Institutional Review Boards (IRBs). Legal and academic requirements for confidentiality are not necessarily the same. In cases of conflict, legal requirements override those set by IRBs. A legal subpoena, for example, will override an agreement between researcher and participant. This is a vital point that must be shared with potential research participants.

Although the importance of fairness, impartiality, objectivity, and confidentiality is beyond dispute, achieving these goals is often easier said than done. To illustrate some of these challenges, in the section that follows the authors present some of the ethical issues they faced when working real-life cases as forensic linguistic consultants.

The Practical Ethics of Being a Researcher and an Expert

One of the authors (Riner) conducted ethnographic research on death penalty cases in Texas, which involved both working as a de facto consultant for defense teams and interviewing jurors after their cases were concluded (Riner 2016). Riner conducted participant observation in four capital trials and worked closely with defense teams on these trials. After the trials concluded, she interviewed jurors who served on the four cases. Her analysis consisted of triangulating on linguistic details of the trials (using court transcripts and her own fieldnotes) and the jurors' use of language in their interviews. Working in these roles consecutively required her to face difficult questions that directly relate to the principles mentioned earlier. What would be the fairest way to present herself to her research participants who, as jurors, had seen her in the courtroom working alongside defense teams? How could she uphold the research standard of impartiality while working with one side of a case?

Riner addressed these questions by conducting a form of "bracketing" (Tufford and Newman 2010). A common qualitative research practice originating in phenomenology, this method involves recognizing and making explicit the researcher's own biases, preconceptions, and other perspectival filters. It requires researchers to be intensely reflexive about their own experiences and understandings and how these may affect research outcomes. During her fieldwork, Riner immersed herself in the trials and her work with the defense teams to achieve ethnographic understanding. Afterward, she bracketed those experiences to approach jurors as an impartial researcher after the trials' conclusion, almost as if she was beginning a new research project. This entailed bringing into focus the perspectives she gained while working on the trials, making them explicit, and then setting them aside as she conducted the second part of her research. This compartmentalization helped to maintain impartiality. Riner's informed consent process with jurors in the second phase of the research involved talking with them about her engagement with the defense teams in their trials. She then assured them that their interview responses would remain confidential and that she would not share anything they said with the attorneys. Most jurors who spoke with her accepted these terms of participation and agreed to be interviewed. There were a couple of occasions on which attorneys asked Riner to share results from her juror interviews. She denied these requests, citing her ethical obligations to her research participants.

This process did not, however, protect Riner from her interview data being potentially subpoenaed in the future. Although rarely discussed, all scholars carry this risk, as US state and federal laws (e.g., Federal Rules of Civil Procedure 45, Federal Rules of Criminal Procedure 17, and similar provisions in most states' laws) may supersede their Institutional Review Board agreements. When this happens, researchers face the conundrum of having to comply with the law while trying to protect the confidentiality of their research participants. There is no way to completely avoid having one's work subpoenaed. Consequently, researchers have the ethical responsibility to make this risk known to their research participants during the informed consent process. An additional suggestion is to delete all identifiable data files after a period of time to avoid their release in the case of subpoena (Black and Riner 2022). When working as a legal consultant, however, this option is not available, as destroying any data or documentation of analysis could be construed as destruction of evidence (Ainsworth 2009: 285).

The following case illustrates just how quickly these problems can occur. Conley recently advised a scientist who was approached by criminal defense lawyers from another state. The lawyers had found a publication by the scientist and thought it might help them in a case. The scientist agreed to have a conversation with the lawyers to explain the work, and did so. Immediately thereafter, the lawyers mailed a subpoena to the scientist, indicating that they expected him to give remote testimony in their case. The scientist had neither the time nor the inclination to become an expert witness. Conley pointed out to the scientist that the mailing of the subpoena from another state was legally inadequate. Conley prepared a response for the scientist stating his reluctance to testify and telling the defense lawyers to direct all further communications to the legal counsel of the scientist's institution. Probably because they did not want to rely on an unwilling expert, the lawyers simply went away. However, had they pressed the issue, the lawyers probably could have legally compelled the scientist to testify about the published work, despite his reluctance. While it is virtually impossible to compel an expert to come into a case and express an opinion, it is possible to compel a scholar to testify about the contents of their writings that are relevant to the case in the role of an ordinary "fact witness".

Practical Challenges of Being a Forensic Expert in the Courtroom

For All Academics

Linguists are not unique among experts in the challenges they face when they enter the adversarial world of litigation. The underlying challenge facing all such experts is the fundamental conflict between the researcher's usual goals of finding truth and advancing knowledge and the lawyer's unswerving duty of zealous advocacy of the client's position. On top of that, the courtroom audience is not a body of fellow academics but a non-scientist fact-finder—a jury or a judge—charged with making a binary choice that leaves no room for "on the one hand, on the other hand" nuance. The expert will be pressed by both sides to give potentially uncomfortable yes/no answers.

A specific area of concern for the academic will be discovery, more extensive on the civil side but nonetheless significant in criminal cases. Though certain things may be immune from discovery (as discussed earlier in this chapter, draft written reports and some communications with the retaining lawyer, for example), the only safe assumption for experts is that everything they write or say in connection with the case can become known to the other side. Moreover, any previous writing the expert has done that is even remotely related to the case will almost certainly be found. The breadth of discovery also means that promises of confidentiality that were made to previous research participants may be in jeopardy.

Discovery necessarily implies use by the opponent, first in a deposition, perhaps again in a *Daubert* hearing, and then at trial, if the case goes that far. At a minimum, the other side will relentlessly probe the expert's qualifications, evidence, and opinions, looking for inconsistencies, weak points, and gaps between evidence and conclusions. Even minor inconsistencies between the expert's current opinions and previously expressed positions are likely to be exploited in ways that a linguist may find unfair.

In facing these challenges, the expert will be working with and against lawyers whose legal and ethical duties are reasonably clear. Pointing out these challenges is not an argument against legal consultancy work. However, in order to be truly prepared, it is essential that any expert thinking of engaging in such work be prepared for what will and can happen.

For Forensic Linguists

The challenges described above are routinely encountered by experts whose disciplines are regularly represented in the courtroom, such as medical doctors, DNA analysts, engineers, and statisticians. In this respect, forensic linguistic experts are no different. However, because linguistic expert testimony is still relatively rare, lawyers and judges may be dealing with it for the first time, and the courts have a very limited body of case law to instruct them on how to treat it.[1] Moreover, judges experienced in dealing with commonly presented expert evidence tend to be skeptical about "hired guns". Once a linguist has been successfully used as an expert, they tend to get used again, as their previous service and capabilities become known within legal circles (a scenario not unique to forensic linguists). Those linguists who are repeatedly retained may reach the status of "professional experts", an arguably problematic position (Butters 2009: 246). There are several reasons why. For example, they may be perceived as trying to profit from testifying (directly, through fees, and indirectly, through enhanced status that can be monetized), which could potentially compromise their scientific integrity and expose them to scrutiny from the opposing side in a case (Tiersma 1993: 122).

Unlike other disciplines like medicine and psychology that have professional standards for practitioners who serve as experts in court proceedings, forensic linguistics has none. Thus, while the legal obligations may be clear—if unwelcome—there is little in the way of ethical guidance. Nor is such guidance likely to be forthcoming, or even feasible. Linguists who benefit from the current lack of standards may also feel no need to change the status quo or even have an incentive to actively resist change. Resolving ethical dilemmas often comes down to doing the right thing in the circumstances. For this reason alone, the experience of a linguist entering the adversary system may be even more uncomfortable than that of other experts.

Experienced forensic linguist Ron Butters argues that establishing professional standards for forensic linguists would have a positive effect, as the "hired gun" perception of professional experts may in fact stem from a "lack of genuine scientific professionalism, not an excess of it" (2009: 246–247). Clearly, some of the qualification issues raised above could be rectified by establishing a set of practical and ethical guidelines that would help establish who should serve as an expert as well as provide standards for their work. Developing a licensure system for forensic linguists who wish to serve as courtroom experts, however, would be impractical, given the variation across legal and education systems, and financially prohibitive for such a relatively small body of practitioners (Stygall 2009). A licensure system could also prove futile, as it could be ignored by judges, who have the ultimate authority to qualify experts.

As yet there has been no resolution to this long-lasting controversy; currently, scant attention has been paid to the issue in publications popular in the discipline. The net effect is that there are currently no official or institutionalized standards that qualify or regulate someone to serve as a forensic linguistics expert. Their status as an expert in any given case is therefore typically at the discretion of the hiring attorney and is subject to qualification by the court. However, the fact that legal professionals rarely have any knowledge about linguistics means that they can be woefully uninformed about whether a person in question has sufficient qualifications to answer the linguistic questions presented in a case (Butters 2009).

Conclusion

As discussed above, the forensic linguist who wishes to serve as an expert witness may face many ethical challenges, above and beyond what experts from other more regulated and historically established disciplines may face. This is not said, however, as discouragement.

On the contrary, we can attest from experience that forensic linguists can and do provide a valuable service not only to individual clients but to the cause of justice generally. In hopes of encouraging others to consider this fascinating area of work, we offer the following five pieces of critical advice:

1. *Stay in your lane*. Do not offer an opinion in an area unless you have full confidence in your expertise. Opposing lawyers are merciless in tormenting experts who seem to have overstepped their qualifications, even modestly.
2. *Assume everything you write or say will become public*. This assumption is overly cautious, but it is far better to be too careful than not careful enough. A corollary of this is that you should never write anything unless the retaining lawyer requests it or has cleared it in advance.
3. *It is better to understate than overstate*. It is far easier to expand upon an understated opinion than to try to back down from an overstatement.
4. *Never allow your retaining lawyer to persuade you to say something you do not fully believe or to phrase something in a way that makes you uncomfortable*. Lawyers are often tempted to get experts to extend their opinions just a little bit further. Just say no. If you violate this rule, you, the lawyer, and the client will almost certainly regret it.
5. *When in doubt about any aspect of your work, ask the retaining lawyer*. You are the forensic linguist, not the lawyer, so make no assumptions about the legal system.

By sharing our insights, cautions, recommendations, and experiences in this chapter, we hope to motivate qualified linguists to consider the possibility of expert engagement. The work itself is usually intellectually stimulating and sometimes provokes new scientific insights. Although it is important to be aware of the potential risks, the practice of forensic linguistics can also offer a linguist significant rewards, both moral and financial.

Note

1 Conley co-edits a law school casebook called *Scientific and Expert Evidence* (Conley and Moriarty 2020). It has entire chapters devoted to each of those other disciplines, but only a few passing references to linguistics, despite his best efforts to find case examples.

References

Ainsworth, Janet. 2009. "A Lawyer's Perspective: Ethical, Technical, and Practical Considerations in the Use of Linguistic Expert Witnesses." *The International Journal of Speech, Language and the Law* 16, no. 2: 279–291.

American Bar Association. 2018. *Model Rules of Professional Conduct*. Retrieved from https://www.americanbar.org/groups/professional_responsibility/publications/model_rules_of_professional_conduct/.

American College of Surgeons. 2004. "Expert Witness Affirmation." Retrieved from https://www.facs.org/media/korhyekd/ewa_certificate.pdf.

Arendt, Hannah. 2006 [1961]. *Between Past and Future*. London: Penguin Classics.

Black, Steven, and Robin Conley Riner. 2022. "Care as a Methodological Stance: Research Ethics in Linguistic Anthropology." In *Research Methods in Linguistic Anthropology*, 97–124. Edited by Sabina M. Perrino and Sonya E. Pritzker. New York: Bloomsbury Publishing.

Butters, Ronald R. 2009. "The Forensic Linguist's Professional Credentials." *The International Journal of Speech, Language and the Law* 16, no. 2: 238–252.
Conley, Robin. 2016. *Confronting the Death Penalty: How Language Influences Jurors in Capital Cases.* Oxford: Oxford University Press.
Conley, John M., and Jane C. Moriarty. 2020. *Scientific and Expert Evidence*. 3rd ed. Boston: Aspen Publishing.
Conley, John M., and David W. Peterson. 1996. "When Ethical Systems Collide: The Social Scientist and the Adversary Process." In *Recent Developments in Forensic Linguistics,* 345–358. Edited by Kniffka, Hannes, Susan Blackwell, and Malcom Coulthard. Bern: Peter Lang Verlag.
Conley, John M., and William M. O'Barr. 1990. *Rules versus Relationships: The Ethnography of Legal Discourse.* Chicago: University of Chicago Press.
Conley, John M., William M. O'Barr, and Robin Conley Riner. 2019. *Just Words: Law, Language, and Power*. 3rd ed. Chicago: University of Chicago Press.
Daubert v. Merrell Dow Pharmaceuticals, Inc. 1993. 509 U.S. 579.
Duranti, Allessandro, Rachel George, and Robin Conley Riner. 2023. *A New Companion to Linguistic Anthropology.* New York: John Wiley & Sons.
Finegan, Edward. 2009. "Expert Linguists and the Whole Truth." *The International Journal of Speech, Language and the Law* 16, no. 2: 267–277.
Finegan, Edward. 1993. "Ethical Considerations for Expert Witnesses in Forensic Linguistics." *Issues in Applied Linguistics* 4, no. 2: 179–187.
Geyh, Charles Gardner. 2013. "The Dimensions of Judicial Impartiality." *Florida Law Review* 65, no. 2: 493–552.
International Association of Forensic Linguists. 2018. *Code of Practice.* Retrieved from https://www.iafl.org/wp-content/uploads/2018/07/IAFL_Code_of_Practice_1-1.pdf.
International Association for Forensic Phonetics and Acoustics. 2020. *Code of Practice.* Retrieved from https://www.iafpa.net/about/code-of-practice/.
Lacey, Hugh. 1999. *Is Science Value Free?: Values and Scientific Understanding*. London: Routledge.
Linguistics Society of America. 2011. *Code of Ethics for Linguists in Forensic Linguistics Consulting.* Retrieved from chrome-extension://efaidnbmnnnibpcajpcglclefindmkaj/https://www.linguisticsociety.org/sites/default/files/code-of-forensic-consulting.pdf.
Maine, Henry J. S. 1861. *Ancient Law.* London: John Murray.
Malinowski, Bronislaw. 1926. *Crime and Custom in Savage Society.* Totowa, NJ: Rowman & Allenheld.
Merry, Sally Engle. 2006. *Human Rights and Gender Violence: Translating International Law into Local Justice.* Chicago: University of Chicago Press.
Morrison, Geoffrey Stewart, and William C. Thompson. 2017. "Assessing the Admissibility of a New Generation of Forensic Voice Comparison Testimony." *Columbia Science and Technology Law Review* 18: 326–352.
Ramseyer, J. Mark, and Eric B. Rasmusen. 2010. "Comparative Litigation Rates." Harvard Law School, John M. Olin Center for Law, Business, and Economics Center Discussion Paper No. 681. Retrieved from chrome-extension://efaidnbmnnnibpcajpcglclefindmkaj/http://www.law.harvard.edu/programs/olin_center/papers/pdf/Ramseyer_681.pdf.
Richland, Justin B. 2008. *Arguing with Tradition: The Language of Law in Hopi Tribal Court.* Chicago: University of Chicago Press.
Riner, Robin Conley. 2016. *Confronting the Death Penalty: How Language Influences Jurors in Capital Cases*. Oxford: Oxford University Press.
Shuman, Daniel W., and Stuart A. Greenberg. 2003. "The Expert Witness, the Adversary System, and the Voice of Reason: Reconciling Impartiality and Advocacy." *Professional Psychology: Research and Practice* 34, no.1: 219–224.
Shuy, Roger. 2009. "Ethical Questions in Forensic Linguistics: Introduction to Papers from a Linguistic Society of America Panel Presentation." *The International Journal of Speech, Language and the Law* 16, no. 2: 219–226.
Shuy, Roger. 2006. *Linguistics in the Courtroom*. Oxford: Oxford University Press.
Solan, Lawrence M. 2012. "Ethics and Method in Forensic Linguistics." In *Proceedings of the International Association of Forensic Linguistics' Tenth Biennial Conference*, 362–368. Retrieved from chrome-extension://efaidnbmnnnibpcajpcglclefindmkaj/https://citeseerx.ist.psu.edu/document?repid=rep1&type=pdf&doi=57e9aa88d33b594a651d9d840c6fd62b336dcb5d.

Solan, Lawrence M., and Peter Tiersma. 2005. *Speaking of Crime.* Chicago: University of Chicago Press.

Solan, Lawrence M., Peter Tiersma, and Tammy Gales. 2024. *Speaking of Crime*. 2nd ed. Chicago: University of Chicago Press.

Stygall, Gail. 2009. "Guiding Principles: Forensic Linguistics and Codes of Ethics in Other Fields and Professions." *The International Journal of Speech, Language and the Law* 16, no. 2: 253–266.

Tiersma, Peter. 1993. "Review Article: Linguistic Issues in the Law." *Language* 69, no. 2: 113–137.

Tufford, Lea, and Peter Newman. 2010. "Bracketing in Qualitative Research." *Qualitative Social Work* 11, no. 1: 80–96.

U.S. Federal Rules of Civil Procedure. 2022. Retrieved from https://www.uscourts.gov/rules-policies/current-rules-practice-procedure.

U.S. Federal Rules of Criminal Procedure. 2022. Retrieved from https://www.uscourts.gov/rules-policies/current-rules-practice-procedure.

U.S. Federal Rules of Evidence. 2020. Retrieved from https://www.uscourts.gov/rules-policies/current-rules-practice-procedure.

13

THE EXPERT WITNESS AND THE ETHICS OF ADVOCACY

Michael O'Laughlin[1]

Introduction

The purpose of a court trial is to resolve disputes and establish responsibility, including degrees of guilt in the case of a crime. Those matters that can be resolved through further investigation, agreement, or a guilty plea do not end up being tried before a judge or a jury. Fundamental to the Anglo-American legal system is the assumption that a just resolution can be reached if parties to the dispute merely present their sides of the story to a competent, unbiased authority. This method for dispute resolution and punishment goes back as far as Greek antiquity (Hazard and Remus 2011; Marsh 2001). A trial, in this system, is an oppositional battle with winners and losers. It is expected that the lawyers for both sides will adopt an adversarial posture and do all in their power to win their case rather than to act as neutral inquirers after truth; this is considered to be the role of the arbiter: the judge or a jury.[3]

The avenue for the presentation of these often-opposite views is witness testimony, and there are two types of witnesses: fact witnesses and experts. A fact witness testifies to what they perceived with their senses, whereas an expert is able to give their opinion, based on their special credentials, and on any reliable science they can cite, or tests that the expert has conducted. Expert testimony is essentially opinion testimony, but that opinion is based on the expert's professional knowledge of a field, such as medicine or engineering, and sometimes on testing (Grimm 2012: 16). A fact witness might be neutral or might be biased, while experts are assumed to be objective.

In part because the adversaries in a US American court are intent on obtaining the most favorable outcome for themselves, rather than on reaching reconciliation or on discovering the truth, legal outcomes can sometimes seem counter-intuitive, unfair, or wrong. One reason that the results of a trial can be surprising is that juries are unpredictable. However, there are also other reasons. The adversarial approach itself can be responsible for many unexpected or unfair outcomes. Trials are often won by the side which is best prepared, or which has the best lawyers, rather than the side which is in the right. This system is easily criticized by those who do not accept the premises upon which it is founded. The system certainly is not perfect. Yet the US system of law has served its ultimate purpose, and many advances in society are due to court decisions.

DOI: 10.4324/9781003391074-15

Experts before the Court

The admissibility of expert testimony in federal court is governed by the Federal Rules of Evidence, which were shaped by the outcomes of three signal cases: 1) *Daubert v. Merrell Dow Pharmaceuticals*, 2) *General Electric Co. v. Joiner*, and 3) *Kumho Tire Co. v. Carmichael*.[4] *Daubert* delineated the criteria by which a test could be considered "scientific", and *General Electric* determined the appellate standard of review of a trial court's admissibility ruling. Finally, after *Kumho Tire*, it is recognized that experts need not be laboratory scientists. From 1993 to 2000, the changes produced by these cases constituted a "revolution", according to Justice Paul Grimm (2012: 13). Under *Kumho Tire*, any person with specialized knowledge of a topic—whatever that topic may be—can give their opinion before a court. Under this broader policy the examination of the credentials of an expert called to the witness stand are even more important than it had been previously. The judge acts as a gatekeeper, and can conduct a voir dire hearing outside the hearing of the jury regarding the education and experience of any purported expert. Under *Kumho Tire*, there can be experts on any number of topics. Experts are critically necessary in a wide range of cases. There are too many topics that come before the court which deal with matters beyond common knowledge for lawyers and judges to proceed without the help of specialists.

Although experts are almost always chosen and paid for by one of the parties to a case, they are expected to remain objective and focused on assisting the court or jury in understanding the matter in dispute. Admonishments to do so include: "An expert witness should never become solely one party's expert advocate nor a 'gun for hire.' Rather, an expert witness should be an advocate of the truth with testimony to help the jury and the Court reach the ultimate truth" (*Van Blargan v. Williams Hospitality Corp.* Puerto Rico. 1991. 754 F. Supp. 246, 249). An expert, in theory, is to act as a neutral source of knowledge and opinion. "While a lawyer is an advocate for his client, an expert witness is supposed to be a source of knowledge and opinions that will aid the Trier of fact" (*English Feedlot, Inc. v. Norden Labs, Inc.* Colorado. 1993. 833 F. Supp. 1498, 1501).[5]

However, these rules are not especially easy to follow. Experts may regard themselves as objective and neutral, but that stance will be quickly challenged once they appear in court. Although experts can write a report laying out the data on which their opinions are based, they cannot present any evidence on the witness stand except by giving answers to questions posed by an attorney. Attorneys ask questions which fit into their strategy of the case. Then opposing counsel asks questions designed to show that the expert is mistaken, or not to be credited. A cross-examination often seeks to twist the statements of an expert, to belittle the expert, or to provoke the expert to anger. In fact, the expert is hardly treated or allowed to be interpreted in a neutral manner. Rather, a judge or jury is given a two-sided look at the expert, one pro, the other con.

The ethical tightrope walked by an expert witness in the United States is not easily understood nor articulated. Rather, experts are more often scorned by commentators, either for being "hired guns", or for coming up with outlandish pseudoscience (Richmond 1997). Why is a judge in a civil case presented with two highly qualified experts who do not agree on a cause of death, or the reason a building has collapsed? The title of a *New York Times* article sums up this perspective: "In U.S., Expert Witnesses Are Partisan" (Liptak 2008). The *Times* article contrasts the US system of dueling experts with that of Australia, where experts communicate with and criticize each other, and then testify together during the trial. This model has a curious name: "hot tubbing" (Butt 2023). Another better-known alternative to the US approach is the practice of Civil Law countries, which employs an Inquisitorial system instead of an Adversarial one. There a *Juge d'instruction* gathers information and calls witnesses. Attorneys, by contrast, play only a minor role, such as suggesting topics of inquiry. In court, it is the judge who questions the witnesses for the simple purpose of arriving at the truth. The rights of the accused are secondary (Merryman Pérez-Perdomo 2007). As then Yale professor John Langbein famously wrote of the advantages of the European system over that of the United States:

> The European jurist who visits the United States and becomes acquainted with our civil procedure typically expresses amazement at our witness practice. His amazement turns to something bordering on disbelief when he discovers that we extend the sphere of partisan control to the selection and preparation of experts. In the Continental tradition experts are selected and commissioned by the court, although with great attention to safeguarding party interests. In the German system, experts are not even called witnesses. They are thought of as "judges' aides". At the American trial bar, those of us who serve as expert witnesses are known as "saxophones". This is a revealing term, as slang often is. The idea is that the lawyer plays the tune, manipulating the expert as though the expert were a musical instrument on which the lawyer sounds the desired notes.
>
> *(Langbein 1985: 835)*[6]

Is it that bad? Opinions will differ, but most will admit there is a grain of truth in Prof. Langbein's scorching criticism of experts under the US adversarial system. The grain of truth is that no expert stands alone, but only speaks in response to an attorney's questions, and to make an attorney's points. Money is another issue for some, but not all experts. Some experts may become infamous for being willing to say anything if the price is right. Other, more ethical experts can be trusted to withdraw from cases where their observations or opinion do not support the position of the client.

To sum up, there is widespread agreement that experts are essential to many cases. The information they provide and their opinions are delivered according to standards of both objectivity and neutrality. Bias or advocacy are reasons for attacking or excluding a witness's testimony. Despite these standards, there is a chorus of criticism of experts focused on the two interrelated problems of bias and advocacy. Bias, often unconscious or inherent, is probably harder to overcome and eliminate than open advocacy (Chin 2019: 21).

However, the very nature of the adversarial system itself may make it difficult for an expert to appear under a flag of neutrality, since the attorneys in the case are invested in their own one-sided presentations, and any expert is easily dragged into the fight. As Justice Grimm (2012: 15–16) notes:

> The role of the expert witness [...] is itself shaped by the demands of the adversary system, rather than some objective search for truth [...]. In such a system, it is naïve to expect that experts will be hired to provide testimony that is not supportive of the case of the party that hired them.

There is therefore always a tension between objectivity and bias, with the expert trying to make his or her opinion clear, while parrying cross-examination attacks on a presentation which may have been one-sided. All experienced experts know that these attacks are coming, and they are therefore careful to say little or nothing that reflects bias or seems to openly advocate for the position of the client.

Where Is Advocacy Often Found?

If advocacy is defined as "the act of pleading for, supporting, or recommending" (Barnett 2017: 1), then advocacy is much easier to avoid than bias. Bias is an attitude, whereas advocacy is an act. Advocacy is within the job description of the attorney, and everything works better if each actor in the court proceeding only does their own job. An expert gives an opinion, the attorney advocates for the client, and the judge (or jury) must then make a decision. However, in cases where an expert exhibits a much deeper knowledge of the question at hand, it is sometimes hard not to simply adopt the expert's conclusion. One area where this can easily occur is that of medical doctors and cases of competence, criminal responsibility, child custody, defenses based on "battered women's syndrome", etc. Too often judges allow a doctor to make what are legal (or moral) decisions that should be made by the court alone. As Paul Appelbaum, MD, writes in "The Medicalization of Judicial Decision-Making":

> In effect, the judge who asks for or tolerates psychiatric testimony about the ultimate issue in a hearing in which a legal-moral determination lies at the core of what the court must do is acquiescing in the medicalization of a profoundly moral question.
>
> *(Applebaum 1999: 2)*

This observation is based on considerable personal experience as an expert witness. Too often Dr. Appelbaum had testified about someone's medical situation, only to be asked by the judge a question such as, "But doctor, is he competent or not?" (Appelbaum 1999: 3). For the doctor, this was a legal, not a medical matter. Non-medical experts are also sometimes in a similar position of advocacy, or near advocacy, often due to their greater knowledge of the matter at hand.

Is There Ever a Place for Advocacy?

The ethical question of advocacy for experts is best addressed within the larger subject of professional ethics in general. One reason is that the ethics of experts has not yet developed beyond the initial stage of establishing basic rules—and the airing of many complaints that experts do not adhere to them: the ethics of experts has actually received relatively little attention. The topic of ethics for many other professionals, as presented in law schools, medical schools, business schools, and other similar training programs, has advanced to a more considered and developed level. Many professionals learn ethics in their schools by studying cases and situations. One learns not only the rules but also how to apply those rules in different contexts. One discovers that one rule may have priority over another and that sometimes an acceptable solution is reached by going beyond the present standard into new territory. Students are taught to become moral agents and are empowered to make their own decisions. Sometimes the only resolution is the least-worse option.

Rules might need to be bent. Students must learn to apply the rules—or not apply the rules—to arrive at a reasonable and equitable solution. Of course, these professionals-in-training must be able to defend their choices. The subject of professional ethics is understood to focus on developing the moral reasoning capability of the student and to prepare that student for a world in which hard choices sometimes need to be made. This training is fundamental and necessary to joining the professional class, because one key difference between an employee and a professional is that if there is a problem, the employee's obligation is to inform their boss or supervisor. By contrast, a professional must very often sort things out and find a solution. Most of the time, one can apply the rules, but if the rules do not fit the case, then some creative thinking may be in order (*Stanford Encyclopedia of Philosophy* 2020; Espen and Schiano 2014; Barnes, Christensen, and Hansen 1994).

This principle lies at the heart of professional ethics today: one must adhere to the rules, but there are times when mere adherence to the rules does not bring one to a practical, reasonable, just position or result. At that point, a well-trained professional must engage in moral reasoning to determine how to properly act. Judges are faced with such scenarios nearly every day, ensuring that outcomes are reasonable, fall within the established legal norms, and are fair.

Just as this type of reasoning is undertaken by judges and attorneys,[7] it should also be part of the work of experts. Most of the time the norms and rules are to be respected, but one cannot exclude the possibility that some rule might need to be set aside in favor of a higher principle. I will illustrate this with a case of my own in which I did not follow the normal rules, but advocated for a solution that probably saved the client's life.

Commonwealth v. al Hussaini

In 2011 an indictment was handed down in Massachusetts against Hussain al Hussaini for Assault and Battery with a Dangerous Weapon, and Mayhem. He was accused of having assaulted another homeless man with a broken bottle and slashed the victim's face. According to the police report, al Hussaini stated that he was getting harassed by several men at Father Bill's Shelter. They called him "Bomber", and "Terrorist", and had beat him up a month earlier. Hussaini said that he was fed up and thought that if he threatened them with a broken bottle, the harassment would stop.

Background and Personal History

I was brought into this case as a cultural expert. Cultural issues can be presented and considered as part of cases involving foreigners or immigrants, in part because otherwise judgments will be made based entirely on the dominant culture's assumptions (Renteln 2004: 6; Einesmann 2010). There is a Ninth Circuit case of Ineffective Assistance of Counsel for not supplying cultural information to the court: *Mak v. Blodgett 970 F2d 614*.

After reading the case materials, I interviewed Mr. al Hussaini. He told me that he was 47 or 48, that he was Iraqi and a Shiite who had grown up in a small town. Mr. al Hussaini's father was a teacher of calligraphy in an art school. The family was not especially religious or different from the rest of the people in their area.

Disaster overtook the family when he was five or six years old. Evidently his father was denounced for having said something critical of Saddam Hussain or the government. The entire family was taken to a prison. He has a clear memory of watching the interrogation of his father there. In order to force him to talk, the interrogators brought his mother over and began to mistreat her in front of her husband. His father then said something to them, for they turned back to him, put him against a wall, and blew his brains out. His mother was then killed, as well. After this horrific event he has no idea what happened to his two brothers and his sister; he never saw them again, but assumes that they died also.

Imprisonment and Torture in Iraq

Mr. al Hussaini was then thrown into an unlit cell that seemed to be four by five meters in size. In this cell there were 20 to 30 men and boys at any given time. There was no room to stretch out flat and no way to know if it was day or night. Food was slid in on a tray under the door. The number of times they were fed each day depended on the mood of the guards. Water came in a bucket. The door was opened only to put people in, take them out, bring in the water bucket, or to empty another bucket that served as a communal toilet. Dead bodies were often left in the cell for three or four days at a time. Sometimes the prisoners made space so that the older ones could lie down flat on the floor. Most often people lay with their back on the floor and their legs up against one of the walls or lay leaning against each other. No one could walk in any direction without stepping on someone else. The prisoners in the cell were infested with lice. When the door was opened, the light from outside momentarily blinded the prisoners. The guards would then enter, holding electrical wires that they used as whips, striking the prisoners at random. If the bucket needed to be emptied, a prisoner was made to carry the bucket out but was whipped if he carried it or emptied it awkwardly because of being unaccustomed to daylight.

Those who were taken out of the cell sometimes never came back. Mr. al Hussaini was removed periodically, either to be tortured or to be raped. During his torture sessions he was sometimes whipped with the electrical wires, sometimes lifted to the ceiling by one leg (which left a broad scar around one of his ankles that I myself saw), sometimes given electrical shocks to his ears or his genitals (which left him without hearing in one ear), and sometimes injected with gasoline (which produced swelling in his extremities, hallucinations, and unbearable hot and cold sensations). During the first years, while he was still one of the younger prisoners, he was also taken out of the cell and raped by the guards both orally and anally. These sexual assaults produced a sense of self-loathing and low self-esteem in him that remained with him in his later life.

Mr. al Hussaini was kept in these inhuman, extreme conditions continually for 12 years. When he was 18 years old, he was taken out to perform military service. He was loaded onto a truck and taken to the border. The Iran-Iraq War was then in progress, and both sides were using untrained youths as combatants. At the front, Mr. al Hussaini was given a Kalashnikov and put in a battle station facing the enemy. Conditions at the front were terrible: fighting raged, and uncertainty reigned. If someone left to get supplies, food, or water, they often did not return. Conscripts died all the time, for all sorts of reasons. Most of his time at the front he spent behind sand shelters listening to the sounds of battle and hoping and praying for death. Dead bodies were everywhere. Some of the conscripts shot themselves so that they would be taken off the battlefield as wounded.

When the conflict drew to a close Mr. al Hussaini was returned to prison and subjected to the same extremely harsh conditions as before. By this point his mental health was failing and he suffered flashbacks of the terrors he had lived through. He remained in prison for two or three more years, until the Gulf War in 1991. His prison was liberated by American forces and he was removed to a refugee camp in Saudi Arabia. He lived in the desert in a tent surrounded by barbed wire for three years. The camp was eventually taken over by the United Nations and a resettlement process began. Mr. al Hussaini was eventually selected for resettlement in the United States.

Thus, in 1994 he was loaded on a plane and taken from Saudi Arabia to New York. From New York he was sent to Boston, where he was to be resettled. The sponsor was a group called the International Rescue Committee (IRC). The Iraqi refugees were put in an apartment, given some money and food stamps, but not a lot of additional help. They had to fend for themselves. The first job Mr. al Hussaini found was as a dishwasher at the Middle Eastern restaurant. While at this job he met an Iraqi who had been in the United States a bit longer than he had. This man suggested that they go together to Oklahoma, where life was very cheap compared to Boston, and where he had a friend who could get them work.

So it was that Mr. al Hussaini went by bus to Oklahoma. They were met at the station by the friend and told that the next day they would meet a Middle Eastern man who owned five hundred houses and was in constant need of workers to clean, paint, and repair his properties. Mr. al Hussaini was hired and found himself living in an apartment owned by this man, Dr. Anwar Abdul. He also worked as a cleaner at a restaurant. Each night he was locked in the restaurant and told that as long as the restaurant was clean in the morning, he could sleep for part of his ten-hour shift. He was a hard worker. He cleaned for six or seven hours and then was able to sleep for the rest of the time. This was a very happy time in Mr. al Hussaini's life—he had work, he had money, and the future seemed promising.

Persecution in the United States

Mr. al Hussaini was thus busily employed in Oklahoma City in 1995. That year, on April 19, Timothy McVeigh and Terry Nichols blew up the Murrah Federal Building. Mr. al Hussaini did not immediately realize that anything unusual had happened that day. He heard another Iraqi say something about a bomb in Arabic, but he was not able to follow any of the excited commentary swirling around Oklahoma City because his English at the time consisted of only the few most necessary words. Then came the announcement that everyone was to go home after only a half day of work because the boss's former wife had been in the building and was among the missing. Sometime later he was told by someone that his picture was on the television. He could not understand the English, but he could see an image on the television screen of himself running. He was then told by someone who could understand the announcement that they were saying that he did the bombing.

His employer met with him later and said that he did not believe that he was involved, but his daughter worked in his office and she was upset by seeing him around, so he would have to leave the job and the apartment. He did leave and began sleeping in his car. The owner at the restaurant job was sympathetic and said that he was willing to help him, but told him that now he could not go in the front door but would have to enter through the rear. Although the police never interviewed him, he was feeling very frightened. As he would eventually learn, before the capture of Timothy McVeigh, a reporter named Jayna Davis from the local television station, KFOR, had begun reporting on a supposed Middle Eastern conspiracy that was behind the bombing. She tied Mr. al Hussaini in because of his resemblance to a drawing being circulated by federal authorities.

He quit the restaurant job. He was now living in his car, and, unfortunately, he began to drink. He had never drunk alcohol previously. Alcohol is forbidden by Islam, but he began to drink, eventually spent all his money, lost his car, and lost his license due to a driving while intoxicated (DWI) charge. Meanwhile, Jayna Davis and her producer continued to expand their investigation, based on the conviction that he was at the heart of the conspiracy. Davis would eventually conclude (this was some years later) that the Oklahoma City bombing was funded by Osama bin Laden. The FBI never took her seriously, and when KFOR was bought by the *New York Times* a year and a half later, she was fired. Despite losing her job, Jayna Davis went on to write *The Third Terrorist: The Middle East Connection to the Oklahoma City Bombing*. In it she concluded that the figure seen with Timothy McVeigh in a brown pickup or at a strip club before the bombing was Hussain al Hussaini, and that the group of Iraqis working for Dr. Anwar Abdul was deeply involved in the bombing.

The official investigation into the Oklahoma City bombing was the largest criminal investigation in American history, involving 28,000 interviews and one billion pieces of information. While Timothy McVeigh and Terry Nichols said in court that they acted on their own, McVeigh's own lawyer, Stephen Jones, later revealed that McVeigh told him there were others involved, and that McVeigh was not the mastermind. The government made very little information available to the defense concerning any other conspirators or any foreign connection, and that led to a plethora of conspiracy theories.

"Al Hussaini as the Third Terrorist" was a story that had found an audience among right-wing elements distrustful of government. On the back cover of *The Third Terrorist* one finds an endorsement from right-wing author Glenn Beck. It reads:

> When you read this book, you are going to be convinced that it is the truth. We need a 9/11 Commission on the Oklahoma City bombing. The evidence is overwhelming that this [terrorist attack] actually has ties to the Middle East and Saddam Hussein. And I want to know what happened!

Likewise, there are stories and commentary on the internet from Fox News and other such sources that loudly endorse the suspicions of Jayna Davis. A Google search by me in 2023 yielded 362,000 hits for the name "Hussain al Hussaini" plus the words "Oklahoma City bombing". Some of these internet articles are alarmist. A representative news item can be found at: http://www.canadafreepress.com/index.php/article/34299: "Homeless" Man Hussain Hashem al-Hussaini is "John Doe #2 in Oklahoma bombing". Here journalist Doug Hagmann hails Jayna Davis's dogged efforts to implicate Hussain al Hussaini in the bombing and describes his own investigative efforts in the Boston area after Mr. al Hussaini was arrested in the case I worked on. In actual fact, Hussein al Hussaini was not a co-conspirator plotting with Timothy McVeigh and Terry Nichols. During one of his hospitalizations, and after the publication of *The Third Terrorist*, he was visited by the FBI and told that he was not a suspect.

As Jayna Davis and her conspiracy-minded associates continued their quest to prove that, among other things, 9/11 could have been prevented if Hussain al Hussaini had been thoroughly investigated, the man himself increasingly found himself adrift. He went briefly to Texas. He contacted the International Rescue Committee there and was given a job at Dulles International Airport making pizza at a restaurant. When he told the IRC that he wanted to go back to Boston, they placed him in the same job at Logan Airport.

Unfortunately, he only lasted there for eight months. His mental health issues were becoming more debilitating, and the story continued to follow him. He began working casual jobs, such as dishwashing and pumping gas, but he would be eventually asked about all the material on the internet. The questions usually asked him were, "Have you been out of state? Have you been to Oklahoma?" His reaction was to deny he was a terrorist, but then he did not return to work the next day. He had trouble facing people who knew his secret, and more and more people knew. He was a hard worker, always ready to work extra hours, and his co-workers asked him if he was sending the money overseas to terrorist organizations. According to him, Jayna Davis began claiming that the box cutters used by the 9/11 hijackers were supplied by him. Increasingly, he saw no way out. Instead of jobs, his life became a series of hospitalizations. He could not function without medication, and his self-medication with alcohol tipped him toward alcoholism.

However, the many medical interventions eventually had a positive effect. For a while it looked like he could have a normal life. Mr. al Hussaini met a Jewish woman named Lisa Cohen at one of his gas station jobs and they began to go out. Eventually he moved in with her and her parents and lived with them for seven or eight years. With their help he was able to get hired doing construction work. He then got a commercial driver's license and drove a truck for a construction and landscaping company. This job lasted two years. He would have stayed longer, but the owner's son began making remarks about him "sending money overseas to kill us". Mr. al Hussaini had a daughter with Lisa, but eventually his relationship with her family began to sour over the lingering suspicions that he was a terrorist. Lisa's older brother was separated or divorced from the mother of his son, and this woman began to complain that her son was not safe around Mr. al Hussaini. She even telephoned Jayna Davis, who told her that al Hussaini was indeed the "baby killer" from Oklahoma. Because these issues came up in the family, Mr. al Hussaini stopped living with the Cohens, although he continued to maintain a friendly relationship with Lisa Cohen and saw his daughter on a regular basis.

Persecution at Father Bill's

After leaving the Cohen home Mr. al Hussaini was homeless and lived in shelters. In Quincy many homeless were served by Father Bill's Place. This shelter was used as a base by Mr. al Hussaini, yet his time there was clouded by friction with a group of homeless men. One of the reasons for this friction was that Mr. al Hussaini was an Arab and a foreigner. He spoke with a heavy accent and his name was similar to that of the hated Saddam Hussein. After the September 11 attacks, many innocent Muslims were persecuted (Barrett 2007), and Mr. al Hussaini was no exception. Another problem was that Mr. al Hussaini had sleep apnea. This condition causes one to be unable to breathe, which then produces snoring and a periodic, loud gasping for breath, so that others often cannot sleep in the same room. Some of the homeless people at the shelter wanted him to leave, voluntarily or involuntarily.

The persecution of Mr. al Hussaini at Father Bill's began with spontaneous acts of cruelty, such as beatings. They would also call him names and spit in his coffee, then force him to drink it. Several times his medicine was stolen, which caused a serious emergency for Mr. al Hussaini, since he was very dependent on his medication to remain stable. These sentiments against Mr. al Hussaini increased once the embittered homeless men at the shelter learned of the suspicions that Mr. al Hussaini was the "Third Terrorist" responsible for the Oklahoma City bombing. His persecutors not only began to call him a bomber and a terrorist, they printed out the stories about him from the internet, using a computer at the public library, and posted the stories on the streets of Quincy.

The campaign against Mr. al Hussaini culminated in false accusations being made against him. In order to get him arrested, his persecutors gave some drugs to a woman at the shelter in exchange for her bringing an accusation that he had "grabbed her ass". At the shelter this accusation was given little credence because the staff make hourly rounds and he was known to be sleeping (and snoring loudly) at the time when he was supposedly molesting the woman, so *she* was ejected from the shelter. However, she then went to the police and he was charged with Indecent Assault and Battery. This charge was eventually dismissed when the woman failed to appear in court, but he was held in jail for four months awaiting trial.

The Department of Mental Health eventually concluded that being at Father Bill's was causing him serious problems and found him a bed at a sober house. This resolved many of the problems that Mr. al Hussaini had. He also began to get counseling at the Boston Medical Center which was geared specifically to victims of torture, and this therapy was of enormous value to Mr. al Hussaini.

Clinical History

According to Mr. al Hussaini, he had been diagnosed with post-traumatic stress disorder, he was bipolar, and suffered from paranoia, flashbacks, depression, and anxiety. He had heard voices and talked to himself. He had been given electric shock therapy. He had been hospitalized at the Arbor Hospital alone 15 to 20 times. He calculated that he has had a total of 53 psychiatric hospitalizations. He had a social worker and a Department of Mental Health worker. He was an alcoholic currently in recovery and living in a Department of Mental Health sober house. He had been sober for eight months when I wrote my report.

Advocacy

In my conclusion I decided to go beyond reporting the tortured and unfortunate history of Mr. al Hussaini. I decided to include a strong recommendation. This was indeed advocacy, but I felt it was warranted.

I wrote that it appeared that the government had a fairly strong case against Mr. al Hussaini. However, I said I hoped that this case would not go to trial, and that there would be no plea nor conviction. I said I believed that this was a unique case, one in which mitigating circumstances warranted prosecutorial discretion on humanitarian grounds. I listed several reasons why I believed such an unusual resolution to this case was warranted:

The most important was that Mr. al Hussaini was mentally ill. He had been hospitalized in psych wards a total of 53 times. He was completely dependent on his medications and had exhibited very troubling behavior when he is not taking his meds. His mental illness could be attributed to his sufferings, which were horrific, and were thus understandable. He had borne unspeakable treatment and torture, and this had left him completely disabled. I argued he should not be held to the same standard as other people.

Secondly, *the consequences of a conviction in this case would go far beyond a jail sentence.* Once convicted of a violent felony, Mr. al Hussaini would then be subject to deportation. At that time anyone who might require special consideration by the Immigration Courts was not likely to get it, as was illustrated by the case of Mark Lyttle, a mentally ill man who was deported in 2008, although he was a US citizen. Mr. Lyttle was deported because it was incorrectly noted in the immigration records that his real name was Jose Thomas and that he was brought to the United States from Mexico when he was three. After arriving in Mexico, Mr. Lyttle was then deported by the Mexican authorities to Honduras. He was finally identified by his family in Guatemala in 2009 and returned to the United States. His story was told in the *New Yorker* (Finnegan 2013). The title of the article, "The Deportation Machine", succinctly captures the danger Mr. al Hussaini faced. Many people who should not be deported were in fact being deported, then and now.

Once in Iraq, Mr. al Hussaini, a man who had not lived in Iraqi society since he was six years old, would be homeless in a country that he would not recognize. He was very dependent upon his medication and therapy for stability. When he arrived in Iraq he would have been off his medications for some time. It was not difficult to imagine what would happen. At best, he would run afoul of the authorities and be imprisoned. At worst, he would offend someone in Iraq and would be injured or killed. As is known and as he himself had told me, mental illness and traumatized people were not well understood in the Middle East (Al-Rawashdeh et al. 2021).

I said in my report that, while the charges against Mr. al Hussaini were serious, and might well warrant jail time, they should not result, domino-like, in his deportation and death, yet these two latter steps were the most likely consequence of a conviction. Mayhem is not normally punished with the death penalty, but death could easily be the outcome of this case. If Mr. al Hussaini's life history illustrated anything, it was that he was not a very lucky man. Because Mr. al Hussaini had suffered some of the worst horrors and privations to have been inflicted on any human being, this case cried out for prosecutorial discretion.

How the Case Was Ultimately Resolved

I am pleased to say that my report was taken seriously. I never testified in the case and in fact I had no further involvement in the matter. My plea that the case be dismissed was not followed, but a reasonable solution was worked out, one that kept Mr. al Hussaini in the United States: he pled guilty to Assault and Battery with a Dangerous Weapon, and the charge of Mayhem was dismissed. He pled to *reckless*, rather than *intentional* Assault and Battery, which greatly improved his chances in the Immigration Court. He was given an 11-month sentence in the county jail. After that, he was transferred to another county and held pending the resolution of his immigration case.

As part of writing this chapter I spoke to Mr. al Hussaini by phone. Many years had passed since I interviewed him. He told me that, all things considered, he is doing well.

Recommendations Regarding Mental Health Matters

Sometimes I am the first person to become aware of a client's possible mental health issues: I then alert the attorney and suggest that a psychologist be hired. I usually recommend someone I know who speaks the same language as the client. I also tell the attorney to give my report to the psychologist, so that they are made aware of a client's life story and any cultural issues that may affect the case.

Culture

When working on cases involving different cultures, one need not be from the same country in order to make comments, although some knowledge of the culture and of how to conduct any research necessary is required. There is a literature on cultural awareness that can be consulted for general principles, but there is no substitute for experience, nor for consultation with educated persons from the country in question. In general, if there is a cultural issue in a case, it will have caused a problem, and one can begin with that problem. For example, I once consulted on a case in which a woman from Africa made strange statements. I was asked to determine if these were due to mental illness or merely to her folk beliefs. In another case, I needed to explain why a young couple, recently arrived from China for study in the United States, would go to the campus police to report that he had raped her after they had consensual sex. He was trying to protect her marriageability.

It is often of great assistance to the attorneys and the court for someone's "foreign-origin" story to be told. Too many people in the dominant culture simply assume that minorities and foreigners are the same. Although they dress alike, they do not think alike nor do they have the same cultural assumptions. The story of the client's life often reveals surprising information. Immigration must be imagined as an hourglass: immigrants are usually from an educated elite, or from the very poor, with many fewer middle-class people leaving their countries. At the bottom of the hourglass, illiteracy, limited education, and hardship should be a focus and can be a mitigating factor. Literacy testing is recommended in both the native language and in English.

Ethics

1. A first principle is that forensic linguists, like all experts, form opinions based on our competency, but do not decide cases. This limited role affects our ethics.
2. Our role is advisory and informational. Ethically, it is essential that we stay within our roles.
3. Since advocacy falls within the job description of the attorney, our role is to supply information and sometimes even an argument that the attorney can adopt and use. There are occasions when we might advocate, and informing the attorney of our suggestions is the proper way to do it.
4. Just as attorneys do not win all their battles, experts are quite often ignored. Ethically, our work is finished when we have submitted our report or finished testifying. Our limited role means that our moral responsibility and our identification with a case is less. While an expert may be quite sure that his or her opinion is correct, rulings and decisions can go against us.
5. Part of being ethical is: *not* taking every case, *not* twisting evidence, *not* allowing attorneys to dictate our conclusions, and *not* overstating our qualifications. There are many *nots*.

6. If an ethically challenging situation presents itself, a linguist must consider not only the rules but also their own stance as a moral agent. Rather than going along, or setting aside doubts, one must weigh the various elements and consider the alternatives. Just as other professionals are expected to think their way through problems, we must do the same.

Conclusion

This chapter has explored the ethics and neutrality of expert witnesses working in the American legal system. There is a chorus of concern that American experts are too often partial and not neutral. I attribute this lack of neutrality primarily to the adversarial system in which experts must work. The ethical issues experts face are little understood and have even been deemed a "dilemma" (Barnett 2017). I agree that experts can sometimes face ethical challenges. My approach is this: rather than framing professional ethics as simple adherence to a checklist of rules, I prefer to engage in the kind of moral reasoning that most professionals learn, and in which individual situations often require a process of prioritizing, contextualizing, and adaptation. Although there may be little discussion or awareness of this approach regarding experts, it is nothing new. All professionals, including judges and attorneys, must resolve difficult problems that defy by-the-book or easy, traditional solutions. I have illustrated my argument with a unique case in which I pleaded, as an expert, for special consideration. I did not do this because I did not know what the norms are or didn't care whether I followed them. Although no advocacy was expected of me, I advocated for a person who had endured extremes of suffering and persecution that left him a broken man. I proposed an unusual resolution of his case to save his life. This was the correct thing to do because virtually all ethical systems hold the respect for life, and its preservation, to be important tenets. While my solution was not the one that was finally negotiated, my explanations and advocacy were what led to the unusual—and I believe ethical—resolution of the matter.

Notes

1 The author's email address is as follows: michaelolaughlin10@gmail.com.

2 For a general statement on ethical issues in forensic linguistics, see Shuy, Roger. 2006. *Linguistics in the Courtroom: A Practical Guide*. Oxford and New York: Oxford University Press: 123–129. Another excellent source of information is Coulthard, Malcolm, and Alison Johnson. 2007. *An Introduction to Forensic Linguistics: Language in Evidence*. London and New York: Routledge.

3 For a thorough examination of the ethical challenges of the advocacy system for attorneys, see Markovits, Daniel. 2008. *A Modern Legal Ethics: Adversary Advocacy in a Democratic Age*. Princeton and Oxford: Princeton University Press.

4 See *Daubert v. Merrell Dow Pharmaceuticals. 1983. 509 U.S. 579*; *General Electric Co. v. Joiner. 1997. 522 U.S. 136*; and *Kumho Tire Co. v. Carmichael. 1999. 526 U.S. 137.*

5 See further, Richmond, D. R. 2000. "Expert Witness Conflicts and Compensation." *Tennessee Law Review* 67, no. 4: 909–948.

6 This article sparked a heated debate. See Reitz, John C. 1990. "Why We Probably Cannot Adopt the German Advantage in Civil Procedure." *Iowa Law Review* 75, no. 4: 987–1010.

7 For examples of this type of moral reasoning in law, see Duarte d'Almeida, Luís. 2015. *Allowing for Exceptions: A Theory of Defenses and Defeasibility in Law.* Oxford and New York: Oxford University.

Case Law

Al-Hussaini Hussain v. Palmer Communications Inc., a Delaware corporation, d/b/a KFOR-TV; Jayna Davis, Brad Edwards, Melissa Klinzing, 2003. No. 00 6366 (W.D. Oklahoma) D.C. No. CIV-97-1535-L.

Commonwealth v. Al-Hussaini, 2011. 1182 CR 00362.

Daubert v. Merrell Dow Pharmaceuticals 1993. 509 U.S. 579.
English Feedlot, Inc. v. Norden Labs., Inc., 1993. 833 F. Supp. 1498, 1501 (D. Colo.).
General Electric Co. v. Joiner 1997 522 U.S. 136.
Kumho Tire Co. v. Carmichael 1999. 526 U.S. 137 (1999).
Mak v. Blodgett 1991. 754 F. Supp. 1490 (9th Cir.).
Van Blargan v. Williams Hospitality Corp. 1991 754 F. Supp. 246,249 (D.P.R.).

References

Al-Rawashdeh, Ahmad B., et al. 2021. "Misconceptions and Public Stigma about Mental Illness." *Indian Journal of Forensic Medicine & Toxicology* 15, no. 2: 1873–1878.

Anderson, Espen, and Bill Schiano. 2014. *Teaching with Cases: A Practical Guide*. Cambridge, MA: Harvard Business School.

Appelbaum, Paul S. 1999, February. "The Medicalization of "Judicial Decision-Making." *The ElderCare Report*.

Barnes, Louis B., C. Roland Christensen, and Abby J. Hansen. 1994. *Teaching and the Case Method: Text, Cases, and Readings*. 3rd. ed. Cambridge, MA: Harvard Business School.

Barnett, Jeff. 2017, March 22. *Advocacy vs. Neutrality: Overcoming the Expert's Dilemma*. Barnett and Leuty, P.C. Retrieved from https://civil-law.com/blog/advocacy-vs-neutrality-overcoming-experts-dilemma/.

Barrett, Paul. 2007. *American Islam: The Struggle for the Soul of a Religion*. New York: Picador.

Boehlert, Eric. 2002, December 3. "Did Timothy McVeigh Have Iraqi Helpers?" *Salon*. Retrieved from https://www.salon.com/2002/12/02/mcveigh_9/.

Butt, Adam. 2023. "Concurrent Expert Evidence in the United States: Is There a Role for Hot Tubbing ?" Civl Jury Project at New York University School of Law. Retrieved from https://civiljuryproject.law.nyu.edu/concurrent-expert-evidence-in-the-united-states-is-there-a-role-for-hot-tubbing/.

Chin, Jason M., et al. 2019. "The Biases of Experts: An Empirical Analysis of Expert Witness Challenges." *Manitoba Law Journal* 42, no. 4: 21–67.

Cole, Richard W., and Laura Maslow-Armand. 1997. "The Role of Counsel and the Courts in Addressing Foreign Language and Cultural Barriers at Different Stages of a Criminal Proceeding." *Western New England Law Review* 19: 193–228.

Coulthard, Malcolm, and Alison Johnson. 2007. *An Introduction to Forensic Linguistics: Language in Evidence*. New York: Routledge.

Davis, Jayna. 2004. *The Third Terrorist: The Middle East Connection to the Oklahoma City Bombing*. Nashville: Nelson Current.

Einesmann, Floralynn. 2010. "Cultural Issues in Motions to Suppress Statements." In *Cultural Issues in Criminal Defense*, 559–628. Edited by Linda Friedman Ramirez. New York: Columbia University.

Finnegan, William. 2013, April 29. "The Deportation Machine." *The New Yorker*.

Grimm, Paul, 2012. "Admissibility of Expert Testimony in Federal Courts: A View from the Bench". In *Ziskin's Coping with Psychiatric and Psychological Testimony*, 6th ed. Edited by David Faust. Retrieved from https://www.google.com/books/edition/_/AZmEbKKjnEUC?hl=en&gbpv=1&pg=PP1.

Hagmann, Doug. 2011, March 11. "'Homeless' Man Hussain Hashem al-Hussaini is 'John Doe #2' in Oklahoma Bombing." *Canada Free Press*.

Hammond, J. S. 2012. "Learning by the Case Method (PDF)." *Harvard Business School*.

Hazard, Geoffrey C., and Dana Remus. 2011. "Advocacy Revalued." *University of Pennsylvania Law Review* 159, no. 3: 751–781.

Langbein, John H. 1985. "The German Advantage in Civil Procedure." *The University of Chicago Law Review* 52: 823–866.

Liptak, John. 2008, August 11. "In U.S., Expert Witnesses Are Partisan." *The New York Times*. Retrieved from https://www.nytimes.com/2008/08/12/us/12experts.html#:~:text=Judge%20Denver%20D.&text=The%20two%20experts%20were%20biased,they%20had%20given%20predictable%20testimony.

Marsh, Charles W. 2001. "Public Relations Ethics: Contrasting Models from the Rhetorics of Plato, Aristotle and Isocrates," *Journal of Mass Media Ethics* 16, nos. 2/3: 79–98.

May, Allyson. 2005. "Advocates and Truth-seeking in the Old Bailey Courtroom." *The Journal of Legal History* 26, no. 1: 83–90.

Merryman, John Henry, and Rogelio Pérez-Perdomo. 2007. *The Civil Law Tradition: An Introduction to the Legal Systems of Europe and Latin America*. Stanford: Stanford University.

O'Laughlin, Michael. 2016. "Addressing Linguistic and Cultural Issues in American Criminal Cases." *Language and Law / Linguagem e Direito* 3: no. 2: 3–44.

O'Laughlin, Michael. 2013. *Expert Report, Commonwealth v Hussain Al-Hussaini.*

Reitz, J. C. 1990. "Why We Probably Cannot Adopt the German Advantage in Civil Procedure." *Iowa Law Review* 75, no. 4: 987–1009.

Renteln, Alison D. 2004. *The Cultural Defense*. Oxford: Oxford University.

Rhodes, Steve. 2002, October "True Believer." *Chicago.* Retrieved from https://www.chicagomag.com/Chicago-Magazine/October-2002/True-Believer/.

Richmond, Douglas R. 2000. "Expert Witness Conflicts and Compensation." *Tennessee Law Review* 67, no. 4: 909–948.

Richmond, Douglas R. 1997. "Regulating Expert Testimony." *Missouri Law Review* 62, no. 3: 485–571.

Ruiz Gonzalez, Carlos. 2013. "Education for Management: Pros & Cons of Using Case Method in the XXI Century." *Journal of Executive Education* 10, no. 1. Retrieved from https://digitalcommons.kennesaw.edu/jee/vol10/iss1/6.

Schlesinger, Rudolf B. 1977. "Comparative Criminal Procedure: A Plea for Utilizing Foreign Experience." *Buffalo Law Review* 26, no. 2: 361–385.

Shuy, Roger. 2006. *Linguistics in the Courtroom: A Practical Guide.* Oxford: Oxford University Press.

Solan, Lawrence M., and Peter M. Tiersma. 2001. "The Linguist on the Witness Stand: Forensic Linguistics in American Courts." *Language* 78, no. 2: 221–239.

Sonenshein, David, and Charles Fitzpatrick. 2013. "The Problem of Partisan Experts and the Potential for Reform Through Concurrent Evidence." *The Review of Litigation* 32, no.1: 1–64.

Stanford Encyclopedia of Philosophy. 2020. "Deontological Ethics." Retrieved from https://plato.stanford.edu/entries/ethics-deontological/.

14

CHALLENGING LEGAL ASSUMPTIONS

Ethics, Due Process, and Forensic Linguistic Research

Mel Greenlee

Introduction

The law is constructed linguistically and socially. However, the ideology of language normally operative within the legal system has often made it difficult for linguists and lawyers to find common ground (Mertz 2016). The study of language in its social context emphasizes that "Language…is key to social justice and can play a powerful role in bringing about positive social change" (Avineri et al. 2019: 3). It is that goal to which both legal and language scholars aspire. While Ainsworth (2008: 11) critiqued Miranda jurisprudence as a product of "hyper-literal parsing" produced by its underlying legal conception of language and communication, Conley, O'Barr, and Conley Riner (2019) have taken forensic linguists to task for seeming to limit their research to linguistic questions mainly of interest to lawyers while ignoring the manner in which the legal system might contribute to the study of language. Rather than advancing questions prioritized by linguists, they suggested, forensic linguists have been in effect allowing "the legal tail to wag[-] the scientific dog" (153).

It is without question that without certain rules of thumb concerning linguistic interpretation, the US legal system would be unable to function.[1] It is equally apparent, however, that some of the prominent and oft-cited legal rules of interpretation are at odds with psychological and linguistic research findings about how language actually works as a contextually determined social medium.[2] As will be shown in criminal case examples to follow, the mismatch between the legal view and social reality may lead to incoherent and skewed results. With a view toward critically exploring the complex nexus between ethics, Due Process, and forensic linguistic research, this chapter examines three core "working assumptions" widespread within the US American legal system.

Working Assumptions of the US Legal System

Assumption 1: Courts Make Accurate Evaluations of Waivers through One-Sided Ritual Questioning

Two contexts may be compared in which persons subject to criminal charges under United States law are informed about their constitutional rights and once informed, agree to forgo them. The

DOI: 10.4324/9781003391074-16

Table 14.1 A Comparison of Miranda and Faretta Waiver Features

Feature	*Miranda*	*Faretta*
Constitutional Right	to Silence, to Counsel	Defense Counsel OR Self-Rep
Administered by	Police	Judge
Setting	Interrogation Room	Open Court
Assistance	None	Defense Counsel
Format	Warning—Do you understand?	Warning—Do you understand?
Evaluation	KIV	KIV
Test	Totality of Circumstances	Totality of Circumstances
Language	Varied but Complex	Generally Complex, Legal
Feedback on Understanding	Yes/No	Yes/No or check box

two waiver contexts are Miranda and Faretta. In *Miranda v. Arizona*. 1966. 384 U.S. 436, the US Supreme Court mandated that prior to questioning, custodial suspects should be informed of their constitutional rights to counsel and against self-incrimination. The "Miranda warnings" are now familiar worldwide in popular media.

Faretta is less well known. Despite the constitutional right of US criminal defendants to defense counsel (*Gideon v. Wainwright*. 1963. 372 US 335), in *Faretta v. California*. 1975. 422 U.S. 806, 835–836, the court recognized a countervailing right to self-representation. It allowed defendants in criminal matters to waive the right to counsel and serve as their own lawyers but cautioned that defendants' choice to do so (as in Miranda) must be an informed decision made with "eyes open". As shown in Table 14.1, these two waivers have many differences in the way they are utilized within the criminal justice system.

As indicated in Table 14.1, the setting and participants in a Faretta waiver differ from those of Miranda, most notably in the presence of defense counsel and a more public locale. Despite their many differences, for both Miranda and Faretta, the legal standard by which waiver validity is judged is the same: the choice must be made "knowingly, intelligently" and voluntarily (Faretta. 422 U.S. at 835) (hereafter referred to as the "KIV standard").

The validity of their decision is judged under a "*totality of the circumstances*" test, which, in theory, takes into account the communicative setting, including the language of the advisement of rights, as well as the waivers' characteristics such as youth or disability (*U.S. v. Garibay*. 9th Cir. 1998. 143 F.3d 534, 536), which may affect their comprehension and responses as well as the inherent power structure of the setting (Garibay. 143 F.3d at 536). This includes elements of the communicative setting such as the language of the advisement of rights. In practice, judicial acceptance of such waivers is very broad (Ainsworth 2008). Importantly, in neither waiver context did the court provide a formula for the required warnings. As examples will show, in both contexts, the usual pattern is a closed response by the waiver (Yes or No) to lengthy admonitions on the part of an authority figure.

Miranda Flaws and an Example

As the Miranda waiver is to be recited in every custodial interrogation, it has received a great deal of attention in empirical research and discussion of justice policy (see, e.g., Rogers et al. 2020, 2010; Pavlenko, Hepford, and Jarvis 2019; Feld 2013a; Domanico, Cicchini, and White 2012; Goldstein et al. 2012). However, many scholars have concluded that the intended protections envisioned by the Miranda court are seldom realized (Ainsworth 2021; Weisselberg 2008). Instead, "court approved formulae have reduced Miranda to a litany that officers read from a card or a waiver form" prior to a suspect's expected statement (Feld 2013a: 424). "Perversely, Miranda allows judges to focus on ritualistic compliance with a procedural formality rather than to assess the voluntariness or reliability of a statement" (Feld 2013b: 24).

According to Rogers et al. (2007), the language of Miranda warnings constitutes a significant and under-rated barrier to valid waiver. Rogers' survey found information on constitutional rights in Miranda warnings "poorly framed, vague and circuitous", with portions so complex that only 11% of Americans would understand it (Ainsworth 2021: 98). Based on several empirical studies, Rogers (2008) estimated that over 300,000 "suspects waive their rights annually while failing to comprehend even 50% of representative Miranda warnings" (777).

Surveys of police administration of the Miranda warnings (e.g., Feld 2013c; Domanico, Cicchini, and White 2012) have shown that the KIV standard is easily satisfied by detainees' minimal responses to rapidly delivered and scripted phrases, which, despite assent to questioning, may be very poorly comprehended. Domanico, Cicchini, and White (2012) demonstrated that the reading level of Miranda warnings exceeded the average educational level of adult detainees; moreover, the warnings were presented more rapidly than surrounding talk. Second language speakers, youth, mentally ill, or intellectually disabled persons are especially vulnerable to misunderstanding (Pavlenko, Hepford, and Jarvis 2019; Ferguson 2012; Goldstein et al. 2012).

The case of Joseph H. may serve to highlight the flaws in evaluating waivers of vulnerable detainees. In this case, a "voluntary" waiver of rights by a ten-year-old boy was accepted as valid, despite arguments that for one so young, the advisement was an empty formalism (*Joseph H. v. California*. 2016. 137 S. Ct. 34, Petition for Writ of Certiorari; Juvenile Law Center 2016).

In 2011, ten-year-old Joseph, who had a recognized learning disability (attention deficit disorder), was arrested for fatally shooting his abusive father in Riverside, California. Joseph was approached by a female law enforcement officer who first Mirandized him before interrogating him in the presence of his stepmother, the victim's wife. According to prosecution evidence, after the rights were read, the boy **k**nowingly, **i**ntelligently and voluntarily "waived" his right to silence and to an attorney and then provided an incriminating recorded statement which was later entered in evidence. Joseph H. was convicted and sentenced for homicide (In re Joseph H. 2015. 237 Cal. App. 4th 517, 522). However, there were many indications that, in fact, Joseph's understanding of the Miranda warnings, as well as his comprehension of the consequences of his actions, were quite limited. In addition to his learning disability, Joseph was determined to have "low-average intelligence" and a history of trauma (In re Joseph H. 237 Cal. App. 4th at 536). These factors may well have influenced his ability to understand his rights under Miranda. For example, although he answered "Yeah" when asked if he understood, Joseph stated that the "right to silence" meant a right to "stay calm". This response suggests that Joseph interpreted this legal phrasing as meaning he should remain quiet while answering the questions.

Despite his counsel's citation to the growing literature on child development and its relevance for legal decisions, Joseph's sentence was upheld on appeal. On petition for review, three of the seven justices on the California Supreme Court filed an unprecedented and strong dissent. They stated that further scrutiny was warranted to determine "whether [...] the Miranda warnings and waiver decision can realistically be made intelligible to very young children", and "whether there is an age below which the concept of a voluntary, knowing and intelligent waiver has no meaningful application". Yet, both the California majority and later the US Supreme Court allowed the judgment to stand.[3]

As the defense counsel argued and the three California Supreme Court justices agreed, the "totality of the circumstances" seriously called into question whether the legal threshold for Miranda had truly been met. There were many factors in the accused's background which may not have been sufficiently considered in the voluntariness calculus. There is much in the linguistic and psychological literature that supports this assertion. As extensive research has shown, cognitive impairment and psychological trauma can diminish an individual's ability to sufficiently understand a Miranda reading (Crane 2017; Cressey 2017; Laird 2016; Goldstein et al. 2012, 2003). Even if Joseph H. had not faced these challenges, the fact that he was underage may have served as an impediment to adequate comprehension.[4] Goldstein and her colleagues, for example, demonstrated that youthful offenders' age and verbal IQ greatly impact their understanding of Miranda, and like the youth earlier surveyed by Grisso (1981), youngsters under about 15 do very poorly in understanding its warnings; their failure correlates with propensity of youth to confess falsely (Tepfer, Nirider, and Tricarico 2009; Goldstein et al. 2003). Even older teenagers may fail to understand the vocabulary of most Miranda warnings. Among the "barriers to comprehension" noted by Rogers and his colleagues (2008) in large-scale studies were "difficult vocabulary, infrequently used words, [and] legalese" (130). Many words commonly used in the warnings required at least a 10th-grade level of education.

In view of this scientific evidence, the courts' decision that Joseph had indeed voluntarily waived his Miranda rights by answering "Yeah" when asked by the police officer whether he had understood, may illustrate what Roger Shuy (1997: 184) has called "one of the odd myths of communication": the notion that words should be believed "without considering the contextual factors" of the communication situation "that may influence the words". Nevertheless, the officials within the criminal justice system and members of the public at large, including detainees, tend to overestimate people's ability to comprehend the warnings, a mistaken impression which contributes to the problem of false confessions (Garrett 2010: 1092). To help avert such miscarriages of justice, researchers have made several concrete suggestions. For example, Goldstein and her colleagues (2003) recommended a pre-test of Miranda understanding for youthful detainees before administration. Ferguson (2012) outlined a dialogue approach to Miranda. Following Domanico, Cicchini, and White (2012), the Communication of Rights Group (CORG) has proposed that police take steps to check suspects' comprehension of the warnings (CORG Guidelines 2015). There has also been broad support for the recommendation that "police should be required to actively assess the degree to which suspects understand their rights" by adopting a more interactive own-words or paraphrase approach (Domanico, Cicchini, and White 2012: 20.) Domanico also advised that "Police should [...] be required to clarify any misconceptions or misunderstandings that the suspect may have" (2012: 20). And the research continues. Recently, Ainsworth (2021) suggested, for example, many ways that linguists could assist in answering legal questions regarding the validity of waivers and the reliability of resulting statements (such as discourse analysis of topics and facts introduced by interrogators versus the accused which could show the cueing of a false confession; examination of suspects' language errors evidencing lack of English proficiency or disability). Despite ongoing scientific research and the many evidence-based recommendations that have come from it, the same flawed Miranda procedures continue to be a part of daily US police practice.

Faretta Flaws and an Example

In contrast to the ubiquity of Miranda, Faretta warnings affect a relatively small proportion of felony cases (Hashimoto 2007). Most US defendants who choose self-representation do so because of "distrust of appointed counsel" (Goldschmidt 2022: 348). In waiving their entitlement to counsel, these defendants, unlike the lone detainees in the interrogation room, present their waiver during an exchange with a judge in the presence of defense counsel. In California procedure, defendants make written Faretta waivers which the judge goes over orally in court at a hearing.

In *Iowa v. Tovar* (2004. 541 U.S. 77, 92), the High Court stated, "We have not...prescribed any...script to be read to a defendant who states that he elects to proceed without counsel. The information a defendant must possess [...] will depend on a range of case-specific factors" (*Iowa v. Tovar*. 541 U.S. at 92). Similarly, the California Supreme Court has held that "general" warnings of the disadvantages attendant on waiver of defense counsel are sufficient (People v. Frederickson. 2020. 8 Cal. 5th 963, 1004; *People v. Riggs*. 2008. 44 Cal 4th 248, 277–278).

The Faretta setting might appear to have more safeguards than Miranda, but these may be illusory if defendants are at odds with counsel and/or so impaired they are unable to accept counsel's advice or realistically assess their own possibility of trial success (Greenlee 2017; White and Gutheil 2016; *People v. Waldon*. 2023. 14 Cal. 5th 288; *People v. Wycoff*. 2021. 12 Cal. 5th 58). Unlike the Miranda context, in which authority figures encourage waiver, in Faretta, judges variously attempt to dissuade pro-se defendants and encourage them, despite their disagreements, to consult with counsel. After these preliminaries, judges ask defendants to fill out a form with questions.[5] The reading level of these forms may be measured by the Flesch-Kincaid test.[6] While the overall reading level required to understand the listed Faretta admonitions is approximately 8th grade (the last grade of primary school in the United States), they vary considerably. The most difficult to comprehend concern the rules of evidence and require college-level skills. The length of written Faretta warnings also varies. In one small sample of 11 California capital cases, the forms ranged from approximately 190 to 1,600 words (Greenlee 2022). Once the form is completed, the judges question defendants in open court, using scripts which closely follow the form's language. For example, a California judicial benchbook provides the following suggested question:

> You have the right to use the process of the court to subpoena any witnesses or records that you may need on your behalf or in your defense. Do you understand that right?
>
> *(Judicial Council of California 2022: 54–46)*

The judge's provision of information on trial rights and related questioning of the defendant regarding them is called a "colloquy" in legal parlance. In these sessions, the majority of talk is by the judge, while defendants' contribution may be minimal, even monosyllabic.

Like the talismanic status of Miranda, if the record discloses that the trial judge collected the form and engaged in a Faretta colloquy, the legal presumption is that the layperson has understood the warnings and is proceeding pro-se despite the known dangers. The Ninth Circuit observed (*Arredondo v. Neven*. 2014. 763 F.3d 1122, at 1130), "We generally presume that defendants seeking to waive their right to counsel understand what they are told regarding that choice". However, given linguistic research on laypersons' understanding of legal language in other contexts, the courts' optimism about how well pro-se defendants understand the Faretta warnings seems unwarranted (e.g., Grieshofer, Gee, and Morton 2022; Martinez, Mollica, and Gibson 2022; Stygall 2021).[7] The complexity of language in Faretta admonitions combined with a generally low level of reading and education among many criminal defendants would undercut the assumption that they comprehend the warnings (Ferguson 2012; Sterngold 2006).

Particularly in the most serious cases, where defendants face the stress of a pending felony or capital trial, lay litigants would appear to be no more likely to understand legal text than, for example, the average layperson who is faced with reading immigration, tax, contract, or credit card documents. In all of these contexts, linguists have shown that the intended audience may, in fact, not comprehend or apply the legal texts as intended (e.g., Martinez, Mollica, and Gibson 2022; Stygall 2021, 2002; Berk-Seligson 2016).

Observations of self-represented capital cases provide an opportunity to examine Faretta procedures in a high-stakes context (Greenlee Forthcoming, 2022, 2017). To illustrate the linguistic complexity of Faretta warnings, an excerpt from a written admonition in a California capital case is shown below.

Case 1: Example of a Faretta Waiver in a California Capital Case

> I UNDERSTAND THAT I HAVE THE RIGHT TO BE REPRESENTED BY A LAWYER AT ALL STAGES OF THE PROCEEDINGS AND, IF I DO NOT HAVE FUNDS TO EMPLOY COUNSEL, ONE WILL BE APPOINTED BY THE COURT FOR ME.
>
> If the Court grants my petition to proceed in propria person, and I am permitted to represent myself, I understand I will have to conduct my own defense without the aid of counsel.
>
> Understanding all of the constitutional rights set forth above, it is my personal desire that I be granted permission by the court to proceed IN PROPRIA PERSONA (acting as my own attorney) and that by making this request I am giving up the right to be represented by a lawyer appointed by the Court.

The above text is taken from the final portion of an over 800-word Faretta advisement (Greenlee 2022). Its lengthy sentences present multiple embedding, phrasal adverbial clauses, conjoined *that* complementizers, a subjunctive phrase ("I be granted") as well as passive verbs—grammatical features which have been flagged by linguists as complicating understanding (Stygall 2021; Randall and Graf 2014; Ferreira 2003; Tiersma 1999). Clearly, the complex admonitions are pitched at a level beyond the educational status of the majority of people to whom they are directed. Thus, as was shown in the discussion of Miranda, the illusion of informed choice under Faretta would seem to rely on unsupported premises concerning defendants' motivation, mental and linguistic competence, as well as an imputed yet improbable talent for processing legalese, all evaluated by a yes/no question. In both settings, the notion of an independent, autonomous actor freely choosing to forgo constitutional rights is a dearly held legal ideal which may in fact be at odds with everyday waivers in the judicial system (Miller 2021). This being the case, one would be forgiven for inferring that the court appears to have greater concern for the ritual of legal language than actual comprehension of that language. The tendency to prioritize form over function is a major impediment to ensuring that waivers such as Miranda and Faretta achieve their goal: namely to safeguard individual rights. To reach this goal, it is essential that the language of the law be understandable by the persons it was designed to protect.

As suggested for Miranda, true understanding of the Faretta language could be measured by a technique which involves actively processing and explaining the warning, rather than solely voicing a monosyllabic affirmation. However, this is not enough. Far more research is needed to evaluate Faretta waivers. For example, forensic linguistic investigations could help to identify the impact that the sheer linguistic complexity of Faretta formulations has upon the level of defendants' comprehension and their ability to exercise their right to self-represent in a court of law. Forensic linguistic studies may also help to ascertain how factors such as education and disability may affect informed choice. Further, examination of talk during the trials of defendants who elect to represent themselves may reveal patterns of misunderstandings and thereby confirm the illusory nature of their purportedly KIV waivers.

Assumption 2: Law Enforcement Transcriptions of Detainee Statements Are Sufficient for Evidentiary Purposes

Assuming that a detainee makes a statement to police during interrogation, additional legal "rules of thumb" govern the manner in which the suspect's statement is recorded and relayed to the court as evidence. In most cases, the initial written transcript of a detainee's statement is prepared by agents of the prosecution. Although some courts appear to treat the recorded and written versions as equivalent, case examples show the entextualization process is far from neutral and may introduce serious alterations to the evidence.

In most cases in the United States and in the United Kingdom, the initial transcript of a detainee's statement is prepared by agents of the prosecution (Haworth 2018; Fishman 2006). Under the evidentiary rules of many US states, the tape-recording is deemed to constitute the evidence, as opposed to any written version. However, in practice, US courts often treat the recording and its transcript as of equivalent evidentiary value. While instructing jurors that the tape is the evidence, courts frequently distribute a written version of the interrogation to "assist" jurors in understanding it, a practice which Solan and Dahmen (2018) characterized as "incoherent".

Ainsworth (2008: 14) has described the legal system's concept of language as a completely transparent conduit for ideas. In line with that ideology, some US courts have treated a written transcript of a recording as analogous to "turning up the volume" of a radio where speech is difficult to decipher.[8] However, a transcript is not a value-less, agenda-free representation of the talk on tape, as Alger (2021), Bucholz (2000), and others have pointed out. What the transcriber chooses to include or leave out may seriously bias the jury's understanding (Haworth 2018; Bucholtz 2000). The transcriber's role, as part of the prosecution team, may also influence the written product. In this respect, producing a written rendition may be seen as a "political act" (Bucholtz 2000: 1453.) Transcribers may also be influenced by confirmation bias (Solan and Dahmen 2018). Aspects of oral speech of interest to linguists, such as "false starts, repairs, repetition, overlap and interruption, although common in spoken language, are routinely omitted from written transcripts" (Haworth 2018: 436). These changes in text are not without effect as such phenomena may contribute crucially to the meaning in a particular interaction. Thus, in the interest of "cleaning up" or paraphrasing talk, transcribers may inadvertently introduce distortions (McMillan 2022; Alger 2021; Jones et al. 2019) and yield "highly significant" alterations to the original meaning of the original exchange (Haworth 2018: 436).

Further, as linguistic research has also repeatedly shown, especially where a recorded statement contains low-volume or garbled passages, the transcriber's version—whether accurate or not—will prime listeners' perception (Solan and Dahmen 2018; Fraser, Stevenson, and Marks 2011). Even when transcript discrepancies are indicated for jurors, the "misinformation effect" may still taint their memory of the recorded evidence (Miller 2016). Psychologists have shown, for example, that memory for evidence is further affected by stereotypes which jurors rely on to fill in evidentiary gaps (Levinson 2007). Despite such compelling scientific evidence, the legal system still has no satisfactory solution to disputed transcripts (Solan and Dahmen 2018).

As discussed in Chapter 21, where the evidence concerns the interrogation of a detainee who does not speak the language of the court (English), jurors' evaluation of recorded interactions is even more filtered and removed from the original (Fishman 2006). The following case illustrates the damage to justice which may originate from the flawed transcript of an interrogation.

Transcript Flaws and an Example: People v. "William Rolando"[9]

The detainee in this case was a 21-year-old high-school dropout with a learning disability. He was questioned by two officers in a lengthy, confrontational, Reid-inspired late-night interrogation session, before eventually being charged with conspiracy and involvement in a multiple homicide. The interrogation was tape-recorded and a 72-page transcript was prepared by the prosecution. The transcript contained numerous annotations of "unintelligible" utterances, most of which were attributed to the suspect. In addition, there were multiple misspellings such as "your" for "you're" and "trader" for "traitor". In addition, overlaps and interruptions between interrogator and suspect were left unmarked; and punctuation was quite inconsistent when included at all. Case 2 displays a linguist's transcription of the original tape, the prosecution's transcription, and the detective's summary of the suspect's statement.

Case 2: A Suspect's Statement, in the Original, the Official Prosecution Transcription, and the Officer's Testimony[10]

Recorded Interrogation

Officer:	You, you're guilty, you, you conspired…
Suspect:	Withholding information.
Officer:	Well, you, uh, not just withholding information, you conspired…
Suspect:	[sniffs]
Officer:	…with these other people…//
Suspect:	// I con- whaddya call it, condoned it?
Officer:	A conspiracy, you planned it with these other people…to go over there….

Prosecution Transcript

Officer:	YOUR GUILTY, YOU CONSPIRED
Suspect:	WITHHOLDING INFORMATION
Officer:	NOT WITHHOLDING INFORMATION, YOUR [SIC] CONSPIRED WITH THESE OTHER PEOPLE
Suspect:	I COMPLETELY CONDONNED [SIC] IT
Officer:	CONSPIRACY YOU PLANNED IT WITH THESE OTHER PEOPLE TO GO OVER THERE…

Officer's Testimony

Toward the end of the interrogation, the detective told the suspect that he was guilty of murder and conspiracy. The suspect told him, "I completely condoned it."

There are several obvious and significant differences between the three renditions of the interrogation.

The jury received information about this speech event solely through the testimony of the principal interrogator, a detective who "refreshed [his] memory" with the written transcript. As shown in Case 2, the officer testified that at the end of the session, the young suspect had admitted that he had "completely condonned" [*sic*] the actions of his co-defendants in the homicide. By recasting the young man's question as a statement, the detective's testimony seriously contaminated the evidence considered by the jurors, as it is well known that jurors consider inculpatory admissions highly persuasive, of both guilt and death-worthiness. Kassin and Neumann (1997) confirmed that confession evidence is "uniquely potent". Such was the case here. After viewing the evidence, the young man was found guilty and sentenced to death.

After his conviction, appellate counsel had the original tape examined by a forensic linguist specializing in pragmatics. The counsel also had the recording processed and analyzed by an audio specialist. Formal linguistic investigation revealed even more discrepancies. Pragmatic analysis, for example, showed that the alleged remark about "condoning" did not fit the topic of the surrounding speech, which addressed another legal concept altogether. In the preceding exchange, the suspect had mistakenly equated "conspiracy" with the notion of "with-holding information". At the moment of the suspect's target utterance, the detective was in the process of explaining the meaning of conspiracy, an explanation which continued after the alleged "condoned" remark. The pragmatic specialist emphasized that the alleged "condone" statement was not only incongruous in the immediate context but did not fit with the stance that the suspect had displayed otherwise toward events throughout the questioning.

Moreover, audio analysis showed that the suspect had uttered this phrase with rising intonation. On this basis, the suspect's remark was not so much a statement but a clarification question concerning yet another legal term he did not understand. Neither the legal meaning of "conspiracy" nor the notion of "condoning" appeared to be in his vocabulary. The audio specialist further determined that the word "completely" was, in fact, never spoken. The expert determined that what the suspect actually said was, "I con-what-a-ya-call it, condoned it?" The suspect had used similar fillers in other parts of the interrogation.

This case constitutes an abject failure of the adversary system. The defense counsel should have challenged the detective's account of the alleged admission by comparing the prosecution's transcript to the original tape. Had this step been taken, the results might have been much different. This failure is below the standard of representation warranted by such a serious case (Gould and Barak 2019; American Bar Association 2003). It serves as a powerful reminder of the critical role that forensic linguistic analysis can play.

There is much that the forensic linguistic community can do to promote the cause of justice. Through their research, they can bring to light the deleterious effects of distortions and contamination in official transcripts. In the interest of helping to avert such problems, forensic linguists have initiated projects to proactively work with professionals in the criminal justice system. (Haworth et al. 2023, for example, have initiated a cooperative project to improve the quality of police transcribers, an innovation to be applauded.) In addition to these large-scale projects, forensic linguists can also make positive contributions by working as expert language consultants on individual cases. As Rickford and King (2016) remind, while working in this capacity, linguists can do much to illuminate and challenge erroneous stereotypes about language and language users. By the same token, linguistic expertise may be brought to bear in countering the notion that written and spoken versions are equivalent, while highlighting those factors which may erode recording reliability along the way, during what Komter (2013: 141) and other linguists have recognized as the "textual travels" of a statement.

Assumption 3: Judicial Instructions to Jurors Erase the Prejudice from Inflammatory Evidence or Argument

One of the most fundamental assumptions of the US criminal jury trial system is that jurors understand and follow the judge's instructions (*Richardson v. Marsh*. 1987. 481 U. S. 200, 211; Simon

2012).[11] Further, there are widespread legal prohibitions against "impeaching the verdict" or challenging the jury's reasoning in arriving at their decision once it has been rendered.[12] The idealized, closed universe of the jury room is one in which decisions are rendered based on "common sense", rationality, and the evidence—theoretically unencumbered by the external society's social prejudices or emotional baggage. Moreover, jury instructions are carefully worded to conform to criminal statutes; and alterations in their presentation are not infrequently a cause for reversal of judgments (Marder 2012; Simon 2012).

Nevertheless, decades of empirical research, using both ethnographic interviews and experimental designs, have shown that the reality of jury decision-making does not conform to this ideal of comprehension and rationality (Simon 2019; Hunt 2015; Simon 2012; Blume 2008; Tiersma and Curtis 2008; Saks 1997; Charrow and Charrow 1979). Rather than treat juries as "oracles", commentators urge that their decision processes should be examined as groups of human beings (Sklansky 2013). This recommendation is made with an eye toward helping to root out the influences of social prejudice which have contributed to inequities in incarceration rates and erroneous convictions. Not insignificantly, a stunning number of convictions that were later overturned involved minorities. African Americans, for example, have long been disproportionately affected by such injustices (Pollis 2022; Nellis 2016). Survey research has also shown long-standing inequalities in the criminal process for Latinx, Indigenous, and other minority community members (see, e.g., Grosso et al. 2019: Nellis 2016; Janisch 2014; Mauer and King 2007).

Problems with jury instructions contribute to this imbalance and injustice. Linguists and psychologists have shown that particular instructions are not understood by jurors (Kramer and Koenig 2014; Stygall 2014; Blume 2008) and research has found that the "mental gymnastics" (Simon 2019; Tiersma 2008) such instructions require may be substituted or supplemented by social stereotypes (Frank 2018; Thompson 2018). At times, it may seem that jurors make decisions *in spite of* their misunderstanding of the instructions, filling in the gaps in legal guidance with popular misconceptions, even of the meaning of legal terms (Marder 2012; Tiersma 2008). Further complicating matters is the fact that jurors may be unaware of their implicit biases and the influence they may have on their decision-making (Bowman 2020). In addition, emotion may play a strong role, despite trial judges' admonitions not to base decisions on it (Kaufman 2020; Bandes 2009). Two types of instructions have proven particularly problematic: (1) limiting instructions, in which jurors must consider evidence solely for a permitted purpose; and (2) judges' cautionary or "curative" admonitions given to jurors when inadmissible evidence or argument is heard in court.

With regard to the question of permissible and impermissible uses of evidence, psychological research has shown that it is nearly "impossible for jurors to forget evidence for one purpose while remembering it for another" (Chortek 2013; Steblay et al. 2006; Eichhorn 1989: 345; see also Spellman and Weaver 2022). This is especially true when the evidence is emotion-laden hearsay, uncharged offenses, allegations of gang membership, or other highly sensitive matters. Trial procedure requires weighing of such evidence prior to its admission, especially where it may lead to a verdict based on the notion of perceived criminal propensity, or an inherent tendency toward criminality (Klein 2020; Frank 2018; Lieberman and Arndt 2000; Saks 1997).

When evidence of other offenses is admitted, "pattern instructions" (standard judge directives based on criminal statutes) are provided to the jury. These instructions are often in the form of a list of permitted uses. They may be given at the time of the evidence's introduction and/or at the end of the evidence presentation. Scholars have criticized this practice as ineffective in preventing verdicts based on notions of a defendant's character defect or tendency toward crime. According to Klein (2020), they are as useful as "waving a magic wand" (153).[13] Judges have also used many colorful metaphors to describe their futility, noting that one "can't unring a bell", or "throw a skunk in the jury box and order jurors not to smell it" (Simon 2012). In part, the ineffectiveness of these instructions may be grounded in jurors' pre-existing social biases. The perceived propensity for crime may be supplied or enhanced by stereotypes about criminal tendencies and race (Frank 2018; Thompson 2018; Goodman 2007).

This issue leads directly to the second type of instructions which have proven particularly problematic: "curative" admonitions. Judicial instructions have been the traditional remedy for improper prosecutorial arguments. The California legislature noted that in the past, such arguments have incorporated metaphors comparing defendants to animals or using other dehumanizing language (California. 2020. Racial Justice Act). Such "flights of oratory", as a Georgia court dubbed them, have generally been met with hand-wringing at most in the appellate courts. This prosecutorial rhetoric has often been assessed as "harmless" to the final verdict, and/or as effectively countered by judges' curative instructions (Bowman 2020; Prasad 2017). Such was the case with *People v. Robinson* (Colorado. 2019. 454 P. 3d 229).

In this criminal case, Mr. Robinson, a Black man, was accused of sexually assaulting a white woman. The prosecutor made "graphic, explicit and repeated reference to the race and skin tone of the defendant and the victim" (Bowman 2020: 71, n. 208). The Colorado Court of Appeal reversed his conviction, observing that the prosecutor's behavior "fundamentally undermine[d] the principle of equal justice" (People v. Robinson. 2017. Colorado Court of Appeal. 459 P.3d 605 at 611), The prosecutor's appeal to bias was so strong, the appellate court held, "the risk that Robinson did not receive a fair trial by unbiased jurors simply is too great to ignore" (459 P.3d at 610–611). The jury had been cautioned by the trial judge against a verdict tainted by bias. However, those warnings, the court of appeal found, had slanted away from race and toward gender. The judge, for example, had instructed the jury "not to allow bias or prejudice, including *gender bias*, or any kind of prejudice based upon gender" to influence their verdict (459 P.3d at 610). On review, these instructions were deemed adequate by the Colorado Supreme Court (People v. Robinson. 2019. 454 P. 3d 229, 235). The Court ruled that any error in the prosecutor's remarks was harmless under the applicable legal standard (Bowman 2020: 107, n. 439). Robinson's conviction was therefore reinstated. Sadly, Robinson's case is by no means exceptional.

In *State v. Zamora* (Washington 2022. 512 P. 3d 512, 516), a Latino defendant was charged with assault on police officers in a confrontation near his relatives' home. In selecting the jury, the prosecutor devoted many questions to issues of "border security, illegal immigration, and crimes committed by undocumented immigrants", even though immigration issues were irrelevant to the assault charges. Zamora was found guilty by the jury and sentenced to prison.

On appeal, Zamora argued that the prosecutor's questions constituted misconduct, as their purpose was to arouse bias against Latinx as a group, denying him a fair trial. The Supreme Court of Washington agreed, holding that "viewed in context, the conduct apparently appealed to the jurors' potential racial or ethnic bias, stereotypes, or prejudice" (Zamora. 512 P.3d at 521). The evident

> purpose of the remarks was to highlight the defendant's perceived ethnicity and invoke stereotypes that Latinx are "criminally" and "wrongly" in the country, are involved in criminal activities such as drug smuggling, and pose a threat to the safety of "Americans".
>
> *(512 P.3d at 521–524)*

The Court further held that the taint from these remarks was so extreme that "instruction would not have cured the [resulting] prejudice" (512 P.3d at 518). Although in this case, the Court agreed there was evidence of bias, in many other cases, appeals on this basis are summarily denied. Thus, remarks grounded in social bias are not evaluated by the same metric in various courts; nor are the remedies equivalent. While some reviewing courts underscore the depth of the potential bias, others decline to require any remedies other than a generic jury admonition.

There is considerable psychological research that calls into question the fairness of such rulings. For example, investigations with mock jurors have shown that evocation of prejudice from such arguments runs very deep: dehumanizing labels may skew verdicts even when "jurors" are unaware of their influence (Bowman 2020; Thompson 2018; Hunt 2015). In mock-jury settings in which evidence depicted a white aggressor, jurors often reported that the aggressor was Black, affected by "racially biased mis-remembering" (Frank 2018). And mock jurors' sentencing decisions in these studies paralleled real-world sentencing statistics with white jurors sentencing Black defendants more harshly, especially for crimes against white victims (Hunt 2015).

In view of such evidence, although change is woefully slow, some progress has been made. For example, legislative action in California now has prohibited arguments appealing to racial stereotypes concerning defendants (California Penal Code 745(a), Racial Justice Act). Along with these innovations, legal scholars have recommended revised jury selection practices and court guidance from more specific jury instructions (Levinson et al. 2021; Ingriselli 2015; Goodman 2007). These proposals emphasize that revised juror questions and instructions must be carefully drafted to avoid suggesting that bias is inevitable or inadvertently endorsing dangerous stereotypes (Levinson et al. 2021.)[14]

Following law professor Mikah Thompson (2018), her colleague Mary Bowman (2020) has proposed that courts should utilize committees to revise or add curative instructions. Relying on bias research, Thompson recommended that in selecting jurors and in instructions, race issues should be made salient, as ignoring them allows the insidious effects of stereotypes to mar jury decisions (Thompson 2018: 1295). While most scholars seem to agree that current generic cautionary language is not enough, some of the newly proposed instructions may suffer from the same defects identified by linguists years ago which make instructions hard for jurors to comprehend or follow. It would be important to avoid, as Tiersma (2008, 1999) observed, instructions that are inexplicably framed in the negative, not specific enough to the jurors' task, which rely on impersonal or passive constructions, offer no rationale for the court's restrictions, and which do not provide an opportunity for feedback on jurors' understanding.

Thus, the collaboration of linguists in drafting new instructions, such as the previous work of Tiersma in revising the California instructions, would be an important contribution to legal improvements and to more equitable jury verdicts. Given the difficulties in accessing some of the more subtle biasing influences, revisions will take a diligent and sustained interdisciplinary collaboration. As Justice Sotomayor wrote over a decade ago (*Calhoun v. U.S.* 2013. 568 U.S. 1206), even in the 21st century, adjudication of criminal offenses in the United States retains a "deep and sorry vein of racial prejudice" that is too often exploited in arguments relying on stereotypes; without specific and understandable guidance from judicial instructions, this pernicious influence may contribute to a miscarriage of justice. Efforts to counter such imbalances will require joint and informed efforts of legal practitioners and many scholars, including forensic linguists. As California's new legislation has recognized, language is a central part of both the "sorry vein" of history but also of efforts to guide juries and courts toward more equitable treatment. It is to be hoped that linguists, whose investigation of jury instructions now spans many decades, will be an important part of improvement endeavors for both types of "cautionary" instructions.

Conclusion

As the discussion above has shown, legal assumptions about the role, importance, and variety of language pervade the criminal adjudication process. Linguists who study legal language have an important role to play in both identifying misconceptions and working to improve the equity of the process. As a matter of ethics, forensic linguists' first allegiance is to linguistics as a science, both in individual cases and policy matters, as the legitimacy of their expert voice depends on grounding and reliable guidance from the accepted methods of the discipline (Ainsworth 2022; Shuy 2009).

When forensic linguists apply their research to challenging basic legal "rules of thumb" such as those above, it is crucial that they continue to test and refine their findings through cross-disciplinary inquiry. Thus, for example, linguists might collaborate with psychologists to create more adequate explanations of operative legal concepts for laypersons, testing the proposed language through controlled experimental methods.

Through such interdisciplinary contact and common endeavors, both legal and linguistic practitioners can gain respect and understanding of the concerns of their colleagues. Along with fundamental research and experimentation, these efforts allow the work of forensic linguists to reach out to affect justice policy. In addition, they advance the depth of knowledge about how language works as a real-world phenomenon.

In light of the strong role that language plays in social bias, among the responsibilities of forensic linguists must be what Rickford and King have characterized as "data-finding and consciousness-raising" regarding observed inequities, as "language lives in society and so must we" (Rickford and King 2016: 982). Forensic linguists have long ventured out from the academic world and will continue to explore and explain language in the complex and changing legal landscape of criminal adjudication.

Notes

1. See Smith (2007).
2. See Rickford and King (2016).
3. Order of the California Supreme Court in case S227929; Chemerinsky 2015; *Joseph H. v. California*, Case No. 15-1086, Brief of Amicus Curiae Juvenile Law Center and The Center on Wrongful Convictions of Youth in Support of Petitioner (March 23, 2016).
4. The California legislature added protection for vulnerable youth subsequent to the Joseph H. case. California Welfare and Institutions Code now requires that prior to any Miranda waiver, a youth aged 17 or under must consult with legal counsel. W&I 625 (2018, 2020).
5. For a recommended list of Faretta admonitions for federal judges, see Goldschmidt 2022: 280–281. Faretta admonitions in the list are similar in their complexity to those in use in California courts.
6. This metric has served as an initial gauge of reading level for many other texts. It is based on the relative length of words and sentences. For more information, see Stockmeyer 2009; Domanico, Cicchini, and White 2012; and Grabeel et al. 2018.
7. Stygall (2021) provides an excellent review of this research.
8. See, for example, *People v. Brown* (1990) 225 Cal. App. 3d 585, 598.
9. A pseudonym is used for the defendant in this case.
10. The transcription conventions used here are as follows: "..." indicates an incomplete utterance with pause, with intonation indicating that speaker intends to continue; and "." indicates end of utterance; and "//" indicates points of overlapping speech.
11. See Tiersma (2008) for a critique of this view.
12. See, for example, California Evidence Code 1150; *Peña-Rodriguez v. Colorado*. 2017. 580 U.S. 206.
13. On this point, see Sklansky (2013: 439) who reviewed experimental data on limiting instructions, concluding that such instructions "sometimes work", "sometimes don't work" (439); and at times could be barely comprehensible. Sklansky contended that the ineffectiveness of such instructions was not universal, but depended on many factors, including the wording of the admonitions. Further, he pointed out that mock-jury research tests individuals' decision-making, unlike that of actual juries, which require group decision-making.

14. A further important goal for improvement is to expand participation in criminal juries, as, regardless of instructions, more diverse juries have been shown to be more meticulous and equitable. Diverse juries examine evidence more carefully, deliberate more, and make fewer errors (Sommers 2006). Further, statistical research has demonstrated that juries are less likely to convict African American defendants when at least one juror is Black (Anwar, Bayer, and Halmarsson 2012). The difference in conviction rates for Black and white defendants was 81% and 66%, respectively. This imbalance disappeared when the jury pool included at least one African American.

Case Law and Statutes

Arredondo v. Neven. 9th Cir. 2014. 763 F.3d 1122.
Calhoun v. U.S. 2013. 568 U.S. 1206
California Penal Code § 745 (2020 Racial Justice Act
California Evidence Code § 1150
California Welfare & Instittions Code § 625
Faretta v. California. 1975. 422 U.S. 806.
Gideon v. Wainwright. 1963. 372 U.S. 335.
In re Joseph H. 2015. 237 Cal. App. 4th 517.
Iowa v. Tovar. 2004. 541 U.S. 77.
Joseph H. v. California. 2016. 137 S. Ct. 34.
Miranda v. Arizona. 1966. 384 U.S. 436.
Peña Rodriguez v. Colorado. 2017. 580 U.S. 206.
People v. Brown. 1990. 225 Cal. App. 3d 585.
People v. Frederickson. 2020. 8 Cal. 5th 963.
People v. Riggs. 2008. 44 Cal. 4th 248.
People v. Robinson. Colorado
Court of Appeal: 2017. 459 P. 3d 605.
Supreme Court: 2019. 454 P. 3d 229.
People v. Waldon. 2023. 14 Cal. 5th 288.
People v. Wycoff. 2021. 12 Cal. 5th 58.
Richardson v. Marsh. 1987. 481 U.S. 200.
State v. Zamora. Washington 2022. 512 P. 3d 512.
U.S. v. Garibay. 9th Cir. 1998. 143 F.3d 534.

References

Ainsworth, Janet. 2022. “Serving Science and Serving Justice: Ethical Issues Faced by Forensic Linguists in Their Role as Expert Witnesses.” In *Language as Evidence: Doing Forensic Linguistics*, 35–53. Edited by Victoria Guillén-Nieto and Dieter Stein. Cham, Switzerland: Springer.

Ainsworth, Janet. 2021. “Miranda Rights: Curtailing Coercion in Police Interrogation: The Failed Promise of Miranda v. Arizona.” In *The Routledge Handbook of Forensic Linguistics*, 2nd ed., 95–111. Edited by Malcolm Coulthard, Alison May, and Rui Sousa-Silva. London: Routledge.

Ainsworth, Janet. 2008. “You Have the Right to Remain Silent…But Only if You Ask for it Just So: The Role of Linguistic Ideology in American Police Interrogation Law-.” *The International Journal of Speech, Language and the Law* 15, no. 2: 1–21.

Alger, Kaitlyn. 2021. “More than what Meets the Ear: Speech Transcription as a Barrier to Justice for African American Vernacular English Speakers.” *Georgetown Journal of Law and Modern Critical Race Perspectives*, 13: 87–104.

American Bar Association. 2003. “ABA Guidelines for the Appointment and Performance of Defense Counsel in Death Penalty Cases.” *Hofstra Law Review* 31, no. 4: 913–1088.

Anwar, Shameena, Patrick Bayer, and Randi Halmarsson. 2012. “The Impact of Jury Race in Criminal Trials.” *Quarterly Journal of Economics* 127: 1017–1048.

Avineri, Netta, et al., eds. 2019. *Language and Social Justice in Practice*. London: Routledge.

Bandes, Susan. 2009. “Victims, ‘Closure,’ and the Sociology of Emotion.” *Law and Contemporary Problems* 72: 1–26.

Berk-Seligson, Susan. 2016. "Totality of Circumstances and Translating the Miranda Warnings." In *Discursive Constructions of Consent in the Legal Process*, 241–264. Edited by Susan Ehrlich, Diana Eades, and Janet Ainsworth. New York: Oxford University Press
Blume, John H. 2008. "An Overview of Significant Findings from the Capital Jury Project and Other Empirical Studies of the Death Penalty Relevant to Jury Selection, Presentation of Evidence and Jury Instructions in Capital Cases." *Cornell Law School*: 5–55.
Bowman, Mary Nicol. 2020. "Confronting Racist Prosecutorial Rhetoric at Trial." *Case Western Reserve Law Review* 71: 39–119.
Bucholtz, Mary. 2000. "The Politics of Transcription." *Journal of Pragmatics* 32: 1439–1465.
Charrow, Robert P., and Veda R. Charrow. 1979. "Making Legal Language Understandable: A Psycholinguistic Study of Jury Instructions." *Columbia Law Review* 79, no. 7: 1306–1374.
Chemerinsky, Erwin. 2015, October 22. "Court Gets it Wrong with Boy who Killed Neo-Nazi Dad." *Orange County Register*. Retrieved from http://www.ocregister.com/articles/%20court-688579-boy-rights.html [https://perma.cc/Q5PL-PZG4].
Chortek, Madelyn. 2013. "The Psychology of Unknowing: Inadmissible Evidence in Jury and Bench Trials." *Review of Litigation* 32: 117–203.
Communication of Rights Group. 2018. "Guidelines for Communicating Rights to Non-native Speakers of English in Australia, England and Wales, and the USA." Accessed April 15, 2021. Retrieved from https://www.aaal.org/guidelines-for-communication-rights##
Conley, John, William O'Barr, and Robin Conley Riner, eds. 2019. *Just Words: Law, Language and Power*. 3d ed. Chicago: University of Chicago Press.
Crane, Megan Glynn. 2017. "Childhood Trauma's Lurking Presence in the Juvenile Interrogation Room and the Need for a Trauma-informed Voluntariness Test for Juvenile Confessions." *South Dakota Law Review* 62: 626–672.
Cressey, Sara. 2017. "Overawed and Overwhelmed: Juvenile Miranda Incomprehension." *Maine Law Review* 70: 87–114.
Domanico, A. J., M. D. Cicchini, and L. T. White. 2012. "Overcoming Miranda: A Content Analysis of the Miranda Portion of Police Interrogations." *Idaho Law Review* 49, no. 1: 1–22.
Eichhorn, L. 1989. "Social Science Findings and the Jury's Ability to Disregard Evidence under the Federal Rules of Evidence." *Law and Contemporary Problems* 42: 341–353.
Feld, Barry C. 2013a. "Behind Closed Doors: What Really Happens When Cops Question Kids." *Cornell Journal of Law and Public Policy* 32: 395–462.
Feld, Barry C. 2013b. "Real Interrogation: What Actually Happens When Cops Question Kids." *Law and Society Review* 47, no. 1: 1–36.
Feld, Barry C. 2013c. *Kids, Cops, and Confessions*. New York: New York University Press.
Ferguson, Andrew Guthrie. 2012. "The Dialogue Approach to Miranda Warnings and Waiver." *American Criminal Law Review* 49: 1437–1491.
Ferreira, Fernanda.2003. "The Misinterpretation of Noncanonical Sentences." *Cognitive Psychology* 47, no. 2: 164–203.
Fishman, Clifford. 2006. "Recordings, Transcripts and Translations as Evidence." *Washington Law Review* 81: 473–524.
Frank, Demetria D. 2018. "The Proof Is in the Prejudice: Implicit Racial Bias, Uncharged Act Evidence and the Colorblind Courtroom." *Harvard Journal of Racial and Ethnic Justice* 32: 1–55.
Fraser, Helen, Bruce Stevenson, and Tony Marks. 2011. "Interpretation of a Crisis Call: Persistence of a Primed Perception of a Disputed Utterance." *The International journal of Speech, Language and the Law* 18: 261–292.
Garrett, Brandon L. 2010. "The Substance of False Confessions." *Stanford Law Review* 62: 1051–1118.
Goldschmidt, Jona. 2022. *Self-Representation: Law, Ethics and Policy*. New York: Lexington Books.
Goldstein, Naomi E., S. M. Kelley, C. L. R. Romaine, and H. Zelle. 2012. "Potential Impact of Juvenile Suspects' Linguistic Abilities on Miranda Understanding and Appreciation." In *The Oxford Handbook of Language and Law*, 299–311. Edited by Peter M. Tiersma and Laurence M. Solan. London: Oxford University Press.
Goldstein, Naomi E. Sevin, et al. 2003. "Juvenile Offenders' Miranda Rights Comprehension and Self-Reported Likelihood of Offering False Confessions." *Assessment* 10, no. 4: 359–369.
Goodman, Chris Chambers. 2007. "The Color of Our Character: Confronting the Racial Character of Rule 404 (b) Evidence." *Law and Inequality* 25: 1–58.

Gould, Jon B., and Maya Pagni Barak. 2019. *Capital Defense: Inside the Lives of America's Death Penalty Lawyers*. New York: New York University Press.

Grabeel, Kelsey L., et al. 2018. "Computerized Versus Hand-Scored Health Literacy Tools: A Comparison of Simple Measure of Gobbledygook (SMOG) and Flesch-Kincaid in Printed Patient Education Materials." *Journal of the Medical Library Association* 106, no. 1: 38–45.

Greenlee, Mel. Forthcoming. "Linguistic Rituals in Faretta Waivers." In *Language and Justice: Communication in Legal Practice*. Edited by Tatiana Grieshofer and Kate Haworth. London: Cambridge.

Greenlee, Mel. 2022, July 15. "It's Not Because I Didn't Warn You." Paper presented to the Law and Society Association Annual Meeting.

Greenlee, Mel. 2017. "Disorder in the Court: Language Use by 'Gray Area' Pro Per Defendants." *Language and Law/Linguagem e Direito* 2, no. 1: 12–31.

Grieshofer née Tkacukova, Tatiana, Matt Gee, and Ralph Morton. 2022. "The Journey to Comprehensibility: Court Forms as the First Barrier to Accessing Justice." *International Journal for the Semiotics of Law-Revue internationale de Sémiotique juridique* 35: 1733–1759.

Grisso, Thomas. 1981. *Juveniles' Waiver of Rights: Legal and Psychological Competence*. New York: Plenum Press.

Grosso, C. M., et al. 2019. "Death by Stereotype: Race, Ethnicity, and California's Failure to Implement Furman's Narrowing Requirement." *UCLA Law Review* 66: 1394–1442.

Hashimoto, Erica. J. 2007. "Defending the Right to Self-Representation: An Empirical Look at the Pro Se Felony Defendant." *North Carolina Law Review* 85: 423–487.

Haworth, Kate. 2018. "Tapes, Transcripts and Trials: The Routine Contamination of Police Interview Evidence " *The International Journal of Evidence and Proof* 22, no. 4: 428–450.

Haworth, K., et al. 2023. "For the Record': Applying Linguistics to Improve Evidential Consistency in Police Investigative Interview Records." *Frontiers in Communication* 8: 1178516.

Hunt, Jennifer S. 2015. "Race, Ethnicity, and Culture in Jury Decision Making." *Annual Review of Law and Social Science* 11: 269–288.

Ingriselli, Elizabeth. 2015. "Note: Mitigating Jurors' Racial Biases: The Effects of Content and Timing of Jury Instructions." *Yale Law Review* 124: 1690–1745.

Janisch, Roy F. 2014. "Native American Incarceration: A Neglected Problem?" In *Punishment and Incarceration: a Global Perspective*, 159–177. Edited by Matthieu Deflem. Leeds, England: Emerald Publishing Group.

Jones, Taylor, Jessica R. Kalbfeld, Ryan Hancock, and Robin Clark. 2019. "Testifying While Black: An Experimental Study of Court Reporter Accuracy in Transcription of African American English." *Language* 95, no. 2: e216–e252.

Judicial Council of California. 2022. *California Judges Benchguide 54: Faretta and Marsden Issues*.

Juvenile Law Center. 2016. "Legal Docket: Joseph H. v. California." Retrieved from https://jlc.org/cases/joseph-h-v-california-re-joseph-h

Kassin, Saul M., and Katherine Neumann. 1997. "On the Power of Confession Evidence: An Experimental Test of the Fundamental Difference Hypothesis." *Law and Human Behavior* 21, no. 5: 469–484.

Kaufman, Sarah Beth. 2020. *American Roulette: The Social Logic of Death Penalty Sentencing Trials*. Oakland, CA: University of California Press.

Klein, Dora W. 2020. "'Obviate!': Addressing Magical Thinking about Limiting Instructions and Character Evidence." *University of Pittsburgh Law Review* 82: 135–172.

Komter, Martha. L. 2013. "Travels of a Suspect's Statement." In *Legal-Lay Communication: Textual Travels in the Law*, 126–146. Edited by Chris Heffer, Frances Rock, and John Conley. London: Oxford University Press.

Kramer, Geoffrey P., and Dorean M. Koenig. 2014. "Do Jurors Understand Criminal Jury Instructions? Analyzing the Results of the Michigan Juror Comprehension Project." In *Instructions, Verdicts, and Judicial Behavior*, 2nd ed., 93–129. Edited by Robert M. Krivroshey. London: Routledge.

Laird, Lorelei. 2016, June 1: "Miranda for Youngsters: Police Routinely Read Juveniles their Miranda Rights, but Do Kids Really Understand Them?" *American Bar Association Journal* 102, no.6. Retrieved from https://www.americanbar.org/groups/public_interest/child_law/resources/child_law_practiceonline/child_law_practice/vol-35/august-2016/police-routinely-read-juveniles-their-miranda-rights--but-do-kid/

Levinson, Ariana, et al. 2021. "Challenging Jurors' Racism." *Gonzaga Law Review* 57, no.3: 365–424.

Levinson, Justin D. 2007. "Forgotten Racial Equality: Implicit Bias, Decisionmaking, and Misremembering." *Duke Law Journal* 57, no. 2: 345–424.

Lieberman, J.D., and J. Arndt. 2000. "Understanding the Limits of Limiting Instructions: Social Psychological Explanations for the Failures of Instructions to Disregard Pretrial Publicity and Other Inadmissible Evidence." *Psychology, Public Policy and Law* 6: 677–711.

Marder, Nancy. 2012. "Instructing the Jury." In *The Oxford Handbook of Language and Law*, 435–445. Edited by Peter M. Tiersma and Laurence M. Solan. London: Oxford University Press.

Martinez, Eric, Francis Mollica, and Edward Gibson. 2022. "Poor Writing, Not Specialized Concepts, Drives Processing Difficulty in Legal Language." *Cognition* 224: 105070.

Mauer, Marc, and Ryan S. King. 2007. "Uneven Justice: State Rates of Incarceration by Race and Ethnicity." Retrieved from https://www.issuelab.org/resources/695/695.pdf

McMillan, Mason. 2022. "Judges *Be* Trippin: A Legal Analysis of Black English in the Courtroom." *Tulsa Law Review* 57, no. 2: 451–488.

Mertz, Elizabeth. 2016. "'Can You Get There from Here?'' Translating Law and Social Science." In *Translating the Social World for Law: Linguistic Tools for a New Legal Realism*, 237–260. Edited by Elizabeth Mertz, William Ford, and Gregory Matoesian. New York: Oxford University Press.

Miller, Amy. 2016. "Jury Suggestibility: The Misinformation Effect and Why Courts Should Care about Inaccuracies in Transcripts that Accompany Recorded Evidence." *Law and Psychology Review* 40: 363–382.

Miller, Kathryn E. 2021. "The Myth of Autonomy Rights." *Cardozo Law Review* 43, no. 2: 375–441.

Nellis, Ashley. 2016. *The Color of Justice: Racial and Ethnic Disparity in State Prisons.* Washington, DC: The Sentencing Project. Retrieved from https://www.sentencingproject.org/publications/color-of-justice-racial-and-ethnic-disparity-in-state-prisons/

Pavlenko, Aneta E. Hepford, and S. Jarvis. 2019. "An Illusion of Understanding: How Native and Non-Native Speakers of English Understand (and Misunderstand) Their Miranda Rights." *International Journal of Speech, Language and the Law* 26, no. 2: 1–21.

Pollis, Andrew. 2022. "The Appellate Judge as the Thirteenth Juror: Combating Implicit Bias in Criminal Convictions." *Temple Law Review* 95, no. 1: 1–43.

Prasad, Praatika. 2017. "Implicit Racial Biases in Prosecutorial Summations: Proposing an Integrated Response." *Fordham Law Review* 86: 3091–3126.

Randall, Janet H., and Lucas R. Graf. 2014. "Linguistics Meets 'Legalese': Syntax, Semantics, and Jury Instruction Reform." Linguistic Society of America Annual Meeting Extended Abstracts, Vol. 5.

Rickford, John R., and Sharese King. 2016. "Language and Linguistics on Trial: Hearing Rachel Jeantel (and Other Vernacular Speakers) in the Courtroom and Beyond." *Language* 92, no. 4: 948–988.

Rogers, R. 2008. "A Little Knowledge is a Dangerous Thing…Emerging Miranda Research and Professional Roles for Psychologists" *American Psychologist* 63, no. 8: 776–787.

Rogers, R., O. Tanveer, E. Y. Drogin, and B. M. Dean. 2020. "Effectiveness of the Miranda Acquiescence Questionnaire for Investigating Impaired Miranda Reasoning." *Journal of the American Academy of Psychiatry and Law* 48: 226–236.

Rogers, Richard, et al. 2010. "Everyone Knows their Miranda Rights: Implicit Assumptions and Countervailing Evidence." *Psychology, Public Policy, and Law* 16: 300–318.

Rogers, Richard, et al. 2008. "The Language of Miranda Warnings in American Jurisdictions: A Replication and Vocabulary Analysis." *Law and Human Behavior* 32: 124–136.

Rogers, Richard, et al. 2007. "An Analysis of Miranda Warnings and Waivers: Comprehension and Coverage" *Law and Human Behavior* 31: 177–192.

Saks, Michael J. 1997. "What do Jury Experiments Tell Us About How Juries (Should) Make Decisions." *Southern California Interdisciplinary Law Journal* 6: 1–54.

Shuy, Roger W. 2009, January 9. "Ethical Questions in Forensic Linguistics: Introduction to Papers From a Linguistic Society of America Panel Presentation, San Francisco, California." *International Journal of Speech, Language & the Law* 16, no. 2: 219–226.

Shuy, Roger W. 1997. "Ten Unanswered Language Questions About Miranda." *Forensic Linguistics/ The International Journal of Speech, Language and the Law* 4, no. 2: 175–196.

Simon, Dan. 2019. "On Juror Decision Making: An Empathic Inquiry." *Annual Review of Law and Social Science* 15: 415–435.

Simon, Dan. 2012. "More Problems with Criminal Trials: The Limited Effectiveness of Legal Mechanisms." *Law and Contemporary Problems* 75: 167–209.

Sklansky, David Alan. 2013. "Evidentiary Instructions and the Jury as Other." *Stanford Law Review* 65: 407–456.

Smith, Peter J. 2007. "New Legal Fictions." *Georgetown Law Journal* 95: 1435–1495.
Solan, Lawrence M., and Silvia Dahmen. 2018. "Legal Indeterminacy in the Spoken Word." In *Meaning and Power in the Language of Law*, 277–302. Edited by Janny Leung and Alan Durante. London: Cambridge University Press.
Sommers, Samuel R. 2006. "On Racial Diversity and Group Decision-Making: Identifying Multiple Effects or Racial Composition on Jury Deliberations." *Journal of Personality and Social Psychology* 90: 597–612.
Spellman, Barbara A., and Charles A. Weaver. III. 2022. "Memory and Law." In Oxford *Handbook of Human Memory*, Vol 2, Edited by M. Kahana and A. Wagner. Oxford: Oxford University Press.
Steblay, Nancy, Harmon M. Hosch, Scott E. Culhane, and Adam McWethy. 2006. "The Impact on Juror Verdicts of Judicial Instruction to Disregard Inadmissible Evidence: A Meta-analysis." *Law and Human Behavior* 30, no. 4: 469–490.
Sterngold, J. 2006, December 27. "Illiteracy Reinforces Prisoners' Captivity." *San Francisco Chronicle.*
Stockmeyer, Norman Otto. 2009. "Using Microsoft Word's Readability Program." *Michigan Bar Journal* 88: 46–47.
Stygall, Gail. 2021. "Legal Writing: Complexity: Complex Documents/Average and Not-so-average Readers." In *The Routledge Handbook of Forensic Linguistics*, 2nd ed., 32–47. Edited by Malcolm Coulthard, Alison May, and Rui Sousa-Silva. London: Routledge.
Stygall, Gail. 2014. "Death Penalty Instructions to Jurors: Still Not Comprehensible After All These Years." *Language and Law/Linguagem e Direito* 1, no. 1: 95–108.
Stygall, Gail. 2002. "Textual Barriers to United States Immigration." In *Language in the Legal Process,* 35–53. Edited by Janet Cotterill. Houndmills: Palgrave.
Tepfer, Joshua A., Laura H. Nirider, and Lynda M. Tricarico. 2009. "Arresting Development: Convictions of Innocent Youth." *Rutgers Law Review* 62: 887–941.
Thompson, Mikah K. 2018. "Bias on Trial: Towards an Open Discussion of Racial Stereotypes in the Courtroom." *Michigan. State Law Review* 5: 1243–1307.
Tiersma, Peter M. 2008. "Asking Jurors to Do the Impossible." *Tennessee Journal of Law and Policy* 5: 105–147.
Tiersma, Peter M. 1999. *Legal Language.* Chicago: University of Chicago Press.
Tiersma, Peter M., and Mathew Curtis. 2008. "Testing the Comprehensibility of Jury Instructions: California's Old and New Instructions on Circumstantial Evidence." *Journal of Court. Innovation* 1: 231–261.
Weisselberg, Charles D. 2008. "Mourning Miranda." *California Law Review* 96: 1519–1602.
White, Mitzi M., and Thomas G. Gutheil. 2016. "A Proposed Model for Assessing Defendant Competence to Self-Represent." *Journal of the American Academy of Psychiatry and the Law* 44, no. 2: 425–436.

15
THE FORENSIC LINGUIST'S ETHICAL DUTY TO EXPLAIN LANGUAGE USE TO TRIERS OF FACT

Dakota Wing

Introduction

As experts in linguistics, forensic linguists are uniquely qualified to analyze and explain language evidence and language use to triers of fact in the legal system. Using their advanced training and experience, forensic linguists have a powerful "toolkit" that is made up of their expert knowledge of the phonetic, phonological, morphological, syntactic, lexical, discursive, textual, and pragmatic mechanics of language (Coulthard, Johnson, and Wright 2016; Shuy 2006). Moreover, forensic linguists are often especially prepared to handle sensitive data, reduce potential biases, ask relevant questions, formulate appropriate research questions, apply accepted methodologies, write expert reports, and explain language use and the language evidence to lay audiences. That having been said, not every case can be handled by any forensic linguist. In order to use their skills effectively and ethically, it is essential that forensic linguists only offer their expertise when their training, knowledge, and psychological constitution are appropriate to help answer the question(s) posed to them. With this caution in mind, when courts are faced with matters relating to language, forensic linguists are not only appropriate and highly valuable experts to consult. As will be demonstrated in this chapter, they also have an ethical duty to explain language use to non-expert triers of fact.

The Forensic Linguist as the Expert Witness

In many common law systems, forensic linguists, like all expert witnesses, are faced with a question of their perceived necessity. When considering language evidence, this may be a specifically difficult task for two (inter-connected) reasons. Firstly, in order to offer their services, a forensic linguist must first be considered an expert whose expertise is needed. This may not always be the case in courtroom settings where expert language users without formal linguistic training routinely may view themselves as qualified and sufficient experts of language. For example, lawyers and judges may often be considered expert language users, specifically in argumentative speech and the use of "legalese", a legal register that they are trained in and regularly use. Mertz (2007) demonstrates that law school trains lawyers to "think like lawyers", which includes socializing them into new ways of using and relating to language. However, such training does not mean that

DOI: 10.4324/9781003391074-17

lawyers (or others trained in specialized types of language use) are trained in the science of language—linguistics. Thus, they may not recognize the value or necessity of the expertise a forensic linguist can offer.

Secondly, a forensic linguist's analysis and expertise must be shown to be outside of the general knowledge of a trier of fact. Tiersma and Solan (2002) note that courts may view jurors as ordinary language users and therefore consider them able to make inferences about language evidence and ordinary language use. They state, "the notion here seems to be: 'Why do we need a language expert? We have twelve jurors who all speak English. Let them decide'" (233). Thus, the domain of language evidence is often viewed to be within the common knowledge of lay people. This widespread perception is based on the misguided belief that users of a language ("language users") are automatically experts in how language is used ("language experts"[1]). However, this is akin to saying that everyone who drives a car understands what is happening under the hood of the car. Just as mechanics have the training to understand how a car works and the tools to analyze a car's performance, trained linguists have the training to understand how language works and the tools to analyze language use.

Descriptive in nature, the study of linguistics is aimed at describing how users of language use language. One of the principal roles of the forensic linguist is, appropriately, to describe language evidence and explain it to triers of fact. Shuy (2002), in discussing his work on trademark cases, describes this role as being a teacher, noting that "[l]inguists teach their audiences" (169). As he further explains:

> it was necessary for the linguists to teach attorneys, judges, and/or juries about the qualities and claims of lexicographical practice, and to convey technical linguistic information about the sounds, grammar, meaning, and discourse of the English language in ways that could be grasped by laypersons.
>
> *(169)*

Likewise, Solan (1998) describes the potential role of the forensic linguist as a "tour guide". This is further reflected by Tiersma and Solan (2002), who note that

> In cases involving complex language about which there is understandable disagreement between the parties, linguists can serve a role by acting as tour guides, walking the judge or jury through the disputed language, and explaining how the disputed language is an example of well-studied linguistic phenomena. The linguist's ultimate interpretation is not very important, and sometimes should not be given. Judges and jurors can use their own intuitions as native speakers. But if the linguist can help the players in the legal system understand in a systematic way the source of these intuitions, she can help jurors to structure their deliberations, and judges to structure their opinions.
>
> *(234–235)*

The authors' point about the forensic linguist being able to offer expert guidance is one well-taken. Although native speakers' intuitions are regularly used by linguists for acceptability judgments, non-linguists lack resources for explaining and testing intuitions. Solan and Gales (2016) caution that intuitions are subject to biases such as confirmation bias and false consensus bias.[2] They further note that "the risk is all the greater when the intuitions are used in legal contexts, where the intuitions of the lawyers are likely to align with their clients' interests, and the intuitions of judges may be outcome-determinative" (256). Thus, forensic linguists trained in scientific theories and methodologies are necessary and appropriate to test and explain intuitions and language use to lay persons in the legal system.

Language Ideologies in Legal Contexts

Language (or linguistic) ideologies are "sets of beliefs about language articulated by users as a rationalization or justification of perceived language use" (Silverstein 1979: 193). Irvine (1989) similarly describes language ideologies as "the cultural (or subcultural) system of ideas about social and linguistic relationships, together with their loading of moral and political interests" (255).

As noted above, a common misperception is that lay "language users" are equivalent to "language experts". This language ideology erroneously assumes that all language users hold the same understanding and perception of language use. Viewing triers of fact as a homogeneous group of language users/experts disregards different peoples' varying experiences with language (which is also at odds with the common law legal systems' aim to seat a heterogeneous jury). For example, people have different cultural backgrounds, socialization experiences, language abilities, and language attitudes which may differently shape how they perceive and relate to language. They may also hold strong ideas about how language is, or should be, used. An excellent illustration of this influence can be seen in the *Florida v. Zimmerman* trial.

In 2013, George Zimmerman was on trial for the killing of Trayvon Marton, a Black teenager. At the trial, Rachel Jeantel testified as a witness for the prosecution, having been on the phone with Trayvon Martin in the moments leading up to his death. Rachel Jeantel, also a Black teenager, spoke African American Vernacular English (AAVE) (with Haitian Creole features). Despite her speaking style being highly systematic, legal actors, the jury, and the media heavily criticized and questioned her credibility as a witness (Rickford and King 2016: also see Sullivan 2017). As a result, her crucial testimony was largely dismissed. Conley, O'Barr, and Conley Riner (2019), in discussing Slobe's (2016) and Rickford and King's (2016) comments on the attitudes toward Rachel Jeantel's testimony, remind us that language ideologies held by legal actors and triers of fact can have significant consequences in the courtroom. They note that

> the mere inability to understand a vernacular of one's own language may trigger a chain of reasoning that can either expose underlying language-based biases or create them: She can't speak English that I can understand, so there must be something wrong with her. Because I can't understand her, I won't believe her. In sociolinguistic terms, the hearer invokes a language ideology that holds that vernacular speakers of the hearer's language are not just nonstandard, but substandard [...and that] there is the problem of preexisting bias against the group associated with either a language or a vernacular.
>
> *(197)*

When triers of fact impose such ideologies on language use, they risk misconstruing language evidence or, as in the case of Rachel Jeantel's testimony, discarding important evidence (Rickford and King 2016). Although this is just one example of triers of fact imposing their understanding of language use on questions posed to them, it highlights the fact that non-linguists often lack an understanding of fundamental linguistic phenomena like accents, dialects, and varieties. Forensic linguists who have been trained in recognizing, analyzing, describing, and explaining language use are uniquely qualified to perform such tasks and help triers of fact better understand language evidence.

Various studies have identified numerous language ideologies in legal systems around the world. For example, in the context of the United States, Shuy (1993) details a number of "misconceptions about language in law cases" (e.g., that meaning is found in individual words, that transcripts are accurate and equivalent to audio recordings, and that discourse participants always have mutual understandings of words being used). Solan and Tiersma (2005) agree and note that "our legal system has embedded within it recurring, but unnoticed, generalizations about how human language works" (180). Haviland (2003) reports a "clash between theoretical and judicial perspectives on language" (764). Further examples include Ainsworth's (2008) discussion of the various language ideologies that underpin interrogation law in the United States requiring unrealistic invocations of rights and a disregard for communicative norms and contextual factors, and Ainsworth's (2012) comments on courts' disregard for a linguistic understanding of how dispreferred responses are typically provided, and the importance of context in understanding silence in evidentiary decisions regarding adoptive/tacit admissions in the United States (and most common law systems). Importantly, these issues have not only been observed in the United States context.

Du (2015) discusses a monolingual language ideology (also see Angermeyer 2008) in Chinese courts in which courts appear unable to accept that speakers of multiple languages may have different degrees of language proficiency in different languages, effectively forcing multilingual speakers to suppress the use of one language or face limiting rights to interpreters. Du (2015) notes that in the studied case, "when the judge finds out that the defendants have certain linguistic competence in the second language, without evaluating the level, he terminates the use of interpretation" (522). Maryns (2005) describes a similar ideology in Belgiam asylum cases.

Also in the context of asylum cases, Jacquemet (2009) comments on the language ideologies at play in transcribing asylum seekers' hearings in Italy, the United Kingdom, and Belgium, including a language ideology linking national identity and ethnolinguistic identity, a monolingual language ideology, and an ideology that "links fluency and truthfulness" (537). These result in transcripts that, through the entextualization process of creating an official record, are based on dominant cultural assumptions, reduce semantic ambiguities and multiple voicing, and distort speakers' original messages. These have important implications influencing asylum decisions and potentially violating asylum seekers' human rights.

In the Commonwealth Caribbean context, Brown-Blake (2014) highlights the ideologies surrounding creole languages that view creoles as ("bad") varieties of English, therefore limiting accused persons' rights relating to the right to be informed of charges in a language they understand and the right to be provided an interpreter at trial if they do not understand the language of the court.

In discussing language ideologies at play in the cross-examination of Aboriginal witnesses in Australia, Eades (2012) emphasizes that language ideologies "often serve the interests of a specific social or cultural group, and in this way they can play an important role in the reproduction of inequality" (474). Indeed, the legal system is a rich context for inequalities to permeate. As this chapter will demonstrate, by offering their professional expertise to triers of fact, forensic linguists can do much to help counteract these inequities. To illustrate how, this chapter addresses an example of a legal doctrine in the United States that disregards professional understandings of language—the way that statements made by a defendant are dealt with in joint trials when a co-defendant's confrontation rights would be violated, what is known as the "Bruton Rule". Although this example is from the US context, the application of linguistic principles and the role (and duty) of the forensic linguist is not limited to this specific law, geographic region, or jurisdiction.

The Bruton Rule

The decision in *Bruton v. United States* (1968) found that in a joint trial, when a defendant's (the "declarant-defendant") out-of-court statement (e.g., a confession) implicates a co-defendant (the "co-defendant"), and the declarant-defendant does not testify at trial, then the admission of the statement violates the co-defendant's right to cross-examine evidence against them. This makes the statement inadmissible. This is known as the "Bruton Rule".

Subsequent judicial decisions found that in order to accommodate both defendants' constitutional rights (i.e., the declarant-defendant's Fifth Amendment right against self-incrimination and the co-defendant's Sixth Amendment confrontation clause rights), the statement must be redacted to eliminate all references to the co-defendant, or the cases must be severed. Dodson (2016) states that "[a]s a practical matter, joint trials are commonplace, and therefore, courts must strike a balance between the efficiency of joint trials and the constitutional rights of the accused" (806). However, since courts are motivated to keep trials joint to save time and money, limit inconveniences to witnesses and victims, and try to ensure consistency in sentencing, redactions are often seen as the preferred solution. Case law, summarized below, has established how such redactions are done:

- *State v. Ferrell (1980)* found that plural pronouns may need to be changed to singular pronouns;
- *Richardson v. Marsh (1987)* found that redactions must eliminate the co-defendant's name as well as all references to their existence;
- *Gray v. Maryland (1998)* found that merely replacing the co-defendant's name with a blank space, a symbol, or the word "deleted" is no different than the unredacted version;
- *State v. Brewington (2000)* found that entire sentences may be removed, and the statement may need to be re-typed.

The following sections demonstrate the role of the forensic linguist in two (non-exclusive) capacities. First, it is shown how a forensic linguist can identify and inform the court or legal counsel about the language ideologies that underlie the proposed solutions to the Bruton Rule. It is recognized that it is rare that an expert witness would now be provided the opportunity to explain this information to triers of fact at trial.[3] Nevertheless, the evolution of the Bruton Rule case law can be seen as a failure to consult with a forensic linguist while establishing such precedents where a linguist could have informed the court of these. And, if faced with an opportunity to overturn solutions to the Bruton Rule, courts should consider the forensic linguist's expertise.

Second, findings will be presented from an analysis regarding a statement in a United States criminal case that underwent redactions to adhere to the Bruton Rule. If forensic linguists are consulted, as they were in this demonstrative case, they can explain to the court (at, for example, evidentiary hearings) how the redactions and edits may change the statement, and the linguistic consequences of such redactions and edits. For example, if a plural nominative pronoun (e.g., "we") is changed to a singular pronoun (e.g., "I"), this has the consequence of changing (and eliminating) the grammatical agent(s). In incriminating statements, the agent(s) of verbs, especially verbs that relate to criminal actions, can be important pieces of information and evidence. Similarly, redacted words and sentences remove context, which can change meanings and affect inferences drawn by readers. It is suggested that, without the expertise of a linguist, potential miscarriages of justice may occur.

Language Ideologies and Redactions

As noted above, in order to adhere to the Bruton Rule, courts have found that statements made by defendants (e.g., confessions) can be redacted (and edited). While courts have and continue to negotiate what types of redactions are or are not permissible,[4] a core linguistic problem underlies these discussions—that redactions, of any type, will contribute to an alteration of the defendants' words and the (linguistic) context. The perception that a statement can be redacted (and edited) yet remain an accurate reflection of the original statement is an assumption based on language ideologies that ignore the role that language has in creating and being dependent on context.

Texts often undergo significant transformations as they "travel" through different institutional settings, such as the legal system (Heffer, Rock, and Conley 2013). Indeed, suspects' statements have received scholarly attention in this matter as they undergo a series of recontextualizations as they are constructed in the police interview room, transformed into a written record, and referred to in court (see, e.g., Komter 2019; Haworth 2018; Rock 2001; Jönsson and Linell 1991). Here, an additional recontextualization process that some statements undergo is discussed.

As Blommaert (2005) notes, "context is potentially everything" (240), and indeed, this includes language. Especially in conversational settings (such as police interviews), language both involves context and provides context for other language. Heritage (1984) describes this as talk being "double contextual" in that it is simultaneously context-shaped and context-renewing. Put another way, Goodwin and Duranti (1992) explain the double contextual nature of talk by stating that "a subsequent utterance not only relies on existing context for its production and interpretation, but that utterance is in its own right an event that shapes a new context for the action that will follow it" (29). In cases regarding the Bruton Rule, courts appear to disregard the dual function of language being both context-creating and context-dependent.

As noted above, scholars have identified various language ideologies throughout the legal system and as guiding judicial decisions regarding language evidence. One that is commonly discussed is referred to as the "referentialist ideology" (Mertz 2007: also see Eades 2012; Ainsworth 2008; Eades 2008; Haviland 2003, for applications and other labels) which Ehrlich (2012) describes as

> a belief in stable, denotational and context-free meaning. According to this idea, meaning resides exclusively in linguistic forms and, as a result, words, phrases, or sentences can be extracted from their original interactional and social context and moved to other contexts without any change in meaning.
>
> *(58)*

It is suggested here that this same ideology undermines decisions relating to the Bruton Rule. This ideology relies on a process referred to as entextualization (Bauman and Briggs 1990). Entextualization is "the process of rendering discourse extractable, of making a stretch of linguistic production into a unit—a text—that can be lifted out of its interactional setting" (73). As Eades (2012) describes, "this process involves DECONTEXTUALIZATION—taking a story (or part of it) out of its context—and RECONTEXTUALIZATION—retelling the story (or part of it) in a new context" (475).

When transcripts are permitted to be redacted and edited, this process of entextualization and the referentialist ideology are apparent in two ways. First, it is clear that a context-free understanding of talk is required to allow for words/phrases/speaking turns to be selectively extracted (redacted) from surrounding talk. Second, unredacted language is also decontextualized from the original surrounding text as it is seemingly understood to stand alone without that surrounding text. As such, the redactions and edits result in a new (linguistic) context. The newly created context recontextualizes the language of the statement. As a consequence, the language in a redacted transcript appears in a different context than the language in the original transcript.

A referentialist ideology warrants an understanding that once language (context) is changed (removed/edited), it does not affect meaning. This understanding undermines an expert linguistic understanding of language. For example, Heritage (1984) notes that utterances "cannot adequately be understood unless except by reference to the context" (242). Changes in context can create changes in meaning. As will be demonstrated below, this is especially true for referential features of language such as pronouns and other deictics that depend on the surrounding language as context. Such edits and redactions result in transformations that can materially change a statement. This includes a defendant's original words and intended meanings.

In cases related to the Bruton Rule in which a statement is required to be redacted and/or edited, the prosecutor in the case is typically the individual responsible for performing the redactions and edits to adequately eliminate all references to the co-defendant. That is, the prosecutor, often the party motivated to have the statement admitted as evidence, makes decisions relating to what language is removed, what language remains, what edits are made, and how language is represented. While defendants may object (and propose their own redactions), and the ultimate decision as to whether the redactions are sufficient is up to a judge, the prosecutor's active role in making redactions and edits situates the prosecutor as a "recontextualizing agent" (Bernstein 1990). The power afforded to the prosecutor to control the language of a suspect's statement should not be overlooked. Bucholtz (2000) points out that transcripts are never neutral and when they are created (or in this case, re-created), certain interests may be promoted over the interests of others. In adversarial systems, prosecutors have specific occupational goals and ideological agendas that are inherently at odds with that of the defendant. This may affect how the redactions are done to the statement. The process of redacting and editing language must therefore be understood as an active process of entextualization entrenched with ideological perspectives. This "forgotten context" (Blommaert 2005) is not a neutral, objective process.

The representation of a suspect's statement is important as it travels into the courtroom as it is often understood to be an accurate representation of what was said in the police interview (Komter 2019; Eades 2012). Thus, despite a prosecutor's influence on a redacted and edited statement, when this statement is presented to the court (including the jury), the statement is held to an assumed standard of reliability and authority. This reflects Bucholtz's (2009) observation that "the trajectory of discourse from speech to recording to transcript or other entextualizing document tends to erase its own tracks, leaving only the final entextualization as the authoritative representation of events" (519). Once a statement is presented to the court, it can be used as a tool to further prosecutorial goals (Komter 2019; Eades 2012), enticing further language ideologies.

One such ideology that Eades (2012) and Trinch (2003) discuss is an ideology of narrative authorship in which the narrative produced through an interview is understood to be the sole product (and the "true story") of the interviewee, therefore "ignor[ing] the collaborative nature of storytelling" (Eades 2012: 478). Eades (2012) demonstrates that this has problematic consequences when the narrative is used to challenge the credibility of the interviewee when faced with inconsistencies. It is suggested here that the same ideology is also problematic when statements are redacted and edited, therefore resulting in a statement that is not solely the suspect's (nor the questioning officer's). Consequently, despite prosecutors' decisions of how and what language is presented in the redacted statement, and sometimes the addition of language, the redacted text is still considered to be solely that of the defendant. In the following section, the effects of such ideologies are discussed. Specific attention is given to the referential ideology at play in the recontextualization of a statement. To make these points, a case study is presented from the author's own work as a consulting forensic linguist.

Redactions in Action: A Case Study

In a recent case, defense counsel for a declarant-defendant in a joint-trial consulted a team of forensic linguists,[5] including myself, to help answer the question as to whether the redactions made by the prosecutor to the declarant-defendant's statement (the "original statement") resulted in a statement (the "redacted statement") that accurately reflected the original statement. Regarding the discussion above, it may seem like there is a clear answer to this—that *any* redaction will result in an inaccurate representation of the original statement—but the role of the forensic linguist here is to identify, describe, and demonstrate what the linguistic changes are, how these changes are made, and what the linguistic effects of the changes are. That is, they need to explain the language evidence to triers of fact. Triers of fact, once presented with this information, can then make an informed decision to answer the ultimate question. Thus, the *linguistic* research question is better understood as: how do the redactions affect the statement?[6] This question is important because the redacted statement would be the only version presented at trial. That is, the redacted statement would be presented as the authoritative representation of the declarant-defendant's account of events and the entire police interview. Dodson (2016) explains that courts understand one of the goals of a "successful" redaction is that the jurors are unaware that the statement has undergone any redactions.

The forensic linguists were provided with the original statement and the redacted statement. These were typed transcripts of the police interview between the declarant-defendant ("SUS") and the questioning detective ("DET"). The police interview lasted approximately two hours and was represented in the original statement in 80 pages. The redacted statement was 40 pages. A substantial amount of language (about half) was redacted. Recall that various court decisions permit/require redactions of all mentions (and the existence of) the co-defendant, including removing entire sentences[7] and editing language (e.g., pronouns). The redactions and edits were done to both the questioning detective's words as in Examples 1 and 2 as well as the declarant-defendant's words as in Examples 3 and 4.

In addition to whole speaking turns (discussed further below), the statement also had individual lexical items removed and sometimes replaced with an alternative word(s). As shown in Examples 1 and 2, lexical items such as the plural referent term "guys" (bolded) were removed from the interviewing officer's question in the original statement.

Example 1

Original statement	*Redacted statement*
DET: "Where did you **guys** go"	DET: "Where did you go"

Example 2

Original statement	*Redacted statement*
DET: "Were you **guys** like drinking and-"	DET: "Were you like drinking and-"

In some cases, the language was also edited (replaced with other language), such as pronouns (bolded) in the suspect's response in Examples 3 and 4.

Example 3

Original statement	*Redacted statement*
SUS: "yeah, **we** were just drinking and stuff"	SUS: "yeah, **I** was just drinking and stuff"

Example 4

Original statement	*Redacted statement*
SUS: "**They** said pull over"	SUS: "**I** pulled over"

Redactions and edits such as these affect the attributions of agency in the declarant-defendant's statement. Since suspect statements have a function of reporting past actions, they are largely constructed through what Ahearn (2010) calls "meta-agentive discourse", or how a person "talk[s] about their own actions and others' actions, how they attribute responsibility for events, how they describe their own and others' decision-making processes" (41). As suspect statements are dialogically constructed (Bucholtz 2009; Jönsson and Linell 1991), agent attributions may be introduced by the questioning officer or the suspect. An analysis of the agency attributions identified two patterns that the redactions had on the declarant-defendant's original agency attributions: agency transfer and agency removal.

When an agent was attributed to an action in the original statement, but the agent changed in the redacted statement, this was considered "agency transfer". For example, in Example 3 above, the pronominal change from "we" to "I" changes who was doing the "drinking and stuff". Likewise, in Examples 1 and 2 above, the change from "you guys" to "you" changes the agency. In the original statement, the referent for "you guys" was defined by the preceding language that included the declarant-defendant and other individuals (including the co-defendant). However, in the redacted version, the removal of "guys" leaves only the pronoun "you". Although "you" can be singular or plural, the surrounding (redacted and edited) context leaves only the declarant-defendant as a referent. As noted above, the removal of language resulted in the removal of the original context.

A similar instance occurs in Example 5 in which the plural pronoun "we" in the original statement refers to the antecedent of one set of individuals, but in the redacted statement, "we" refers to a different (smaller) set of individuals. Often, the meanings of pronouns are defined by their anaphoric referents. The shift in meaning of "we" in this example is due to the redactions of the surrounding language (that eliminate the existence of the co-defendant) in the redacted statement. As examples such as this demonstrate, when language is removed (or edited), anaphoric references can also change. When context changes, meanings can change.

Example 5

Original statement	*Redacted statement*
SUS: "**We** was in his mom's van"	SUS: "**We** was in a van"

"Agency removal" consisted of removing an agent (and most often also the verbal action the agent was responsible for). This typically occurred when parts of, or whole, speaking turns were redacted. For example, when the questioning officer was trying to determine where an event happened, the declarant-defendant stated that he had no idea and that he thought they were in a different part of town. This answer prompted the officer to state "so you were really turned around". The declarant-defendant provided a 166-word narrative beginning with "yeah, you know" and included 16 agency attributions. Notably, these included significant events, including the initial interactions with the victim(s), an escalation of violence, and the production of the murder weapon. Of the 166 words (and the accompanying agency attributions), 165 were redacted, leaving only the initial word "yeah" as a response. This results in the redacted statement not including information such as who initiated contact with the victims, who escalated the violence, and who had the murder weapon. Likewise, in other redacted speaking turns, the individual whom the declarant-defendant identified as being the individual/agent responsible for robbing and murdering the victim(s) was also removed.

In the agency analysis, we identified 14 different agents (individuals or different groups of individuals) in the original statement. The declarant-defendant was the proclaimed agent in 36.9% of the total agency attributions. In the redacted statement, there were only four agents (again, either individuals or different groups of people) and the declarant-defendant was the agent in 81.6% of the total agency attributions. Notably, the other agents in the redacted statement included the victims and two other witnesses—individuals who were not suggested to have been responsible for committing the accused crimes. This left the declarant-defendant as the only other party present.

In addition to redactions and edits affecting agency attributions, it also changed verbal actions, as in Example 4. The redaction of the reporting verb ("said") used to report the speech of others in the original version left only the verbal action ("pulled over") in the redacted version. In this case, an important issue surrounded who was driving the vehicle, as this person would be, at the very least, considered an accomplice in the alleged acts that followed. Therefore, the redactions and edits in this example may have significant incriminating import for the declarant-defendant. The redactions and edits also affected definiteness, as in Example 5 where a possessive form implying definiteness in the original version ("his mom's") is replaced with an indefinite article ("a") in the redacted version. The lack of specificity created by the indefinite article may produce a sense of uncertainty and unreliability which could potentially affect the perceived credibility of the supposed speaker—the declarant-defendant.

Additionally, the redactions and edits altered conversational practices and therefore affected pragmatic and inferential meanings. For example, in Example 6, the redaction creates ambiguity as the detective's second, unrelated question is redacted but the corresponding answer is not. Responses to two originally separate questions are presented in the redacted transcript as responses to one question as a single adjacency pair.

Example 6

Original statement	*Redacted statement*
DET: You got any kids? SUS: No DET: Will your phone like, make a noise or something? SUS: Yeah	DET: You got any kids? SUS: No Yeah

Although it is likely that this was done in error, without the engagement of a forensic linguist this issue would probably not have been addressed. Without an understanding of the recontextualization process that the statement has undergone, future audiences (e.g., juries) are left with a statement, as in this case, that misrepresents the declarant-defendant's words and intended meaning.

In Example 6, the declarant-defendant's response in the redacted statement to the question of "You got any kids?" appears as "No Yeah". This is ambiguous. While people may have different family dynamics in which this question could elicit an unclear answer (e.g., an unborn child, a blended family, a deceased child, etc.), there is no further clarification—be it provided by the declarant-defendant or requested by the questioning detective—to explain this ambiguity. Thus, a reader is left to make an inference about the intended meaning. In a criminal trial in which a person's credibility is in question, ambiguity may affect triers of fact's perceptions of credibility.

In Example 7, a response is likewise presented in the redacted transcript as part of a response to a question that appears many speaking turns and questions prior in the original transcript.

Example 7

Original statement	*Redacted statement*
DET. Um, so um, ok. Um, alright. So we can go through your phone, we can go through the video and look through both of those things, right? Ok.	DET: [...] Um, so um, ok. Um, alright. So we can go through your phone, we can go through the video and look through both of those things, right? Ok.
[*11 questions/29 turns omitted*]	DET: Ok. Cool. Alright.
DET: If he gets cornered by the police, is he gonna try to hurt himself, or hurt the police?	SUS: **That'** s what I' m saying, man. It' s like, what the fuck? You know, like–
SUS: I don' t think he will.	
DET: Ok. Is he soft?	
SUS: Yes sir. That' s what I' m saying.	
DET: Ok. Cool. Alright.	
SUS: **That'** s what I' m saying, man. It' s like, what the fuck? You know, like–	

As shown in Example 7, the redactions change the surrounding context and the referents of the deictic term "that" (bolded). In the original transcript, "that" refers to the declarant-defendant's claim that he does not think that the co-defendant will hurt himself or the police (if they were to try to arrest him) and that the co-defendant is "soft". However, in the redacted statement, "that" is referring to the detective's question about going through his phone and a video. Thus, the redactions result in a change in meaning.

In addition to the overtly linguistic consequences of the redactions, other important interactional information may be distorted when the language of a statement is redacted or edited. For example, in this demonstrative case, the declarant-defendant was cooperative with the police, both conversationally and practically. Thus, in general adherence to Grice's (1975) maxims of the cooperative principle, in addition to providing responses to the questioning detective's questions, the declarant-defendant also expressly agreed to aid the detective in their investigation by providing access to his phone and agreeing to help set up an opportunity to arrest the co-defendant. Such actions contained in this language evidence demonstrate cooperation. However, this information is altered or not included in the redacted version. Thus, the redactions contribute to an obscured sense of cooperation by the declarant-defendant. In the US legal system, cooperation with authorities can be considered a mitigating factor during sentencing decisions in certain cases.

Conclusion

This chapter suggests that judicial decisions regarding the admissibility of some language evidence, such as matters relating to the Bruton Rule in the US legal system, are guided by language ideologies. The demonstrative case presented above illustrates the possible effects of such ideologies—in this case, that edits and redactions to suspects' statements can change meanings. The law underlying the Bruton Rule provides a situation in which language evidence is permitted to be changed (edited and redacted); this is a form of evidence contamination. As Haworth (2018) notes, such recontextualizations of police interviews is "in stark contrast to the strict principles of preservation applied to physical evidence" (429). While prosecutors, defense counsel, and judges may do their best to comply with the Bruton Rule in their roles as actors in an adversarial system, when working within the Bruton context—or any law in any jurisdiction regarding language evidence—all actors can potentially benefit from the expertise of a forensic linguist.

In matters such as this, forensic linguists therefore have a responsibility to explain to non-experts the linguistic concepts at issue with the language evidence, such as the language ideologies underlying certain practices and the accompanying linguistic consequences that result from them. This responsibility includes consulting on cases, conducting analyses, writing reports, and offering testimony. If such expertise is not provided, the linguistic consequences can have potential legal consequences.

Ainsworth (2022) notes that, contrary to attorneys who have ethical duties to cases and clients, "the [forensic] linguist's only duty of loyalty is to the science of linguistics" (36). It is suggested here that this duty begins with linguists recognizing that they have a responsibility to consult on matters regarding language evidence, when their training, knowledge, and constitution allow.[8] This assertion aligns with other suggestions for linguists to provide their expertise in the legal system, such as Rickford and King's (2016) recommendation for linguists to "[s]ay yes to invitations to work on cases, projects involving AAVE speakers in court, police encounters, and so on" (981), and for social justice more generally. They note,

> as we think of the millions of vernacular dialect speakers who could benefit from the expertise and experience of linguists, we feel that it is not just their languages and dialects that are on trial, but LINGUISTICS itself, and that we can and should do more. [...] more of us need to get out of our offices, labs, or libraries and make a difference in the world.
>
> *(981)*

As Leonard, Ford, and Christensen (2017) note, "although forensic linguistics has been used to advantage in a multiplicity of cases, it is still underutilized in both criminal and civil matters" (897). The underutilization of (forensic) linguistics in matters regarding language evidence is concerning as language evidence, and judicial decisions affecting language evidence, are, and have been, central to many cases. Furthermore, in precedent-based (common law) legal systems, the risk is that when decisions about language evidence are based on language ideologies and without the expertise of a linguist, future cases are also at risk of exploiting and reinforcing such ideologies.

Acting as an expert witness also benefits the academic community with the advancement of scholarly knowledge. Much academic work in forensic linguistics stems from casework (e.g., Shuy 2013; Grant 2012; Bucholtz 2009; Coulthard 2002; Bucholtz 2000; Shuy 1998, 1993; Labov 1982) and journals such as the *International Journal of Speech, Language and the Law* continue to welcome case reports for publication. Fulfilling a duty to consult on cases can also help to highlight and fill research gaps relating to linguistic issues in the legal system. Serving as an expert consultant for triers of fact can therefore simultaneously assist in furthering the field of forensic linguistics and social justice.

Notes

1 The term "language expert" is used here to refer to those with education and training in analyzing and describing language use. As Bucholtz et al. (2014) point out, all language users are indeed linguistic experts. The distinction I aim to make here relies on the tools that trained linguists can apply in order to test intuitions and explain language use.
2 Indeed, forensic linguists may also be susceptible to such biases, but forensic linguists are often trained to be aware of them and employ methodological steps to reduce biases.
3 However, see Haviland (2003) in which his role as an expert witness involved identifying language ideologies in various cases.
4 See Dodson (2016) for an overview.
5 Hofstra University's Forensic Linguistics Capital Case Innocence Project.
6 Importantly, the question here is not whether the statements are accurate as to the actual spoken interaction.
7 Sentences here refer to the original transcriber's imposed punctuation choices appearing in the transcript.
8 Of course, a forensic linguist also has a right to say "no" to cases for a variety of reasons (see Shuy 2002; and Ainsworth 2022), including the mental well-being of the forensic linguist (see Elstein 2019; and Giménez, Elstein, and Queralt 2021).

Case Law

Bruton v. United States. 1968. 391 U.S. 123.
Gray v. State of Maryland. 1998. 523 U.S. 185.
Richardson v. Marsh. 1987. 481 U.S. 200.
State of North Carolina v. Brewington. 2000. 352 N.C. 489.
State of North Carolina v. Ferrell. 1980. 46. N.C. App. 52.
State of Florida v. *Zimmerman*. 2013. 114 So. 3d 446.

References

Ahearn, Laura M. 2010. "Agency and Language." In *Society and Language Use*, 28–48. Edited by Jürgen Jaspers, Jan-Ola Östman, and Jef Verschueren. Amsterdam: Top of Form John Benjamins Publishing.

Ainsworth, Janet. 2022. "Serving Science and Serving Justice: Ethical Issues Faced by Forensic Linguists in Their Role as Expert Witnesses." In *Language as Evidence: Doing Forensic Linguistics*, 35–53. Edited by Victoria Guillén-Nieto and Dieter Stein, Cham, Switzerland: Springer International Publishing.

Ainsworth, Janet. 2012. "The Meaning of Silence in The Right to Remain Silent." In *The Oxford Handbook of Language and Law,* 288–298. Edited by Lawrence M. Solan and Peter M. Tiersma. Oxford: Oxford University Press.

Ainsworth, Janet. 2008. "You Have the Right to Remain Silent…But Only if You Ask for it Just So: The Role of Linguistic Ideology in American Police Interrogation Law." *International Journal of Speech, Language and the Law* 15, no. 1: 1–21.

Angermeyer, Philipp Sebastian. 2008. "Creating Monolingualism in the Multilingual Courtroom." *Sociolinguistic Studies* 2 no. 3: 385–404.

Bauman, Richard, and Charles L. Briggs. 1990. "Poetics and Performances as Critical Perspectives on Language and Social Life." *Annual Review of Anthropology* 19, no. 1: 59–88.
Bernstein, Basil. 1990. *Class, Codes and Control: The Structuring of Pedagogic Discourse*. New York, NY: Routledge.
Blommaert, Jan. 2005. *Discourse: A Critical Introduction*. New York, NY: Cambridge University Press.
Brown-Blake, Celia. 2014. "Expanding the Use of Non-Dominant Caribbean Languages: Can the Law Help?" *The International Journal of Speech, Language and the Law* 21, no. 1: 51–82.
Bucholtz, Mary. 2009. "Captured on Tape: Professional Hearing and Competing Entextualizations in the Criminal Justice System." *Text & Talk* 29, no. 5: 503–523.
Bucholtz, Mary. 2000. "The Politics of Transcription." *Journal of Pragmatics* 32, no. 10: 1439–1465.
Bucholtz, Mary, Audrey Lopez, et al. 2014. "Sociolinguistic Justice in the Schools: Student Researchers as Linguistic Experts." *Language and Linguistics Compass* 8, no. 4: 144–157.
Conley, John M., William M. O'Barr, and Robin Conley Riner. 2019. *Just Words: Law, Language, and Power*. Chicago: University of Chicago Press.
Coulthard, Malcolm. 2002. "Whose Voice is It? Invented and Concealed Dialogue in Written Records of Verbal Evidence Produced by the Police." In *Language in the Legal Process*, 19–34. Edited by Janet Cotterill. London: Palgrave Macmillan.
Coulthard, Malcolm, Alison Johnson, and David Wright. 2016. *An Introduction to Forensic Linguistics: Language in Evidence*. New York, NY: Routledge.
Dodson, Margret. 2016. "Bruton on Balance: Standardizing Redacted Codefendant Confessions Through Federal Rule of Evidence 403." *Vanderbilt Law Review* 69, no. 3: 803–844.
Du, Biyu. 2015. "The Silenced Interpreter: A Case Study of Language and Ideology in the Chinese Criminal Court." *International Journal for the Semiotics of Law-Revue internationale de Sémiotique juridique* 28, no. 3: 507–524.
Duranti, Alessandro, and Charles Goodwin, eds. 1992. *Rethinking Context: Language as an Interactive Phenomenon*. Cambridge: Cambridge University Press.
Eades, Diana. 2012. "The Social Consequences of Language Ideologies in Courtroom Cross-Examination." *Language in Society* 41, no. 4: 471–497.
Eades, Diana. 2008. *Courtroom Talk and Neocolonial Control*. Berlin: Mouton de Gruyter.
Ehrlich, Susan. 2012. "Text Trajectories, Legal Discourse and Gendered Inequalities." *Applied Linguistics Review* 3, no. 1: 47–73.
Fadden, Lorna, and Solan, Lawrence M. 2015. "Expert Witness Communication." In *Communication in Investigative and Legal Contexts: Integrated Approaches from Forensic Psychology, Linguistics and Law Enforcement*, 209–228. Edited by Gavin Oxburgh, Trond Myklebust, Tim Grant, and Rebecca Milne. Malden, MA: Wiley Blackwell.
Giménez, Roser, Solly Elstein, and Sheila Queralt. 2021. "The Pandemic and the Forensic Linguistics Caseworker's Wellbeing: Effects and Recommendations." *International Journal of Speech, Language & the Law* 27, no. 2: 233–254.
Grant, Tim. 2012. "TXT 4N6: Method, Consistency, and Distinctiveness in the Analysis of SMS Text Messages." *Journal of Law and Policy* 21, no. 2: 467–494.
Grice, H. Paul. 1975. "Logic and Conversation." In *Speech Acts*, 41–58. Edited by Peter Cole and Jerry L. Morgan. Leiden, The Netherlands: Brill.
Haviland, John B. 2003. "Ideologies of Language: Some Reflections on Language and US Law." *American Anthropologist* 105, no. 4: 764–774.
Haworth, Kate. 2018. "Tapes, Transcripts and Trials: The Routine Contamination of Police Interview Evidence." *The International Journal of Evidence & Proof* 22, no. 4: 428–450.
Heffer, Chris, Francis Rock, and John M. Conley, eds. 2013. *Legal-lay Communication: Textual Travels in the Law*. Oxford: Oxford University Press.
Heritage, John. 1984. "Garfinkel and Ethnomethodology." In *Structures of Social Action: Studies in Conversation Analysis*, 346–369. Edited by J. Maxwell Atkinson and John Heritage. Cambridge: Cambridge University Press.
Heritage, John, and Steven Clayman. 2011. *Talk in Action: Interactions, Identities, and Institutions*. Chichester, UK: John Wiley & Sons.
Hirsch, Susan F. 1998. *Pronouncing and Persevering: Gender and the Discourses of Disputing in an African Islamic Court*. Chicago: University of Chicago Press.

Irvine, Judith T. 1989. "When Talk isn't Cheap: Language and Political Economy". *American Ethnologist* 16, no. 2: 248–267.

Jacquemet, Marco. 2009. "Transcribing Refugees: The Entextualization of Asylum Seekers' Hearings in a Transidiomatic Environment." *Text & Talk* 29, no. 5: 525–546.

Jönsson, Linda, and Per Linell. 1991. "Story Generations: From Dialogical Interviews to Written Reports in Police Interrogations." *Text-Interdisciplinary Journal for the Study of Discourse* 11, no. 3: 419–440.

Komter, Martha. 2019. *The Suspect's Statement: Talk and Text in the Criminal Process*. New York, NY: Cambridge University Press.

Labov, William. 1982. "Objectivity and Commitment in Linguistic Science: The Case of the Black English Trial in Ann Arbor." *Language in Society* 11, no. 2: 165–201.

Leonard, Robert A., Juliane E. Ford, and Tanya K. Christensen. 2017. "Forensic Linguistics: Applying the Science of Linguistics to Issues of the Law." *Hofstra Law Review* 45, no. 3: 881–897.

Lippi-Green, Rosina. 1997. *English with an Accent: Language, Ideology, and Discrimination in the United States*. New York, NY: Routledge.

Maryns, Katrijn. 2005. "Monolingual Language Ideologies and Code Choice in the Belgian Asylum Procedure." *Language & Communication* 25, no. 3: 299–314.

Mertz, Elizabeth. 2007. *The Language of Law School: Learning to "Think Like a Lawyer"*. New York, NY: Oxford University Press.

Rickford, John R., and Sharese King. 2016. "Language and Linguistics on Trial: Hearing Rachel Jeantel (and Other Vernacular Speakers) in the Courtroom and Beyond." *Language* 92, no. 4: 948–988.

Rock, Francis. 2001. "The Genesis of a Witness Statement." *Forensic Linguistics* 8, no. 2: 44–72.

Shuy, Roger W. 2013. *The Language of Bribery Cases*. New York, NY: Oxford University Press.

Shuy, Roger W. 2006. *Linguistics in the Courtroom: A Practical Guide*. New York, NY: Oxford University Press.

Shuy, Roger W. 2002. *Linguistic Battles in Trademark Disputes*. London: Palgrave Macmillan.

Shuy, Roger W. 1998. *The Language of Confession, Interrogation, and Deception*. Thousand Oaks, CA: Sage Publications.

Shuy, Roger W. 1993. *Language Crimes: The Use and Abuse of Language Evidence in the Courtroom*. Cambridge, MA: Blackwell.

Silverstein, Michael. 1979. "Language Structure and Linguistic Ideology." In *The Elements: A Parasession on Linguistic Units and Levels*, 193–247. Edited by Paul R. Clyne, William F. Hanks, and Carol L. Hofbauer. Chicago: Chicago Linguistic Society

Slobe, Tyanna. 2016. "Creepy-Ass Cracker in Post-Racial America: Don West's Examination of Rachel Jeantel in the George Zimmerman Murder Trial." *Text & Talk* 36, no. 5: 613–635.

Solan, Lawrence M. 1998. "Linguistic Experts as Semantic Tour Guides." *Forensic Linguistics* 5, no. 2: 87–106.

Solan, Lawrence M., and Tammy Gales. 2016. "Finding Ordinary Meaning in Law: The Judge, the Dictionary or the Corpus?" *International Journal of Legal Discourse* 1, no. 2: 253–276.

Solan, Lawrence M., and Peter M. Tiersma. 2005. *Speaking of Crime: The Language of Criminal Justice*. Chicago: University of Chicago Press.

Sullivan, Grace. 2017. *Problematizing Minority Voices: Intertextuality and Ideology in the Court Reporter's Representation of Rachel Jeantel's Voice in the State of Florida v. George Zimmerman Murder Trial*. Doctoral dissertation, Georgetown University.

Tiersma, Peter M., and Lawrence M. Solan. 2002. "The Linguist on the Witness Stand: Forensic Linguistics in American Courts." *Language*, 78, no. 2: 221–239.

Trinch, Shonna L. 2003. *Latinas' Narratives of Domestic Abuse: Discrepant Versions of Violence*. Amsterdam: John Benjamins Publishing.

16

THE GOLDEN RULE AND THE GOLDEN THREAD

The Ethics of Communicating the Criminal Standard of Proof

Chris Heffer

Introduction

The criminal standard of proof in common law legal systems is the degree of evidential persuasion required by the prosecution to convict a defendant in a criminal case. It is part of the fundamental legal-ethical principle "that a person may not be convicted of a crime unless the government proves guilt beyond a reasonable doubt" (Garner 2009: 1306). This umbrella notion of the presumption of innocence actually contains within it three different foundational ethical principles of proof: (1) that the defendant is to be presumed innocent until proven guilty (the presumption of innocence proper); (2) that it is the prosecution rather than the defense who have to prove the case (the burden of proof); and (3) that they have to prove the case to a very high standard if the defendant is to be convicted (the standard of proof). The societal significance of conveying the criminal standard of proof successfully to the jury cannot be overestimated. If jurors misunderstand the standard, they might convict or acquit defendants unjustly, and inconsistent application of the standard may undermine the rule of law. The reasonable doubt standard has been described in the United States as "a constitutional cornerstone of the criminal justice system" (*United States v. Pepe 1974*: 1142–1143), and, in the United Kingdom, Viscount Sankey famously described the burden and standard of proof as the "golden thread" that runs through "the web of the English Criminal Law" (*Woolmington v. DPP* 1935: 481–482).

Most work on standard of proof instruction has focused on issues of linguistic comprehensibility, or the complexity of linguistic features in the proof instructions (Tiersma 2001; Dumas 2000; Charrow and Charrow 1979), whether mock jurors in psychological testing can actually comprehend current instructions (e.g., Woody and Greene 2012; Saxton 1998; Kramer and Koenig 1990)—they mostly cannot—or, from a legal perspective, what type of approach to instruction would be most effective (e.g., Cicchini 2019; Picinali 2015; Horowitz and Kirkpatrick 1996). In terms of the effectiveness of instruction, while there is a general consensus across jurisdictions that current instruction is not fit for purpose, there is currently no consensus on the optimal approach to instructing on the standard; and, as with many academic topics, arguments tend to go back and forth without resolution: Should quantification be part of the instruction (Woody and Greene 2012; Slovic, Monahan, and MacGregor 2000; Kagehiro and Stanton 1985; Tribe 1971)? Should the

DOI: 10.4324/9781003391074-18

definitional focus be on reasonable doubt or beyond reasonable doubt (Solan 1999)? Should the phrasing "beyond reasonable doubt"[1] be used or should it be replaced with an ordinary language term or no formula at all (Keane and McKeown 2019)? Should the wording of the instruction be fixed or flexible (Heffer 2006)?

Stepping back from these "local" arguments on practice, this article takes a highly original approach to the question by focusing on the *ethics* of communicating the standard of proof to the jury. Taking an ethical perspective on communication means asking what we owe others when we speak, or, in this case, what "judicial authorities" (those responsible for the instructions) owe others in designing and delivering the instructions. This approach is also a deliberate subversion of normative expectations within linguistics. Ethics have long been eschewed in linguistics since the discipline is meant to be a descriptive rather than normative or prescriptive discipline. There has, accordingly, been some pushback against a research program on the ethics of speech that includes exploration of our commitments to sincerity and epistemic responsibility in discourse (Heffer 2020), empirical investigation of arrogance and humility in debate (Hanel et al. 2023), and discussion of the ethics of persuasion in the legal process (Heffer In Prep).[2] Yet ethics have recently sneaked in through the back door into linguistics through the field of research ethics, or what we as researchers owe others (participants, stakeholders) when we conduct linguistic research. In recognition of the intended focus of this volume, I end with reflections on the ethics of researching the criminal standard of proof instructions, but my primary intention in the chapter is to show that ethics should be more central to linguistic analysis and not simply a fundamental part of linguistic methods.

Beginning with a philosophically more robust version of the ancient Golden Rule that we should treat others as we would wish to be treated ourselves, the chapter delineates four features (accuracy, consistency, comprehensibility, and persuasive engagement) that judicial authorities owe their audiences when they communicate the criminal standard to juries. The chapter then goes on to show how this ethical template can help resolve many of the key issues involved with conveying the standard, even in the absence of a consensus on optimal instruction. It will be argued that while some of the ethical requirements are largely met, others—particularly those that are owed to jurors—have been largely ignored. However, as this chapter will demonstrate, although Viscount Sankey's golden thread may have become frayed, it can still be mended.

What Judicial Authorities Owe Others When Conveying the Criminal Standard of Proof

Let us start by considering who owes what to whom when conveying the criminal standard of proof to juries. The participation framework (Goffman 1981) of jury instruction is quite complex and can only be sketched in outline here. On the addresser side (the party who owes), the trial judges are the "animators" (sound boxes) who deliver the instructions orally to the jury. However, while English and Welsh judges have considerable discretion in the wording of their standard of proof instructions and so become at least partial authors (Heffer 2006, 2005, 2002), US judges often have a minimal authorship role, if any.[3] Instead, proof instructions, like other jury instructions in the United States, are mostly authored by judicial drafting committees. Finally, the "principals", those who are ultimately responsible for what is said (Goffman 1981), are primarily the drafting committees and the appeals courts that oversee them in the United States. While English and Welsh judges have more of an authoring role, they are still subject to legal authorities such as the Judicial College and the Appeals Courts. To simplify matters, particularly given the significant jurisdictional differences, this chapter will operationally conflate the animators, authors, and principals involved into the umbrella term "judicial authorities". On the addressee side (the party to whom is owed), Goffman's distinction between "ratified participants" (those who are ratified to speak in the communication event) and "overhearers" is not particularly useful because even the direct addressees, the jurors, are not allowed to speak except to deliver their verdict. Instead, the distinct addressees of the instruction are considered. These addressees are the jurors who are directly addressed and who are responsible for applying the standard to the evidence; the legal parties (prosecution and defense) who have a vested interest in what is said; and the appeal court justices who might eventually adjudicate what is said.

What, then, do judicial authorities owe these direct and indirect addressees when they communicate the criminal standard of proof? To answer that question without embarking on a treatise on ethics, it is useful to turn to the philosophical golden rule. The Golden Rule should not be confused with the "golden rule argument" in trial advocacy in which the trial lawyer asks the jury to imagine themselves in the position of the complainant or defendant in the crime story being narrated by the lawyer and then to consider how they would feel in that position. This type of "argumentation" is rightly frowned on as manipulative: the lawyer is trying to sway the emotions of the jurors rather than appealing to reason. This is a gross distortion of the philosophical golden rule, which is actually designed to make our thinking more impartial by forcing us to consider the situation from other perspectives. In its simplest forms (e.g., "Do unto others as you would have them do unto you"), it is open to the types of objections made by Kant, who considered it "trivial" in comparison to his surprisingly similar "Categorical Imperative" (Kant 2012: 4:430). For example, Kant claimed that if a judge were to follow the golden rule, they should not convict a murderer because they would not want to be convicted of murder if they were in the defendant's shoes. Or, in Parfit's (2011) example, a White racist hotel manager could claim they were following the golden rule by not admitting Black guests because they would be happy not to be admitted to a Black-run hotel that did not admit White guests. Such objections arguably follow the letter of the rule rather than its spirit. However, it is possible to convey the rule in a more philosophically rigorous form, as in Derek Parfit's reformulation:

> We ought to treat everyone as we would rationally be willing to be treated if we were going to be in all of these people's positions, and would be relevantly like them.
>
> *(Parfit 2011: 327)*

Put in this way, it is clear that one needs to put oneself in the position of not just one person but the positions of all those affected by our actions. Thus, in a murder case, the judge needs to consider not just the defendant's perspective and imagine themselves as guilty. The judge would also have to consider the perspective of the victim's family, the murderer's family, and society at large, as none of us wants to see murderers acquitted. Similarly, the White racist hotel manager needs to consider not the position of a fellow Black racist, but to put themselves in the position of the non-racist Black guest *wanting* to stay in their hotel.

The Golden Rule, as reformulated by Parfit, works well in the context of conveying the criminal standard. Effectively, it asks the judicial authorities to rationally imagine themselves in the psychosocial position of each of the people affected by the instruction. Thus, if you were the defendant, or their legal representative, you would not want the judge to inadvertently shift the burden of proof to the defense or to convey too low a threshold for conviction. If you were the complainant, or the prosecution, you would not want the judge to inadvertently set the threshold for conviction too high. Combining those perspectives, you would want the threshold to be set as accurately as possible. If you were a Court of Appeal Justice, you would want the instruction to convey the standard accurately but would also want it to be consistent with instructional practice. So far these are easy thought experiments because they only involve considering institutional positions: complainant, defendant, prosecutor, higher court Justice.

The more difficult thought experiment is for the judicial authorities to imagine themselves in the psychosocial position of the juror. To be "relevantly like" an average juror, for example, one has to imagine having an average level of intelligence and education; possessing little understanding of the law; having little if any experience of trials; being familiar with the phrase "beyond reasonable doubt" but not having a real understanding of what it means in a legal context; and possessing no experience of applying legal rules to decide cases. In that psychosocial position, how would we want to be treated? Well, firstly, we would want to be able to understand what the legal formula conveying the standard means and to be able to easily understand the words surrounding that formula. But, secondly, we would also want help, through explanation and exemplification, with trying to apply that standard to the evidence in a legal case. Finally, we would want to be treated with dignity: to not be treated insincerely or as stupid or undeserving of explanation. Where jury instructions tend to go wrong is when they often show little sign of judicial authorities having put themselves in the position of the juror.

Applying the Golden Rule, then, we can establish that the judicial authorities owe significant others the following:

(1) **Accuracy** in conveying the standard: the burden of proof needs to remain on the prosecution and the threshold for conviction needs to be set very high. Accuracy (Williams 2002) or, in virtue epistemological terms "epistemic responsibility" (Code 1987), is an essential part of professional ethics.

(2) **Consistency** in conveying the standard: jurors across trials need to be receiving the same notion of the standard. Consistency of application is an ethical cornerstone of the rule of law.

(3) **Comprehensibility** of the criminal standard of proof instruction. It must be comprehensible to the average juror so that the jury as a group can apply the standard effectively.[4] Both the legal formula conveying the standard and the wording surrounding that formula need to be understandable to the average juror. This is the ethical principle of accommodation to the other.

(4) **Persuasive engagement** with the jury. It is not sufficient for the jury to understand the instructions: they need to be persuaded to apply them carefully to the evidence. Furthermore, in doing so, judicial authorities should not be insincerely withholding information or dogmatically claiming against the evidence that a certain approach works. Ethically, we are concerned here with the dignity of the juror.

However, judicial authorities do not owe stakeholders perlocutionary success or the successful uptake of instructions such that juries apply the standard accurately and consistently. There is currently a lack of expert consensus on what the most effective type of instruction would be to guarantee perlocutionary success, and evidence of successful uptake has remained elusive. At the same time, though, so much instruction falls short of giving stakeholders what they are owed that there is plenty that can be done before the perfect means of instruction can be found. In what follows, accuracy and consistency are first considered. Then comprehensibility of both legal formulas and instructional language is examined before finally the issues relating to persuasive engagement with the jury are addressed.

Accuracy and Consistency

Judicial authorities need to have both an accurate idea of the criminal standard and be able to apply it accurately and consistently.

The Concept of the Criminal Standard of Proof

Standards of proof generally reflect the subjective confidence level that is required of fact-finders in order to convict.[5] The appropriate confidence level depends on the risks involved and the relative gravity of convicting the innocent versus acquitting the guilty. In criminal cases, it is generally held that wrongful conviction is far worse than wrongful acquittal, because the wrongfully convicted are deprived of their liberty or, in capital punishment jurisdictions such as China, Iran, Saudi Arabia, and some US states, even of their life. Consequently, there is a strong international consensus both that the criminal standard of proof must be a very high standard and that it must maintain the burden of proof on the prosecution (Keane and McKeown 2019). This is true not only in common law jurisdictions but also in civil law systems (Clermont and Sherwin 2002). While the criminal standard of proof is not set in stone, Keane and McKeown (2019: 509) note that "there are no indications in the international jurisprudence that in the foreseeable future there is likely to be any move towards lowering the very high standard currently required".

The current criminal standard emerged not as a way of setting a high threshold for conviction, but as a way of persuading reluctant jurors, fearful of making a theologically "false judgment": rather than make an objectively true judgment, they *only* had to be "convinced in reason" (Whitman 2008). It thus constituted a lowering of the standard from "beyond *any* doubt" to "beyond *reasonable* doubt" (Morano 1975: 515). By the mid-19th century, jurists were associating the criminal standard with the highest level of theological certainty, or moral certainty. Moral certainty was "the kind of certainty on which it is sensible to rely for all practical purposes, even though cast-iron guarantees are lacking" (Dear 1992: 622). The highest legal authorities across the common law world show remarkable consensus in identifying the evidential threshold of the standard as the modern equivalent of moral certainty (i.e., "subjective certainty", e.g., *Victor v. Nebraska*).[6] This focus on a subjective mental state of conviction is demonstrated by the directions recommended by national judicial advisory boards in both the United States and England and Wales. For example, the US Federal Judicial Center recommends the following definition of "beyond reasonable doubt": "Proof beyond a reasonable doubt is proof that *leaves you firmly convinced* of the defendant's guilt" (Federal Judicial Center 1988, emphasis added). And the Judicial College for England and Wales advises judges to tell the jury that: "The prosecution will only succeed in proving that D[name of defendant][7] is guilty if *you have been made sure* of D[name of defendant]'s guilt" (Picton, Ormerod, and Tayton 2023: 5-3 Ex 1, emphasis added). Trials inevitably take place under conditions of uncertainty: if the evidence against a defendant were beyond the shadow of a doubt, the case would never go to trial. So the criminal standard is conceived in terms of the highest subjective conviction possible within that condition of uncertainty, whether we call that moral, subjective, practical, or functional certainty.

However, in the 20th century, the standard also started being conceived in terms of probabilities rather than subjective conviction. For example, in a much-cited passage from *Miller v. Pensions* (1947), Lord Denning conceived the standard very much in terms of possibility and probability:

> It need not reach certainty, but it must carry *a high degree of probability*. Proof beyond reasonable doubt does not mean proof beyond the shadow of a doubt. The law would fail to protect the community if it admitted *fanciful possibilities* to deflect the course of justice. If the evidence is so strong against a man as to leave only a *remote possibility* in his favour which can be dismissed with the sentence "of course it is *possible*, but *not in the least probable*", the case is proved, but nothing short of that will suffice.[8]

There is some consensus within the legal community that, if conceived in probabilistic terms, the criminal standard corresponds to roughly 85–90% probability. There have always been advocates of quantifying *beyond reasonable doubt* for the jury, and there has been a renewed interest in this (e.g., Smith 2022; Pi, Parisi, and Luppi 2019; Newman 2007). Traditionally, though, quantification has been rejected by legal scholars and philosophers (e.g., Laudan 2006; Cohen 1977; Tribe 1971) partly because of what it seems to suggest about sentencing. The very notion that one could put away a murderer for life because they are *probably* guilty is generally considered unacceptable, just as it is generally considered unacceptable for judges to deliver their sentences or judgments in probabilistic terms (e.g., "You have been found guilty because you most likely murdered your wife"). Furthermore, it needs to be remembered that when talking about probabilities in the context of jury decision-making, we are talking about subjective rather than objective probabilities. As Laudan (2006) points out, it is doubtful whether most jurors would be able to relate their degree of belief, or confidence level, in the defendant's guilt, to a numerical probability:

> While I could perhaps tell you that I believe strongly in the guilt of the accused, that I found the prosecution's case powerful and that of the defense weak, I would be hard pressed to tell you whether I would assign a probability of 95, 85, or 75 percent to the prosecution's hypothesis that the defendant committed the crime. And even if I did pull some specific number out of a hat, it would be a stretch to imagine that it accurately represented something like my "degree of belief" in the defendant's guilt.
>
> *(77–78)*

These are important observations because they also throw doubt on psychological literature in this area (e.g., Magnussen et al. 2014; Slovic, Monahan, and MacGregor 2000) that depends precisely on mock jurors assigning probabilities to their degree of belief in guilt or to the degree of belief required by the standard.

Accurate Communication of the Criminal Standard

From the perspective of rational decision-making, the legal term "beyond reasonable doubt" is arguably the most accurate formulation of the agreed standard since it shows how we can arrive at a state of subjective certainty under conditions of objective uncertainty. Namely, it addresses the fact that when considering multifarious elements from two opposing perspectives, we need to set aside trivial anomalies in the evidence if we are to come to a firm decision. Anomaly hunting is one of the central facets of conspiracy thinking. Conspiracy theorists hunt for and find hundreds of apparent anomalies in the evidence to "prove", for example, that COVID is caused by 5G masts or that there was widespread voter fraud in the US presidential elections. We avoid the conspiracy thinking trap by putting aside such anomalies as trivial, irrelevant, or highly speculative, and by going with the overwhelming evidence in the case that firmly points to COVID being caused by a virus and the 2020 US elections being won by Biden. In Denning's terms, the conspiracy theories are technically possible but not in the least probable. One recent proposal for explaining "beyond reasonable doubt" is in terms of surprise (Cheng and Ginther 2018; Risinger 2018): we (the jury) would be very surprised if it turned out that the defendant was not guilty after all, just as we would be very surprised if it turned out that a conspiracy theory were true. So pointing to the need to put aside trivial or hairsplitting doubts when the evidence points very strongly in favor of the prosecution is crucial, particularly for those jurors who might lean slightly toward conspiracy thinking.

At the same time, the ethical obligation to accuracy can conflict with the other obligations. With increasing recognition that jurors had trouble understanding the legal term "beyond reasonable doubt", from the mid-20th century in the jurisdiction of England and Wales, judges started being recommended to use a paraphrase of "beyond reasonable doubt": "the prosecution must make you *sure*" or "You must be satisfied so that you are *sure*" (Heffer and Coulthard In Press). *Sure* can certainly represent a state of subjective certainty and is clearly an ordinary word rather than a technical term and so is much more comprehensible (Heffer 2007, 2006). However, there is a particular problem with this qualitative adjective in terms of accuracy: namely, in everyday life, *sure* is scalar but, when used in the proof direction, *sure* is meant to be fixed at the very high level required by the criminal standard (Heffer and Coulthard In Press). Analysis of concordances from large corpora of the language shows that we can be somewhat sure or reasonably sure or absolutely sure and that the modifiers are not necessarily used to explicitly convey that degree of sureness. For example, we often say "I'm sure you're right" not to mean "I'm genuinely convinced you are right" but to concede that you *may* be right. It is not at all unusual to ask, "How sure are you?", and thus it is to be expected that jurors will ask, "How sure do we have to be?" Yet that is not a legally permissible question because the standard is meant to be a binary, like "guilty": sure or not sure. And so the correct legal response is the absurd "*sure* cannot be modified". Accordingly, *sure*, if stipulated to have the legal meaning of "subjective certainty", is accurate. However, if it is understood as a scalar ordinary language term, it is not accurate.

Message Consistency

Clearly, the less discretion the judge has in wording their instruction, the greater the degree of message consistency. While the US Constitution does not require specific wording beyond the formula "proof beyond a reasonable doubt", most federal and state jurisdictions have a pattern or model instruction that is prepared carefully by drafting committees and then is either read verbatim to juries or is only slightly adapted by the judge. Such an approach clearly scores highly for consistency: juries in any criminal trial in a given jurisdiction will be receiving effectively the same instruction, which guarantees the rule of law. However, before pattern instructions were introduced, there was a vicious circle of Appeal Court approval and disapproval of specific wordings. This had a chilling effect on judge-jury communication and severely tied the hands of drafting committees, who could not disregard statutory-approved language (Tiersma 2001). Thus legal precedent meant that wordings became fossilized in time and became archaic and difficult for contemporary juries to understand (see below). That having been said, giving judges much greater authorship rights, as in England and Wales, can potentially lead to inconsistencies across trials but tends to increase comprehensibility[9] and persuasive engagement with the jury.

Comprehensibility

Key ethical issues relating to the comprehensibility of the criminal standard of proof instruction include: the claim that "beyond reasonable doubt" is an ordinary language term; the comprehensibility of overly complex definitions of the legal term; and the use of *sure* as an ordinary language alternative.

Beyond Reasonable Doubt as "Ordinary Language"

According to the plain meaning rule, terms used in legal writing are either given stipulated meanings (which recognize their legal status) or they are said to have the same meaning as in ordinary language. When the "beyond reasonable doubt" formula emerged, it would have been considered ordinary language. "Beyond reasonable doubt" simply meant "beyond reason to doubt" and "reasonable" would have been inevitably understood as "according to reason" (Whitman 2008; Morano 1975). Moreover, the term "moral certainty", or "subjective certainty" with which it was linked, was well understood by ordinary churchgoers (i.e., most of the population) of the time (Whitman 2008). However, as the religious context of "beyond reasonable doubt" waned, it ceased to be part of ordinary language, jurors started having problems understanding it, and the majority of jurisdictions moved to treating the term as a legal one by stipulating its meaning in jury instructions. However, these definitions in turn became antiquated, with complex syntax and terms such as "moral certainty" and "captious [i.e., trivial] doubt" becoming archaic or rare. In coping with juror misunderstanding created by these complex definitions, many courts of appeal in the United States have discouraged definition altogether. Pi, Parisi, and Luppi (2019: 456–457 n.9) provide a long list of such decisions eschewing definition by judges. Some have simply shied away from the problem: "As a matter of Illinois common law, trial judges are instructed not to attempt to define reasonable doubt for the jury" (*Adorno v. Melvin* 2017). Others implicitly suggest that the term is ordinary: "Our decisions hold that 'reasonable doubt does not *require* definition'" (emphasis added) (*United States v. Fields* 2011). While others still are fairly explicit: "We have consistently and repeatedly held that reasonable doubt is *self-explanatory*" (*Postelle v. State* 2011, emphasis added). Some Australian jurisdictions (e.g., New South Wales, Victoria) are even more explicit about *beyond reasonable doubt* being ordinary language that jurors can understand. Thus, while the New South Wales Law Reform Commission recognized in 2008 that it "commonly occurs in criminal trials [that] the jury asks for some further explanation of the expression" (NSWLRC 2008: 69), this evidence of lack of understanding was and still is trumped by the authoritative voice of a man often described as "Australia's greatest jurist": "Since 1961, when Chief Justice Dixon referred to the formula 'beyond reasonable doubt' as 'time honoured,' appellate courts in Australia have consistently held that it is *an expression well understood by ordinary people*" (NSWLRC 2008: 67, emphasis added). Yet the appellate courts hold that the expression is "well understood by ordinary people" precisely by referring to Chief Justice Dixon's authoritative word. The argumentative circularity here is typical of dogma, or "unwarranted discursive closure" (Heffer 2020: 183).

There are, in fact, multiple forms of evidence that show that "beyond reasonable doubt" is no longer an ordinary language term but a legal term requiring explanation. There is the linguistic evidence of language change. Today, "according to reason" is the less common of the two main contemporary meanings of *reasonable*, the more common one being *moderate*, *fair*, *average*. But then the term can be construed as something like "beyond a fair amount of doubt", which clearly lowers the standard (Heffer 2007, 2006). Moreover, the link with "moral certainty" has been lost: now that term is more likely to be understood as meaning certainty about one's morals. Then there is the lexicographical evidence. The term is invariably included in specialist legal dictionaries such as *Black's* (Garner 2009) and often tagged with the label "legal" in general dictionaries, including the Australian Macquarie dictionary (Macquarie 2010). Furthermore, where it *is* included as an ordinary language word, the definition does not correspond to the legal one. For example, the COBUILD dictionary, the first to be based on analysis of actual language use, gives "beyond doubt" and "beyond reasonable doubt" as synonyms (COBUILD 2001). The empirical evidence on the comprehension of the term also shows that there is poor understanding of its meaning (Solan 1999). For example, Horwowitz and Kirkpatrick (1996: 668) found that juries given undefined *beyond reasonable doubt* instructions, in comparison to those given defined instructions, were less sure of their decisions, less likely to concentrate on the evidence, and more likely to set lower standards of proof. Moreover, in terms of professional practice, the emergence in the 19th century of the judicial practice of defining the term (or, at least, "reasonable doubt") strongly suggests that jurists no longer expected that jurors would understand the term without explanation. The fact that the majority of jurisdictions still define the term further suggests an understanding that the term is a legal one requiring explanation. While it is *possible* that the term has returned to being ordinary language in the interim, and the majority jurisdictions are simply wrong, it is *not in the least probable*. Arguably, jurists who hold the ordinary language line are confusing passive recognition of a term, which in all probability most people will have heard of, with an active correct contextual understanding of a term. The fact that people recognize the term does not preclude them from understanding it to mean "beyond all doubt" or "beyond a fair amount of doubt", which in the trial context are both clearly inaccurate. Similarly, shortly before C. J. Dixon made his influential obiter in Australia, Lord Goddard suggested replacing the term with the ordinary language term "sure", precisely because "beyond reasonable doubt" was no longer part of ordinary language. Finally, there is evidence from those jurisdictions that do not define the term that juries frequently seek a definition or explanation from the trial judge and there is a recognition by judges that this is the case (Keane and McKeown 2019; *The Queen v. Dookheea* [2018] at [27]; NSWLRC 2009: 69; Young, Cameron, and Tinsley 1999; R v. Lifchus [1997]).

Given these multiple forms of evidence that "beyond reasonable doubt" is a legal term, it is right, not only legally but also ethically, that it should be defined for the jury: we owe it to the jurors to explain this legal term to them so that they can carry out their vital task properly. We have a general epistemic responsibility to take care that what we say is true (Heffer 2020). Some speakers/authors, though, including journalists, academics, lawyers, and experts of all types, have a particular "duty of epistemic care" (Heffer 2020: 203–205) that derives from their role responsibilities (Hart 2008: 212–213). In such cases, there is a strong warrant that speakers will have taken care to get their facts right and that, if they are not certain of them, they will hedge their claims. Failure to take epistemic care in such circumstances is not merely irresponsible but can be described as "epistemically negligent" (Heffer 2020: 202–207). Judges and, in the United States in particular, the drafting committees who are responsible for the words uttered in jury instruction, have a particular duty of epistemic care to make sure that what they are saying with respect to the criminal standard of proof is true. Yet judicial authorities in "non-defining" jurisdictions such as New South Wales and Victoria in Australia and Oklahoma and Kentucky in the United States effectively make false statements about the term being ordinary language. There is no suggestion that these judicial authorities are lying: they may well sincerely believe that "beyond reasonable doubt" is indeed ordinary language. However, they are clearly in the grips of dogma, the main feature of which is a closure to alternative viewpoints (Heffer 2020:183–188; Christensen 2011). For example, in the face of understandable jury questions on the meaning of "beyond reasonable doubt", judges and justices may impute a lack of intelligence to the jurors. In *R v. Chatzidimitriou* (in the State of Victoria), the jury, thwarted in their attempt to get the trial judge to explain the formula, asked for a dictionary. The judge, arguing the case with counsel, made the following statement:

> HIS HONOUR: Mr O'Doherty, I know that the courts have taken a very strong view that judges ought not give "beyond reasonable doubt" a legal definition and in fact the only direction that's really permissible is that which I've given – they're plain English words and that it connotes a very high standard of proof in the law. But when one tells lay people that it is a matter of plain language and they want to understand the language better, why should they not have recourse to an authorised dictionary?
>
> (R v. Chatzidimitriou *2000, at [36])*

There is an evident contradiction here between asserting on the one hand that the formula is "plain language" and acceding on the other hand to a demand for a dictionary. The judge tries to resolve it by imputing linguistic ignorance to the jurors who need to "understand the language better". As one of the justices put it even more condescendingly at appeal: "to give them a dictionary is, on one view, to let the jurors do by reading what any competent English teacher might be able to do by recollection" (Heffer 2013: 220).

While one might ideally hope for practicing judges or justices to consult academic work on the subject, in practice it is unusual for legal professionals to go beyond legal cases and materials. However, it takes some effort not to listen to the many juries asking for judges to explain a term that, as ordinary people, they clearly do not understand, and which is understood as a legal term in many other jurisdictions and clearly defined as a legal term in legal dictionaries. Given the role responsibility of the judicial authorities in these jurisdictions and their expert status, this refusal to recognize the legal status of "beyond reasonable doubt" could be described as not only being epistemically negligent but also blatantly unethical.

The Comprehensibility of Beyond Reasonable Doubt Instructions

We have established that "beyond reasonable doubt" is a legal term with a specific legal meaning that jurors very often do not understand. Accordingly, it is ethically right to try to explain what is meant by the term. However, there is plenty of evidence that jurors are not understanding the proof instructions. This is not surprising if we consider as an example the proof instructions delivered by Judge Ito in the O. J. Simpson criminal trial (Tiersma 2001):

> A defendant in a criminal action is presumed to be innocent until the contrary is proved, and in case of a reasonable doubt whether his guilt is satisfactorily shown, he is entitled to a verdict of not guilty. This presumption places upon the prosecution the burden of proving him guilty beyond a reasonable doubt. Reasonable doubt is defined as follows. It is not a mere possible doubt; because everything relating to human affairs is open to some possible or imaginary doubt. It is that state of the case which, after the entire comparison and consideration of all the evidence, leaves the minds of the jurors in that condition that they cannot say they feel an abiding conviction of the truth of the charge.

There are numerous linguistic features here that are known to cause comprehension difficulties: complex syntactic embedding (particularly the last sentence) (Charrow and Charrow 1979; Felker et al. 1981: 43); excessive use of the passive ("is presumed", "is proved", "is entitled", "is defined") (Tiersma 1999: 207); nominalizations ("presumption", "the contrary") (Felker et al. 1981: 38); abstract subjects ("a defendant in a criminal action", "this presumption") (Heffer 2006); rare or archaic words ("abiding") (Tiersma 1999: 204–205). Most of these features are common to jury instructions in general (Charrow and Charrow 1979).

An explanation often given for the difficult language in jury instructions is that legal language is required to convey the instructions correctly. Yet, unlike explanations of the criminal offenses, there is no need for legal language in the proof instructions beyond the legal formula itself. This can be seen by comparing US American Judge Ito's instruction with an English judge's attempt at instruction:

> [T]he prosecution must prove the defendant's guilty. He does not have to prove anything. How do they succeed in doing that? Well, that is quite simple. You have already heard it, but I emphasise it because it is of fundamental importance. And that is that they must make you *sure* he is guilty. Nothing less will do. It is not enough that, "Oh, we think he may have acted badly", or "I suppose he did it". That is just not enough. You must be *sure* of his guilt before you can convict in respect of any of these counts.
>
> *(Heffer 2002: 240)*

This instruction is far from ideal, not just because of the issues relating to the accuracy of *sure* as representing the criminal standard but also because the hypothetical speech ("Oh, we think he may have acted badly") is potentially condescending to the jurors. However, all the sentences are very simple and the only legal language consists in the terms "prosecution", "defendant" and "counts", which should be well understood by the jurors by the end of the trial.

Accordingly, there is an ethical obligation to try to ensure that the instructions given to the jury are as clear and comprehensible as possible. This ethical obligation often comes into conflict, particularly in US jurisdictions, with legal precedent that establishes certain wordings as required. Tiersma (2021, 2001) found, for example, in working on a drafting committee in California, that he was severely restricted in the sorts of changes he could make because words and phrases had been authorized by precedent. Yet comprehensible instructions are a major part of what is owed to jurors.

The question of *how best* to convey the concept of the criminal standard itself, on the other hand, is more difficult to construe in ethical terms. This is because, while the question has occupied lawyers and scholars for decades (Keane and McKeown 2019; Picinali 2015), we are still not close to reaching an academic or legal consensus. Choosing a method where the overwhelming expert consensus is that it is the most effective method is an ethical choice: an ethical doctor owes it to the cancer patient to prescribe chemotherapy rather than vitamins. However, if there is not an overwhelming expert consensus, then this cannot be an ethical choice. Defining or explaining "beyond reasonable doubt" belongs more to this second category than the first. Almost all the proposals made have had supporters and detractors but none have come with such overwhelming proof that to ignore the conclusions could be considered epistemically negligent. In contrast, we have seen that it is epistemically negligent to ignore the fact that "beyond reasonable doubt" is a legal term requiring explanation.

At most one can say that certain definitions might "distort" the standard. Distortion (Gilovich, Griffin et al. 2002) can be seen as a form of "discourse pathology" (Heffer 2020: 80–81) in which the speaker unintentionally misrepresents a state of affairs because they have not taken sufficient care in establishing and/or conveying the facts (2020: 188–195). For example, Solan (1999) argues convincingly that definitions that focus solely on the concept of "reasonable doubt" inadvertently switch the burden of proof from the prosecution to the defense: the jurors are encouraged to search for real doubts that would exonerate the defendant rather than considering the overall strength of the prosecution case and whether that leaves them firmly convinced about the guilt of the defendant. Similarly, there is a growing consensus (e.g., *R v. Lifchus*; *R v. Wanhalla* 2007) that the well-meaning analogy that the level of conviction required is what one would expect when making important decisions in one's life is actually very misleading (in the distortion sense) since people can actually be quite cavalier about such decisions (e.g., buying a house on a hunch or marrying someone they are currently infatuated with). Such decisions involve subjective perceptions, which can be notoriously inaccurate. Jurors, by comparison, have the advantage of being independent adjudicators who can consider the case more objectively and very often in more detail than they would if it were an important matter in their own life. However, while there might be a growing consensus on a few things that should not be said, there is little agreement on what should be said.

In conclusion, one can say that it is ethically right to define or explain "beyond reasonable doubt" and to *try* to do so in a way that will be both comprehensible to and actionable by the jurors. On the other hand, while certain wordings might distort the standard, there is a lack of sufficient expert consensus to say that a certain approach to definition or explanation is the ethically right one.

"Sure" as an Ordinary Language Alternative

Replacing "beyond reasonable doubt" with an ordinary language formula like "the prosecution must *make you sure*" or "the evidence must leave you *convinced*" appears to avoid some of the pitfalls in attempting to define the standard legal term. However, there are two potential ethical problems with this approach.

Firstly, given that words used frequently in legal contexts tend to take on legal meanings, there is always the risk that the ordinary language term will eventually become a legal term with a specific legal meaning that is potentially just as alien to the lay juror as "beyond reasonable doubt" but without recognition that it is, in fact, a legal term. As indicated above, this is arguably what has happened with the formula *make you sure* used in England and Wales (Heffer and Coulthard, In Press). While dictionary definitions seem to suggest a meaning that is close to the intended legal meaning of *sure*, e.g., "free from hesitancy or uncertainty (with regard to a belief, conviction, etc.)" (CED 2003), corpus linguistic evidence suggests that *sure* is actually used in a wide variety of contexts where the meaning could not be construed as equivalent to "beyond reasonable doubt" (Heffer and Coulthard, In Press). Yet precedent has now set clearly legal parameters to the meaning of *sure*. Given that this is now the legal meaning of *sure*, it is unethical to withhold these legal meanings from the jury, but that would mean defining and explaining the legal meaning of *sure* in the proof instructions, and there is less to hang a definition on than with the standard formula.

Secondly, in the case of *sure* used in England and Wales, it effectively co-exists as a standard with "beyond reasonable doubt". While the recommended formula has for many decades been based on *sure*, barristers are still permitted to use "beyond reasonable doubt" in their speeches to the jury so that the judge then has to explain to jurors that the terms "sure" and "beyond reasonable doubt" mean the same thing, which must sound counter-intuitive to them. Furthermore, judges can unwittingly introduce some serious distortions when juries inevitably ask, "How sure do we have to be?" The correct legal response, as we have seen, is that the legal term *sure* cannot be modified in any way, even if it is very frequently modified as an ordinary language term. This, in itself, is misleading for jurors who are not told specifically that *sure* has a legal sense. However, in trying to respond to this question, judges will sometimes respond that the jurors need to be "sure beyond reasonable doubt". Yet in this case, the traditional formula is being used as a scalar postmodifier to indicate the degree of sureness that is required, which is impermissible in terms of the legal meaning of *sure* (Heffer and Coulthard, In Press).

Persuasive Engagement

Conceptual accuracy, consistency of delivery, and textual comprehensibility are necessary conditions for successful jury uptake of a criminal standard of proof instruction, but they are not sufficient. In order to comply with instructions, jurors need to have both the capability and willingness to do so (Heffer 2008). In terms of capability, research shows that jurors are sufficiently competent to undertake their task but that they are hampered by incomprehensible instructions. In terms of willingness, nothing is done to extrinsically motivate the jurors: they are not paid, they are not rewarded for performing well, and they are not required to motivate their verdict. Therefore, the system relies on intrinsic motivation, and that needs to be nurtured through persuasive engagement with the jury. The jury needs to be respected for playing their vital role and to be treated with dignity. But, as "ordinary citizens" surrounded by experts and not ratified to speak, jurors can easily be treated as mere means to a verdict rather than as individuals with their own dignity. The late Bethany Dumas (2000a) argued that jurors should be treated as "expert" fact-finders, and while that superficially seems to jar with the fact that they are chosen precisely for their role as "ordinary" citizens, it makes very good sense in terms of treating the jury with respect.

One of the most important acts of respect is to recognize that a specific set of individual jurors have come together for this specific trial with this set of witnesses and defendants to deliberate on and reach a specific verdict. Yet if, as is often the case in US jurisdictions, the focus is on delivering pattern legal instructions accurately and consistently, without considering whether or not the jurors will be able to understand and apply them but thinking primarily of whether the instructions will avoid risk of appeal, then the jurors are not being considered as ends in themselves and so are not being treated with dignity. And if they are not treated with dignity, they will not be motivated to comply carefully with the instructions. Jury instructions, including the criminal standard of proof, are often delivered to the jury as a driver delivers an Amazon parcel: routinely and without explanation. And the parcel contains instructions like a set of terms and conditions that have been carefully crafted to comply with the law but without thinking of the end user.

Take again Judge Ito's proof instructions in the O. J. Simpson criminal trial. The proof instructions are generally considered to be the most crucial jury instructions of all and, in recognition of this, they tend to be covered very early in an English judge's summing-up. Yet Judge Ito's presumption of innocence instructions occurred after 3,580 words and were followed by a further 1,980. Nothing was done to flag up the instructions as crucial, as in the sample English instruction above: "You have already heard it, but I emphasise it because it is of fundamental importance". It would have been difficult for the jurors to sustain the attention to even get to the proof instructions. Indeed, because the instructions are seen as a set of terms and conditions, there is often no attempt to structure the instructions in a meaningful way for the jury. Then, although the judge is instructing the specific jury seated in the trial, there is nothing in the wording to suggest that. Indeed, the only mention of "jurors" is as a postmodifier of "minds" deeply embedded in the abstract definition of reasonable doubt: "It is that state of the case which, after the entire comparison and consideration of all the evidence, leaves *the minds of the jurors* in that condition that they cannot say they feel an abiding conviction of the truth of the charge", which repeats verbatim an instruction given over 170 years ago (*Commonwealth v. Webster* 1850, 59 Mass.: 295, 320). This is apparently not something that the jurors need to pay attention to: it is simply an abstract definition of reasonable doubt included as part of the terms and conditions that come with the jury's deliberation task.

There are many ways in which judges can engage more persuasively with the jury. English and Welsh judges are now encouraged to integrate the legal instructions with the case[10] (Auld 2001). One might wonder how one can do that with a generic standard of proof instruction, but one way is to "narrativise" the instruction (Heffer 2005, 2002) (i.e., to provide human subjects, agency and action). Thus, the Judicial College encourages judges to use the defendant's name in the proof instructions rather than the abstract "the defendant" and some even personalize the lawyers and, if very straightforward, specify the offense. So, for example: "Ms Smith has to prove that Mr Jones murdered his wife. Mr Jones and his barrister Mr Doughty don't have to prove anything". Nelson (2013) found that integrating the proof instructions with the specific case details improved comprehension and that narrativizing the instruction further improved comprehension. Other means of engaging actively with the jury include encouraging jurors to take notes and ask questions of witnesses during the trial and the judge during deliberation; allowing jurors to discuss the case during breaks; and summing up the case for the jury, including the types of questions the jury should be asking of the evidence. Each of these measures has its costs. For example, judges in England and Wales complain that very "keen" jurors can slow down the trial process considerably by continually wanting to ask questions (personal communication); discussion of the case can lead to premature deliberation (Kerr and Jung 2018); and the judge summing up the case is often seen as infringing too far into the jury's role (Robertshaw 1998). However, focusing solely on accuracy and consistency also has very considerable costs if the jury then fail to comply with the instructions. There are no easy answers, but the questions need to be asked if jury instruction is to be an ethical enterprise.

The Ethics of Researching Communication of the Standard of Proof

Most of this article has been concerned with the central ethical issue of what judicial authorities owe others when communicating the criminal standard of proof. In this final section, I shall briefly consider what researchers into jury instructions owe stakeholders (judges and judicial authorities, jurors, the public) in conducting their research. I shall focus on just two issues that are particularly pertinent to jury instruction research: deference and deception.

With respect to deference, just as one can have too much deference to scientific authority, which can lead to anti-democratic views of decision-making (Howell et al. 2020), so one can have too much deference toward legal authority when researching legal sites. There has been a tendency in applied linguistics to take the institutional status quo as given and then consider how linguistic analysis might improve communication within the already highly constrained institutional context. Working with institutional stakeholders generally means working within those strict constraints. In the case of jury instruction, that means accepting the general approach taken within the jurisdiction and then tinkering with the wording of the instructions. There certainly can be great value in doing this. The late Peter Tiersma (2021, 2001) and Bethany Dumas (2000) undoubtedly improved pattern jury instructions in California and Tennessee respectively. However, where judicial authorities in a given jurisdiction hold, against all evidence, that *beyond reasonable doubt* is an ordinary language term that all jurors understand (Heffer 2013), or that everyday *sure* cannot be qualified (Heffer and Coulthard In Press), there is a moral obligation on the researcher to point out that these beliefs are wrong. Similarly, where judges do not respect the autonomy or intelligence of jurors, it is arguable that this needs to be pointed out. Although jurors *decide* the criminal case, they have no *voice* (Hymes 1996) in the process. Accordingly, they are frequently misunderstood by judicial authorities. As a researcher into jury instruction, I have always seen the need to give voice to jurors who are otherwise silenced and left unheard. This desire to give jurors voice can lead to an ethical clash with the need to respect the professionalism of individual judges and justices. However, this would seem to be necessary where those institutionally powerful judges are attacking the jury unfairly as showing "absolutely fundamental deficits in understanding" (Heffer 2019: 248), or as not being able to do without a dictionary "what any competent English teacher might be able to do by recollection" (Heffer 2013: 220).

Turning to deception, one needs to balance respect for the juror against the urgent need for more ecologically valid research into how juries deliberate. Most empirical studies of jury instruction are very weak in terms of ecological validity. Often mock jurors are simply presented with a short written vignette describing a case. Researchers often do not consider how the choice of that case might affect the results. The fact is that possible solutions to the evident problems (better explanation, paraphrase, integration with the evidence, etc.) have not been sufficiently tested in valid and reliable environments to be confident that they will improve matters significantly. The most valid and reliable jury studies require "alternate" juries who are following the same case as the real jury and who deliberate on the case, just like the real jury. Ideally, from a research perspective, the alternate jury, consisting in participants who, like those on a real jury, have been called for jury duty, would not know that they are not actually making a consequential decision since it is only when jurors believe that their decision has real consequence that we can be sure that they will be deliberating like a real jury. However, this is a very significant deception and deception is a red flag in research ethics. But the research ethical issue with deception is not the deceiving itself but the "reasonableness of withholding the information from the person who is deceived" (Athanassoulis and Wilson 2009: 44). Given the gravity of jurors not understanding the instructions they are meant to apply to the evidence in a case, it *might* be reasonable to deceive alternate jurors who have already been called for jury duty, though this would need to be done exceptionally carefully.

An example of why such research is desperately needed can be seen in the series *The Jury: Murder Trial*, which first aired on the United Kingdom's Channel 4 network in 2024. It was advertised with the following strapline:

> How much can we trust our justice system? In this landmark experiment, a real-life murder trial is restaged in front of two juries of ordinary people.
>
> *(https://www.channel4.com/programmes/the-jury-murder-trial-signed)*

There was clear deception going on here because the two juries were unaware of the existence of the other one during the trial and consequent deliberations. Yet there was a greater framing deception: that this was a "landmark experiment". Far from being "a forensic look at the inner workings of justice", like all reality TV (the series seemed to be modeled on *The Traitors*), the program went for sensationalism over reality, presenting "a carefully curated selection of 'characters', with distinct personality traits or 'interesting back-stories' to spice up the deliberations…a process made worse by the 'cut-away' confessional interviews" (Foxsmith 2024). Widely criticized by UK legal professionals (e.g., Baksi 2024; Foxsmith 2024), the case of spousal murder was deliberately chosen as one that could go either way and the jurors carefully selected to increase the odds that the two juries would deliver different verdicts (as they did). The judge's summing-up to the jury (if it existed at all in the trial restaged by actors) did not even make it onto the show at all and we have no idea if or how the criminal standard of proof was conveyed. As Baksi put it, the series was "so flawed, with authenticity sacrificed on the altar of drama and ratings, that it was potentially damaging to the public's respect for the justice system" (2024). Indeed, journalists writing for both the left-wing *Guardian* and the right-wing *Telegraph* erroneously concluded that the series showed a justice system in a parlous state (Mangan 2024; Singh 2024).

Conclusion

Perlocutionary uptake by the jury of the criminal standard of proof is essential to guarantee a fair criminal justice system. We do not yet have all the answers as to how this can best be achieved, and more robust empirical research is required to find the optimal form of instruction. By applying the Golden Rule, though, we can establish what judicial authorities owe those affected by their instruction. These speech ethical principles are accuracy (in communicating the standard of proof), consistency (in delivering the instruction), comprehensibility (of both the legal formula used to represent the standard and the wording of the instruction as a whole), and persuasive engagement with the jury (by motivating them and treating them with respect). Judicial authorities have tended to focus on accuracy and consistency, which are certainly owed the defendant and the legal parties, but this has generally come at the expense of comprehensibility and persuasive engagement with the jury. But without these latter two elements, accuracy and consistency are pointless because there is no way of forcing jurors to comply with the instructions.

In terms of accuracy, "beyond reasonable doubt" captures well the sense of subjective certainty but only if it is understood in the correct legal sense. "Sure" can also convey subjective certainty, but only if understood in its legal sense, with the added confusion that jurors are very unlikely to recognize "sure" as a legal term. In terms of comprehensibility, to claim that "beyond reasonable doubt" is an ordinary language term, and thus in no need of definition or explanation, is epistemically negligent given the overwhelming weight of evidence against this view. To claim that "sure" is an ordinary language term in the English legal context is more excusable even if the accumulation of jury questions should be raising doubts in judges' minds. Clearly, the most consistent way of delivering the instruction is by repeating verbatim the same words but this can lead to fossilization of the wording of the instruction, which can then become incomprehensible over time. Detached, abstract instructions also fail to persuasively engage the jury and fail to treat them with respect.

This chapter has considered the question purely from an ethical perspective. It is recognized that it is very difficult to make fundamental changes in instructional practice. One issue is the fear that by explicitly recognizing that a jurisdiction has got instruction on the criminal standard of proof wrong, this will open the floodgates to appeals from those convicted on the basis of the previous instruction (McKeown 2022). However, radical changes have been made historically to the proof instructions and so it is certainly possible for this to happen again. Thus, as this chapter has shown, the Golden Rule can help mend the Golden Thread.

Notes

1 In this chapter, this form is used as a phrase lemma. Variations on the base form "beyond reasonable doubt" include "beyond a reasonable doubt", "beyond any reasonable doubt", and "beyond all reasonable doubt". The base form tends to be used in Commonwealth countries. The other forms, and particularly the one with the indefinite article, tend to be used in US jurisdictions.

2 So far this reaction has been informal (anonymous reviews, oral reaction to conference papers, more interest shown from philosophy than linguistics) rather than published.

3 Legal jurisdictions are complex and do not correspond to nation-states. For example, the United States has 50 State jurisdictions and 13 Federal circuits. The United Kingdom has three jurisdictions: England and Wales, Scotland, and Northern Ireland. Surprisingly different procedures can be found across jurisdictions within a single nation-state, and I do not claim to know the procedures across all these jurisdictions.

4 Unlike warning labels, it need not be intelligible to a juror of lower than average intelligence because it is assumed that there will be a sufficient number of jurors of average or above-average intelligence who will be in a position to help others with the meaning of the instruction. This assumption is probably less reliable in jurisdictions that exclude legal and other professionals and that allow for peremptory challenges to jurors.

5 Objective standards also exist in some jurisdictions (e.g., requirements for corroboration).

6 Victor was convicted of first-degree murder and was sentenced to death. He appealed partly on the grounds that the judge, following Nebraskan pattern instructions, had used the "archaic" term "moral certainty". The Supreme Court agreed that the term was archaic but held that the standard was sufficiently conveyed by the surrounding language. In doing so, they reaffirmed the meaning of *beyond reasonable doubt* as a moral or subjective certainty.

7 The Judicial College recommends using the name of the defendant rather than keeping things abstract.

8 *Miller v. Minister of Pensions* [1947] 2 All E.R. 372 at 373; 63 T.L.R. 474. Emphasis added.

9 I am not aware of any empirical work comparing comprehension of pattern proof instructions with comprehension of judge-authored instructions, so this is based on linguistic analysis of the language of the text (see below) rather than jury comprehension tests.

10 They are also required to review the evidence in the case but that usually follows the legal instructions themselves.

Case Law

Adorno v. Melvin [2017] 876 F.3d 917
Commonwealth v. Webster [1850], 59 Mass. 295
Miller v. Minister of Pensions [1947] 2 All E.R. 372 at 373; 63 T.L.R. 474.
Postelle v. State [2011] 267 P.3d 114
R v. Chatzidimitriou [2000] VSCA 91 – 1 VR 493
R v. Lifchus [1997] 3 SCR 320
R v. Wanhalla [2007] 2 NZLR 573
The Queen v. Dookheea [2018] *ILRC 476*
United States v. Fields [2011]
United States v. Pepe [1974] 501 F.2d 1142, 1143 (10th Cir.)
Victor v. Nebraska [1994] 511 U.S. 1
Woolmington v. DPP AC462

References

Athanassoulis, N., and J. Wilson. 2009. "When is Deception in Research Ethical?" *Clinical Ethics* 4, no. 1: 44–49.

Auld, R. E. 2001. *Review of the Criminal Courts of England and Wales: Report*. London: Stationery Office. Retrieved from http://www.criminal-courts-review.org.uk.

Baksi, C. 2024, March 11. "Trial by Television Got Juries All Wrong." *The Law Society Gazette*.

CED. 2003. *Collins English Dictionary*. Glasgow: HarperCollins.

Charrow, R. P., and V. R. Charrow. 1979. "Making Legal Language Understandable: A Psycholinguistic Study of Jury Instruction." *Columbia Law Review* 79, no. 5: 1306–1374.

Cheng, E. K., and M. Ginther. 2018. "Surprise vs. Probability as a Metric for Proof." *Seton Hall Law Review* 48: 1081–1094.

Christensen, D. 2011. "Disagreement, Question-begging and Epistemic Self-criticism." *Philosophers' Imprint* 11, no. 6: 1–22.

Cicchini, M. D. 2019. "Reasonable Doubt and Relativity." *Washington and Lee Law Review* 76: 1443–1449.

Clermont, K. M., and E. Sherwin. 2002. "A Comparative View of Standards of Proof." *The American Journal of Comparative Law* 50: 243–276.

COBUILD. 2001. *Collins COBUILD English Dictionary for Advanced Learners*. Glasgow: Harper Collins.

Code, L. 1987. *Epistemic Responsibility*. Hanover, NH: Brown University Press.

Cohen, L. J. 1977. *The Probable and the Provable*. Oxford: Clarendon Press.

Dear, P. 1992. "From Truth to Disinterestedness in the Seventeenth Century." *Social Studies of Science* 22, no. 4: 619–631.

Dumas, B. 2000a. "Jury Trials: Lay Jurors, Pattern Jury Instructions, and Comprehension Issues." *Tennessee Law Review* 67: 701–742.

Dumas, B. 2000b. "US Pattern Jury Instructions: Problems and Proposals." *Forensic Linguistics* 7, no. 1: 49–71.

Federal Judicial Center. 1988. Federal Judicial Center Pattern Criminal Jury Instructions. Washington, DC: Federal Judicial Center.

Felker, D. B., et al. 1981. *Guidelines for Document Designers*. Washington, DC: American Institutes for Research.

Foxsmith, G. 2024. March 7. "Judging the Jury." *The Law Society Gazette*.

Garner, B., ed. 2009. *Black's Law Dictionary*. St. Paul, Minnesota: West Publishing Company.

Gilovich, T., D. Griffin, and D. Kahneman. 2002. *Heuristics and Biases: The Psychology of Intuitive Judgment*. Cambridge: Cambridge University Press.

Goffman, E. 1981. *Forms of Talk*. Philadelphia: University of Pennsylvania Press.

Hanel, H. P., et al. 2023. "Using Self-affirmation to Increase Intellectual Humility in Debate." *Royal Society Open Science* 10: 220958.

Hart, H. L. A. 2008. *Punishment and Responsibility: Essays in the Philosophy of Law*. Oxford: Oxford University Press.

Heffer, C. In Prep. *Rhetoric and Rights: A Theory of Forensic Discourse*. New York: Oxford University Press.

Heffer, C. 2020. *All Bullshit and Lies? Insincerity, Irresponsibility and the Judgment of Untruthfulness*. New York: Oxford University Press.

Heffer, C. 2019. "Suppression, Silencing, and Failure to Project: Ways of Losing Voice While Using It." In *Rethinking Language, Text and Context*, 237–253. Edited by R. Page, B. Busse, and N. Nørgaard. London: Routledge.

Heffer, C. 2013. "Communication and Magic: Authorized Voice, Legal-linguistic Habitus and the Recontextualization of 'Beyond Reasonable Doubt'." In *Legal-Lay Communication: Textual Travels in the Law*, 206–225 Edited by C. Heffer, F. Rock, and J. M. Conley. New York: Oxford University Press.

Heffer, C. 2008. "The Language and Communication of Jury instruction." In *Dimensions of Forensic Linguistics*, 47–65. Edited by J. Gibbons and M. T. Turell. Amsterdam: John Benjamins:

Heffer, C. 2007. "The Language of Conviction and the Convictions of Certainty: Is 'Sure' An Impossible Standard of Proof?" *International Commentary on Evidence* 5, no. 1: Article 5.

Heffer, C. 2006. "Beyond 'Reasonable Doubt': The Criminal Standard of Proof Instruction as Communicative Act." *The International Journal of Speech, Language and the Law* 13, no. 2: 159–188.

Heffer, C. 2005. *The Language of Jury Trial: A Corpus-aided Analysis of Legal-Lay Discourse*. Basingstoke: Palgrave Macmillan.

Heffer, C. 2002. "'If You Were Standing in Marks and Spencers': Narrativisation and Comprehension in the English Summing-up." In *Language in the Legal Process*, 228–245. Edited by J. Cotterill. Houndmills: Palgrave.

Heffer, C., and M. Coulthard. In press. "'Being Sure': Judicial Practices in Directing Juries on the Criminal Standard of Proof." In *Communication and Legal Practice*. Edited by T. Grieshofer and K. Haworth. Cambridge: Cambridge University Press.

Horowitz, I. A., and L. C. Kirkpatrick. 1996. "A Concept in Search of a Definition: The Effects of Reasonable Doubt Instructions on Certainty of Guilt Standards and Jury Verdicts." *Law and Human Behavior* 20, no. 6: 655–670.

Howell, E. L., et al. 2020. "Deference and Decision-making in Science and Society: How Deference to Scientific Authority Goes Beyond Confidence in Science and Scientists to become Authoritarianism." *Public Understanding of Science* 29, no. 8: 800–818.

Hymes, D. 1996. *Ethnography, Linguistics, Narrative Inequality: Toward an Understanding of Voice*. London: Taylor and Francis.

Kagehiro, D. K., and W. C. Stanton. 1985. "Legal vs. Quantified Definitions of Standards of Proof." *Law and Human Behavior* 9, no. 2: 159–178.

Kant, I. 2012. *Groundwork of the Metaphysics of Morals*. Cambridge: Cambridge University Press.

Keane, A., and P. McKeown. 2019. "Time to Abandon 'Beyond Reasonable Doubt' and 'Sure': The Case for a New Direction on the Criminal Standard and Why and How it Should be Used." *Criminal Law Review* no. 6: 505–527.

Kerr, N. L., and J. Jung. 2018. "Should Jurors be Allowed to Discuss Ttrial Evidence before Deliberation?: New Research Evidence." *Law and Human Behavior* 42, no. 5: 413–426.

Kramer, G. P., and D. M. Koenig. 1990. "Do Jurors Understand Criminal Jury Instructions? Analyzing the Results of the Michigan Juror Comprehension Project." *Journal of Law Reform* 23, no. 3: 401–437.

Laudan, L. 2006. *Truth, Error, and Criminal Law: An Essay in Legal Epistemology*. Cambridge: Cambridge University Press.

Macquarie. 2010. *The Macquarie Dictionary Online. Sydney: Macquarie Dictionary.* Sydney: Macquarie Dictionary.

Magnussen, S., D. E. Eilertsen, K. H. Teigen, and E. Wessel. 2014. "The Probability of Guilt in Criminal Cases: Are People Aware of Being 'Beyond Reasonable Doubt'?" *Applied Cognitive Psychology* 28, no. 2: 196–203.

Mangan, L. 2024, February 26. "The Jury: Murder Trial review – Surely this is the End of the UK Legal System as We Know It." *The Guardian*.

McKeown, P. 2022, July 29. "Dreaded Questions, Doubtful Answers and the Trouble with 'Sure'." *New Law Journal*: 13–14

Morano, A. A. 1975. "A Reexamination of the Development of the Reasonable Doubt Rule." *Boston University Law Review* 55: 507–528.

Nelson, S. 2013. *Directing Jurors in England and Wales: The Effect of Narrativization on Comprehension.* PhD Thesis, Cardiff University.

Newman, J. 2007. "Quantifying the Standard of Proof Beyond a Reasonable Doubt." *Law, Probability, and Risk* 5: 267–269.

New South Wales Law Reform Commission. 2008. Consultation Paper 4—Jury Instructions. Sydney: New South Wales Law Reform Commission.

Parfit, D. 2011. *On What Matters*. Oxford: Oxford University Press.

Pi, D., F. Parisi, and B. Luppi. 2019. "Quantifying Reasonable Doubt." *Rutgers University Law Review* 72, no. 2: 455–508.

Picinali, F. 2015. "The Threshold Lies in the Method: Instructing Jurors about Reasoning beyond Reasonable Doubt." *The International Journal of Evidence and Proof* 19, no. 3: 139–153.

Picton, M., D. Ormerod, and E. A. Tayton. 2023. *The Crown Court Compendium, Part I: Jury and Trial Management and Summing Up.* Judicial College.

Risinger, D. M. 2018. "Leveraging Surprise: What Standards of Proof Iimply that We Want from Jurors, and What We Should Say to Them to Get It." *Seton Hall Law Review* 48, no. 4: 965–994.

Robertshaw, P. 1998. *Summary Justice*. London: Cassell.

Saxton, B. 1998. "How Well Do Jurors Understand Jury Instructions? A Field Test Using Real Juries and Real Trials in Wyoming." *Land and Water Law Review* 33, no. 1: 59–189.

Singh, A. 2024, February 26. The Jury: Murder Trial Will Make You Lose all Faith in the British Justice System." *The Daily Telegraph.*

Slovic, P., J. Monahan, and D. MacGregor. 2000. "Violence Risk Assessment and Risk Communication: The Effects of Using Actual Cases, Providing Instruction, and Employing Probability versus Frequency Formats." *Law and Human Behavior* 24: 271–296.

Smith, O. K. H. 2022. "The Beyond a Reasonable Doubt Standard of Proof: Juror Understanding and Reform." *The International Journal of Evidence and Proof* 26, no. 4: 291–308.

Solan, L. M. 1999. "Refocusing the Burden of Proof in Criminal Cases: Some Doubt about Reasonable Doubt." *Texas Law Review* 78, no. 1: 105–147.

Tiersma, P. M. 2021. "Redrafting California's Jury Instructions." In *The Routledge Handbook of Forensic Linguistics*, 267–280. Edited by M. Coulthard, A. May, and R. Sousa-Silva. Abingdon: Routledge.

Tiersma, P. M. 2001. "The Rocky Road to Legal Reform: Improving the Language of Jury Instructions." *Brooklyn Law Review* 66, no. 4: 1081–1119.

Tiersma, P. M. 1999. *Legal Language*. Chicago: University of Chicago Press.

Tribe, L. H. 1971. "Trial by Mathematics: Precision and Ritual in the Legal Process." *Harvard Law Review* 84, no. 6: 1329–1393.

Whitman, J. Q. 2008. *The Origins of Reasonable Doubt: Theological Roots of the Criminal Trial.* New Haven: Yale University Press.

Williams, B. 2002. *Truth and Truthfulness: An Essay in Genealogy*. Princeton, NJ: Princeton University Press.

Woody, W. D., and E. Greene. 2012. "Jurors' Use of Standards of Proof in Decisions about Punitive Damages." *Behavioral Sciences & the Law* 30, no. 6: 856–872.

Young, W., N. Cameron, and Y. Tinsley. 1999. *Juries in Criminal Trials Part 2: A Summary of the Research Findings*. Wellington, New Zealand: New Zealand Law Commission. Preliminary Paper 37, vol. 2: 83–83.

17

ETHICAL UNCERTAINTIES AND INCONSISTENCIES IN DIMINISHED RESPONSIBILITY RULINGS

The Case for Fictionalism

Felicity Deamer and Sam Wilkinson

Introduction

In England and Wales, "diminished responsibility" is a partial defense to murder, which is heavily dependent on psychiatric evidence. If the defendant is successful in proving the defense on the balance of probabilities, the result is a conviction for manslaughter rather than murder. As such, a diminished responsibility plea is an extremely important process for the defendant, the families of the victims, and society. When changes to section 2 of the Homicide Act 1957 (HA 1957) under section 52 of the Coroners and Justice Act 2009 were implemented over a decade ago (April 10, 2010), Parliament's intention was clearly to enhance the role of the psychiatrist in diminished responsibility cases and to "medicalise" the defense. However, since the implementation of these reforms, there have been more "contested" diminished responsibility trials (Mackay and Mitchell 2017).

The existing literature on tensions between psychiatry and law has primarily focused on the risk factors and causes underlying the ethical issues and biases that are considered to be at play within Forensic Psychiatry (Eastman and Rix 2021; Eastman et al. 2018; Murrie 2015; Eastman et al 2012; Appelbaum 2008; Commons 2004) and has generally been directed toward individual psychiatrists in the form of practical guidance. For example, Eastman and Rix (2021) focus on biases in expert witness practice. We intend to go beyond this by bringing in a contemporary debate within philosophy of psychiatry, namely, the fundamental rationale behind the very idea that psychiatric conditions exculpate.

In this chapter, we will examine the ways in which uncertainties in the conceptual underpinnings of diminished responsibility are reflected in inconsistencies in psychiatric evidence and how that evidence is reflected on by judges in their sentencing remarks. We will demonstrate how linguistic analysis can unearth these irregularities and contradictions by using two contrasting cases.

We start by examining the rationale behind the exculpatory force of psychiatric conditions, which fundamentally turns on answers to deep, philosophical questions about the nature of responsibility, and impediments to it. We then look closely at judges' sentencing remarks to show how there is an inconsistent flitting between different implicit philosophical views about the role that

DOI: 10.4324/9781003391074-19

psychiatric conditions play in the exculpatory process. We end by suggesting that this oscillation between implicit views can't be made sense of, unless we adopt a fictionalist approach to mental illness and moral responsibility, as per Wilkinson (2023).

Introducing the Diminished Responsibility Plea

In this section, we start by looking at how psychiatric exculpation plays out in the criminal justice system within England and Wales, through the legislation of diminished responsibility. We then discuss the philosophical rationale behind this legislation.

The Legislation

Here is how the diminished responsibility legislation is presented in the government legislation, in the Homicide Act (1957) (as amended by the Coroners and Justice Act (2009)):

> The defendant must be suffering from an abnormality of mental functioning which:
>
> - arose from a recognised medical condition;
> - substantially impaired the defendant's ability to do one or more of the following
>   - understand the nature of their conduct;
>   - form a rational judgment;
>   - to exercise self-control.
> - provides an explanation for defendant's acts and omissions in doing or being a party to the killing.

Of course, almost every element of this legislation is ambiguous and open to interpretation. But what is the underlying rationale for this exculpation in general?

What seems to be captured by the second bullet point is that, in order for someone to be fully responsible for their behavior, and hence fully legally culpable, certain criteria need to hold (in this legislation, the three listed as "understanding the nature of their conduct; form a rational judgment; to exercise self-control"). The rationale behind the plea of diminished responsibility is that psychiatric illness, in some cases at least, seems to prevent these criteria from holding. This immediately raises the following questions. First: what are these criteria, precisely? Second: is psychiatric illness special in how it interferes with them or is it simply one context among countless others where these criteria are interfered with, and hence a reduced sentence is warranted? Unfortunately, neither of these two questions has been conclusively answered. Nor is it entirely clear which answers to these philosophical questions legal theorists and legislators are basing their legislation on.

The first bullet point, referring to a "medically recognized condition" is perhaps especially ambiguous. Is the main point about it being recognized as "medical"? Or that it's a "recognized condition"? In other words, does it point to the idea that something must be deemed *medically* wrong with someone, even if it is a unique condition that defies diagnostic categorization (i.e., is unlike any other condition, but is surely medical)? Or does it point to the (more demanding) requirement of someone to get a clear diagnostic label? The latter is clearly arbitrary. Why should a medical condition have to be a recognized condition in order for it to exculpate? The former is arguably, but less obviously, problematic too. Why does a condition need to be recognized *as* medical in order to exculpate? Why does the distinctly medical status of a problem hold such sway in the court of law?

Crucially, there is a tension between this first bullet point and the second. The second tells us that there are criteria that need to hold. The first tells us that those criteria have to be interfered with by a medically recognized condition (in either of the two senses) if the plea of diminished responsibility is to be granted. But what if something that isn't a medically recognized condition interferes with these criteria? Do we not have a viable diminished responsibility plea? And conversely, what if we have a "medically recognized condition" that doesn't interfere with these conditions? It is not clear what the exact relationship between these two bullet points really is. They seem to both be thrown in for good measure. More worryingly, they embody opposing and incompatible approaches to the issue, what Wilkinson (2023, 2022) calls "intrinsic and incidental approaches". We will explain what this phrase means shortly. Let's move away from the legislation itself, and move on to the conceptual underpinnings and rationale.

The Necessary Criteria of Responsibility

Let us start with a trivial case of responsibility. Jane is hungry and believes that there is food in the fridge. Jane goes to the fridge, finds food, eats the food, and is no longer hungry. Jane has behaved in a completely rational manner and is responsible for her actions. As currently described, it is odd to talk about culpability, though, since nothing blameworthy has taken place. So let's suppose the fridge is a shared office appliance, and the food Jane ate was not hers. That's theft! Jane is both responsible for her actions and, at first glance, culpable of wrongdoing.

What makes Jane responsible for her actions? A standard philosophical answer is as follows: she fulfills an "informational" criterion and a "control" criterion. She has an accurate informational perspective on what is going on, and she has control over what she is doing.

Now, these criteria can fail to hold in fairly mundane circumstances. Suppose she genuinely believes that the food in the fridge is hers. Her culpability would be reduced, and we would expect her to use that in an excuse of her actions. She is acting on the basis of inaccurate information. She thought she was eating her own food. Now for the control criterion. Suppose she was acting under coercion, at gunpoint. Her culpability would be similarly reduced. In more everyday parlance, if you are misled, or coerced (or otherwise not in control, like sleepwalking), you are less culpable of wrongdoing (indeed, you may not be culpable at all).

What psychiatric conditions might mislead or coerce? Delusion might mislead. Addiction might coerce. Psychosis might do a bit of both: you have an inaccurate take on the world, but are being driven toward certain actions, too. Notice, though, that in these cases, the fact these are medically recognized conditions (in either the first or second sense) is irrelevant. They interfere with the necessary conditions of responsibility in the very same way that other, more mundane things might, like being lied to, or held at gunpoint.

Complicating the Picture

So, it is clear that these criteria can fail to hold in non-medical contexts. But do medical conditions only exculpate by interfering with these criteria? Many philosophers think not. They opine that medical conditions can reduce culpability without affecting the informational or control criteria. For example, somebody who has a personality disorder as a result of an abusive childhood, Foetal Alcohol Spectrum Disorder (FASD), or certain types of acquired brain injury (ABI), doesn't have an inaccurate take on the world, nor do they necessarily feel coerced. And yet we may feel that their condition should be taken into account when they behave reprehensibly. Furthermore, the exculpation does not attach to a particular action, but to actions more generally. They lack, as it were, more of a background *capacity* for responsibility. This points to a distinction between two different senses of responsibility: "you are responsible for this action" vs. "you are a responsible adult".

How is this seemingly different (and more general) sense of "lack of responsibility", which is not about interfering with an informational or control criterion for a particular action, accounted for philosophically? According to some views, these individuals may, for example, lack an appreciation of right and wrong. Moral realists, like Susan Wolf (1987), can then fold this into informational criteria: there is an exculpatory ignorance of facts here, somewhat akin to that of the delusional individual, it's just an ignorance of "moral facts", of right and wrong. Of course "ignorance of right and wrong" applies more generally across many potential actions than a short-lived psychotic episode where you think that your spouse has been replaced by an evil robot. Notice that this ignorance of right and wrong is what exculpates young children. We do not think of children as addicts or deluded: they just don't yet "know right and wrong".

Of course, many philosophers find this level of moral realism (the postulating of moral facts) unappealing. And yet, they accept a need to exculpate those with personality disorders, brain injury, etc. So how is this to be done? One way is via an intrinsic approach to the relationship between medical and moral discourse. But this is in opposition to the incidental approach.

Incidental and Intrinsic Approaches to Psychiatric Exculpation

Incidental approaches to psychiatric exculpation claim that some mental health conditions *happen to* (i.e., "incidentally") interfere with these criteria of responsibility. So, a delusion or hallucination will interfere with the informational criterion. Addiction (or some other form of compulsion, like Tourettes) will interfere with the control criterion. Something like psychopathy, or FASD, will interfere with the agent's knowledge of right and wrong, and so on.

This incidental approach has two interesting features. First, there is nothing special about mental illness. It can interfere with these conditions, but so can other things that aren't mental disorders. So, being delusional, or hallucinating, is a bit like being lied to; being in the grip of a strong, addictive urge is like being forced to act at gunpoint, and so on. And while certain psychiatric or neurodevelopmental disorders can remove your ability to understand right and wrong, so does being a young child, or having had a particularly problematic upbringing.

Intrinsic approaches to psychiatric exculpation are very different. They claim that an intrinsic part of considering someone to have a mental illness is removing responsibility for relevant actions. This makes psychiatric exculpation special.

It should now be clear why the first bullet point in the legislation, with its reference to a "recognized medical condition", embodies an intrinsic approach, while the reference to the "impaired ability of the defendant" to fulfill the conditions of responsibility reflects an incidental approach.

On the Linguistic Analysis of Sentencing Remarks

Linguistic analysis is likely to yield insights at many different stages of the entire process of a case of diminished responsibility to unearth how psychiatric exculpation actually works. This could include analysis of (1) the written expert's report; (2) the process by which that report is presented and interpreted in the courtroom; and (3) the process by which the psychiatric evidence presented to the jury is shaped by the dyadic nature of the adversarial setting. However, for the purposes of this chapter, in which we primarily set out to motivate such an endeavor by appealing to philosophical and conceptual arguments, we limit our focus to exploring insights which, from a preliminary examination of sentencing remarks, contain extensive reference to the psychiatric reports themselves, while also affording the judge an opportunity to offer their stance on the case and to explain and justify their sentencing decisions.

What Sentencing Remarks Are

This sub-genre of courtroom language data is especially interesting for our purposes because it is where the judge establishes the "official" narrative of the case, or the "judicial truth" as Campos-Pardillos (2022) puts it. The judge's remarks offer an account not only of the trial itself, but also of why a specific sentence has been imposed given the circumstances of the case and the evidence that has been considered, as established by the court, and as interpreted by the judge. It might appear at first as though sentencing remarks function solely to explain the sentence imposed to the convict (not least because the offender is addressed in the 2nd person throughout). However, as Campos-Pardillos (2020: 148) notes, sentencing remarks have an exemplary role, and as such the audience extends way beyond those present in the courtroom, "reinforc[ing] the connection between the judiciary and society by showing that justice has been made". The judge seeks both to explain how the sentencing guidelines are being implemented and to justify their decisions while seeking legitimation from their audience (Campos-Pardillos 2022). Heffer (2010: 208) describes the sentencing phase as where "finally the constraints on the evaluative dimension of narrative are lifted and…the violence of punishment that the words of sentencing represent needs to be motivated with more than a purely referential factual account of the crime events".

Given that murder trials in which a plea of diminished responsibility is granted typically present examples in which the sentence imposed is significantly lighter than would be expected otherwise, the sentencing remarks become all the more important. Often, the judge will seek to preempt the perception that the punishment may not be severe enough, particularly in high-profile cases in which the public interest has been peaked, and unratified hearers are highly present, and the judge finds themselves seeking legitimation from the whole of society. As such, sentencing remarks in diminished responsibility cases are often characterized by persuasive language, where the judge is attempting to convince the audience that justice has indeed been done: "the punishment must not only fit the crime, but also be 'seen' to fit the crime" (Campos-Pardillos 2022: 164).

What we aim to demonstrate is that psychiatric evidence is likely to be compromised by uncertainties and inconsistencies in the conceptual underpinnings of forensic psychiatry, but also that those same misconceptions are behind the principles guiding the plea of diminished responsibility in the first place. Forensic psychiatric reports look the way they do because they are shaped by the needs and procedures of the court. Purported facts are woven into persuasive stories that fit our moral frameworks, which are then filtered back through legal frameworks; and this tension between narrative and anti-narrative results in a messy discourse around the relationship between mental disorder and culpability.

The Narrative Structure of Sentencing Remarks

The structure of sentencing remarks, much like the trial as a whole, is somewhat dictated by the needs and procedures of the court, and this is all the more marked in cases featuring a plea of diminished responsibility, as various principles and guidelines must be followed. The judge themselves must evaluate the case, including the strength of the psychiatric evidence, and scrutinize it in relation to the principles to be considered, including the culpability of the defendant, the seriousness of the offense, and the dangerousness of the offender.

Heffer (2005: 101) draws an important distinction between the "crime story"—the skeleton structure of events as conceived of and/or established within the court—and instances of narrative discourse in which that story is narrated or told. Heffer (2010: 193) refers to "The crime story [being] pieced together during the investigatory stages of a case and [it] not [being] linked to any one specific telling", whereas the opening and closing speeches are instances of that version of events being told or narrated by the prosecution and defense, each weaving a slightly different story through the use of narrative discourse features (Harris 2005; Heffer 2005; Cotterill 2003).

Sentencing remarks present an opportunity for that crime story to be picked up again by the judge and re-told one final time, post-conviction, when the jurors have, as Heffer (2020: 206) describes, "integrate[d] the trial evidence with their pre-existing scripts for event sequences similar to those in dispute and their generic expectations about what makes a coherent story" and come down on the side of either the prosecution or the defense. In this final telling by the judge, the story is told in a way that explains and justifies the sentence that is about to be passed. This is where we see evidence of the judge using a variety of lexical, grammatical, and discoursal means to convince the audience that the sentence being imposed is just.

Below are the central elements of the sentencing remarks for both Case 1 and Case 2. They appear in different orders and are not always clearly delineated, but the central elements are the same in both, largely because they are dictated by the needs and procedures of the court.

1) *Context*: A short summary of the case
2) *Procedural history*: Details when a plea of diminished responsibility was suggested/established
3) *Description of offender and the killing(s)*: Discusses and evaluates the offender's background, context, mental state at the time of the killing(s)
4) *Psychiatric reports*: Extracts from the psychiatric reports
5) *Rules and procedures*: Discusses rules and procedures which guide sentencing
6) *Psychiatric evidence in relation to rules and procedures*: One in light of the other
7) *The sentence*: A final decision and course of action in relation to sentencing

There is a general consensus that the trial constitutes something of a "hybrid discourse" which is neither entirely narrative nor completely non-narrative. A verdict is sought, and the decision in relation to that verdict is made on the basis of two necessarily coupled types of reasoning: (1) logico-scientific reasoning which involves the search for context-independent descriptions and explanations; and (2) narrative reasoning which entails the search to understand people's actions and intentions in a given context. The result is a continuous tension between truth-seeking and narrative (Heffer 2010, 2005).

Sentencing remarks in diminished responsibility cases offer a microcosm of that tension except perhaps with an additional challenge coming from the logico-scientific reasoning pulling from the two very different domains of medicine and law. The judge is attempting to convince the audience of a course of action which has been made on the basis of some ground (scientific) truths, but those supposed ground truths (from the domain of psychiatry, a branch of medicine) have been shaped and manipulated to fit the needs and procedures of the court. As discussed above, there is no clear or agreed link between psychiatric diagnostics and culpability, but one must be asserted for the diminished responsibility process to work.

Heffer (2010) discusses how throughout the trial there is a shift in goals from decision-making (guilty or not guilty) to fact determination, and back to decision-making. Heffer writes:

> the discursive work performed in earlier [parts of the trial] will necessarily influence later ones. For example, both the legal framework and the story constructions will influence the type of facts that counsel will focus on in witness testimony. Similarly, the closing speeches will weave together the facts determined in the evidential phase into convincing narratives but they will also anticipate the legal charge or summing up by fitting these stories into a clear legal framework.
>
> *(Heffer 2010: 194)*

We propose that something similar is going on within sentencing remarks in diminished responsibility cases. The initial context setting and the section of the remarks in which the judge describes the offender and the details surrounding the killing(s) are the clearest instances of narrative mode in sentencing remarks. They present an opportunity for the judge to re-tell the prosecution's story (the narrative that the jury has accepted in arriving at their guilty verdict) but with additional persuasive techniques in order to justify the punishment being imposed. A narrative is established in which numerous rhetorical devices come into play to cast the events within a moral framework, which sets up the sentence to be imposed. The constraints on the use of evaluative language imposed by courtroom rules and procedures are lifted at the sentencing stage, and as Heffer (2010: 207) observes, "the verbal intensity of judgement appears to be calibrated with the severity of the crime and the length of the sentence". This is particularly interesting when considering sentencing in diminished responsibility cases as there is resistance in this calibration process; the severity of the crime and the nature of the sentence imposed are typically precisely not well aligned. The more heinous the killing, the more victims' families and society at large seek retribution and protection from the perpetrator, but in cases in which the offender is deemed to have been mentally unwell at the time of the killing (not uncommon in the most shocking of murders), they are more likely to be given a hospital order than a lengthy prison sentence. It is this negative correlation that potentially results in linguistic patterns in these highly persuasive sections of sentencing remarks.

We argue that the specifics of any given case are likely to trigger an initial moral reaction, either a strong desire for retribution (i.e., in virtue perhaps of the shocking and horrific nature of the index offense) even in the face of a relatively clear "medically recognized condition", or the opposite: namely, there is a prima facie leaning toward exculpation in the absence of such a clear diagnosis. Moreover, that the initial stance taken toward the facts of the case will shape the sentencing remarks throughout, from the initial sentencing remarks through to the discussion of the psychiatric evidence itself. Take the following two contrasting cases by way of example.

Context Setting, Description of Offender, Context, and Actions

Case 1

The facts of Case 1 demonstrate from the outset that the defendant was in the throws of a serious mental disorder at the time of the killing. A diagnosis of schizophrenia had preceded the killings, and his behavior and actions on that day were consistent with that diagnosis. In a situation such as this, it is often the case that the crime is so horrific in nature that it "shocks the nation", and the judge, along with everybody else, must set aside their extremely strong desire for retribution and seek a psychiatrist's expert opinion to confirm the ground-truth about the defendant's mental state at the time of the killing(s). We see this reflected in the context setting.

Extract 1. Case 1 judge's context setting

> In the early hours of 13th June last year, you committed a series of atrocities in this city which ended the lives of three innocent people. You went on to attack three more; fully intending, but failing, to kill them too. Your sickening crimes both shocked the nation and wrecked the lives of your surviving victims and the families of them all.

Case 2

In contrast, in Case 2, the judge congratulates the prosecution on their "characteristically fair and balanced opening" during the trial, and quotes them as having said that in this case, "Context is everything", which sets the narrative mode of the whole set of sentencing remarks. The judge states that the facts are "extremely sad" and "tragic" and the judge, along with everybody else, seeks to find an explanation or a justification for the events that unfolded that day. They perceive the defendant to be moral, and the killing to be the result of a "perfect storm" of unfortunate happenings. They seek a way to shape the principles and guidelines around the notion of diminished responsibility to allow the defendant to escape the full thrust of the law. They see that the defendant was certainly in a difficult context, was certainly stressed and exhausted, and they think it's plausible that they were not of sound mind when the killing took place due to those external stressors. We see this reflected in the context setting and description of events.

Extract 2. Case 2 judge's context setting

> On 20th June 2014, the Defendant, Mr Smith (aged 78) stabbed his wife, Mrs Smith (also aged 78), in the heart and killed her. They had been happily married 57 years and were a devoted and loving couple. Mrs Smith had been suffering for several years with complex medical problems and increasing dementia. Mr Smith had been her constant and devoted carer. Her condition had considerably worsened in the previous days and weeks. She did not want to return to hospital or to suffer any more. He was under intolerable strain at the time and acted under impulse. He called the emergency services and admitted what he had done.

In the context setting of these sentencing remarks, the evaluation is woven into the core narrative clauses through extensive and strategic use of attitude markers appearing in the form of evocative nouns, verbs, adverbs, and adjectives.

Additionally, just as Cotterill (2003) demonstrated to be the case in opening statements, in the Case 1 sentencing remarks, we see evidence of apparently more neutral words such as "committed" contributing to a negative semantic prosody due to its frequent collocation with negative words such as "atrocities" and "crime". The term "commit" also suggests a controlled intentionality to the offender's actions that day. The word "commit" and all its associations are notably absent from the Case 2 sentencing remarks. Also notable is the use of the transitive verb "wreck" in the Case 1 sentencing remarks, which typically takes a more intentional agent in the active role (e.g., "the player wrecked his chances of being selected" or "the kids wrecked the playroom"), and as such arguably connotes a degree of willful negligence. This is in contrast to apparent synonyms such as "destroy", or "ruin", which can often take non-agentive forces in the active role (e.g., "the storm destroyed the fence" or "the relationship ruined his life"), thus connoting less intentionality. In fact, intentionality is also very explicitly marked in the Case 1 sentencing remarks with "fully intending, but failing, to kill...", and the opposite is very clearly marked in the Case 2 remarks with “acted under impulse”, “under intolerable strain”, and the factual non-narrative description, “stabbed his wife...and killed her”.

Judge’s Obligation to Rules and Procedures

This element of sentencing remarks is where the judge asserts their stance on the (prosecution’s) acceptance of the plea of diminished responsibility, and the language used in relation to that acceptance again demonstrates that calibration that the judge is having to make between the severity of the crime and the nature of the sentence being imposed. In Case 1, the narrative constructed by the judge in the earlier parts of the remarks is one of heinous crimes featuring incomprehensible amounts of violence inflicted by an individual who fully intended to kill. So when it comes to the judge expressing their deontic stance on the acceptance of the plea and the subsequent impact on sentencing, we can see in Extract 3 that there is something of a reluctant obligation to proper procedure expressed through the modal verb “must”. An example from Case 1 is shown in the judge’s following remark:

Extract 3. Case 1 judge’s stance taking

> I must, of course, factor [the psychiatrists’] opinions into the process of sentencing although I am not bound by them...judges do not sentence in a vacuum....I must have regard to the sentencing guidelines.

It is clear from even this one extract that the judge is acutely aware that the severity of the crime and the nature of the sentence imposed are going to be particularly poorly calibrated in this case, and that society at large as well as the victim’s families will be seeking retribution. With this in mind, he is keen to emphasize the lack of calibration in cases of this kind. This point is exemplified from the following remark extracted from Case 1: “none of the evidence relating to your mental state detracts from the horror of your actions or the disastrous impact which they have had upon many people”. In contrast, in the Case 2 remarks, the judge expresses their affective attitude toward the acceptance of the plea through the adverbial phrase “entirely right and appropriate”.

Extract 4. Case 2 judge’s stance taking

> in view of the context and evidence in this case...it is entirely right and appropriate that the prosecution should [accept a plea to manslaughter] and I sentence on this basis.

In contrast to Case 1, in Case 2, the judge quite clearly expresses a positive stance toward the acceptance of the plea. The judge positions himself in support of the rules and procedures here, not at the mercy of them as in Case 1.

Psychiatric Reports

Psychiatrists provide evidence, and that evidence is supposed to make up part of the non-narrative, truth-seeking, logico-scientific aspects of the trial. The primary function of the report is to answer the question: Does the defendant have a recognized mental health condition that reduces their culpability for the killing?

However, we see in the following few extracts from the psychiatric reports read out as part of the sentencing remarks that the motivation to seek expert psychiatric evidence, the details of the case, and how they sit within our moral frameworks, shape the way the evidence (the reports themselves) looks. In Case 1, the evidence is more logico-scientific. The offender clearly meets the criteria of a diagnosis of a serious mental disorder, and little to no persuasive language or narrative structure is required to convince the court of this fact. In Case 2, the psychiatric evidence takes on a narrative mode, with persuasive discourse features, attempting to convince the court (and wider society at sentencing) that it is "ok" for him to be almost entirely let off, despite the fact that he doesn't really meet the criteria.

In Case 1, when the psychiatric diagnosis is referred to, the prognosis is explained, and adjectives such as "chronic" and "enduring" are applied to it, but at no point do the doctors feel it necessary to emphasize or draw attention to the legitimacy of such a diagnosis within the Mental Health Act (1983; 2007). It is assumed to be common ground among all interested parties.

Extract 5. Case 1, report 1 (Dr Greeves)

> [The defendant] has an established diagnosis of paranoid schizophrenia. This is a chronic and enduring psychotic mental illness. His illness will never be "cured", and he will require long term, very cautious management.

The psychiatrist makes a series of assertions relating to the symptoms the defendant was experiencing at the time, and these assertions are made with a high degree of certainty.

Extract 6. Case 1, report 1 (Dr Greeves)

> His history of mental illness and his own explanation of the offences indicate that he was experiencing hallucinatory voices, persecutory delusional beliefs, among other psychotic symptoms.

In stark contrast, the first psychiatric report in Case 2 is written in narrative mode. It encompasses a series of linguistic and content choices through which the psychiatrist endeavors to convince the audience that the facts of the case fit the criteria for granting a plea of diminished responsibility. The psychiatrist first attempts to fit the facts with the diminished responsibility criteria but there is an awareness that the proposed medical diagnosis is not typically understood to impact an individual's ability to form rational judgments and or maintain self-control to the extent that they might stab their wife, so the psychiatrist quickly switches to narrative mode, bringing in contextual information which can be combined with the medical "facts" to create a persuasive case for why the full force of the law should not be applied to the defendant in this case.

In contrast to Case 1, as we can see in Extract 7, the first of the psychiatric reports elicited in Case 2 the psychiatrist seeks to establish the legitimacy of a potentially lesser-known psychiatric diagnosis, namely that of a *moderate depressive disorder*.

Extract 7. Case 2, report 1 (Dr Hallow)

> Mr Smith…was suffering with a moderate depressive episode, which is a mental disorder within the meaning of the Mental Health Act 1983 (amended in 2007)…A moderate depressive disorder is a recognized medical condition.

The term "moderate depressive" is repeated four more times during the same paragraph of the report, either as a disorder or an episode. The switching between disorder and episode is significant in itself as disorder picks out a significant and enduring state, whereas episode denotes a specific moment, a temporary experience, or a defined period of time. Given that one of the requirements for a plea of diminished responsibility to be granted is that the defendant is considered to have been suffering from a serious mental disorder at the time of the indexed crime, the psychiatrist is ticking that box via the use of the term "disorder", whereas he then switches to referring to an episode when talking about the occasion of the killing because stabbing one's wife in the heart is not necessarily something one would associate with a relatively mild psychiatric disorder such as a moderate depressive disorder. In referring to a specific episode, he is setting that episode up to be the focus of the report, a specific extraordinary episode in which a multitude of factors came together and led to the killing (which is critically important if the audience are to determine that Mr. Smith no longer presents a risk to society).

The second psychiatric report also asserts with a high degree of certainty that the defendant was suffering from an abnormal state of mind (a part of the definition of a recognized mental disorder as stipulated in the legal framework for diminished responsibility) and a moderate depressive episode or adjustment disorder.

Extract 8. Case 2, report 2 (Dr James)

I think it is highly likely he was suffering from an abnormal mental state at the time of the alleged indexed offence.

Extract 9. Case 2, report 2 (Dr James)

This may present a moderate depressive episode; it could also be described as an adjustment disorder.

Environmental Factors and Good Character

In contrast to Case 1, in Case 2 there is little to no mention of any symptoms associated with the defendant's proposed psychiatric diagnosis. Instead, unlike in Case 1 (as seen in Extract 10), there is reference to environmental factors that are purported to have combined with the mental disorder and/or abnormal mental state to create a "perfect storm" (a metaphor used by the judge elsewhere in the remarks) which led to the *episode* in which the killing took place, as well as (in Extract 11) his positive character and relationship with his wife (something which plays no role in a psychiatric evaluation).

Extract 10. Case 2, report 2 (Dr Hallow)

He had not slept for days and this was with a background of chronic poor sleep. He had lost weight and described to [his doctor at the time] feeling in low mood.

Extract 11. Case 2, report 2 (Dr Hallow)

Mr Smith was a devoted husband and cared and supported his wife, Mrs Smith, on a round the clock basis for years.

Discussion of Psychiatric Reports in Relation to Rules and Procedures

Then there is the judge's evaluation of the psychiatric evidence. Remember, the judge is having to calibrate the seriousness of the crime with the sentence to be passed. In Case 1 there is a reverse correlation. The logico-scientific evidence interrupts the narrative, offering something thought of as a ground-truth. The judge's use of the modal verb "must" demonstrates how he reluctantly defers to this. He demonstrates that the outcome is predetermined; the needs and procedures of the court dictate that they "must" consider the psychiatric evidence. But to comfort the audience, the judge emphasizes that this doesn't impact our representation of these crimes as horrific and that the desire for retribution must be met.

Extract 12. Case 1 judge's obligation to needs and procedures of the court

> The expert evidence is to the effect that but for your mental illness you would not have committed these appalling crimes. I must further consider the extent to which punishment is required. In this context, I must, of course, bear fully in mind the devastating impact which your offending has had upon your victims and their families. I note in passing, however, that the diagnosis of treatment resistant schizophrenia means that, in any event, it is very likely you will never be released.

In effect, this is saying, explicitly to the defendant, but obliquely to society and more pertinently to the families and loved ones of the victims: due to rules and procedures in place, my hands are tied in terms of the official sentence that I can give. However, don't worry; he's not getting off, it still means detention for the rest of his days. We cannot officially see this as punishment, but it effectively is.

Case 2 is very different. The judge is grateful for the narrative nature of the psychiatrist's evidence. They are thanked for their "lucid and powerful" contributions to the narrative. In direct opposition to the constraints that the rules and procedures place upon the judge in Case 1, they provide invaluable support for what the judge wants to do.

While both defendants are granted their pleas of diminished responsibility, and charged with manslaughter rather than murder, the defendant in Case 2 is free to go.

In short, the practical outcomes for the defendant, and the signaling to the stakeholders in the two cases, could hardly be more different.

Conclusions

What is the relationship between these sentencing remarks and the underlying rationale behind diminished responsibility pleas? Here are some tentative conclusions. There is talk of generic interferences to conditions of responsibility. But there are also attempts to talk about problems both in medical terms and with appeal to distinctly medical categories.

By the letter of the law, Case 1 fits the diminished responsibility plea more clearly than Case 2, under both the weaker and more demanding readings of "medically recognized condition". And yet, what we see in the sentencing remarks in Case 1 is a judge who feels reluctantly compelled to grant diminished responsibility. (Interestingly, even given the scathing nature of the sentencing remarks, the sentence itself was appealed on the grounds that it was "unduly lenient". The appeal was overturned.) We see, in addition to this, that canonical psychiatric status of the defendant is played down and eclipsed by agentive and moralizing language.

In stark contrast, Case 2 does not as clearly and unambiguously fulfill the criteria of diminished responsibility. What we see in those sentencing remarks is a contextually informed intuition of leniency. The psychiatric "facts" are then marshaled in support of the already-arrived-at conclusion. One can easily imagine that if the defendant hadn't killed his wife, he would never have received a psychiatric diagnosis (for a "moderate depressive episode" or "adjustment disorder"). An episode is not in itself a diagnosis of a disorder, and "adjustment disorder" is, unlike the severe psychosis of the defendant of Case 1, a highly contested category. Within psychiatry, it is defined as "a problematic reaction to stressful events". This is not only a very vague definition but also one that risks medicalizing all manner of "problematic reaction". Many crimes are arguably "problematic reactions". What needs articulating is what makes a reaction problematic in a way that warrants psychiatric status.

What we clearly see from examining these two cases is the following. The psychiatric "facts", as presented by the psychiatric expert witnesses, are supposed to wield enough authority to ground the judgment, whether murder or manslaughter. And yet, what we really see is that there is a pre-existing intuition based on many contextual factors which remains firm and independent of a rather nebulous and malleable "psychiatric expertise". In Case 1, an unsympathetic judge is reluctant to truly accept the idea that the, rather abundant, psychiatric evidence fully exculpates. In Case 2, the judge is very keen to accentuate the rather threadbare psychiatric evidence in order to push for an exculpatory outcome.

These patterns fit with psychiatric fictionalism. The psychiatric "facts" are presented as causal and authoritative justifications of legal conclusions. However, they are shaped in myriad ways by pre-judged intuitions about the case. And so, they do not, as they purport to, factually and incidentally, interfere with the criteria of moral responsibility. Rather, they serve to intrinsically "up-medicalize" and "down-moralize" in line with the judge's intuitions about the case.

Forensic linguistics can be used to unearth inconsistencies in how psychiatric evidence is used in court. What needs to be examined, in particular, is whether there is a problematic, pervasive, but unacknowledged gap between the purported, factual exculpatory force of psychiatry and psychiatric evidence, and the way in which it is actually wielded as a post hoc instrument to fulfill already intuited conclusions.

References

Appelbaum, Paul. 2008. "Ethics and Forensic Psychiatry: Translating Principles into Practice." *Journal of the American Academy of Psychiatry and the Law* 36, no 2: 195–200.

Campos Pardillos, Miguel Ángel. 2022. "Metaphors as a Foundation for Judges' Reasoning and Narratives in Sentencing Remarks." In *Metaphor in Legal Discourse,* 146–168. Edited by I. Seskauskienė. Newcastle: Cambridge Scholars.

Campos Pardillos, Miguel Ángel. 2020. "Sentencing Remarks as a Legal Subgenre: 'R v Darren Osborne.'" Estudios de Traducción 10: 17–33. https://doi.org/10.5209/estr.68057.

Commons, Michael Lamport, Patrice Marie Miller, and Thomas Gutheil. 2004. "Expert Witness Perceptions of Bias in Experts." *Journal of the American Academy of Psychiatry and the Law* 32: 70–75.

Cotterill, Janet. 2003. *Language and Power in Court: A Linguistic Analysis of the O.J. Simpson Trial.* Basingstoke: Palgrave.

Eastman, Nigel, Sanya Krljes, Richard Latham, and Marc Lyall. 2018. *Handbook of Forensic Psychiatric Practice in Capital Cases.* London: The Death Penalty Project.

Eastman, Nigel, and Keith Rix. 2021. "Bias in Expert Witness Practice: Sources, Routes to Expression and How to Minimise it." *British Journal of Psychiatric Advances* 28, no 1: 35–45.

Eastman, Nigel, et al. 2012. *Oxford Specialist Handbook of Forensic Psychiatry*. Oxford University Press.

Harris, Sandra. 2005. "Telling Stories and Giving Evidence: The Hybridisation of Narrative and Nonnarrative Modes of Discourse in a Sexual Assault Trial." In *The Sociolinguistics of Narrative*, 215–237. Edited by Joanna Thornborrow and Jennifer Coates. Amsterdam: Benjamins.

Heffer, Chris. 2010. "Narrative in the Trial: Constructing Crime Stories in Court." In *The Routledge Handbook of Forensic Linguistics*, 199–217. Edited by Malcolm Coulthard, Alison May, and Rui Sousa-Silva. London: Routledge:

Heffer, Chris. 2005. *The Language of Jury Trial: A Corpus-Aided Analysis of Legal-Lay Discourse*. New York: Springer.

Mackay, Ronnie, and Barry Mitchell. 2017. "The New Diminished Responsibility Plea in Operation: Some Initial Findings." *Criminal Law Review*, 18–23.

Murrie, Daniel, and Marcus Boccaccini. 2015. "Adversarial Allegiance Among Expert Witnesses." *Annual Review of Law and Social Science*, no 11: 37–55.

Wilkinson, Sam. 2023. "Psychiatric Fictionalism and Narratives of Responsibility." *Philosophical Explorations* 26, no 2: 91–109.

Wilkinson, Sam. 2022. *Philosopy of Psychiatry: A Contemporary Introduction.* New York: Routledge.

Wolf, Susan. 1987. "Sanity and the Metaphysics of Responsibility." In *Responsibility, Character, and the Emotions: New Essays in Moral Psychology*, 46–62. Edited by Ferdinand David Schoeman. Cambridge: Cambridge University Press.

18

TIGHTENING THE REIN ON EXPERT EVIDENCE

Ethical Reasons for Judicial Gatekeeping in the Canadian Context and Implications for Forensic Linguistics

Jennifer Glougie

Introduction

The vast majority of scholarship in the area of ethics and expert witnesses in the context of forensic linguistics has been done by the linguists who act themselves as experts in forensic linguistics matters (see, e.g., Finegan 2021; MacLeod 2021; Clarke 2018; Butters 2010; Finegan 2010; Shuy 2010; Stygall 2010; Siegel 2005). This literature includes many examples of cases where expert forensic linguistic evidence helped to deliver justice. There are significantly fewer publications in this area that provide perspectives from the lawyers who retain forensic linguistics experts on behalf of their clients (Ainsworth 2010). This chapter complements this body of work and presents the perspective of the trier of fact who is tasked with evaluating proposed expert evidence to decide whether or not it should be included. In this context, the chapter explores a recent example from Canada that shows the devastating consequences that can occur when courts admit expert evidence resulting in a miscarriage of justice.

The case in point involves an official Inquiry into Pediatric Forensic Pathology in Ontario (henceforth the "Inquiry"). Commissioned in 2007, the Inquiry had been prompted by a number of highly publicized cases in which innocent people were criminally charged and, in many cases, convicted on the basis of flawed expert evidence. In 2008, the Inquiry's commissioner, Mr. Justice Stephen T. Goudge, released his report (henceforth the "Report" or the "Goudge Report"). The Report exposed serious flaws in the way pediatric forensic pathology had been practiced in Ontario throughout the last decades of the 20th century. It also brought to light many disturbing insights into how the criminal justice system had relied on and sanctified these practices in pediatric forensic pathology. The combined result was over a dozen miscarriages of justice.

While the Inquiry did not recommend reforming the law of evidence, it did state that the court should exercise greater vigilance when evaluating expert evidence, albeit within the existing framework. Jurisprudence over the following decade and a half demonstrates that Canadian courts have taken up the gauntlet. The result has been the most significant shift in the law surrounding

DOI: 10.4324/9781003391074-20

expert evidence in Canada since the test for the admissibility of expert evidence was first articulated in the mid-1990s.

This shift has, hopefully, reduced the potential damage done by charlatan experts in Canada. Under the new approach, Canadian courts now evaluate proposed evidence separate and apart from assessing experts' qualifications. In doing so, the court assesses whether the matters on which experts purport to give evidence legitimately fall within the scope of their expertise. In cases where this threshold is not met, an expert's opinion and/or evidence is excluded. As a part of this review process, the court also evaluates whether there are realistic concerns about the capacity of experts to give evidence in a manner consistent with their duty to the court. Finally, the court more vigilantly evaluates the expert evidence for (un)conscious bias. In cases where it cannot be determined that said evidence is free from bias, it is formally excluded.

As this chapter will show, these developments have implications for forensic linguists who undertake the role of expert witness in Canada and elsewhere. Importantly, as this chapter will demonstrate, the issues raised regarding the quality control and management of expert witnesses and their evidence is not only relevant to the Canadian legal system. It also has implications for the professional ethics of expert forensic linguistic witnesses in other jurisdictions which might not require the same kind of evaluation.

The Inquiry

There is a popular running joke among law students who study in Commonwealth countries where the legal systems are founded on the British common law. You can always tell the result of a decision written by Lord Denning, Master of the Rolls, by reading its introduction. To the extent that principle is true of the 2008 Goudge Report, the first paragraph of the executive summary contains a pretty significant spoiler. The Report begins:

> The sudden, unexpected death of child is a devastating event for parents, for family, and for the entire community. If something suggests that a criminal act may have been involved, the devastation takes on a further tragic dimension. This reality lies at the core of the work of this Inquiry.
>
> *(2008: 1)*

The scientific expert who was involved in providing the evidence for the "devastating event" Goudge refers to, Dr. Charles Smith, was the preeminent pediatric forensic pathologist in Canada's largest province of Ontario. Dr. Smith received his fellowship in pediatric pathology in 1981 and immediately began working full time at Toronto's world-renowned Hospital for Sick Kids ("Sick Kids Hospital"). Throughout much of the 1980s and 1990s and into the early 2000s, Dr. Smith provided forensic evidence for numerous cases.

During Dr. Smith's tenure, autopsies were conducted on a fee-for-service basis. These services were often provided by the pathologists employed at hospitals, but these post-mortem examinations were not considered part of their regular hospital duties. Not all pathologists felt comfortable with performing these examinations, however. Many of the pediatric pathologists at Sick Kids Hospital, the institution involved in the Inquiry, refused in fact to perform this work, arguing that they did not have the requisite expertise skills. Dr. Smith was an exception, however. Although like most pathologists in Ontario at the time, Dr. Smith had no formal training or certification in forensic pathology, he and a handful of his colleagues were willing to take on this work.

Almost as soon as Dr. Smith began performing pediatric forensic pathology, concerns began to emerge about his competence to do so. Initially, these concerns were raised by those he had helped to accuse as well as their friends and family. Likely for that reason, these complaints were largely ignored. However, even as early as the late 1980s, courts were expressing wariness about the nature of Dr. Smith's evidence and the dogmatic confidence with which he presented it.

For example, in a 1989 decision, [*R. v. M(S)* [1991], O.J. No. 1383, the court summarily rejected Dr. Smith's expert evidence that a child, Amber, had died of shaken baby syndrome. In this case, Dr. Smith had in fact conducted the autopsy in question. Nonetheless the court preferred the evidence of other experts who had not participated in the death investigation. The court expressed significant reservations about Dr. Smith's expert evidence, ultimately concluding he lacked objectivity; failed to properly investigate or consider all of the relevant facts before arriving at his conclusion; failed to keep an adequate record of his work; and lacked familiarity with the relevant scientific literature. The court found Amber had not died of shaken baby syndrome as Dr. Smith opined and entered an acquittal.

As in the case above, many of these official concerns about Dr. Smith's evidence were lodged even before he was made director of the Ontario Pediatric Forensic Pathology Unit (OPFPU). The first of its kind in Ontario or anywhere else in Canada, the OPFPU was created in 1991 and operated out of Sick Kids Hospital. Despite the court's indictment of his approach, his conclusions, and his evidence, Dr. Smith was appointed as the OPFPU's first director.

During his tenure as OPFPU director, between 1991 and 2005, Dr. Smith conducted autopsies in or provided consultative opinions on 45 "criminally suspicious" child or infant deaths. Where criminal charges flowed from his consultation, he regularly provided expert evidence, both in the form of reports and testimony. Dr. Smith also regularly taught seminars and gave lectures on pediatric forensic pathology to coroners, counsel from the Crown prosecution service and criminal defense bar, and law enforcement.

By the time he was removed from his directorship in 2005, Dr. Smith's expert evidence was instrumental in securing the conviction of 12 caregivers whose charges died unexpectedly while in their care. An additional caregiver was found liable for the child's death, though not held criminally responsible by reason of mental disorder. These were the parents, uncles, aunts, and babysitters of otherwise healthy young children who died in what Dr. Smith believed, and the Crown agreed, to be criminally suspicious circumstances.

However, Dr. Smith's prominence and concerns about his competence grew in tandem. By the early 2000s, Dr. Smith's work was starting to make the news, and the reporting was cause for significant alarm. For example, in 2001, Canadian news media reported on the case of a mother who was charged with the 1997 murder of her daughter Sharon. Dr. Smith's expert opinion was that Sharon had died after being stabbed 88 times. He rejected the mother's position that Sharon died after being attacked by her grandfather's pit bull terrier. He claimed, pathologically speaking, to be able to distinguish the difference between a stab wound and a dog-bite wound. Sharon's file was reviewed by another pathologist who disputed Dr. Smith's conclusions. Her remains were ultimately exhumed and examined by Ontario's chief pathologist who found evidence of numerous dog-bite wounds and no evidence of a stabbing. Later, in his evidence before the Inquiry, Dr. Smith conceded that he had actually never worked a case involving a dog bite before Sharon's case. The Crown opted not to pursue the murder charges against Sharon's mother in 2001. By then, however, she had spent two years in jail, mostly in an isolation cell, and another two years living in a shelter with strict bail conditions awaiting trial. Her other four children had been placed into foster care throughout that entire period.[1]

This was not the only miscarriage of justice that relied on Dr. Smith's expertise. Another case that came to light in the early 2000s involved the death of 11-month-old Jenna. Dr. Smith's investigation uncovered, but ignored, a pubic hair near the child's vagina. The pubic hair was collected, but not initially submitted for testing and not included in Dr. Smith's report. While Dr. Smith was convinced that Jenna had died at a time when only her mother could have killed her, testing (when it was eventually done) established the pubic hair was not Jenna's mother's. In 1999, the Crown decided not to pursue the murder charges against Jenna's mother. By this time, Jenna's sister had spent two years in foster care. In the meantime, the police continued their investigation into Jenna's death. Subsequent experts they consulted provided grounds to believe that Dr. Smith's estimate as to her time of death was incorrect and that it was likely she received her fatal injuries while in the care of a 14-year-old male babysitter. In 2005, the babysitter confessed to an undercover police officer to having punched, poked, sexually assaulted, and killed Jenna.

By the mid-2000s, Ontario's Chief Coroner, who ostensibly oversaw the OPFPU and Dr. Smith's performance as its director, could no longer ignore the mounting concerns and criticism. He retained five experienced pathologists unaffiliated with Sick Kids Hospital (henceforth the "Panel") and commissioned them to perform an independent review of Dr. Smith's work. The Panel reviewed all 45 of the "criminally suspicious" cases in which Dr. Smith was involved as director of the OPFPU and identified 20 of them as being "problematic" (2008: 7). In those 20 cases, which amount to almost half of the total cases in which Dr. Smith had been involved, the Panel took issue with his findings of fact; the opinions he drew from those facts; and/or the testimony he subsequently gave in support of the charges that were filed, usually on the basis of his professional opinion.

The Inquiry was established six days after the Chief Coroner's Review published its results. The goal of the Inquiry was to review what had gone so horribly wrong in the practice of pediatric forensic pathology—from the practice itself to its use in the criminal justice system. The goal of the Inquiry was to make recommendations that would protect against such miscarriages of justice happening again. While the Inquiry did not focus expressly on Dr. Smith, the reality is that he was the preeminent pediatric forensic pathologist operating in Ontario throughout the period in question, and it was his professional competency in death investigations that came under specific scrutiny.

The Inquiry, like the Chief Coroner's Review, found significant scientific flaws in Dr. Smith's practices in forensic pathology. The Inquiry found that Dr. Smith had failed to follow basic principles and accepted methodologies. He almost never attended the death scene. Before conducting an autopsy, he had failed to ensure he always had relevant information that was widely accepted as necessary within the field. He had also been indiscriminate about accepting irrelevant information—particularly about the social history of those allegedly involved with the deceased child.

However, the deficits identified did not end there. From a forensic perspective, the Inquiry found that Dr. Smith had failed to understand the role of an expert and had neglected to properly undertake the duties associated with that role. Expert witnesses must be neutral and independent, yet Dr. Smith clearly saw himself as being on the same "team" as law enforcement and the prosecution. In one particularly egregious example, he had even allowed police to intercept phone calls he conducted with suspect family members in which he discussed the contents of his autopsy report. His early informal expressions of opinion to police were found to be too categorical and often skewed the investigation from its earliest stages, as had happened in both Jenna's and Sharon's cases. He was also found to have regularly engaged in speculation on matters about what could have happened even in the absence of any pathological evidence to support his ideas. The Inquiry also determined that he had routinely failed to acknowledge when his opinion was controversial within the scientific field. Instead, he presented his viewpoint as dogmatic. Finally, he failed to acknowledge limitations of his expertise and he testified on matters well outside of them. Not all of the failings that the Inquiry identified were isolated to Dr. Smith, however.

The Inquiry also found serious flaws with how he had been allowed to operate as an expert within the criminal justice system. The lack of regulation had ultimately permitted Dr. Smith's expert opinion to be used to charge and convict innocent people of murder. In the end, the Inquiry found a miscarriage of justice had occurred in all of the 13 cases where Dr. Smith's participation resulted in a finding of liability, whether it was a criminal conviction or the declaration of no criminal responsibility by reason of mental disorder.

Dr. Smith had his medical license revoked in 2011. This was no condolence to the individuals who, on the basis of his word, spent years and in some cases decades in jail or on bail for the most heinous acts which either never happened or for which they were not responsible. Nor was it consolation to the families who had other children taken from them and put in foster care or up for adoption without their consent. The fact that Dr. Smith lost his medical license also did nothing to compensate people who had spent hundreds of thousands of dollars, losing their homes and life savings or sinking into crippling debt, in their attempt to defend themselves against the weight of Dr. Smith's opinion. These individuals and families faced these devastating hurdles, all while trying to grieve the loss of a child.

The Shift in the Canadian Approach to Expert Evidence

Expert evidence has always played an uneasy role in courtroom processes. Witnesses are generally required to give evidence only about events they witnessed without giving their opinions about those events. Expert opinion evidence is an exception to this rule. That is, experts are generally permitted to provide opinion evidence in a manner no other witness can, provided their opinion is about a matter that falls within the scope of their expertise. In doing so, expert witnesses are permitted to opine on areas that would normally be reserved for the trier of fact. In that regard, expert witnesses walk a fine line; they must provide their evidence in a way that provides assistance to the court—which has no expertise in the subject matter at issue—while not overstepping their role by usurping the trier of fact.

Other than to serve as a cautionary tale, the Inquiry resulted in the most consequential shift in the law around expert evidence in Canada since the legal test itself was established in *R. v. Mohan* [1994], 2 S.C.R. 9 ("*Mohan*"). *Mohan* is a 1994 decision of the Supreme Court of Canada, which sets out the threshold criteria against which expert evidence must assessed. The expert evidence must be (a) relevant (b) necessary (c) not excluded by operation of some other evidentiary rule, and (d) given by a properly qualified expert. Even where the *Mohan* criteria are met, the court has the discretion to exclude evidence where its probative value is outweighed by its prejudicial effect. While this "gatekeeping function" has risen to prominence in the aftermath of the Goudge Report, it has always been available to courts.

The Inquiry made 169 recommendations that, in its view, would put better safeguards in place to ensure that the evidence of self-proclaimed "experts" like Dr. Smith would not be routinely and indiscriminately accepted, and to avoid or at least limit the amount of damage people like him might cause. Most of the Inquiry recommendations focus on implementing a system of oversight and accountability, and regulating how law enforcement are allowed to work with experts involved in death investigations.

Twelve of the Inquiry's 169 recommendations concerned the role of the judiciary and pertain specifically to the admissibility of expert evidence at court. Remarkably, the recommendations did not call for a full reform of the law of evidence. As Paciocco (2009a) notes, the Report is "a measured reaction to the incompetence it exposed" (139). Rather, the recommendations encourage the court to exercise greater vigilance over the evaluation process already set out in the *Mohan* framework.

Evidence-Based Approach to the Evaluation of Expert Evidence

American practitioners and scholars will almost surely have noticed that the *Mohan* test for the admissibility of expert evidence does not codify any particular criteria for evaluating the evidence itself. In this way, the *Mohan* criteria differ from those set out in *Daubert v. Merrell Dow Pharmaceuticals Inc.* (1993), 509 U.S. 579. This is not to say that Canadian courts do not consider threshold reliability, but rather that scientific validity is not a condition precedent to the admissibility of expert opinion evidence (*Abbey #2*, para 108). As Professor Paciocco, now Justice Paciocco of the Ontario Court of Appeal, noted,

> Clearly it is inappropriate to consider all expertise as science, or to require all expertise to attain the scientific method. Some expert witnesses rely on science only in a loose sense. Actuaries apply probability theory and mathematics to produce decidedly unscientific results. Appraisers make subjective assessments of objective data, as do family assessment experts. Professionals testifying to standards of care within their profession are doing nothing scientific. Yet Daubert spawned a jurisprudence that was fixated for a long time with science. This led lower courts to commit two kinds of error. First, it caused some lower courts to hold that the Daubert test and gatekeeping role is confined to scientific expertise. Experts who were not scientists would not be subjected to the reliability inquiry prescribed by Daubert. Second, it caused other courts to apply the criteria listed in Daubert in a wooden fashion, even to non-scientific forms of expertise. Each of these two kinds of errors was caused by the failure to take context into account.
>
> *(2001: 57)*

The Goudge Report makes the same observation in the context of forensic pathology, when it noted the following:

> Forensic pathology provides a good example of a discipline that has not traditionally engaged in random testing or determining rates of error. The reasons are obvious: testing and reproducibility cannot be used to verify a cause of death. The forensic pathologist's opinion must instead rely on specialized training, accepted standards and protocols within the forensic pathology community, accurate gathering of empirical evidence, attention to the limits of the discipline and the possibility of alternative explanations or error, knowledge derived from established peer-reviewed medical literature, and sound professional judgment.
>
> *(2008: 493)*

Thus, while expert evidence must always demonstrate a sufficient level of reliability to warrant its admission, a flexible approach to the determination of that reliability is essential (*Abbey #2*, para 111; *Kumho Tire Co. v. Carmichael,* 526 U.S. 137, 119 S. Cr. 1167 (1999)).

Of course, a court cannot assess threshold reliability if it does not understand how the expert arrived at the opinion they seek to offer in evidence. To this extent, the Canadian courts have adopted an evidence-based approach to the evaluation of expert witness evidence, which requires the following:

1. the theory or technique used by the expert must be reliable, and so too must the *use* of that theory or technique by the expert;
2. the expert must not be biased (the expert must "keep an open mind to a broad menu of possibilities");
3. the expert must be objective and complete in collective evidence, must reject information that is not germane to the theory or technique being used, and must be transparent about all of the information and influences they have been exposed to; and
4. the expert must clearly express not only the opinion, but also the complete reasoning process that led to it, and must be candid about the shortcomings of the theory or technique employed and the opinion reached, offering fair guidance on the level of confidence that can be placed in the opinion expressed (Paciocco 2009a: 146–147).

This evidence-based approach was expressly endorsed by the Ontario Superior Court of Justice in *R. v. France*, 2017 ONSC 2040 ("*France*"). Since then, courts across Canada have regularly rejected expert evidence not because they take issue with the expert's qualifications, but because the expert did not satisfy the court that their opinion was consistent with this approach. In *Moffitt (Litigation Guardian of) v. TD Canada Trust*, 2023 OJ No. 2181, for example, the expert's evidence was excluded, among other things, because their report failed to explain the methodology they used to arrive at their opinion, or how that methodology was used in their analysis.

An evidence-based approach requires the expert witness to show their work, which better positions the court to assess threshold reliability. In other words, as a result of the Inquiry's recommendations, Canadian courts have shifted away from a "trust me" model of expert evidence to a "persuade me" model (Paciocco 2009a: 146). Moreover, adopting an evidence-based approach also puts the court in a better position to scrutinize the expert's ability to fulfill their duty to the court and ensure the expert's opinion evidence does not stray beyond the bounds of their expertise. It is the courts' renewed and reinvigorated commitment to undertaking this level of scrutiny that has caused the shift in how expert opinion evidence is admitted or excluded.

White Burgess and the Expert's Duty to the Court

One area where the Inquiry did make recommendations that arguably required an amendment, or addition, to the law of evidence was with respect to the duty of an expert witness to the court. Expert witnesses have always had a duty to give opinion evidence that is impartial, independent, and free from bias; and courts have exercised their discretion to exclude expert opinion evidence where there are concerns about the expert's ability to fulfill that duty. In *White Burgess Langille Inman*, 2015 SCC 23 ("*White Burgess*"), however, the Supreme Court of Canada not only imposed a requirement that courts consider the expert's capacity to fulfill that duty as part of their assessment of admissibility. It also developed the framework for doing so.

The issue in *White Burgess* was whether an otherwise properly qualified expert witness's evidence should be excluded because of concerns about their ability to fulfill their duty to the court. The court acknowledged that the duty of an expert involves three related concepts: impartiality, independence, and neutrality. Impartiality reflects an objective assessment of the question at hand. Independence means the opinion is the product of the expert's independent judgment, uninfluenced by the potential outcome of the litigation or which party retained them. Neutrality means their opinion is unbiased and does not unfairly favor one party's position over the other.

The plaintiffs in *White Burgess* were company shareholders who brought an action against their former auditors for failing to apply generally accepted auditing and accounting principles, causing them financial loss. The plaintiffs relied, in part, on the expert testimony of an employee of another accounting firm. The defendants sought to have the expert evidence excluded on the grounds she was not impartial. They argued she could be exposed to liability if her report was not accepted by the court and, as a partner in her firm, she could be personally liable as a result.

The Supreme Court of Canada had to decide whether the elements of the expert's duty go to admissibility (i.e., the threshold requirements from *Mohan*) rather than simply weight. In other words, can a court exclude an expert's evidence if it is not convinced they are capable of fulfilling their duty to the court, or should those concerns be addressed as a matter of deciding how much weight to give that evidence after it is admitted? The court concluded that an expert's lack of impartiality or independence can go to the admissibility of their evidence at the threshold stage. In that respect, it reiterated that

> expert witnesses have a duty to assist the court that overrides their obligation to the party calling them. If an expert is unable to fulfill that duty, they do not qualify to perform the role of an expert and should be excluded.
>
> *(para 46)*

However, this does not prevent the court from also considering the expert's impartiality and independence in deciding how much weight to give their evidence, if the evidence is otherwise admitted.

Given that a lack of impartiality or independence could disqualify an expert's evidence at the threshold stage, the court went on to determine what the threshold admissibility requirement should be in relation to the expert's capacity to fulfill their duty. The threshold for establishing an expert's neutrality is not particularly an onerous one. If an expert swears an oath or affirms that their opinion is impartial, independent, and free from bias, the court will generally be satisfied that the expert is capable of performing their duty in a way that doesn't impede their qualifications to do so. Once the expert has given such an affirmation, the burden falls to the party seeking to oppose the admission of their evidence to raise a "realistic" concern about the expert's capacity to act in accordance with their duty.

Since *White Burgess*, the Rules of Court in British Columbia (BC) and other Canadian jurisdictions have imposed a requirement that experts include such an attestation as part of an expert report before it can be accepted at court. In BC, the Supreme Court Civil Rules require the following:

> **Rule 11-2 – Duty of Expert Witnesses**
>
> **Duty of expert witness**
>
> (1) In giving an opinion to the court, an expert appointed under this Part by one or more parties or by the court has a duty to assist the court and is not to be an advocate for any party.
>
> **Advice and certification**
>
> (2) If an expert is appointed under this Part by one or more parties or by the court, the expert must, in any report he or she prepares under this Part, certify that he or she
>   a. is aware of the duty referred to in subrule (1),
>   b. has made the report in conformity with that duty, and
>   c. will, if call on to give oral or written testimony, give that testimony in conformity with that duty.

The court will generally consider the first branch of the *White Burgess* test to be satisfied if the expert provides a report that conforms to the requirements of a rule that is in Rule 11-2.

None of this is to say that, before *White Burgess*, courts did not or could not consider witness partiality as part of the inquiry into proper qualifications. On the contrary, in *White Burgess*, the court reviewed a number of decisions where an expert witness's evidence was excluded because of a lack of impartiality or independence, including where the proposed expert had too personal a connection to the litigation. For example, a proposed expert was excluded where they were a party to the litigation (*Ocean v. Economical Mutual Insurance Co.* 2010 NSSC 315), the party's lawyer in another related matter (*Fellowes* (1998), 40 OR (3d) 456; *Royal Trust Corp.* (2000), 49 O.R. (3d) 187), or the father of the lawyer attempting to call him (*R. v. Docherty* 2010 ONSC 3628). Proposed experts have also been excluded where they had an interest in the outcome of the litigation, such as where their remuneration depended on the trial's success (*Bank of Montreal* [2001] O.J. NO. 1096), where they stood to incur liability depending on the result of the trial (*Hutchingame* 2006 BCSC 271), or where their retainer was simply inappropriate (*Dean Construction* 2011; ONSC 4629). Similarly, the court reviewed a number of decisions where the expert's stance or behavior as an advocate was found to justify their exclusion (e.g., *Carmen Alfano Family Trust (Trustee of) v. Piersanti* 2012 ONCA 297; *Kirby Lowbed* 2003 BCSC 617; *Gould* 2012; ONSC 5184). However, while prior to *White Burgess*, courts *could* undertake an evaluation of the expert's capacity to fulfill their duty (and exclude their evidence where concerns arose), after *White Burgess*, they *must* undertake such an evaluation, at least at a rudimentary level.

The Scope of an Expert's Expertise

In addition to the requirement to evaluate the expert's capacity to fulfill their duty to the court, the Canadian courts have become more diligent about distinguishing between an evaluation of the expert and an evaluation of the scope of their expertise. In other words, even in cases when an expert is properly qualified, the court will go on to determine whether the evidence they propose to give falls within their specialized area of knowledge.

In Dr. Smith's case, there was no doubt he had the training and the credentials to be qualified as an expert in medicine, with a specialization in pediatric pathology. The problem—separate and apart from the manner in which he practiced—was that he regularly gave evidence about *forensic* pathology, a field in which he had no training, no qualifications, and no experience other than which he had gained "on the job".

The Inquiry found that, having been seduced by Dr. Smith's credentials and qualifications, many courts had given him free rein to testify on any topic and accepted his testimony as gospel even when he testified about matters well outside of his expertise. To combat this going forward, the Inquiry suggested the following: "When a witness is put forward to give expert scientific evidence, the court should clearly define the subject area of the witness's expertise and vigorously confine the witness's testimony to it" (Recommendation 129).

In response to the Inquiry, courts have actively sought to confine the expert's testimony in accordance with Recommendation 129. In *McLeod v. BC (Superintendent of Motor Vehicles)*, 2023 BCJ No. 407, for example, the court excluded the expert's opinion evidence because they refused to identify the scope of their expertise, despite being ordered to do so.

Of course, the expert's capacity to perform their duty and the scope of expertise rarely arise independently of each other. For the most part, where a court provides reasons justifying the exclusion of an expert's evidence, those reasons demonstrate significant overlap between these concerns. This was the case in *R. v. Scott*, 2018 BCSC 1739 ("*Scott*"), where the court excluded the evidence of Dr. Gabor Maté, a renowned medical doctor with specializations in addiction and childhood trauma. The defense in *Scott* proposed to call Dr. Maté to give expert opinion evidence about the origins of the accused's addiction in early life trauma and the psychological effect of addiction and relapse (para 2). Dr. Maté is a prominent public figure, with recognized expertise and qualifications. He is a sought-after public speaker nationally and internationally who has published extensively on the areas of addiction and childhood trauma. He was cited as a member of the Order of Canada in 2018 for contributions to the field of addictions and mental health. Despite these qualifications, Dr. Maté was not found qualified by the court to give expert evidence in support of the accused in *Scott*. In making its decision, the court relied on the fact that Dr. Maté:

- had not done independent research in these areas, relying on the research of others and coming to his own conclusions about it. In that sense, his expertise developed through personal experience, namely his many years of researching the literature, finding a gap in medical thinking and writing about it in his books that seek to explain the issues to the public;
- publishes regularly, but his books are bestsellers that are not subject to peer-review; and
- his expertise did not extend to forensic psychology or neuropsychology, yet he opined on those issues for the purposes of his report

For these reasons, the court determined that Dr. Maté was not a qualified expert in the areas in which he purported to give evidence.

The court went on to find the Crown had established reasonable concerns that Dr. Maté would be incapable of performing his duty to the court. The court found that, in voir dire, he became defensive and argumentative. He testified that he had "little faith in the ability of the courts to 'come to terms with human psychology as it is not one of their strengths'" (para 92). The court found Dr. Maté's comments in this regard to contain a

> strong element of grandiosity that lends weight to the concern over a lack of impartiality. Indeed, such an attitude blinds him to his obligation of impartiality and his duty to the Court as an expert witness. Moreover such a view tends to suggest that Dr. Maté believes his opinion is the correct view and the courts are simply unable to get it right because they do not understand human psychology.
>
> *(para 93)*

Finally, the court was troubled by the transcript of Dr. Maté's interview with the accused which, in its view, read "like an individual approaching an interview with a particular thesis in mind and then asking questions to confirm that thesis" (para 121). Dr. Maté's interview style raised the court's concerns not only about the potential for bias, but also that the interview itself might have contaminated the accused's memory about his childhood (para 121). For all of these reasons, Dr. Maté's expert opinion evidence was also excluded in its entirety. As this case illustrates, the changes in the law of expert evidence in Canada appear to have helped in ensuring that the job of giving expert testimony is limited to those who are properly qualified to give it.

Implications for Forensic Linguists Serving as Expert Witnesses

The majority of forensic linguists who engage in a consultancy or expert witness practice do not work within Canada. Nonetheless, the Inquiry and the recommendations that flow from it make a persuasive case for the development of best practices, regardless of what the law in other jurisdictions requires. If forensic linguistics, as a field of scholarship, is susceptible to the types of concerns the Inquiry unearthed in pediatric forensic pathology, then in all likelihood the law in other jurisdictions will, eventually, be forced to adopt an approach as Canada has done.

It is important to acknowledge that the field of forensic linguistics suffers from many of the same challenges—and is open to the same criticism—as pediatric forensic pathology as it was practiced in Ontario until the early 21th century. Like pediatric forensic pathology, forensic linguistics is unregulated and has no mechanism for oversight or accountability. The Inquiry made numerous recommendations aimed at providing regulatory oversight to promote transparency and accountability in the selection and use of forensic expert witnesses in Canada. However, there is no easy method by which the entire practice of forensic linguistics could be regulated, even if there were good reason to want to do so. As a result, it is for the most part left to individual scholars to ensure they are using best practices in performing their consultancy work. In that respect, the shift in Canadian law can help guide individual practitioners as to what those best practices might look like.

Gaining Appropriate Training

A scholar of law and language, including forensic linguistics, should undertake specific training about the obligations and duties of acting as an expert witness before taking on a consultancy or expert witness practice. Even a scholar who engages only in consultancy work has the capacity to contribute to a miscarriage of justice if they undertake that role without a proper understanding of their role as an opinion provider. Recall the cases of Jenna and Sharon discussed above.

One of the most crucial flaws identified in the Inquiry was Dr. Smith's obvious belief that having established expertise in one specialization was qualification enough to allow him to provide expert testimony in another. Dr. Smith also clearly and quite falsely believed that he honed his skill as an expert over the course of the decades he spent performing that role. The Inquiry pointed out just how flawed those beliefs were. It is not enough to have expertise within a particular subject matter. An expert should also have specific training required for the forensic analysis and know how to accomplish this analysis in a manner consistent with their duty to the court. Put simply, the role of the expert witness is not something a scholar can or should learn on the job.

While forensic linguistics is the study of language in the legal context, including the use of language as evidence in criminal or other courtroom matters, practitioners of forensic linguistics do not necessarily have any specific competence in what their role as a consultant or expert requires. A forensic linguist who practices or intends to practice as a consultant should undertake specific training and certification in that regard.

Limiting Evidence to the Scope of Your Expertise

Expert witnesses are given rein to provide opinion evidence in ways that other witnesses are not, provided it falls within the scope of their expertise. This is because the expert is retained specifically to assist the trier of fact understand evidence that the expert has specialized knowledge about and the judge or jury do not. An expert has no greater knowledge than the judge or jury on matters that don't fall within their specific area of study. An expert that attempts to give evidence beyond the scope of their expertise has usurped the role of the trier of fact.

The fact a scholar has established expertise within a field does not mean they have expertise in each of the field's sub-areas of study. For example, a forensic linguist may have expertise in authorship attribution, but not forensic phonetics, and vice versa. A forensic linguist acting as an expert witness must therefore limit their opinions to only those sub-areas of the field over which they can legitimately claim scholarship and expertise.

This requirement applies equally when the expert witness is preparing their report and when they are giving evidence. In Canada, the court must assess the *Mohan* criteria on an ongoing basis: *R. v. Sekhon*, 2014 SCC 15. A witness who is properly qualified at voir dire on the basis of their report will find their evidence excluded at trial if they venture beyond the scope of their expertise in testimony.

Although other jurisdictions may have other regulations, the practitioner in forensic linguistics should clearly explain the exact scope of their expertise and provide the scholarship that supports that assertion either in their expert report or in their attached curriculum vitae. They should be aware of and resist counsel's attempts to have them testify on matters beyond that scope.

Showing Your Work

The evidence-based approach to expert evidence in Canada requires that the expert clearly express not only their opinion but also the complete reasoning process that led to it. To the extent there are shortcomings in the theory or approach the expert witness used in arriving at their opinion, those shortcomings should be identified and acknowledged. Experts must also be transparent about what information and influences they have been exposed to in preparing their report. To this extent, Grant and Grieve (2022) provide helpful advice on how to avoid confirmation basis when providing consultancy work.

The case of *De Angelis v. Siermy*, 2022 BCSC 31 is relevant here. The case involved an authorship dispute concerning how a testator intended to distribute her estate on her passing. In 2002, the testator executed a will that would leave the bulk of her estate to one of her nieces. In 2011, however, she executed a new will changing the primary beneficiary to a different niece. The original beneficiary sued, claiming the amendment to the will breached the contract that had been created when the first will was executed in 2002. In doing so, the plaintiff relied on three purported "letters from the grave": three letters the testator wrote, to be distributed after her death, explaining why she distributed her estate as she did. There was a dispute as to the authenticity and authorship of two of the three letters from the grave. The plaintiff retained a forensic linguist who performed a forensic stylistic analysis of the three letters, comparing them to known writings from the testator, and concluded that she likely wrote all of them. The plaintiff's rebuttal expert did not provide an opinion on the authorship question but pointed out the shortcomings with the forensic stylistics approach the initial expert had used and argued why it did not meet the benchmark of threshold reliability. The court accepted the rebuttal expert's evidence on that issue and, while it admitted both reports, it gave no weight to any of the linguistic evidence.

The court relies on experts not only to fill in knowledge gaps on the subject matter of their expertise but also to satisfy itself that the evidence meets the standard of threshold reliability. A party to a court proceeding, whether that is an accused person or a party to litigation, should not need to retain a rebuttal expert to challenge the initial expert's qualifications, methodology, or approach where there are concerns about threshold reliability The expert witnesses should identify those limitations themselves in their report and let the court decide how to address them.

Making Decisions Available

Forensic linguists who engage in consultancy and expert witness work should consider making publicly available copies of all decisions in which the court either admitted or excluded their evidence. One of the main criticisms of Dr. Smith's practice was that, while he sometimes shared with his colleagues and superiors when he "helped the prosecution win" a case, he was rarely so forthcoming when he helped them lose. While providing access to the expert's report itself is potentially fraught with privacy and privilege concerns, there is no such issue providing a copy of the court's decision on whether or not to accept the report into evidence, since court decisions are public documents. Moreover, it is the court's rationale for including or excluding the expert's evidence that is of most interest. There are many reasons a court might exclude an expert witness's evidence that are in no way an indictment of their work or their reputation: this is particularly true where the exclusion is made as part of the court's gatekeeping function. However, it was this lack of transparency that allowed Dr. Smith to perpetuate miscarriages of justice for as long as he did. By making decisions publicly available to other scholars, forensic linguists who engage in this type of work promote transparency that allows others to learn from their experiences, and ultimately promote the development of best practices on the basis of real casework.

Conclusion

The Inquiry exposed serious and long-standing flaws in the practice of pediatric forensic pathology as it was practiced in Canada into the beginning of the 21th century. These flaws were perpetuated in a judicial system which failed to scrutinize the "expertise" of so-called experts and ultimately brought the justice system into disrepute. The horror that the Inquiry exposed properly caused Canadian courts to rethink their approach to evaluating expert evidence within the *Mohan* framework. Courts in Canada must now evaluate threshold reliability on the basis of an evidence-based approach, must explicitly consider the expert's capacity to fulfill their duty to the court, and must consider the expert separate and apart from the evidence they propose to give. In doing so, the courts have tightened the reins on experts and, hopefully, ended the reign of the charlatan expert.

Forensic linguists who act as expert witnesses or consultants would benefit from implementing in their own practice certain safeguards to help convince courts they are not such charlatans. They can, and should, do this by ensuring they are properly trained to give expert evidence in the first place; by explicitly explaining their methods, approaches, and scope of their expertise, and then acting within it; and by engaging in practices to promote transparency in the field.

It is crucial that forensic linguists who want to pursue a consultancy or professional practice first establish a strong foundational background in both forensics and linguistics. One of the lessons of Dr. Smith's key failures was assuming his expertise in one area automatically granted him expertise in others. He also entirely failed to recognize that the art of acting as an expert witness itself requires its own training and expertise. By learning from the harms caused by these misconceptions, forensic linguists can do much to raise the ethical standards of the field and help to avert future miscarriages of justice.

Notes

1 Sharon's mother sued Dr. Smith, as well as the Province of Ontario, for damages. Her claim was ultimately resolved out of court.

Case Law

Bank of Montreal v. Citak, [2001] O.J. No. 1096

Carmen Alfano Family Trust (Trustee of) v. Piersanti, 2012 ONCA 297

Daubert v. Merrell Dow Pharmaceuticals Inc..1993. 509 U.S. 579

De Angelis v. Siermy, 2022 BCSC 31
Dean Construction Co. v. M.J. Dixon Construction Ltd., 2011 ONSC 4629
Dussiaume v. Sandoz Canada Inc., [2023] BCJ No. 890
Fellowes, McNeil v. Kansa General International Insurance Co. (1998), 40 OR (3d) 456
Gould v. Western Coal Corp., 2012 ONSC 5184
Hutchingame v. Johnstone, 2006 BCSC 271
Kirby Lowbed Services Ltd. v. Bank of Nova Scotia, 2003 BCSC 617
Kumho Tire Co. v. Carmichael. 1999. 526 U.S. 137
Moffat v. Canada (Minister of Citizenship and Immigration), 2019 FCJ 896
Moffitt (Litigation Guardian of) v. TD Canada Trust, 2023 OJ No. 2181
Ocean v. Economical Mutual Insurance Co., 2010 NSSC 315
R. v. Abbey, [1982] S.C.R. 24
R. v. Abbey (2009), 97 O.R. (3d) 330
R. v. Bouvette, 2023 BCJ No. 634
R. v. Docherty, 2010 ONSC 3628
R. v. France, 2017 ONSC 2040
R. v Millard, 2018 ONSC 4410
R. v. Mohan [1994], 2 S.C.R. 9
R. v. M(S) [1991], O.J. No. 1383
R. v. Scott, 2018 BCSC 1739
R. v. Sekhon, 2014 SCC 15
Royal Trust Corp. of Canada v. Fisherman (2000), 49 O.R. (3d) 187
White Burgess Langille Inman v. Abbott and Haliburton Co., 2015 SCC 23

References

Ainsworth, Janet. 2010. "A Lawyer's Perspective: Ethical, Technical, and Practical Considerations in the Use of Linguistic Expert Witnesses." *International Journal of Speech Language and the Law* 16, no. 2: 279–291.

Butters, Ronald. 2010. "The Linguistic Expert's Professional Credentials." *International Journal of Speech Language and the Law* 16, no. 2: 237–252.

Chin, Jason M., Michael Lutsky, and Itiel E. Dror. 2019. "The Bias of Experts: An Empirical Analysis of Expert Witness Challenges." *Manitoba Law Journal* 42, no. 4: 21–67.

Chin, Jason M., and Rory McFadden. 2020. "Expert Witness Codes of Conduct for Forensic Practitioners: A Review and Proposal for Reform." *Canadian Journal of Law and Justice* 2: 23.

Clarke, Isobelle. 2018. "'I Consider Myself to be a Service Provider': Discursive Identity Construction of the Forensic Linguistic Expert." *International Journal of Speech Language and the Law* 25 no. 1: 79–101.

Finegan, Edward. 2021. "Expert Testimony by Linguists in US Courts." *Language and Law / Linguagem e Direito* 8, no. 1: 22–42.

Finegan, Edward. 2010. "Expert linguists and the Whole Truth." *International Journal of Speech Language and the Law* 16 no. 2: 267–277.

Goudge, Hon Stephen. 2008. *Inquiry into Pediatric Forensic Pathology in Ontario Report*. Toronto: Ministry of the Attorney General.

Grant, Tim, and Jack Grieve. 2022. "The Starbucks Case: Methods for Addressing Confirmation Bias in Forensic Authorship Analysis." In *Methodologies and Challenges in Forensic Linguistic Casework*, 113–128. Edited by Isabel Picornell, Ria Perkins, and Malcolm Coulthard. NJ: Wiley-Blackwell.

MacLeod, Nicci. 2021. "Art vs. Craft; Expert Evidence in the England and Wales Criminal Justice System." *Language and Law / Linguagem e Direito* 8, no. 1: 43–58.

Paciocco, David M. 2009a. "Taking a 'Goudge' Out of Bluster and Blarney: An 'Evidence-based Approach' to Expert Testimony." *Canadian Criminal Law Review* 13: 135–157.

Paciocco, David M. 2009b. "Unplugging Jukebox Testimony in an Adversarial System: Strategies for Changing the Tune on Partial Experts." *Queen's Law Journal* 34: 565.

Paciocco, David M. 2001. "Context, Culture, and the Law of Expert Evidence." *Advocates' Quarterly* 24, no. 1: 42–63.

Paciocco, David M., and Lee Stuesser. 2015. *The Law of Evidence*, 7th ed. Toronto: Irwin.

Shuy, Roger. 2010. “Ethical Questions in Forensic Linguistics: Introduction to Papers from a Linguistic Society of America Panel Presentation, San Francisco, California, January 9, 2009.” *International Journal of Speech Language and the Law* 16, no. 2: 219–226.

Siegel, M. 2005. “Finding Conversational Facts; A Role for Linguistics in Court.” *International Journal of Speech Language and Law* 12, no. 2: 255–278.

Stygall, Gail. 2010. “Guiding Principles: Forensic Linguistics and Codes of Ethics in Other Fields and Professions.” *International Journal of Speech Language and Law* 16, no. 2: 253–266.

19
ETHICAL EXPERT PRACTICE
Engaging with Everyday Linguistic Harms

Kelly Elizabeth Wright

Introduction

Positionality and Terminology Statements

I am a Black biracial cisgender woman from Appalachia and I identify as a member of the Queer community. I am currently 37 years old. I am working class. I am an experimental sociolinguist, lexicographer, and scholar-activist who is occasionally called upon to offer expert testimony. My expertise is most often applied in consultation on the origins of words and their usage in a given context. I neither analyze nor report the issues of oppression highlighted in this chapter from an impartial stance; I work to secure linguistic justice for all people.

Generally, I will employ *language user* in place of *speaker* and *language use* in place of *speech* because the issues covered in this chapter are relevant to all modalities of communication. Readers can assume when the terms *speech/speaker* are present, I intend to be modality specific. However, they cannot make the same assumption when such terms appear in quoted material. Additionally, for consistency's sake, I will refer to the contemporary standardized variety of English as *Mainstream US English* (MUSE), at times making a distinction between this Standard variety and that which is often anticipated in the hyperformal and hyperanalytic context of the courtroom. The word *Standard* will be capitalized when referring to the contemporary standardized variety, as names of other languages like Kwéyòl and Tagalog are capitalized.

Primary Concerns

This volume is concerned with the (more) ethical application of linguistic knowledge to matters of law. The purpose of this chapter is to bridge an institutional divide between the formal linguistic literature highlighting linguistic oppression and the forensic linguistic literature highlighting linguistic justice. At the intersection of these literatures an overarching ethical question echoes: what is the obligation of our field to help others? First we will encounter the ways in which forensic linguistics solidified as a discipline, shifting focus from protecting the coerced to identifying the radical. Next we will engage with methodological equity issues and cautions raised across recent forensic linguistic literature to frame a conversation on linguistic justice. Finally, experimental outcomes are presented which demonstrate that all language users perceive and apply sociolin-

 DOI: 10.4324/9781003391074-21

guistic information. This consideration will upturn some soil for forensic linguists seeking to sow equity in their daily practice as experts.

The directionality of expert work is another primary concern of this chapter, stemming from a recognition that a majority of forensic linguists apply their expertise in service of the state or corporations and not in service of individual citizens seeking protection of their civil rights. This is not a moral judgment; the forensic linguist is often hired only to interpret evidence or identify a user or their intent. And in general the formal linguist is concerned with further elaborating models of linguistic structure and function so that linguistic theory may better predict observed user behavior. The oppressive linguistic phenomena which manifest or are relevant in legal cases also manifest and are relevant for situations of everyday use. Such situations occur when embodiment (Bucholtz and Hall 2016) and linguistic production intersect with personal prejudices and institutional traditions (Craft et al. 2020; Levon 2014) to create opportunities for perceivers to—both passively and actively—reify oppressive language ideologies with their behavior (Wright 2017; Hill 2008). In the combined literatures presented below, there is no explication of applying formal or forensic linguistic findings to aid an individual or group experiencing linguistic discrimination. This chapter—in asking what is the fields' obligation to help others—intends to motivate new research providing such an explication.

Becoming socially responsible scientists requires relocating our inquiry into veins which are not commensurate with value in an academic meritocracy (Wolfram et al. 2008), and this chapter asks unincentivized professionals to communicate and to confront the global, institutional pressures which have heretofore separated formal and forensic linguistic disciplines. I must echo Watt and Brown's (2020: 411) assertion that "Forensic speech scientists are much the best-placed experts to offer" institutional solutions for linguistic oppression; if linguistic justice is to be actively approximated in any institution, we must begin with courts of justice. In general, linguistic justice as project confronts a preference for standardized ways of languaging (Wright 2022; Baker-Bell 2020; Lippi-Green 2012) which determine the ways in which our bodies are capable of externalizing thought (Flores and Rosa 2015; Labov et al. 2011). A general preference for Standard language use creates environments where not only free speech but equitable perception of any kind is diminished or threatened (King and Sumner 2014).

Many would describe free speech as a situation in which they are free from consequence stemming from the content or interpretation of an utterance. Entirely free language use implies a scenario where one's intended meaning freely appears in another language user's mind in any context. This is ostensibly telepathy: a direct, physiological transfer of information from one user to another resulting in total understanding. Such a state of being is so elusive that the most advanced language models cannot approximate it and indeed no species has achieved it (yet). Taking one step back from telepathy then, let's consider linguistic production within the scope of a courtroom. In a just scenario, any language user would be able to address the court in their preferred variety, style, or modality. In just scenario, all parties involved in a legal proceeding would know and use the same language varieties, styles, and modalities and respect the variation between them. Each step removed from this ideal creates space where a language user's civil rights can become suppressed because there exists a general preference for certain ways of languaging baked in to legal proceedings. Each degree of linguistic mismatch between judge and jury, between officer and suspect, between the people and the accused, between a defendant and a prosecutor, must be accounted for in the equitable application of linguistic knowledge to matters of law. Such an accounting rarely occurs.

The forensic linguistics literature presented in this chapter draws attention to the ways in which linguistic perception is necessarily "attenuated" (Schilling and Marsters 2015: 197) by the measurable and observable properties of an utterance which condition the way in which a user is perceived. The authors below raise red flags by exemplifying consequences related to inequitable or discriminatory percepts and practices resulting in distressing outcomes—particularly in capital cases. My intention is to emphasize that the same perceptual mechanisms which underlie such condemning attention function continuously—lensing the everyday linguistic actions of all (Babel 2015). What is crucial to recognize—and what summarily calls us to action—is that completely equitable usage does not exist. Thus, considerations of inequity are always relevant no matter how language is being used by whom or where (Fishman 1965). In this light, language experts cannot remain ethical while applying the majority of their expertise in service of the state or corporations. This chapter calls for work that actively builds out accessible resources for the public's use in securing their linguistic human rights, principally their freedom of expression.

Origins of Forensic Linguistic Practice in the United States

Early on, much of forensic linguistics in the United States was practiced by government-contracted experts assessing oral admissions of guilt for the presence of police coercion. Many of these assessments were inconclusive because many experts were unable to compare instances of a suspect's usage outside of police custody. In the mid-1960s, a number of cases were taken up by the US Supreme Court where so-called "custodial interrogations"[1] (Elek 2016: 123, *Miranda v. Arizona*) had taken place. The court's deliberations determined that violations of the Fifth Amendment (which protects US citizens from self-incrimination) had occurred and the original rulings were overturned on appeal. These cases establish what are now collectively known as Miranda rights, which ensure an individual is aware of their right against self-incrimination. Police officers were issued cards with the exact wording of *Miranda*, ordered to carry them at all times, and to read those words aloud to any and all persons of interest. The standardized Miranda rights statement was uttered so ubiquitously that even today, over a half a century since their introduction, mediated portrayals of US law enforcement routinely display this procedure. After implementation of the Supreme Court's rulings, the focus of many forensic linguistic researchers shifted to other issues such as witness questioning and trademark disputes.[2] Today, forensic linguistics has expanded to include a wide variety of legal questions surrounding language user verification and identification. Contemporary research is also applied to eliminating suspects (Jessen 2020), determining the severity of an incident (like a hate crime) (Elek 2016), and identifying abuse of process (Greenlee 2020).

Forensic linguistics confronts issues of linguistic justice most often in the literature (not in the courtroom) when researchers cite specific cases where translation errors or negative social ideologies have corrupted an individual's rights to due process or equal representation. For example, Schilling and Marsters (2015) consider a case in which a German railway worker was identified as a suspected blackmailer through using a specialized term: *Langsamfahrstrecke*. The authors claim to have defined the suspect's profession through their so-called "sociolect" (Jessen 2020: 384). They also cite the inherent fuzziness involved in grouping an unidentified speaker (or suspect) merely by their use of jargon or other standardized forms. They therefore echo forensic linguist and criminal profiler James R. Fitzgerald's caution to not "render opinions as to individuals' degree of facility with professional, standard and/or literary registers" (Schilling and Marsters 2015: 198). However, Schilling and Marsters (2015) do not take the next step of connecting such professional, Standard, and literary registers with the centuries of oppression which created them (Wright 2021; Wolf 2010; Hill 2008; Bonfiglio 2002).

Linguistic Profiling

Linguistic Profiling Theory and Expert Work

Linguistic and stylistic variables gain meaning in areal and sociocultural contexts through habitual association with particular social groups, lifeways, and character types—associations which are shaped through the personal beliefs and practices of language users (Wright 2019). This is so because "[p]articipation in discourse involves a continual interpretation of forms in context, an in-the-moment assigning of indexical values to linguistic forms" (Eckert 2008; see also van Berkem 2018; Summer and Samuel 2009; Silverteiin 1981). Linguistic Profiling can be understood as speaker-specific characteristic identification and it operates similarly to racial profiling, where people are perceived to possess certain characteristics that are socially indexed to their racial type, such as aggression with Blackness (Clemons and Grieser 2023; Wright 2023; Weissler 2021). Linguistic Profiling indexes certain characteristics to language use (versus to skin color or style of dress) and is cued through "a patchwork of perceptions that vary along…region, class, education, race, and religion, among other traits including sexual orientation and speech impediments" (Baugh 2003: 171). For the remainder of this chapter, I will use the term *linguistic profiling* when considering voice-based protected class identification which leads to inequitable outcomes and the cognitive processes underlying this phenomenon. I will use the term *user profiling* when considering speaker-specific characteristics which are systematically defined and measured for the purpose of identifying a suspect or determining the severity of a crime. Linguistic Profiling rises to the level of linguistic oppression when in-the-moment assigning of perceived speech characteristics marks a language user in some way and discrimination ensues.

The case-by-case nature of expert work limits the comparability of forensic linguistic findings and each trained listener must, at some stage of their analysis, "rely on previous experience with the voice or language variety in question" (Schilling and Marsters 2015: 200) to craft their informed decision. We can understand an analyst's experience as conditioned by norms of practice (which derive from personal, embodied experience; Bucholtz and Hall 2016) and norms of performance (which derive from mediated characterizations, ideologies, and second-hand knowledge (Heaton 2018; Johnstone 2016; Silverstein 1998). Schilling and Marsters (2015: 200) contend that non-experts appear to "listen only holistically" because they do not describe the phonetic features which lead to their percepts,[3] stating that the subtlety of certain indexed productions escape those "who may be familiar with stereotypical pronunciations but not their authentic linguistic patterning". This statement exemplifies the distinction between norms of practice and norms of performance but assumes the expert is free from the undue influence of knowledge derived through ideology and characterization. This is surely not the case as Schilling and Marsters (2015: 199)—while appealing for consideration of socialization factors when defining user ethnicity from speech—mention that a "vast body" of literature exists which exemplifies "variation by ethnicity, by localized communities of practice such as gangs"; this vast body is not cited. Schilling and Marsters (2015) are able to shade their appeal with essentialized characterizations (Namboodiripad and Henner 2022) of Black language use due to the strong and uncontested nature of the same linguistic ideologies they caution readers into resisting. The regular and predictable aspects of linguistic production which inform how an individual is perceived also inform the expert's sociological and demographic decisions when establishing a suspect's likely identity or their intent to harm. The expert's linguistic perceptions are a more refined (Sumner et al. 2014), aware (see Wright 2022), and informed (Schilling and Masters 2015) operation of Linguistic Profiling. However, forensic linguistic literature, even the most engaged, rarely considers the effects Linguistic Profiling may have on User Profiling.

Cross-Disciplinary Applications of Linguistic Profiling Theory

Greenlee (2020) puts LINGUISTIC PROFILING into perspective by considering the process of death qualification. In US capital cases, if a defendant is found guilty, the jury immediately proceeds to a sentencing phase where they must choose between life without the possibility of parole or death. Setting all other issues of equitable representation aside, Greenlee (2020) highlights linguistic concerns arising from the ways in which jurors are selected. Many of these processes are conditioned by "the legal system's unspoken, normative view of how language *should* work" (Greenlee 2020: 229, emphasis original), or what Greenlee calls "language shading", which he asserts conditions legal "proceedings from start to finish" (242). These unspoken views are not exclusive to the courtroom—or even to contemporary US American English usage. In fact, they are an exemplification of what Lippi-Green (2012) terms the "Standard language ideology", or:

> a bias toward an abstracted, idealized homogenous spoken language which is imposed and maintained by dominant bloc institutions and which names as its model the written language, but which is drawn primarily from the spoken language of the upper middle class.
>
> *(64)*

Greenlee (2020) highlights that courtroom registers necessarily influence the production of all language users and that no one's everyday language is what is expected in legal arenae. Greenlee (2020) presents the coercive power of language use as deliberate manipulation which effectively exploits near-universal discourse practices (i.e., saving face) along with "readily accepted yet erroneous ideas about minority communities and speakers" (233, see also Rosa 2019). Such manipulation on the part of legal professionals reveals systematic knowledge of the sociolinguistic aspects of language use and the ways in which they shape perception of an individual's character in real time and over time.

We need not know the history or evolution of a given variety to have impressions about what types of people may use language that way or what contexts and registers seem appropriate for a given variety's usage. Intuitions about what language is proper or appropriate or civil or clear are very strongly and commonly held (Weldon 2021; Flores and Rosa 2015; King and Summer 2014). At a given point in time, a population can be assumed to have near-similar schema for extant linguistic ideologies, no matter which individual acts of being trigger recognition of a given social type (Wright 2023, 2022). While the interactional exigency for LINGUISTIC PROFILING is decidedly variable, it is dependent on where and when users come into contact—as each context activates different, culturally relevant behaviors that embed ideology in linguistic forms (Babel 2015; Kroskrity 2010; Silverstein 1998). Moreover, like all types of profiling, LINGUISTIC PROFILING operates bidirectionally—amplifying desirable outcomes for certain individuals and dampening them for others (see *Hud v Blackwell*). More positive instantiations of LINGUISTIC PROFILING are particularly insidious because in this Standard-language-ideology-infused sea we are swimming in (Lippi-Green 2012), many language users have always experienced an amplification of desirable outcomes, giving them no means of developing expectations outside of the status quo (Wright 2022). Positive instances of discrimination are extremely difficult to prove, if for no other reason than a lack of will to actively deny oneself access. The existence of such outcomes must not be overlooked as experts consider the full scope of LINGUISTIC PROFILING and the ways in which it shapes access and opportunities for individuals.

In the US housing market, mismatches in interactional exigency can result in positive or negative outcomes in the voice-only context of a rental property inquiry phone call (Scharinger et al. 2011; Massey and Lundy 2001; Purnell et al. 1999). Apart from its importance to listening landlords, trustworthiness represents a characteristic which is less likely to be perceptible through indexed features in speech, but which listeners nonetheless make rapid decisions about as they form a profile of *who* is speaking (Wright 2023). Such cognitive impressions are based on the constellation of features a given listener hears; their enregisterment (Johnstone 2016) in a given context; the listener's personal ideologies of race, space, and personhood; and the discourse context. Evidence of the indexation of voices to social groups or character types is important to establish because (as Bonfiglio (2002) discusses) as the Standard variety developed "its racial and ethnic unmarkedness, it also became conventionalized to index information about a speaker's moral purity" (Wright 2023). Wright (2023) reports that a female speaker of African American Language receives up to a 10-point decrease in TRUSTWORTHINESS ratings as compared to a MUSE or Southern American speaker because Black voices are correlated (both experimentally and impressionistically; see Boucaichith et al. 2019; Grosse 2015; Levon 2014; Podesva and Szakay 2013; Mendoza-Denton 2004) to masculinity. These correlations demonstrate that perceiving only one socially indexed trait in a voice may in turn produce a generally negative overall impression of a speaker's character. More importantly, these results demonstrate that a listener does not need to accurately identify a speaker's membership in a protected class or specific phonetic correlates for consistent, discriminatory phenomena to arise from LINGUISTIC PROFILING.

Confronting Solutions

To address discriminatory linguistic phenomena, some in the field have advocated applying computer technology (Jessen 2020; Watt and Brown 2020) and with recent developments in artificial intelligence, enthusiasm for and acceptance of automated forms of forensic linguistic analysis will no doubt increase. Automated Speech Recognition (ASR) systems, for example, now make "assessments on the acoustics of...speech samples, and have proven to be very good at this one job" (Watt and Brown 2020: 407) when analyzing individuals producing MUSE or another normative variety of a globally recognized language (Martin and Wright 2022). Despite these successes ASRs operate far from optimally. Similarity and typicality analyses are only as explanatory as their optimal output; our models do not approach optimality because the data needed *to make them just* simply does not exist. Our best option moving forward as socially responsible scientists—as ethical practitioners—is to employ equitable and representative models of description and experimentation[4] (Wright 2022: 166).

Crucially, only the most well-resourced forensic linguist is positioned to develop the "bespoke corpora" (Watt and Brown 2020: 405) needed to minimize the error rate of machine-based methods. Semiautonomous approaches may indeed possess heightened objectivity in some applications. Thus, the repeated suggestion that forensic linguistics cannot take a meaningful step toward accuracy without more data and data-processing tools is a sound one (Jessen 2020). But it must not be forgotten that such tools may also possess a heightened capability of increasing discriminatory outcomes precisely because of their assumed objectivity. Even the explanatory capacity of models trained on reference corpora from well-documented languages of wider communication (such as Arabic or Mandarin) are limited by their size and generally coarse-grained nature. Giving machines the ability to make finer-grained judgments without first addressing underlying imbalances in training datasets (Martin and Wright 2022) creates nothing more than a faster processor (not a smarter or more objective one), reifying all of the same harmful and uninformed linguistic ideologies of the individuals who created the texts they've learned from (Caliskan et al. 2017; Wright 2017).

The key to solving the problem of perceptual bias in forensic linguistic assessments does not lie exclusively in collecting larger and more varied amounts of data to feed faster and more sophisticated computer algorithms. Moreover, although the urgency of developing "greater quantities of up-to-date and broad-ranging empirical data" (Watt and Brown 2020: 411) for more efficient machine-based forensic linguistic methods is understood, it seems such entreaties are all too frequently made with the aim of improving the ways forensic linguists might better serve the state. The dire consequences of uninformed percepts (Greenlee 2020), potential mistaken identities (Jessen 2020), and justice deferred through misunderstanding (Greenlee 2020) are all cited to highlight ways in which the government's time may be wasted or space for appeal is created or a guilty party is allowed to go free. Even Watt and Brown's (2020) passion seems to stem from a desire to punish hate crime perpetrators appropriately rather than a desire to secure linguistic human rights. A greater ethical imperative is being overlooked.

Through decades of linguistic research, we have established the following: the way written language is presented in a courtroom significantly conditions outcomes, and legal practitioners actively manipulate language to exclude certain social groups from US juries (*Miller-El vs Dretke*, Grrenlee 2020; Rickford and King 2016; Hanson and Hawley 2011). Standard language ideologies are so strong, using a language other than the mainstream often translates to perceived untrustworthiness (Wright 2022; Greenlee 2020; Rosa 2019; Rickford and King 2016; Flores and Rosa 2015). Using accented speech can have the same outcome (Wright 2023; Craft et al. 2020; Chakraborty 2017; Rickford and King 2016; Cambell-Kibler 2010). Transcriptionists often encode errors in translating MUSE and accented speech which result in both more severe sentencing and overturns on appeal (Jones et al. 2019). Human observers—both trained and untrained—encode qualitative impressions of an individual when perceiving language use (Andraszewicz et al. 2011; Feldstein 2011). Manipulating a speech signal to seem sadder, faster, louder, deeper, gendered, or foreign (Bouavichith et al. 2019; D'Onofrio 2019; Babel and Russel 2015; Kim 2015; Levon 2014) can toggle those qualitative impressions even when the lexical content remains unchanged or a listener is unfamiliar with the language being used (Watt and Burns 2012). Machine-based analyses are only as strong as the representativeness of their underlying datasets (Martin and Wright 2022).

Courts value forensic linguistic analysis because it is ordered, incorporated, and presented as evidence at trial, but the aspects of usage that forensic linguistics is primarily concerned with—identity, indexation, comprehension, and sincerity—bear no straightforward link to physical properties which can be measured in the speech signal. The behavior of courts makes clear their awareness that language use can be variously perceived and actively manipulated. More work is needed to develop protections to mitigate potentially negative outcomes from LINGUISTIC PROFILING phenomena (Angwin et al. 2016). It seems sufficient evidence for and understanding of linguistic realities exists such that it could be brought to bear in service of individuals who experience linguistic oppression daily in nearly all aspects of social navigation. Thus, the global community of forensic linguists must develop regularized means of enacting linguistic justice.

Conclusion

Enacting Engaged Scholarship

For contemporary language users, inequity is inescapable and embodiment is the largest conditioning factor of language use. The universality of linguistic oppression reflects the universality of the underlying perceptual processes which maintain it and highlights that the specific identities being perceived matter less than the social superstructure valuing produced varieties (Zhang 2005;

Bourdieu 1977). The universality of Linguistic Profiling also builds a so-called "deniability" (Baugh 2023: 169) into institutional processes which regularly employ it publicly as a proxy for private hatreds and stereotypes. Forensic linguists are not removed from the influence of language ideology by their expertise, but rather apply their practice under such influence. Our everyday, passive application of language ideology can warp outcomes for at-risk populations.

In his treatise on Linguistic Profiling across international geopolitical landscapes, Baugh (2023) draws the following conclusion:

> we have come to recognize that the relief needed to advance linguistic human rights and increase access to equal opportunities throughout the world may also rely on the goodwill of those who are in positions of political influence and power within their respective societies.
> *(174)*

We formal and forensic linguists must exercise such goodwill in our respective societies. Engagement, according to Wolfram et al. (2008), is not centered around a final product—it is a state of being which can infuse our processes, can allow the ways we already do things to reflect our personal—and radical—recognition of mutual equity between linguistic forms and their users. So allow this chapter to be both the ethical axe swinging over your head and the friendly voice encouraging you to incorporate the more well-described consequences of cross-modal, cross-identity, and cross-linguistic institutionalized Linguistic Profiling into your expert work. The potential effects of Linguistic Profiling, be they positive or negative, are not confined to a particular region, language, group of people, or variety being produced. Rather, they extend beyond questions of language use and amplify extant patterns of privilege and oppression. Contextual and intra-speaker variation is relevant for the everyday language user—who possesses many styles (or distinct ways of languaging (Wright 2022; Eckert 2001; Bourdieu 1991; Bell 1984) which develop through exposure to norms of appropriateness (Flores and Rosa 2015) and personal navigation of them (King 2021; D'Onofrio 2015). What is crucial for the forensic linguist to apply—both in terms of User Profiling and in terms of serving those experiencing Linguistic Profiling—is that "sociolinguistic perceptions of speech...are inevitably relative psychological constructs" (Baugh 2023: 171). One person's, a few researchers', a given litigator's or judge's, one corpus' perspective is but that—a single facet of the full scope of languaging that is being analyzed or weaponized. Imposing one way of reading speech features is professionally inappropriate and ethically remiss because the indexicality of a given feature is fluid, controlled by who is using what language with whom and when (Silverstein 1998; Fishman 1965). Embedding such an understanding in the world's organized courts—where the social guarantees of fundamental human rights are defined and upheld—should be a driving engine of scholarly production across both formal and forensic linguistics.

What is the obligation of our field to help other people? Developing equitable and representative methods matters because all people deserve the respect of being acknowledged, of having their observations, motivations, and positions appreciated as issuing from a full person, with a full and functional linguistic system. Additionally, every language user who contributes to our surveys and experiments deserves having their body, mind, and being be knowable behind their data points (Wright 2023). The language user is a receiver, perceiving and categorizing linguistic data as they experience the discrete moments of life, and using it to contribute not only to an ever-evolving, internal, conception of self, but also to an ever-evolving, external, collectively held, known, and usable entity we refer to as *the language*. Every person deserves accessible means of enacting their linguistic human rights. The far-reaching—and often invisible—effects of Linguistic Profiling will continue to be felt as long as discrimination legally refers only to acts that occur in physical proximity. I challenge formal and forensic linguists to develop research which seeks to fill gaps in access, particularly for groups underrepresented in the phonocentric, colonized, and heteronormative LINGUISTIC PROFILING literature such as the Deaf, the migrant, the incarcerated, and those with intersectional identities. Individuals should be confident in exercising their freedoms without working against negative stereotypes of language and being, indeed without suffering discrimination in any form. Language users deserve the right to legally challenge discrimination they experience when linguistic freedoms do not manifest; may the combined work of our fields going forward create various means of manifesting those freedoms for generations to come.

Notes

1 Custodial interrogations occur when a suspect or accused individual (and sometimes witnesses and associates) are questioned while in police custody of that of other authorities.Custodial interrogations occur when a suspect or accused individual (and sometimes witnesses and associates) are questioned while in police custody of that of other authorities.

2 This chapter will not consider corporate forensic linguistic work, but all of the discriminatory phenomena explicated here are relevant to such applications and future work should seek to invigorate more ethical practice of experts in corporate arenae.

3 (A point which directly contradicts the authors' warning not to judge the value of user performance based on an absence of standardized jargon, but I digress…)

4 Equitable and representative models allow participants the agency to self-describe their identities while also affording researchers the opportunity to transparently report on the nature of the populations who provide their data, which may be subsequently analyzed and critiqued.

Case Law

Miller- El v. Dretke (2005) 545 U.S. 231, 256.
Miranda v Arizona (1966) 384 U.S. 436.
R v. Slade and Ors (2015) EWCA Crim 71.
Secretary, HUD vs Blackwell (1990) 908 F.2d 864.

References

Abdul Karim, M. F. 2021. "The Effect of Resume Whitening on African Americans Ingroup Members' Perceived Likability, Hireability, Future Encounters, and Emotional Reactions: The Role of Perceived Racial Identity." Doctoral dissertation, Indiana University.

Alexiadou, Artemis and Terje Lohndal. 2018. "Units of Language Mixing: A Cross-Linguistic Perspective." *Frontiers in Psychology* 9: 1719.

Andraszewicz, Sandra, Junichi Yamagishi, and Simon King. 2011. "Vocal Attractiveness of Statistical Speech Synthesisers." IEEE International Conference on Acoustics, Speech and Signal Processing (ICASSP), 5368–5371. doi:10.1109/ICASSP.2011.5947571.

Angwin, Anthony J., et al. 2016. "Searching for the Trace: The Influencer of Age, Lexical Activation and Working Memory on Sentence Processing." *Journal of Psycholinguistic Research* 35, no. 1: 101–117.

Babel, Anna. 2015. *Awareness and Control in Sociolinguistic Research.* Cambridge: Cambridge University Press.

Babel, M., and J. Russell. 2015. "Expectations and Speech Intelligibility." *Journal of the Acoustical Society of America*, 137, no. 5: 2823–2833.

Baker-Bell, April. 2020. *Linguistic Justice: Black Language, Literacy, Identity, and Pedagogy*. NCTE-Routledge Research Series. Milton: Taylor and Francis.

Baugh, John. 2023. "Linguistic Profiling across International Geopolitical Landscapes." *Daedalus* 152, no. 3: 167–177.

Baugh, John. 2003. "Linguistic Profiling. Black Linguistics: Language, Society, and Politics." In *Africa and the Americas*, 155–68. Edited by Sinfree Makoni, Geneva Smitherman, Arnetha F. Ball, and Arthur K. Spears. New York: Routledge.

Bell, Allan. 1984. "Language Style as Audience Design." *Language in Society* 13, no. 2: 145–204. New York: Cambridge University Press.

van Berkum, Jos J. A., et al. 2008. "The Neural Integration of Speaker and Message." *Journal of Cognitive Neuroscience* 20, no. 4: 580–591.

Bonfiglio, Thomas Paul. 2002. *Race and the Rise of Standard American: Language, Power and Social Process*. Berlin: De Gruyter.

Bouavichith, Dominique A., et al. 2019. "Perceptual Influences of Social and Linguistic Priming are Bidirectional." In *Proceedings of the 19th International Congress of Phonetic Sciences*, 1039–1043. Edited by S. Calhoun, P. Escudero, M. Tabain, and P. Warren. Melbourne, Australia: Australasian Speech Science and Technology Association Inc., and International Phonetic Association.

Bourdieu, Pierre. 1991. *Language and Symbolic Power*. Cambridge: Harvard University Press.

Bourdieu, Pierre. 1977. "The Economics of Linguistic Exchanges." *Information (International Social Science Council)* 16: 645–668.

Bucholtz, Mary, and Kira Hall. 2016. "Embodied Sociolinguistics." *Sociolinguistics: Theoretical Debates* 1: 173–200.

Caliskan, A., J. J. Bryson and A. Narayanan. 2017. "Semantics Derived Automatically from Language Corpora Contain Human-like Biases." *Science* 356, no. 6334: 183–6. doi: 10.1126/science. aal4230.

Campbell-Kibler, Katherine. 2010. "The Effect of Speaker Information on Attitudes Toward (ING)." *Journal of Language and Social Psychology* 29, no. 2: 214–223.

Chakraborty, Rahul. 2017. "A Short Note on Accent-bias, Social identity and Ethnocentrism." *Advances in Language and Literary Studies* 8, no. 4: 57–64.

Clemons, Aris Moreno, and Jessica A. Grieser. 2023. "Black Womanhood: Raciolinguistic Intersections of Gender, Sexuality and Social Status in the Aftermaths of Colonization." *Daedalus* 152, no. 3: 115–129.

Craft, Justin T., et al. 2020. "Language and Discrimination: Generating Meaning, Perceiving Identities, and Discriminating Outcomes." *Annual Review of Linguistics* 6, no. 1: 389–407.

D'Onofrio, Annette. 2019. "Complicating Categories: Personae Mediate Racialized Expectations of Non-native Speech." *Journal of Sociolinguistics* 23: 346–366.

D'Onofrio, Annette. 2015. "Persona-based Information Shapes Linguistic Perception: Valley Girls and California Vowels." *Journal of Sociolinguistics* 19, no. 2: 241–256.

Eckert, Penelope. 2008. "Variation and the Indexical Field." *Journal of Sociolinguistics* 12, no. 4: 453–476.

Eckert, Penelope. 2001. "Style and Social Meaning." In *Style and Sociolinguistic Variation,* 119–126. Edited by Penelope Eckert and John Rickford. Cambridge: Cambridge University Press.

Elek, Balázs. 2016. "To the Recommendation of Using 'Linguistic Fingerprints' in the Criminal Procedure." *Comparative Legilinguistics* 28: 120–136.

Feldstein, Stanley, Faith-Anne Dohm, and Cynthia L. Crown. 2001. "Gender and Speech Rate in the Perception of Competence and Social Attractiveness." *The Journal of Social Psychology* 141, no. 6: 785–806.

Fishman, Joshua A. 1965. "Who Speaks what Language to Whom and When?" *La Linguistique* 1, no. 2: 67–88.

Flores, Nelson, and Jonathan J. Rosa. 2015. "Undoing Appropriateness: Raciolinguistic Ideologies and Language Diversity in Education." *Harvard Educational Review* 85, no. 2: 149–171.

Fought, Carmen. 2011. "Language and Ethnicity." In *The Cambridge Handbook of Sociolinguistics*, 238–258. Edited by Rajend Mesthrie. Cambridge: Cambridge University Press.

Greenlee, Mel. 2020. "Capitally Speaking: Language and Bias in Capital Trials." In *The Routledge Handbook of Forensic Linguistics*, 228–244. Edited by Malcolm Coulthard, Alison May, and Rui Sousa-Silva. New York: Routledge.

Grosse, Terry. 2015. "From Upspeak To Vocal Fry: Are We 'Policing' Young Women's Voices?" *NPR*. Retrieved from https://www.npr.org/transcripts/425608745?storyId=425608745&ft=nprml&f=425607817.

Hanson, Andrew, and Zackary Hawley. 2011. "Do Landlords Discriminate in the Rental Housing Market? Evidence from an Internet Field Experiment in U.S. Cities." *Journal of Urban Economics* 70, no. 2–3: 99–114.

Heaton, Hailey. 2018. "Media Influence on Implicit and Explicit Language Attitudes." *DeepBlue*. Accessed April 5, 2022. Retrieved from http://hdl.handle.net/2027.42/146019.

Hill, Jane H. 2008. *The Everyday Language of White Racism*. 3rd ed. Chichester: Wiley- Blackwell.

Hillier, Amy E. 2005. "Residential Security Maps and Neighborhood Appraisals: The Home Owners' Loan Corporation and the Case of Philadelphia." *Social Science History* 29, no. 2: 207–33.
HUD.gov. n.d. "The Fair Housing Act." In *US Department of Housing and Urban Development*. Washington, DC. Retrieved from https://www.hud.gov/program_offices/fair_housing_equal_ opp/fair_housing_act _overview#_The_Fair_Housing.
Jessen, Michael 2020. "Speaker Profiling and Forensic Voice Comparison." In *The Routledge Handbook of Forensic Linguistics*, 382–399. Edited by Malcolm Coulthard, Alison May, and Rui Sousa-Silva. New York: Routledge.
Jessen, Michael. 2007. "Speaker Classification in Forensic Phonetics and Acoustics." In *Speaker classification I: Fundamentals, Features, and Methods*, 180–204. Edited by Christian Müller. Berlin: Springer.
Johnstone, B. 2016. "Enregisterment: How Linguistic Items Become Linked with Ways of Speaking, Language and Linguistics." *Compass*, 10: 632–643. doi:10.1111/lnc3.12210.
Jones, Taylor, Jessica Rose Kalbfeld, Ryan Hancock, and Robin Clark. 2019. "Testifying While Black: An Experimental Study of Court Reporter Accuracy in Transcription of African American English." *Language* 95, no. 2: e216–e252.
Kim, Seung Kyung. 2015. "Speech, Variation, and Meaning: The Effects of Eotional Prosody on Word Recognition." Doctoral dissertation, Stanford University.
King, Sharese. 2021. "Rethinking Race and Place: The Role of Persona in Sound Change Reversal." *Journal of Sociolinguistics* 25: 159–178.
King, Sharese, and Meghan Sumner. 2014. "Voices and Variants: Effects of Voice on the Form-based Processing of Words with Different Phonological Variants." *Proceedings of the Annual Meeting of the Cognitive Science Society* 36. Accessed April 4, 2022. Retrieved from https://escholarship.org/uc/item /26g0009q.
Kroskrity, Paul V. 2010. "Language ideologies: Evolving perspectives." In *Society and Language Use,* 192–211. Edited by Jürgen Jaspers, Jan-Ola Östman, and Jef Verschueren. Amsterdam: John Benjamins.
Labov, William. 1994. *Principles of Linguistic Change*. Oxford: Blackwell Publishers.
Labov, William, et al. 2011. "Properties of the Sociolinguistic Monitor." *Journal of Sociolinguistics* 15, no. 4: 431–463.
Levon, Erez. 2014. "Categories, Stereotypes, and the Linguistic Perception of Sexuality." *Language in Society* 43, no. 5: 539–566.
Lippi-Green, Rosina. 2012. *English with an Accent: Language, Ideology, and Discrimination in the United States*. 2nd ed. New York: Routledge.
Martin, Joshua, and Kelly Elizabeth Wright. 2022. "Bias in Automatic Speech Recognition: The Case of African American Language." *Applied Linguistics* 44, no. 4: 613–630.
Massey, Douglas S., and Garvey Lundy. 2001. "Use of Black English and Racial Discrimination in Urban Housing Markets." *Urban Affairs Review* 36, no. 4: 452–469.
Mendoza-Denton, Norma. 2004. "Language and Identity." In *The Handbook of Language Variation and Change,* 475–499. Edited by J. K. Chambers, Peter Trudgill, and Natalie Schilling-Estes. London: Blackwell.
Merriam-Webster. n.d. "Accent." In *Merriam-Webster.com dictionary*. Accessed November 9, 2023. Retrieved from https://www.merriam-webster.com/dictionary/accent.
Namboodiripad, Savithry, and Johnathan Henner, J. 2022. "Rejecting Competence–Essentialist Constructs Reproduce Ableism and White Supremacy in Linguistic Theory: A Commentary on 'Undoing Competence': Coloniality, Homogeneity, and the Overrepresentation of Whiteness in Applied Linguistics." *Language Learning*, 73. no. S2: 321–324.
Podesva, Robert J., and Anita Szakay. 2013. "Gender Differences in the Acoustic Realization of Creaky Voice: Evidence from Conversational Data Collected in Northern California." *The Journal of the Acoustical Society of America* 134: 4238.
Purnell, Thomas, William J. Idsardi, and John Baugh. 1999. "Perceptual and Phonetic Experiments on American English Dialect Identification." *Journal of Language and Social Psychology* 18: 10–30.
Reed, Paul E. 2018. "The Importance of Appalachian Identity: A Case Study in Rootedness." *American Speech* 93, no. 3–4: 409–424.
Rickford, John R., and Sharese King. 2016. "Language and Linguistics on Trial: Hearing Rachel Jeantel (and other Vernacular Speakers) in the Courtroom and Beyond." *Language* 92, no. 4, 948–988.
Rosa, Jonathan. 2019. " 'They're Bilingual…that Means they Don't Know the Language': The Ideology of Languagelessness in Practice, Policy, and Theory." In *Looking Like a Language, Sounding like a Race:*

Raciolinguistic Ideologies and the Learning of Latinidad, 125–143. Edited by Johnathan Rosa. Oxford: Oxford University Press.

Scharinger, Mathias, Philip J. Monahan, and William J. Idsardi. 2011. "You Had Me at 'Hello': Rapid Extraction of Dialect Information from Spoken Words." *NeuroImage* 56, no. 4: 2329–2338.

Schilling, Natalie, and Alexandria Marsters. 2015. "Unmasking Identity: Speaker Profiling for Forensic Linguistic Purposes." *Annual Review of Applied Linguistics* 35: 195–214.

Secretary, HUD v Blackwell. n.d. "Casemine." Accessed October 1, 2023. Retrieved from https://www.casemine.com/judgement/us/5914c015add7b049347b1ac5.

Silverstein, Michael. 1998. "The Uses and Utility of Ideology: A Commentary." In *Language Ideologies,* 123–145. Edited by Bambi B. Schieffelin, Kathryn A. Woolard, and Paul Kroskrity. Oxford: Oxford University Press.

Silverstein, Michael. 1981. "The Lmits of Awareness." *Texas Working Papers in Linguistics* 84: 1–30.

Sumner, Meghan, Seung Kyung Kim, Ed King, and Kevin B. McGowan. 2014. "The Socially Weighted Encoding of Spoken Words: A Dual-route Approach to Speech Perception." *Frontiers in Psychology* 4: 1.

Sumner, Meghan, and Arthur G. Samuel. 2009. "The Effect of Experience on the Perception and Representation of Dialect Variants." *Journal of Memory and Language* 6, no. 4: 487–501.

Watt, Dominic, and Georgina Brown. 2020. "Forensic Phonetics and Automatic Speaker Recognition." In *The Routledge Handbook of Forensic Linguistics*, 400–415. Edited by Malcolm Coulthard, Alison May, and Rui Sousa-Silva. New York: Routledge.

Watt, Dominic, and Juliet Burns. 2012. "Verbal Descriptions of Voice Quality Differences Among Untrained Listeners." *York Papers in Linguistics* (Series 2), 12a: 1 – 28.

Weissler, Rachel Elizabeth. 2021. "Leveraging African American English Knowledge: Cognition and Multidialectal Processing." *DeepBlue*. Accessed on April 4, 2022. https://dx.doi.org/10.7302/2706.

Weldon, Tracey. 2021. *Middle-Class African American English*. Cambridge: Cambridge University Press.

Winford, Donald. 2015. "Origins of African American Vernacular English: Beginnings." In *The Oxford Handbook of African American Language*, 85–104. Edited by Jennifer Bloomquist, Lisa J. Green, and Sonja J. Lanehart, Oxford: Oxford University Press.

Wolf, Eric R. 2010. *Europe and the People Without History*. 2nd ed. Berkeley: University of California Press.

Wolfram, Walt, Jeffrey Reaser, and Charlotte Vaughn. 2008. *Languages and Linguistics Compass* 2, no. 6: 1109–1134.

Wright, Kelly Elizabeth. 2023. "Housing Policy and Linguistic Profiling: An Audit Study of Three American Dialects." *Language* 99, no. 2: e58-e85.

Wright, Kelly Elizabeth. 2022. "Black Professionalism: Perception and Metalinguistic Assessment of Black American Speakers' Sociolinguistic Labor." *ProQuest Dissertations and Theses Global*.

Wright, Kelly Elizabeth. 2021. "Race and Language: Considering Black linguistic experiences." Chicago Linguistics Society (Plenary). *YouTube*. Accessed April 5, 2022. Retrieved from https://www.youtube.com/watch?v=np46elWCaKc.

Wright, Kelly Elizabeth. 2019. "Experiments on Linguistic Profiling of Three American Dialects." (Qualifying Thesis) University of Michigan. Accessed May 2, 2022. Retrieved from http://hdl.handle.net/2027.42/163508.

Wright, Kelly Elizabeth. 2017. "Accounting for Vox Populi: Adjusting the Cost-benefit Model of Language Planning by Incorporating Network Analysis in the Ghanaian Context." In *Proceedings of the ATRA Closing Workshop -Trieste, May 25–26, 2016*. Trieste: Edizioni Universtià di Trieste.

Zhang, Qing. 2005. "A Chinese Yuppie in Beijing: Phonological Variation and the Construction of a New Professional Identity." *Language in Society* 34, no. 3: 431–466.

20

A CRITICAL EVALUATION OF ADJUSTMENTS AND ACCOMMODATIONS IN THE LEGAL SYSTEM IN ENGLAND AND WALES FOR SEX ABUSE CRIMES INVOLVING ADULT WITNESSES WITH AN INTELLECTUAL DISABILITY

Michelle Aldridge and Tina Pereira

Introduction

There is a legal presumption, in England and Wales, that all witnesses are competent to give evidence unless information is presented to suggest otherwise (the Youth Justice and Criminal Evidence Act (YJCEA) 1999: 53). An early ethical question then, is how the Justice System can be confident that an adult who has an intellectual disability, that is, an individual with a lower intellectual ability (i.e., typically an IQ below 70) and an impairment of adaptive and social functioning (NICE 2024), is provided the opportunity to make decisions and give their evidence effectively and in a manner that best aligns with their communication abilities. To answer that question, it is necessary to first evaluate what procedures are taken to determine competence to take part in a trial and achieve consent to participate from these witnesses. How can it be determined, with certainty, that witnesses who want to exercise their fundamental legal right to give evidence can do so without influence. An equally important question is how it can be reliably ascertained that those who do not give permission to be interviewed are not forced to participate in the legal process against their will. It is crucial, whenever possible, that it is the witness who makes the decision and exerts their epistemic knowledge of themselves and their future. It is ethically essential that no professional be allowed to usurp the witness's right to make this decision. The wishes of individuals with an intellectual disability are particularly susceptible to being ignored or overridden (Antaki et al. 2008), and this may be particularly so in the Justice System, when the professionals around them are driven by their deontic responsibilities or stereotypical assumptions of a witness's ability and credibility (or lack of) to participate in the legal process and give an accurate, complete, and coherent account of events.

 DOI: 10.4324/9781003391074-22

For some adult witnesses with an intellectual disability (AWIDs) in England and Wales, it is often the police who make this decision in conjunction with AWIDs' carers and/or legal guardians, and for AWIDs with complex communication needs an intermediary should be consulted. The question is whether this is the optimum way to determine the ability to participate in an interview for those AWIDs, as we are aware (e.g., Baroness Newlove of Warrington 2018: ii) that there is inconsistency in practice amongst police forces in their interactions with AWIDs partly because "there is no set process of needs assessment for special measures" (Victims' Commissioner 2021: 6). Furthermore, to make an informed decision, AWIDs need to understand the pros and cons of participating in a legal process. This understanding must include the longer-term impact such a decision may have on them, if any. To facilitate such understanding, the assistance of trained professionals can be invaluable. As in medical decision-making where a negotiation of epistemic and deontic rights and responsibilities between a patient and a health care provider is required (Landmark 2015), the same may be true for facilitating informed decision-making amongst AWIDs. Consent to participate in an investigative interview should, for example, involve a similar negotiation between the complainant and a specially trained police representative (e.g., an interviewing officer), with the assistance of an intermediary, to help ensure that AWIDs have fully understood the process and its potential impact on their future.

Once it is determined that an AWID is competent to participate in the English and Welsh criminal justice system, AWIDs are officially defined as "vulnerable" by virtue of a disorder or disability (section 16[4] of the 1999 Act) (Achieving Best Evidence (ABE) 2022: 12) and as such are eligible to avail of Special Measures as detailed in sections 23 to 30 of the YJCEA 1999 Act (ABE 2022: 21). These "measures" consist of procedural adjustments intended to eliminate specific barriers to the effective participation of witnesses in the legal system. There are three measures that are particularly relevant to the AWID population:

- Section 27: The use of video-recorded interviews as evidence-in-chief. The police investigative interview will be visually recorded and subject to approval by the Crown Prosecution Service (CPS), will be played in court as the witness's evidence-in-chief, thus eliminating the need for the AWID to retell their account in court again.
- Section 29: Communication through intermediaries. An intermediary is a trained communication specialist whose professional role is to facilitate two-way communication between a vulnerable witness and the legal professionals. They are instructed by the court and have a neutral role.
- Section 30: The use of communication aids such as pictures and objects. The intermediary can introduce the use of such augmentative and alternative communication to replace or complement speech.

These must be explained to AWIDs along with the opportunity to opt out of all or some of them (Aldridge 2019). Once such decisions are made, it is recommended in the ABE guidelines (2022: 78) that the police investigative interview will progress through the following four phases:

1. Rapport
2. Free-narrative account

3(a) Questioning: Topic division followed by clarification and development of the account

3(b) Questioning: Systematic introduction and probing of any wider investigative material that may be important to the investigation

4. Closure

It is likely that the interview will be visually recorded (ABE 2022: 61) and once the interview process has been completed, the interview will be digitally passed to the Crown Prosecution Service to determine whether the case should go forward to court. Once in court, AWIDs will often experience similar cross-examination strategies as any witness would (see Conley and O'Barr 2005 for an overview), albeit with the benefit of Special Measures if a judge so directs. Much research has been conducted on issues of fairness and equity for vulnerable witnesses, both those with and without intellectual disabilities, along the legal path (see Morrison et al. (2019) for an overview) but comparatively little attention has been given to the procedural steps that take place even before the trial process begins. In the following, the chapter examines the ethical decisions that need to be made in each one of those steps.

Determining AWID Competence and Obtaining Consent

In English and Welsh law, "all people, whatever their age, are competent to act as witnesses unless they cannot understand questions asked of them in court or cannot answer them in a way that can be understood with, if necessary, the assistance of special measures" (sections 53 and 54 of the 1999 Act and R v B [2010] EWCA Crim 4) such as the involvement of an intermediary and communication aids (sections 29 and 30). Indeed, the Mental Capacity Act (2005) states that everybody should be assumed to have capacity unless established otherwise, stating as summarized in ABE (2022: 46) "that a communication issue should not be confused with a capacity issue and that every effort should be made to communicate with people, using whatever methods are necessary". An intermediary therefore may be instructed in these circumstances (see section 29 of the 1999 Act) along with adopting an alternative and/or augmentative means of communication. The importance of differentiating competence and communication is also made in ABE (2022: 78) where it advises that the fact that the phased investigative interview approach may not be appropriate for interviewing some witnesses with the most challenging communication skills (e.g., those only able to respond "yes" or "no" to a question) should not mean that the most vulnerable of witnesses are denied access to justice.

Sometimes the AWID's disability will be obvious and the possibility of applying for certain Special Measures will be straightforward. At other times, the witness's needs may not be readily apparent. Such is the case with AWIDs who have "hidden disabilities" such as autism or attention deficit hyperactivity disorder (ADHD). Often neurodiverse witnesses will attempt to hide their potential differences and perform with a cloak of competence. Identifying vulnerability can be hugely difficult in such cases. It is therefore essential that the investigative team have "no presumptions or preconceptions" (ABE 2022: 20). Time must be taken to thoroughly establish the witness's abilities as well as disabilities at the pre-interview stage. Intermediaries can assist the police at this stage, assessing an AWID's communication and interaction abilities to determine effective ways in which best evidence can be facilitated in interview. People who know the witness well can also be consulted to determine best practice for each individual AWID (ABE 2022). It is appreciated that planning can be curtailed by the urgency of the case (ABE 2022); however, it is truly a false economy to rush preparation in a case involving AWIDs. When procedural shortcuts are taken, discrimination may occur, prejudices may dominate, and rights may be violated. Therefore, it is critical that law enforcement have both the time and skill set to accurately identify witness needs (Roberts, Cooper, and Judge 2005).

Once a trained professional has determined whether an AWID is indeed competent to participate in a police interview—regardless of any communicative challenges they may have—it is essential to ascertain whether the individual freely consents to participate in this interview process.

This step presents its own challenges. For example, an AWID's world knowledge may not be typical. Consequently, they may not appreciate that what happened is outside the normal boundaries of intimacy. Even if they know something inappropriate has taken place, they may be fearful of disclosing the events. This fear may be particularly great if, for example, the person involved is influential in their living arrangements or career opportunities. They may benefit from the companionship or support (e.g., financial, logistical, physical, etc.) offered by the alleged assailant and fear losing that relationship. Alternatively, they may simply fear the consequences of "telling on" someone (Disability Abuse (lwa.org.uk) Sequeira, Howlin, and Hollins 2003). Another possibility is that they may experience irrational guilt and believe (or have been made to believe) that they were complicit in some way. Given these complex reactions, it is important that those who have a duty of care for AWIDs are educated and trained in understanding these different perspectives so they can empower an AWID who has potentially undergone sexual assault to disclose what they have experienced and enable them to communicate with honesty and integrity.

It is important to note that not all cases of suspected sexual assault are reported by the alleged victim but by a third party to whom a disclosure has been made. When this happens, the AWID may not understand why they are being asked to speak with the police. They may well be fond of the alleged offender and may not recognize them as the described perpetrator. They may be fearful of authority and assume that the reason they are being brought to the police is because they have done something wrong. In such cases, it is morally imperative to ensure that sufficient time is given to appropriately explain what is going on and the roles of all the people involved. Part of this information-giving process involves helping AWIDs to understand the many possible consequences of consenting to a police interview: e.g., the case may go to court; they may be required to attend the proceedings; their personal life experiences may be made public; if the case goes to trial, they may need to be cross-examined; and they may not agree with or even understand the verdict. Only after AWIDs are assisted in understanding these issues is it possible to enable them to give their informed consent to a police interview, if they choose. In brief, the AWID's needs and wishes toward giving evidence must be "ascertained, not assumed" (Burton, Evans, and Sanders 2006: 14).

AWIDs and the Investigative Interview

Once it is determined that informed consent has been obtained, a decision must be made about where to conduct the investigative interview. In order to achieve an acceptable standard of recording that meets the requirements of court rules, it is recommended that the interview be conducted at a specifically designed "interview suite" (ABE 2022: 225). Such suites have the advantage of possessing recording equipment like strategically located cameras, typically one that can pan-tilt-zoom and another that has a wide-angle lens to record the entire room. They also usually contain specific furniture, lighting, color schemes, etc. which have been carefully chosen to offer a safe and effective environment for facilitating witnesses to talk with the police. Furthermore, there is sometimes a medical room onsite should input be required by medical personnel. Such suites also routinely have a monitoring room so the interview can be observed and monitored by trained equipment operators, normally other police officers whose responsibility it is to ensure the quality of the recording (ABE 2022: 226). Additionally, these officers serve as a second pair of ears and eyes, ensuring all relevant investigation-relevant information has been captured. Their role is independent from that of the interviewer, but they can speak with the interviewer and highlight, for example, topics that need further inquiry before the interview closes. Despite these special features, the above advantages offered by an interview suite may not be so apparent for AWIDs and they may express a preference to be interviewed at their residential address or some other familiar location instead. This may be because, for example, they find it disconcerting to be in an unfamiliar location with unregulated sensory stimuli. Having a set routine can be extremely stabilizing for some AWIDs and the disruption to their routine caused by the interview appointment may present an additional challenge and thereby potentially undermine their ability to give evidence.

This point leads to the importance of the issues of comfort surrounding the interview process. Accessibility can pose a major problem for AWIDs who have been scheduled for a police interview. For example, can the witness reach the room, use toilet facilities, and take refreshments independently or will access issues disempower them further? To avoid discrimination and potential human rights violations, it is important that information be gathered about the specific communication and interaction idiosyncrasies of each AWID, and persons familiar with the AWID (and not involved in the case) should be consulted to determine the most appropriate time, location, and "Special Measures" recommended for holding the interview.

By using portable video-recording equipment (ABE 2022: 225), police can interview AWIDs in environments that are more familiar to them. This decision, however, also comes with its own logistical and ethical challenges. Once the interview takes place away from a purpose-built location, it becomes more vulnerable to challenge—questions may be asked why a different building is being used and what the police hope to achieve through this alternative venue; this may lead to allegations that there might be people or environmental triggers coaching the witness there; queries will be raised as to whether the recording will be so crisp or could visual and audible detail be lost. Furthermore, it is possible that the non-neutral location could be problematic for the witness and the professionals. The familiar location may create stress for the witness by their fearing being seen or overheard during the interview; it is possible then that being interviewed there may be intrusive and cause additional stress, and for the interviewer, the unfamiliar location, equipment, and recording set-up may unsettle them, causing them to underperform. The specifically designed suites have been modified, over the years, to create the best environment for all parties and once one is not used, perhaps for an unavoidable reason, a different set of issues will emerge which may add to the jury being encouraged to question the credibility of the case. Of course, none of these issues make the case less credible, in themselves, but it is possible to demonstrate that equity in the legal system is fraught with many dilemmas at each stage of the process.

Special Measures

In this section we consider the challenges involved in employing Special Measures and reflect on AWIDs' likely understanding of them. If it is agreed that Special Measures are granted (which is likely in alleged cases of sexual abuse with AWIDs), it is crucial that these adjustments are fully explained to the witness. This requirement is specified both in Right 2 of the Statutory Code of Practice for Victims of Crime (Ministry of Justice 2020) and in Charter Standard 4 of the Witness Charter (Ministry of Justice 2013). According to these regulations, AWIDs must be given the chance to "opt out" of using Special Measures if they wish. AWIDs may reject the use of Special Measures because they do not wish to be treated differently and feel they can manage without them. However, simply telling an AWID that they have options is ethically insufficient. The police must be sure that AWIDs understand what their options are to facilitate their choice-making and to ensure that all the Special Measures desired by the AWIDs are applied for consideration by the judge. Once communicated, the wishes of AWIDs must be respected, but it is important to understand that while these measures do often facilitate the giving of evidence, they are not without complication as we now go on to illustrate.

The Use of Video Recordings

From the perspective of many investigators, and indeed the main recommendation of ABE (2022), visually recorded evidence-in-chief (the police investigative interview) is described as best practice for the AWID to achieve their evidence-giving. In cases, however, where an AWID has a history of being video-recorded or photographed against their will, this measure could not only trigger secondary trauma but might also undermine the investigative process. Being recorded without informed consent or without full consideration of their history could cause the AWID to re-experience the emotions associated with the allegation and result in their giving qualitatively and quantitatively poor evidence. Complying with these wishes may also pose immediate ethical dilemmas and subsequent legal challenges, however. If, for example, on the AWID's request the police forgo recording and opt for taking a written statement from an AWID with a compromised ability to read and write, questions may be raised later, in relation to the quality and credibility of their evidence. Is the written statement (albeit a statutory right the AWID can assert) in that witness's best interests? Are the sentiments expressed and the events recounted in the written statement genuinely authored by the AWID in their "own words" as stipulated in the ABE 2022 guidelines? If Special Measures are declined for the investigative interview, but later used for the court trial and discrepancies in the testimony occur, how would the court view these apparently conflicting processes? To avoid these and other procedural questions, it is not unusual for law enforcement to prefer that investigative interviews conducted with AWIDs are recorded, provided that the court determines that this modality is likely to maximize the quality of the witness's evidence and is not against the interests of justice to admit such video-recorded evidence (section 27[2]).

If the decision is made to video the interview, it is essential that the recording be of good quality as it is the "key element underpinning a prosecution" (HMCPSI and HMIC 2014: 4) and that the AWID's account comes across as being complete, coherent, and accurate (Hansard 1999). Understandably, these high standards are not always met. It sometimes happens, for example, that ABE recordings are considered less reliable because of problems with technology (Plotnikoff and Woolfson 2019: 113); indeed, some judges have also expressed concern about the quality of some ABE interviews and poor technology (Victims' Commissioner 2021: 8), and it can be argued that such concerns are ultimately detrimental to fair access to justice for AWIDs. Interestingly, the quality of a non-vulnerable complainant's written statement is typically not scrutinized to the same extent: a barrister usually elicits oral evidence from such a complainant by using their written statement as a guide. It is what a non-vulnerable complainant says that is heard by the jury, not necessarily their written statement. Therefore, a witness's oral evidence is controlled and regulated by their barrister's type and manner of questioning. By comparison, the visual recording of the AWID's interview may not even reach the jury as the CPS may intervene on the grounds that it is not in the public interest to proceed.

Communication through Intermediaries

In planning an interview, the ABE (2022) guidelines state that the witness's race, culture, ethnicity, and first language should be given due consideration by the interviewing team. AWIDs who are not native English or Welsh speakers can utilize the services of an interpreter to translate to/from their native language but some AWIDs may also use alternative and augmentative communication (AAC) as their first "go-to" tool to supplement or replace spoken language, and therefore they will also require the assistance of a registered intermediary (RI) to introduce AAC in the legal process as required (The Statutory Code of Practice for Victims (Ministry of Justice 2020).

Once selected and instructed, it is hoped that the same RI is involved from the planning phase of the interview until the close of the case. RIs are communication specialists with additional training in legal processes (UK Parliament 1998) who assess the witness's communicative abilities, including their use of AAC, and advise the interviewing officer on the types of questions and language use that is appropriate. These recommendations include suggestions regarding the sensory and interactional modifications, accommodations, and adaptations needed to ensure the AWID gives the best evidence possible including adopting communication aids such as drawings, wooden figures, miniature furniture, and sticky notes.

The aim of a police investigative interview is to elicit the best possible account of the alleged incident (ABE 2022: 19) and so the "quality of a witness's evidence" is paramount. Quality is defined as the "completeness, coherence and accuracy" of the evidence (ABE 2022: 172 (section 16[5] of the 1999 Act). While "completeness" is self-explanatory, "coherence" is somewhat more complicated. Legally, it refers to a witness's ability to provide evidence by giving answers to the questions put to them; and those answers are expressed in a way such that they can be understood both individually and collectively (section 16[5] of the 1999 Act) (ABE 2022: 12). In assessing the coherence of a witness's account, the issue of accuracy is germane as there is no complete agreement over what this means. For some, it could be interpreted to mean "truthfulness" but this cannot be measured as no one other than the witness and accused knows what happened, while for others, accuracy is better interpreted as the ability of the AWID to communicate what they intend to communicate, as faithfully as possible. In other words, their evidence is communicated in a manner that most closely replicates their communicative intent (Pereira 2021). Whichever interpretation one prefers, in the end, the accuracy of an AWID's statement must be checked through thorough questioning as it would be for any witness.

Importantly, AWIDs are not expected to take the oath in police investigative interviews. However, they may be asked to demonstrate that they know the difference between telling the truth and lying. This is sometimes achieved through scenarios (see appendix G ABE 2022: 198 as examples). The effectiveness and appropriateness of this technique for AWIDs are questionable as scenarios require the ability to imagine and understand a range of possible options, which some AWIDs struggle with. Another common strategy to encourage AWIDs to provide truthful statements is to "simply" remind them to tell the truth by giving instructions like "Please tell me all you can remember about what happened. Don't make anything up or leave anything out. It is very important to tell the truth" (ABE 2022: 83).

Initially, this directive may appear innocuous. But although aimed at simplifying language to match the linguistic abilities of an AWID, the above instruction could be taken to imply: "I think you might make things up or forget things". Such lines can be potentially disempowering and may reinforce the stereotype that AWIDs are unreliable witnesses. Interviewing officers who are sensitive to these problems often qualify the above directive by saying: "This does not mean I think you are lying. This is just something that we have to do. It's something that we say with everyone". The problem with such disclaimers is that they are syntactically complex and may not be understood by AWIDs. Furthermore, neuro-typical witnesses do not receive such reminders when they give the oath—the oath may be repeated parrot fashion, but its sincerity goes unchallenged, not so the intentions of AWIDs. Although designed to assuage worry and upset, such lines could also potentially create doubt with the jury that an AWID's credibility is being questioned even when they are advised in court this is not how they should be interpreted. The ethical dilemma officers and interview trainers face here is how best to consistently balance the conflicting deontic responsibility of ensuring a witness is telling the truth with jury perceptions and a witness's potential negative emotional response.

The Use of Communication Aids

It is well known that AWIDs are likely to have cognitive limitations which might restrict their memory, give them limited speech abilities, and reduce their understanding (Belva et al. 2012; Milne and Bull 2006). Consequently, it is likely that they will give incomplete information about "what happened". While this may also be true of non-vulnerable witnesses, AWIDs may also have a reduced linguistic ability to give their evidence. This lack of specificity and detail does not mean they have not been abused or that they do not remember the abuse. Instead, it means that they need support to give their evidence. RIs can offer AWIDs alternative ways of giving evidence. Rather than telling what happened, AWIDs may show it using communication aids provided by the RI (Pereira and Aldridge 2022; Pereira 2021). Communication aids can be employed to enable the AWID to demonstrate what happened by using high-tech objects like iPads or low-tech objects like pictures, wooden figures, and miniature furniture, for example (The Advocate's Gateway 2015). There is a growing body of research on their use and relevance in police investigative interviews (Pereira and Aldridge 2022; Pereira 2021; Mattison and Dando 2020; Henry et al. 2017) but the ongoing challenge is to demonstrate that they are as relevant as the spoken word.

Another potential challenge is that spoken language is very rich and nuanced in comparison to visual communication. In spoken language, there are numerous words to describe and differentiate between activities, including the intensity, speed, and degree to which they are completed. For example, there is a clear difference between "he *put* it in" and "he *pushed* it in" (implying greater intensity and perhaps seriousness of allegation). Communicating and understanding this difference using communication aids may not be as easy for an unfamiliar user. This is why recruiting an intermediary who has the necessary specialized skills in using aids can be critical to the success of the investigative interview. Intermediaries are proficient in different professional areas, but aid introduction and management require specialist training and skill that perhaps not all intermediaries are competent in, and thus, AWIDs may miss out because a relevantly trained intermediary is not available (Plotnikoff and Woolfson 2015). This barrier where one's native "language" cannot be used would not exist for witnesses with competent verbal abilities showing the unwitting prejudices some vulnerable witnesses may experience.

Attending Court

If a case goes to court, ideally, the same RI who worked with the AWID during the investigative interview should be used for the trial, for consistency and effectiveness, since rapport would have been established. Additionally, that RI would have specific knowledge of the AWID's idiosyncratic communication strengths and difficulties, enabling communication strategies to be tailored to their individual needs precisely and promptly. While retaining the same RI is desirable, it is not a guaranteed right for AWIDs and practical and logistical difficulties in ensuring consistency may cause uncertainty and potential change in personnel. This could result in unsettling the AWID which will not enhance their evidence-giving. The ethical dilemma is to balance the need for consistency with ensuring that cases progress through the court system speedily.

There are other Special Measures that are a matter of official policy for the court case; these include:

- Using movable physical barriers such as portable partitions to screen the witness and defendant from each other, ensuring they cannot be seen or see each other (section 23);
- Allowing the witness to give evidence by way of live video link (section 24);
- Permitting evidence to be given in private (the public gallery is closed, and non-case essential persons are excluded (section 25);
- Removing the wigs and gowns of court officials (section 26);
- Allowing cross-examination to be video-recorded in advance of the trial (section 28).

Giving evidence live in court is very challenging for everyone (Jacobson, Hunter, and Kirby 2015) but especially for AWIDs. The implementation of these Special Measures is a huge advance in facilitating equal access to justice as AWIDs. Specifically, memory of the allegation over time, reliving the original traumatic event(s) because of cancelled court dates, unfamiliarity with the court environment and process, waiting for months before the trial is listed in court, and emotional factors can all negatively impact the quality of an AWID's evidence. Pre-recorded cross-examination (S 28) can often assist a vulnerable individual in minimizing those barriers to effective communication.

However, looking at the process and use of Special Measures more holistically, we conclude that the consequences may not always be entirely positive. More specifically, we currently do not have known research on how the video recording may be received by juries even though judges routinely explain that this manner of presenting their evidence-in-chief is now not uncommon. We might also wonder whether some jurors are less attentive to evidence received on a video monitor compared with watching evidence being given live in court and whether they give the evidence less weight because of the distancing effect. Research does not support this concern (Ellison and Munro 2011), but we must be mindful that such results come from mock court cases as it is not permissible to talk with the jury in England and Wales.

We have a growing body of research on the efficacy and usefulness of communication aids in assisting AWIDs communicate their account but as yet we do not know how the jury responds to communication aids. Research must be conducted to determine if their relevance, potential, and function are being sufficiently and accurately explained to the jury. Jurors need clarification and direction on how to interpret and attach meaning to communication aids in ways that the words of a non-vulnerable witness may not. While great strides have been made giving AWIDs a "voice" using communication aids, we suggest pushing the boundaries further in our goal of mainstreaming communication aids.

Turning to cross-examination, we note that Special Measures have begun to level the playing field facilitating AWIDs' ability to answer a barrage of difficult questions. Without doubt, this is a linguistically challenging experience for all witnesses, and much has been written about the problems of cross-examination (Kassin, Williams, and Saunders 2020; Zajac, O'Neill, and Hayne 2012; O'Barr 1982) and the adversarial techniques used to advance the case of the defense (see Zydervelt, Zajac, Kaladelfos, and Westera 2016 for a review). The rigor of the questions during cross-examination is a legal requirement of the adversarial system as the accused is innocent until proven guilty. For that reason, AWIDs are often likely to be subject to robust cross-examination techniques, despite significant, clear, guidance from the intermediary, judge, and cross-examination toolkits that recommend that questions be asked using modified language and tone. If these guidelines are not followed, AWIDs are potentially at a disadvantage. For that reason, it is now common practice for advocates to send their proposed cross-examination questions to the instructed intermediary for comments prior to asking them (The Advocate's Gateway 2019). Those comments may suggest amendments to the linguistic structure of the questions, such that they are in keeping with the AWID's linguistic profile.

During cross-examination, non-vulnerable witnesses traditionally stand in the witness box alone and answer the questions directed at them. This means that the words heard by the jury and their emotional paralinguistic features such as intonation, tone, and gaze can draw the jury into listening and believing their account. In contrast, vulnerable AWIDs are likely to be in a different room from the jury if they avail of S24 (evidence via video link). The physical distance may make it more difficult (although still possible) for the jury to view their immediate emotions and facial expressions, along with accompanying changes in language and voice. The potential dilution of impact was noted by some judges and witnesses in recent surveys reported by the Victims' Commissioner (2021: 34) where it was summarized that there seems to be a prevailing view among criminal practitioners, and some witnesses, that evidence given by video link is less impactful on jurors and so should be avoided. This is despite any empirical evidence which supports this view (Fairclough 2020 in the Victims' commissioner review 2021). The dilemma the legal system faces at this point is what is in the best interests of justice.

Irrespective of whether AWIDs appear in the courtroom or on video link, they will often not appear on their own. Frequently, they will be accompanied by an intermediary who may move aids into the witness's space or repeat what the witness or others in court have said. The intermediary is an impartial officer of the court, present to assist communication between both parties (the vulnerable person and the court), but we need further research to understand what the jury may perceive of the co-constructed evidence. Historically, minimally verbal individuals have been thought of as incapable of presenting their evidence in a plausible manner (Perlman et al. 1994), but we are optimistic that the jury sees intermediary input as minimizing those preconceived notions of credibility and linguistic competence. It is the responsibility of the judge to protect the interests of vulnerable witnesses (ABE 2022: 142), and we hope their input can override the jury's potential preconceived expectations and stereotypes.

Conclusions: What Can Forensic Linguistics Offer?

According to Article 6 of the Human Rights Act (UK Parliament 1998) everyone in England and Wales is entitled to a fair hearing in a court of law. Respect for this humanitarian principle is a matter of considerable pride and rightly so. As this chapter has shown, the social vulnerabilities of people with intellectual disabilities puts them at greater risk of being victims of crime (Henry and Wilcock 2013) while potentially having less ability to give a full account of what happened to them. Pleasingly, in recent years, the Justice System in England and Wales has gone some way in accommodating their differences and offering equitable justice to AWIDs.

Given the different communication styles and needs of AWIDs, we acknowledge that the current legislation in England and Wales has made strides in overcoming possible discrimination in the police investigative interview and in the courtroom. Speaking about Special Measures, the Victims' Commissioner (2021: 1) concluded that

> many people who would not have been able to give their evidence have been enabled to do so because of these brilliant but simple devices. Many more witnesses have been saved unnecessary stress. And the courts have functioned more effectively as a consequence.

And she goes on to write, "Special measures are a success story" (Victims' Commissioner 2021: 7) and "the introduction of Registered Intermediaries is the single biggest improvement in the criminal justice system over the last thirty years" (Victims' Commissioner 2021: 27). Indeed, minimally verbal witnesses are now coming forward to give evidence in ways that simply were not possible before the YJCEA 1999. Clearly, research has shown that communication aids can enable evidence-giving in non-leading ways (Pereira and Aldridge 2022; Pereira 2021; Mattison and Dando 2020; Henry et al. 2017), and training for professionals is ongoing, and will become mandatory, through the Inns of Court College of Advocacy (ICCA) and other training providers, for all involved in the legal path to work with the Special Measures in achieving justice for all. As yet, however, the system is not perfect. Plotnikoff and Woolfson (2019) give a powerful and sobering review of the shortcomings of the current system citing lack of available and specialized intermediaries, patchy police interview styles and adherence to ABE recommendations, and post-code lotteries for resources as some of the ongoing problems. Thus, it could be posited that Special Measures level the playing field given the resources we currently have, but what can the field of forensic linguistics offer to help secure further the rights of AWIDs who have experienced sexual abuse and wish to have their "voice" heard?

This is a difficult question to answer given our adversarial system but we can campaign for fast(er) tracking of cases for vulnerable witnesses as we know episodic memory decreases over time (Tulving 1972); we can campaign for more resources so that there are more skilled RIs available to support AWIDs; and we can raise awareness so that there is greater appreciation of how intermediaries' skill set can help in the delivery of justice (Baroness Newlove of Warrington (2018: ii and Victims' Commissioner 2021: 8). We can campaign for more resources so that all police interview locations and court rooms have state-of-the-art recording equipment and are accessible for all. In relation to communication aids and the relative emphasis given them in interviews, if AWIDs are to be afforded equitable access to the justice system, it is essential that greater attention be given to the way their evidence is recorded. Currently, while one camera focuses on the AWID's face, a second pans out to capture the entire room. If aids are used, it is recommended that one camera be focused on the communication aids, while another camera is positioned to record the simultaneous facial non-verbal motions of the AWID so that the jury can fully appreciate the intent of the communication aids.

We can campaign for more resources so that all relevant professionals are trained in identifying and working with vulnerable witnesses. More recently, in piecemeal fashion, accommodations have become available for defendants, though there is currently no government provision of communication support for them (Victims' Commissioner 2021: 12) and we can campaign that parity is achieved in the magistrate's court as well as in Crown court (Victims' Commissioner 2021 :8). We can raise awareness about the credibility of Special Measures to overcome potential biases, so as noted, for example, in the Victims' Commissioner's report (2021: 37),

> the belief that a witness' evidence loses impact and perhaps even authenticity because they are not in the room may well be unfounded but if this is, as we understand, a myth, it needs to be busted as it seems to be widely held.

Importantly, too, forensic linguists could consider broadening their professional views on communication to include the visuo-spatial-gestural modality in addition to the aural-oral modality. Communication is more than words. Finally, where this population is concerned, while accepting individual differences, we might aim for conventional use of regularly used signs and symbols so that their introduction within a case is less idiosyncratic and arbitrary. As noted by Fairclough (2021: 168), "many other jurisdictions, such as Scotland, Northern Ireland, New Zealand and Australia are looking to us for a blueprint on how to adapt their criminal proceedings for vulnerable participants", and thus we must continue to strive to improve ethical opportunities for all.

Case Law

Charter Standard 4 of the Witness Charter (Ministry of Justice 2013).
House of Commons - Hansard. 1999. Explanatory Notes - YJCEA. UK Parliament. London.
Human Rights Act 1998. UK Parliament (Article 6).
Mental Capacity Act (2005).
R v B [2010] EWCA Crim 4 (England and Wales Court of Appeal (Criminal Division).
Youth Justice and Criminal Evidence Act (YJCEA). 1999.
The Statutory Code of Practice for Victims of Crime (Ministry of Justice 2020)

References

Aldridge, Michelle. 2019. "Police Delivery of the Opt-out Procedure for Children's Court Evidence: Evidence of Inadequate Language Awareness." *Journal of Language Awareness* Special Edition, 28, no. 3: 166–185.

Antaki, Charles, Warren Finlay, Chris Walton, and Louise Pate. 2008. "Offering Choices to People with Intellectual Disabilities: An Interactional Study." *Journal of Intellectual Disability Research* 52, no. 12: 1165–1175.

Baroness Newlove of Warrington. 2018. *A Voice for the Voiceless: The Victims' Commissioner's Review into the Provision of Registered Intermediaries for Children and Vulnerable Victims and Witnesses*. Victims' Commissioner: Positive Change for Victims.

Belva, Brian, Johnny Matson, Megan Sipes, and Jay Bamburg. 2012. "An Examination of Specific Communication Deficits in Adults with Profound Intellectual Disabilities." *Research In Developmental Disabilities* 33: 525–529.

Burton, Mandy, Roger Evans, and Andrew Sanders. 2006. "Implementing Special Measures for Vulnerable and Intimidated Witnesses: The Problem of Identification." *Criminal Law Review* 11, no. 1: 229–240.

Conley, John, and William O'Barr. 2005. *Just Words: Law, Language and Power*. University of Chicago Press.

Disability Abuse: Iwa.org.uk understanding-abuse/who-is-affected-by-abuse/disability-abuse/.

Ellison, Louise, and Vanessa Munro. 2011. "Special Measures in Rape Trials: Exploring the Impact of Screens, Live Links and Video-recorded Evidence on Mock Juror Deliberations." Economic and Social Research Council (ESRC RE-000-22-4277).

Ewin, Robert. 2015. "The Vulnerable and Intimidated Witness: A Socio-Legal Analysis of Special Measures." *Journal of Applied Psychology and Social Science* 1, no. 2: 31–54.

Fairclough, Samantha. 2021. "The Consequences of Unenthusiastic Criminal Justice Reform: A Special Measures Case Study." *Criminology & Criminal Justice* 21, no. 2: 151–168.

Fairclough, Samantha. 2020. "Special Measures Literature Review." *Commissioned Report*.

Fairclough, Samantha. 2018. "Speaking up for Injustice: Reconsidering the Provision of Special Measures through the Lens of Equality." *Criminal Law Review* 1: 4–19.

Henry, Lucy, and Rachel Wilcock. 2013. "Witnesses with Intellectual Disabilities." *International Journal of Disability, Development and Education* 60, no. 1: 1–2.

Henry, Lucy, et al. 2017. "Verbal, Visual, and Intermediary Support for Child Witnesses with Autism During Investigative Interviews." *Journal of Autism and Developmental Disorders* 47, no. 8: 2348–2362.

HMCPSI and HMIC. 2014. *Achieving Best Evidence in Child Sexual Abuse Cases - A Joint Inspection*. London: HM Crown Prosecution Service Inspectorate HM Inspectorate of Constabulary.

Jacobson, Jessica, Gillian Hunter, and Amy Kirby. 2015. *Inside Crown Court: Personal Experiences and Questions of Legitimacy*. Bristol University Press.

Kassin, Saul, Lorri Williams, and Courtney Saunders. 2020. "Dirty Tricks of Cross-Examination: The Influence of Conjectural Evidence on the Jury." *Law and Human Behavior* 14: 373–384.

Landmark, Anne Marie Dalby, Pal Gulbrandsen, and Jan Svennevig. 2015. "Whose Decision? Negotiating Epistemic and Deontic Rights in Medical Treatment Decisions." *Journal of Pragmatics* 78: 54–69.

Mattison, Michelle, and Coral Dando. 2020. "Police Officers' and Registered Intermediaries' Use of Drawing During Investigative Interviews with Vulnerable Witnesses." *Psychology, Crime and Law* 26, no. 2: 167–185.

Milne, Becky, and Raymond Bull. 2006. "Interviewing Victims of Crime, including Children and People with Intellectual Disabilities." In *Practical Psychology for Forensic Investigations and Prosecutions*, 7–24. Edited by Mark Kebell and Graham Davies. John Wiley and Sons Ltd.

Ministry of Justice. 2022. *Achieving Best Evidence Criminal Proceedings: Guidance on Interviewing Victims and Witnesses and Guidance on using Special Measures*. London: Ministry of Justice.

Ministry of Justice. 2020. *Registered Intermediary Procedural Guidance*. London: Ministry of Justice: London.

Ministry of Justice. 2013. *The Witness Charter: Standards of Care for Witnesses in the Criminal Justice System*. London: Ministry of Justice.

Morrison, Joanne, Rachel Forrester-Jones, Jill Bradshaw, and Murphy Glynis. 2019. "Communication and Cross-Examination in Court for Children and Adults with Intellectual Disabilities: A Systematic Review." *The International Journal of Evidence & Proof* 23, no. 4: 366–398.

NICE. 2024. *National Institute for Health and Care Excellence. Learning Disabilities: What is it?* Retrieved from https://cks.nice.org.uk/topics/learning-disabilities/background-information/definition/#:~:text='A%20significantly%20reduced%20ability%20to,%2C%20which%20started%20before%20adulthood.

O'Barr, William. 1982. *Linguistic Evidence: Language, Power, and Strategy in the Courtroom*. New York: Academic Press.

Perlman, Nitza, Kristine Ericson, Victoria Esses, Barry Isaacs, and Ronald Roesch. 1994. "The Developmentally Handicapped Witness: Competency as a Function of Question Format." *Law and Human Behavior* 18, no. 2: 171–187.

Pereira, Tina. 2021. "Critical Evaluation of the Impact of Low Technology Communication Aids on the Quality of Evidence Elicited from Witnesses with a Learning Disability in Registered Intermediary-Mediated Achieving Best Evidence Police Investigative Interviews." Unpublished PhD dissertation.

Pereira, Tina, and Michelle Aldridge. 2022. "Show Me What Happened: Low Technology Communication Aids Used in Intermediary Mediated Police Investigative Interviews with Vulnerable Witnesses with an Intellectual Disability." *The International Journal of Evidence & Proof* 27, no. 1: 83–104.

Plotnikoff, Joyce, and Richard Woolfson. 2019. *Falling Short? A Snapshot of Young Witness Policy and Practice: A Report for the NSPCC*. https://learning.nspcc.org.uk/media/1672/falling-short-snapshot-young-witness-policy-practice-full-report.pdf.

Plotnikoff, Joyce, and Richard Woolfson. 2015. *Intermediaries in the Criminal Justice System*. Policy Press.

Roberts, Paul, Debbie Cooper, and Sheelagh Judge. 2005. "Monitoring Success, Accounting for Failure: The Outcome of Prosecutors' Applications for Special Measures Directions under The Youth Justice and Criminal Evidence Act 1999." *International Journal of Evidence & Proof* 9, no. 4: 269–290.

Sequeira, Heather., Patricia Howlin, and Sheila Hollins. 2003. "Psychological Disturbances Associated with Sexual Abuse in People with Learning Disabilities." *British Journal of Psychiatry* 183: 451–456.

The Advocate's Gateway. 2019. *Toolkit 1. Ground Rules Hearing and the Fair Treatment of Vulnerable People in Court*. London: The Advocate's Gateway. https://www.theadvocatesgateway.org/_files/ugd/1074f0_846f9ab1f1e94dd7bd58bcc62f76ddb8.pdf.

The Advocate's Gateway. 2015. *Toolkit 14. Using Communication Aids in the Justice System* London: The Advocate's Gateway. https://www.theadvocatesgateway.org/_files/ugd/1074f0_f5d07af401574f289401e2c0df981c4d.pdf.

Tulving, Endel. 1972. "Episodic and Semantic Memory." In *Organization of Memory*, 381–403. Edited by Endel Tulving and Wayne Donaldson. New York: Academic Press.

Victims' Commissioner. 2021. *Next Steps for Special Measures: A Review of the Provision of Special Measures to Vulnerable and Intimidated Witnesses*. Victims' Commissioner for England and Wales.

Zajac, Rachel, Sarah O'Neill, and Harlene Hayne. 2012. "Disorder in the Court Room? Child Witnesses under Cross Examination." *Developmental Review* 32: 181–204.

Zydervelt, Sarah, Rachel Zajac, Andy Kaladelfos, and Nina Westera. 2016. "Lawyers' Strategies for Cross Examining Rape Complainants: Have We Moved Beyond the 1950s?" *British Journal of Criminology* 57: 551–569.

21

THE PROBLEMATIC NATURE OF L1 AND TRANSLATION IN US CRIMINAL COURTS

Access, Ethics, and Errors

Mel Greenlee

Introduction

Neuropsychologists Artiola i Fortuny and Mullaney (1998) once asked, "Can the absurd be ethical?" referring to clinical evaluations without adequate attention to language differences. These could imperil patients as well as violate clinicians' ethical duties, they explained. A similar ethical challenge persists in a parallel realm, the legal system, where inadequate attention to language may also have disastrous consequences. This chapter examines some of the most prominent of these complex issues.

In the US criminal justice system, persons who do not speak the language of the courts, or whose command of the language is limited, face formidable obstacles to justice. These hurdles exist at every stage, from initial contact with the police (*U.S. v. Silva-Arzeta*. 2010. 10th Cir. 602 F.3d 1208), through questioning or interrogation (Filipović 2022; Berk-Seligson 2000), to confrontation of evidence at trial and its consideration by a jury (Miller et al. 2011). In some cases, these obstacles continue to post-conviction challenges where a miscarriage of justice may have occurred (O'Brien et al. 2019).

Similarly, when victims or witnesses seek help from law enforcement, their pleas may be ignored or misunderstood when not in English (CRLA 2022; Boyette and Santana 2014; Turkewitz 2014). In the most extreme instances of misunderstanding, those needing help may even be harmed by police (see, e.g., Nasser 2021).

Under federal law in the United States, public service providers, such as federal courts and police departments, are obligated to provide meaningful access to services regardless of users' lack of fluency in English; this obligation is known as "Language Access" (see, e.g., Migration Policy Institute 2024). But legal practice has not kept pace with the need for certified language interpreters (Gentry 2020; Davis 2016; Miller et al. 2011). There is also an ongoing need for reliable and competent written translation (Killman 2020; NAJIT 2019).

Language access difficulties are particularly acute for Indigenous language speakers (Gentry 2020; Medina 2019). Cobian, Melchor, and Cazares (2024) cite a census of over 20,000 Latin Americans in Los Angeles who speak neither English nor Spanish. CIELO (2022) identified 17

DOI: 10.4324/9781003391074-23

Indigenous languages in that area, with speakers also in 23 US states. Without state-certified interpreters, Indigenous speakers may be forced to rely on Spanish, a second language many do not speak well, with further impediments when courts rely on relay interpretation from Spanish to English (CIELO 2022; Lehwalder 2021).

An Indigenous resident of rural California recounted her frustration in trying to tell police details of an incident she witnessed (CRLA 2022: 11, 45):

> I saw my neighbor and her husband having a fight.... He got mad and...he ran her over... I saw how my neighbor's bones had been broken, and she was dying. She died minutes later...About one hour later, the policeman came to my house to ask what had happened. He spoke English, and I couldn't talk to him, so I asked my [minor] daughter to interpret for me, and she explained to him what had happened.

While such a crime is hopefully rare, impediments and miscommunication can be common. This chapter will explore some conundrums and questions concerning language in US criminal courts, which draw in ethical issues of legal fairness. The chapter begins with an examination of courts' view of linguistic experts themselves. Part 1 then turns to major themes confronting justice in translation. Using case examples, central issues for limited English speakers and speakers of languages other than English are addressed. Part 2 presents examples of cases where access to justice may have been hampered by linguistic impediments. The chapter concludes with lessons learned from instances where legal language practices have gone wrong. Concrete suggestions for improvement, with a collaborative role for linguists, are also offered.

Part 1: Overview of Enduring Issues Regarding L1 in Legal Proceedings

The Definition and Role of Language Experts

Most linguists are accustomed to being questioned about how many languages they speak, when questioners confuse "linguists" with "polyglots". In much the same way, language experts who work in legal contexts are accustomed to judges using the word "translators" to refer to professionals who provide oral interpretation into English from another language. As explained by Benmaman and Framer (2015: 131):

> The general term translation...[has]...two distinct categories, *translation* and *interpretation*. The two terms should not be used interchangeably. *Translation* refers to the written form of this process while *interpretation* denotes the oral form. [Italics added]

A further long-standing misconception is that in either rendition, conversion from the source language (L1) into the target one (L2) is a word-for-word process. One state court analogized language interpreters to court reporters, viewing both as performing a relatively mechanical task (*State v. Lopez Ramos*. 2019. Minn. 929 N.W. 2d 414). But legal scholars and language researchers recognize that overly literal interpretation may produce nonsense (Davis 2016; Fishman 2006; Mikkelson 1999). Professionals who serve as language experts and trainers for court interpretation and translation emphasize the complex skills and education required (Benmaman and Framer 2015; González, Dueñas, Vásquez, and Mikkelson 2012; https://najit.org/position-papers/).

Nevertheless, under the prevailing law of many jurisdictions in the United States, interpreters at interrogations are regarded as implied agents of the speaker, or language conduits, rather than witnesses who can be confronted and cross-examined at trial (Bolitho 2019; Kracum 2014; Ross 2014). The role of interpreters and translators is the subject of considerable controversy, as their function in the justice system is not uniformly regulated (See Tamura 2019a; *Ye v. United States*. 2016. WL 860966. Petition for Writ of Certiorari and related briefs of amici curiae). Thus, as argued by amici in *Ye*, interpreters are uncertain about whether they could be called to testify, as it may depend on their jurisdiction. California adheres to the "language conduit" rule, treating the interpreter at interrogation as a relay, with their words (in English) attributed to the suspect (see *People v. Torres*. 1989. 213 Cal. App. 3d 1248). In contrast, appellate courts in Washington, for example, have rejected the conduit notion, holding that interpreters' failure to testify is error when other witnesses merely report the defendant's statements in English via the interpreter's words; such hearsay reports by non-testifying witnesses are generally prohibited by defendants' constitutional right to confrontation. (See, e.g., *State v. Gonzalez-Hernandez*. 2004. 122 Wash. App. 53. 92 P. 3d 789; *People v. Garcia Trujillo*. 1997. 89 Wash. App.23. 948 P. 2d 390.)

Because non-English speaking persons' words may undergo more than one transformation on their journey through the legal process, it is important to review the levels at which distortions may occur. The first starts well outside the courtroom. The second level concerns how the transformed language is cast into evidence in court.

Level 1: Witness Statements and Suspect Interrogations

At the first level, a crucial issue concerns the competence and possible conflict of interest of the "interpreter" who assists in eliciting statements from witnesses, or in interrogating a suspect (Tamura 2019a; Filipović and Vergara 2018; Berk-Seligson 2000). Berk-Seligson (2000: 232) found that these "interpreters" were ad hoc, a "highly heterogenous" group, often lacking professional language skills or ethical standards. There are many legal rules for interpreters in court, but no corresponding binding guidelines for interpreters during police interrogations of suspects (Filipović 2022). While police officers must provide Miranda warnings to suspects prior to interrogation, the unknown L1 competence of the interpreter (often another officer) and the way suspects' waivers of rights are gathered may leave little assurance that detainees have understood. Pavlenko (2023) found that Miranda rights were administered to limited English speakers solely in English in over one-third of cases she reviewed, and experimental research showed that even intermediate English learners failed to understand parts of the warnings (Pavlenko, Hepford, and Jarvis 2019).

In interviewing witnesses, police may rely on other officers, bystanders, or children, both at the scene and at the police station (CRLA 2022; Davis 2016). California Rural Legal Assistance (CRLA) research interviewers found that when family members were used, these most often were minor children, as in the homicide witness's experience noted above. Reporting practices vary among the 18,000 police jurisdictions in the United States. In some instances, witnesses may go to the police station with a friend or relative. Where officers speak Spanish, reports might be made in that language. However, a failure to translate the information can be devastating. In a notorious 2014 case in New York, failure to translate and follow up on a Spanish report concerning domestic violence contributed to a triple homicide (Boyette and Santana 2014; Turkewitz 2014).

Level 2: Language as Courtroom Evidence and Courtroom Interpretation

The problems in interpretation outside of court and frequent use of ad hoc interpreters (at Level 1) lead to the second problem—the verification of accuracy for any transcript or translation of police interaction outside of court. Where none of the legal actors controls the source language, such verification can be very problematic.

Fishman (2006) has provided a comprehensive analysis of legal cases wrestling with the problems of recorded evidence where translations are at issue. For example, what if the witness verifying the English version is a confidential informant? How should the court resolve disputes where recordings are only partially intelligible? May summaries of L1 evidence guide testimony? In Australia, Fraser (2022) has set out guidelines for the preparation of appropriate transcripts, noting the many problems that arise even within a single language.

In live courtroom interpretation, case law includes instances where trial judges appear to merely throw up their hands, lacking sufficient evidence to determine whether the interpretation was accurate. In *U.S. v. Long*. 9th Cir. 2002. 301 F.3d 1095, both the trial judge and defense counsel expressed doubt about the interpreter's Lao and Hmong skills, but a defense motion for mistrial was denied. The appellate court found no error, simply noting "there is no way to know what really happened" as neither party spoke the witness's languages. Rather, such "occasional lapses" did not require a remedy (*Long*, 301 F.3d 1105).

Legal issues involved in challenging conversion into English, either oral or written, involve a complex of evidentiary rules. To properly raise an error in courtroom interpretation, an attorney must promptly object, but how to object when no one in the courtroom besides the interpreter controls the two languages? By relegating interpreter/translator issues to evidentiary jurisprudence, reviewing courts have set a nearly impossible standard which requires both L1 and L2 competence on the part of counsel. It also subjects any errors to a very rigorous legal hurdle on review (Santaniello 2018; Turner 2013; Benmaman 2000).

For written transcripts, case law encourages parties to attempt agreement pre-trial on translated statements (see, e.g., *U.S. v. Onori*.1976. 5th Cir. 535 F.2d 938; Fishman 2006). This may occur even when the original recording contains lacunae or confusing passages, such as in the account of a dubious Spanish/English telephonic interpretation detailed by Angelelli (2015). Her analysis showed that communication breakdowns and misunderstandings during questioning by law enforcement officers caused much confusion, but these problems were not noted in the written account of the session; the English transcript was nevertheless accepted by the court.

The California Evidence Code (see § 751 et seq.) and the Rules of Court require a certification by the translator. However, where a recorded L1 interrogation is converted to a written English translation, prosecutors may skirt this requirement by presenting a witness (such as a police interrogator) who merely uses the translation to "refresh their recollection" while the translation itself is not introduced as evidence. This may mean that defense counsel lacks an opportunity to thoroughly dispute the validity of the translation.

While one very old California case (*People v. Lee Fat*. 1880. 54 Cal. 527) rejected a translation without the translator's certification, more recent cases deem such omissions harmless error (*People v. Garcia*. 2022. WL1468074; *People v. Torres*.1985. 164 Cal. App. 3d 326; *People v. Zamora*. 2012. WL 6621741). Federal courts have accepted even summaries of translations as evidence. For example, in *U.S. v. Pena-Espinoza*. 1995. 9th Cir. 47 F.3d 356, "condensation" and summaries by Federal Bureau of Investigation (FBI) officers were accepted, despite one justice's strong dissent (47 F. 3d at 363, Reinhardt J., dissenting). Justice Reinhardt maintained that the FBI witnesses were "describing conversations about which they had no factual knowledge based solely on their opinions that the transcripts were accurate", so their accounts lacked essential evidentiary foundation. Acceptance of either oral or written renditions of L1 into English without adequate verification or professional credentials is worrisome and may be, in fact, far from harmless. (See

Leal v. State 1989. Tex. Crim. App. 782 S.W. 2d 844, finding the trial court erred in admitting unsworn prosecution translation in a murder case.)

For oral interpretation, reliance on telephonic and other remote language services has grown since the pandemic and continues to expand in use and acceptance (California Judicial Council 2021; National Center for State Courts (NSCS) 2018). Nevertheless, interpreters who appear remotely should have proper certification, which is not always clear. For example, in *State v. Lopez Ramos* 2019. Minn. 929 N.W. 2d 414, the native Mam-speaking defendant's conviction (Neuman 2021) rested on a recorded Spanish/English interrogation interpreted remotely by an interpreter whose name and credentials were never stated. (Mam is an Indigenous language of Guatemala.)

Remote interpreters may also face numerous technological obstacles. Video "imposes constraints on the perception of non-verbal clues and general visual perception, and...makes direct eye contact virtually impossible" (Braun 2011: 271). Early, large-scale research on remote interpreting in Europe found that it tended to "entail increased stress and fatigue, a decline in interpreting quality and a change in working conditions" for language specialists (Braun 2019: 11). Experimental research in Australia confirmed that telephone interpretation, especially, may lead to significant detriments in accuracy. Nevertheless, investigators found that under ideal conditions, video interpreting (VRI) accuracy was not significantly worse than in-person performance (Hale et al. 2022). Ideal conditions, though, are rarely the pattern in individual US courtrooms, even where virtual proceedings do not involve L1 to English interpretation. Many difficulties in remote proceedings have been detailed in criminal matters and in immigration proceedings (Benninger et al. 2021; Bradley and Farber 2021; Turner 2021a).

In a large state study of remote guilty pleas, over a quarter were marred by technical problems (Turner 2021b). In immigration court, problems include "pixelated screens, sound quality issues, and dropped Internet reception". Immigration judges reported "extremely problematic issues with interpretation and coordination between telephonic interpreters and the [video] units" (Bradley and Farber 2021: 554). Courts have endeavored to make improvements through upgraded technology as well as practice guidelines and uniform standards (see, e.g., California Judicial Council 2021; National Center for State Courts 2018).

Nevertheless, fundamental questions remain concerning both interpreter accuracy and evaluations of remote proceedings by those affected. The extent to which witness credibility is affected by remote interpretation is a vital research area. Hale et al. (2024) found that credibility could be influenced by interpreters' English language ability. Setting is also an important factor (Kostroff 2023). Technical problems with interpretation would magnify these difficulties.

Moreover, perception of fairness in the courts relies on a litigant's or witness's assessment that he or she has "the opportunity to be heard" (Bandes and Feigenson 2020: 1320; Clair 2020). Participants in virtual hearings may report a feeling of unreality (Benninger et al. 2021), which could be exacerbated by the isolation of VRI technology. Courts have stressed the advantages of VRI in making interpreters available for witnesses speaking less common languages, but too little attention has been paid to L1 litigants' or witnesses' assessment of how well they can "tell their story" by this means. These issues deserve detailed ethnographic research with L1 speakers in their own preferred languages.

Proposed Improvements

Level 1 Solutions

Many solutions have been proposed for interpretation problems during police interrogation or questioning. One recommendation would require professional interpreters at these sessions

(Filipović 2022). Triano-López (2015) called for implementing legislation, highlighting the errors of inept police "interpreters" and their repercussions. But legislatures have not responded and the proposal was previously rejected by the courts. See, e.g., *Baltazar-Monterossa v. State*. 2006. 137 P.3d 1137, where the defendant claimed error because the police "interpreter" at his interrogation was neither certified nor independent; the Nevada court rejected his claim, considering any dispute merely an evidentiary quibble.

Another proposed improvement would require recording of police interactions with suspects, witnesses, and victims. Bang et al. (2018: 15) observe that many US states require recording of interrogations, which "promotes…transparency, integrity, and accuracy in the criminal justice system". Conversely Berk-Seligson (2009) observed heightened dangers to justice where police made no recordings of interpreted interrogations (see also Victor Rosario case, *post*). Tamura (2019b) also maintains recording is a must, highlighting that Massachusetts requires recording of all interrogations and interviews using an interpreter (*Commonwealth v. AdonSoto*. 2016. 475 Mass. 497). Video recording is preferable. In *Taylor v. State*. 2016. Maryland. 130 A. 3d 509, the court examined a video recording and took special notice of the interpreter's role. It found many interpretation problems in a lengthy interrogation of a hearing-impaired defendant, in which the interpreter had provided both oral and sign-language communication.

Level 2 Solutions

Various suggestions have also been made to address linguistic obstacles to fair and equal treatment within the courtroom. For instance, a recording requirement has been proposed for interpreted proceedings to preserve the source language (Santaniello 2018; National Center for State Courts 2006). California courts have consistently denied motions for audio recording of both L1 and English interpretation in criminal trials (see, e.g., *People v. Coriano*, RT 2031).[1] In denying such a motion, the court stressed the role of the certified court interpreter, pointed to unavailability of appropriate equipment, and voiced worry over "issues cropping up that were never addressed at trial". A Maryland judge in *Wallace v. State*. 2020. WL 2731228 gave a similar rationale, denying a motion to videotape sign language interpretation. These decisions render interpreter errors unrecoverable absent counsel's contemporary objection (Turner 2013).

In view of the poor quality of ad hoc interpreters for interrogations (at Level 1), Larsen (2020) has proposed that when a tape of a non-English interrogation is presented as evidence, the original interpreter's version should be excised and only a version prepared by a forensic transcription specialist used when the tape is played in court, reasoning that the expert's version would be more accurate. But this proposal would short-circuit an important part of the language conversion. The National Association of Judiciary Interpreters and Translators (NAJIT) has provided very clear guidelines for the work of forensic transcription, which require a two-step process; only after a detailed transcript is prepared in L1 should the work of translation into L2 be undertaken (NAJIT 2019; Benmaman and Framer 2015).

Bolitho (2019) would go even further, urging that for recorded evidence, the court interpreter should provide live interpretation into English for jurors. But NAJIT (2019) cautions against this practice, as posing too great a risk of inaccuracy. Live interpretation of tape recordings would put inordinate stress on the court interpreter. It might also eliminate the possibility for detailed research on the variety of L1 in the evidence, an issue which Rainof (1991) underscored as essential. For many reasons, Bolitho's proposed solution appears impractical.

How jurors are told to evaluate interpreted or translated evidence is also critical and is a matter of some controversy. In many jurisdictions, such as California, when an audio recording of an out-of-court statement is played in court, jurors are given a transcript to follow. However, the jury is instructed that the tape, rather than the written version, is the primary evidence (*People v. Brown.* 1990. 225 Cal.App.3d 585, 598–599). For recordings in English, jurors are instructed that the transcript is merely an aid. Even where the prosecution's taped evidence includes many inaudible portions, courts have sanctioned jurors' use of a disputed transcript (*People v. Polk.*1996. 47 Cal. App. 4th 944). The instruction to use a transcript is modified when taped evidence is in another language. Then jurors are instructed according to the following model, drawn from *People v. Arancibia.* 2015. WL 5095991 [ordered depublished].

> You are about to [hear] [watch] a recording in the [specify the foreign language] language. A transcript of the recording has been admitted into evidence. The transcript is an official English-language translation of the recording, [...] Although some of you may know the [specify the foreign language] language, it is important that all jurors consider the same evidence. Therefore, you must accept the English translation contained in the transcript even if you would translate it differently.

Some legal scholars maintain that playing an audio or video recording jurors may not understand is confusing (Richter 2022; Fishman 2006). Richter proposed amending the federal Best Evidence Rule to allow admission of translated transcripts without the recordings. Fishman (2006) observed that eliminating the recording would prevent problems, such as jurors attempting to divine speaker emphases or attitudes in an unfamiliar language and cultural context.

While the avoidance of these complications is understandable, there is another possible motivation for dispensing with the recordings that is extremely troubling, namely thwarting jurors who may understand the source language (Moya 2021; Perea 1992). According to some commentators (and judges), if individual jurors understand the original recording, they may try to serve as second-guessing "underground" interpreters who introduce unreliable and untested evidence into the jury's decision-making (see *People v. Quinonez.* 2014. WL 7263745). In *People v. Arancibia.* 2013. 213 Cal. App. 4th 1465 (ordered depublished), the appellate court explicitly disapproved of using both L1 and English in a transcript given to jurors. At trial, the judge had provided a bilingual transcript to jurors along with the Spanish tape. Jurors should receive any such transcript only in English, the reviewing court decided.

But such jurisprudence goes counter to research on bilingual communities. As observed in the amicus brief in *Hernandez v. New York.* 1991. 500 U.S. 352, 1990. WL 10013132: 17, comprehension of a language one speaks is not optional, "Bilinguals cannot switch English on and Spanish off. Both language systems are constantly switched on. Requiring bilinguals to act otherwise is contrary to human mental processes". Nevertheless, in *Hernandez*, a plurality of the Supreme Court sanctioned dismissal of two jurors for giving "hesitant" responses (saying they would "try") when asked to listen solely to the official English (500 U.S.356). Legal commentary on the decision has argued that rather than regard L1 competence as a cause for suspicion, courts should treat bilingual jurors as possessing unique and valuable expertise (Valentine and Lee 2015).

Persisting doubts about jurors who speak and understand a minority language may go along with long-standing biases against minority community members and immigrants, a malady long plaguing the judicial system (Moya 2021; Rose 2020; Bribriesco 2010). For example, surveys have shown that defendants using an interpreter receive less favorable bail decisions (Rengifo, Rouzbahani, and Peirce 2020); accented witnesses are perceived as less credible (Cantone et al. 2019); and Spanish-speaking Latinx plaintiffs receive less compensation than those who speak English in personal injury suits (Ewing, Reyes III, and Wetherbe 2008).

Clearly, unanimity and jurors' consideration of the same evidence is fundamental. However, instructions which ask L1-speaking jurors to inform the court privately if they contest the interpretation are to be preferred over blanket orders to ignore any such defects, treating L1 competence as a disadvantage.

Minimal Attention to Language Issues

Despite widespread acknowledgement of significant problems, there has been comparatively little attention to language issues from advocates. Recent research on criminal proceedings and their speed may help explain why. It is estimated that only 5% of US criminal cases go to trial, with the rest resolved by plea-bargains (Refo 2004). Review of a large sample showed the average such proceeding lasted less than 30 minutes (Redlich et al. 2022). The lack of time spent in plea-bargaining may raise unique and serious challenges for limited English speakers (Chang 2008).

Fraser (2022) observed that often busy defense attorneys in Australia have little time to review transcripts of recorded statements against the original. And in the United States, the overloaded case list of public defenders is a well-known problem (Gottlieb 2023; Pace et al. 2023). Budget constraints, in addition, may limit defense attorneys' use of check-interpreters and language experts who could monitor the quality of language conversion.

Lawyers may give only minimal attention to language issues because they do not consider such challenges as potentially advantageous to their clients. In practice, language issues are seldom a cause for reversal (Benmaman 2000). Legal scholars (e.g., Santaniello 2018; Turner 2013) have proposed changing the appellate standard to allow language defects, such as interpreter errors at trial, to receive more attention. If the rule were changed and trials recorded, Santaniello argues, a language expert could be consulted by the court and appellate counsel to assist in identifying problems and errors, to more directly and efficiently seek a remedy.

For challenges to written translation errors, the legal standard is a very difficult hurdle as well. California courts have held that "transcripts of admissible tape recordings are only prejudicial if it is shown they are so inaccurate that the jury might be misled into convicting an innocent [person]" (*People v. Brown*. 1990. 225 Cal. App. 3d 585, 599). This criterion has been applied to both written translations and English-only transcripts. At present, undoing the effects of erroneous interpretation or translation is extremely difficult in the United States. In the following section, examples of challenges are presented. However, as will be shown, these cases have mixed success, can take a very long time, and usually involve issues beyond disputed translation and interpretation.

Part 2: Case Examples of Language Gone Wrong

Successful challenges to translation and interpretation do occur. However, they most often take place when the reviewing court finds other errors, beyond those of language, in the proceedings. O'Brien et al. (2019: 1693–1694) reviewed 29 cases of Latinx exonerees, finding that in at least 12 of their cases, English language proficiency was an important factor, as defendants did not speak English well or "fully understand spoken English". However, the authors conceded that in challenging these convictions, the following was also true (2019: 1705):

> Even if defense counsel has doubts about an interpreter's competence, expending the substantial time and resources to identify mistranslations is probably not cost-effective, as it is unlikely to convince a court that the defendant was prejudiced. [citation omitted] As a result, we cannot know how often inadequate translation prejudices defendants.

The importance of language in these challenges is shown in the cases below.

*Case 1: Victor Rosario (*Comm. v. Rosario. *2017. Mass. 74 N.E. 2d 599)*

Victor Rosario was a Spanish-speaking 24-year-old falsely found guilty of arson-murder. His erroneous conviction was ultimately undone, in part, by scientific expertise debunking flawed forensic evidence. Fire science experts' re-examination invalidated the prosecution's theory of Molotov cocktails as the fire's origin. Another factor in overturning his conviction was the defendant's communication during a very long interrogation (O'Brien et al. 2019).

A mental health expert reviewing Mr. Rosario's interrogation and history found that during his interrogation, he was suffering from severe psychological symptoms such as hallucinations which made his speech incoherent. A further key witness was the interpreter for the interrogation. The interpreter provided a sworn affidavit declaring that an admission to setting the fire with a Molotov cocktail, attributed to Mr. Rosario, was in fact inserted into his signed statement by a detective. That inculpatory statement, prepared in English by the police, was not translated into Spanish before Rosario signed it (Brief of Amicus Curiae 2016 WL 6468398). The interpreter also described Mr. Rosario's mental deterioration over the course of the all-night questioning. Mr. Rosario finally was released in 2017 (Comm. v. Rosario. Mass. 2017. 74 N.E. 2d 599) and the charges were dismissed. He was compensated for his 1983 conviction by the state of Massachusetts in 2021 (see https://www.law.umich.edu/special/exoneration/Pages/casedetail.aspx?caseid=5196). Despite this positive ending, Victor Rosario endured prison for over 30 years before his wrongful conviction was finally overturned.

Case 2: People v. Velasco-Palacios. *2015. 235 Cal.App.4th 439*

In a California case, review of the prosecution's translation of the defendant's Spanish statement prevented a serious error which otherwise would have tainted the outcome of a plea bargain. Mr. Velasco-Palacios, age 21, was accused of a sexual offense. He had been interrogated in Spanish and two English translations were prepared—one by the defense and one by the prosecution.

Defense counsel discovered that the prosecution's version contained an admission not in the defense transcript and sought the original recording. After reviewing it, he determined that the prosecutor had inserted an inculpatory remark by the suspect into the English version, apparently to influence the plea-bargain being negotiated. The defendant denied making the statement, which would have significantly enhanced his risk of receiving a punishment of life in prison.

The prosecutor later admitted that he had inserted the statement into the English translation but claimed he had added this as a joke. The trial court was not amused and dismissed the case, with strong remarks concerning "outrageous" prosecutorial misconduct in creating such a fraud (*Velasco-Palacios*. 235 Cal.App.4th 439, at 444).

Case 3: People v. Guerra. *2006. 37 Cal.4th 1067*

The case of Jose Guerra involved a monolingual defendant. Unfortunately, he died in prison while post-conviction challenges related to intellectual disability were still pending. His case illustrates the problem of conflicting interest when a police officer with inadequate language skills acts as an interpreter for an interrogation (Filipović and Vergara 2018; Berk-Seligson 2000) as well as the great difficulty of convincing courts of its enduring taint to legal rights.

Mr. Guerra was a Guatemalan national who had immigrated a year before his arrest. He had no experience with the US justice system. He was questioned regarding a homicide which occurred near the jobsite where he had been working as a manual laborer. He had significant mental limitations, with an overall intellectual level later assessed as equivalent to that of a 12-year-old child. Those who knew him relayed that he was "mentally slow and easily confused" (*In re Guerra.* 2007. S134332, Amended Petition for Habeas Corpus:178).

Upon his detention, a uniformed police officer served as the interpreter for administration of the Miranda rights, which Guerra had great trouble understanding. This difficulty was not only due to Mr. Guerra's cognitive limitations, but also the officer's inept Spanish interpretation. Mr. Guerra asked many questions of the officers. But it soon became clear that the Spanish-speaking officer's primary role was to facilitate his interrogation as a suspect. This officer was instrumental in convincing Mr. Guerra to retract his request for counsel, which otherwise would have stopped the questioning. After Guerra finally appeared to grasp the concept of free defense counsel, a long exchange ensued between Mr. Guerra, the interpreting officer (Int.), and another officer present. A relevant portion of the officers' exchange regarding counsel is shown below (*In re Guerra*, Amended Habeas Petition, 3819–3821).[2]

Mr. Guerra: Pues, está bien que me lo nombren también un abogado, verdad?[*Well, it's fine for an attorney to be appointed for me also, right?*]**Int.**: Entonces quiere un abogado Vd.?[*So then you want an attorney?*]**Mr. Guerra**: Sí, quiero un abogado.[*Yes, I want an attorney.*]Int **to Officer 1**:He says, "Yes, I want an attorney".**Officer 1**: Okay, understand this. If he wants the attorney, he doesn't wish to speak to us…then from the information that we have, he's going to be arrested for murder. We'll book him in the jail right now.**Int**: Okey, quiero que, dice el detectivo que quiero que, que sepas esto. Entonces, si quieres el abogado aquí, y no quieres hablar con los detectives del caso de ayer, entonces quiero que, que ellos detoctives, tienen suficiente, eh, in- información de lo que oc- ocurrió ayer que ahorita te van a arrestar a ti y te van a meter a la cárcel por, por la, por la muerte de la señora[-][*Okay, I want to, the detective says, that I want you to, to know this. So then, if you want the attorney here, and you don't want to talk with the detoctives about the case yesterday, then I want to, that, that they have enough, uh, in- information about what ha- happened yesterday that they're going to arrest you right now and they're gonna put you in jail for, for, the death of the*[-] *lady*]**Mr. Guerra**: O sea, es que, es que no te entendí, ¿verdad?[*In other words, the thing, the thing is that I didn't understand, right?*]***[Later in same exchange]**Int**: …y si quiere al abogado aquí presente, entonces ellos te van a lle- a meter a la cárcel ahorita.[*…and if you want the attorney present here, then they're going to ta- put you in jail right now.*]

When the evidence was presented as part of a motion to suppress the resulting statement, the trial court acknowledged problems in interpretation but admitted the statement anyway. The trial court's assessment is provided below:

> The officer doesn't act as an interpreter. He filters things and changes things and filters it back instead of *acting like a machine* and repeating what the officers say in English and getting back to the officers in Spanish what the officers say in English. He is supposed to translate into Spanish for the defendant or the interviewer, then take whatever the interviewee says and translate it from Spanish into English. But instead of that, you get the officer participating and it leads to great problems. [emphasis added]
>
> *(2 RT 525)*

The trial court's remarks align with the "conduit" view of interpretation which envisions language conversion as a mechanical process (see *Correa v. Superior Court.* 2002. 27 Cal. 4th 444). The court's overly simplistic concept of interpretation clearly underestimates the sophisticated and complex nature of language experts' contribution.

Furthermore, the judge's comments on the officer/interpreter's "participating" allude to the inherent conflict of interest in his dual role. Nevertheless, the trial court determined Guerra's statement was voluntarily made, and his (translated) words were used as powerful evidence for conviction (20 RT 5373-91). Despite the many misunderstandings in the Miranda exchange and multiple threats of jailing, on appeal the California Supreme Court found "no causal link" between the officers' (including the "interpreter's") statements and the defendant's decision to forgo counsel and talk to detectives (*People v. Guerra*. 2006. 37 Cal.4th 1067, 1096).

Post-conviction counsel unsuccessfully challenged this bland view of the interaction with additional relevant facts, including Mr. Guerra's cognitive limitations and a detailed analysis of the Spanish and English at his interrogation by a qualified language expert (*In re Guerra*, Amended Habeas Petition, Ex. 194.) The expert found that the officer/"interpreter" not only had inadequate language skills, with errors in vocabulary, grammar, and meaning, but also failed to translate the defendant's questions for his fellow officers. The officer's performance, she opined, was well below the standards of professional interpretation.

Habeas counsel argued that these defects tainted the voluntariness of Guerra's statement, especially in light of his mental disability, an argument the California court rejected without elaboration. Nevertheless, the expert's analysis revealed that like defendants in cases examined by Pavlenko (2023), Mr. Guerra received problematic admonitions of his rights. Moreover, the dual role of the police "interpreter" allowed law enforcement to take advantage of Mr. Guerra's inexperience and confusion. The officers evidently relied on a shared language to persuade Mr. Guerra to give up these protections. As Berk-Seligson (2000) found over two decades ago, police interpreters' flawed language rendition was no bar to admission of the defendant's statement. Like the case study of Filipović and Vergara (2018), in Guerra, the officer's attempt to serve two roles simultaneously was impossible and as counsel argued, caused serious defects. But in considering post-conviction challenges, the reviewing court limited its hearing to intellectual disability, which could nevertheless have aided in clarifying the improper linguistic influences, had Mr. Guerra not died while it was still in progress.

Case 4: People v. Hobart and Konard

It is not unusual in the United States for police officers with some second language skills to also assist in questioning witnesses. In *People v. Hobart and Konard*,[3] the damaging ripple effects from inadequate witness interpretation are dramatically shown in a California capital trial.

In this case, police questioned a middle-aged Mandarin-speaking woman concerning two intruders at her home. Although she understood some English, she often made mistakes, so police initially received confusing and ungrammatical answers. To assist, they enlisted another officer, a heritage speaker of Mandarin, whose language skills were also limited. When police asked the witness if one of the defendants (who spoke English) had threatened her, the police "interpreter" gave suggestive answers back, partly because he did not know how to translate the word "threat" in Mandarin. This led to even more confusion. In the repeated questioning, the witness adopted some of the officer/interpreter's suggestions.

But all these complexities and instances of confusion were subsequently smoothed over in the English police report of the interview. At the time of the defendants' trial, a few years after the police interviews, the witness was allowed to refresh her recollection by using the prosecution's flawed report, further influencing her testimony. The prosecutor later used the witness's wording to underscore the defendants' allegedly dangerous intentions in the incident.

In post-conviction investigation, defense counsel worked with a competent and certified interpreter who examined the original Mandarin exchange. This linguistic analysis revealed many interpretation errors and showed the tainted origin of the exaggerated threat evidence which had infected and influenced the trial jurors' penalty decision (Petition for Habeas Corpus at 118, 3198). As the *Hobart and Konard* case illustrates, enduring harm can result from inept ad hoc police interpreters. Further, as Berk-Seligson (2000) confirms, these lasting flaws can also affect the memory and subsequent testimony of trial witnesses on which the jury bases its verdict.

Case 5: In re Benavides Figueroa. *2018. 4 Cal. 5th 576*

As this case shows, inept interpretation by unqualified police officers may be disguised in written English translations used by witnesses and considered by jurors, lending credence to arguments attributing any inconsistencies in defendants' accounts to attempts to avoid guilt (Greenlee 2021; O'Brien et al. 2019). Mr. Benavides Figueroa's conviction and sentence were eventually reversed when a reviewing court determined that forensic evidence was fatally flawed. The reviewing court found that the prosecution's theory, alleging the victim had died as a result of child sexual abuse, was in fact false and unsupported. However, the defendant's release from Death Row came only after a quarter century of diligent legal advocacy on post-conviction.

At his original trial, the prosecution had argued that jurors should not credit Mr. Benavides Figueroa's denials of child abuse, in both interrogation and in-court testimony, because his answers to questions were inconsistent and avoidant. However, when his statements were closely analyzed, it became clear that much of the alleged inconsistency was attributable to poor interpretation compounded by inept written translation.

Mr. Benavides Figueroa, a working-class Spanish speaker, had been interrogated by one officer while another assisted in the role of "interpreter". The latter's facility in Spanish was quite limited, resulting in multiple unintelligible passages and numerous misunderstandings. Expert interpreter and translator, Haydée Claus, who later examined the recorded interrogation, set forth a lengthy analysis of the police "interpreter's" errors. Mistakes were found in Spanish vocabulary, syntax, and even pronunciation. The officer relied on English words when his Spanish failed which created even more confusion (*In re Figueroa*. 2008. Amended Petition for Habeas Corpus, Exh. 63).

The expert further found that interpreter errors were sanitized and disguised in the English translation jurors reviewed. The English translation introduced additional errors. The communication considered by the jurors thus not only distorted the competence of the interrogator's Spanish, but also gave an erroneous impression of the defendant's mental state, credibility, and voluntariness (*In re Figueroa*. 2008. Amended Petition for Habeas Corpus).

Perhaps the most obvious example of Spanish misinterpretation and its erasure in the English transcript was the officer's use of the Spanish word "rapada", "shaved clean", for the intended "violada", which means "raped". This error reflected the impoverished Spanish vocabulary of the officer, who relied on a false cognate. For example, when the officer/interpreter intended to say, "It looks like she was RAPED [Sp. Violada] from behind", he actually said "parece que fue rapada por atrás, por el rectal" which means "it looks like she was shaved clean from behind, from the 'rectal'". The transcriber's English version totally obscured this error and rendered the detective's utterance as "it seems that she was raped from behind. From the rectal". The transcriber's version also omitted Mr. Benavides Figueroa's immediate question "¿Cómo?" meaning "What?" (Amended Habeas Corpus Petition, Exh. 63: 5214).

More subtle alterations, which further affected the quality of the evidence from his statement, were also "cleaned up" in the English transcript. For example, at one point, the officer/interpreter states, "No entiendes por qué". [*You don't understand why.*] However, in the transcript the passage appears with a change in syntax: [*You don't understand. Why?*] In another instance, the defendant says, "Ya la encontré tirada afuera". [*I found her lying outside.*] This statement is translated as "*I found her laying on floor*", manifesting a large shift in meaning. This last mistranslation seriously affected the credibility of Mr. Benavides Figueroa's statement, as the prosecution contended that he had lied about where he found the injured child. In post-conviction counsel's challenge to his case, the linguistic expert showed that, from the very beginning, the interactions between Mr. Benavides Figueroa and the police officers were crucially misrepresented and errors cleaned up in translation. The contaminated evidence the jurors considered was thus doubly distorted from the defendant's original statement, lending false credence to the prosecutor's argument for guilt.

O'Brien et al. (2019) provide an insightful analysis of the Benavides Figueroa case and underscore the cumulative harm done by the poor interpretation and translation which combined to give a false impression of the defendant and his statements. The prosecutor's arguments, grounded in language errors, contributed to an innocent man spending decades of his life falsely imprisoned. A final case shows how cultural myopia and language can combine to bring about serious legal error.

Case 6: US v. Hayat. *2013. 9th Cir. 710 F. 3d 875*

In 2006, Hamir Hayat, a young man of Pakistani origin living in rural California, was convicted of supporting terrorism. The principal evidence came from a lengthy interrogation. Based on that evidence, the government alleged that Hayat had attended a terrorist training camp for three to six months during a family visit to Pakistan. Additional evidence was an Arabic saying found in the defendant's wallet that a government witness testified was both uncommon and "not peaceful", averring that the type of person who would carry it was "[a] person who perceives him or herself as being engaged in war for God against an enemy" (*U.S. v. Hayat*. 2019 WL 3423538, E.D. Cal.). Hayat was convicted and sentenced to over 20 years in prison.

Post-conviction habeas investigation found numerous alibi witnesses who would have contradicted the government's story of Hayat's attendance at a jihadist camp, as Hayat was with family during the vast majority of his time in Pakistan, without the extended absences posited by the prosecution (*U.S. v. Hayat*. 2019. WL 3423538). Moreover, a cultural expert explained that the Arabic saying in the defendant's wallet was "part of a common prayer, with no sinister meaning" (Edelko 2020).

Hayat was released after 14 years in prison (*U.S. v. Hayat*. 2019 WL 3423538). The court hearing habeas evidence found that Hayat's original defense counsel had prejudicially erred in failing to present the alibi witnesses, and further missed the mark by failing to call an appropriate Arabic language expert to explain the prayer that the prosecutor had claimed showed Hayat's jihadist leanings. The Hayat case illustrates the enormous importance of situating language in its appropriate cultural context.

Conclusion

Review of issues in language conversion and concrete cases has illustrated the complex obstacles to justice faced by persons who do not speak the language of the court. How could such unethical and unequal treatment be revised? Many research-based suggestions have been mentioned above. In collaboration with courts, linguists have assisted in training and development of materials for focused Language Access and served as amici curiae when questions concerning language arise in the courts; forensic linguists' analyses have provided critical foundation for individual case challenges to erroneous and tainted convictions.

But a more ethical procedure would seek to prevent such a language divide from the outset, emphasizing the crucial importance of language issues to due process. A first step to improvement requires accountability and transparency: universal recording of interpreted proceedings at both levels 1 and 2 as well as augmented resources for an informed defense would go a long way toward these goals. Improved VRI technology may make more professional interpreters available for interrogations and interviews, but investigation is needed on the accuracy of remote methods as well as their provision of perceived procedural justice for witnesses, victims, and defendants. Ethnographic interviews such as CRLA's with "end-users" are vital for improvements in public services. As technology advances and populations of L1 speakers needing services change, collaborative efforts between language experts and legal specialists can serve to set priorities for policy implementation. It is to be hoped that more attention to language issues, and more involvement of linguistic expertise from the outset, will prevent the long wait for justice that defendants endured in the cases discussed here and will mean that victims of crimes no longer fear that their reports will be misunderstood or ignored if not in English.

Notes

1 Cases in post-conviction posture are identified by pseudonym. RT refers to the reporter's transcript of trial.
2 In this excerpt, text in italics represents the post-conviction language expert's translation of Spanish utterances from the tape of the interrogation. Misspellings of English words, such as "detoctive", are intended to show the police "interpreter's" mistakes in Spanish pronunciation of that word.
3 Defendants are identified by pseudonyms.

Case Law

Baltazar-Monterrosa v. State. 2006. 122 Nev. Adv. Op. 56, 137 P.3d 1137.
Commonwealth v. AdonSoto. 2016. 475 Mass. 497, 58 N.E.3d 305.
Commonwealth v. Rosario. Mass. 2017. 47 Mass. 69, 74 N.E.3d 599.
- Brief of Amici Curiae, New England Innocence Project, the Innocence Project, Inc., and the Boston College Innocence Program. 2016. WL 6468398
- https://www.law.umich.edu/special/exoneration/Pages/casedetail.aspx?caseid=5196

Correa v. Superior Court. 2002. 27 Cal. 4th 444.
Hernandez v. New York. 1991. 500 U.S. 352.
- Brief for the Mexican American Legal Defense and Educational Fund and the Commonwealth of Puerto Rico, Department of Puerto Rican Community Affairs in the United States, as Amici Curiae in Support of Petitioner, filed Nov. 28, 1990. 1990 WL 100131132.

In re (Benavides) Figueroa. 2018. 4 Cal. 5th 576.
In re J.F. Guerra. 2007. Amended Petition for Habeas Corpus, Case no. S134322.
Leal v. State. 1989. Tex. Crim. App. 782 S.W. 2d 844.
People. v. Arancibia. 2013. 213 Cal. App. 4th 1465 [ordered depublished].
People v. Brown. 1990. 225 Cal.App.3d 585.
People v. Garcia. 2022. WL 1468074.
People v. Garcia Trujillo. 1997. 89 Wash. App.203, 948 P. 2d 390.
People v. Guerra. 2006. 37 Cal. 4th 1067.
People v. Lee Fat. 1880. 54 Cal. 527.
People v. Polk. 1996. 47 Cal. App. 4th 944.
People v. Quinonez 2014. WL 7263745.
People v. Torres. 1985. 164 Cal.App.3d 266.
People v. Torres. 1989. 213 Cal. App. 3d 1248.
People v. Velasco-Palacios. 2015. 235 Cal. App. 4th 439.
People v. Zamora. 2012. WL 6621741.

State v. Gonzalez-Hernandez. 2004.122 Wash. App. 53, 92 P. 3d 789.
State v. Lopez Ramos. 2019. Minn. 929 N.W. 2d 414.
Taylor v, State. 2016. Maryland. 226 Md. App. 317, 130 A.3d 509.
US v. Hayat. 2013. 9th Cir. 710 F.3d 875.
US v. Hayat. 2019. WL 3423538 (U.S. District Court No. 2: 05-cr-240-GEB (E.D. Cal. July 29, 2019)). [order granting relief]
US v. Long. 2002. 9th. Cir. 301 F.3d 1095.
US v. Onori. 1976. 5th Cir. 535 F.2d 938.
US v. Pena-Espinosa. 1995. 9th Cir. 47 F. 3d 356.
US v. Silva-Arzeta. 2010. 10th Cir. 602 F.3d 1208
Wallace v. State. 2020. WL 2731228.
Ye v. US. 2016. WL 860966. Petition for Writ of Certiorari and Amicus Briefs.

- Brief of Interpreting and Translation Professors as Amici Curiae in Support of Petitioner.
- Brief for the Massachusetts Association of Court Interpreters, Inc. as Amicus Curiae in Support of Petitioner.

Statutes

California Evidence Code

§ 751. Oath required of interpreters and translators

(a) An interpreter shall take an oath that he or she will make a true interpretation to the witness in a language that the witness understands and that he or she will make a true interpretation of the witness' answers to questions to counsel, court, or jury, in the English language, with his or her best skill and judgment.

(c) A translator shall take an oath that he or she will make a true translation in the English language of any writing he or she is to decipher or translate.

§ 752. Interpreters for witnesses

(a) When a witness is incapable of understanding the English language or is incapable of expressing himself or herself in the English language so as to be understood directly by counsel, court, and jury, an interpreter whom he or she can understand and who can understand him or her shall be sworn to interpret for him or her.

(b) The record shall identify the interpreter who may be appointed and compensated as provided in Article 2 (commencing with Section 730) of Chapter 3.

§ 753. Translators of writings ["writings" includes recordings]

(a) When the written characters in a writing offered in evidence are incapable of being deciphered or understood directly, a translator who can decipher the characters or understand the language shall be sworn to decipher or translate the writing.

(b) The record shall identify the translator who may be appointed and compensated as provided in Article 2 (commencing with Section 730) of Chapter 3.

Source: Mendez, Miguel A. 2023. "Evidence-A Concise Comparison of the Federal Rules with the California Evidence Code (West 2014)."

California Rules of Court, Rule 3.1110(g): "Exhibits written in a foreign language must be accompanied by an English translation, certified under oath by a qualified interpreter."

References

Angelelli, Claudia V. 2015. "Justice for all? Issues Faced by Linguistic Minorities and Border Patrol Agents during Interpreted Arraignment Interviews." (U. Alicante MonTI) *Monografías de Traducción e Interpretación* 7: 181–205.

Artiola i Fortuny, Lidia, and Helene A. Mullaney. 1998. "Assessing Patients Whose Language You do not Know: Can the Absurd be Ethical?" *The Clinical Neuropsychologist* 12, no.1: 113–126.

Bandes, Susan A., and Neal Feigenson. 2020. "Virtual Trials: Necessity, Invention, and the Evolution of the Courtroom." *Buffalo Law Review* 68, no. 5: 1275–1352.

Bang, Brandon L., Duane Stanton, Craig Hemmens, and Mary K. Stohr. 2018. "Police Recording of Custodial Interrogations: A State-by-state Legal Inquiry." *International Journal of Police Science and Management* 20, no. 1: 3–18.

Benmaman, Virginia. 2000. "Interpreter Issues on Appeal." *Proteus* 9, no. 4: 1–9.

Benmaman, Virginia, and Isabel Framer. 2015. "Foreign Language Interpreters and the Judicial System." In *Cultural Issues in Criminal Defense*, 4th ed., 101–188. Edited by Ama N. Appiah. New York: Juris. Publishing Company.

Benninger, Taylor, Courtney Colwell, Debbie Mukamal, and Leah Plachinski, Stanford Law School Criminal Justice Center. 2021. *Virtual Justice? A National Study Analyzing the Transition to Remote Criminal Court.* Retrieved from https://law.stanford.edu/wp-content/uploads/2021/08/Virtual-Justice-Final-Aug-2021.pdf.

Berk-Seligson, Susan. 2009. *Coerced Confessions: The Discourse of Bilingual Police Interrogations.* New York: Mouton de Gruyter.

Berk-Seligson, Susan. 2000. "Interpreting for the Police: Issues in Pre-trial Phases of the Judicial Process." *International Journal of Speech, Language and the Law* 7, no. 2: 212–237.

Bolitho, A. 2019. "The Hearsay and Confrontation Clause Problems Caused by Admitting what a Non-Testifying Interpreter Said the Criminal Defendant Said." *New Mexico Law Review* 49: 193–235.

Boyette, Chris, and Maria Santana. 2014, February 20. "A Woman's Plea in her Native Language goes Untranslated; Three Lives are Lost." Retrieved from https://www.cnn.com/2014/02/19/us/new-york-domestic-killing-warnings/index.html.

Bradley, Liz, and Hillary Farber. 2021. "Virtually Incredible: Rethinking Deference to Demeanor When Assessing Credibility in Asylum Cases Conducted by Video Teleconference." *Georgetown Immigration Law Journal* 36, no. 16: 515–570.

Braun, Sabine. 2019. "Technology and Interpreting." In *The Routledge Handbook of Translation and Technology*, 271–288. Edited by M. O'Hagan. London: Routledge.

Braun, Sabine. 2011. "Recommendations for The Use of Video-mediated Interpreting in Criminal Proceedings." In *Videoconference and Remote Interpreting in Criminal Proceedings*, 265–287. Edited by Sabine Braun and J. L. Taylor. Guildford: University of Surrey.

Bribriesco, Andrew W. 2010. "Latino/a Plaintiffs and the Intersection of Stereotypes, Unconscious Bias, Race-Neutral Policies, and Personal Injury." *Journal of Gender, Race and Justice* 13, no 2: 373–412.

California Judicial Council. 2021. *Language Access Metrics Report.* Retrieved from https://www.courts.ca.gov/documents/LAP-Metrics-Report-2021-Fall.pdf.

California Rural Legal Assistance (CRLA). 2022. *Language Access and Law Enforcement in Kern County, California.* Retrieved from https://www.pewtrusts.org/-/media/assets/external-sites/health-impact-project/2022/06/language-interpretation-services-hia-report.pdf.

Cantone, Jason A., Leslie N. Martinez, Cynthia Willis-Esqueda, and Taija Miller. 2019. "Sounding Guilty: How Accent Bias Affects Juror Judgments of Culpability." *Journal of Ethnicity in Criminal Justice* 17 no. 3: 228–253.

Chang, Annabel R. 2008. "Lost in Interpretation: The Problem of Plea Bargains and Court Interpretation for Non-English-speaking Defendants." *Washington University Law Review* 86: 445–480.

CIELO, UCLA AISC, UCLA Promise Institute for Human Rights, UCLA Bunche Center. 2022, January 31. *We are Here: Indigenous Diaspora in Los Angeles.* Retrieved from https://mycielo.org.

Clair, Matthew. 2020. *Privilege and Punishment: How Race and Class Matter in Criminal Court.* Princeton: Princeton University Press.

Cobian, Jessica, Heidy Melchor, and Sebastian Cazares. 2024, May 6. "Opinion: How LA can stop Excluding Latin American Indigenous Language Speakers." *Los Angeles Times.* Retrieved from https://www.latimes.com/opinion/story/2024-05-06/los-angeles-latino-indigenous-languages-population.

Davis, Wendy M. 2016, March. "Language Barriers: Justice Moves Slowly for Those Who Need Interpreters." *American Bar Association Journal* 102, no. 3: 56–63.

Edelko, Bob. 2020, February 15. "Federal case against Lodi terror Suspect Hamid Hayat Comes to an End." *San Francisco Chronicle* Retrieved from https://www.sfchronicle.com/news/article/Federal-case-against-Lodi-terror-suspect-Hamid-15057833.php.

Ewing, Bradley T., Angel L. Reyes III, and James C. Wetherbe. 2008. "Estimating the Effect of Non-English Speaking Hispanic on Personal Injury Jury Trial Outcomes." Texas Tech University, Rawls College of Business, ISQS Working Paper.

Filipović, Luna. 2022. "The Tale of Two Countries: Police interpreting in the UK vs in the US." *Interpreting* 24, no. 2: 254–278.

Filipović, Luna, and Suzanne Abad Vergara. 2018. "Juggling Investigation and Interpretation: The Problematic Dual Role of Police Officer-interpreter." *Language and Law: Linguagem e Direito* 5, no. 1: 62–79.

Fishman, Clifford S. 2006. "Recordings, Transcripts, and Translations as Evidence." *Washington Law Review* 81, no. 3: 473–523.

Fraser, Helen. 2022. "A Framework for Deciding How to Create and Evaluate Transcripts for Forensic and Other Purposes." *Frontiers in Communication* 7: 898410.

Gentry, Blake. 2020. "O'odham Niok? In Indigenous Languages, US 'Jurisprudence' Means Nothing." *Chicanx-Latinx Law Review* 37, no. 1: 29–64.

González, Roseann Dueñas, Victoria F. Vásquez, and Holly Mikkelson. 2012. *Fundamentals of Court Interpretation. Theory, Policy, and Practice*. 2nd ed. Durham, NC: Carolina Academic Press.

Gottlieb, Aaron. 2023. "Making Gideon Count? Public Defender Resources and Felony Case Outcomes for Black, White, and Latinx Individuals." *Race and Justice* 13, no. 4: 506–529.

Greenlee, Mel. 2021. "Capitally Speaking: Language and Bias in Capital Trials." *The Routledge Handbook of Forensic Linguistics, Second Edition*, 228-244. Edited by: Malcolm Coulthard, Alison May, and Rui Sousa-Silva. London: Routledge.

Hale, Sandra, Jane Goodman-Delahunty, Natalie Martschuk, and Julie Lim. 2022. "Does Interpreter Location Make a Difference? A Study of Remote vs. Face-to-face Interpreting in Simulated Police Interviews." *Interpreting* 24, no. 2: 221–253.

Hale, Sandra, Natalie Martschuk, Jane Goodman-Delahunty and Julie Lim. 2024, April. "Juror Perceptions in Bilingual Interpreted Trials." *Perspectives: Studies in Translation Theory and Practice*, 1–24.

Killman, Jeffrey. 2020. "Translation in the Shadows of Interpreting in US Court Systems: Standards, Guidelines and Practice." In *Institutional Translation and Interpreting,* 62–83. Edited by Fernando Prieto Ramos. London: Routledge.

Kostroff, Mallory. 2023. "Behind the Screen: Examining the Human Consequences and Constitutional Ramifications of the Virtual Criminal Defendant." *Washington and Lee Journal of Civil Rights and Social Justice* 29, no. 3: 245–289.

Kracum, John. 2014. "The Validity of *United States v. Nazemian* following *Crawford* and Its Progeny: Do Criminal Defendants Have the Right to Face Their Interpreters at Trial." *Journal of Criminal Law and Criminology* 104, no. 2: 431–456.

Larsen, Maureen. 2020. "Why Defendants Should Have the Opportunity to Challenge the Accuracy of Interpreted Statements by Cross-Examination." *Criminal Law Bulletin* 56, no. 2: 59.

Lehwalder, Ashton. 2021. "How Indigenous-Language Court Interpreters and Clients Navigate the U.S. Court System Under Strict Court-Interpreting Guidelines." Honors Thesis, Portland State University.

Medina, Jennifer. 2019, March 22. "Anyone Speak K'iche' or Mam? Immigration Courts Overwhelmed by Indigenous Languages." *New York Times*. Retrieved from https://www.nytimes.com/2019/03/19/us/translators-border-wall-immigration.html.

Migration Policy Institute. 2024. *Frequently Asked Questions on Legal Requirements to Provide Language Access Services*. Retrieved from https://www.migrationpolicy.org/programs/language%C2%A0access-translation-and-interpretation-policies-and-practices/.

Mikkelson, Holly. 1999. "Verbatim Interpretation: An Oxymoron." Unpublished paper: 1–11.

Miller, M. A., L.W. Davis, A. Prestidge, and W.G. Eggington. 2011. "Finding Justice in Translation: American jurisprudence Affecting Due Process for People with Limited English Proficiency Together with Practical Suggestions." *Harvard Latino Law Review* 14: 117–153.

Moya, Sarah. 2021. "Language Barriers and Cultural Incompetency in the Criminal Legal System: The Prejudicial Impacts on LEP Criminal Defendants." *Fordham Urban Law Journal* 49: 401–433.

Nasser, Sharifa. 2021, April 6. "No Charges in Death of Ejaz Choudry, Shot and Killed by Police While in Crisis." *CBC News* (Toronto, Canada).

National Association of Judiciary Interpreters and Translators (NAJIT). 2019. "General Guidelines and Minimum Requirements for Transcript Translation in Legal Settings." *NAJIT Position Papers on Issues Affecting Court Interpreters and Translators*. Accessed April 15, 2024. Retrieved from https://najit.org/position- papers/.

National Center for State Courts (NCSC). 2018. *Remote Interpreting Guide for Courts, Court Staff, and Justice Partners*. Retrieved from https://www.ncsc.org/__data/assets/pdf_file/0021/18705/remote_interpreting_-guide.pdf.

National Center for State Courts (NCSC). 2006. *Judges' Guide to Standards for Interpreted Proceedings*. Retrieved from https://nmcenterforlanguageaccess.org/cms/images/JudgesGuideToStandardsForInterpretedProceedings.pdf.

National Registry of Exonerations. 2023. Retrieved from https://www.law.umich.edu/special/exoneration/Pages/about.aspx.

Neuman, Alicia. 2021. "Criminal Law: Incompatible Approaches to Interpreters' Translations: Protecting Defendants' Right to Confront—*State v. Lopez-Ramos*, 929 N.W.2D 414 (Minn. 2019)." *Mitchell Hamline Law Review* 47, no. 2: 751–781.

O'Brien, Barbara, Klara Stephens, Maurice Possley, and Catherine M. Grosso. 2019. "Latinx Defendants, False Convictions, and the Difficult Road to Exoneration." *UCLA Law Review* 66: 1682–1715.

Pace, Nicholas M., Malia N. Brink, Cynthia G. Lee, and Stephen F. Hanlon. 2023. *National Public Defense Workload Study*. Santa Monica, CA: Rand Corporation. Retrieved from https://www.rand.org/pubs/research_reports/RRA2559–1.html.

Pavlenko, Aneta. 2023. "Language Proficiency as a Matter of Law: Judicial Reasoning on Miranda Waivers by Speakers with Limited English Proficiency (LEP)." *International Journal for the Semiotics of Law-Revue Internationale de Sémiotique Juridique*: 1–29.

Pavlenko, Aneta, Elizabeth Hepford, and Scott Jarvis.2019. "An Illusion of Understanding: How Native and Non-native Speakers of English Understand (and Misunderstand) their Miranda Rights." *International Journal of Speech Language and the Law* 26, no. 2: 181–207.

Perea, Juan F. 1992. "Hernandez v. New York: Courts, Prosecutors, and the Fear of Spanish." *Hofstra Law Review* 21, no. 1: 1–61.

Rainof, A. 1991-, July/August. "Where Showing the Finger Points to the Truth." *California Attorneys for Criminal Justice (CACJ) Forum* 18: 50.

Redlich, Allison D., Kirsten Domagalski, Skye A. Woestehoff, Amy Dezember, and Jodi A. Quas. 2022. "Guilty Plea Hearings in Juvenile and Criminal Court." *Law and Human Behavior* 46, no. 5: 337–352.

Refo, Patricia Lee. 2004. "The Vanishing Trial." *Journal of Empirical Legal Studies* 1, no. 3: v–vii.

Rengifo, Andres F., Diba Rouzbahani, and Jennifer Peirce. 2020. "Court Interpreters and the Political Economy of Bail in Three Arraignment Courts." *Law and Policy* 42. no. 3: 236–260.

Richter, Liesa L. 2022. "Lost in Translation: The Best Evidence Rule and Foreign-Language Recordings in Federal Court." *Iowa Law Review* 108: 1839–1883.

Rose, Jasmine B. Gonzáles. 2020. "Color-Blind but Not Color-Deaf: Accent Discrimination in Jury Selection." *New York University Review of Law and Social Change* 44: 309–354.

Ross, Casen B. 2014. "Clogged conduits: A Defendant's Right to Confront his Translated Statements." *The University of Chicago Law Review* 81, no.4: 1931–1989.

Santaniello, Lisa. 2018. "If an Interpreter Mistranslates in a Courtroom and There is no Recording, Does Anyone Care?: The Case for Protecting LEP Defendants' Constitutional Rights." *Northwestern Journal of Law and Social Policy* 14, no. 1: 91–124.

Tamura, Tomoko. 2019a. "Requiring the Police Interpreter's In-court Testimony: The Best Way to Ensure Accuracy." *Invitation to Interpreting and Translation Studies* 21: 1–21.

Tamura, Tomoko. 2019b. "Qualification and Translation Accuracy Issues of Law-Enforcement Officers Serving as Interpreters in the US: Quantitative Analysis of Court-Ruling-Derived Data." *Interpreting and Translation Studies: The Journal of the Japan Association for Interpreting and Translation Studies* 19: 23–44.

Triano-López, Manuel. 2015. "The Pre-trial Stages of Arrest and Police Questioning: Implications for Interpreters and Translators in the United States." *Procedia-Social and Behavioral Sciences* 212: 256–260.

Turkewitz, Julie. 2014, May 11. "Language Barriers Continue to Thwart Victims of Crime." *New York Times*. Retrieved from https://www.legalservicesnyc.org/news/language-barrier-continues-to-thwart-victims-of-crimes/.

Turner, Jenia. 2021a. "Remote Criminal Justice." *Texas Tech Law Review* 53: 197–271.

Turner, Jenia. 2021b. "Virtual Guilty Pleas." *University of Pennsylvania Journal of Constitutional Law* 24: 211–275.

Turner, Will. 2013. "¿Qué Dijo? The Plain Error Rule's Effective Denial of Due Process to Non-English-Speaking Criminal Defendants." *Alabama Criminal Civil Rights and Civil Liberties Law Review* 3: 141–152.

Valentine, A., and I. I. Lee. 2015. "My Ears Hear More than English: Granting Multilingual Jurors Accommodations and Treating Multilingualism as a Common Type of Juror Expertise." *Boston College Law Review* 56: 1249–1285.

SECTION III

Researching, Teaching, Publishing, and Providing for the Next Generation

22

"TOO CLOSE TO HOME"

A Dialogue on the Ethics of Forensic Linguists' Proximity to Their Data

Grace Sullivan Buker

Introduction

Our task as forensic linguists is to examine data as it relates to language, crime, and the law. We are social scientists. Therefore, we must execute objective, unbiased, empirical research that will aid in advancing the field and providing justice for those involved. This chapter explores the balancing act faced by forensic linguists in the professional quest for impartiality, objectivity, and justice in the world of forensic linguistic data, while also acknowledging the impact that this work may have on our own personal lives.

Many of us analyze language data from crimes so heinous that we must take extraordinary measures to protect not only the victims or survivors but also ourselves. Some examples of potentially disturbing topics that forensic linguists may encounter include suicide (see Nick 2014; Prokofyeva 2013), human trafficking (see Olsson 2016), sexual child abuse (see Grant and Macleod 2020), ransom, kidnapping, death threats, and more. In the face of such difficult subject-matter, it is important to ask whether it is possible, or even beneficial, to conduct the necessary forensic analyses from a detached, unbiased position. It is not unheard of for forensic linguists to find themselves feeling incredibly partial, subjective, and even connected to the cases they are presented, and the fates of the people involved. Although understandable, when such emotions arise, the researcher has become too biased to continue, and the most ethical decision is to decline further involvement.

While the answer to this ethical challenge is obvious in theory, in practice this is not always the case. This chapter explores professional strategies for managing personal connectedness with data and offers concrete suggestions for conducting forensic linguistic research with objectivity and justice in mind. These suggestions are based upon not only a review of relevant scientific literature but also anonymized interviews conducted with senior investigators and practitioners in forensic linguistics and allied disciplines.

The roots of this ethics chapter began over a decade ago, when I was speaking with a friend. She was working with INTERPOL on cases involving missing and abducted children and human trafficking. I remember her saying that she knew her time there was limited, and she would most likely leave before she had children of her own. That interaction deeply affected me and my approach to forensic linguistic research. At that time, I was a doctoral student. I had not been exposed to any

DOI: 10.4324/9781003391074-25

cases or data that impacted me in the way she described. However, as I began to expand my own research and read more cases, I realized there were certain communities of vulnerable populations that I would feel "too close to" and that emotional proximity would turn me into the dreaded "biased scientist". I heeded my friend's advice and decided not to pursue research involving these populations. That decision meant though that I was faced with the equally distasteful prospect of becoming one of those "armchair researchers" who only study countries they have never visited and analyze the languages and cultures of people they see no reason to understand. Faced with this dilemma, I sought advice from respected colleagues who, I found, utilized similar strategies to avoid becoming "too close to" or "too removed from" the forensic data they work with. Before those strategies are presented, the next section will focus on the scientific literature that has been published on the topic of professional detachment.

In the conceptualization of this chapter, I reached out to my friend again and asked her about her experience and why she predicted her departure from INTERPOL once she became a mother. She said the following:

> truth be told, I feel like if I'd had a better working environment, I would have been able to handle the stress of the subject matter much better. However, my workplace was incredibly toxic and unsupportive so I knew I wouldn't receive any help from them.

This quote inspired me to investigate the existing research on coping strategies for dealing with traumatic subject matter, particularly as a researcher. My suspicions were confirmed that there is a lack of institutional support for many professions that are exposed to traumatic content, as described in the quote above, and most practitioners are left to come up with their own strategies to cope with stress. The following section discusses the existing literature and coping strategies for handling workplace trauma, and also poses questions about the ethical nature of the practitioner distancing oneself from the content, perhaps at the expense of their own psychological well-being.

Reviewing Vicarious Trauma and Practitioners' Burnout

In the field of forensic sciences, there has been extensive work on the impact of regularly being exposed to data involving trauma, violence, graphic imagery, and other disturbing content. Much of this work, however, is dedicated to forensic practitioners, such as those who are victim-facing, as in psychiatrists or law enforcement, rather than those behind the scenes (Pirelli, Formon, and Maloney 2020: 455).

In their research on social workers, Newell and MacNeil (2010) write about the emotional and psychological risks of this profession. They introduce the notion of "vicarious trauma", or a secondary trauma response, most commonly researched as a work-related trauma exposure in which a practitioner is exposed to the trauma of others (sometimes repeatedly) and then experiences a secondary trauma response as a result.

During my interviews, I received a response from a forensic linguist who reported a powerful personal experience of vicarious trauma. The situation came about while conducting their own interview with prosecutors about their decision-making. As the linguist reported:

> The most stark example comes from an interview I did with a prosecutor who specialized in investigating internet crimes against children (generally child pornography cases). In her interview, she went into detail about the horrifying evidence she sees daily, how the perpetrators hide in plain sight, and even when they are caught rarely face significant con-

> sequences. As a mother of two young children, this was particularly difficult to hear. I felt incredibly distracted the next few days, and even though this was now months ago, I still think about that interview at random times.

This response echoed much of what my INTERPOL friend had said. It also powerfully illustrates the fact that forensic linguistic researchers may experience vicarious trauma, second or even third hand. The respondent also emphasized something that I found to be true early on in my own forensic linguistics research:

> It's not going to get easier. There's a lot about Forensic Linguistics that's hard, it's not just the data. It could be the deposition or testimony when providing expert testimony.

The ability of professionals to deal with this occupation hazard varies greatly. Without intervention, it is not unusual for people who work in high trauma environments to develop what researchers call "compassion fatigue", or an experience of physical or emotional depletion that is associated with caring for individuals in significant physical or emotional pain (Newell and MacNeil 2010). The literature on vicarious trauma and compassion fatigue is largely geared toward social workers, and the recommendations within the literature focus on institutionalizing preventative measures during training. Similarly, attention is paid to the inclusion of self-care on a macro (systemic) and micro (personal) level to combat compassion fatigue and vicarious trauma.

These concepts are different than "burnout" which is associated with a sense of being chronically overworked and/or overwhelmed from the quality and quantity of work. While burnout has been documented across a wide variety of occupations, vicarious trauma and compassion fatigue are primarily seen in fields where professionals are exposed to serial, intense, and prolonged trauma. Such is the case in the forensic sciences, including forensic linguistics.

To date, there has been extensive research on the impact of regularly being exposed to data involving trauma, violence, graphic imagery, and other disturbing content (Elstein and Kredens 2020; Pirelli, Formon, and Maloney 2020; Clarke and Kredens 2018; Nick 2014; among others). Much of this work, however, has focused on "victim-facing" forensic practitioners, such as psychiatrists and police officers who interact directly with traumatized populations. Research has also been conducted to investigate the experiences of professionals who work behind the scenes (Pirelli, Formon, and Maloney 2020). The authors write: "Those working in the forensic subfields of psychology, psychiatry, social work, and nursing may be even more susceptible to developing problems associated with [vicarious trauma, compassion fatigue, and burnout] given the intensive nature of their work" (2). As mentioned above, much of the existing literature focuses on front-line practitioners—law enforcement, military, health care professional—therefore, extrapolation from these professional areas was helpful in Pirelli, Formon, and Maloney (2020) work on forensic professionals behind the scenes, as they echo that empirical literature on this matter is "essentially nonexistent in forensic psychology, specifically" (2).

One could assume, and the testimonial from my friend at INTERPOL illustrates, that mental load of working with traumatic populations naturally extends to researchers, as well, although there is a massive gap in the literature about this phenomenon. Elstein and Kredens' (2020) work fills the dearth of literature in this field as it relates to researchers and examines occupational stress in forensic linguistic practice, including many coping mechanisms that forensic linguistic researchers undertake when working on traumatic data. In their survey, they found that researchers claim to rarely be affected by secondary trauma, to which the authors credit the many coping strategies undertaken by them: desensitization, talking to others, mentally preparing themselves for the data, seeing their work's impact to a larger societal need, and putting a distance between themselves and the data. The respondents surveyed discuss employing many coping strategies to protect themselves against secondary trauma on a more micro level, such as talking to others, mentally preparing themselves for the data, and focusing on the potential of their work to address larger societal needs. However, none of the informants quoted in their survey discussed macro level, i.e., an institutionalized support system, wherein they were trained on the possibilities of vicarious trauma, compassion fatigue, or burnout (as recommended by Pirelli, Formon, and Maloney 2020; Newell and MacNeil 2010); or provided systematized counseling or group discussions to discuss effective coping strategies.

The Ethics of Detaching from Stress

This returns us to the larger question: is this an ethical double-bind (à la Tannen 1994) wherein you acknowledge the potential for experiencing vicarious trauma and a potential for being too subjective in your analysis, so you consciously pursue strategies toward distancing yourself from the data? This also begs the question: is detachment even possible? Elstein and Kredens (2020: 58) hint to detachment as a strategy used by many of the forensic linguistic researchers they interviewed, but the authors also indicate the following: "in a nutshell, [detachment] can be effective in the short-term, but maladaptive if relied on consistently" and then cite work by Snyder and Mann Pulvers (2001).

Upon further review of Snyder and Mann Pulvers' (2001) work on coping with stress, they discuss several parameters of how stress can be "weighted" or measured, thus how they impact an individual experiencing said stressor. For example, they acknowledge that individuals do not respond to the same stressor in the exact same way; nor are some stressors "bigger" or "smaller" than other stressors. Rather, these stressors are related to how much they impact the individual. The authors give an example of an undergraduate student trying to get into a competitive graduate program receiving a C grade in a class as compared to a retired person auditing the class for fun receiving the same grade. This stressor has more impact in the long run for the student, not just for their future prospects, but potentially their identity as being a "good student".

Snyder and Mann Pulvers (2001) include a diagram of an avoidance/approach continuum when dealing with a stressor, where an individual may selectively detach themselves from the stressor (avoid) or act on the stressor (approach). The authors state that regularly, when a person pursues the avoidance path to a stressor, they may also feel that they are lacking in productive resources to cope with said stressor. They write that this person may then feel that they themselves are lacking in the characteristic to effectively deal with a stressor, and thus the cycle of avoidance and inability to regulate one's own emotions. On the approach side of the continuum, the authors examine strategies that are in direct opposition to the avoidance strategy. Therefore, instead of ignoring the stressor and detaching from the stressor, the person *leans in* to the stressor with an understanding of "I have been here before and handled it OK" (15) and problem solves the stressor with their agency in mind.

Don't Press the Button

One section in Snyder and Mann Pulvers' (2001) work illustrates an issue that appears to be counter to the ethos of empirical research: one must detach themselves from their data to be an objective researcher. The authors write about a seminal work on stress by Wegner et al. (1987) where participants are instructed "to not think about a white bear". However, the participants thought even more about this white bear; they became consumed with images of the white bear. The conclusion, therefore, was "attempted avoidance or suppression of thoughts about a given stimulus results in even greater cognitive attention to that stimulus" (Snyder and Mann Pulvers 2001: 12). The authors call this "paradoxical hyperfixation" and write that contrary to the belief that avoidance and detachment from the stressor is a useful self-regulator tool, this often backfires and makes the person hyperfixate on the stressor. As seen in much of the survey-based research on participants' own micro-strategies for handling stress, many individuals discuss utilizing these *don't press the button* strategies for detachment—however, Snyder and Mann Pulvers' (2001) work, as well as Wegner et al.'s (1987), suggest this may be a maladaptive strategy and backfire.

Coping Mechanisms for Being "Too Close" to Your Data

Just Say No

As mentioned above, when asked to take on an emotionally charged case that risks compromising the researcher's ability to work objectively, one ethical option is simply to decline. As one interviewee aptly stated:

> You have a few options: you can say no, you don't want the case, or you can write your report saying what the linguistics says and if I don't tell you what the linguistics says about this case, I'd be doing a disservice to justice.

However, in everyday practice, it is not always easy to turn down an offer to take on a case. For example, a common concern I once experienced myself and now often hear my students mention is the worry that they feel compelled to take any opportunity that comes. The reason for this feeling is simple: they worry that if they pass on one opportunity because they know the content is disturbing or triggering, then they will not be asked to participate again in the future. The reminder that we are under no obligation to pursue every case we are offered may be true, but it also reflects a certain type of privilege and job security not felt by all practitioners.

Many emerging practitioners may not feel empowered to turn down professional opportunities, which suggests the need for a strong support system of advisers or mentors—some who have already achieved the ability to say no professionally—who could be instrumental in bolstering the practitioners' confidence. This chapter will reiterate throughout the text the need for more institutionalized support systems not only to cope with the material as a forensic linguistics practitioner but also to aid in professional development and protection against retaliation.

Give Yourself Time to Process

When you do decide to take on a case that may be emotionally challenging, it is important to take all the time you need to truly process the information. As the interviewees below explain, this step is crucial to protecting your well-being.

> I have to remind myself that these stories are important for people to hear. So, I have to find a way, for my own mental health, to disengage from the work so that I can keep doing it. Allowing myself first to really sit and feel a horrible case, and then talking about it with my husband or a friend has been one strategy that helps me let it go.

> I also made sure from a self-protection point of view that I gave myself time to recognize and respect the nastiness that I was often dealing with. I never wanted to become fully immune to it.

As these seasoned forensic linguists stress, when dealing with confronting material, it is critical that we give ourselves time to acknowledge its effect on us and to resist the urge to avoid addressing this dynamic. Even if psychological disengagement can be temporarily effective, ultimately you can never "unsee" or "unhear" some details. By denying the fact that our data may trigger us, we reduce our ability to identify the ways our work may be affected and thereby compromise our effectivity. Thus, we have an ethical responsibility not to become disassociated from our work. Thus, while it is important not to become overwhelmed by the difficult details of our data, complete dissociation is also undesirable. As one respondent explained: "I never wanted to detach completely and act as if the details weren't there…[if I did that] then I would not be working on the case". A very similar point was made by another one of the senior forensic linguists who served as an interviewee for this chapter:

> If a case relates to specific abuse then we can't completely detach from it—it's likely the reason we're looking at the data. And also we are humans, to deny that data might affect us would be dangerous (at an individual level, and a professional). In order to truly make sure that you and your analysis aren't being adversely impacted by the data, you have to recognize the severity of it. It becomes somewhat cyclical, and you end up having to take a holistic approach.

This response beautifully articulated my own internal debate about how the data impacts us and whether that may in fact be a good thing! I appreciated this respondent's unapologetic approach to this type of social science analysis, since we as forensic linguists do have a unique task in dealing with the "nasty" cases and we have a unique skill set to try and help aid in the delivery of justice.

Talk About It…Safely

I conducted an in-person interview with a federal law enforcement officer who is trained in high threat response tactics and emergency medical services (EMS). This respondent shared their EMS training textbook *Emergency Care and Transportation of the Sick and Injured* (Gulli, Ciatolla, and Barnes 2011). This reference provides detailed information about Critical Incident Stress Management (CISM) which "was developed to address acute stress situations and potentially decrease the likelihood that PTSD will develop after such an incident" (46). The CISM debriefing sessions occur within 24 to 72 hours after the acute stressful incident, and may have to be repeated at a later time. These programs are offered throughout the United States and CISM teams can be requested through the EMS's employer or called directly by the personnel.

Of course, law enforcement is not alone in providing institutional fora for emotional processing. In interviews conducted with licensed social workers and therapists, I was told that many also have formal strategies for talking through confronting material. One respondent mentioned, for example, that she has a weekly meeting with her supervisor and other therapists on her team. Called a "consultation group", this supervisory session is dedicated to discussing not only how their clients are faring but also how the therapists themselves are handling their casework. Consultation groups are also spaces where therapists can discuss personal matters that may impact the care they provide their clients.

Sadly, in comparison to some of these other professions, there is a wide gulf in the amount of resources, training, and support that forensic linguists are given to take care of our own mental health and occupational stressors. Some of the respondents I interviewed indicated that they sometimes turned to friends and family to help them process difficult material, as seen with the respondent above who shared she needed to sit with the case before sharing it with loved ones.

This raises a question about the ethics of exposing people, particularly those untrained in the matters of traumatic content, à la "offloading" your trauma onto them. If we as ethicists are advocating for our practitioners to talk to someone to release themselves from secondary trauma, we must also advise these practitioners to avoid tertiary-quaternary-etc. trauma infliction. This anecdote reinforces the need for a systemic support system wherein the practitioners' field provides an opportunity for individuals to reach out to trained professionals to "offload" their concerns in a safe and productive way.

Similarly, we can also rely on each other in the profession to compare notes on how our colleagues cope with the stressors we experience in the field. However, it is important to note that when coping strategies are left up to personnel without sufficient oversight or assistance from professionals trained in stress and trauma management, people may resort to potentially maladaptive coping strategies. To avoid that negative spiral, it is imperative that we not shy away from seeking professional help whenever we need it, particularly if it is not provided by our professional network.

Pitfalls of Extreme Distancing

Returning to the notion of proximity to one's data, many of my respondents mention the need to distance oneself from extraneous information.

> As a forensic linguist and someone who works with digital discourse data, I also follow a set of principles when looking at all types of sensitive data, such as the need to keep participants' identities as anonymous as possible, which is one of my primary concerns.
>
> I don't think it's possible to detach or disengage completely. I think when people attempt it, it's called avoidance, and though that can bring some short-term relief, it generally takes a negative toll in the long-term.

This is echoed in Elstein and Kredens (2020: 58), in which two respondents described how they compartmentalized the data they examined in certain cases:

> (Extract 9) I focus very much on the language part and on the linguistic analysis and I try not to read the details of the case more than once.
>
> (Extract 10) When I look at a piece of text it's—you don't really read it. You are looking at the individual features and so you are able, or I am able, to separate myself from what I am actually reading.

I asked my respondents if there has ever been an instance in which they avoid the details of the case or try to distance themselves from the data through compartmentalizing, as described above. One respondent said:

> That's hard as a sociolinguist, we know that language gets meaning from context. It's hard, there's a tricky balance. As much as you like to think 'I'm keeping it in the four corners of this paper and that's it', it's a lot harder than that.

Focus on the Positive Impact of Research

According to Elstein and Kredens (2020), one of the principal internal motivations of forensic linguists is to aid in the delivery of justice, which is a major coping strategy for dealing with stressful data. This point was reflected in commentary made by the interviewees. For example, during my interview, one forensic linguist offered the following insight:

> I have found myself drawn to data that many would consider hard. I think that is because I wanted to help, at least if only through witnessing that these things were happening. I'm aware that this need to bear witness to the unjustness/abuse someone else is suffering comes from my own background. I also felt a responsibility to help while I was able, knowing that I might not always be able to deal with difficult data.

The desire or "draw" to work with data that one finds personally significant was also mentioned by another interviewee who stated: "I encourage my students to work on data that has real-world implications and also a personal connection, or 'mesearch'". In some circles, *mesearch* has been belittled by critics as "academic narcissism" and "diary-writing for the over-educated" (Pickles 2017). In this chapter, however, a neutral definition offered by Visonà (2023) is used: "data that has real-world implications and a personal connection" (*personal communication*). Like my colleague above, I have similarly advised students to seek out research that appealed to them or that they had a (positive!) emotional connection to, because few people enjoy writing (or reading) about subject matters that do not resonate with them on a personal level.

By electing to work with data that has personal significance, forensic linguistic work can be enormously rewarding, even when the data is difficult. The feeling of real-world implications and internal motivation of aiding in justice is sometimes termed "compassion satisfaction" (Stamm 2005). Compassion satisfaction is frequently studied in front-line trauma practitioners. Consider Letson et al. (2020) who studied a total of 885 child advocates for children who experienced abuse. In this study, they found the child welfare workers who had significantly lower compassion satisfaction studies also had high levels of burnout scores. These findings contributed to research around preventative measures toward burnout and the correlation of compassion satisfaction and professional burnout.

These findings are echoed by Clarke and Kredens (2018) who argue that the sense of social justice and duty is an integral component of a forensic linguist. This is evidenced not only by the notion of pursuing "mesearch" to aid in the identity construction of the researcher's own career pursuits but also as a coping mechanism. Seeing the positive aspects of one's research is not the same as making light of the situation.

Gallows humor is prevalent in professions where workers frequently have to deal with traumatic situations, such as death, injuries, victims, and working with vulnerable populations, including police officers, emergency medical services, military, funeral workers, and more (see Christopher 2015; Maier, Bernstein, and Musholt 1989; Muklay 1989; Haig 1986; Kubie 1971). The research on whether this is an effective coping mechanism is split, however. On the one hand, there are studies that have found that gallows humor may be effective in mitigating stress (as seen with Maier, Bernstein, and Musholt 1989). On the other hand, there is research that warns that humor may be used as a suppression tool for anxiety rather than a coping tool (Snyder and Pulvers 2001). Although more research is needed to determine the efficacy of using this type of humor as a coping mechanism, it is clear that gallows humor can be viewed as inappropriate outside the professional community of practice (Sultanoff 1995). As Moran and Massam (1997) also warn, gallows humor may actually transfer the stress of the situation from the individual experiencing the stress to become a source of stress for others.

Focus on Your Job as a Linguist Pursuing Justice

Although forensic linguists can and should take pride in being a part of a profession that can make important societal contributions, it is ethically imperative that they not overstep their professional boundaries. As one interview stated:

> The role of an expert in the case is to be true to your expertise. As an expert linguist, it's not our job to be a crime fighter or an advocate. Our job is to say, alright I'll look at the evidence and say, ok what can I say about the language.

I received a similar response during an online interview I conducted with a forensic linguistics professor, researcher, and practitioner. During our conversation, I asked about researchers' proximity to their data and whether it is advisable to pick the data your research consciously to either avoid or pursue data you feel close to. Her response beautifully laid out what our "purpose" as forensic linguists is:

> We are the people that are supposed to be neutral, the law enforcement officers' job is to get the bad guys, a social worker's job is to be an advocate.
>
> [She continued] There may be a case you work on where you worry 'what if the defendant actually did the crime?' You have a few options: you can say no, you don't want the case, or you can write your report saying what the linguistics says and if I don't tell you what the linguistics says about this case, I'd be doing a *disservice to justice*.

This is reminiscent of the findings seen with Clarke and Kredens' (2018: 11) work on identity construction of forensic linguists, who write, "Even an informant who construes their identity in matter-of-fact, detached terms does evoke the justice dimension":

> (Excerpt 20) I consider myself to be a service provider erm yes basically that's the thing I am

> (Excerpt 21) first of all I'm a service provider to law enforcement er to justice er and my job is to assist their work erm by doing forensic analysis with the help of linguistics

This connects the ideas of identity construction and compassion satisfaction with that of pursuit of justice.

Overall, the conversations I had with forensic linguistic practitioners had commonalities with the findings of Elstein and Kredens (2020) who found a throughline of coping strategies in their respondents: detachment, compassion satisfaction, pursuit of justice, and talking to others. However, my conversations do not appear as aligned with Elstein and Kredens (2020: 65) who write, "Our findings indicate that the forensic linguists in our sample are generally not affected by the graphic content involved in their work but acknowledge the potentially traumatic nature of it". The graphic content of many cases seems to haunt some of the practitioners I spoke with, in addition to the occupational pressure of researching, teaching, and providing expert analysis. My findings support Elstein and Kredens (2020) who found that there is a wide gulf in the amount of resources, training, and support that *forensic linguists are given to take care of our own mental health and occupational stressors*. We have had to look to each other and compare notes on how our colleagues cope with the stressors of the profession.

A Call for Institutional Support

Inspired by these findings, I reached out to the victim-facing practitioners who regularly work with the victims of trauma to discuss their approaches to coping with vicarious trauma. The respondents discussed the micro- or self-directed strategies for coping with stress explored above; however, they also mentioned more institutional support.

I conducted an in-person interview with a federal law enforcement officer who is trained in high threat response tactics and emergency medical services. The respondent stated that they have general training on recognizing symptoms of stress in the victims of trauma they are caring for, but also in themselves, e.g., high heart rate, high blood pressure, irritability, difficulty sleeping. The respondent mentioned a training textbook that lays out the strategies for coping with occupational stress as an emergency medical technician (EMT), which includes many of the self-directed micro-strategies discussed so far (e.g., improve quality of sleep, minimize exposure to the stressor, increase exercise). However, employers of EMTs are also instructed to provide *critical incident stress management*, or a process to debrief the first responder after a traumatic event. It is important to note, this support is an acute response to one (or few) instances of trauma, i.e., a mass-casualty event, and the onus is on the employer to provide this for their employees. This is a good start; however, this is not a regularly occurring support system to manage chronic stress and vicarious trauma throughout an individual's career.

An improvement of institutionalized support for first-line practitioners is seen with social workers and therapists. I interviewed two licensed therapists (one via email and another via Zoom) and began our conversations to clarify what they advise their patients, and in turn themselves, regarding coping strategies that actually work—not just what are common coping strategies (e.g., gallows humor). The rationale was analogous to this overarching discussion of *proximity*, and whether one could be too close or too far away from their patients' trauma (see "avoidance and approach" cycle from Snyder and Mann Pulvers 2001).

One respondent said the following regarding detachment:

> I don't think it's possible to detach or disengage completely. I think when people attempt it it's called avoidance, and though that can bring some short-term relief, it generally takes a negative toll in the long-term. [She continued by saying:] There's a lot of awareness and conversation in my field around burnout and self-care, so again, really focusing on acknowledging and identifying when things are feeling heavy and taking a break, taking care of the self, intentionally noticing and letting go.

Both respondents discussed the notion of a consultation group, or a supervisory session, dedicated to discussing not only how their clients are faring but also how the therapists themselves are handling their casework. Consultation groups are a space where therapists can discuss other matters, such as anything personal going on at home or anything else that may impact how the therapist can provide care for their clients.

The second respondent sang the praises of her consultation group as a way for her to talk through, with trained professionals, her hesitation to take on new clients who have experienced traumas that, in her words, are "too close to home", as in sexual assault. Through the consultation group, her colleagues and mentors were able to talk through her misgivings and advise her on the proper way, for her as a practitioner and a human being, to take on the client (or not).

These supervision or consultation groups are a mandatory aspect of their professions and for these two therapists are weekly sessions with their peers and supervisors. I contacted the law enforcement officer respondent to ask if there was anything similar for them, and he reiterated what was seen in their training textbooks, that the onus was on the individual to reach out for help. The law enforcement officer echoed the feeling that seeking professional help was seen as a last-ditch effort and very much stigmatized in the profession as it indicates that you are unable to do your job effectively; furthermore, much of what he experiences in his professional socialization is the maintenance of a certain air of masculinity and having it "together".

These anecdotes return to the overarching question of this chapter that one's own proximity to their data should be considered and evaluated with the practitioner's professional and personal well-being in mind. The existing literature provides a number of recommendations for more micro-strategies for coping with vicarious trauma; however, this last section illustrates the need for institutionalized support for practitioners of all kinds.

Conclusion

At the end of this chapter, I return you to the original ethical dilemma that sparked the conversation taking place here: can you be "too close" to your data? And if so, should you try to detach yourself from it to be a better, more objective scientist? If you're too distant from your data, are you running the risk of becoming an armchair researcher who examines data at a "safe" distance? Let's then say you decide to seek out data that you find a personal connection to that has real-world implications (e.g. "mesearch" [Visonà 2023]), will you then be too invested in the data or analysis for you to conduct scientifically sound, empirical research? The answer to that question depends on whether you take adequate precautions before and interventions during your work. As this chapter has illustrated, the ethical dilemma of whether or not to work with data of personal interest is not unique to the field of forensic linguistics. It is very common in allied fields such as social work, psychology, and law enforcement.

These interviews I conducted illustrated that when coping strategies are left up to the personnel (i.e., forensic linguistic researchers and law enforcement), it appears people resort to potentially maladaptive coping strategies, such as unhealthily detaching from the data and other avoidance strategies. These interviews also made me wonder about the stigma around seeking help since the stress is more than you can handle. This worry is discussed explicitly in the Snyder and Mann Pulvers (2001: 12) text on coping strategies for stress, which states: "the person on the avoidance path perceives that he will be unable to regulate his emotions effectively in the face of the given stressor. Very real concern that he will be controlled by his emotions".

I recommend we take cues from these disciplines. We need to develop a culture where emerging scholars and storied forensic linguists alike are free to discuss issues like vicarious trauma and professional burnout, and share our experiences with coping mechanisms. By expanding the agency for creating these spaces to move beyond private individuals to include institutions such as universities or forensic linguistic organizations, we can improve the communication around how this work impacts our lives and focus on addressing the stressors in a healthier way. Furthermore, a more codified training early on in forensic linguistic training would be beneficial as well, in addition to continued institutional discussions on ethics in forensic linguistic research (LSA 2020; Shuy 2009; Stygall 2009; Finegan 1993; Hamilton 1993). Similar to how many allied professions have already acknowledged the job-related emotional tolls on personnel, we can learn from our colleagues and begin to teach effective coping strategies that many of us have had to figure out through professional socialization with colleagues or through trial and error. In sum, more institutional support is needed to assist forensic linguists deal with the inevitable connections they will form with their data and their subsequent need for dialogue regarding this connection. This all reminds me of a quote by American poet Robert Frost, who wrote “the best way out is always through”. Upon reflecting on this quote, author John Green adds, “and the only way through is together”.

References

Alexander, David A., and Susan F. Atcheson. 1998. “Psychiatric Aspects of Trauma Care: Survey of Nurses and Doctors.” *Psychiatric Bulletin* 22, no. 3: 132–136.

Alexander, David A., and Andrew Wells. 1991. “Reactions of Police Officers to Body-Handling after a Major Disaster: A Before-and-After Comparison.” *The British Journal of Psychiatry* 159, no. 4: 547–555.

Christopher, Sarah. 2015. “An Introduction to Black Humour as a Coping Mechanism for Student Paramedics.” *Journal of Paramedic Practice* 7, no. 12: 610–617.

Clarke, Isobelle, and Krzysztof J. Kredens. 2018. “I Consider Myself to Be a Service Provider: Discursive Identity Construction of the Forensic Linguistic Expert.” *International Journal of Speech, Language and the Law* 25, no. 1: 79–107.

Elstein, Solly, and Krzysztof Kredens. 2020. “Occupational Stress in Forensic Linguistic Practice.” *Journal of Applied Linguistics & Professional Practice* 17, no. 1: 50–72.

Finegan, Edward. 1993. “Ethical Considerations for Expert Witnesses in Forensic Linguistics.” *Issues in Applied Linguistics* 4, no. 2: 179–187.

Grant, Tim, and Nicci Macleod. 2020. *Language and Online Identities: The Undercover Policing of Internet Sexual Crime.* London: Cambridge University Press.

Gulli, Benjamin, Joseph A. Ciatolla, and Leaugeay Barnes, eds. 2011. *Emergency Care and Transportation of the Sick and Injured.* Jones & Bartlett Learning.

Haig, Robin A. 1986. “Therapeutic Uses of Humor.” *American Journal of Psychotherapy* 40, no. 4: 543–553.

Hamilton, Heidi, E. 1993. “Ethical Issues for Applying Linguistics to Clinical Contexts: The Case of Speech-Language Pathology.” *Issues in Applied Linguistics* 4, no. 2: 207–223.

Kubie, Lawrence S. 1971. “The Destructive Potential of Humor in Psychotherapy.” *American Journal of Psychiatry* 127, no. 7: 861–866.

Letson, Megan M., et al. 2020. “Identifying Compassion Satisfaction, burnout, and Traumatic Stress in Children’s Advocacy Centers.” *Child Abuse & Neglect* 110, no. 3: 104–240.

Lefcourt, Herbert M. 2001. “The Humor Solution.” In *Coping with Stress: Effective People and Processes*, 68–92. Edited by C. R. Snyder. New York, NY: Oxford University Press.

LSA. 2020, August 13. “Code of Ethics for Linguistics in Forensic Linguistic Consulting.” Retrieved from https://www.linguisticsociety.org/sites/default/files/code-of-forensic-consulting.pdf.

Maier, Gary J., Michael J. Bernstein, and Edmund A. Musholt. 1989. “Personal Coping Mechanisms for Prison Clinicians: Toward Transformation.” *Journal of Prison & Jail Health* 8, no. 1: 29–39.

Moran, Carmen, and Margaret Massam. 1997. “An Evaluation of Humour in Emergency Work.” *The Australasian Journal of Disaster and Trauma Studies* 3, no. 3: 1–11.

Mulkay, Michael. 1989 *On Humour, Its Nature and Place in Modern Society.* Polity Press.

Newell, Jason M., and Gordon A. MacNeil. 2010. “Professional Burnout, Vicarious Trauma, Secondary Traumatic Stress, and Compassion Fatigue.” *Best Practices in Mental Health* 6, no. 2: 57–68.

Nick, Iman. 2014. “First Responder Suicide Notes: What Do They Tell Us?” *2nd Annual Roundtable of the Germanic Society of Forensic Linguistics* [Roundtable presentation]. Mainz, Germany.

Pickles, Matt. 2017. “‘Mesearch’ – When Study Really Is All about Me.” *BBC News*. Retrieved from https://www.bbc.com/news/business-39856894.

Pirelli, Gianni, Dana L. Formon, and Kelsey Maloney. 2020. “Preventing Vicarious Trauma (VT), Compassion Fatigue (CF), and Burnout (BO) in Forensic Mental Health: Forensic Psychology as Exemplar.” *Professional Psychology: Research and Practice* 51, no. 5: 454–466.

Prokofyeva, Tatiana. 2013. *Language Use in Two Types of Suicide Texts.* Dissertation from Linköping University, Department of Culture and Communication, Language and Culture. Retrieved from http://urn.kb.se/resolve?urn=urn:nbn:se:liu:diva-102099.

Shuy, Roger. 2009. “Ethical Questions in Forensic Linguistics: Introduction to Papers from a Linguistic Society of America Panel Presentation, San Francisco, California, 9 January 2009.” *International Journal of Speech, Language, and the Law* 16, no. 2: 219–226.

Snyder, C. R., and Kimberley Mann Pulvers. 2001. "Dr. Seuss, the Coping Machine, and 'Oh, the Places You'll Go'." In *Coping with Stress: Effective People and Processes*, 3–29. Edited by C. R. Snyder. New York, NY: Oxford University Press.

Stamm, B. Hudnall. 2005. *The ProQOL Manual.* Idaho: Idaho State University.

Stygall, Gail. 2009. "Guiding Principles: Forensic Linguistics and Codes of Ethics in Other Fields and Professions." *International Journal of Speech, Language, and the Law* 16, no. 2: 253–248.

Sultanoff, Steven. 1995. "Levity Defies Gravity: Using Humor in Crisis Situations." *Therapeutic Humor* 9, no. 3: 1–2.

Tannen, Deborah. 1994. *Talking from 9 to 5: Women and Men at Work.* New York, NY: Harper Collins.

Visonà, Mark. 2023. Personal Correspondence.

Wegner, Daniel M., David J. Schneider, Samuel R. Carter, and Teri L. White. 1987. "Paradoxical Effects of Thought Suppression." *Journal of Personality and Social Psychology* 53, no. 1: 5–13.

23

NAVIGATING LEGALITY, LINGUISTIC RIGHTS, AND SOCIAL JUSTICE IN LINGUISTIC RESEARCH WITH DEAF REFUGEES

Nina Sivunen and Johanna Ennser-Kananen

Introduction

Located at the intersection of forensic linguistics and critical applied linguistics (Pennycook 2001), this chapter focuses on the ethical questions and principles that arise when scholars in applied linguistics work with so-called "vulnerable" populations, such as deaf adults with refugee backgrounds, particularly as legal issues surface.[1]

For this chapter, "legality" is defined as compliance with applicable legal rules and regulations. This operationalization follows the common definition of social justice as "principles of 'fairness' and 'equality' for all people" (Sensoy and DiAngelo 2017: xix). This chapter draws on a conceptualization of linguistic justice that stems from the field of language policy where "legality" is understood to be a realization of all linguistic human rights (Skutnabb-Kangas and Philippson 1994). This chapter investigates how the above three principles can be enacted together, particularly in contexts where participants and researchers may feel accountable to several competing authorities of justice. These standards include national regulations and guidelines; international human rights agreements; disciplinary and institutional expectations; and their own social positions. With a few notable exceptions (e.g., Nick 2018; Bigelow and Pettitt 2015) research in this area is scarce. This is especially true when it comes to work with deaf refugees.

To help fill this significant gap in the literature, this chapter will present original data collected during a two-year ethnographic study with deaf refugees who, at the time, had little formal education and emergent print literacy. The participants' proficiency in (named) sign languages of origin varied considerably, from home signing to master level (Sivunen and Tapio 2020; Sivunen 2019). The chapter also offers data from interviews with four reception center workers who have worked with deaf asylum seekers. The larger study, from which three vignettes were taken for presentation in this chapter, examined the participants' communication and interaction strategies and used a linguistic ethnographic approach (Hou and Kusters 2019; Copland and Creese 2015).

As this data will show, language barriers and deficit views pose a threat not only to the constructed legitimacy of refugees' stories (e.g., Smith-Khan 2017) but also to their ability to understand and claim their "political subjectivity", which Behrman (2014) defines as "ability to assert one's own identity and needs in ways that do not fit the paradigm of nation-state-capital, a construct

 DOI: 10.4324/9781003391074-26

that gains its normative power through law" (21). This threat is especially striking in high-stakes situations where access to vital information (e.g., about asylum processes) is either unavailable or inaccessible to refugees because their linguistic and semiotic practices are not captured by what is commonly understood as "language". Theorizing Sivunen's experience of communicating with research participants via different linguistic repertoires and semiotic resources, this chapter discusses how deaf refugees who communicate in ways not officially recognized, such as home signing, may experience threats to their linguistic legitimacy (Ennser-Kananen 2018, 2014) and political subjectivity. The chapter also addresses how these threats may be minimized. In addition, a detailed discussion is provided about how ethical academic positionalities can unfold in work with marginalized populations. Finally, several recommendations are made for rethinking and reorienting our work as researchers in these contexts.

Data and Context

The case study profiled in this chapter is based on Sivunen's two-year ethnographic study with deaf refugees in Finland. Conducted from 2015 to 2017 (see Sivunen 2021; Sivunen and Tapio 2020; Sivunen 2019), this investigation included Sivunen's fieldnotes and interviews with refugee-background adults and reception center workers. In this chapter, the term "refugee" is used for people who left their home seeking safety and/or opportunities, regardless of their official status. This chapter rejects the highly problematic categorization of refugees by reasons for their escape.[2] In some instances, the term "asylum seeker" is used to indicate that the participants of this study were in the process of applying for asylum. Following a linguistic ethnographic approach (Hou and Kusters 2019; Copland and Creese 2015), the original study examined deaf refugees' communication and interaction strategies in six different reception centers, at a folk high school for the deaf, and in their homes. Sivunen found and recruited participants by sending emails to different reception centers.

In Finland, deaf asylum seekers are placed in different reception centers around the country because there is no one facility specialized in serving deaf clients. In response to the relatively large number (32,477) of refugees who arrived in Finland between 2015 and 2016, many new reception centers were hastily established, for instance by converting old hotels, hospitals, and apartment buildings into temporary homes. Ten deaf asylum seekers (six men and four women) from countries in West Asia participated in the original study. They all reported seeking safety outside of their countries of origin and having interrupted formal education backgrounds and emergent print literacy skills. In addition to these commonalities, they exhibited a wide spectrum of proficiency in the sign languages of their former home countries. While some reported using home signs, others knew a few official signs, and still others had high levels of the sign language of their prior home countries (e.g., Arab Sign) (see Sivunen 2019). Sivunen interviewed each of the research participants up to three times during a two-year study, as the circumstances allowed. The interviews were conducted in either International Sign (IS) or Finnish Sign Language (FinSL).[3] Accompanied by FinSL interpreters, Sivunen also interviewed four reception center workers. Informed consent was obtained by using a combination of gestures, pictures, signing, and printed Finnish texts. Before more details about the research of this fieldwork are shared, the chapter provides some background information about sign languages in the context of the asylum process.

International Sign: An Inclusive Language in Asylum Procedures?

Given its widespread use and importance amongst deaf communities around the world, International Sign has been identified as global deaf lingua franca (Kusters 2021). However, IS does not fulfill some commonly used and highly problematic criteria of "language" (Hansen 2016; Whynot 2016). For example, it has no so-called "native speakers" and is not tied to a national territory (Green 2014: 446). For these reasons, some scholars have argued that IS is a contact language (Kusters 2020; Adam 2012). Currently, no consensus has been reached in the linguistic community about the most appropriate classification of IS. As Kusters (2020) notes, it remains "an ephemeral and variable linguistic phenomenon of which the nature is difficult to pin down" (52).

Even without matching levels of proficiency in IS or another (sign) language, signing deaf people around the world typically have the ability to communicate with each other by using linguistic and semiotic strategies such as *calibrating* (Moriarty and Kusters 2021) and *cross-signing* (Byun 2023; Zeshan 2015). Calibrating and cross-signing among signers at international encounters means that they learn and use features of each other's signed and written languages such as signs, written words, and fingerspelling for building mutual understanding. These translingual practices (Canagarajah 2013) make use of shared resources and features of sign languages, such as lexicon, grammar, and iconic signs, and include mouthing, fingerspelling, pointing, and gesturing (Moriarty and Justers 2021; Zeshan 2015; Hiddinga and Crasborn 2011). Importantly, along with these common multimodal and semiotic resources (Kusters et al. 2017), the shared experience of being deaf (Crasborn and Hiddinga 2015) further facilitates the building of mutual understanding via IS. As Green (2014) argues, "communicating in IS relies on and produces mutual moral orientation among signers" (445). In other words, there may be shared values embedded in and reinforced by the use of IS. Nevertheless, such (assumed) commonalities are also problematic. Like with any other language, there is wide inter-group variation in its use; and these variations relate to underlying sociocultural and political differences. Signers rely on shared resources and strategies while communicating with each other (Kusters 2021) and are thus context-dependent (Sivunen and Tapio 2020; Hiddinga and Crasborn 2011).

In the context of deaf international gatherings, such as at the World Federation of the Deaf Congress, IS has been found to include many signs from American Sign Language (ASL) (Kusters 2021, 2020; Whynot 2016) as well as features of other sign languages commonly used in the Global North (Whynot 2016; Green 2014: 452). Kusters (2020) found that there is a linguistic continuum of the amount of ASL features that surface in IS, which is evaluated differently by different signers. While some are in favor of the use of ASL in IS, there are many others who find this practice counterproductive to communication and community-building.

These findings relate to Whynot's (2016) investigation that demonstrated that expository IS does not provide full access to information for monolingual signers who do not communicate in ASL and English, have interrupted formal education, and have no opportunities to travel to deaf international gatherings. These differences of opinion and experience underscore the fact that the conventionalized form of IS is tied to a certain amount of privilege, mobility, and proficiency in literacies (İlkbaşaran 2015) and signs with iconic motivations. These, in turn, are often related to white, European, so-called "Western" cultural practices which are not always transparent or familiar to those who come from other backgrounds (Kusters 2021: 405; see Sivunen 2019; Whynot 2016). As Moriarty and Kusters (2021) point out, the "moral ideals and the ability to draw on diverse semiotic repertoires are not equally available to all deaf people because of differential access to socio-economic capital, and to language(s)" (287). Thus, at the same time that IS has played a critical role in building connections between members of the deaf community across the globe, these connections have also systematically excluded certain groups of deaf people whose linguistic practices have not been acknowledged or accepted in "Western" countries. This includes Nordic countries.

As will be described in the following sections, the participants in Sivunen's study did not have access to formal education in signed languages, and the fact that they only had a few years of formal education in spoken and written language(s) in their original home countries posed significant linguistic and legal challenges when faced with the widespread use of (conventionalized) IS in the asylum processes of Finland. To illustrate these obstacles, the following sections present three vignettes from Sivunen's fieldnotes and personal reflections. As will be shown, these narratives not only demonstrate the equity issues that arise in the work with deaf refugees. They also point to the ethical imperative for reorientation and reform in legal and academic policies and practices followed by professionals who work with these vulnerable communities.

When Legal Issues Become Personal

As we show, Vignette 1 highlights the mismatch between deaf refugees' life realities and the existing policies and practices for asylum seekers. It also illustrates the conflicting goals of all agents in this situation.

Nabil (pseudonym) is a middle-aged deaf asylum seeker from West Asia. He arrived in Finland in the autumn of 2015. During this year, part of what was later problematically termed the "refugee-crisis", 32,477 refugees arrived in Finland (Maahanmuuttovirasto, henceforth Migri 2023). Nabil was moved to a reception center, an old hotel in the middle of nowhere. The interaction with Nina took place in a big dining room with many tables and chairs surrounding them. There were big windows on one side of the dining room and the view was toward the forest.

VIGNETTE 1: Please Come In—and Interpret

I arrived at the reception center in the middle of nowhere in Finland and brought a sign language interpreter with me so the staff would be able to communicate with me. I met a deaf refugee in the lounge. Nabil [pseudonym] seemed confused. He stared at me and the sign language interpreter. It was obvious that he was not sure who I was and I had no chance to explain and introduce myself as a researcher, because when members of the staff joined us, I unexpectedly became a relay interpreter for Nabil. I could not refuse that request. The social worker wanted to ask him some questions and inform him about some vital issues regarding the asylum procedure and living at the reception center. As I learned later, he came from West Asia and was the only deaf person in that reception center. None of the staff or other clients there could sign and communicate with him. I started to interpret for him in International Sign. It seemed that he did not understand me well and was still confused about who I was. I did not fully understand him either. Nabil used Arab Sign, which was unfamiliar to me. He started to visually express his traumatic experiences of violence in his home country to me. Although this part was easy to understand because of the shared resources and features of sign languages and the use of gestures (e.g., movements of body, facial expressions and pointings), it was hard for me to grasp all of his whole story, translate, respond, and ask more questions. Because we did not share a signed or a spoken language, nor did we have a common alphabet (he was an emergent reader in Arabic but not familiar with the Latin alphabet, and I do not read Arabic), we could not rely on strategies like fingerspelling or writing for communicating names of people and places, for example. After our first encounter, I asked the staff at the reception center to order him a sign language interpreter who knows International Sign for future meetings. (Fieldnotes from October 2015)

Vignette 1 illustrates some common misconceptions non-signing officials in the asylum process have about communication in sign language and deaf people in general. It is generally assumed that there is one sign language that is universal. Operating on this false assumption, the employees at the reception center saw the visit of a researcher (Nina), who was a deaf sign language user and an accompanying IS interpreter, as an opportunity to establish communication with Nabil. They were either not aware of or simply disregarded the fact that the respective sign languages (and also their alphabets and fingerspelling) might not be mutually intelligible. Furthermore, in expecting the researcher to serve as an interpreter without any preparation or negotiation, the staff at the reception center also dismissed her professional background and role in the situation. As a result, the professional identity of the researcher was relegated to the background. Not only did this interfere with the stage of rapport building between study participants and researcher that is crucial in ethnographic research. It also put the researcher in the difficult position of having to decide on the spot which course of action would be most legally appropriate, most conducive to her research, and the most ethical, while simultaneously being expected to translate to authorities in FinSL and process stories of violence and trauma being recounted by the study participant.

It is important here to point out some of the local policies that were at play in this situation. In Finland, it is the legal responsibility of the reception center to provide refugees with relevant information about asylum procedures "in written form in a person's mother tongue or in a language the person can reasonably be expected to understand" (746/2011, §15). The Finnish Act on the Reception of Persons Applying for International Protection and on the Identification of and Assistance to Victims of Trafficking in Human Beings (746/2011, §27) also specifies that "[a]n issue can be interpreted or translated into a language the customer can *reasonably be expected* to understand" (emphasis added). This policy raises the questions of what constitutes a "reasonable expectation" and from whose point of view. From the perspective of a Finnish official, knowing (Finnish) sign language or IS might seem "reasonable". This kind of presumptive thinking is commonplace amongst government officials who consider it "reasonable" to conflate nationality with dominant language competence. For example, a person with Finnish nationality is simply presumed to be Finnish-speaking and someone who is Iranian is assumed to be Farsi-speaking. The fact that such assumptions may be proven correct in some situations does not mean that they are correct in every situation. The problem here is that so-called "reasonable expectations" tend to be informed by hegemonic discourses and ideologies. As Vignette 1 demonstrated, the consequences of these faulty expectations were to the detriment of deaf refugees with emergent print literacy like Nabil. This is not to say that the social workers in the reception center were insensitive to the communicative needs of the asylum seekers. In fact, as Vignette 1 also illustrated, they were acutely aware of the communicative urgency but lacked the proper resources to respond.

In Finland, staff working at reception centers are permitted to book interpreters for their clients for a variety of occasions, including visits to a nurse, a social worker, or a social counselor. However, this does not apply to everyday situations. According to the Finnish Immigration Office, as a general rule, refugees "should make an effort to make do without an interpreter" (Migri 2021). Such policies are based on the premise that "making an effort" is enough to overcome linguistic boundaries—a supposition that is highly problematic even for hearing language users. As a large body of research within applied linguistics has documented, language acquisition or successful communication is not (mainly) a matter of intrinsic motivation (or "effort") but largely depends on the societal opportunities for formal learning and social interaction (Ennser-Kananen 2018; Darvin and Norton 2015; Norton 2013). The expectation that asylum seekers should simply "make an effort" seems almost cynical in the context of deaf refugees, particularly those who have only emergent print literacy. No amount of effort on their part could erase their deafness or make up for the center's blatant unpreparedness for it.

Ironically, with regard to its legal policies, Finland is commonly considered a pioneer country in sign language interpretation for the deaf, hard-of-hearing, deafblind, as well as for the speech-impaired. Unlike many other countries both inside and outside of Europe, in Finland, these services are guaranteed by the Act on Interpretation Services for Persons with Disabilities (133/2010). This legislation means deaf people have the right to a sign language interpreter and the use of interpretation services is free. However, this right only applies to official residents of Finland, i.e., any person who has a place of domicile (Municipality of Residence Act 201/1994) officially listed in the Finnish population registry. Asylum seekers, both deaf and hearing, do not fall into this category and are thus not covered by public social and health services (Ministry of Social Affairs and Health 2022). Since deaf asylum seekers can't book or access interpretation services themselves, they are dependent on the reception centers and the Finnish Immigration Service to do this. It is entirely up to the officials in these institutions to decide when, where, and if interpretation services are needed. This practice would seem to be in conflict with the United Nations' Convention Rights of Persons with Disabilities (CRPD). Specifically Article 21 states the following:

> States Parties shall take all appropriate measures to ensure that persons with disabilities can exercise the right to freedom of expression and opinion, including the freedom to seek, receive and impart information and ideas on an equal basis with others and through all forms of communication of their choice.
>
> *(paragraph 1)*

It could be reasonably argued that the right of refugees to receive information "on an equal basis with others" is ultimately violated by predicating it on their inclusion in the Finnish population registry. A similarly strong case could be made for the argument that the Finnish policy also critically impinges upon their eligibility for necessary public social and health services.

Given the high-stakes nature of asylum processes, the actions taken by the reception center staff to use the researcher to establish a communication channel with their client is understandable. As one reception worker conceded: "Actually we have had no contact with him [Nabil], we only contact him if we absolutely have to" (Interview with Nina, May 2016). As the interviewee went on to explain, there had been attempts made by co-workers to communicate with Nabil through written Arabic. However, these efforts had been unsuccessful:

> We have one employee who uses Arabic and he tried to communicate with Nabil by writing notes and the deaf reads from the notes. We wrote detailed information like times and dates on notes. But it did not work, because his literacy skills are bad. [...] We wished that it had worked, but it didn't.
>
> *(Interview, May 2016)*

Nabil's case and his linguistic isolation reflect severe shortcomings of the Finnish legal, health, and social care systems. By and large, these systems seem to be designed for people who are equipped with the mental, social, and linguistic abilities to claim their rights and advocate for themselves. Such self-advocacy is very difficult if not entirely impossible for people like Nabil. Not only do they lack access to information, they are also unfamiliar with the notion of having a language with a name (Gal 2006). As Sivunen noted in her larger study, the research participants typically referred to their languages generically as simply "signing" or "Arab Sign" (Sivunen

2021, 2019). Many of them had never had the opportunity to learn about the value and importance of (sign) language(s) and resources they could potentially tap into as signers. They were completely unaware of the important role (named) languages play in high-stakes situations, such as asylum hearings (Smith-Khan 2022; Maryns 2014 [2006]).

These study participants are unfortunately not an exception. To this day, only 10% of deaf people around the world receive an education and only 2% of these in sign language (Haualand and Allen 2009). Although the World Federation of the Deaf (WFD) and national deaf associations worldwide have continued to work toward the official recognition, promotion, and protection of sign languages around the world, there is still a long way to go before real equity or justice are achieved (WFD 2016). With its restrictive access to interpretation services, the Finnish system can be said to facilitate the linguistic and social isolation of deaf refugees. This threat may be particularly great for those with interrupted formal schooling and emergent print literacy (Sivunen 2019). Thus, rather than working toward equity and inclusion, these policies can perpetuate a sense of languagelessness and deprive refugees of their legal rights and necessary resources. In addition, these policies put staff at reception centers under terrific pressure to try to compensate for the scarcity of communicative resources. As a consequence of these legal inequities, individual, relational, ethical, and scientific concerns are overshadowed. As Vignette 1 made painfully clear, the presence of a deaf researcher cannot fix these systemic problems, not even momentarily. And, although attempts to draw on any available strategies and resources are understandable against this backdrop, the re-positioning of a researcher into the role that should be filled by a professional interpreter only serves to further marginalize the deaf person in need of assistance.

What Is Legally Legitimate Language?

Rami (pseudonym) is a young adult deaf asylum seeker from West Asia. He also arrived in Finland in the autumn of 2015 and was moved to a reception center. Rami received a negative asylum decision. Around the time of his asylum hearing he acquired some Finnish Sign Language skills, so that the interaction in Vignette 2 took place mainly in FinSL. In the following vignette, Rami shares his frustration with the impact his deafness has in the asylum process.

VIGNETTE 2: No Language, No Asylum?

At the reception center there was another [asylum seeker] who came from the same hometown in X [country of origin] as me. He got asylum fast! I was surprised and asked and wondered why. [...] He is hearing! He can speak. He can speak easily and powerfully. I can't [speak]. I am deaf and I have to repeat my signing many times and it is hard. Now it [signing] is going better. I hope for the best. (Interview 3, 2017)

Rami's experience is sadly not unusual. In many industrialized countries, asylum hearings are an important part of the immigration policy and are used for controlling and deciding on asylum seekers' futures (Gill and Good 2019). Given the centrality of the interview to this decision-making process, access to information in preparation for these hearings is vital (see Smith-Khan 2022). In an interview, a reception center worker expressed concern over another deaf client's case and his communication skills:

> Is his knowledge good enough? Does he know he should tell a lot? Does he understand enough to make it through his story positively? I am really worried about this. I have tried to talk about these issues when the interpreter was here. I tried to talk about these issues and I tried to emphasize those things you [a deaf person] should tell the police and immigration service in hearings. I mean that he should talk about those issues there, because it has significance. Hearing asylum seekers talk with each other about their experiences, about what happened in asylum hearings…what was special there, and many other issues. A deaf person misses out on all of that [information].
>
> *(Interview with reception center worker, May 2016)*

As required by law (746/2011, 15§), all of the study participants received information about asylum procedures and hearings. However, that information was typically offered in a written language and/or through sign language interpreters in language(s) the participants were not familiar with. As the social worker commented above, the refugees' opportunities for receiving information about complex legal processes, asking informed questions, or discussing the procedures with their peers at the reception center are severely limited. In fact, none of the study participants had ever actually used legal sign language interpreters in their countries of origin. Instead their hearing family members and children had acted and continued to act as language brokers. The resulting lack of necessary information lies at the roots of the personal frustration and legal disadvantage that refugees like Rami routinely experience.

It is important to mention here that at his asylum hearing, Rami was indeed provided with sign language interpreters. The fact he was given this service was not a function of his having made a formal request, however. In Finland, the information that police or border guard officials enter into the electronic case management system ("Ulkomaalaisasioiden asiankäsittelyjärjestelmä", or UMA) determines the type of interpreting services refugees receive (Migri 2018: 14). Although officials at the Finnish Immigration Service may have some experience in using remote telecommunication services to enlist the services of (spoken language) interpreters who reside abroad and are proficient in less commonly spoken languages or dialects (Migri 2018: 15), this kind of service is not, to our knowledge, routinely used in the case of deaf applicants. Instead, as the research participants reported, during their hearings, two sign language interpreters who knew FinSL and IS were present. They also mentioned that only one of these interpreters was deaf. Using a deaf sign language interpreter as a relay interpreter is important, because deaf signers are typically very skillful at calibrating and switching the level of communication in order to seek and negotiate mutual understanding (Moriarty and Kusters 2021).

Although such interpretation strategies are an essential part of effective communicative processes, they may not be appreciated or even allowed in asylum hearings. For instance, according to the guidelines of the Finnish Immigration Service (Migri 2020), the interpreter is not allowed to change the form of the question of the authority, stop the applicant, or repeat the question if the applicant doesn't reply directly (Migri 2020: 6; see also UNHCR 2017). This restrictive view of communication exacerbates the linguistic barriers of language learners (or users with low proficiency), deaf people communicating across several sign languages, and people with emergent print literacy who often rely heavily on oral dialogue, repetition, extralinguistic cues, calibrating, and linguistic redundancy. This means that asylum hearings are critically undermined, even when interpretation is available. The negative repercussions were described by a Finnish social worker at one of the reception centers. According to the interviewee, such obstacles can ruin a deaf refugee's chances during the hearing:

> Really, from A to Z, everything has to run smoothly because it all has an effect on the asylum decision. Everything depends on how the hearings go with the police and the immigration service. [...In this case]...the police forgot to order [sign language] interpreters. It is annoying.
>
> *(Interview with Sivunen, May 2016)*

In addition to potential absence of services, proper training and preparation is also frequently lacking. Especially between 2015 and 2016, many Finnish officials who participated in asylum hearings were new and inexperienced, and the available trainings for asylum hearings were short and insufficient (Migri 2018). To make matters worse, seated across from these underprepared migration officials were deaf applicants who not only lacked access to relevant information and exchange but also often struggled to tell their story, let alone present their arguments. As Rami explained above, he had to repeat his signing many times in order to communicate his message, either because he was unsure how to express himself or his message was not understood. Difficulties in offering a convincing "performance" at the asylum hearing because of trauma and language barriers further compound these disadvantages (UNHCR 2017). In his comment, Rami made a connection between being deaf/having limited sign language proficiency in FinSL and IS and receiving a negative decision. There is evidence that buttresses Rami's assertion.

In the larger study that contextualized this paper, Sivunen (2019) found that 70% of the study participants were denied asylum, including two who were returned to another country based on the Dublin Agreement. Although comparison is difficult, larger national statistics contextualize this number: In 2015, 7,466 applicants were granted international protection in Finland: 1,874 persons received positive and 1,225 received negative asylum decisions (4,367 cases were dismissed or expired). In 2016, 28,208 decisions were made, out of which 7,745 were positive and 14,282 negative (and 6,181 were "other") (Migri 2023). Although it is impossible to say for sure that deafness led to the negative asylum decisions in Rami's and the other study participants' cases, it is important to keep in mind that "[i]nconsistencies between the information (names, dates, places) provided during successive interview in the different stages of investigation can—in combination with other criteria—be used as grounds for rejection" (Maryns 2014: 1). For many deaf asylum seekers, providing such information often poses a great challenge, particularly if they have emerging literacy skills and do not use fingerspelling, which severely limits, or completely negates, their opportunity to communicate effectively with interpreters, who in turn may not know or understand the formal or home signs for the names and places deaf asylum seekers are signing. By law, authorities should take into consideration a person's circumstances such as age, the state of health (including if an applicant is deaf or blind), as well as how prior experiences may affect their ability to provide a coherent argument for their asylum application (Migri 2018; EASO 2015). However, in practice, it is difficult to say to what degree this actually happens. Given the rejection rates mentioned above, it is important to ask whether or even if an applicant's deafness can be disentangled from other factors that inform the asylum decision such as the actual story being told.

According to Migri (2018), an applicant's disabilities should even be considered a powerful argument in favor of approving an asylum application because such restrictions may negatively impact a person's ability to protect themselves in active conflict or from arbitrary violence (e.g., mines, bombs) (Migri 2018: 34). What is important to note in the discourse around deafness in the context of asylum procedures is its consistent labeling as "personal issue", which is in stark contrast to much of today's research in applied linguistics which considers successful communication to be the responsibility of all the agents involved. Most recently, Canagarajah (2022) entirely rejects viewing language competence as internalized personal knowledge, and sees it instead as a manifestation of social relationships. From this perspective, language competence is a question of sociomaterial embeddedness that is essential not only for meaningful communication but also for community building.[4] Seen through this lens, Rami's (in)ability to tell a compelling story in the asylum hearing is not merely due to his deafness, his identity as "foreign" language user, or his emergent literacy; rather it also hinges on his environment's (in)ability to create a space where his communicative resources have traction and validity.

While some (albeit severely lacking) services exist in Finland that have been designed to accommodate users of different languages, people with disabilities, and deaf refugees, the process does not support applicants who fall into several or all of those categories and thus need multiple layers of coordinated support. However, the problems discussed thus far are not merely an issue of lacking or insufficient support but reflect deeper societally dominant understandings of language (and) rights. When language rights are at stake, what is seen as a language is critical. As García et al. (2021) recently argued, "the sociopolitically imposed concept of a named language has little to do with how racialized bilinguals language or, for that matter, how any bilingual languages" (6). This assertion, of course, also applies to experiences of the deaf refugees featured in this chapter. The failure to recognize IS as a language in most countries around the world and the inability of the current support system to recognize multilingual and multimodal communication (see Kusters et al. 2017), including, for example, different sign languages, home signs, and writing/drawing, as a vital and valid communicative repertoire, leaves deaf refugees with emerging print literacy and interrupted/no formal education in a state of socially imposed communicative inability, or rather communicative being-disabledness. Relatedly, Moriarty Harrelson (2017) criticizes the linguistic orthodoxy that still permeates sociolinguistics and applied linguistics and portrays deaf people as not having a language. In her research with deaf people in Cambodia, she shows that their participants were drawing on a plethora of linguistic elements, which were different from those of named spoken or signed languages. As in that study, it is argued here that such communication practices should be recognized, promoted, and used. In the end, a society that declares a person as "languageless" reveals itself to be unwilling to recognize basic human rights.

From Social Justice to Anticoloniality

Deaf people around the world share not only the experience of being deaf but also the challenge of negotiating differences in their situations and backgrounds (Friedner and Kusters 2014). As has been shown in the two previous vignettes, differences in the social status of deaf asylum applicant are particularly pronounced. As a citizen of the country where the study participants sought refuge, in the third and final vignette presented in this chapter, the researcher reflected upon her own experiences in relation to theirs:

VIGNETTE 3: What Do We Share?

As a deaf white person, I am in some ways in a privileged position. After my birth and finding out that I was deaf, my hearing parents immediately started to learn FinSL in order to communicate with me. A few years later I got a deaf sibling, and (along with Finnish), FinSL became one of the main languages in my family. I was educated bilingually in FinSL and Finnish. I know several European and North American signed and written languages. IS is familiar to me because I have traveled, studied, worked, and lived in different countries around the world. Having access to sign language interpreters and using interpreting services in different areas is a natural part of my life. Advocating for human rights of the deaf both nationally and internationally is a big part of my professional and personal life. The situation of my research participants is quite different. As deaf people, they received very little formal education in their prior home countries. When I met them, they were emergent readers of Arabic, and there was great variety among them in terms of their proficiency levels in the sign language of their prior home country. As refugees escaping war and terror, they had experienced physical violence in their home countries and during their flight to Finland. Although we share the identity of deafness, it is impossible to gloss over the existing differences in languages, mobilities, and freedom between me and my research participants. Those existing injustices make me feel uncomfortable. Although I felt helpless at times, I tried to be more than a "researcher only" by connecting with them and supporting them emotionally during their asylum procedure. I also acted as a friend, social worker, a psychologist, and a relay interpreter in some situations. (personal reflection, November 2021)

Along with their deafness, the challenges faced by the refugees profiled here included their positions as emergent readers and learners of national sign language(s). The normal communicative hurdles these factors involved were amplified by the importance these skills have for the asylum process. In other words, for some of these refugees, language learning was vital in the most literal sense, as they were under immense pressure to acquire the communication strategies that were officially recognized by their new environment. As Vignette 3 described above, this experience was clearly different from that of the researcher, despite their shared deafness.

Another obvious difference between researcher and participants was their educational backgrounds and opportunities they had been given or denied throughout their lives. Coming from a national context with a history of promoting bilingual education in sign languages for the deaf (Tang, Adam, and Simpson 2020), Nina was granted a right that is only afforded to 2% of deaf children and youth around the world (Haualand and Allen 2009). By comparison, the research participants' education was typically short, interrupted, and offered only orally.

In the face of such glaring inequities which sometimes are not always immediately apparent to academics who do not have these challenges, the role and responsibility of the researcher and its institutional environment have to be revisited. While we may learn in research courses on ethics (which are generally few and far between) about the importance of obtaining informed consent and protecting the identities of research participants who come from so-called "vulnerable" communities, a greater shift in how we position ourselves as researchers is needed. This may be particularly the case for scientists who come from the Global North.

The focus of this chapter has been on the extremely complex case of deaf refugees, who come to interact with academics, social workers, and migration officials in Europe. As this chapter has shown, research and practice require greater recognition of the sociohistorical relations in these interactions. For this, a so-called "social justice" approach may be insufficient. As Adam (2020) has pointed out, in academic spaces of the Global South, there is increasing reluctance to use social justice frameworks that have been developed in the Global North. Instead, Adam notices "a decolonial turn" (para 6) that implies a "contestation with the universalisation of Euro-centric frameworks of human values" and a focus on "*the sources* of systematic oppression and dominance" (para 9). Also in the field of applied linguistics, lines of research have gained traction that are motivated by anticolonial efforts.[5] For instance, work by Anya (2020), Meighan (2020), Von Esch, Motha, and Kubota (2020), Kubota (2020), Anya (2016), Motha (2014), and Kubota and Lin (2009), as well as scholars in raciolinguistics (e.g., Rosa and Flores 2017; Flores and Rosa 2015) have made a strong case for centralizing antiracist, anticolonial work in language education, and applied linguistics as a whole. Relatedly, Cushing-Leubner et al. (2022) have documented how even anticolonial approaches to language education research can end up perpetuating inequities and colonial power structures. Nevertheless, an anticolonial lens may be helpful in discussing work with deaf, racialized refugees, and the role of the researcher in it. Rather than a naive approach to social justice that addresses differences merely through "aid" and "support" structures, what is needed is an honest examination of the role nations, officials, institutions, and academics play in producing and prolonging the existing inequities in our society. This work has already begun (e.g., Keskinen 2019; Keskinen et al. 2016), but there is still much more that needs to be done.

Conclusions

In offering the above three vignettes, several points have come to the fore that need to be considered further and more deeply. First, as this chapter has shown, law and institutional policies can be in direct conflict with interpersonal, ethical, and scientific needs. Where vulnerable communities and individuals are concerned, the consequence of these conflicts can be linguistic and social isolation. Further, as this chapter has also demonstrated, researchers who seek to investigate these inequities may quickly find themselves confronted with ethical questions that university courses rarely if ever address.

Based on our experiences as scientists, we contend that applied linguists should revist the question, "What is a language?", particularly in contexts where language rights are at stake or legal matters are tied to high-stakes communicative processes, such as asylum hearings. Given that many named languages have roots in colonial efforts to administer and oppress stolen territories (Heller and McElhinny 2017), the definition of language and by extension the recognition of linguistic rights and resources has to be approached from a sociohistorically sensitive perspective that acknowledges linguistic oppression as a tool of epistemic superiority and the position of colonized peoples as "eternal second-language speaker(s)" and thus second-class citizens (Mitova 2020: 202). Such (and other) power dynamics have been transferred from colonial to contemporary migration and asylum processes (Mayblin and Turner 2020; Schinkel 2018) and surface with multiple layers of discrimination, including linguistic marginalization, ableism, white supremacy, and educationalism.

It is also recommended that researchers critically consider their positionality. This reflection should include the positionality of their institutions within the larger higher education landscape and societal context. Such reflection can help bring to the surface some of the ways that social identities intersect and interact with one another. For instance, as researchers and institutions, we need more insights into how constructs of gender, race, ability, and class are intertwined and amplify each other in our work, individually and collectively. Although it is agreed that more activist research that addresses societal inequities is needed, this chapter also contends that this work needs to extend also (and maybe especially) to practices that intend to be inclusive but still have the potential to perpetuate discriminatory patterns of social stratification (e.g., by constructs of class, race, gender, nationality, and (dis)ability). This chapter is then a call for all of us to rethink our research approaches and premises. However, an ethical approach to research that consists merely of "do's and don't's" will not suffice. Rather, what is needed is a fundamental rethinking of how we understand and enact our role in our professions and the larger society.

Notes

1 In this chapter we use "deaf" with a lower-case "d" because we do not refer to deaf people as belonging to a particular linguistic-cultural group, as members of a deaf signing community (as big D), or as people with hearing loss as a medical and audiological condition, because d/D distinction is problematic and has caused the experiences of exclusion (see also Kusters, De Meulder, and O'Brien 2017, 13–15; Pudans-

Smith et al. 2019). Rather, "deaf" here means all kinds of deaf and hard of hearing people who have sensory asymmetries with proficiency variation in written and signed languages (see also De Meulder, Kusters, Moriarty and Murray 2019).

2 Refugees are often categorized by the reason for their flight, such as war, persecution, or economic hardship. Reasons for displacement are usually complex and dynamic, and refuse an overly simplistic categorization, particularly a clear-cut distinction between forced displacement and voluntary migration (see also Nayeri 2019).

3 Nina's IS repertoire includes many conventionalized IS strategies, such as English mouthing, fingerspelling (one-handed), and signs from European and North American national sign languages such as American Sign Language (ASL). Thanks to her frequent travels around the world, her knowledge also includes several European written languages as well as European and North American signed languages. As illustrated by Whynot's (2016) experimental data from the World Federation of the Deaf (WFD), in academic settings where conventionalized IS is largely used, such a mix of influences in signers' IS repertoires is quite typical.

4 Canagarajah's (2022) theory was based on his own experience as a cancer patient.

5 We use anticoloniality to describe a deliberate thinking and acting against continued colonial domination that is woven into the fabric of our neoliberal societies, i.e., what Quijano (2007) termed "coloniality". Drawing on Zembylas (2022), we use anti- rather than decolonial to account for the fact that decolonial efforts always include the repatriation of land, whereas this chapter article is concerned with refusing and resisting the coloniality of policies and practices related to research, language, migration, and asylum. The term "anticolonial" is used here to refer to an intentional and systematic effort of dismantling contemporary colonial processes. The terms "decolonial" and "decolonization" are not used here as they are inextricably linked to Indigenous rights and land reclamation (Tuck and Yang 2012).

References

Adam, Robert. 2012. "Language Contact and Borrowing." In *Sign Language: An International Handbook*, 841–860. Edited by Roland Pfau, Markus Steinbach, and Bencie Woll. Berlin: De Gruyter Mouton.

Anya, Uju. 2020. "African Americans in World Language Study: The Forged Path and Future Directions." *Annual Review of Applied Linguistics* 40: 97–112.

Anya, Uju. 2016. *Racialized Identities in Second Language Learning: Speaking Blackness in Brazil.* New York: Routledge.

Behrman, Simon. 2014. "Legal Subjectivity and the Refugee." *International Journal of Refugee Law* 26, no. 1: 1–21.

Bigelow, Martha, and Nicole Pettitt. 2015. "Narrative of Ethical Dilemmas in Research with Immigrants with Limited Formal Schooling." In *Ethics in Applied Linguistics Research*, 841–860. Edited by Peter I. De Costa. New York: Routledge.

Byun, Kang-Suk. 2023. *Establishing Intersubjectivity in Cross-signing.* Doctoral dissertation. Radboud University. Max Planck Institute for Psycholinguistics.

Canagarajah, Suresh. 2022. *Language Incompetence: Learning to Communicate Through Cancer, Disability, and Anomalous Embodiment.* London: Routledge.

Canagarajah, Suresh. 2013. *Translingual Practice: Global Englishes and Cosmopolitan Relations.* London: Routledge.

Copland, Fiona, and Angela Creese. 2015. *Linguistic Ethnography: Collecting, Analysing and Presenting Data.* London: SAGE Publications.

Crasborn, Onno, and Anja Hiddinga. 2015. "The Paradox of International Sign: The Importance of Deaf-Hearing Encounters for Deaf-Deaf Communication Across Sign Language Borders." In *It's a Small World: International Deaf Spaces and Encounters*, 59–69. Edited by Michele Friedner and Annelies Kusters. Washington, DC: Gallaudet University Press.

Cushing-Leubner, Jenna, Mel Engman, Johanna Ennser-Kananen, and Nicole Pettitt. 2022. "Imperial Straightening Devices in Disciplinary Choices of Academic Knowledge Production." *Language, Culture and Society* 3, no. 2: 201–230.

Darvin, Ron, and Bonny Norton. 2015. "Identity and a Model of Investment in Applied Linguistics." *Annual Review of Applied Linguistics* 35: 36–56.

De Meulder, Maartje, Annelies Kusters, Erin Moriarty, and Joseph. J. Murray. 2019. "Describe, Don't Prescribe. The Practice and Politics of Translanguaging in the Context of Deaf Signers." *Journal of Multilingual and Multicultural Development* 40, no. 10: 892–906.

EASO 2015. *EASO Practical Guide: Evidence Assessment. European Asylum Support Office. Publications Office of the European Union.* Accessed December 21, 2021. Retrieved from https://easo.europa.eu/sites/default/files/public/EASO-Practical-Guide_-Evidence-Assessment.pdf.

Ennser-Kananen, Johanna. 2018. "'That German Stuff': Negotiating Linguistic Legitimacy in a Foreign Language Classroom." *Journal of Language and Education* 4, no. 1: 18–30.

Ennser-Kananen, Johanna. 2014. "The Right to Be Multilingual: How Two Trilingual Students Construct Their Linguistic Legitimacy in a German Classroom." Doctoral dissertation. University of Minnesota, Digital Conservancy.

Finlex 746/2011. *Laki kansainvälistä suojelua hakevan vastaanotosta sekä ihmiskaupan uhrin tunnistamisesta ja auttamisesta [Finnish Act on the Reception of Persons Applying International Protection and on the Identification of and Assistance of Victims of Trafficking in Human Beings].* Retrieved from https://www.finlex.fi/fi/laki/ajantasa/2011/20110746.

Finlex 133/2010. *Laki vammaisten henkilöiden tulkkauspalvelusta [Finnish Act on Disabled People's Interpretation Service].* Retrieved from https://www.finlex.fi/fi/laki/alkup/2010/20100133.

Finlex 201/1994. *Kotikuntalaki [Municipality of Residence Act].* Retrieved from https://finlex.fi/fi/laki/ajantasa/1994/19940201.

Flores, Nelson, and Jonathan Rosa. 2015. "Undoing Appropriateness: Raciolinguistic Ideologies and Language Diversity in Education." *Harvard Educational Review* 85, no. 2: 149–171.

Friedner, Michele I., and Annelies Kusters. 2014. "On the Possibilities and Limits of 'DEAF DEAF SAME': Tourism and Empowerment Camps in Adamorobe (Ghana), Bangalore and Mumbai (India)." *Disability Studies Quarterly* 34, no. 3. Retrieved from https://dsq-sds.org/article/view/4246/3649.

Gal, Susan. 2006. "Migration, Minorities and Multilingualism: Language Ideologies in Europe." In *Language Ideologies, Policies and Practices: Language and the Future of Europe*, 13–27. Edited by Clare Mar-Molinero and Patrick Stevenson. Basingstoke: Palgrave MacMillan.

García, Ofelia, et al. 2021. "Rejecting Abyssal Thinking in the Language and Education of Racialized Bilinguals: A Manifesto." *Critical Inquiry in Language Studies*18, no. 3: 203–228.

Green, E. Mara. 2014. "Building the Tower of Babel: International Sign, Linguistic Commensuration, and Moral Orientation." *Language in Society* 43, no. 4: 445–465.

Gill, Nick, and Anthony Good. 2019. *Asylum Determination in Europe: Ethnographic Perspectives.* Cham, Switzerland: Springer Nature.

Hansen, Martje. 2016. "What Is International Sign? The Linguistic Status of a Visual Transborder Communication Mode." In *International Sign. Linguistic, Usage, and Status Issues*, 15–31. Edited by Rachel Rosenstock and Jemina Napier. Gallaudet University Press.

Haualand, Hilde, and Colin Allen. 2009. *Deaf People and Human Rights.* World Federation of the Deaf and Swedish National Association of the Deaf. Retrieved from https://www.rasit.org/files/Deaf-People-and-Human-Rights-Report.pdf.

Heller, Monica, and Bonnie McElhinny. 2017. *Language, Capitalism, Colonialism: Toward a Critical History.* University of Toronto Press.

Hiddinga, Anja, and Onno Crasborn. 2011. "Signed Languages and Globalization." *Language in Society* 40, no. 4: 483–505.

Hou, Lynn, and Annelies Kusters. 2019. "Sign Languages." In *The Routledge Handbook of Linguistic Ethnography,* 340–355. Edited by Karin Tusting. London: Routledge.

İlkbaşaran, Deniz. 2015. "Social Media Practices of Deaf Youth in Turkey: Emerging Mobilities and Language Choice." In *It's a Small World: International Deaf Spaces and Encounters,* 112–126. Edited by Michele Friedner and Annelies Kusters. Washington, DC: Gallaudet University Press.

Keskinen, Suvi. 2019. "Intra-Nordic Differences, Colonial/Racial Histories, and National Narratives: Rewriting Finnish History." *Scandinavian Studies* 91, nos. 1/2: 163–181.

Keskinen, Suvi, Salla Tuori, Sari Irni, and Diana Mulinari, eds. 2016. *Complying with Colonialism: Gender, Race and Ethnicity in the Nordic Region.* London: Routledge.

Kubota, Ryoko, and Angel M. Y. Lin, eds. 2009. *Race, Culture, and Identities in Second Language Education: Exploring Critically Engaged Practice.* New York: Routledge.

Kusters, Annelies. 2021. "International Sign and American Sign Language as Different Types of Global Deaf Lingua Francas." *Sign Language Studies* 21, no. 4: 391–426.

Kusters, Annelies. 2020. "The Tipping Point: On the Use of Signs from American Sign Language in International Sign." *Language and Communication* 75: 51–66.

Kusters, Annelies, Maartje De Meulder, and Dai O'Brien. 2017. Critically Mapping the Field. In *Innovations in Deaf Studies. The Role of Deaf Scholars,* 1–53. Edited by Annelies Kusters, Maartje De Meulder, and Dai O'Brien. New York: Oxford University Press.

Kusters, Annelies, Max Spotti, Ruth Swanwick, and Elina Tapio. 2017. "Beyond Languages, Beyond Modalities: Transforming the Study of Semiotic Repertoires." *International Journal of Multilingualism* 14, no. 3: 219–232.

Macedo, Donaldo, ed. 2019. *Decolonizing Foreign Language Education: The Misteaching of English and Other Colonial Languages*. New York: Routledge.

Maahanmuuttovirasto (Migri). 2023. Statistics. Accessed December 11, 2023. Retrieved from https://tilastot.migri.fi/index.html#applications?l=en&start=540&end=551.

Maahanmuuttovirasto (Migri). 2021. *Interpreting*. Accessed December 21, 2021. Retrieved from https://migri.fi/en/interpreting.

Maahanmuuttovirasto (Migri). 2020. *Opas tulkeille.* Accessed December 21, 2021. Retrieved from https://migri.fi/documents/5202425/6164491/Opas+tulkeille+(fi).pdf/6a44b5a3-48d8-433e-778f-dd668d1d1f80/Opas+tulkeille+(fi).pdf?t=1598360531351.

Maahanmuuttovirasto (Migri). 2018, June 13. *Maahanmuuttoviraston selvitys sisäministerille turvapaikkapäätöksentekoon ja -menettelyyn liittyen.* Accessed December 21, 2021. Retrieved from https://migri.fi/documents/5202425/8701378/Maahanmuuttoviraston+selvitys+sis%C3%A4ministerille+turvapaikkap%C3%A4%C3%A4t%C3%B6ksentekoon_valmis.pdf.

Maryns, Katrijn. 2014 [2006]. *The Asylum Speaker: Language in the Belgian Asylum Procedure.* New York: Routledge.

Mayblin, Lucy, and Joe Turner. 2020. *Migration Studies and Colonialism.* Cambridge: Polity.

Meighan, Paul J. 2020. "A Case for Decolonizing English Language Instruction." *ELT Journal* 74, no. 1: 83–85.

Ministry of Social Affairs and Health. 2022. *Social Security and Services of Asylum Seekers in Finland.* Accessed January 28, 2022. Retrieved from https://stm.fi/en/social-security-and-services-of-asylum-seekers-in-finland (last accessed 28.1.2022).

Mitova, Veli. 2020. "Decolonising Knowledge Here and Now." *Philosophical Papers* 49, no. 2: 191–212.

Moriarty, Erin and Annelies Kusters. 2021. "Deaf Cosmopolitanism: Calibrating as a Moral Process." *International Journal of Multilingualism* 18, no. 2: 285–302.

Moriarty Harrelson, Erin. 2017. "Deaf People with 'No Language': Mobility and Flexible Accumulation in Languaging Practices of Deaf People in Cambodia." *Applied Linguistic Review* 10, no. 1: 55–72.

Motha, Suhanthie. 2014. *Race, Empire, and English Language Teaching: Creating Responsible and Ethical Anti-Racist Practice.* New York: Teachers College Press.

Nayeri, Dina. 2019. *The Ungrateful Refugee: What Immigrants Never Tell You.* New York: Catapult.

Nick, Iman M., ed. 2018. *Forensic Linguistics: Asylum-seekers, Refugees and Immigrants.* Wilmington: Vernon Press.

Norton, Bonny. 2013. *Identity and Language Learning.* Bristol, Blue Ridge Summit: Multilingual Matters.

Pennycook, Alastair. 2001. *Critical Applied Linguistics: A Critical Introduction.* New York: Routledge. https://doi.org/10.4324/9781410600790.

Pudans-Smith, Kimberly K., Katrina R. Cue, Ju-Lee A. Wolsey, and Diane M. Clark. 2019. "To Deaf or Not to Deaf: That Is the Question." *Psychology* 10: 2091–2114.

Quijano, Aníbal. 2007. "Coloniality and Modernity/rationality." *Cultural Studies* 21, no. 2/3: 168–178.

Rosa, Jonathan, and Nelson Flores. 2017. "Unsettling Race and Language: Toward a Raciolinguistic Perspective." *Language in Society* 46, no. 5: 621–647.

Schinkel, Willem. 2018. "Against 'Immigrant Integration': For an End to Neocolonial Knowledge Production." *Comparative Migration Studies* 6, no. 1: 1–17.

Sensoy, Özlem, and Robin DiAngelo. 2017. *Is Everyone Really Equal?: An Introduction to Key Concepts in Social Justice Education.* New York: Teachers College Press.

Shin, Hyunjung. 2006. "Rethinking TESOL from a SOL's Perspective: Indigenous Epistemology and Decolonizing Praxis in TESOL." *Critical Inquiry in Language Studies* 3, nos. 2/3: 147–167.

Sivunen, Nina. 2021. "Kuurojen turvapaikanhakijoiden kielikäsityksien muokkautuminen turvapaikkaprosessin aikana". [Deaf Asylum Seekers' Transformation of Language Beliefs During the Asylum Procedure] *Puhe ja Kieli* 41, no. 2: 159–178.

Sivunen, Nina. 2019. "An Ethnographic Study of Deaf Refugees Seeking Asylum in Finland." *Societies* 9, no. 1/ 2: 1–16.

Sivunen, Nina, and Elina Tapio. 2020. "Do You Understand (Me)?" Negotiating Mutual Understanding by Using Gaze and Environmentally Coupled Gestures Between Two Deaf Signing Participants. *Applied Linguistics Review* 13, no. 6: 983–1004.

Skutnabb-Kangas, Tove, and Robert Phillipson. 1994. *Linguistic Human Rights: Overcoming Linguistic Discrimination.* Berlin: Mouton de Gruyter.

Smith-Khan, Laura. 2022. "Inclusive Processes for Refugees with Disabilities: Improving Communication for Deaf Forced Migrants." In *Handbook of Disability. Critical Thought and Social Change in a Globalizing World*, 1–26. Edited by Marcia H. Rioux, José Viera, Alexis Buettgen, and Ezra Zubrow. Singapore: Springer.

Smith-Khan, Laura. 2017. "Telling Stories: Credibility and the Representation of Social Actors in Australian Asylum Appeals." *Discourse and Society* 28, no. 5: 512–534.

Tang, Gladys, Robert Adam, and Karen Simpson. 2020. "Educating Bilingual and Multilingual Deaf Children in the 21st Century." In *Understanding Deafness, Language and Cognitive Development. Essays in Honor of Bencie Woll,* 183–204. Edited by Gary Morgan. Amsterdam: John Benjamins Publishing Company.

Tuck, Eve, and K. Wayne Yang. 2012. "Decolonization Is Not a Metaphor." *Decolonization: Indigeneity, Education and Society* 1, no. 1: 1–40.

UNHCR 2017. *Handbook for Interpreters in Asylum Procedures.* Vienna: UNHCR Austria. Accessed December 10, 2021. Retrieved from https://www.unhcr.org/dach/wp-content/uploads/sites/27/2017/09/AUT_Handbook-Asylum-Interpreting_en.pdf.

United Nations, Human Rights. *Convention on the Rights of Persons with Disabilities* (entry into force: 3 May 2008, in Accordance with Article, vol. 45, no. 1). Retrieved from https://www.ohchr.org/en/instruments-mechanisms/instruments/convention-rights-persons-disabilities.

Von Esch, Kerry S., Suhanthie Motha, and Ryuko Kubota. 2020. "Race and Language Teaching." *Language Teaching* 53, no. 4: 391–421.

WFD. 2016. *Know and Achieve Your Human Rights Toolkit.* Helsinki: World Federation of the Deaf. Retrieved from https://wfdeaf.org/our-work/human-rights-of-the-deaf/.

Whynot, Lori A. 2016. *Understanding International Sign: A Sociolinguistic Study.* Washington, DC: Gallaudet University Press.

Zembylas, Michalinos. 2022. "Affective and Biopolitical Dimensions of Hope: From Critical Hope to Anticolonial Hope in Pedagogy." *Journal of Curriculum and Pedagogy* 19, no. 1: 28–48.

Zeshan, Ulrike. 2015. "'Making Meaning': Communication Between Sign Language Users Without a Shared Language." *Cognitive Linguistics* 26, no. 2: 211–260.

24

ETHICAL RESEARCH PRACTICES WITH HIGHLY STIGMATIZED POPULATIONS

Engaging with Individuals Convicted of Sexual Offenses

Stina Lindegren

Introduction

Sexual violence is a widespread threat to the health, well-being, and basic human rights of a significant portion of the population (World Health Organization 2021; Macdowall et al. 2013). Women and children are particularly affected. The issue of sexual abuse understandably evokes strong negative emotions in most of us. Perceptions of those who commit sexual abuse are nearly universally negative. The stigma is exceptional (Tewksbury 2012) and entails numerous obstacles when conducting research on this population.

Professionals working with individuals who have sexually offended may face a range of challenges (Shrim and Baum 2022; Frost 2016; Hackett 2000). Difficult moral emotions and secondary trauma are common negative side-effects and this can also affect analytical professionals, such as researchers (Woodhams and Duran 2024). While most people want to end sexual violence, few are willing to engage with those who commit such acts or are at risk of doing so (cf. Jahnke 2018). For those who do choose to engage, their efforts are sometimes met with resistance and even hate (Walker 2023). Researchers can be accused of "siding with the perpetrator" instead of the victim. Moreover, a study by Rydberg et al. (2023) shows that the public tends to view experts as less credible when they discuss individuals who have sexually offended, compared to topics like drunk driving. Credibility appears to decline, particularly when their message challenges widely held but empirically unsupported beliefs such as the necessity of punitive policies for handling individuals convicted of sexual offending.

Accordingly, the stigma associated with being a "sex offender" appears to affect researchers as well. Additionally, various inner moral conflicts or issues may arise for researchers in this field. Researching highly stigmatized populations, such as individuals committing sexual crimes, is also associated with several ethical challenges when recruiting research subjects, collecting data, and presenting results. I discuss some of these challenges in this chapter and offer potential solutions.

 DOI: 10.4324/9781003391074-27

Methodology

In this chapter, I draw primarily on the interview research I conducted at Uppsala University, Sweden, in 2020–2024, and my dissertation titled "Support and Treatment for Men Convicted of Sexual Offending: Readiness, Change, and Previous Help-Seeking" (Lindegren 2024). I also use my experience as a practitioner when I previously worked with individuals convicted of sexual offenses as a probation officer and sex offender treatment therapist in the Swedish Prison and Probation Service.

The dissertation is comprised of four papers or studies. The aim was to investigate the support and treatment experiences of adult men convicted of sexual offenses in Sweden (Lindegren 2024). The dissertation examines their experiences of change following participation in sex offender treatment and how they engage with risk-reducing interventions. The data includes pre- and post-tests (n = 99) from the first cohort of participants in a fairly new sex offender treatment program in the Swedish Prison and Probation Service (Lindegren 2022), as well as in-depth interviews with men convicted of sexual offenses (n = 19) (Lindegren 2025a, 2025b). The interviews were conducted with both participants (n = 13) and non-participants (n = 6) of the sex offender treatment program. Ethical approval was granted by the Swedish Prison and Probation Service and the Swedish Ethical Review Authority (no. 2020-04038), in compliance with the Swedish Ethical Review Act (SFS 2003: 460).

Interviews and Interviewees

The interview studies used purposive sampling to select participants, including both individuals who had completed a sex offender treatment program and those who had not participated, either due to negative attitudes or hesitation toward treatment (Lindegren 2024). Exclusions were made for individuals under 18, those considered high-risk for violence toward professionals, or those with serious mental disorders. The sample size was planned using the concept of information power (Malterud, Siersma, and Guassora 2016). Prison staff handled participant recruitment, and participation was voluntary, with no financial compensation offered. Written informed consent was obtained, and it was clearly stated that participation or non-participation would not affect participants' sentence conditions (Lindegren 2024).

The sample comprised 19 adult men incarcerated for sexual offenses, representing a range of prison security levels (low, medium, and high) (Lindegren 2024). With ages ranging from 26 to 82 years, most of the interviewees had committed serious sexual offenses such as rape. Sentences ranged from six months to nine years. There were equal numbers of interviewees whose victims were children and adults. A majority of the victims were closely related to the interviewee.

In-depth interviews were carried out in four prisons, some by telephone due to COVID-19, between June 2021 and July 2022 (Lindegren 2024). A semi-structured interview guide was followed, using an experiential/phenomenological approach (Aspers 2009). The interview approach was influenced by teller-focused interviews, suitable for sensitive topics, emphasizing the interviewee's narrative (Lindegren 2024; Hydén 2014). Data were analyzed mainly inductively using reflexive thematic analysis (Braun and Clarke 2019, 2006).

The Scandinavian Context

This chapter discusses research conducted in, and practitioner experiences from, Sweden. Sweden can be seen as an outlier in some aspects, which are worth taking into consideration. Sweden is considered a wealthy, highly gender-equal country with strong egalitarian values. The state-run Prison and Probation Service is known for its humane prison conditions and a rehabilitation-focused approach rather than one centered on punishment, a characteristic often referred to as Scandinavian exceptionalism (Pratt 2007). However, some scholars contend that Scandinavian criminal policy has recently shifted toward a more punitive approach (Storgaard 2022; Tham 2019). Swedish sexual offense legislation is extensive, and the definition of "rape" is quite broad, encompassing all non-consensual acts comparable to sexual intercourse (Bladini and Svedberg Andersson 2020; Holmberg and Lewenhagen 2020). Sweden does not have a public registry for individuals convicted of sexual offenses or laws requiring notifications about their status, as is common in the United States.

While participation in sex offender treatment programs is voluntary and free of charge in Sweden, inmates are strongly incentivized to take part in exchange for various privileges. The interviewees who participated in the research presented in this chapter were in a program with a flexible manualized approach. Grounded in cognitive behavioral therapy, and utilizing the risk-, need-, and responsivity model (Lindegren 2024; Bonta and Andrews 2017), this program aims to reduce recidivism risk. Those who deny guilt may still be included, provided they acknowledge relevant risk factors addressed by the program (denial itself is not a treatment focus). Program therapists are mainly psychologists or social workers employed by the Prison and Probation Service.

Macro-Ethical Considerations

Person-First Language

Similar to the debate about how to name people who have been subjected to abuse ("victim" or "survivor"?) (Schwark and Bohner 2019), there has been discussion about appropriate terms for individuals convicted of sexual offending. While there has been a shift in other areas where labeling terms like "addict" have been replaced by person-first language, a clear trend is yet to be seen regarding those who have sexually offended (D. A. Harris 2021). Traditionally, they have been referred to as "sex offenders" and this is still often the case.

I propose that in research or any work involving individuals convicted of sexual offenses, it is essential to condemn the actions that constitute the offense, while simultaneously respecting the individual as a person and recognizing their inherent dignity and human rights (Braithwaite 2020; Mullins and Kirkwood 2019). As such, the term "sexual offender" is not used in this chapter, as it suggests a persistent tendency to commit sexual offenses and reduces the person to their crime. Instead descriptors like "individual convicted of a sexual offense" are used throughout this chapter.

Why is such person-first language preferable? Empirical research has shown that using the label "sexual offender" rather than person-first language influences people's perceptions of the likelihood of reoffending and their attitudes toward restrictive public policies (Harris and Socia 2016). In addition, such labels can make it harder for individuals to seek help from preventive services (Piché et al. 2018; Levenson, Willis, and Vicencio 2017). Willis (2018: 727) points out a crucial problem with the term: "Why call someone by what we don't want them to be?" Moreover, person-first language is in accordance with the 2021 recommendations from the Council of Europe (McCartan 2022) and North American organizations like the Association for the Treatment and Prevention of Sexual Abuse (2023). Nonetheless, it is understood that terms like "individuals who have sexually offended" are comparatively long and can impede readability. For that reason, abbreviations and acronyms like "ICSO" are often used in scientific literature. Unfortunately, these shortened forms can become new stigmatizing labels.

Avoiding Reinforcement of Stereotypes and Myths

There are many myths and stereotypes about the characteristics of individuals who have committed sex offenses, which do not reflect the typical person who perpetrates sexual abuse. These include notions like the "stranger danger" myth. According to this misconception, a person who commits sexual violence is often cast as a deviant stranger—a monster lurking in the shadows (Harper, Hogue, and Bartels 2017; Levenson et al. 2007; Cowburn and Dominelli 2001). In real life, however, sexual violence is often perpetrated by someone the victim already knows and trusts, such as a parent, relative, or a romantic partner (Brå 2023; Andersson, Pettersson, and Jacobsson 2020). The stereotypes and intense stigma associated with sexual violence perpetration can help explain the emergence of extreme legal measures and social movements, such as "pedophile hunting", even in Scandinavian countries where the approach toward sexual offenses is less punitive than in other regions of the world (Jørgensen, Olesen, and Lentz 2023; Lomell 2020).

In many societies, there is often a strong desire to distance oneself from individuals who commit sexual violence. Research with an intersectional perspective has explored the stereotypes present in the ongoing societal discourse on sexual and intimate partner violence. Men with immigrant backgrounds are often seen as representatives of a violent, domineering, and patriarchal masculinity (Christensen and Jensen 2014; Ranta-Tyrkkö 2011). This negative stereotype can feed directly into the distancing desire, where individuals who use sexual violence often are consequently viewed as being foreign and different from the "modern man", or where the victim-blaming of women who have been raped is seen as something that only happens in remote, rural, conservative environments (Berggren and Gottzén 2021; Gottzén and Franzén 2020).

Accordingly, those who perpetrate sexual violence may frequently be seen as "the other", distinct from people who are familiar or loved. This perception has several consequences. First, it risks silencing the voices of those who have been abused by someone who does not fit the stereotype (Cowburn and Dominelli 2001). In the worst-case scenario, the victimized person, whether an adult or a child, is not believed and might be accused of lying. Second, the stranger danger myth may obscure the role played by structural power dynamics such as the gender-based dimensions of domestic sexual violence. A third risk is that individuals who are at risk of perpetrating sexual violence, or who have already done so, may have difficulties identifying with this category and thus refrain from seeking help to change their behavior (Piché et al. 2018; Levenson, Willis, and Vicencio 2017).

For these reasons, I argue that researchers should avoid reinforcing such stereotypes and myths. It is important to recognize that the extreme cases that are typically seen in the newspaper headlines and in true crime media are usually not very representative. In order to help provide data that is maximally useful, it is imperative that scientists study individuals who represent the majority of people who commit sexual offenses. Presenting research using nuanced illustrations of these individuals, including both their harmful actions and their non-pathological sides, offers a more accurate picture and is also important to combat myths and stereotypes.

There are additional important ethical reasons for avoiding research that exclusively focuses on the most extreme cases. Highlighting sadistic acts and psychopathic personalities may actually pose certain risks. Individuals who commit sexual offenses may have difficulties seeing their sexual transgression as abuse in comparison to these extreme cases, which can prevent them from changing their behavior or seeking help. Furthermore, those who are victimized may also minimize the abuse they have been subjected to because they think that what happened to them was not "as bad" as in these extreme cases, or they conclude that the person who abused them was "not evil". Lastly, in the mistaken belief that the extreme cases accurately represent what is typical for sexual violence, the general public, as well as those close to people who have abused and/or been abused, might also add to this minimization.

A Victim's Perspective

Another macro-ethical challenge lies in striking a delicate balance between, on the one hand, respecting individuals who have committed a sexual offense and acknowledging their human rights and inherent value, while on the other hand, recognizing the gravity of sexual violence and the trauma experienced by victims of sexual abuse (Lindegren 2024). This complexity is further heightened by the occasional overlap between these two groups (Glasser et al. 2001). However, well-conducted research and a broader understanding of those who have perpetrated sexual violence arguably will ultimately benefit both groups. Studying those who perpetrate sexual violence is crucial to improving prevention. To do so, it is important to take into account the perspective, opinions, experiences, thoughts, and emotions of the person who has committed a sexual offense.

However, individuals who have sexually offended often deny or express justifications for their offending behavior (Ware and Mann 2012). Research that uncritically presents denial, justifications, and victim-blaming narratives may be offensive or harmful to those who have been victimized. It is important to acknowledge this risk in research and ensure that such accounts are only included when essential to the analysis. In such cases, they should always be critically examined and/or contextualized.

Even when this precaution is taken, researchers involved in sexual perpetration research may find themselves accused of siding with the perpetrator instead of the victim. They may also encounter additional misconceptions, such as the belief that they are mentally ill or even harbor pedophilic interests themselves (Walker 2023). Personally, in social media, I have been accused of neglecting the power dynamics of violence against women. I have also received hostile remarks, including conspiracy theories suggesting that I am "spreading the narrative" of certain political parties. Additionally, some perceive me as entirely lacking credibility simply because I focus on perpetration rather than victimization. These misperceptions have ethical implications, as they can erode trust in researchers and lead people to dismiss scientific facts and hold on to myths and stereotypes (Rydberg et al. 2023). While it can feel exhausting to state the obvious, my experience has led me to the conclusion that making positional statements, where the victim's perspective is clearly spelled out, is necessary whenever the word limit allows. For example, in my dissertation (Lindegren 2024), I formulated it like this:

> It should go without saying that I firmly condemn all acts constituting sexual harassment, abuse, or violence. I want to contribute to improved understanding of these phenomena. One of my primary aims is to reduce sexual crimes and the harms associated with them. I fully stand with and support all victims and survivors of sexual abuse. I want to combat, not only men's violence against women, but all kinds of sexually violent or abusive acts, regardless of age, gender, or any other social category. Improving prevention is one important reason why I have professionally dedicated the last decade to this topic.
>
> *(17)*

Such declarations might help prevent others from drawing the wrong conclusions.

Interviewing the Most Hated

Scholars like Brinkmann and Kvale (2005) challenge the assumption that qualitative research is inherently more ethical or superior to quantitative methods that are supposedly uncaring and dehumanizing. When conducting qualitative research, it is equally important to abide by ethical standards to avoid causing harm to the interviewee. At the same time, simply following ethical rules or principles may not be enough; researchers must also understand when and how to apply them. Brinkmann and Kvale (2005) accordingly argue for employing Aristotelian "practical wisdom", where researchers develop the ability to recognize events in their context and make judgments based on that understanding. In this section, I discuss some aspects of practical wisdom that are important for conducting ethical interviews with individuals convicted of sexual offenses.

Asymmetrical Power Relations

One main ethical challenge involves the power imbalances between researchers and study participants. A key ethical consideration when researching incarcerated individuals is the potential for informants to feel coerced into participating due to asymmetrical power relations. While it is challenging to completely eliminate these concerns, in my research, I addressed this issue by clearly informing prospective interviewees, both in written and oral forms, that they could withdraw at any time (Lindegren 2024). I also emphasized that their sentence conditions would not be affected by their decision for or against participating in the interviews. When working with people who are incarcerated, it is essential to provide clear information about the independence of the research and the researcher from the criminal justice system.

On a few occasions, interviewees asked me for advice or information regarding matters related to their sentences, such as how to influence risk assessments. In line with feminist research approaches like the one proposed by Oakley (2016), I consider it unethical to expect the interviewees to share information with me without offering anything in return. Hence, when they specifically sought guidance, I chose to offer information or my professional opinion on the issues they raised, provided I felt I had relevant knowledge (Lindegren 2024). This approach demonstrated respect for the interviewees as human beings, which was the polar opposite of what many of them routinely experienced.

Individuals convicted of sex offenses are not only a highly stigmatized group in the general society. They are the lowest in the hierarchy in prisons (Ugelvik 2015). Furthermore, those who have abused children are the lowest within the group of those convicted of sexual offenses. This circumstance is something that can affect the interview situation and interactional processes between the researcher and the interviewee. At the end of each session, I asked participants if they had felt comfortable with the interview situation (Lindegren 2024). Two interviewees, somewhat unexpectedly, asked how I felt interviewing them—"sex offenders". This question demonstrated their awareness of the negative emotions people associate with both their crimes and themselves as individuals.

Consistent with the earlier discussion on feminist ethics, I believed it was important to respond honestly. I explained that this was not my first encounter with people who had been sexually abusive and that I found the interviews interesting and meaningful because I believe such knowledge can help prevent sexual offenses and improve society. I also clarified that, while I condemn sexual offenses like most people, I think it is crucial to understand these crimes and to gain insights to improve treatment experiences. I asserted that my intention was not to excuse the offenses but rather to enhance the effectiveness of rehabilitation. Both interviewees nodded and expressed agreement with my response. These events could be viewed as a form of stigma management on the interviewees' part, providing an opportunity for the researcher to support such processes. I believe this is an important ethical responsibility, as it can help prepare the convicted individual for future reintegration and an offense-free future.

Recruiting Hesitant Research Participants in a Non-Intrusive Way

The population of individuals convicted of sexual offenses is characterized by low levels of trust (Mann et al. 2013). As a result, it can be a challenge to find individuals who are willing to volunteer for research. Additionally, issues like shame, stigma, and taboos can increase the risk of potential interviewees declining to participate or refusing to continue their participation later on. It can be difficult to navigate effective recruitment strategies without being intrusive.

In my research, at one point, an interviewee, after initially agreeing to participate, decided he no longer wanted to continue (Lindegren 2024). He informed me via telephone that he was in a bad mood and had lost interest in taking part. This situation raised ethical concerns, particularly about balancing respect for informed consent with the need for novel knowledge, which sometimes requires researchers to be somewhat persistent (Lindberg and Sjöström 2019). This was especially important in this case, as the interviewee had shared perspectives that differed from the existing data, meaning his participation could have added valuable insights. During our phone conversation, I reiterated that participation was entirely voluntary, and I expressed my genuine interest in hearing his perspective (Lindegren 2024). I also made it clear that he was welcome to change his mind and join the study at any time. After consulting with my supervisors, we considered options such as calling him again or sending a letter to express my continued interest and remind him that he was welcome to participate. The letter was chosen as a less intrusive option. He called me shortly after receiving it and expressed his desire to participate. After the interview, he mentioned that he was pleased to have contributed to the research. Thus, he did not seem to have perceived the overall recruitment strategy as intrusive.

This experience also makes another important point about the value of engaging people who have sexually offended in interview research. In addition to providing crucial knowledge that can help improve preventive measures, participating in interviews can serve potentially redemptive functions for the individual. For example, the majority of the interviewees in my research stated that they participated in the research project because they wanted to give something back to the community, to do good (Lindegren 2025a). As such, research can be seen as a way to promote accountability and rehabilitation, which may benefit both the interviewees themselves and the people who have been victimized.

A Trauma-Informed and Holistic Approach

Practical wisdom involves researchers using their skill in noticing to respond effectively to important events and reactions during interviews (Brinkmann and Kvale 2005). This skill set requires emotional reflexivity, a key aspect of high-quality research (Bergman Blix and Wettergren 2015), in particular when researching complex topics where research participants are stigmatized. Concretely, emotional reflexivity includes being sensitive to emotional cues during interviews and managing both the interviewees' and one's own emotions and potential distress, so-called "inward-directed emotion work" (Bergman Blix and Wettergren 2015). This aspect is particularly important when conducting interviews with people who have committed sexual offenses or other violent crimes. Interviews with these individuals may provoke negative emotions or recollections of past traumatic experiences, which are common among people who have perpetrated sexual abuse (Jespersen, Lalumière, and Seto 2009).

For that reason, in addition to information letters that provide contact numbers for support services, interviews need to include inquiries into any adverse effects in the end. In my research, I adhered to the basic principles of trauma-informed interviewing (Lindegren 2024; Todd-Kvam and Goyes 2023). This involved closely observing facial expressions or other distress signals, especially from those who disclosed past trauma or post-traumatic stress disorder diagnoses, and responding appropriately to avoid re-traumatization, such as changing topics or refraining from further probing. At a few points, two interviewees showed signs of intense distress, guilt, and anxiety, possibly indicating they were emotionally overwhelmed or re-experiencing their own trauma. Some of the interviewees also cried. During these moments, I evaluated the situation and applied various de-escalation strategies such as validating verbal and non-verbal cues, offering comments to normalize their reactions, and shifting the interview to less sensitive topics or questions. In instances where interviewees were emotionally affected, I made a conscious effort at the end of the interview to assess their mental state and reactions. I also reminded them of the support contacts listed in the research participant information. Additionally, I asked prison staff to be mindful of the interviewees' emotional states after the sessions. None of the interviewees reported or displayed signs of significant distress when the interview concluded. On the contrary, nearly all participants expressed satisfaction with their contribution to the research. They also mentioned that the reflections they had had during the interviews were beneficial for them. Their newly won insights reportedly served as an additional form of emotional processing of their experiences.

In my opinion, researchers can benefit from a holistic and person-centered approach that pays special attention to recognizing the person behind the offense. One of the aspects the interviewees in my research highlighted as being the most crucial for their experience in the sex offender treatment program was the way that the therapist treated them like human beings and not "sex offenders" (Lindegren 2025a). Being recognized as a person seems to have improved their engagement in treatment and outcomes that are in line with treatment goals. At the same time, the desire to distance oneself from one's offenses can also take counterproductive forms.

Jacobsson and Åkerström (2013) warn that interviewees may attempt to redefine their assigned social category to avoid stigma. Denial and justifications may also serve as a defense mechanism (Maruna and Mann 2006). The psychological drive to protect oneself from feelings of guilt and shame can be particularly great in relation to the incarceration process for a sex offense, as statistics show that this group is at high risk of suicide (Jeglic, Spada, and Mercado 2013) and that this risk is heightened particularly during remand and release from prison (Haglund et al. 2014; Fazel et al. 2008). Even after these stages end, interviewees who have committed violent offenses may still try to manage the impression they make upon the researcher either by giving responses they feel will enhance their image or by avoiding responses that might have the opposite effect. This phenomenon, called the "social desirability bias", may skew the results obtained when investigating morally sensitive research topics like violent offending (Presser 2004). In my own research, as a strategy to anticipate this obstacle, I decided to begin each interview by allowing the interviewees to share something about themselves or their background that was unrelated to their offense (Lindegren 2024). By doing this, I let them know that I acknowledged them as persons and not simply as "sex offenders". I also decided against gathering background information at the start of the interview, as is often done. Instead, I waited until the end of the interview to ask very sensitive, specific, and detailed questions about the offense and atypical sexual interests. My perception was that this was an effective strategy, as I noticed many interviewees who were initially hesitant and withdrawn at the start of the interview had become noticeably more relaxed by the end. Finally, I made a point of concluding the interviews on a positive and non-stigmatizing note. To do so, I asked them what they were most proud of while participating in the sex offender treatment or serving their sentence.

The Researcher's Role—An Emotional and Moral Balance Act

Inner Conflicts and Complex Emotions

Engaging with individuals who have committed serious sexual offenses is not without personal, moral, and emotional challenges for the investigator (Shrim and Baum 2022; Frost 2016; Hackett 2000). The researcher may experience strong negative emotions, typically related to the offense, such as anger, disgust, and resentment. Moreover, there is a risk of developing secondary trauma responses including anxiety, increased suspicion, and loneliness (Shrim and Baum 2022). Importantly, the source of this distress may not only come from working with people who have committed violent offenses but also from the negative attitudes others may have about that work. For example, a study of probation officers who worked with individuals convicted of sexual offending showed that they lacked support from several of their colleagues and their partners (Shrim and Baum 2022). They felt isolated and faced attacks both "from within and from outside". These attacks were commonly manifested as criticism, contempt, and disgust toward their work. According to the researchers, there was a link between the exclusion and loneliness the probation officers reported and their limited ability to share their experiences with colleagues and family members. Negative emotions are not the only potential source of inner conflict for professionals who work with people who have committed violent crimes.

Positive emotions can also produce inner conflicts and feelings of guilt. This phenomenon is well described by Ievins (2023), who conducted an ethnographic study about individuals convicted of sexual offending in the UK prison system. She purposefully did not look up the individuals' offenses while conducting the participatory observation to reduce the potential for bias. However, when she concluded her field observations, she could not resist the urge to find out about the criminal records of some of the inmates she had met and liked. She found out that their offenses were far more serious than she had thought. In the passage below, Ievins describes her shock:

> Sitting in that Stafford McDonald's, my confusion stemmed from the fact that not only did I not know whose side I had been on, I also did not know whose side I should be on. Wherever I placed my sympathy, it cast a shadow—over the victims whose stories had been obscured or over the prisoners whose moral identity had been stained—and I couldn't even decide who was placed lowest in the hierarchy of credibility. I didn't want to betray my participants by allowing their offenses to shape how I saw them, but I also worried that liking someone who had been convicted of such serious sex offenses implied that I thought their offenses didn't matter.
>
> *(147)*

As Ievins describes, engaging with those who have sexually offended is a difficult task that often gives rise to various inner moral battles. It can be a lonely journey, as most people cannot relate to these experiences. Supervision, or even therapy, can help to cope with these inner conflicts (Andersson 2022; Shrim and Baum 2022). Additionally, separating the person from the abusive act is essential to sustainable and ethical practices with individuals who have sexually offended (Braithwaite 2020). In practice, this is, however, not an easy task. In the next sections, I elaborate on these issues.

Critical but Not Confrontational

It can be challenging and provocative to listen to excuses and victim-blaming from people who have sexually offended. Research demonstrates that professionals working with individuals convicted of sexual offenses may also feel moral pressure to hold the person accountable for what they have done (Youssef 2022; Ware and Mann 2012). Moral conflicts and negative emotions may result in what psychotherapists call "countertransference" (Andersson 2022). This is when the negative emotions are directed at the client, which may result in a confrontational approach.

Within psychotherapeutic settings, confrontational approaches can increase the client's resistance and disengagement from treatment (Marshall et al. 2009; Marshall 2005). Confrontation typically triggers defensive reactions such as denial and other justifications, and thus interferes with rehabilitative goals and processes. Where sexual offenses are concerned, the negative repercussions of confrontational approaches are not limited to the person who has committed the offense. Encounters that enhance denial and justification may be counterproductive for people who have been the victim of a sexual offense, as they often value accountability (McGlynn and Westmarland 2019).

Although empirical research on this issue is scarce, it is highly likely that these phenomena are similar in a research interview situation, where the establishment of rapport is very similar to the therapeutic alliance between a therapist and a client. While some scholars propose confrontation in research interviews (Brinkmann and Kvale 2005), I argue, in line with Blagden and Pemberton (2010), that in order to adhere to ethical principles and receive truthful and rich accounts from highly stigmatized research participants, it is advisable to adopt a non-confrontational approach. This position is supported by research that has shown the non-confrontational approach to be effective in sex offender treatment research (Lindegren 2024; Mullins and Kirkwood 2022; Marshall 2005; Serran et al. 2003). In my research, I decided early on not to judge or challenge the participants. I also chose not to engage in debates about whether or not they had committed a crime.

A non-confrontational approach, however, does not mean that the researcher should not adopt a critical stance when interviewing participants, analyzing their responses, or later reporting the results. It is crucial that the researcher, like the therapist, avoids falling into what is known as "collusion"—an unconscious dynamic where they unintentionally support or reinforce excuses, victim-blaming, or violence-justifying attitudes (Ware and Mann 2012; Hackett 2000). Avoiding collusion is essential as the reinforcement of offense-supportive cognitions may increase the recidivism risk. To avoid collusion it is important to name abusive acts for what they are—violence or abuse—while simultaneously showing genuine empathy and positive regard for the individual as a person (Andersson 2022). In other words, disapprove of the act but not the person (Braithwaite 2020). Taking responsibility for one's abusive acts is typically a relational process that takes time (Andersson 2022; Kruse 2020). However, it is not the researcher's role to facilitate this process when conducting research.

Self-Awareness and Self-Care

While the sections above have highlighted various inner conflicts and emotions that can affect the researcher, the question remains how to counteract these processes and potential pitfalls that may create barriers to ethical practices. I propose that continuous self-reflection and self-care in the form of social and emotional support, as well as mindfulness techniques, can be valuable tools. They can prevent secondary trauma symptoms or other emotional reactions that might impede a researcher's ability to act ethically when encountering individuals convicted of serious offenses.

One way to enhance self-awareness and reflexivity is to keep a reflective journal throughout the research process (Braun and Clarke 2019). In this journal, the investigator can record their reflections, insights, emotions, impressions, and evolving ideas. It is recommended to document reflections immediately after each interview. Another suggestion is to have psychotherapeutic supervision throughout the research process. It can also be helpful to use mindfulness strategies such as acknowledging, accepting, and then releasing negative thoughts and feelings, like clouds passing by. As mindfulness theory suggests (Academic Mindfulness Interest Group and Academic Mindfulness Interest Group 2006), we should not avoid or ignore challenging emotions but rather accept them while not letting them interfere with or influence the task.

Additionally, I believe it is essential for professionals to find a supportive community of colleagues who share similar experiences. Reflecting on and processing emotional strain is often difficult with friends and family, not only because of confidentiality concerns but also because many people are unwilling to engage with the dark and complex aspects of this work (Shrim and Baum 2022). Those who have not interacted with individuals convicted of sexual offenses might struggle to relate to the internal conflicts as earlier described by Ievins (2023), and may also be reluctant to hear about the harrowing details of sexual abuse. As professionals in various roles, I believe it is important for us to support one another, as I have personally found such collegial support to be invaluable. Although employing self-awareness and self-care strategies cannot prevent the difficulties that can come with working on confronting issues, it can help to mitigate the short- and long-term negative effects.

Conclusion

The contextual aspects of Sweden might differ from those in many other countries, making it difficult to directly transfer the findings from this chapter. It is reasonable to assume that some ethical issues involved in conducting research on individuals who have sexually offended may pose different challenges in countries with other legal systems. However, there are some basic insights that can nevertheless be generalized from the work presented here. Adhering to principles of human rights, respecting individuals' integrity, and treating others with dignity are generally uncomplicated when dealing with people who have been victims of violent crimes. These responses are based on values that society typically agrees upon, at least in theory. However, the real challenge arises when we engage with individuals who have caused significant harm, such as those who have committed serious sexual offenses. In this chapter, I argue that it is crucial to uphold these very same principles of ethical behavior. Moreover, I would argue that it is in such complex situations that society's commitment to these ethical standards is truly tested.

In this chapter, I have highlighted the ethical obstacles, dilemmas, and challenges involved in conducting research with individuals convicted of sexual offenses. I have discussed potential solutions and emphasized the importance of using person-first language, adopting a trauma-informed and non-confrontational approach, practicing continuous self-reflection, and seeking support from a professional community with shared experiences. As this chapter has also shown, employing ethically responsible research practices with populations of people who have committed crimes not only helps to safeguard their human rights but also facilitates the generation of knowledge that can improve prevention, promote therapeutic intervention, and challenge stereotypes and myths—ultimately benefiting people who have been sexually abused and potentially reducing sexual violence in general. With those goals in mind, there may be great benefit in transferring the approaches and insights described in this chapter to work involving other highly stigmatized populations.

Acknowledgements

I want to express my gratitude to all the interviewees and staff at the Swedish Prison and Probation Service for their support, as well as to Stefan Sjöström, Maritha Jacobsson, Cecilia Kjellgren, and colleagues for their insightful feedback on the research that provides the foundation for this chapter. The research did not receive any external funding.

Legal Sources

Lag om etikprövning av forskning som avser människor [the Swedish Ethical Review Act]. SFS 2003:460. Utbildningsdepartementet [Ministry of Education].

References

Academic Mindfulness Interest Group, Melbourne, and Melbourne Academic Mindfulness Interest Group. 2006. "Mindfulness-Based Psychotherapies: A Review of Conceptual Foundations, Empirical Evidence and Practical Considerations." *Australian & New Zealand Journal of Psychiatry* 40, no. 4: 285–294.

Andersson, Peter. 2022. "Countertransference and Therapeutic Holding: Working with a Perpetrator of Domestic Violence." *Psychodynamic Practice* 28, no. 3: 275–295.

Andersson, Tommy, Rickard Pettersson, and Maritha Jacobsson. 2020, June. "Women and Men Sexually Violated by Closely Related Perpetrators over a Lifespan. Prevalence, Revictimization, and Association to Adverse Childhood Conditions and Experiences." *Children and Youth Services Review* 113: 104897.

Aspers, Patrik. 2009. "Empirical Phenomenology: A Qualitative Research Approach (The Cologne Seminars)." *Indo-Pacific Journal of Phenomenology* 9, no. 2: 1–12.

Association for the Treatment & Prevention of Sexual Abuse. 2023. *Children with Sexual Behavior Problems*. 2nd ed. Association for the Treatment & Prevention of Sexual Abuse. Retrieved from https://members.atsa.com/ap/CloudFile/Download/pgGxjO4p.

Berggren, Kalle, and Lucas Gottzén. 2021. "The Rape Capital or the Most Gender-Equal Country in the World? Masculinity, Hybridity and Young Men's Intimate Partner Violence in Sweden." In *Men, Masculinities and Intimate Partner Violence*, 66–80. Edited by Lucas Gottzén, Margunn Bjørnholt, and F Boonzaier. New York: Routledge.

Bergman Blix, Stina, and Åsa Wettergren. 2015. "The Emotional Labour of Gaining and Maintaining Access to the Field." *Qualitative Research* 15, no. 6: 688–704.

Bladini, Moa, and Wanna Svedberg Andersson. 2020. "Swedish Rape Legislation from Use of Force to Voluntariness - Critical Reflections from an Everyday Life Perspective." *Bergen Journal of Criminal Law and Criminal Justice* 8, no. 2: 95–125.

Blagden, Nicholas, and Sarah Pemberton. 2010. "The Challenge in Conducting Qualitative Research With Convicted Sex Offenders." *The Howard Journal of Criminal Justice* 49, no. 3: 269–281.

Bonta, James, and Donald. A. Andrews. 2017. *The Psychology of Criminal Conduct*. 6th ed. New York: Routledge, Taylor & Francis Group.

Brå [Swedish National Council for Crime Prevention]. 2023. "Anmälda Brott 2022. Slutlig Statistik. [Reported Crimes 2022. Final statistics.]" *Brottsförebyggande Rådet* [Swedish National Council for Crime Prevention]. Retrieved from https://bra.se/download/18.6bb714d018722cf91a6b4b3/1680615035046/Statistikrapport_anmalda_2022.pdf.

Braithwaite, John. 2020. "Restorative Justice and Reintegrative Shaming." In *Criminal Justice Theory*, 1st ed., 281–308. Edited by Cecilia Chouhy, Joshua C. Chocran, and Cheryl Lero Jonson. New York: Routledge.

Braun, Virginia, and Victoria Clarke. 2006. "Using Thematic Analysis in Psychology." *Qualitative Research in Psychology* 3, no. 2: 77–101.

Braun, Virginia, and Victoria Clarke. 2019. "Reflecting on Reflexive Thematic Analysis." *Qualitative Research in Sport, Exercise and Health* 11, no. 4: 589–597.

Brinkmann, Svend, and Steinar Kvale. 2005. "Confronting the Ethics of Qualitative Research." *Journal of Constructivist Psychology* 18, no. 2: 157–181.

Christensen, Ann-Dorte, and Sune Qvotrup Jensen. 2014. "Combining Hegemonic Masculinity and Intersectionality." *NORMA* 9, no. 1: 60–75.

Cowburn, Malcolm, and Lena Dominelli. 2001. "Masking Hegemonic Masculinity: Reconstructing the Paedophile as the Dangerous Stranger." *British Journal of Social Work* 31, no. 3: 399–415.

Fazel, Seena, Julia Cartwright, Arabella Norman-Nott, and Keith Hawton. 2008. "Suicide in Prisoners: A Systematic Review of Risk Factors." *The Journal of Clinical Psychiatry* 69, no. 11: 1721–1731.

Frost, Liz. 2016. "Exploring the Concepts of Recognition and Shame for Social Work." *Journal of Social Work Practice* 30, no. 4: 431–446.

Glasser, M., I. Kolvin, D. Campbell, A. Glasser, I. Leitch, and S. Farrelly. 2001. "Cycle of Child Sexual Abuse: Links between Being a Victim and Becoming a Perpetrator." *British Journal of Psychiatry* 179, no. 6: 482–494.

Gottzén, Lucas, and Anna G. Franzén. 2020. "Othering the Rapist: Rurality, Sexual Violence and the Bjästa Case." In *Rape in the Nordic Countries*, 155–170. Edited by Marie Bruvik Heinskou, May-Len Skilbrei, and Kari Stefansen, London: Routledge.

Hackett, Simon. 2000. "Sexual Aggression, Diversity and the Challenge of Anti-Oppressive Practice." *Journal of Sexual Aggression* 5, no. 1: 4–20.

Haglund, Axel, et al. 2014. "Suicide after Release from Prison: A Population-Based Cohort Study from Sweden." *The Journal of Clinical Psychiatry* 75, no. 10: 1047–1053.

Harper, Craig A., Todd E. Hogue, and Ross M. Bartels. 2017, May. "Attitudes towards Sexual Offenders: What Do We Know, and Why Are They Important?" *Aggression and Violent Behavior* 34: 201–213.

Harris, Andrew J., and Kelly M. Socia. 2016. "What's in a Name? Evaluating the Effects of the 'Sex Offender' Label on Public Opinions and Beliefs:" *Sexual Abuse* 28, no. 7: 660–678.

Harris, Danielle Arlanda. 2021. "Desistance From Sexual Offending." *Current Psychiatry Reports* 23: 1–7.

Holmberg, Stina, and Lars Lewenhagen. 2020. *Reported and cleared rapes in Europe. Difficulties of international comparisons. Report 2020:13.* Swedish National Council for Crime Prevention. Retrieved from https://bra.se/download/18.7d27ebd916ea64de5306c65f/1601393665407/2020_13_Reported_and_cleared_rapes_in_Europe.pdf.

Hydén, Margareta. 2014. "The Teller-Focused Interview: Interviewing as a Relational Practice." *Qualitative Social Work* 13, no. 6: 795–812.

Ievins, Alice. 2023. *The Stains of Imprisonment: Moral Communication and Men Convicted of Sex Offenses.* 1st ed. Vol. 10. Oakland: University of California Press.

Jacobsson, Katarina, and Malin Åkerström. 2013. "Interviewees with an Agenda: Learning from a 'Failed' Interview." *Qualitative Research* 13, no. 6: 717–734.

Jahnke, Sara. 2018. "The Stigma of Pedophilia: Clinical and Forensic Implications." *European Psychologist, Special series: Recent Advances in Criminal Psychology* 23, no. 2: 144–153.

Jeglic, Elizabeth L., Ashley Spada, and Cynthia Calkins Mercado. 2013. "An Examination of Suicide Attempts Among Incarcerated Sex Offenders." *Sexual Abuse* 25, no. 1: 21–40.

Jespersen, Ashley F., Martin L. Lalumière, and Michael C. Seto. 2009. "Sexual Abuse History among Adult Sex Offenders and Non-Sex Offenders: A Meta-Analysis." *Child Abuse & Neglect* 33, no. 3: 179–192.

Jørgensen, Kasper, Annette Olesen, and Lene Wacher Lentz. 2023. "'Det Er Ligesom at Bo i et Glashus'—Politiets Tilsynsbesøg Hos Seksualforbrydere Dømt for Overgreb Mod Børn. ['It's Like Living in a Glass House'—the Police Supervision Home Visits at Sex Offenders Convicted of Sexual Abuse against Children]." *Retfærd: Nordisk Juridisk Tidsskrift* 2, no. 176: 51–67.

Kruse, Anja Emilie. 2020. "The Why, the Who and the Wherefore. Explanations, Self-Change and Social Responses in Men's Narratives of Sexual Violations." Dissertation, copy for defence ceremony, University of Oslo, Norway.

Levenson, Jill S., Yolanda N. Brannon, Timothy Fortney, and Juanita Baker. 2007. "Public Perceptions about Sex Offenders and Community Protection Policies." *Analyses of Social Issues and Public Policy* 7, no. 1: 137–161.

Levenson, Jill S., Gwenda M. Willis, and Claudia P. Vicencio. 2017. "Obstacles to Help-Seeking for Sexual Offenders: Implications for Prevention of Sexual Abuse." *Journal of Child Sexual Abuse* 26, no. 2: 99–120.

Lindberg, Jens, and Stefan Sjöström. 2019. "Påstridigt fältarbete: om rekrytering av svårtillgängliga forskningspersoner. [Persistent Field Work: About Recruitment of Hard-to-reach Research Subjects]." In *Etiska dilemman: forskningsdeltagande, utsatthet och samtycke* [Ethical Dilemmas: Research Participation, Vulnerability and Consent], 119–131. Gleerups Utbildning AB.

Lindegren, Stina. 2025a. "Participants' Experiences of Sex Offender Treatment: Searching for Desisting Narrative Identities." *International Journal of Offender Therapy and Comparative Criminology* 69, nos. 10/11: 1426–1445.

Lindegren, Stina. 2025b. "Looping Disruption: A Relational Mechanism Enhancing Treatment Readiness among Individuals Convicted of Sexual Offending?" *Sexual Abuse* 37, no. 3: 282–308.

Lindegren, Stina. 2024. "Support and Treatment for Men Convicted of Sexual Offending: Readiness, Change, and Previous Help-Seeking." [Dissertation, Uppsala University]. Retrieved from https://uu.diva-portal.org/smash/record.jsf?pid=diva2%3A1849844&dswid=1899.

Lindegren, Stina. 2022. "A Pilot Study of the Swedish Sexual Offender Treatment Program." *Research on Social Work Practice* 32, no. 3: 328–344.

Lomell, Heidi Mork. 2020. "Selvoppnevnte rettshåndhevere: Om fremveksten av 'pedojegere' på nett. [Self-announced Upholders of Justice: About the Development of 'Paedophile Hunters' on the Internet]." *Tidsskrift for Rettsvitenskap* 133, no. 5: 660–690.

Macdowall, Wendy, et al. 2013. "Lifetime Prevalence, Associated Factors, and Circumstances of Non-Volitional Sex in Women and Men in Britain: Findings from the Third National Survey of Sexual Attitudes and Lifestyles (Natsal-3)." *The Lancet* 382, no. 9907: 1845–1855.

Malterud, Kirsti, Volkert Dirk Siersma, and Ann Dorrit Guassora. 2016. "Sample Size in Qualitative Interview Studies: Guided by Information Power." *Qualitative Health Research* 26, no. 13: 1753–1760.

Mann, Ruth E., Stephen D. Webster, Helen C. Wakeling, and Helen Keylock. 2013. "Why Do Sexual Offenders Refuse Treatment?" *Journal of Sexual Aggression* 19, no. 2: 191–206.

Marshall, William L. 2005. "Therapist Style in Sexual Offender Treatment: Influence on Indices of Change." *Sexual Abuse* 17, no. 2: 109–116.

Marshall, William L., Liam. E. Marshall, G. A. Serran, and M. D. O'Brien. 2009. "Self-Esteem, Shame, Cognitive Distortions and Empathy in Sexual Offenders: Their Integration and Treatment Implications." *Psychology, Crime & Law* 15, nos. 2/3: 217–234.

Maruna, Shadd, and Ruth E. Mann. 2006. "A Fundamental Attribution Error? Rethinking Cognitive Distortions†." *Legal and Criminological Psychology* 11, no. 2: 155–177.

McCartan, Kieran F. 2022. "Refining Processes in Policy and Practice in Working with People Accused or Convicted of a Sexual Offense." *HM Inspectorate of Probation Academic Insights* 01: 1–16.

McGlynn, Clare, and Nicole Westmarland. 2019. "Kaleidoscopic Justice: Sexual Violence and Victim-Survivors' Perceptions of Justice." *Social & Legal Studies* 28, no. 2: 179–201.

Mullins, Eve, and Steve Kirkwood. 2022. "Unpicking Social Work Practice Skills: Warmth and Respect in Practice." *Qualitative Social Work* 21, no. 6: 1063–1083.

Mullins, Eve, and Steve Kirkwood. 2019. "Dams, Barriers and Beating Yourself Up: Shame in Groupwork for Addressing Sexual Offending." *Journal of Social Work Practice* 33, no. 4: 369–384.

Oakley, Ann. 2016. "Interviewing Women Again: Power, Time and the Gift." *Sociology* 50, no. 1: 195–213.

Piché, Lyne, Jeffrey Mathesius, Patrick Lussier, and Anton Schweighofer. 2018. "Preventative Services for Sexual Offenders." *Sexual Abuse* 30, no. 1: 63–81.

Pratt, John. 2007. "Scandinavian Exceptionalism in an Era of Penal Excess: Part I: The Nature and Roots of Scandinavian Exceptionalism." *British Journal of Criminology* 48, no. 2: 119–137.

Presser, Lois. 2004. "Violent Offenders, Moral Selves: Constructing Identities and Accounts in the Research Interview." *Social Problems* 51, no. 1: 82–101.

Ranta-Tyrkkö, Satu. 2011. "High Time for Postcolonial Analysis in Social Work." *Nordic Social Work Research* 1, no. 1: 25–41.

Rydberg, Jason, Kelly M. Socia, Christopher P. Dum, and Katrina Cole. 2023. "Shooting the Messenger: How Expert Statements on Stigmatized Populations Negatively Impact Perceived Credibility." *Justice Quarterly* 42, no. 1: 64–89.

Schwark, Sandra, and Gerd Bohner. 2019. "Sexual Violence—'Victim' or 'Survivor': News Images Affect Explicit and Implicit Judgments of Blame." *Violence Against Women* 25, no. 12: 1491–1509.

Serran, G., Y. Fernandez, W. L. Marshall, and R. E. Mann. 2003. "Process Issues in Treatment: Application to Sexual Offender Programs." *Professional Psychology: Research and Practice* 34, no. 4: 368–374.

Shrim, Elad, and Nehami Baum. 2022. "Treating Sex Offenders: Effects on Male and Female Therapists." *Journal of Interpersonal Violence* 37, nos. 3/4: NP1733–NP1758.

Storgaard, Asbjørn. 2022. "Democratic Emancipation in Scandinavian Offender Rehabilitation – Between Policy Discourse and Actual Practice." *Nordisk Tidsskrift for Kriminalvidenskab* 109, no. 1: 228–235.

Tewksbury, Richard. 2012. "Stigmatization of Sex Offenders." *Deviant Behavior* 33, no. 8: 606–623.

Tham, Henrik. 2019. "Har det verkligen inte skett en straffande vändning? En kommentar av Henrik Tham, professor em., Kriminologiska institutionen, Stockholms universitet." [Has It Not Been A Punitive Turn? A Comment from Henrik Tham, Professor Em., Department of Criminology, Stockholm University]. *Nordisk Tidsskrift for Kriminalvidenskab* 106, no. 2: 259–265.

Todd-Kvam, Mari, and David R. Goyes. 2023. "Dreams and Nightmares: Interviewing Research Participants Who Have Experienced Psychological Trauma." In *Ethical Dilemmas in International Criminological Research*, 130–148. Edited by Michael Adorjan, and Rosemary Ricciardelli. London: Taylor & Francis Group.

Ugelvik, Thomas. 2015. "The Rapist and the Proper Criminal." In *Narrative Criminology Understanding the Stories of Crime*, 23–41. Edited by Lois Presser and Sveinung Sandberg. New York University Press.

Walker, Allyn. 2023. "Transphobic Discourse and Moral Panic Convergence: A Content Analysis of My Hate Mail." *Criminology* 61, no. 4: 994–1021.

Ware, Jayson, and Ruth E. Mann. 2012. "How Should 'Acceptance of Responsibility' Be Addressed in Sexual Offending Treatment Programs?" *Aggression and Violent Behavior* 17, no. 4: 279–288.

Willis, Gwenda M. 2018. "Why Call Someone by What We Don't Want Them to Be? The Ethics of Labeling in Forensic/Correctional Psychology." *Psychology, Crime & Law* 24, no. 7: 727–743.

Woodhams, Jessica, and Fazeelat Duran. 2024. "A Model for Secondary Traumatic Stress Following Workplace Exposure to Traumatic Material in Analytical Staff." *Communications Psychology* 2, no. 1: 1–10.

World Health Organization. 2021. *Violence against Women Prevalence Estimates, 2018: Global, Regional and National Prevalence Estimates for Intimate Partner Violence against Women and Global and Regional Prevalence Estimates for Non-partner Sexual Violence against Women: Executive Summary*. World Health Organization. Retrieved from https://iris.who.int/bitstream/handle/10665/341338/9789240026681-eng.pdf?sequence=1.

Youssef, Carollyne. 2022. "Therapeutic Alliance and Its Setting with Those Who Have Sexually Offended and Implications for Community Maintenance Programs." *Journal of Offender Rehabilitation* 61, no. 6: 310–336.

25

ETHICAL CONSIDERATIONS WHEN WORKING WITH POLITICAL EXTREMIST THREAT COMMUNICATIONS

I. M. Nick and Ulrike Preiss

Introduction

According to international law enforcement authorities, recent years have seen a palpable increase in criminal activities perpetrated by political extremist organizations. For example, according to data gathered by the European Union's Agency for Law Enforcement Cooperation or "Europol", in 2021, there were 18 terrorist attacks and 388 arrests made. Just two years later, in 2023, there were 120 attacks and 426 arrests made (European Union Terrorism Situation Report 2024: 10). In the 2024 Global Terrorism Index, it was reported that in 2023, the number of deaths caused by terrorist acts had increased by 22% over 2022, representing the highest level of lethality reached since 2017 (Institute for Economics and Peace 2024: 2). These developments have significantly elevated the demand for research across multiple forensic sciences disciplines.

Within forensic linguistics (FL), there is a pressing need for investigations which can help investigators gather clues about the potential identity of author(s) of political extremist threat communications (PETCs); determine the authenticity of PETCs recovered; estimate the propensity, immediacy, and severity of future criminal actions; and inform strategic decision-making to eradicate or mitigate potential harm to persons and/or property. Moreover, forensic linguistic research into PETCs can also help to increase our collective understanding of the etiology of political extremism so that we are in a better position to safeguard democratic principles and human rights. Despite the many potential benefits of and the obvious need for FL research on PETC, far more work is needed (Ebner, Kavanagh, and Whitehouse 2022).

One of the hindrances to meeting this scientific demand is the difficulty in finding a sufficient supply of authentic material. However, as this chapter explores, the challenge is not merely pragmatic. There are also numerous ethical issues surrounding the procurement and preparation of PETC even before the analysis can commence. And, as this chapter explains, once the data has been obtained, the ethical dilemmas do not end. After providing a brief introduction to political extremism and threat communications, this chapter discusses these ethical issues with regard to right-wing and left-wing PETCs, offering insights from analysis of contemporary and historical radical movements. In addition, practical suggestions are given to help safeguard the FL inves-

DOI: 10.4324/9781003391074-28

tigator's professional and personal boundaries. Despite the specialized focus of this chapter, it is hoped that the ethical questions raised and experience-based recommendations offered here will be applicable and useful for others working with highly sensitive, potentially disturbing forensic linguistic data.

Political Extremism: Forensic Linguistic Research into Extremist Threat Communication

The investigation of political extremism is exceedingly complex. A reflection of this complexity is the lack of unanimity in the terminology and classifications used. There are many definitional questions that have not yet been resolved (Freilich et al. 2024; Duga and Distler 2016; Young 2016; Weinberg, Pedahzur, and Hirsch-Hoefler 2004). Can the terms "terrorism" and "extremism" be used interchangeably, or should they be used to describe different yet overlapping criminological phenomena? If a mass casualty attack was unplanned, is it best to categorize it as a hate crime or is it an act of terrorism? Are crimes committed by lone wolves (i.e., persons who self-radicalize without having direct contact with an ideological group) somehow different than offenses committed by political extremist group members? Despite these and other expert-level debates, it is generally agreed that political extremism is far from being a marginal phenomenon. According to statistics gathered by (inter)national government agencies and watchdog organizations alike, in recent years, there has been a troubling increase in political extremism (e.g., Europol 2024; Bundesamt für Verfassungschutz 2023; Australian Government Preventing and Countering Terrorism and Violent Extremism Report 2022; Bundesamt für Verfassungsschutz 2020). Around the world, individuals and groups with extreme political ideologies are forming and perpetrating acts of violence to impose their political views upon others (Institute for Economics and Peace 2024). Interestingly, although the global incidence of violent acts of political extremism has decreased, the number of victims who have lost their lives in these acts has increased (Institute for Economics and Peace 2024: 2). This means that the lethality of political extremism has risen as perpetrators succeed in finding methods that maximize the number of injured and dead. No country or society is completely immune from experiencing these mass casualty events. Nevertheless, as much terrorist research has determined, there are certain factors which can substantially elevate the potential for violent political extremism. These society-level aggravators include restricted opportunities for socio-economic advancement; widespread marginalization and discrimination; poor governmental enforcement of the rule of law; and long-standing, unresolved inter-group conflicts. If left unchecked, the resulting political fault lines can quickly lead to critical ruptures in the structures which traditionally hold families, communities, and nations together.

From vaccination to immigration, environmental protection, and religious affiliation, the number and variety of political issues which can become weaponized along political party lines of identity demarcation are seemingly endless. Debate, even when contentious, in and of itself is not a threat. What becomes problematic is when tolerance ceases to exist for differing opinions and the people who hold them. Once that happens, segments of ideological groups may resort to the use of violence to exert control over their opponents. The challenge for the criminal justice system is to anticipate when communication has crossed the threshold between debate and threat. Though always difficult, this differentiation has become especially arduous in a period of time when it has become increasingly commonplace to make political comments in a form or manner which violates long-standing norms of polite, civil, respectful conversation. In such an altered communicative landscape, where language is regularly used as a weapon to cause injury, it is difficult to determine which menacing messages may be harbingers of violent action. Which one of the countless online and offline communications was produced by that small set of authors who pose a serious and immediate danger of escalating from threat to execution, be it spontaneous or with malice aforethought? This is where FL examinations of threatening communications can be of invaluable service.

Described briefly, this body of work can be divided into several different investigatory foci. These include but are not limited to the following: (1) seeking to identify linguistic and paralinguistic features that correlate significantly with an increased propensity for committing physical acts of violence (Baele et al. 2023; Scrivens et al. 2022; Kleinberg, van der Vegt, and Gill 2021; Baele 2017; Glasgow and Schouten 2014; Smith 2008); (2) comparing the language lone-actor and terrorist groups (Baele 2017; Kaati, Shrestha, and Sardella 2016); (3) investigating questions concerning the potential authorship of threat communications (Windsor 2020; Nini 2018; Abbasi and Chen 2005; Smith and Shuy 2002; Miron and O'Douglas 1979; Frind 1966); (4) seeking to determine overarching linguistic characteristics of (political) extremism, terrorism, and threatening communications in general (Tompkinson 2023; Ebner, Kavanagh, and Whitehouse 2022; Bojsen-Møller et al. 2020; Muschalik 2018; Nick 2018; Ebling et al. 2013; Fraser 1998; Storey 1995; Klemperer 1947); (5) exploring historical continuities between the rhetorical strategies and language policies of contemporary political extremist groups with their ideological antecedents (Nick 2019; Lanzke 2016; Maas 1984); (6) testing automated systems for detecting threatening communications in online discourse (Jaki et al. 2019; Kaati, Shrestha, and Sardella 2016; Neuman et al. 2015); (7) investigating the ways in which political extremist groups use language to perpetuate and/or activate widespread prejudice, discrimination, and acts of violence against targeted "others" (Williams and Tzani 2024; Nick 2019; Schuppener 2010; Baugut and Neuman 2022; Koehler 2014; Abbasi and Chen 2005); (8) researching social media and threatening communications (Hiaeshutter-Rice and Hawkins 2022; Torregrosa et al. 2022; van der Vegt et al. 2019); (9) investigating the linguistic mechanisms used to first recruit and then radicalize individuals to the extent that they are ready to take their own lives and the lives of others (Hollewell and Longpre 2021; Windsor 2020; Koehler 2014); and (10) researching threatening communications and the law (Gales 2019; Karoli Christensen 2019; Rohr 2017; Solan and Tiersma 2015).

Given the rising risks to public safety, there is a pressing need for scientific investigation into political extremist threat communications. However, the amount of FL research conducted in this area remains disappointingly low. One of the main reasons for this discrepancy is the notorious difficulty in obtaining authentic samples for investigation. Interestingly, the problems in gaining access come from two opposing parties. On the one side, there are the offenders who communicate in clandestine networks in order to protect both their members and their political objectives. On the other side, there are the law enforcement and government decision-makers who are loathe to share highly sensitive data, save for a small set of highly vetted individuals. As a result, those researchers who elect to study this area must first overcome the formidable two-sided obstacle of identifying and accessing appropriate text materials.

As this chapter points out, however, the initial challenge facing the PETC researcher is not solely pragmatic: i.e., how to get their hands on these materials. There are also many difficult ethical questions around how to handle such materials once they have been gathered. The focus of this current chapter is examining many of these ethical concerns. More specifically, this chapter presents insights and recommendations from the authors based on their separate, yet complementary, PETC research. While Ulrike's work involves the in-depth examination of a corpus of contemporary right-wing and left-wing PETCs provided by the federal police authorities, Iman's research involves the examination of historical PETCs produced during the Third Reich and gathered from archival records. Although the time period and the data recovery methods of the two different researchers differed, the ethical issues they faced and the solutions they found were much the same. Importantly, the commonality of these strategies was not limited to the treatment of highly sensitive PETC documentation. It also extended to the ethical imperative to protect oneself and others, be they members of the FL community or beyond. The chapter ends with a brief summary of the primary points made and suggestions for other researchers interested in exploring this important area of forensic linguistics.

The Ethical Challenges of Obtaining a Sufficient Supply of Authentic PETC Data

Today, thanks in no small measure to the internet, it is very easy to locate messages with threatening content. Examples of hate-filled speech can be readily found on most, if not all, social media platforms, chat forums, internet news channels, etc. In addition, as a by-product of the general public's seemingly insatiable interest in true crime, there are countless internet articles, blogs, and posts that provide facsimiles of evidentiary material, such as authentic threat letters from past forensic cases. FL researchers interested in examining PETCs, therefore, may decide to use an expanding number of tools to scrape[1] data from these internet sources for use in their investigations. Without doubt, these tools make it possible to gather large amounts of information which would otherwise be difficult to impossible for the forensic linguist to gather manually. However, data scraping comes with risks. Aside from serious questions surrounding the accuracy and quality of the materials gathered, there are important ethical issues involved.

First and foremost is the issue of informed consent. It is imperative to first determine who and/or what entity or entities is the legal owner of the language in question and then to obtain the necessary permissions before scraping. In many cases, the owner may be the legal proprietor of the hosting platform. Failure to obtain permission may result in prosecution for a wide assortment of legal infractions, including the illegal access and use of data, violation of terms of use, plagiarism, copyright infringement, and even electronic trespassing (Krotov, Johnson, and Silva 2020; Chiauzzi and Wicks 2019; Gold and Latonero 2018; Hirschey 2014; Bassett and O'Riordan 2002).[2] Obtaining official permission of the platform owners may satisfy legal requirements, but it may not be enough ethically if the persons who produced the original texts did not consent to having their communications seen by a third party (Hudson and Bruckman 2004), let alone used for a forensic investigation. Ethically speaking, it is highly advisable to also obtain permission from the platform users before their data is extracted. Therein lies the practical problem, particularly when large sets of data from unknown, sometimes even untraceable, groups of users are involved.

In such cases, it may not be feasible or even possible to obtain all users' permission. There are those who may contend that, under such circumstances, asking individual user permission is not necessary as the sheer size of the dataset realistically precludes the identity of any one user being revealed. Aside from the fact that today's technology has made it shockingly easy to obtain the identity of a single individual on the basis of very little information (Chiauzzi and Wicks 2019; Gold and Latonero 2018; Zook et al. 2017), that conveniently ignores the fact that, regardless of whether the identity of the unwitting research "participant" is revealed, the users' expectations of privacy have nevertheless been violated and the ethical scientific norm of informed consent has been breached (Ess 2018; Hirschey 2014). And this is to say nothing of the complex of intersecting data protection laws which may be triggered by the varying nationalities of the platforms and users involved. Complicating the issue even further is the potential that a portion of users may be made up of "vulnerable persons" which then necessitates a significantly heightened level of care. The fact that data is available in the public domain does not guarantee lack of harm when it is used by a third party (Zook et al. 2017), no matter how noble their objectives may be. Moreover, it is also important to keep in mind that the data yielded from web scraping may very well contain illegal material, the transfer or possession of which may be considered illegal (Brewer et al. 2021).

So, as convenient and simple as large-scale data scraping may appear at first, in reality, the practice involves many complex ethical and legal challenges. As seductive as these large datasets may initially seem, they come with heightened responsibilities. Just as in a small-scale study, researchers who elect to work with large datasets have an ethical responsibility to do as much as possible to protect each and every one of the study participants. This responsibility of care does not diminish with each participant obtained. We must always remember that behind every data point is "a person somewhere in that process that could be directly or indirectly impacted by the research" (Markham and Buchanan 2012: 7). As will be described in the following segments, immediately after the hurdles surrounding informed consent have been resolved, the next obstacle is often not far behind: namely, finding enough verifiably authentic data that fits the specific linguistic parameters of the investigation.

Ulrike's Investigation of Contemporary PETC

Ulrike's empirical research centered on investigating author-specific style features of PETC (Preiss 2024). It was therefore necessary to minimize the possibility that the different text characteristics found were actually a function of text-type variation. This meant controlling several variables. The basic content of the extremist communication was restricted to two sub-groups: (1) threatening and defamatory messages; and (2) responsibility claims. The PETC dataset was limited to analog texts. Digital communications were not included. The type of political ideology examined was either right-wing or left-wing extremism. In addition, as the analysis was to be synchronous, the texts had to have been produced within the same period of time, i.e., between 1999 and 2011. Given all of these selection criteria, it was rather challenging to gather a PETC corpus large enough to run the scheduled statistical tests. The difficulties of independently compiling such a specialized corpus are formidable.

The time needed and the resources required can easily exceed those available to an individual researcher. An excellent solution is requesting PETC materials from cooperating law enforcement agencies. As Freilichet al. (2024) caution, as a general rule, "police agencies usually share data only with select researchers" (190). This reticence is understood given the sensitivity of the information. In addition, given their own limitations, securing assistance from local police departments can also become prohibitively time-consuming for the researcher. Perseverance is necessary. Luckily, there are some larger-scale law enforcement institutions that have their own collection of forensic materials already on-hand that forensic investigators can apply to use for research purposes. In Ulrike's case, that institution was the Federal Criminal Police Office of Germany or the "Bundeskriminalamt" (BKA) which had already compiled PETC corpora for their own investigatory purposes. After application and official review, Ulrike was given permission to access the agency's store of PETC.

When using information gathered by law enforcement agencies, it is important to bear in mind possible biases that exist in the dataset. There are groups which may be over-represented or under-represented, given professional biases or governmental policies that (dis)favor the arrest and/or prosecution of particular segments of the population. The procedures involved in the collection may be radically different than those of the academic researcher, as a result of the differing professional needs for the dataset (Bowen 2018). The data may also contain information that is pertinent for policing but tangential or entirely irrelevant for an FL investigation. To avoid the potential of this extralinguistic information biasing the researcher, a proactive protocol for cleaning the data must be devised in advance. Failure to consider these points can inadvertently taint the findings. Ultimately, devising and implementing a solid plan of action for maintaining proper data hygiene is an ethical requirement of any research project. However, when working with particularly sensitive data provided by an official agency, mistakes or lapses in data management and security protocols can have legal repercussions for the researcher, colleagues, and their institutional affiliation, as well as for the official agency which graciously provided access to the data. Breaches in security, no matter how innocent, can also negatively impact the reputation of all parties involved in the eyes of the general public. Given these risks, officials tend to exercise extreme caution in selecting the researchers with whom they will share data. The fact that one has been granted access is indeed a privilege. Nevertheless, researchers are ethically obligated to refrain from engaging in any activity which may compromise their ability to render independent, fact-based assessments of the linguistic evidence provided to them. Being appreciative of the opportunity to work with official data must not be confused with feelings of being beholden. If there is any allegiance which the researcher may have in the handling of forensic data, it is to the facts.

Iman's Investigation of Historical PETC

One of Iman's areas of linguistic specialization is forensic onomastics which can be basically described as the investigation of names and naming within the legal sphere. Much of her work on right-extremism has involved examining the ways in which naming policies were instituted by law makers during the Third Reich to systematically identify, isolate, and implement the mass extermination of targeted victims. Her research also examines how Shoah survivors used these discriminatory onomastics policies to help engineer their escape from persecution during the War; and how, after the fall of Hitler's Reich, Nazi offenders constructed aliases to escape prosecution for their crimes against humanity after the War (Nick 2019). As the basis for this historical research, Iman not only interviewed survivors, eyewitnesses, and their relatives, but also conducted years of archival research, visiting numerous museums and archives with special collections devoted to the Holocaust and National Socialism.

The materials contained in these holdings are scrupulously vetted by historical experts for their authenticity. This means that researchers who are able to gain permission to use these collections have an invaluable store of historical documentation at their disposal as well as a team of archivists and research librarians who can offer expert assistance. Moreover, most modern archives, large and small, now have computerized systems that allow for quick and easy digital searches of their collections. These automated catalog systems can certainly save investigators considerable time and money when attempting to locate relevant documentation. However, as any investigator who uses archival documentation can attest, there is often terrific variability in the way automated search services operate.

While some institutes organize their materials across nationally or even locally relevant time periods, others organize their holdings primarily by subject-matter or language, and still others use a combination. Alongside these organizational differences, there can also be considerable variation in the terms used as keywords to tag the materials in the database as well as radical inconsistencies in the accompanying content descriptions provided for each file. This means that researchers must invest the time to become intimately familiar with the idiosyncrasies of each archive to successfully identify file holdings which might potentially contain useful information for their investigation.

Once a set of potentially useful files has been identified, it is not unusual to find, however, that some of the most interesting and important documents have been placed under a temporary seal. In some cases, these seals have been imposed at the request of a private citizen who donated the materials under the condition that they not be made available for public view for a predetermined period of time. In other cases, it is a government agency that has sealed the documents in the interest of protecting (inter)national security and relations. Immediately after World War II, for example, a number of these seals were placed on particularly sensitive military documents, either completely or partially restricting their use for several decades—typically until the person(s) involved were assured to have died. Even now, nearly eight decades after the War's end, there are still some records that still remain sealed. However, progressively, scientists whose credentials and objectives have passed review, are being granted access to those once highly top-secret materials. Although many of the actors who were involved in the activities and events described in these records have long since passed, this is not always the case; and even if the actors themselves have died, their relatives may still be alive. For that reason, extreme care and sensitivity are required in the way the information gained from these once censored materials is handled.

Ethical issues surrounding the use of historical PETC documentation are not limited to the responsible use of information, however. They also subsume the physical handling of the archival materials. Among those documents that have been de-classified and are now available for scholarly use, a goodly number have not been digitalized. From yellowing handwritten notes to brittle typewritten reports, these documents are extremely fragile and often difficult to read thanks to a combination of factors. These include the gradual erosion of ink and paper, the introduction of unwanted stains, and changes in handwriting conventions. In the 1970s, in an effort to protect such delicate documents, there was a general movement to create microfiche reproductions. Once hailed as a major step forward, these reproductions are notoriously difficult to read and devilishly expensive to print onto paper. In the face of these problems, many modern document users may be frustrated to discover that cameras are still strictly prohibited by museums and archives. While the logic behind restricting the use of flash cameras on light sensitive documents is generally understood, with the introduction of the modern smartphone, such restrictions may seem ridiculous. However, there are a number of reasons why strict prohibitions against unauthorized reproductions exist.

For example, as it is impossible to check whether the cameras of private users truly pose no threat to sensitive documents, a general prohibition is the only sensible answer. It must also be remembered that payments made to the official reproduction centers are a source of income for the chronically underfinanced repositories. The reproduction orders received may serve as important sources of data regarding the frequency and type of document use provided by the institution. These records, by extension, can then be used by administrators when it comes time to apply for new or continued funding. Finally, reproducing the materials without the express permission of the holding institution may, in fact, constitute copyright infringement which can have serious consequences for individual researchers, their projects, and any institution with which they are affiliated.

In addition to all these points, it is important to keep in mind that violating the use protocols placed on archival collections ultimately violates the trust that is essential to the relationship between archivists and researchers. At the end of the day, respecting the rules is an issue of common courtesy: when you are a guest in someone's house, it is incumbent upon you to respect their rules. With that adage in mind, it should not be assumed that the absence of a prohibition is tantamount to permission. When in doubt, ask.

A final ethical point to keep in mind at all times: the materials held in archives represent real people's lives. Oftentimes, the materials that are held in these collections have been donated to the institution or put on loan by individuals whose families were directly affected. It is not unusual to find that in addition to researchers, these archives are visited by survivors and/or family members who are trying desperately to determine what happened to a loved one. The strict policies regulating the use of archival materials were also developed out of respect for these visitors. For researchers, these special collections are of immeasurable scientific value. However, for survivors and their families, they may be the only tangible direct tie to their individual, familial, and national heritage. This aspect is yet another reason for strict restrictions against making unauthorized photos or reproductions within archives. While scientists may only see archives as information repositories, others may consider them as hallowed spaces for reflection and remembrance.

Therein lies yet another reason why it is ethically imperative to assiduously comply with the institutions' regulations of use. This includes being sure to return materials as you found them. Be sure to place all documents in the same file, in the exact same order. This is not only an issue of politeness. Documents that are misfiled are essentially lost. Accordingly, when you locate misplaced information or discover that a document is missing, immediately let someone on staff know. They will appreciate it and so will the next user who comes in after you.

The Ethical Challenges of Handling and Securing PETC Data

In many ways, PETCs must be handled just like any other forensic data: namely, very carefully. Once permission for use is secured, it is important to understand that this authorization rarely extends to other individuals. Depending on the sensitivity of the data, those prohibitions may even extend to supervisors and colleagues. Limitations may also be placed on presenting or publishing findings. Even when use of short extracts and examples from the original PETC sources is permitted in scientific arenas such as academic conferences, researchers may be required to ensure that audience members do not record any of the material presented. Even if these stipulations are not included in the original contract of use, it is crucial that researchers specifically ask about the confines of use and have those terms included in the written contract. That way it is clear to all involved parties what the data is to be used for, how it is to be handled, and who is permitted access.

Data storage requirements should also be addressed. Password-protected hard drives, locked rooms, privacy filter on your screen—all these are examples of how to keep sensitive data safe. However, these precautions may not be enough. It may also be recommended and even required that all PETC data be deleted and/or destroyed as soon as the research project and its associated publications have been completed. In instances where the destruction requirement is not imposed, it is essential that data be maintained securely. Respecting these regulations is a question of ethical responsibility not only for the persons, groups, and organizations represented in the data, be it directly or indirectly. It is also out of ethical consideration to persons who have no direct relationship to the data and may be harmed by it. In an age when working from home has become increasingly common, extra care must be taken to protect family members from exposure to forensic material. This warning is especially great in households where minors are present. As many a parent can attest, it is sometimes the youngest members of the family who are the most tech-savvy when it comes to breaching security systems. Even when data is exclusively kept in the workspace, extreme caution is still advised to protect against unauthorized access by colleagues, or worse yet, "bad actors". Sadly, there are many examples in which PETCs have served as criminal inspiration for recruiting and radicalizing others to take criminal action that ends in destruction, injury, or death. Preventing this risk of such copy-cat crimes is yet another reason to keep PETC data securely.

Confidentiality must be maintained as well. This frequently entails properly filtering the data to protect the privacy and safety of all persons involved, be they victim or victimizer, however directly or indirectly. There are many different strategies which can be utilized for anonymization.

Many involve the removal or substitution of names that can lead to identification. However, names are not the only linguistic element that can serve as personal identifiers. In Ulrike's contemporary PETC corpus, several letters made indirect reference to a particular politician or public figure, by describing political actions, certain events, or public statements. With such information—which is often impossible to entirely anonymize without losing the necessary context—any reasonably well-informed reader would probably be able to correctly identify the person referenced, either immediately or after consulting some outside source. It is impossible to completely avoid such a chain of associations being made, but being aware of this potential is critical. Only then can appropriate preventative steps be taken.

Although Iman's PETC research into Nazism often involves archival data, this ethical recommendation is the same. Even without specifically mentioning an individual's name, descriptions about a location and/or event can serve as important identification clues. This potential for discovery is heightened by the number and variety of often digitalized resources available where complete names are provided. In view of that fact, extreme caution must be exercised to protect people's identities. One of the primary methods of achieving this goal is, of course, anonymization. As straightforward and simple as that might sound, in practice, there are many ethical questions that need to be answered before implementation.

Who Should Anonymize the Data?

The question of who should anonymize PETC data depends not only on the above-mentioned use restrictions but also on the study design. At times, it may be determined that having an authorized third party perform the anonymization is the most advisable option to reduce the risk of biasing team members responsible for analyzing and/or interpreting the dataset. However, this strategy is not without drawbacks. Increasing the number of people who are responsible for anonymizing the data can increase the potential for errors, misinterpretations, and/or inconsistencies. To guard against this happening, it is imperative that a policy of procedures be devised, training be provided, and a protocol be developed for resolving potential inconsistencies. These procedural precautions can help ensure consistency and reliability of anonymization across all persons involved. That being said, for many projects, there simply are not sufficient financial or staff resources to allow for third-party assistance. When this is the case, the PETC researcher can still develop a set of procedures for anonymization and periodically spot-checking the data for consistency.

If automatized anonymization is used, special care must be taken to perform quality checks. For all of their sophistication, computer programs are often still unable to reliably and consistently differentiate between what should and should not be anonymized, how, and to what extent. Added to potential inconsistencies is the very real possibility that certain items have been missed by the automatic sweep, due to, for example, problems in the legibility of the text, errors or alterations in spelling, and the appearance of nameless descriptions which nevertheless clearly identify an individual. Such unwanted variations can skew the data, confound the findings, and impact the inferences drawn. In FL, this chain of errors can then have serious, long-reaching impact for the society at large, when such tainted FL evidence is used in judicial decision-making, law enforcement practices, and governmental policy development.

Therefore, regardless of whether automatic or manual anonymization methods are used, quality checks must be performed to protect the integrity of the data and any subsequent findings. Quality checks are necessary even when the PETC data given to the researcher has already been anonymized by, for example, a police authority or administrative official during the archiving process. When such pre-screened data is provided, the challenge of anonymization is reduced but not entirely eliminated. It is essential that time and care be invested in reviewing the material to make sure that all sensitive data has in fact been anonymized and possible errors have been corrected. Depending on the diligence of the data provider, performing these checks can be just as challenging and time-consuming as implementing one's own anonymization from scratch. For that reason, although it may be tempting to take the data provider at their word and assume that the data is clean, it is ultimately, however, the professional and ethical responsibility of the data user to guarantee that the PETC material is truly "clean".

What Should be Anonymized?

Although it might sound contra-intuitive when talking about anonymization, there might be certain items in the data that need not or should not be deleted or substituted. A very simple example would be the name "Hitler". Even if one were to replace it, along with all the other surnames in the dataset, with "LastName", the anonymized version of "Heil LastName" would still be readily identifiable as "Heil Hitler". Most would agree, this strategy would not suffice. At the same time, one must ask whether there is truly a need to replace the names of (in)famous historical figures (and places) whose actions and identities are well known. In such instances, unless there is a need to protect a third person, anonymization might be wholly unnecessary. It is sometimes argued that names that have already appeared in the public domain no longer require anonymization. However, the question we should ask ourselves in such cases is what is the motivation for revealing an actor's identity? Is it truly necessary to provide contextualization or clarification? If not, then it is always advisable to err on the side of caution which means protecting people's identities. Although this moral imperative is generally unquestioned when the person involved is considered "vulnerable" (e.g., a victim or a witness of a crime), the situation may appear much less clear in the case of victimizers. There are some, for example, who argue that it is ethically defensible to provide the names of persons who have committed crimes, past or present.

This argument is fraught for many reasons. No matter how heinous an offender's crimes may be, they in no way diminish the responsibility of the researcher to act ethically. The role of the FL scientist is not then to seek retribution but to provide objective evidence for the fair and equitable dispensation of justice. That responsibility is not restricted to those with whom we can easily identify or sympathize, nor does it gradually expire or terminate once the original person(s) involved pass away. Our personal and professional ethical responsibility remains in perpetuity; and it extends not only to the living, but also to the dead. Thus, it is incumbent upon ourselves as both individuals and as a field to exercise extreme caution when handling and releasing data. The guiding principle is "to reveal as little as possible, and as much as necessary". The decision about whether or not to anonymize a certain name or place requires serious thought and must be based on sound, objective reasoning.

The Ethical Challenges of Reporting PETC Data

Throughout academia, it has now become standard practice to issue a trigger warning[3] when a presentation contains materials that some audience members might find objectionable. During linguistics conferences, such warnings are commonly given when a presentation contains insulting, defamatory, or discriminatory language. However, there are some researchers who feel that these warnings are insufficient and that more stringent steps should be taken, such as refraining from reading aloud excerpts from PETC data or masking hate speech with circumscriptions such as "n*" or "nxxxxx".

It is agreed that great care must be exercised when selecting PETC material for presentation. It is not necessary to use excerpts that feature the most extreme insults and swear words when less extreme material can be used to illustrate a linguistic feature. It is also important to remember that the more one works with PETC material, or any forensic text type for that matter, the more de-sensitized one may become. Audiences that have never or rarely encountered such material, such as students, may have fewer or less robust affective shields. One of the tasks of an effective public speaker is to modify the content and style of a presentation to fit the experience, training, and general constitution of a specific audience.

Still, when presenting findings to other linguists, we would argue against altering the original language found in the dataset. In our opinion, replacing letters with "x"s or using an abbreviation with an asterisk reduces the authenticity of the data, which is to be avoided when presenting to other scientists. Yes, it is important not to re-traumatize vulnerable colleagues; and no, we do not want to reinforce the injurious use of language. However, a distinction must be made between accurately reporting inflammatory speech for scientific purposes and personally producing such injurious speech oneself.

In fact, it is because of our acute awareness of and sensitivity to the harm that language can cause that we argue against masking, reducing, or otherwise obfuscating just how radical, vile, and outrageous PETC can be. In fact, we would contend that the inaccurate presentation of PETC language in professional settings, no matter how well meaning, is ethically fraught in that it risks giving a false picture of the true level of threat involved. Further exacerbating the problem is the tendency to use these "masking strategies" rather arbitrarily. In scientific papers and presentations, for example, it is not unusual for circumventions like the "n-word" to appear right next to other epithets that are arguably just as potentially hurtful and insulting. To be completely consistent, one would have to mask any and all demonstrably or perhaps even potentially objectionable language which would, in our opinion, be wholly undesirable for the scientific presentation, discussion, and investigation of threatening language. Just as a medical examiner, crime scene photographer, or forensic odontologist would not only be permitted, but indeed expected, to present completely accurate, highly detailed, maximally unfiltered evidence—which, understandably, lay audiences might find deeply disturbing— so too must the FL researcher be permitted to do the same within formal, scientific, professional contexts.

Does this mean that we believe issuing content notifications given before presenting potentially upsetting PETC examples in professional settings is superfluous? Absolutely not. Even in an FL conference, it is clear that not everyone in attendance works on the same crime type. Just as we appreciate receiving a trigger warning when listening to a presentation by a colleague working on, say, sexual abuse, we do the same when we present our PETC data. In our opinion, offering these alerts is an ethical requirement of creating and maintaining safe work spaces. It is imperative that forensic linguistic gatherings not be turned into harmful environments ruled by the ethos, "If you can't take the heat, get out of the kitchen". That kind of toxic mentality risks driving out highly competent compassionate scientists from our field. It is also deeply ethically fraught to treat forensic data as a kind of professional litmus test.

The Ethical Mandate to Protect Oneself and Others Exposed to PETC Data

The internet and darknet are filled with political extremist content. On the one hand, it can be useful for researchers to surreptitiously observe or even join these criminal circles to gather intelligence about their structure, origins, objectives, recruitment strategies, and criminal activities. FL experts who have used this method have gathered critical data about the linguistic practices of the groups and individuals within these radicalized political circles. Gaining an understanding of the underlying mindset of these organizations and individuals is extremely important. Thanks to the within-group knowledge they have been able to obtain, both scientists and law enforcement officials alike are in a better position to understand the meanings and significance of linguistic features that appear in PETCs.

On the other hand, it is essential not to underestimate the potential danger of these organizations, both online and off. Extreme caution must be exercised before signing up for newsletters, joining chats groups, attending meetings (online or off), or contacting people who are or have been active in extremist organizations. Just as the researcher can secretly gain access to these illicit organizations, they can do the same in the reverse. Offenders in online networks are well known for "hacking back" (Décary-Hétu and Aldridge 2015: 132). It is strongly advised that researchers who are considering using these methods of data collection consult with law enforcement authorities first and maintain contacts throughout the project for advisement and protection (Brewer et al. 2021). It is important to stress that the dangers are limited to those investigating contemporary political extremist organizations.

While working with archival documents, it is not unusual to find that someone has slipped in contact information about an upcoming meeting for an illegal extremist group; or after ordering publications on extremism, researchers can find themselves placed on contact lists for active radical organizations, without their permission. When this happens, it is paramount that the FL investigator take immediate steps to protect their safety and that of their loved ones and associates by reporting the incident to all responsible authorities.

The mandate for self-care is not limited to one's physical safety, however. The prolonged exposure to the disturbing language and content of PETC can also take a mental and psychological toll. Working effectively and responsibly with such materials requires developing situational self-awareness of one's own personal needs and limits, and taking decisive corrective steps to restore one's balance whenever needed. Practicing such self-care is not, as some might criticize, an exercise in self-indulgence. As much research has shown, safeguarding one's own health increases the forensic professional's ability to think critically and act effectively. According to Bryan and Blackman (2011), self-care may even be seen as a "social responsibility", "a proactive process" that allows for "informed balanced processing" (17–19). Therefore, practicing self-care is both a personal and a professional necessity when working with confronting material. In the following sub-sections, the co-authors reflect on their individual experiences with processing their PETC research.

Ulrike's Experiences with PETC and Self-Care

The BKA data I worked with consisted of 114 threat messages with right-wing extremist content and 50 responsibility claims linked to left-wing extremist crimes. There was considerable variability in the texts I examined. Some had only a few words, others were more than ten pages long. Many were handwritten but some were composed on a computer or typewriter. The style ranged from abnormally formal to shockingly vulgar.

In the responsibility claims, the authors frequently gave detailed descriptions of the crimes they had allegedly committed, along with eerily calm, almost indifferent, justifications for why it had been necessary to burn cars, smash windows, block trains, or injure and kill people. By comparison, the threat messages were more often extremely emotional, filled with offensive rantings about and promises to cause harm to targeted victims such as Jews and immigrants. Reading this material initially left me feeling quite powerless and frightened, even though I am not in a group that was being directly threatened.

As a White, Christian, German woman from a well-situated, educated family, I do not have the profile typically targeted in the PETCs I investigated. Still, the fact that many of my dearest friends were members of these threatened groups was more than enough to personally affect me. In fact, I was horrified by the texts I read. The brutality of the language and the resoluteness of the authors was deeply disturbing. Having had the luck of growing up in a very peaceful environment, I had never encountered this kind of aggression before and I admit, confronting it initially scared me.

I very quickly realized that I needed to create some professional distance, not only to be able to conduct a neutral, scientific study, but also to protect myself. Between readings, I made sure to take regular breaks to do something enjoyable and put my mind at ease. I also made a point of speaking with close friends who were also FL colleagues that helped me process what I was investigating. With them, I was able to not only discuss my research findings which were fascinating but also work through my initial discomfort and fear.

Thanks to these self-help strategies, I was able to work on the PETC corpus for several years while remaining happy and healthy. I learned a lot about myself and my limits during this time. I realized that there are many forensic topics I can handle, but there are others I can't and don't want to deal with at all. As an FL researcher, it is important to remain aware of one's personal limits and to never exceed them. Part of that self-protection is accepting that one's personal limits may be very different from those of your colleagues. Some of my colleagues are not able to work with PETC data. I personally draw the line with child abuse cases. Resisting the urge to compare yourself with others who can seemingly endure more is imperative. It is only when we know and respect our own personal borders that we do our best professional work. In a very real sense, personal self-care is integral to being an ethically responsible and effective forensic scientist.

Iman's Experiences with PETC and Self-Care

As an African American woman and the daughter of two academicians who spent their lives fighting for Civil Rights, I was introduced very early on to the dangers of politicized racial hatred. My awareness was not limited to intellectual theories, but was informed by direct personal experience. I will never forget the first time I saw the Ku Klux Klan march down Main Street in full regalia or felt the sting of being called the n-word by teachers in school. It is against that background that I took my first course on the Holocaust as an undergraduate with a double major in German and Psychology. That course literally changed my life. After I graduated, I moved to Germany, where I have continued to research hatred and the Holocaust ever since. As a visible foreigner in Germany, I have been personally attacked by Neo-Nazis many times. I have been spat on, shoved, punched, had my hair pulled, interrogated by police and so-called "well-meaning citizens"; I have been refused service and medical attention; and over the years I have been called an assortment of filthy names to my face and in anonymous messages. Most of these assaults have been from complete strangers, but some have been from people I knew. These experiences are not peculiar to me. Over the many years of university teaching in Germany, I have witnessed several of my students go through similarly painful events.

I will always remember the day one of my students from Poland came into my office hour, crying. She explained that she had lost her way in the city and had asked a friendly looking pair of elderly women if they could help her find her way back to the university. They stopped and smiled and one of them said, "You have an interesting accent. Where do you come from?" My student explained that she was a Polish exchange student. Hearing that, the women laughed out loud before one proclaimed, "Slavs make excellent slaves!" and the other chimed in, "And soap!" I consoled my student, in the same way that my own professors had comforted me so many years before. I write all this to say that I am certainly no stranger to hate-fueled extremism. Still, the first time I laid my hands upon original archival records documenting the annihilation of millions of people was quite disturbing. The shocking level of planning contained in those records was only matched by the depth of the perpetrators' relentless depravity. There are many FL experts who maintain that the only way to properly deal with such data is to focus on providing a factually accurate analysis. I agree.

As a forensic scientist, it is imperative not to allow one's personal feelings to impede one's professional responsibility to provide accurate, factual, evidence, as devoid as possible of cognitive bias. Indeed, I would argue that the more egregious the crime, the greater the public need is for accurate and reliable scientific evidence that is as free as possible from personal prejudice. At the same time, I also believe strongly that striving for professional excellence must not be confused with becoming callous, being inconsiderate, or devoid of personal empathy or conscience. Ultimately, I do what I do *because* I care and not in spite of that fact. In my opinion, one of the reasons why FL is such a societal asset is that its methodological tools and theories can help provide the evidence and insights needed to reveal past injustices and prevent more from occurring in the future. For that potential to be reached, it is, in my opinion, essential that we nurture the humanity within ourselves and our discipline. To do this, we must prioritize taking care of ourselves and taking care of those around us. When we are exhausted from extreme levels of stress thanks to heavy workloads, confronting materials, and competing responsibilities in our professional and private lives, we are unable to perform at our highest level. In other words, failing to take care of ourselves increases the propensity for errors. While this is never desirable, it is particularly problematic when working with data of such potential import. For that reason, taking time out to take care for one's self and those we love is not a luxury, it is an ethical necessity.

Conclusion

On the 20th of December 2024, in the German city of Magdeburg, a 50-year-old psychiatrist drove to the annual Christmas market in a black BMW. Once there, he maneuvered his vehicle around the protective barriers and drove directly into an unsuspecting crowd of visitors, injuring more than 200 people before police successfully apprehended him. In the investigation that followed, it was determined that the assailant had sent a flurry of increasingly disturbed PETCs in which the self-described Islamophobe and Elon Musk fan praised right-wing extremists' efforts to stop immigration from Middle Eastern countries. It was further revealed that, for more than a decade, the assailant had sent PETCs to a variety of German officials, news agencies, and private citizens. For example, in February 2023, he used a contact form to send the following message to officials in the German Ministry of the Interior: "Is it necessary to kill 20 people on the street, in order to get justice in Germany?"[4] (*Die Zeit* 2025). Then, in May, he posted: "German terrorism will be brought to justice. It's very likely that I will die this year in order to bring justice" (*Die Zeit* 2025). Less than eight months after that PETC, he committed one of the largest mass casualty assaults in German postwar history, leaving officials and the general public asking, "Why didn't the authorities take more decisive action to prevent this tragedy from happening?" (*Die Zeit* 2025).

Exactly 12 days later, a white Ford pickup truck plowed into a crowd of revelers celebrating the New Year on New Orleans' famous Bourbon Street. After crashing the vehicle, the 42-year-old offender opened fire upon police officers before he was finally subdued. The FBI's search of the offender's vehicle yielded an ISIS flag and two homemade explosives. Subsequent review of CCTV images taken before the attack revealed images of the assailant walking through the French Quarter where he placed other explosive devices. The offender was identified as a US Army veteran who had become politically radicalized (Sarnoff 2025). According to law enforcement, hours before launching his attack, the perpetrator had searched online for information about the Christmas market attack in Germany. At last count, the New Orleans attack resulted in the deaths of 15 people and the injury of 57 others (Frazier 2025). Although, during the writing of this chapter, investigations are still underway, authorities have indicated that preceding his attack, the assailant made a series of video-recorded PETCs (Frazier 2025).

These two mass casualty terrorist attacks took place in two different countries and were committed by offenders with diametrically opposing political ideologies. Nevertheless, the actions of these two violent offenders have much in common. In both cases, after a period of radicalization, the assailants used violence to seek revenge for perceived injustices in their personal lives and publicly declare their allegiance to an extremist ideology. While such acts were once exceedingly rare, that is no longer the case. As mentioned at the start of this chapter, in countries around the world today, we are seeing a disturbing increase in violent political extremism. This trend appears to be due to a proliferation of factors that provide fertile ground for these acts of terror from radicalized actors. Some of the factors that research has determined may precipitate a dangerous shift toward extremism include personal grievances, a lack of social supports, social isolation, and prolonged exposure to hate-messages that target others as the source for personal and social evils (Pilkington 2022; Bilewicz and Soral 2020; McCauley and Moskalenko 2008). Sadly, the increase in these factors in nations around the world has been accompanied by dramatic increases in violent political extremism. According to the 2024 Global Terrorism Index report of the Institute for Economics and Peace (2024): "Globally, higher levels of political terror, lower respect for human rights, high levels of militarization, the existence of policies targeting religious freedom, group grievance, and political instability all correlate with higher levels of terrorism" (48).

One factor that is making identifying and apprehending these offenders so challenging is the fact that their demographic backgrounds and ideologies are becoming so mixed and variable that they deftly defy traditional methods of profiling (Brace, Baele, and Ging 2023; Roberts and Wallner 2023). During Congressional Testimony, former FBI Director Christopher Wray described this emerging form of "mixed, unclear, and unstable" ideological extremism as "salad bar extremism": an analogy based on this emerging trend for offenders to pick and choose seemingly contradictory ideologies in unexpected and difficult to predict ways (Kidwai 2024). One behavioral factor which political extremist perpetrators do still seem to share, however, is their tendency to produce violent PETCs in advance of their attacks (i.e., "leakage") (Brown et al. 2024; Rose and Morrison 2023; Ebner, Kavanagh, and Whitehouse 2022; Dudenhoefer et al. 2021; Ellis et al. 2016; Cohen et al. 2013; Meloy et al. 2012; Meloy and O'Toole 2011). It is here that FL researchers can provide an invaluable service in helping to combat this escalating threat to domestic and international security. There are so many questions which FL is in a unique position to help answer. For example, there is a pressing need for FL investigations that compare the PETCs of so-called "lone wolf" perpetrators and extremist groups. Also needed are FL comparisons across the political spectrum as well as between differing languages, cultures, nations, time periods, and communicative media (e.g., gaming vs. traditional online social media platforms). And, given the recent technological developments, there is also a pressing need for FL research that examines AI-generated PETCs. There is also a critical need for more research on the victims of PETCs. These are just a few of the issues that FL is particularly well primed to address. Without doubt, the examination of PETCs is a truly fruitful area of FL research with the potential to greatly expand our collective scientific understanding of this growing criminological phenomenon.

As aggression, hostility, belligerence, and general intolerance continue to permeate everyday parlance, both online and offline, FL investigations into PETC can be applied to assist law enforcement to ferret through that noise and identify credible and imminent risks to property and lives. Work in this area can also shed light on the weaponization of language with an eye toward developing effective safeguards against the misinformation and disinformation that can destabilize vital democratic institutions. As Fontana, Nannicini, and Tabellini (2023) assert, the danger represented by these extremist activities must not be underestimated, for "[i]t is well understood that the presence of large extremist parties can hinder the functioning of democratic institutions. In the most severe cases, an extremist party can directly threaten the viability of democratic rule" (723). As important as it is to recognize this danger, in order to help guard against it, it is imperative that forensic linguists exercise the utmost care to ensure that the manner in which we procure, process, and present data does not violate the very ethical principles we are duty bound to help protect.

Notes

1 Web scraping is sometimes referred to as "mirroring", "spoofing", or "cloning". It involves the extraction of data from websites. The term "web crawling" refers to the automated process of locating, scanning, and indexing internet sources. In big data investigations, both web scraping and web crawling are commonly used in tandem. As mentioned in this chapter, there are an increasing number of artificial intelligence (AI) tools which can be used for these processes. However, it is ethically problematic to use these or any other automated tools without understanding in depth precisely how they operate. Without that knowledge, it is difficult to impossible to ensure that their operation does not involve functions which are in violation of established norms for protecting research subjects and the society at large. Tools, for example, that store data retrieved in cloud services, however temporarily, may violate confidentiality regulations. In addition, care must be taken to ensure that the data gathered does not allow backdoor access by unauthorized persons, or create copies for use outside of the control of the researcher for purposes beyond those expressly permitted by the initial Ethics Review Boards.

2 For more on the (il)legalities of web scraping and web crawling, interested readers are encouraged to consult the list of legal cases provided below.

3 Although trigger warnings have become widespread practice, the empirical psychological research conducted to date has uncovered compelling evidence that they may not be nearly as effective in protecting audiences against experiencing negative emotions or re-traumatization as is generally believed. (For more see Bridgland, Jones, and Bellet 2024; Kimble et al. 2021; Jones, Bellet, and McNally 2020; Dickman-Burnett and Geaman 2019; Sanson, Strange, and Gary 2019.) Given the importance of this issue for instruction and presentation in the forensic sciences and allied disciplines, the quantitative research that is beginning to emerge deserves special attention.

4 The English translations of the German originals were provided by the co-authors.

Legal Cases

eBay v. Bidder's Edge, Inc., 100 F. Supp. 2d 1058, 54 USPQ 2D (BNA) 1798 (ND Cal. May 23, 2000)
Ticketmaster Corp. v. Tickets-com, Inc. US District Court, CD California March 7, 2003
Facebook v. Power Ventures, 844 F.3d 1058 (9th Cir. 2016)
Associated Press v. Meltwater, 931 F. Supp. 2d 537 (2013)

References

Abbassi, Ahmed and Hsinchun Chen. 2005. "Applying Authorship Analysis to Extremist-group Web Forum Messages". *IEEE Intelligent Systems*. 20, no. 5: 567–75.

Australian Government. 2022. "Preventing and Countering Terrorism and Violent Extremism Report." Accessed March 15, 2025. Retrieved from https://www.dfat.gov.au/sites/default/files/Preventing

-Countering-Terrorism-Violent-Extremism-2022-26-Australias-International-Engagement-Update-Way-Ahead.pdf.

Baele, Stephane 2017. "Lone Actor Terrorist Emotions and Cognition: An Evaluation Beyond Stereotypes." *Political Psychology* 38, no. 3: 449–468.

Baele, Stephane, Katherine Boyd, Travis Coan, and Elahe Naserian. 2023. "Lethal Words: An Integrated Model of Violent Extremists' Language." *Studies in Conflict and Terrorism.* Retrieved from https://www.tandfonline.com/doi/epdf/10.1080/1057610X.2023.2213963?needAccess=true.

Bassett, Elizabeth, and Kate O'Riordan. 2002. "Ethics of Internet Research: Contesting the Human Subjects Research Model." *Ethics and Informative Technology* 4, no. 3: 233–249.

Baugut, Philip, and Katharina Neuman. 2022. "Online Propaganda Use Driving Islamist Radicalization." *Information, Communication, and Society* 23, no. 11: 1570–1592.

Bilewicz, Michał, and Wiktor Soral. 2020. "Hate Speech Epidemic: The Dynamic Effects of Derogatory Language in Intergroup Relations and Political Radicalization." *Political Psychology* 41 (Supplement 1): 3–33.

Bojsen-Møller, Marie, Sune Auken, Amy Devitt, and Tanya Karoli Christensen. 2020. "Illicit Genres. The Case of Threatening Communications." *Tidsskriftet Sakprosa* 12, no. 1: 1–53.

Bowen, Robin. 2018. *Ethics and the Practice of Forensic Science*. London: CRC Press.

Brace, Lewys, Stephane Baele, and Debbie Ging. 2023. "Where Do 'Mixed, Unclear, and Unstable' Ideologies Come From? A Data-Driven Answer Centered on the Incleosphere." *Journal of Policing, Intelligence and Counter Terrorism* 19, no. 2: 103–124.

Braun, Christian A. 2007. *Nationalsozialistischer Sprachstil. Theoretischer Zugang und praktische Analysen auf der Grundlage einer pragmatisch-textlinguistisch orientierten Stilistik.* Heidelberg: Winter.

Brewer, Russell, Bryce Westlake, Tahlia Hart, and Omar Arauza. 2021. "The Ethics of Web Crawling and Web Scraping in Cybercrime Research: Navigating Issues of Consent, Privacy, and Other Potential Harms Associated with Automated Data Collection." In *Researching Cybercrimes*, 435–456. Edited by Anita Lavorgna and Thomas J. Holt. Berlin: Springer.

Brigland, Victoria, Payton Jones, and Benjamin Bellet. 2024. "A Meta-analysis of the Efficacy of Trigger Warnings, Content Warnings, and Content Notes." *Clinical Psychological Science* 12, no. 4: 751–771.

Brown, Olivia, et al. 2024. "Online Signals of Extremist Mobilization." *Personality and Social Psychology Bulletin.* https://doi.org/10.1177/01461672241266866.

Bryan, Venise, and Galicia Blackman. 2011. "The Ethics of Self-care in Higher Education." *Emerging Perspectives* 32, no. 2: 14–34.

Bundesamt für Verfassungsschutz. 2023. "Brief Summary. 2023 Report on the Protection of the Constitution." Accessed March 14, 2025. Retrieved from https://www.verfassungsschutz.de/SharedDocs/publikationen/EN/reports-on-the-protection-of-the-constitution/2024-06-brief-summary-2023-report-on-the-protection-of-the-constitution.pdf?__blob=publicationFile&v=4.

Bundesamt für Verfassungsschutz. 2020. "Antisemitismus und Rechtsextremismus." Accessed March 15, 2025. Retrieved from https://www.verfassungsschutz.de/SharedDocs/hintergruende/DE/rechtsextremismus/antisemitismus-im-rechtsextremismus-3.pdf?__blob=publicationFile&v=5.

Chiauzzi, Emil, and Paul Wicks. 2019. "Digital Trespass: Ethical and Terms of Use Violations by Researchers Accessing Data from an Online Patient Community." *Journal of Medical Internet Research* 21, no. 2: e11985.

Cohen, Katie et al. 2013. "Detecting Linguistic Markers for Radical Violence in Social Media." *Terrorism and Political Violence* 26, no. 1: 246–256.

Décary-Hétu, David, and Judith Aldridge. 2015. "Sifting Through the Net: Monitoring of Online Offenders by Researchers." *The European Review of Organised Crime* 2, no. 2: 122–141.

Dickmann-Burnett, Victoria, and Maribeth Geaman, 2019. "Untangling the Trigger-Warning Debate: Curating a Complete Toolkit for Compassionate Praxis in the Classroom." *Journal of Thought* 53, nos. 3/4: 35–52.

Die Zeit. 2025, January 16. "BKA listet laut Bericht 105 Vorgänge zu Taleb al-Abdulmohsen auf" [According to a report by the BKA Taleb al-Abdulmohsen had 105 legal charges pending] Accessed January 25, 2025. Retrieved from https://www.zeit.de/politik/uropeand/2025-01/uropean-anschlag-todesfahrer-bka-bericht-105-vorgaenge.

Dudenhoefer, Anne-Lynn, et al. 2021. "Leaking in Terrorist Attacks: A Review." *Aggression and Violent Behavior* 58: 101582.

Duga, Laura, and Michael Distler. 2016. "Measuring Terrorism." In *The Handbook of the Criminology of Terrorism*, 187–205. Edited by Gary La Free and Joshua D. Freilich. New York: Wiley.

Ebling, Sarah, Joachim Scharloth, Tobias Dussa, and Noah Bubenhofer. 2013."Gibt es eine Sprache des politischen Extremismus?".[Is there a language of political extremism?] In *Die da oben. Texte, Medien, Partizipation*, 3–67. Edited by Frank Liedtke. Bremen: Hempen.

Ebner, Julia, Chris Kavanagh, and Harvey Whitehouse. 2022. "Is There a Language of Terrorists? A Comparative Manifesto Analysis." *Studies in Conflict and Terrorism*. https://doi.org/10.1080/1057610X.2022.2109244.

Ellis, C., et al. 2016. "Analysing th Processes of Lone-Actor Terrorism: Research Findings" *Perspectives on Terrorism* 10, no. 2: 33–41.

Ess, Charles. 2018. "Internet Research Ethics." In *The Oxford Handbook of Internet Psychology*, 487–502. Edited by Adam Joinson, Katelyn McKenna, Tom Postmes, and Ulf-Dietrich Reips. Oxford: Oxford University Press.

Europol. 2024. *European Union Terrorism Situation Report 2024*. Retrieved from https://www.europol.europa.eu/publication-events/main-reports/uropean-union-terrorism-situation-and-trend-report-2024-eu-te-sat.

Fraser, Bruce. 1998. "Threatening Revisted." *International Journal of Speech, Language, and the Law* 5, no. 2: 159–173.

Frazier, Kierra. 2025, January 15. "FBI Reveals New Orleans Attacker's Online Search History, Releases New Image." *CBS News*. Accessed January 16, 2025. Retrieved from https://www.cbsnews.com/news/fbi-new-orleans-attacker-online-searches-new-image/.

Freilich, Joshua, Steven Chernak, Rachel Arietti, and Noah Turner. 2024. "Terrorism, Political Extremism, and Crime and Criminal Justice." *Annual Review of Criminology* 7: 187–209.

Frind, Siegrid. 1966. "Die Sprache als Propagandainstrument des Nationalsozialismus." [Language as Propaganda Instrument of National Socialism]. *Muttersprache* 76: 129–135.

Fontana, Nicola, Tommaso Nannicini, and Guido Tabellini. 2023. "Historical Roots of Political Extremism: The Effects of Nazi Occupation of Italy." *Journal of Comparative Economics* 51, no. 3: 723–743.

Gales, Tammy. 2019. "Threatening Contexts: An Examination of Threatening Language from Linguistic, Legal, and Law Enforcement Perspectives." In *The Routledge Handbook of Language in Conflict*, 472–490. Edited by Matthew Evans, Lesley Jeffries, and Jim O'Driscoll. New York: Routledge.

Glasgow, Kimberly, and Ronald Schouten. 2014. "Assessing Violence Risk in Threatening Communications." *Proceedings of the Workship on Computational Linguistics and Clinical Psychology: From Linguistic Signal to Clinical Reality*, 38–45.

Gold, Zachary, and Mark Latonero. 2018. "Robots Welcome? Ethical and Legal Considerations for Webcrawling and Scraping." *Washington Journal of Law, Technology, and Arts* 13, no. 3: 275–312.

Hiaeshutter-Rice, Dan, and Ian Hawkins. 2022. "The Language of Extremism on Social Media: An Examination of Posts, Comments, and Themes on Reddit." *Frontiers in Political Science* 4: 805008.

Hirschey, Jeffrey. 2014. "Symbiotic Relationships: Pragmatic Acceptance of Data Scraping." *Berkeley Technology Law Journal* 29: 1–38. Retrieved from https://papers.ssrn.com/sol3/papers.cfm?abstract_id=2419167.

Hollewell, Georgia, and Nicolas Longpré. 2021. "Radicalization in the Social Media Era: Understanding the Relationship between Self-radicalization and the Internet." *International Journal of Offender Therapy and Comparative Criminology* 66, no. 8: 896–913.

Hudson, James, and Amy Bruckman. 2004. "'Go Away': Participant Objections to Being Studied and the Ethics of Chatroom Research." *Information and Society* 20, no. 2: 127–139.

Institute for Economics and Peace. 2024. *Global Terrorism Index 2024*. Retrieved from https://www.economicsandpeace.org/wp-content/uploads/2024/02/GTI-2024-web-290224.pdf.

Jaki, Sylvia, et al. 2019. "Online Hatred of Women in the Incels.me Forum: Linguistic Analysis and Automatic Detection." *Journal of Language, Aggression, and Conflict* 7, no. 2: 240–268.

Jones, Payton, Benjamin Bellet, and Richard McNally. 2020. "Helping or Harming? The Effect of Trigger Warnings on Individuals with Trauma Histories." *Clinical Psychological Science* 8, no. 5: 905–917.

Kaati, Lisa, Amendra Shrestha, and Tony Sardella. 2016. "Identifying Warning Behaviors of Violent Lone Offenders in Written Communication." IEEE 16th International Conference on Data Mining, 1053–1060.

Karoli Christensen, Tanya. 2019. "Indirect Threats as an Illegal Speech Act." *The Sign of the V: Papers in Honour of Sten Vikner*, 113–130. Edited by Ken Ramshøj Christensen, Johanna Wood, and Henrik Jørgensen. Århus: Aarhus University.

Kidwai, Saman. 2024, September 26. "Salad Bar Extremism." *Modern Diplomacy*. Accessed March 15, 2025. Retrieved from https://moderndiplomacy.eu/2024/09/26/salad-bar-extremism/.

Kimble, Matthew et al., 2021. "Student Reactions to Traumatic Material in Literature: Implications for Trigger Warnings." *PLoS* 16, no. 3: e0247579.
Kleinberg, Bennett, Isabelle van der Vegt, and Paul Gill. 2021. "The Temporal Evolution of a Far-right Forum." *The Journal of Computational Social Science* 4, no. 1: 1–23.
Klemperer, Victor. 1947. *LTI. Notizbuch eines Philologen*. Berlin: Aufbau-Verlag.
Koehler, Daniel. 2014. "The Radical Online: Individual Radicalisation Processes and the Role of the Internet." *Journal for De-radicalisation* 1, no. 1: 116–134.
Krotov, Vlad, Leigh Johnson, and Leiser Silva. 2020. "Tutorial: Legality and Ethics of Web Scraping." *Communications of the Association for Information Systems* 47: 555–581.
Lanzke, Alice. 2016. "Viraler Hass: Rechtsextreme Wortergreifungsstrategien im Web 2.0" [Virtual Hate: Right-extremist linguistic strategies for capturing control in the web 2.0] In *Strategien der extremen Rechten. Hintergründe, Analysen, Antworten*, 623–630. Edited by Stephan Braun, Alexander Geisler, and Martin Gerster. Wiesbaden: Springer.
Maas, Utz. 1984. *"Als der Geist der Gemeinschaft eine Sprache fand": Sprache im Nationalsozialismus. Versuch einer historischen Argumentationsanalyse*. Opladen: Westdeutscher Verlag.
Markham, Annette, and Elizabeth Buchanan. 2012. "Ethical Decision-making and Internet Research: Recommendations from the AoIR Ethics Working Committee." *Readings in Virtual Research Ethics*, 27–44. Edited by Charles Ess and Steven Jones. New York: IGI Global Scientific Publishing.
McCauley, Clark, and Sophia Moskalenko. 2008. "Mechanisms of Political Radicalization: Pathways toward Terrorism." *Terrorism and Political Violence* 20: 415–433.
Meloy, J., and O'Toole, M. 2011. "The Concept of Leakage in Threat Assessment." *Behavioral Sciences and the Law* 29, no. 4: 513–527.
Meloy, J., et al. 2012. "The Role of Warning Behaviors In Threat Assessment: An Exploration and Suggested Typology." *Behavioral Sciences and the Law* 30, no. 3: 256–279.
Miron, Murray, and John O'Douglas. 1979, September. "Threat Analysis: The Psycholinguistics Approach." *The F.B.I. Law Enforcement Bulletin* 48, no. 9: 5–9.
Muschalik, Julia. 2018. *Threatening in English. A Mixed Method Approach*. Amsterdam: John Benjamins Publishing Company.
Neuman, Yair, Dan Assaf, Yochai Cohen, and James Knoll. 2015. "Profiling school Shooters: Automatic Text-based Analysis." *Frontiers in Psychiatry* 6. Retrieved from https://www.frontiersin.org/journals/psychiatry/articles/10.3389/fpsyt.2015.00086/full.
Nick, I. M. 2019. *Personal Names, Hitler, and the Holocaust: A Socio-onomastic Study of Genocide in Nazi Germany*. Lanham, MD: Lexington Books.
Nick, I. M. 2018. "In the Wake of Hate. A Mixed-Method Analysis of Anonymous Threatening Communications Sent During the 2016 US Presidential Election". *Nordic Journal of Linguistics* 2, no. 41: 183–203.
Nini, Andrea. 2018. "An Authorship Analysis of the Jack the Ripper Letters." *Digital Scholarship in the Humanities* 33, no. 3: 621–636.
Pickard, Harry, Georgios Efthyvoulou, and Vincenzo Bove. 2022. "What's Left After Right-wing Extremism? The Effects on Political Orientation." *European Journal of Political Research* 62, no. 1: 338–351.
Pilkington, Hillary. 2022. "Radicalization As and In Process: Tracing Journeys Through an 'Extreme-Right' Milieu." *Studies in Conflict and Terrorism*: https://doi.org/10.1080/1057610X.2023.2169896.
Preiss, Ulrike. 2024. *Stile extremistischer Tatschreiben. Charakteristika rechter und linker inkriminierter Texte*. Berlin: de Gruyter.
Roberts, Karl, and Claudia Wallner. 2023. *Beyond Ideological Categories: Mixed Ideologies and What They Mean for Preventing and Countering Violent Extremism*. Luxembourg: European Union.
Rohr, Marc. 2017. "'Threatening' Speech: The Thin Line Between Implicit Threats, Solicitation, and Advocacy of Crime." *Rutgers Journal of Law and Public Policy* 13, no. 2: 150–174.
Rose, Miss, and John Morrison. 2023. "An Exploratory Analysis of Leakage Warning Behavior in Lone-Actor Terrorists." *Behavioral Sciences of Terrorism and Political Aggression* 15, no. 2: 179–214.
Sanson, Mevagh, Deryn Strange, and Maryanne Garny. 2019. "Trigger Warnings are Trivially Helpful at Reducing Negative Affect, Intrusive Thoughts, and Avoidance." *Clinical Psychological Science* 7, no. 4: 778–798.
Sarnoff, Leah. 2025, January 5. FBI Releases Timeline of Suspect Shamsud-Dim Jabbar's New Orleans Attack." *ABC News*. Accessed January 5, 2025. Retrieved from https://abcnews.go.com/US/fbi-releases-timeline-suspect-shamsud-dim-jabbars-new/story?id=117280639.
Schuppener, Georg. 2010. *Sprache des Rechtsextremismus. Spezifika der Sprache rechtsextremistischer Publikationen und rechter Musik*. Leipzig: Edition Hamouda.

Scrivens, Ryan, Garth Davies, Tina Gaudette, and Richard Frank. 2022. "Comparing Online Posting Typologies Among Violent and Nonviolent Right-wing Extremists:" *Studies in Conflict and Terrorism*: 1–23. https://doi.org/10.1080/1057610X.2022.2099269.

Scrivens, Ryan, Garth Davies, and Richard Frank. 2018. "Searching for Signs of Extremism on the Web: An Introduction to Sentiment-based Identification of Radical Authors." *Behavioral Sciences of Terrorism and Political Aggression* 10, no. 1: 39–59.

Smith, Sharon, 2008. "From Violent Words to Violent Deeds? Assessing Risk from Threatening Communications." *The International Journal of Speech, Language, and the Law* 15, no. 1: 105–107.

Smith, Sharon, and Roger Shuy. 2002. "Forensic Psycholinguistics: Using Language Analysis for Identifying and Assessing Offenders." *F.B.I. Bulletin* 71, no. 416–21.

Solan, Lawrence, and Peter Tiersma. 2015. "Threats." In *Speaking of Language and Law*, 223–229. Edited by Lawrence Solan, Janet Ainsworth, and Roger Shuy. Oxford: Oxford University Press.

Storey, Kate.1995. "The Language of Threats." *International Journal of Speech, Language, and the Law* 2, no. 1: 74–80.

Tompkinson, James. 2023. *Spoken Threats from Production to Perception*. Cambridge: Cambridge University Press.

Torregrosa, Javier, et al. 2022. "A Survey on Extremism Analysis Using Natural Language Processing." *Journal of Ambient Intelligence and Humanized Computing* 14: 9869–9905.

van der Vegt, Isabelle, Paul Gill, Stuart Macdonald, and Bennett Kleinberg. 2019. *Shedding Light on Terrorist and Extremist Content Removal*. Global Research Network on Terrorism and Technology. Retrieved from https://static.rusi.org/20190703_grntt_paper_3.pdf.

Völkner, Paula. 2024, December 23. "Rechtsextremist, Linksextremist oder Islamist? Nach Magdeburg-Anschlag ordnet Experte das Täterprofil ein. In: Frankfurter Rundschau." Accessed January 13, 2025. Retrieved from https://www.fr.de/politik/magdeburg-anschlag-weihnachtsmarkt-attentat-taeter-profil-motiv-ermittlungen-zr-93484200.html.

Weinberg, Leonard, Ami Pedahzur, and Sivan Hirsch-Hoefler. 2004. "The Challenges of Conceptualizing Terrorism." *Terrorism and Political Violence* 16, no. 4: 777–794.

Williams, Thomas and Calli Tzani. 2024. "How Does Language Influence the Radicalization Process? A Systematic Review of Research Exploring Online Extremist Communication and Discussion." *Behavioral Sciences of Terrorism and Political Aggression* 16, no. 3: 310–330.

Windsor, Leah. 2020. "The Language of Radicalization: Female Internet Recruitment to Participation in ISIS Activities." *Terrorism and Political Violence* 32, no. 3: 506–538.

Young, Joseph. 2016. "Measuring Terrorism." *Terrorism and Political Violence* 31, no. 2: 323–345.

Zook, M., et al. 2017. "Ten Simple Rules for Responsible Big Data Research." *PLoS Computational Biology* 13, no. 3: e1005399.

26

ETHICAL CONSIDERATIONS IN CORPUS-BASED LEGAL INTERPRETATION ANALYSIS

Brett Hashimoto and Derek Christian Haderlie

Introduction

In the United States, judges who are required to interpret legal texts rely heavily on the Ordinary Meaning Doctrine, which dictates that "[w]ords are to be understood in their ordinary, everyday meanings—unless the contexts indicates that they bear a technical sense" (Scalia and Garner 2012: 69). This constraint can be challenging when considering the types of linguistic questions that judges face. Consider, for instance, the case of *United States v. Robinson* (2024), which revolved around the ordinary meaning of the word "strangulation" and whether that word ordinarily entailed violent intent or whether instead it could be unintentional. Traditional approaches for determining the ordinary meaning of a term, such as appeals to intuition, etymology, and dictionaries, are often unreliable, irrelevant, or limited (Mouritsen 2010). As a result, many have advocated that judges use corpus-based legal interpretation analysis (CLIA) for determining ordinary meaning (recently summarized in Römer-Barron and Cunningham 2024). CLIA promises to be more reliable and empirically grounded than previously used methods (Keller and Egbert 2021; Lee and Mouritsen 2021; Gries 2020). But its detractors argue that in addition to a wide range of other criticisms, its use in legal interpretation is not sufficiently deterministic or scientific to be useful or that it is commonly misapplied or executed substandardly[1] (Bernstein 2021; Ehrett 2019; Klapper 2019; Zoldan 2019; Tankersley 2018; Herenstein 2017; Ramer 2017). The fact that there are so many varied criticisms of CLIA indicates a need for a robust theoretical framework that outlines what kind of methods would be most appropriate for understanding legal texts. Proponents and critics alike need a common ground with common vocabulary for assessing methods of linguistic analysis, including CLIA.

In light of this need, we propose a modest philosophical framework that provides principles for evaluating the appropriateness of CLIA. This framework is built on the notions of responsibility: epistemic responsibility and moral responsibility. In developing the framework, we focus on the responsibilities that jurists have to various stakeholders, for example, the parties involved in particular cases, those affected by their setting of precedent, the general citizenry who put their trust in the justice system, and so forth. The resulting framework can then be directly applied by legal theorists, researchers, and practitioners to better understand how to evaluate the appropriate use of CLIA and to address the quality of criticisms of CLIA in general and specific applications.

DOI: 10.4324/9781003391074-29

Literature Review

As earlier mentioned, CLIA and its applications have incurred a variety of criticisms (Bernstein 2021; Tobia 2020; Ehrett 2019; Klapper 2019; Zoldan 2019; Hessick 2018; Tankersley 2018; Herenstein 2017; Ramer 2017). These criticisms strike at multiple levels of the implementation of CLIA as seen in Table 26.1, including the proposition that language can be studied empirically, data analysis methods, and the ability of stakeholders to understand CLIA. Table 26.1 provides a non-exhaustive list of criticisms that have been levied against CLIA in each of the respective categories.

What is apparent from systematically reviewing these various criticisms is that many detractors do not appear to understand either the nature of scientific research broadly, or of linguistic analysis and corpus research specifically. We acknowledge that a number of these criticisms, and their attendant misunderstandings, are due to looking at specific applications of CLIA where insufficient detail or explanation was provided.

In response to these many criticisms, Gries (2020) attempts to systematically address a variety of these concerns by, for instance, pointing out that many of them are "straw man" arguments that were never made by CLIA proponents and by showing how some of these arguments have already proven to be empirically false. Similar work has been performed by several others to logically and empirically argue the usefulness of corpus linguistics (Gries and Gales 2024; Lee and Mouritsen 2021; Tobia 2021). Other CLIA advocacy research has worked to refine or build upon CLIA methods to make them more reliable and rigorous as well as to expand their application (Egbert and Lee 2024; Gries, Slocum, and Tobia 2024; Keller and Egbert 2021; Phillips and Egbert 2018). However, what is apparent from the literature is that detractors and advocates of CLIA do not share a framework for evaluating CLIA applications that both groups can agree upon. We hope to remedy this situation by introducing a framework around the core concept of responsibility that will provide critics and defenders of corpus linguistics alike with a shared background of philosophical assumptions by which to assess certain uses of corpus linguistics.

Methods

The framework we provide has two components. First, background motivations for the central organizing concept: the concept of epistemic and moral responsibility. Second, a structured set of principles that judges, advocates, and others can use in determining morally and epistemically responsible use of corpus linguistics within the law. We develop a set of principles by way of conceptual analysis. This is the process of analyzing a concept and developing a set of inferences that we might make when applying the concept in a particular case.[2]

Motivating the Central Concept

The central organizing concept for our philosophical framework is the concept of responsibility. Specifically, we shall focus on *moral* and *epistemic* responsibility. Here we use "moral" in a broad way, to capture both the moral and the ethical. This is to go beyond mere codes of conduct and to talk about the more general application of moral and ethical principles. We use "epistemic" broadly as well, using it to capture issues related to knowledge, belief, evidence, and justification.

Moreover, we can distinguish between two notions of responsibility. On the one hand, a person can be responsible in the sense that they can be appropriately praised or blamed for their action. In

Table 26.1 Categories of Criticisms of CLIA

Category	*Examples of Critiques*
Theoretical justification of the use of CLIA	• Ordinary meaning is not connected to frequency (Klapper 2019; Hessick 2018) • CLIA is not consistent with the canons of interpretation (Tankersley 2018)
Empirical study of language	• Only human intuition can be useful for abstracting the optimal term from the facts for a particular case (Klapper 2019) • Frequency of occurrence from a corpus does not indicate anything useful (Hessick 2018; Herenstein 2017)
Data selection/corpus design	• The corpora that CLIA analyses have used are not appropriate (Bernstein 2021; Ehrett 2019; Zoldan 2019; Hessick 2018) • The sample sizes are too small (Klapper 2019)
Data analysis methods	• There is a lack of consistency between analyses (Hessick 2018; Tankersley 2018) • There is a lack of transparency of methods (Tankersley 2018Ramer 2017)
Manner of interpreting results	• There is excessive analyst discretion in interpreting results (Zoldan 2019; Tankersley 2018) • Because different kinds of analyses can be conducted, results can be different, leading to opposing outcomes from different corpus analyses (Bernstein 2021; Tankersley 2018; Ramer 2017)
Analyst background expertise	• Non-experts are doing complex linguistic analysis but cannot be relied upon to perform such analyses adequately (Ehrett 2019; Tankersley 2018) • Using corpora does not require experts, but judges lack the training to use corpora (Ramer 2017)
Ability of the relevant stakeholders to understand CLIA	• The public would not be able to use corpora, which threatens the principle of notice (Ehrett 2019; Zoldan 2019; Hessick 2018; Herenstein 2017) • The legislature would not be able to use corpora, which presents challenges to drafting laws (Herenstein 2017)
Manner that CLIA makes it into the court room	• If each side produces corpus research, then it would result in competing analyses (Hessick 2018) • CLIA use of sua sponte (judicial action taken without request from parties) involves too much discretion on the part of the judge (Tankersley 2018)
Practicality of CLIA	• It is impossible for a single person to analyze enough data for it to be meaningful (Klapper 2019) • Hiring experts to perform CLIA unnecessarily increases the cost for parties and decreases law's accessibility (Jennejohn, Nelson, and Nunez 2021; Hessick 2018)
Alternative methods are superior to CLIA	• Dictionaries are better because they are more principled (Tankersley 2018) • Dictionaries are easier to use (Ehrett 2019)

other words, a person is responsible when *she did it* (Talbert 2024). On the other hand, a person can be responsible in the sense that they perform an action in light of and appropriate to the demands and requirements made upon them by their moral and epistemic duties and obligations. In other words, a person is responsible when she performs her action in a way that accords with her duties and obligations; she is irresponsible when she fails to do so. It is this latter notion of responsibility that we will focus on in the remainder of the chapter. And we judge that our readers will be able to distinguish by context any use of the former notion of responsibility.

We think that what has been at issue in debates over corpus linguistics in the law is whether users make *responsible* use of it. Criticisms of the tools often claim to show that judges or others have used corpus linguistics in morally or epistemically irresponsible ways (Ehrett 2019; Tankersley 2018). Defenses have often responded in kind by attempting to show how easily one can responsibly use corpus linguistic analysis, for instance, by setting critical values or doing double-coding qualitative data (Keller and Egbert 2021; Phillips and White 2017). For this reason, we think that developing a framework around the core concept of responsibility will provide critics and defenders of corpus linguistics alike with a shared background of philosophical assumptions by which to assess certain uses of corpus linguistics. We can see why responsibility looms so large by considering a toy case.

> *Alfie's Bike*: Two boys are having a dispute. Alfie claims that his bike was "stolen" by Bobby. Bobby claims that he did not "steal" the bike; he only borrowed it. Carrie, the adult on the scene, is asked to adjudicate. Her question: how should she understand "steal" in this context? She agrees that stealing is wrong and should be addressed, but she is less clear about its attribution in the initial accusation. She must determine the meaning of "steal".

There are several ways that Carrie could proceed to make her judgment. Here are just three.

A. **Carrie could take a vote on the meaning**. The problem is that this approach needlessly defers to public opinion, which is notably unreliable about the truth of these sorts of things. Public opinion is often swayed by bias and misconception. A second problem is that a survey may only reveal the children's biases and misconceptions, instead of telling her something about what "steal" means.
B. **Carrie could unilaterally decide a meaning for stealing**. The problem is that although this would surely expedite her adjudication, it would put Carrie into the position of arbitrarily deciding whether Bobby stole the bike. Perhaps to avoid arbitrariness, Carrie could consult her intuitions about the meaning of "steal". But such an approach is liable to personal bias, lacks transparency, and is otherwise unreliable.
C. **Carrie could consult an expert on language analysis**. In contrast with options A and B, this avoids the worry about unreliable methods: she is seeking guidance from someone who is an expert about analyzing the relevant language. She also avoids the arbitrariness or partiality worries that come with unilaterally deciding on the meaning.

Although these three alternatives do not exhaustively represent all of Carrie's options, they are sufficient to illustrate the importance of ethical and epistemological responsibility in how Carrie proceeds. For now, notice that were Carrie to take option A, she would be acting epistemically irresponsibly, since she would be relying on well-known unreliable (and invalid) methods for discovering the meaning of "steal". Were Carrie to take option B, she would be acting morally irresponsibly, since it appeals to no standard of justice, but to arbitrary choice on Carrie's part.

Thus, of the three options surveyed, Carrie would be most responsible if she were to take option C. For only in this way would she (a) seek an epistemically reliable method of determining the meaning of "steal" and (b) seek to avoid arbitrary choice or whichever choice she was partial to.

Principles of the Framework

Given the importance of responsibility for legal interpretation, the question remains how we determine whether a particular use of CLIA is responsible. That is, what makes one more or less responsible in their use of corpus linguistics in particular cases. Our proposal is to answer this question with a set of interconnected and gradable principles. We state each principle schematically, with the intention that when evaluating a particular use of CLIA, evaluators will fill in the details as demanded by the particular case at hand. We can state our first principle, where *S* is the person making a judgment and *p* is a proposition:

> **Validity**: *S* is epistemically responsible in judging that *p* only if and to the extent that *S's* methods for determining whether *p* are valid.

A method *M* is valid for determining whether *p* is valid only if it gives us results about our intended target of inquiry, namely *p*. We can further distinguish between internal and external validity. A method is *internally valid* whenever the method gives us accurate results. And the method is *externally valid* whenever it generalizes or can be applied to other situations. As we mentioned earlier, in Carrie's case, surveying the children on the meaning of "steal" would have dubious validity. Suppose that the children have a bias for Bobby over Alfie. It is likely that in attempting to inquire about the meaning of "steal", Carrie's survey would yield results not about the meaning of the term but about who is most liked between Bobby and Alfie. In this way, Carrie's methods would be invalid, since her method would fail to give her results about her intended target of inquiry or the meaning of "steal".

Not only do we want our methods to be valid, but we also want them to give us information that we can trust.

> **Reliability**: *S* is epistemically responsible in judging that *p* only if and to the extent that *S* uses the most reliable methods available to her to judge that *p*.

Note that we specifically focus on the *most reliable method* in our formulation of Reliability. This is because it may be that, for certain domains of investigation, the methods available to use are all unreliable. We can understand reliability either in terms of actual frequency or terms of possible and actual frequency (or propensity). We say that a process is reliable if it has a high truth-ratio. We leave just how high a truth-ratio is required vague. It need not be 1.0, but should be higher than 0.5 (Goldman and Beddor 2021). That is, given a method *M* assigning a truth-value to a proposition *p*, the probability that *M* will assign the correct truth-value can be less than 1.0 but must be greater than 0.5.

Suppose that *S* has access to two competing methods, M_1 and M_2, for determining whether *p*. And further suppose that M_1 has a truth-ratio of 0.8 and M_2 has a truth-ratio of 0.9. On these assumptions Reliability demands that if *S* is responsible, she will use the most reliable available to her: M_2. Now, it should be obvious why Reliability plays an important role in our framework: epistemically responsible agents are aiming at truth. And using the most reliable methods available to us to figure out what is true puts us in a better position to get true judgments. But Reliability and Validity are not, by themselves, enough. Consider the worry raised by BonJour (1980), concerning a hypothetical clairvoyant named Norman. Norman has a *perfectly reliable* and *valid* faculty of clairvoyance. Whenever he is asked about the ordinary meaning of a word, he immediately forms a perfectly accurate belief every time about the very word of interest. He does not know *why* his belief is true; it is just that he always forms the correct belief. In terms of reliability and validity, Norman's clairvoyant faculty is perfect. His clairvoyance has a truth-ratio of 1.0. The problem, Bonjour points out, is that while Norman will form true judgments every time he uses his clairvoyance, he is not *justified*. He has no reasons for believing what he does. He just believes it. So, even though Norman might judge that *p* as a result of using his perfectly reliable method of clairvoyance, he will not be *justified* in believing that *p*. These reflections suggest a third principle:

> **Justification**: *S* is epistemically responsible in judging that *p* only if and to the extent that *S* understands why *p* is true.

Norman fails the requirement of Justification, and thus fails to be an epistemically reliable agent. Furthermore, if another agent were to rely on Norman's testimony alone in forming a belief that *p* they would also fail Justification, since they would in a sense be alienated from the reasons why *p* is true by virtue of only coming to believe *p* on the basis of Norman's testimony. And because of this, even though Norman is *in an important sense* an expert, relying on his testimony alone would be irresponsible.

Not only should practitioners apply these principles, but they must also provide evidence of their application in a way that is accessible to others. This is especially important in the law, given the practical significance and import of the results of CLIA for individuals affected by those results. With that in mind, we believe that judges, attorneys, and others who use CLIA when interpreting statutes have a responsibility to "leave a paper trail" for how they achieved those results. Otherwise, it is impossible for others to know whether their use of CLIA was morally and epistemically responsible. This suggests a fourth principle:

> **Transparency**: *S* is responsible in asserting a judgment that *p* (in a legal context) only if and to the extent *S* is transparent in her application of the first three principles.

If a judge or attorney fails to adhere to Transparency, it makes it impossible to evaluate their particular use of CLIA in the law, since they provide us with insufficient evidence that they have adhered to the first three principles. This is critical in an adversarial legal system, such as in the United States, because an opposing party (especially if they utilize an expert witness on linguistics) can evaluate the validity and reliability of an analysis only insofar as they could replicate the same results. If methods are not transparent (such as in reliance on one's intuition), they cannot be reproduced and, therefore, cannot be falsified. Transparency is also critical because decisions can be appealed to higher courts, where judges can review interpretative evidence.

These four principles, together and without qualifications, can be extremely burdensome for individual agents. One might not always be able to tell whether a method is the most reliable and valid one or not. Or one may simply not have access to all the reasons why *p* is true, even when she has judged that it is by way of a reliable method. This may be due to in-principle limitations on her access to information. These and other limitations suggest a fifth principle that constrains applications of the first four principles:

> **Practicality**: When seeking to determine whether *p*, *S* must take into account practical constraints such as timeliness, expense, and comprehensibility.

Let us consider these in turn. In the famous novel, *Hitchhiker's Guide to the Galaxy*, Douglas Adams describes a supercomputer called Deep Thought. Deep Thought is the most powerful computer ever created, and capable of answering any question. Now, let us suppose (as is plausible) that Deep Thought can satisfy both Reliability and Justification, and would therefore be a responsible source of information. Nevertheless, Deep Thought would fail Practicality. Why?

First, Deep Thought may take centuries to determine whether a given proposition is true or not and why. In the book, it takes 7.5 million years to determine the answer to the question, "What is the meaning of life?", failing a timeliness constraint. Second, Deep Thought would be exorbitantly expensive, requiring enormous amounts of resources. Finally, although Deep Thought may be able to determine why some proposition is true, satisfying Transparency, it may be that those reasons that make the proposition true and their relationship to the proposition will not be comprehensible to the average jurist or jury member. This would undermine the value of their transparency, since it would effectively make a proposition *p* transparent to *S* only if *S* was unable to comprehend why *p*. For these reasons, in some cases there may be practical constraints that make the epistemic ideal inaccessible. Practicality gives us a principled way of knowing when it is admissible to proceed despite falling short of the epistemic ideal.

It should be plain that Practicality has a moral dimension. Sometimes whether *p* is true or not makes a morally significant difference in a person's life. Whether it is true that Albert killed Bob or not really does matter to Albert when he is on trial. He may be unable to pay bail, and detaining him indefinitely dramatically impacts his life. This is especially salient in the case that Albert is innocent.

There is a fifth principle that we include as part of our framework:

> **Lawfulness**: *S* is responsible in judging that *p* only if and to the extent that *S* administers the proper application of the law.

This principle is related to the concept of The Rule of Law. As Waldron (2023) points out,

> The most important demand of the Rule of Law is that people in positions of authority should exercise their power within a constraining framework of well-established public norms rather than in an arbitrary, ad hoc, or purely discretionary manner on the basis of their own preferences or ideology.

In line with this, administering the proper application of the law requires attorneys and judges to adhere to the demands and rigors of the legal system. This can mean that they must act consistent with the law, even when this conflicts with their own personal opinions. Practicality and Lawfulness raise the issue of pragmatic encroachment. Pragmatic encroachment occurs when practical and/or moral considerations affect what you are justified in believing and doing. Important considerations for pragmatic encroachment might be whether or not including/excluding information is fair or whether an inquiry into the meaning of the law has been sufficiently comprehensive in its thoroughness. Whether practical/moral considerations should determine what we believe or not is controversial. But defenders of pragmatic encroachment, such as Jason Stanley, use contrasting cases like the following to motivate it.

> *Low Stakes*: Hannah and her wife Sarah are driving home on a Friday afternoon. They plan to stop at the bank on the way home to deposit their paychecks. It is not important that they do so, as they have no impending bills. But as they drive past the bank, they notice that the

> lines inside are very long, as they often are on Friday afternoons. Realizing that it was not very important that their paychecks are deposited right away, Hannah says, "I know the bank will be open tomorrow, since I was there just two weeks ago on Saturday morning. So, we can deposit our paychecks tomorrow morning".
>
> *High Stakes*: Hannah and her wife Sarah are driving home on a Friday afternoon. They plan to stop at the bank on the way home to deposit their paychecks. Since they have an impending bill coming due, and very little in their account, it is very important that they deposit their paychecks by Saturday. Hannah notes that she was at the bank two weeks before on a Saturday morning, and it was open. But, as Sarah points out, banks do change their hours. Hannah says, "I guess you're right. I don't know that the bank will be open tomorrow".
>
> *(Stanley 2005: 3–4)*

It should also be clear how our more specific interest to applying this framework to *the law* provides some reason to be less doubtful about pragmatic encroachment, given the practical and moral weight that legal decisions bear on the lives of those to whom they apply.

So far, we have presented a framework that centers on the notion of epistemic and moral responsibility. We have proposed six principles that can help us identify responsible and irresponsible ways of forming judgments.

> **Validity**: *S* is epistemically responsible in judging that *p* only if and to the extent *S*'s methods for determining whether *p* are valid.
>
> **Reliability**: *S* is epistemically responsible in judging that *p* only if and to the extent *S* uses the most reliable methods available to her to judge that *p*.
>
> **Justification**: *S* is epistemically responsible in judging that *p* only if and to the extent *S* understands why *p* is true.
>
> **Transparency**: *S* is responsible in asserting a judgment that *p* (in a legal context) only if and to the extent *S* is transparent in her application of the first three principles.
>
> **Practicality**: When seeking to determine whether *p*, *S* must take into account (a) timeliness (b) expense, and (c) comprehensibility.
>
> **Lawfulness**: When seeking to determine whether *p*, *S* must weigh a judgment that *p* against the justice of *p*.

These principles provide a balanced approach to determining responsibility. It merits emphasis that applying these six principles may not be sufficient on its own to ensure moral and epistemically responsible use of CLIA. Figure 26.1 visually displays how these principles may work together in helping evaluate whether an analysis is a sufficiently responsible application of interpretative methods.

Objections

In the next section, we will apply our framework to two real-world cases, but there are two objections that can help us clarify parts of our framework.

Objection 1: We initially motivated the centrality of responsibility by appeal to the case with Alfie's bike. We considered three approaches Carrie might take. But why think that this has

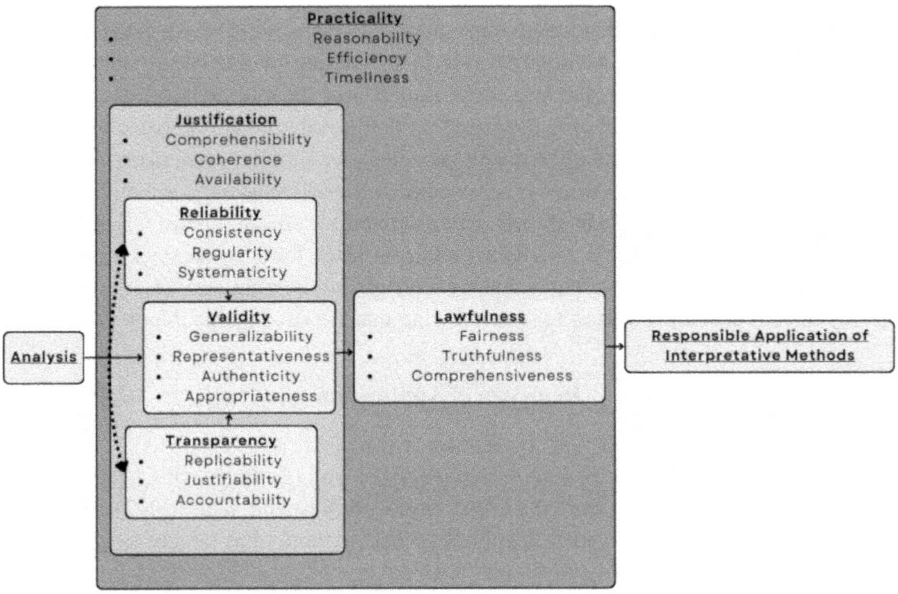

Figure 26.1 Framework of Responsible Application of Interpretative Methods

exhausted all of Carrie's options? Perhaps instead she could look at the *Oxford English Dictionary*, or she could scry a crystal ball. The problem is that by limiting the cases, we have inadvertently steered the Alfie case in a way that inappropriately serves our aims.

Response: This objection fails to note that the number of kinds of cases are not what is relevant to our discussion. We could have used 2 or 20 cases. Regardless, it would have been reasonable to describe them as more or less responsible and irresponsible solutions to Carrie's interpretative problem. What matters is just that there are enough cases to illustrate the rough difference between a responsible and an irresponsible response.

Objection 2: Why think the TRANSPARENCY of reasons is needed? Here are two reasons to think all we need is RELIABILITY:

a) All we care about is avoiding errors. Often in a judicial setting, the stakes are incredibly high. Therefore, getting things right is more important than knowing why we got them right.
b) We motivated the addition of JUSTIFICATION in our framework by appeal to the need to have justification. But, if we think that justification is not about what evidence we have access to, but is rather about getting things right, then reliably getting the right outcome will give us everything necessary for justification, since justification will correspond to the objectively right outcome. Consider a case where someone provides very bad reasons for their belief that *p*. If *p* is true, then one can search for justification and slowly recover reasons for *p*. In a slogan, if you have the right answer, then you can know that there is a really good in-principle justification for the answer you give. And an in-principle justification is the kind of justification we were after in the first place.

Response: Let us be concessive and say that JUSTIFICATION is *not* always a requirement for being epistemically responsible. Nevertheless, the legal context is a special case that makes JUSTIFICATION necessary. So, in response to (a), we do care about more than just avoiding errors. For example, plausibly due process demands that judges make their judgments transparent, in the relevant sense. Moreover, transparency plays an important role in the setting of precedent, since transparency will allow others to better understand how they might follow the interpretative decisions made by judges. So, in fact, we do care not only about getting things right but also in being able to replicate that. And unless we want to constantly reinvent the wheel, we should be transparent in our reasons.

And, in response to (b), RELIABILITY is an externalist friendly epistemic principle. The addition of JUSTIFICATION simply strengthens the epistemic demands made on agents. Given the presumption of innocence, JUSTIFICATION is a necessary addition to the framework, since we not only want to get the right answer, we want to know that we have gotten the right answer. JUSTIFICATION goes a long way toward providing a basis for our knowing that we have gotten things right.

Real-World Applications

Here we will apply our framework to evaluating two real-life uses of CLIA: (1) the Michigan Supreme Court's use of CLIA in *People v. Harris* (2016), and (2) the use of CLIA in the analysis of the three emoluments clauses from the Constitution (Phillips and White 2017). Together, these two cases are instructive, for the latter exemplifies what we think of as largely responsible uses of corpus linguistics in the law, while the former does not.

People v. Harris

First, *People v. Harris*. In this case, the defendants (three police officers) had been called to testify at a Garrity hearing. During the hearing, the defendants made false statements. It later became clear that they had made false statements when a video recording of the incident under investigation surfaced. In *People v. Harris*, the question before the court was whether the defendants' false statements counted as "information" or not. The statute the court was asked to interpret was Disclosures by Law Enforcement Officers Act: "An involuntary statement made by a law enforcement officer, and any *information* derived from that involuntary statement, shall not be used against the law enforcement officer in a criminal proceeding" [emphasis added] (Michigan Compiled Laws 15.393). If the false statements counted as information, then they could not be used in criminal proceedings. If they did not count, then they could. Ultimately, the majority opinion was that the false statements did count as information. Both the majority and the minority opinions used CLIA in their attempts to determine whether the false statements counted as information or not, but we will focus our analysis on the majority's use of CLIA.

Because the majority provides very little evidence about how they used CLIA, they fail TRANSPARENCY. This makes evaluating their use difficult. But given what information we do have, we can ask questions relevant to the other principles of the framework. To what extent was the majority's use of CLIA valid? We think it likely that the majority's use of CLIA was largely invalid. The majority used the Corpus of Contemporary American English (COCA) to analyze "information". The majority provides no justification of their use of CLIA as a method for interpreting other than citing Mouritsen (2012: 839), "(COCA) allows users to 'analyze ordinary meaning through a method that is quantifiable and verifiable'". But, without further justification being offered, it is unclear that COCA is itself the right corpus for analyzing the meaning of "information" in the statute.

To what extent was the majority's use of CLIA reliable? Given information gathered from the minority's response, we have reason to believe that the majority's use of CLIA was unreliable. For example, the minority suggests that the collocates that the majority's argument hinges upon are only a small fraction of a percent of the collocates, suggesting that non-factive uses of information are rare, exotic, or aberrant. Ignoring the majority of collocates without comment makes judgments about meaning highly unreliable.

Did the majority's use of CLIA adhere to JUSTIFICATION? We think not. Suppose they were correct that the false statements made by the defendants counted as information. Even on this assumption, we have reason to think the majority did not understand why their claim was correct. The majority used adjective collocates of "information", but additional details about basic and important parameters, such as the collocation directionality or window size, were not provided. They likewise provided no information about the number of collocates or concordance lines they examined. Since they apparently failed to systematically or carefully use CLIA, it would not be true that they understand *why* the false statements counted as information.

So, the majority were irresponsible, having failed to satisfy the first three principles of the framework. But we might wonder if there were practical constraints that made it difficult (if not impossible) to use CLIA in a valid, reliable, justified, and transparent way. We think not. As a matter of practicality, it would have been relatively easy and inexpensive for the majority to improve their use of CLIA greatly by simply reporting exactly what analyses they performed and how they performed them. Because all evidence that we can ascertain suggests that the CLIA application was not thorough or reflective of the truth, it does not seem to meet any reasonable standard of fairness or LAWFULNESS. So, the majority fails every principle of the framework. We therefore conclude that the majority was irresponsible in their use of CLIA.

Emoluments Clauses

The second case comes from Phillips and White (2017), where they address a question over the meaning of "emoluments" in the US Constitution, which states that "no Person holding any Office of Profit or Trust under them, shall, without the Consent of the Congress, accept of any present, Emolument, Office, or Title, of any kind whatever, from any King, Prince, or foreign State" (US Constitution, art. 1, sec. 9). Various lawsuits were made against Donald Trump that purport that Trump was the recipient of emoluments while he served as president. The question before the court was how we should interpret "emoluments" for determining whether Trump violated his constitutional duties:

> If the broad, general sense is the correct sense of "emolument(s)" in these two constitutional clauses, and assuming the Foreign Emoluments Clause applies to the President, then they are likely correct that the President has violated the Constitution. But if the Constitution uses the narrow, public office, employment, or service sense of "emolument(s)," then based on the facts alleged it would appear that the President has not violated these constitutional clauses since no one has claimed that he is in the official employ or an officer of a foreign state.
>
> *(Phillips and White 2017: 185)*

What made the use of CLIA in this case difficult for Phillips and White was that there were no relevant pre-existing corpora for them to use that adequately represented the relevant time period in the relevant variety: Founding Era American English (Repo, Hashimoto, and Laippala 2023). Thus, Phillips and White assembled their own corpora and applied some systematic corpus analysis within those datasets.

Was Phillips and White's use of CLIA highly valid? Yes, at least as valid as is reasonable given constraints of practicality and existing methodology. They extensively describe and motivate the use of their corpora, specifically aimed at English usage before and during the framing of the constitution (three Corpus of Founding Era American English subcorpora, Hashimoto 2023: 1. Evans Early American Imprint Series, 2. National Archives Founders Papers Online, and 3. Hein Online). They include details about the sampling of the data, types of texts, and number of texts and words in the sample. They also specify that each is designed to represent somewhat different groups of people, and they acknowledge various limitations of their sample (Phillips and White 2017: 204). All of this evidences that Phillips and White made reasonable efforts to ensure their use of CLIA was valid.

Was their use of CLIA highly reliable? Again, yes. In their study they checked the frequency of "emolument" as a flemma;[3] gave a list of the top 15 collocates ranked by the collocation statistic, Mutual Information (MI) of "emolument" in a 5L/5R context; 3- to 5-grams with "emolument" in them ordered by frequency; 784 randomly sampled concordance lines of "emolument" (Phillips and White 2017). They also looked at how "emolument(s)" was used in context, by checking the language around the term, "usually about 150–200 words surrounding the term" and independently double-coding each for semantic information using a grounded coding scheme after which they calculated interrater reliability (Phillips and White 2017: 207). These efforts all improve the probability that their conclusion is true, therefore making their analysis reliable. Did they adhere to JUSTIFICATION and TRANSPARENCY in their use of CLIA? In other words, do they understand and explain why "emoluments" means what it means? Yes. They suggest that there are two plausible interpretations of the meaning of "emoluments" and provide credible evidence in both cases that these are reasonable interpretations based on empirical evidence obtained through replicable methods.

Perhaps the biggest drawback of Phillips and White's use of CLIA is that their analysis would be less high in PRACTICALITY, given the amount of work they spent manually double-coding hundreds of coding concordance lines. This much work may not be timely or cost-effective in some cases. Nevertheless, we think that Phillips and White's analysis is highly responsible and can serve as an exemplar of responsible use of CLIA.

Conclusions

The proposed framework articulates an approach to both the evaluation of CLIA research as well as a set of heuristics to evaluate the quality of criticisms to the CLIA applications generally. These advances push forward CLIA research in its pursuit of high ethical standards in ensuring this kind of high-stakes analysis is performed appropriately and assessed fairly. However, this framework is merely a first step in highlighting the major areas of consideration in CLIA research. The proposed framework itself is neither perfect nor complete. Specifically, we expect additional work to flesh out subcomponents of each major facet of the framework as well as refinement to more particularly articulate its major principles. Future research should also apply this framework to specific areas of application (e.g., in intellectual property law or contract law) to evaluate the extensibility of the framework beyond statutory and Constitutional applications. As illustrated by our evaluation of previous analyses, this framework also highlights weaknesses in previous CLIA research, especially in providing adequate explanations of methodological procedures and justifications for critical decisions in analyses. This framework also highlights major areas where CLIA can and should make improvements to its methodology (e.g., a need for a greater need for standardization in approaches for how critical, subjective decisions are made as part of analysis). In short, this framework can play a major role in the ethical application of CLIA and has the potential to facilitate CLIA becoming much more widely adopted by making it more comprehensible and intellectually accessible to relevant stakeholders, while also maintaining a high degree of empirical rigor.

We see this framework being used in primarily two ways by practitioners. Firstly, in cases where detractors of CLIA question the appropriateness of the use of CLIA altogether, this framework can be used by practitioners to demonstrate when CLIA can be a useful tool in providing evidence for answers to legal interpretation questions. In other words, by applying the framework, the kind of illogical or otherwise poor critiques of CLIA (of the kind that were discussed in Table 26.1) can be easily dismissed. Secondly, practitioners who use CLIA can show by using the framework the extent to which analyses that they perform are appropriate. This is especially useful in the increasingly common scenario where two dueling corpus analyses are submitted by opposing parties. In other words, lawyers, judges, and expert linguists can use this framework as a basis for a rubric to demonstrate how their application of CLIA may be maximally ethically and epistemically responsible.

Notes

1 It is worth noting that critics of the use of corpus applications tend to have a different jurisprudential theoretical background than its proponents. However, empirical evidence demonstrates that regardless of one's theoretical orientation, corpus linguistics can be useful in legal interpretation (Hashimoto and Heilpern 2023).

2 A detailed description of conceptual analysis can be found in Bennett (2017).

3 A flemma includes all inflections of words whose base forms are spelled the same way, regardless of part of speech or meaning (Pinchbeck 2014; Jarvis and Hashimoto 2021).

Case Law

People v. Harris. 2016. 499 Mich. 332, 885 N.W.2d 832.
United States v. Robinson. 2024. 3:24-cr-00117 (E.D. Va.).

References

Bennett, Karen. 2017. "Conceptual Analysis and its Limits." *Philosophic Exchange*: 46, no. 1: 1–12.

Bernstein, Anya. 2021. "Legal Corpus Linguistics and the Half-Empirical Attitude." *Cornell Law Review* 106, no. 6: 1397–1456.

Bonjour, Laurence. 1980. "Externalist Theories of Empirical Knowledge." *Midwest Studies in Philosophy* 5: 53–73.

Egbert, Jesse, and Thomas R. Lee. 2024. "Prototype-by-Component Analysis: A Corpus-Based, Intensional Approach to Ordinary Meaning in Statutory Interpretation." *Applied Corpus Linguistics* 4, no. 1: 100078, 1–9.

Ehrett, John S. 2019–2020. "Against Corpus Linguistics." *Georgetown Law Journal Online* 108: 50–73.

Goldman, Alvin, and Bob Beddor. 2021, May 21. "Reliabilist Epistemology." *Stanford Encyclopedia of Philosophy*. Retrieved from https://plato.stanford.edu/entries/reliabilism/.

Gries, Stefan Th. 2020. "Corpora and Legal Interpretation." In *The Routledge Handbook of Forensic Linguistics*, 628–643. Abingdon, Oxon: Routledge, Taylor & Francis Group.

Gries, Stefan Th., and Tammy Gales. 2024. "Talking Across the Interdisciplinary Aisle: A Guide for Legal and Corpus-Linguistic Scholars and Practitioners." *Applied Corpus Linguistics* 4, no. 1: 100086, 1–9.

Gries, Stefan Th., Brian G. Slocum, and Kevin Tobia. 2024. "Corpus-Linguistic Approaches to Lexical Statutory Meaning: Extensionalist vs. Intensionalist Approaches." *Applied Corpus Linguistics* 4, no. 1: 100079, 1–14.

Hashimoto, Brett. 2023. "Corpus of Founding Era American English: Designing a Corpus for Interpreting the United States Constitution." *Corpora* 18, no. 1: 1–14.

Hashimoto, Brett, and James Heilpern. 2023. "Solving the Cherry-Picking Problem in Legislative History Use: A Corpus-Based Approach for Empirical Intentionalist Legal Interpretation Analysis." *International Journal of Language and Law* 12: 48–72.

Herenstein, Ethan J. 2017–2018. "The Faulty Frequency Hypothesis: Difficulties in Operationalizing Ordinary Meaning through Corpus Linguistics." *Stanford Law Review Online* 70: 112–122.

Hessick, Carissa Byrne. 2018. "Corpus Linguistics and the Criminal Law." *Brigham Young University Law Review* 2017: 1503–1530.

Jarvis, Scott, and Brett Hashimoto. 2021. "How Operationalizations of Word Types Affect Measures of Lexical Diversity." *International Journal of Learner Corpus Research* 7, no. 1: 163–194.

Jennejohn, Matthew, Samuel Nelson, and D. Carolina Núñez. 2021. "Hidden Bias in Empirical Textualism." *Georgetown Law Journal* 109: 767–807. BYU Law Research Paper No. 21–07.

Keller, Daniel, and Jesse Egbert. 2021. "Hypothesis Testing Ordinary Meaning." *Brooklyn Law Review* 86: 489–532.

Klapper, Shlomo. 2019. "(Mis)judging Ordinary Meaning?: Corpus Linguistics, the Frequency Fallacy, and the Extension-Abstraction Distinction in 'Ordinary Meaning' Textualism." *British Journal of American Legal Studies* 8: 327–369.

Lee, Thomas R., and Stephen C. Mouritsen. 2021. "The Corpus and the Critics." *The University of Chicago Law Review* 88, no. 2: 275–366.

McLean, Stuart. 2018. "Evidence for the adoption of the Flemma as an Appropriate Word Counting unit." *Applied Linguistics* 39, no. 6: 823–845.

Mouritsen, Stephen C. 2012. "Hard Cases and Hard Data: Assessing Corpus Linguistics as an Empirical Path to Plain Meaning." *Columbia Science and Technology Law Review* 13: 156–205.

Mouritsen, Stephen C. 2010. "The Dictionary Is Not a Fortress: Definitional Fallacies and a Corpus-Based Approach to Plain Meaning." *Brigham Young University Law Review* 2010: 1915–1980.

Pinchbeck, Geoffrey. G. 2014. Lexical frequency profiling of a large sample of Canadian high school diploma exam expository writing: L1 and L2 academic English [Roundtable presentation]. American Association of Applied Linguistics, Portland, OR, USA.

Phillips, James C., and Jesse Egbert. 2018. "Advancing Law and Corpus Linguistics: Importing Principles and Practices from Survey and Content Analysis Methodologies to Improve Corpus Design and Analysis." *Brigham Young University Law Review* 2017: 1589–1619.

Phillips, James Cleith, and Sara White. 2017. "The Meaning of the Three Emoluments Clauses in the U.S. Constitution: A Corpus Linguistic Analysis of American English from 1760–1799." *South Texas Law Review* 59, no. 2: 181–236.

Ramer, John D. 2017. "Corpus Linguistics: Misfire or More Ammo for the Ordinary-Meaning Canon." *Michigan Law Review* 116, no. 2: 303–328. https://doi.org/10.36644/mlr.116.2.corpus.

Repo, Liina, Brett Hashimoto, and Veronika Laippala. 2023. "In Search of Founding era Registers: Automatic Modeling of Registers from the Corpus of Founding Era American English." *Digital Scholarship in the Humanities* 38: 1659–1677.

Römer-Barron, Ute, and Clark D. Cunningham. 2024. "Applied Corpus Linguistics and Legal Interpretation: A Rapidly Developing Field of Interdisciplinary Scholarship." *Applied Corpus Linguistics* 4, no 1: 100080, 1–7.

Scalia, Antonin, and Bryan A. Garner. 2012. *Reading Law: The Interpretation of Legal Texts*. St. Paul, MN: Thomson West.

Stanley, Jason. 2005. *Knowledge and Practical Interests*. New York: Oxford University Press.

Talbert, Matthew. 2024. "Moral Responsibility." In *The Stanford Encyclopedia of Philosophy* (Fall 2024 Edition). Edited by Edward N. Zalta and Uri Nodelman. Retrieved from https://plato.stanford.edu/archives/fall2024/entries/moral-responsibility/.

Tankersley, Daniel C. 2018. "Beyond the Dictionary: Why Sua Sponte Judicial Use of Corpus Linguistics Is Not Appropriate for Statutory Interpretation." *Mississippi Law Journal* 87, no. 4: 641–678.

Tobia, Kevin. 2021. "The Corpus and the Courts." *University of Chicago Law Review Online* 2021: 1–36.

Tobia, Kevin. 2020. "Testing Ordinary Meaning." *Harvard Law Review* 134: 1027–1063.

Waldron, Jeremy. 2023. "The Rule of Law." In *The Stanford Encyclopedia of Philosophy* (Fall 2023 Edition). Edited by Edward N. Zalta and Uri Nodelman. Retrieved from https://plato.stanford.edu/archives/fall2023/entries/rule-of-law/.

Zoldan, Evan C. 2019. "Corpus Linguistics and the Dream of Objectivity." *Seton Hall Law Review* 50, no. 2: 401–448.

27

GET YOUR HANDS OFF MY IDIOLECT! A SPECTRUM MODEL OF MISUSE OF IDIOLECTAL INFORMATION AS A HEURISTIC TOOL

Peter S. Harrison and Dominic Watt

Introduction

With the highly successful aim of bringing attention to the potential opportunities and dangers of the technology, on the 16th of May 2023 Senator Richard Blumenthal, Chair of the US Senate Judiciary Committee's subcommittee on Privacy, Technology and the Law, opened a congressional hearing on the regulation of artificial intelligence (AI) using a speech written by OpenAI's ChatGPT Large Language Model (LLM) and voiced through a text-to-speech (TTS) program designed to mimic the Senator's voice. The rapid development and implementation of such artificial intelligence and machine learning (ML) systems, in particular where such systems are linked to the operation and development of TTS programs, will revolutionize both the practice of forensic linguistics and the problems to which forensic linguistics seeks solutions. The changes this revolution will bring will almost certainly occur in ways that are as yet difficult to accurately predict. These developments amplify and make urgent the need for a proper understanding of the ethical foundations of the use of idiolectal information within forensic linguistics and more broadly. What they do not do is change the ethical playing field. For the authors of the present chapter, those ethical foundations are, and will continue to be, based upon a respect for accuracy, privacy, and property; the decrease of conscious and unconscious bias and the increase in impartiality; an understanding of analytical limits; and an appreciation of cross-cultural and multilingual issues. The rapid advancement of AI/ML will, however, potentially make analyzing this ethical landscape quite difficult. This is particularly true where the internal operation of these systems becomes ever more opaque, even to the creators of the system (Romero 2023; Sison et al. 2023; Yampolskiy 2020), and where the involvement of AI in (perhaps all) analytical processes becomes ubiquitous and non-apparent to the recipients of the analytical results.

Although there will be important questions with regard to the impact of LLM and TTS technologies on the development of analytical tools *within* the practice of forensic linguistics, the key focus of this chapter will be the impact of LLMs and TTS on the existing and developing problems

DOI: 10.4324/9781003391074-30

upon which forensic linguistics will be brought to bear. To this end, the chapter will examine the ethical questions around the use, and in particular "mis"-use, of idiolectal information (envisaged in its broadest form). In particular, the chapter will look to situate uses (and misuses) of idiolectal information on a spectrum of "degree of misuse". It should be stated from the outset that this allocation is not intended to be definitive. There are many unavoidable subjectivities in allocating a position on such a spectrum. There are also many opportunities for arguing what constitutes a "misuse". However, this approach is intended as a heuristic tool to aid analysis of, and provoke debate about, those things that we should consider important when looking at the use/misuse of idiolectal information. In this way, this chapter aims at providing a structure for further analysis of this rapidly developing field. Indeed, readers are invited to consider where they themselves would place the identified "uses/misuses" on the provided spectra and, of course, consider where yet further uses and misuses might fall. Although this chapter will have a particular focus on misuses of idiolectal information which are enabled, or enhanced, by the use of LLMs and TTS programs, it is important to examine examples of "non-AI-/TTS-enabled" misuse of idiolectal information, both to tease out fundamental principles and to situate them within the spectrum of misuse alongside AI-/TTS-enabled misuse.

What Is "Idiolectal Information"?

An "idiolect" is a person's unique sub-variety of a language. Sometimes likened to a "linguistic fingerprint" (e.g., Shim 2022; Coulthard 2004), the notion of an idiolect has a long history, and in a chapter of this length we cannot adequately review the literature documenting its development (see Barber and Garcia Ramirez 2021; Heck 2006; Higginbotham 2006; George 1990). We start here, however, by examining Bernard Bloch's original definition:

> The totality of the possible utterances of one speaker at one time in using a language to interact with one other speaker is an idiolect. [...] Our definition implies (a) that an idiolect is peculiar to one speaker (b), that a given speaker may have different idiolects at successive stages of his career, and (c) that he may have two or more idiolects at the same time.
>
> *(1948: 7)*

Bloch's definition is not free of conceptual problems. For one thing, the set of possible utterances that an individual speaker could produce during an interaction at a given time is almost infinite, for all practical purposes. So a definition of idiolect as unconstrained as this is not a very useful one. We could focus instead on what a speaker would be likely to say or write under a given set of circumstances, but our prediction could only be based on what the speaker had previously said or written. That might constrain our model of that person's idiolect too tightly: we would not want to rule out the capacity of a speaker's grammar to generate entirely novel—or at least hitherto unattested—utterances, as they in fact might constitute a very large proportion of the speaker's language productions. So far as we can ascertain, the uniqueness of written strings of a length greater than half a dozen words or so is almost guaranteed, so long as the speaker is not using formulaic constructions with a fixed wording (Culwin and Child 2010; Coulthard 2004). And uniqueness in speech is a certainty, given that even a single speaker can never utter the same string in exactly the same way twice. This is not a self-evident aspect of speech production that we can dismiss just because it is trivially true. After all, how would one go about classifying which phonetic properties of an utterance are accidental by-products of the coordinated movements of the speech organs, and which are under the speaker's "deliberate" control (Nolan 1999; Ohala 1989)? The speaker's idiolect will, presumably, be specified by both accidental and deliberate features, since both will contribute to the uniqueness of the individual's voice.

The implications (a–c) in the Bloch quote above do not necessarily follow from the first part of his definition, but let us accept that implication (b) follows from (a), and (c) follows from (b). Implication (a) is automatically true if an idiolect is to be defined as the unique language habits of an individual speaker. However, much of a speaker's language competence must coincide with that of other speakers of the same language by virtue of the fact that they can be grouped together *as* speakers of the same language. We could therefore talk of idiolect as the unique combination of language features possessed by a speaker even if none of the regular features contained within that set is actually idiosyncratic to that person (Chaski 2013). The second implication (b), allows for changes to an individual's idiolect over time (see de Paiva, Duarte, and Guy 2021, and other entries in Beaman and Buchstaller 2021, such as Bowie 2021; Reubold and Harrington 2021; Rickford 2021; or Standing and Petré 2021). While the idiolects of children change very rapidly as they mature, changes to an adult's relatively fixed language competence across the lifespan can nevertheless come about because of changes to that person's knowledge of the language, perhaps through illness or injury to the brain, or via education, contact with other speakers, choice of reading, media consumption, social conformity pressures, and so on (Peng 2020; Sankoff 2018; Harrington, Palethorpe and Watson 2000). In line with Bloch's (c), speakers may be bi- or multi-idiolectal if they have competence in more than one dialect or language. We might also include additional codes such as registers or styles under this heading. An advocate or politician, for instance, might adopt an idiosyncratic, personalized style when engaging in formal legal discourse as well as when they are communicating with friends or family members, while a novelist might choose to switch between authorial styles for artistic effect or for obfuscatory purposes (to disguise his/her true identity, as in the cases of Mary Ann Evans/George Eliot or J. K. Rowling/Robert Galbraith; see Cafiero and Camps 2023; Bevendorff et al. 2022; Juola 2015; Burrows 1987).

Attempts to operationalize the concept of idiolect—providing objective evidence that individual language users do in fact possess unique language systems manifested by something analogous to a fingerprint—are immediately beset by practical difficulties. As Barlow (2010) points out:

> [t]he term idiolect is well-known in linguistics and is distinguished by the fact that there is probably no other linguistic term in which there is such a large gap between the familiarity of the concept and lack of empirical data on the phenomenon.
>
> *(1)*

If we were to confine our effort just to an individual's written language, the task would be challenging enough. However, in the domain of speech, in which the number of degrees of freedom is very much larger than in writing, our problems would be significantly greater. As we discussed in an earlier paper (Watt, Harrison, and Cabot-King 2019), it is extremely difficult to characterize what it is about a speaker's voice that makes it unique. Since we cannot capture every utterance the speaker has produced, will produce, or could potentially produce, we must make do with an analysis based on a representative sample of that person's speech. But how big a sample would we need to establish distinctness of the speech of person X from that of everyone else, supposing that that is even a feasible goal in principle (McLachlan 2016)? What parameters would we choose for the purpose, since we cannot describe or measure every possible phonetic or acoustic property of our samples? Even if it were feasible, would it be worthwhile? For all practical purposes, it is accepted that the voice model we derive for individual speakers, no matter how richly detailed, is still only partial. So while a voice model's incompleteness increases the possibility that we will find that two speaker models are identically matched and are "vocal twins", so to speak, the missing information would generally not raise the likelihood of finding such matches to such an extent that it would give us major cause for worry.

For the purposes of this chapter, we will define "idiolectal information" to be information that is sufficient to identify an individual from others within a particular circumstance. It need not encompass all the information that would define or represent every element of a person's idiolect or indeed be required to definitively identify that individual from all others—particularly as that is impossible to achieve. This definition is therefore contextual. Where one is looking at a narrow situation the amount of information that is sufficient to identify an individual from others will likely be less than that required in a more complex situation.

For the analysis in this chapter, we have split idiolectal information into three broad categories:

(1) *Solely verbal information* in which an attempt is made to determine the speaker's identity through analysis of their word use alone;
(2) *Solely vocal information* in which an attempt is made to determine the speaker's identity through analysis of their vocal characteristics alone; and
(3) Combined *verbal and vocal information* in which an attempt is made to determine the speaker's identity through analysis of both their pattern of word use and vocal characteristics, i.e., utterance content and form are considered simultaneously.

Under some circumstances it may be necessary only to consider one channel—the linguistic content but not the phonetic form, or vice versa—while under others we need to take account of both simultaneously. When attempting to determine what makes an author's writing style distinctive from that of another author, for instance, all we are interested in is the words and syntactic structures exemplified in the text.

On other occasions, we may only wish to consider what makes a speaker's voice sound different from the voice of some other person/people. It might not matter much which language is being spoken, or whether we understand what is being said. It should be borne in mind that the acoustic phonetic properties peculiar to a speaker's voice are the mixed product of physical factors—the anatomy and physiology of his/her vocal tract—combined with behavioral ones, i.e., learned habits combined with speaker-specific tendencies that could not be predicted on the basis of the speaker's biological characteristics alone.

In the third scenario, it may be crucial to consider both the message and the medium at once: capturing a speaker's distinctive linguistic profile may be possible only if verbal and vocal information is combined with the speaker's lexical and grammatical habits. The latest generation of powerful AI tools makes it feasible to derive convincing models of an individual's idiolect in terms of both speech and language with great rapidity, and to generate linguistic artifacts based upon them which may be used to deceive observers. We discuss the ethical dimensions of this possibility below.

LLM and TTS and Idiolectal Information

An LLM is a type of AI model that is capable of processing and generating human language. These models are typically based on deep neural networks and are trained on enormous amounts of text (Tang, Chuang, and Hu 2023; Brown et al. 2020). OpenAI's GPT-3.5 chatbot was trained on 570GB of internet text data, which is the equivalent of hundreds of billions of words (Romero 2021). LLMs use this training data to first learn the patterns and structure of human language and then to generate (mostly) coherent and grammatical sentences. LLMs have the potential to significantly impact forensic linguistics by providing new methods for analyzing and interpreting linguistic data. LLMs may be able to analyze text in ways that were previously difficult or impossible. This could allow forensic linguists to identify patterns and linguistic features that were previously overlooked, leading to more accurate and faster analyses (Boratyn et al. 2023; Desaire et al. 2023a, 2023b; Nagel 2023; Sánchez, Moreno, and Jiménez López 2022).

To what extent can LLMs analyze idiolects? Since LLMs are trained on vast amounts of text data from various sources and can learn to recognize and predict patterns in language use, they are capable of identifying idiolectal idiosyncrasies in vocabulary and syntax and using those to mimic individual authorial style (e.g., that of H. P. Lovecraft, as per Garrido-Merchán, Arroyo-Barrigüete, and Gozalo-Brizuela 2023). However, the accuracy and effectiveness of LLMs in analyzing idiolects will depend on the amount, and nature, of data available for training, as well as the complexity and variability of the idiolect in question. Crucially, an LLM that has the capacity to write text in an author's "style" may be used to write unauthorized text for a wide variety of reasons, from pastiche through to disinformation and fraud (Augenstein et al. 2023; Jones, Nurse, and Li 2022; Jones et al. 2022).

Voice sampling is the process of collecting and analyzing a person's voice in order to extract useful information from the signal, such as the language being spoken and the linguistic content of the sampling for use in, for example, automatic transcription. Voice sampling may also be used for classifying a speaker according to regional accent, gender, and age group; or detecting the speaker's emotional states. Such sampling has long given rise to multiple ethical considerations (e.g., obtaining informed consent; ensuring privacy and security; promoting accuracy and fairness; restricting use and purpose; promoting transparency and accountability; and respecting cultural sensitivity) (Bohm et al. 2017; Cowling 2015). There have also been major concerns over the use of illicitly obtained samples of our speech for fraudulent or other criminal purposes (e.g., Khatsenkova 2023; Müller, Pizzi, and Williams 2022; Perot and Mostert 2020; Shirvanian, Saxena, and Mukhopadhyay 2018; Chen et al. 2017; Feng, Fawaz, and Shin 2017). However, the dramatic development of TTS programs has added a further layer of concern. Allied to LLM-generated text, these programs can be utilized for convincing deepfake voice generation that in turn can be used in disinformation and fraud (de Rancourt-Raymond and Smaili 2023; Vaccari and Chadwick 2020).

LLMs also have the potential to significantly impact TTS programs by improving their accuracy, naturalness, and fluency. The result is the generation of more realistic-sounding speech that is much more convincing and increasingly necessitates less human supervision. Speech synthesis algorithms such as those developed by ElevenLabs (elevenlabs.io) are already capable of generating synthetic speech that is indistinguishable from human speech, especially when it is presented to listeners under less than optimal conditions, such as over a telephone line (Mai et al. 2023). As Senator Blumenthal sought to highlight, these new technologies are not only cheaply or even freely available, they are also already being used for malicious purposes such as creating deepfake audio recordings for fraud, impersonation, disinformation campaigns, or the generation of fake evidence. Online interaction with the Microsoft Bing ChatGPT chatbot is, according to some observers, becoming uncannily similar to interaction with humans (e.g., Chomsky, Roberts, and Watumull 2023; Roose 2023), even if ChatGPT "hallucinates and produces 'bullshit'" (Sison et al. 2023: 8). We are thus entering a new phase in the arms race between the security services and cybercriminals, with an as yet only limited ability to identify whether deceptive or defamatory communications are created by machines under close human supervision, or by machines creating material semi- or fully independently (*Walters v. OpenAI LLC*, Ga. Super. Ct., 23-A-04860-2).

Misappropriation of writing and voice samples from individual language users is easy in many cases, and "voiceprint" models based upon those samples can with great rapidity be generated for deployment in spoofing attacks. Notable recent examples from the United Kingdom include a fake advertisement video purporting to be of British financial expert Martin Lewis, which appeared on the Facebook social media platform in July 2023 (BBC News, https://www.bbc.co.uk/news/uk-66130785), and two audio files which were claimed to be covert recordings of the Labour Party leader Sir Keir Starmer berating an assistant and complaining about their conference venue (Cuthbertson 2023).

How Should We Define "Misuse" of Idiolectal Information?

In formulating a "spectrum of misuse" as a heuristic tool to aid analysis of the misuse of idiolectal information, the authors are aware that the concept of misuse is, for legal scholars, something of a minefield which encompasses many legal philosophical principles. As stated, the intention here is not to create a definitive spectrum. However, we believe it is an exercise worth attempting, at least to stimulate discussion around what misuse of an idiolect actually is (and is not) and what elements constitute such misuse.

Whilst the specific definition of misuse obviously depends on the broader context of the use and the nature of the thing being used/misused, for this chapter the term "misuse" broadly refers to the harmful use of a thing in a way that is either contrary to its intended purpose or is inappropriate in its application. In what way can a use be said to be "contrary to its intended purpose"; or, where there is arguably no intended purpose per se, how might an application be "inappropriate"? The answers to these questions clearly depend on what "appropriate" use means in relation to a particular thing. For the most part, an individual's idiolect is arguably not something which has been "created" for a "purpose", although one can very much think of many ways in which individuals develop or switch their idiolect according to social situation or surroundings.[1]

We suggest that what fundamentally determines the "inappropriateness" of the way idiolectal information is used rests on a series of fundamental principles: confidentiality/privacy (including data protection); ownership; and protection against defamation, misrepresentation, or fraud. In the following sections, we will deal with each in turn. As it is impossible to review relevant jurisprudence in all jurisdictions, we have identified some helpful examples from legislation from the European Union (EU), United States, and United Kingdom.

Fundamental Principles That Determine the (In)appropriate Use of Idiolect

Confidentiality/Privacy

It is not clear whether principles of privacy or confidentiality give rise to any type of control over the idiolect (Watt, Harrison, and Cabot-King 2019). For example, in England and Wales there is no common law right to privacy as such (see *Kaye v. Robertson* ([1991] FSR 62)). However, the law in England and Wales has long protected confidential information from improper disclosure (see for example *Coco v. A N Clark (Engineers) Ltd* ([1969] RPC 41). Claimants have used this right to protect commercially sensitive information as well as confidential information which is of personal significance. However, where individuals merely have their voice (or other elements of their idiolect) recorded, no confidential "information" is conveyed as such. Nevertheless, the English law of Breach of Confidence needs to be interpreted and applied to give effect to Article 8(1) of the European Convention on Human Rights (ECHR) (*Campbell v. Mirror Group Newspapers Ltd*

([2004] UKHL 22). This article states that "[e]veryone has the right to respect for his private and family life, his home and his correspondence". This proclamation very much follows the provisions of Article 12 of the Universal Declaration of Human Rights (UDHR).

If an individual, unknowingly, has their voice recorded to create a "voiceprint", then, following *Campbell*, it is not inconceivable that a court might consider that the mandate to respect their private life extends a cloak of "confidentiality" over their voice. This decision may be made even though the individual's voice may not otherwise be of a confidential nature. However, it is likely that any such right is highly dependent upon the intent of the person recording the voiceprint. No matter the intent, it is unlikely that it would create any right of control over another person's idiolect per se (Watt, Harrison, and Cabot-King 2019). Though arising from a fundamental human right, as will be examined in the following sections, the right of privacy has taken different evolutionary routes in different jurisdictions.

Data Protection

The holding and processing of personal information by all types of organization is an increasing concern for the people whose data is being held. Legislatures across the globe are developing laws controlling the way in which such data can be held and processed (Kontargyris 2018). One of the most influential is the EU General Data Protection Regulation (EU 2016/679) or GDPR which imposes a strict set of rules on those holding personal data. Under GDPR Article 9, especially stringent rules are applied to so-called "special categories" of personal data which are deemed "particularly sensitive in relation to fundamental rights and freedoms" (GDPR Recital 51). Such special categories include "biometric data for the purpose of uniquely identifying a natural person" (GDPR Art 9(1)). Biometric data is defined in GDPR Article 4 as "personal data resulting from specific technical processing relating to the physical, physiological or behavioral characteristics of a natural person, which allow or confirm the unique identification of that natural person, such as facial images or dactyloscopic [fingerprint] data". So long as a voiceprint is being used as a unique identifier of a physical person, it is hard to imagine that it would fall outside the ambit of Article 9 (Watt, Harrison, and Cabot-King 2019).

"Ownership"

The ownership or possession of a property right in (or over) idiolectal information is a complex, and in many ways unclear, area (Melzner, Bonezzi, and Meyvis 2023; Gibson 2022; Scott et al. 2019; Watt, Harrison, and Cabot-King 2019). There are a number of intellectual property rights that may, to a greater or lesser extent, provide some protection over the elements of the idiolect. This chapter will focus on four: (1) copyright; (2) passing-off; (3) right in privacy; and (4) right of (sonorous) image. In the following sub-sections, each of these rights is examined in turn with reference to key elements of an idiolect. The examination starts with the "physical" vocal elements and progresses to the "verbal" elements.

Copyright

THE VOCAL ELEMENT OF AN IDIOLECT

Building on the 1971 Convention for the Protection of Producers of Phonograms Against Unauthorized Duplication of Their Phonograms—the "Geneva Phonograms Convention"—the majority of jurisdictions worldwide provide protection for sound recordings.[2] Therefore, where a voice has been "captured" in a fixed form through a recording, that recording will attract copyright protection. Importantly, that recording will be owned by the person *making* the recording, *not* the person who possesses the recorded voice. However, what of the voice alone? Does copyright protect a voice which has not been "fixed" in a recording? Many jurisdictions provide a "right in performance", which is separate from copyright, to control the recording of a live performance.[3] This is a right to prevent fixation by others but applies only to dramatic or musical works, the reading or recitation of a literary work, or a variety act (see, for example, UK Copyright Designs and Patents Act 1988 ("CDPA") Section 180(2)). It will not protect a voice outside of that context. Performing artists also have a "moral right" to prevent "derogatory treatment" of their performance (CDPA Section 205F) and the right to be identified as the performer of a live performance or on a recording of their live performance (CDPA Section 205C).[4] Again, this is a right to control a fixation by others, but it also only applies to dramatic or musical works, the reading or recitation of a literary work, or a variety act, and does not protect a voice outside of that context. Within the United States, the "unfixed" voice alone is not protected.[5]

Not all jurisdictions are as certain on this question of fixation. The Berne Convention for the Protection of Literary and Artistic Works of September 9, 1886, states in Article 2(2) that it is for signatories to prescribe that works in general or any specified category shall not be protected unless they have been fixed in some material form. However, UK law states that copyright does "not subsist in a literary, dramatic or musical work unless and until it is recorded, in writing or otherwise" (UK CDPA Section 3(2)); and fixation in relation to other works is implicit in the wording of other sections of the Act.

This requirement for fixation may, at first blush, seem somewhat surprising. However, the key point here is that it is broadly understood that copyright protects the particular *expression* of an idea, not the idea itself (see for example, *Baigent and Leigh v. The Random House Group Ltd* [2007] EWCA Civ 247). A painter's idea for a painting is not protectable under copyright until it is rendered in a *tangible* form, after which it is protected by copyright. This requirement for tangibility arguably derives from two, closely related, principles. Firstly, establishing the subsistence of an intellectual property right such as copyright and enforcing it against third parties fundamentally requires a degree of certainty for third parties, such that they may fairly navigate around such rights and be protected against spurious claims (Harrison 2020). That certainty is, arguably, not provided by an idea in an artist's head until the idea is rendered into a tangible expression of that idea. Secondly, since intellectual property rights create extremely powerful ways to restrict third-party activity, their subsistence must be philosophically justified (Harrison 2020). If the purpose of copyright is to protect created "works" from being copied by others so as to reward the authors of those works for their creativity and to incentivize others to put their creative efforts into the creation of such works (Shiffrin 2008), something must have to be done in terms of producing a tangible work to deserve the reward, beyond merely coming up with an amorphous, ill-defined, concept. Although their voice may very much be a part of the artistic persona of a vocal artist, with regard to copyright law at least, the mere concept or "idea" of that voice will not be protected until it is captured in a tangible record of an expression of that idea. However, as will be explored in this chapter, there are many other areas of the law that can provide protections of the intangible idea of a voice. Readers seeking further analysis of the question of copyright and the spoken word are directed to the review by MacQueen (2005).

THE VERBAL ELEMENT OF AN IDIOLECT

Although verbal elements of an idiolect may, when subjected to linguistic analysis, be sufficient to identify an individual or useful in the identification of an individual in the presence of other evidence, the identifying phrases may themselves be short, be scattered through speech, and/or only be definitive when in combination with other phrases which are equally scattered through speech. As such, the verbal element of an idiolect presents two key problems for copyright subsistence: (1) is the work sufficiently large? and (2) is it original? Copyright law across many jurisdictions recognizes a *de minimis* principle that if a work is too small, it will not be granted protection.

Within the EU, the question of whether a work was sufficiently large to attract copyright protection was addressed in *Infopaq International A/S v. Danske Dagblades Forening* ([19 July 2009] ECR I-6569). For the Court of Justice of the EU, the question was not about size *as such*. Instead, it held that the question was actually one of originality of the work in question. Subsistence of copyright across jurisdictions depends on a work being "original", and the question of originality has preoccupied courts across the world.[6] In being asked to address the *de minimis* question the court in *Infopaq* developed what has turned out to be a seminal test for originality: the reproduction of 11-word extracts from newspaper articles would constitute reproduction of a copyright work provided that the reproduced extracts were subject-matter which was original in the sense that the extracts were their "author's own intellectual creation" (*Infopaq* para 37). This, and subsequent decisions of the Court (Rosati 2018), suggest that provided a work is identifiable as distinct and original in the sense that it demonstrates the "author's own intellectual creation", it will then attract copyright protection, regardless of its size.

To what extent can the verbal elements of an idiolect be seen to represent an "expression of the author's own intellectual creation"? As stated above, the individual identifying phrases may themselves be short, imitative of others, scattered throughout speech, and only definitive when in combination with other phrases which are equally scattered through speech. Although we are aware of no case law on the point, it is difficult to imagine a "mosaic" of phrases as constituting something sufficient to be an "expression of the author's own intellectual creation". For these authors, short vocal samples are also likely to fall foul of the *de minimis* and originality rules set out above; and the more that a voice sample is fractionated, the less likely it is to attract copyright protection. Importantly, even if such a sample were to attract such protection, the first ownership of the copyright would sit with the person that had made the recording of the voice, rather than with the person from whom the samples were taken (unless, of course, that person had recorded their own voice).

It is also difficult to see why a combination of both verbal and vocal elements of an idiolect would be treated significantly differently under copyright law than would be each of the vocal and verbal elements alone. For the present authors, it is not inconceivable that a combination of vocal and verbal elements could combine to create an impression of an "expression of the author's own intellectual creation". However, given that the fixation problems discussed above will still apply in relation to the vocal element, it is likely that courts would take something of a "reductionist" approach to each element of the idiolect.

Even where elements of an idiolect *could* combine to allow for copyright protection, it is worth mentioning that various jurisdictions provide a specific "fair use" or "fair dealing" exemption from copyright infringement where the third-party use of the copyright work is a parody or pastiche. For example, Article 5(3)(k) of EU Directive 2001/29/EC allows for "caricature, parody or pastiche" of a work provided that it does not conflict with a normal exploitation of the work and does not unreasonably prejudice the legitimate interests of the right holder. In *Johan Deckmyn and Vrijheidsfonds VZW v. Helena Vandersteen and Others*, the Court of Justice of the European Union ruled that for a parody exemption to apply, the otherwise infringing work must: (i) "evoke an existing work, while being noticeably different from it", and (ii) "constitute an expression of humour or mockery", and that a court considering such a parody defense should seek to strike a fair balance between the rights of the rights holders of the original work and the maker of the parody. In the United States, the parodies of a work may potentially be permitted as fair use of a copyright work subject to application of the tests for fair use set out under the US Copyright Act of 1976 §107 (see *Campbell v. Acuff-Rose Music, Inc.*, 510 U.S. 569 (1994)). It remains moot how a caricature, parody, or pastiche exemption would exactly apply to copyright protection of elements of an idiolect, although for the present authors there appears no fundamental reason why such elements would be treated differently from other copyright works.

Passing-Off

The law of passing-off, or "misappropriation" as it is more commonly called in the United States, is found in most common law jurisdictions. It is based upon the idea that it is wrong to mislead consumers into thinking that your product is actually somebody else's.[7] In theory, there is no reason why the use of a very distinctive voice in a commercial context (say as a voiceover used in a marketing campaign) could not be protected under the law of passing-off. In the English Court of Appeal case of *Irvine v. Talksport Ltd* ([2003] EWCA Civ 423); [2003] 2 All ER 881; [2003] EMLR 538), the court agreed with the claimant that although he was not a traditional trader per se, he did have commercial goodwill arising from his celebrity status and his ability to use that fame to endorse products.

To be protected under passing-off, a voice would have to be very distinctive in itself—so much so that in the absence of any other identifying clues, the listener would be able correctly to put a name to the voice. Ideally, the voice would have become so strongly associated with a product that when consumers heard it, they would bring that product to mind.[8] In addition, any unauthorized mimicry would need to be so good as to mislead the vast majority of listeners into believing that the voiceover really was being provided by the original celebrity. It is also hard to divorce the voice from what is being said. Messages that explicitly revealed an "irreverent" nature of a sound-alike would perhaps take the voiceover further away from being passing-off. A further hurdle is that passing-off requires the existence of *commercial* goodwill and misrepresentation within a commercial or endorsement context. It does not create any right over a celebrity voice which is outwith this commercial sphere, nor does it create a right in non-famous voices.

Right in Publicity

The protection of celebrity endorsement in the United States has taken a different evolutionary path from passing-off. The principle that underlies the approach taken in the United States following *Haelen* is to prevent the undue enrichment of the person misusing a celebrity's right, regardless of whether that use damages that celebrity's commercial "goodwill".[9] Although the court in *Midler v. Ford* did not think that copyright could exist in a voice alone, it did find that Midler's "rights in publicity" could as equally sit with her voice as with her photographic image.[10] As in the United Kingdom, in the United States, for a right in based upon the concept of personality to exist in a voice alone, the voice would need to be very distinctive. However, the US test appears not to require confusion on the part of the public. This is because the cause of action is not based on harm to the celebrity claimant, but on unfair gain by the defendant.[11] However, like the English and Welsh law of endorsement passing-off under *Irvine*, the US right in publicity only applies where a celebrity has some degree of recognition in the mind of the public.

Right of (Sonorous) Image

In the *Papillon* case (TGI Paris ord., réf., Feb. 27, 1970), the Tribunal de Grande Instance in Paris held that the claimant had a "right of image", and suggested that the unauthorized use of the celebrity claimant's photograph on a book published by the defendant had given improper endorsement to the defendant's book. At first examination, this looks something like "endorsement passing-off" in the case in *Irvine*, or the US "right of publicity" established in *Haelan*—both of which require a degree of celebrity to be triggered. However, the French right to control an image actually arises out of Article 9 of the Civil Code of France, which provides that everyone has the right to respect for their private life ("Chacun a droit au respect de sa vie privée"). This declaration is, of course, itself related to Article 8(1) ECHR which, as mentioned above, stipulates that "Everyone has the right to respect for his private and family life, his home and his correspondence".

Given that Article 9 refers to "everyone", the question is whether this "right of image" is more akin to a right of privacy which would protect those who have no celebrity. The French case law is inconsistent (Caron 2006; Logeais and Schroeder 1998). French case law has extended to granting a right of image over a voice (TGI Paris, 3e ch., 1re sect., 27 sept. 2004: préc.), the voice being described as a "sonorous image" [*image sonore*] (Lepage 2005: 2). However, although this applies on the basis of relatively little fame and, it would appear, no requirement for distinctiveness, some element of exposure to the public realm seems to be required (Tafforeau 2007).

Defamation

Many jurisdictions will have mechanisms to protect a person's character from serious damage. However, the interaction between the protection of character and the fundamental right to freedom of opinion and expression (including freedom to hold opinions without interference) (Article 19 UDHR and Article 10 ECHR) leads to a complex balance. In England and Wales, for example, defamation requires a publication of a written or verbal statement that has caused, or is likely to cause, *serious* harm to a person's reputation. For a statement to be defamatory, it must refer to the claimant; and it must either be untrue or incapable of being excused by any other legal defense (Defamation Act 2013 sections 1–3). Whether an LLM-/TTS-created deepfake can be considered defamatory will hinge on whether it could be considered to be a statement that would cause serious reputational damage but which should be protected by "freedom of expression" (Kugler and Pace 2021; Ray 2021; Harris 2019; Ice 2019).

Fraud and Fraudulent Misrepresentation

Where voice samples are used for voice cloning for fraudulent purposes such as gaining access to a bank account, that action will very likely constitute the crime of fraud. Although the treatment of fraud across jurisdictions is variable (Crentsil 2015), fraud is broadly concerned with the dishonest making of a false representation with the intention of obtaining some gain for yourself or another, causing loss to another, or exposing another to a risk of loss. Clearly, use of voice cloning to access another person's bank account would involve false representation and would (all other elements being present) constitute a fraudulent act.

Outside of criminal law, in some jurisdictions, the fraudulent use of a voiceprint or "digital spoofing" could give rise to a civil law claim under the tort of fraudulent misrepresentation. This is where an untrue statement of fact or law made by one party induces another to enter into a contract which thereby causes that party loss. The deceived party's remedies are rescission of the contract. This would involve the cancellation of the contract and the parties put back to where they were before the contact was entered into. It would also involve the award of damages for loss or harms arising out of having entered into the contract. It should be noted that this tort will protect the recipient of the fraudulent voiceprint, rather than the person whose voiceprint is being copied. Neither the crime of fraud nor the tort of fraudulent misrepresentation creates any sort of "ownership" in a voiceprint by the person from whom the voiceprint was created.

What Is, and Is Not, "Consent"?

The question of consent to use of idiolectal information is a key element in any analysis of use/misuse. One might simply assume that the giving of consent immediately turns what would constitute a potential misuse into a consensual use. However, we need to ask what consent means in this context. Consent is the voluntary and informed agreement given by a person who has the legal capacity to understand and make decisions about a specific action or activity. It involves a person providing their explicit permission for a particular action or activity to take place. This permission must be given freely (i.e., without coercion, pressure, manipulation, or undue influence); and should be capable of being withdrawn at any time. As such, consent requires clear communication and understanding of what is being agreed to. This includes the nature of the activity, the risks involved, and any relevant terms or conditions. Accordingly, essential to defining the use or misuse of idiolectal information in any particular situation is determining whether the consent given was truly free, prior, and informed.

Proposed Spectra of Idiolect Misuse

For all the proposed spectra set out below, for ease of reference within the figures and text, we have given an alphanumeric code to each level of misuse. We have used the following code:

a) VER ("verbal element spectrum") to denote misuse of the verbal elements of idiolect without a vocal element;
b) VOS ("vocal element spectrum") to denote misuse of the purely vocal elements of an idiolect (without a verbal element); and
c) VEVOS ("verbal and vocal elements combined spectrum") to denote misuse of both the verbal and vocal elements of an idiolect used together.

The lowest number on each scale denotes the lowest level of misuse and the highest number the highest. Please note, however, that this numbering scheme is only intended as a guide as to qualitative *order*, not as a representation of any quantitative scale or relative degree of misuse.

Verbal Elements

Figure 27.1 shows a proposed spectrum of misuse (VER) of the verbal elements of idiolect without a vocal element. The proposed order of misuses is as follows:

VER1. At the base of the inverted triangle, at the lowest level of misuse, is a situation where a subject has given their unequivocal prior informed consent to the use of their idiolectal information. This might occur, for example, as part of a research project into the idiolect, though one can imagine many other scenarios. Almost by definition, this cannot be misuse. Nevertheless, users of the information must be certain they have gained informed consent and that the information is used only for those purposes for which consent was granted.

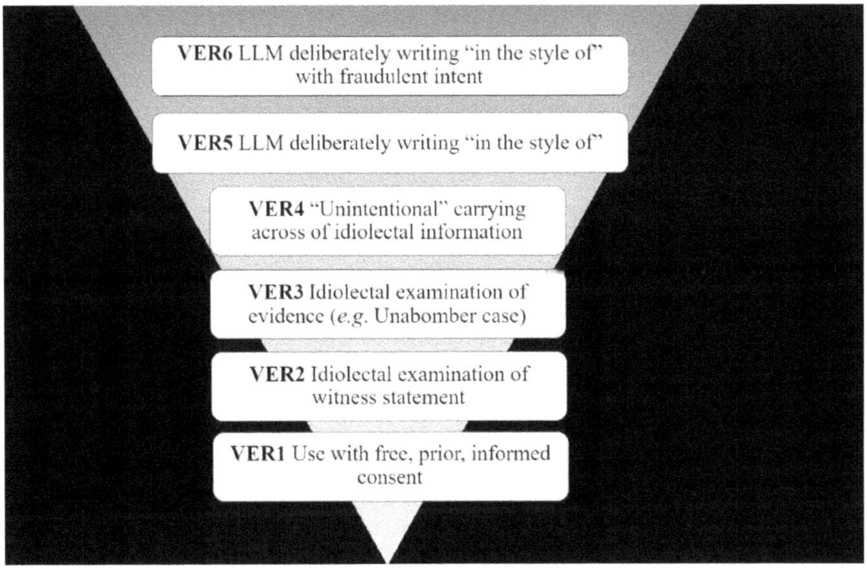

Figure 27.1 Spectrum of Idiolect Misuse (Without Vocal Element)
Degree of misuse decreases from top to bottom of inverted triangle.

VER2. This is the idiolectal examination of a witness statement and/or court-adduced evidence to assist in determining the veracity of the evidence. In the near future, convincing deepfakes produced by LLM/TTS may start to undermine the reliability of court evidence (Breen 2021). Examination of the verbal elements of an idiolect may allow a veracity check. Although the adducer of evidence may not have *expressly* consented to such an analysis, there is an arguable case that such consent is implied. Here the user needs to be extremely sensitive to the limits of such implied consent, however. Idiolectal examination may also apply in relation to determining what a witness actually means, for example, in a statement given to the police.[12] Again, a user will need to be very careful in relation to the extent of implied consent, notably where the analysis is being used against the subject.

VER3. There will also be cases where non-statement evidence used in court or a police investigation has not been formally adduced by the subject but has nevertheless been produced by the subject. The idiolectal examination of the manuscript "Industrial Society and Its Future" released by the "Unabomber" Ted Kaczynski in 1995 is an interesting example here (Coulthard 2006). For 17 years Kaczynski mailed, or hand-delivered, bombs to universities and other targets across the United States. The US Federal Bureau of Investigation (FBI) had failed to identify the bomber until he posted a 35,000-word essay ("Industrial Society and Its Future") claiming to explain his motives and his views of the failures of modern society. After the essay was published in the *Washington Post*, Kaczynski's brother recognized various distinctive turns of phrase in the work which caused him to suspect that Kaczynski was the author of the work. Further linguistic analysis by the FBI confirmed Kaczynski as a suspect and led to his eventual tracking down to a mountain cabin in Montana. Clearly, in this type of case the subject has not expressly or impliedly consented to such forensic examination, but here the requirement for consent is surely overridden by the public interest in securing a conviction. Once again, however, the extent of such public interest will be a limited exception.

VER4. This is the "misuse" of idiolectal information which perhaps inadvertently occurs during an act of plagiarism. Although the misuse here is strictly speaking coincidental and may not be intended to create harm to the subject (at least in terms of their idiolect), it is unlikely that there is any implied consent from the subject in having their work plagiarized.

VER5 & VER6. Toward the upper end of our "misuse" spectrum, we have an LLM deliberately writing "in the style of an individual" without the subject's consent. Here there is the misuse of the verbal elements of the subject's idiolect but the degree of misuse will, we propose, depend on the *intent* behind the act and harm to the subject. If the fake use is clearly satirical, then the degree of misuse will likely be less, termed in this spectrum as VER5.

The situation is different when the intent behind the deepfake is fraudulent (termed in this spectrum as VER6). This might occur, for example, where the use is intended to defraud the subject of money. Such cases would be at the far end of the spectrum. A real-life UK example is presented by the scam advertisements featuring the deepfaked face and voice of the respected British financial adviser Martin Lewis (Mostert and Cruz 2022). Alongside these examples are deepfakes used for disinformation purposes which may or may not cause serious reputational damage. We propose situating such misuse between satire and fraud. Use that is defamatory would sit above that which was not defamatory. However, teasing out disinformation from satire, and determining the degree of reputational damage (if any), will be problematic and would touch upon jurisdictionally determined case law/legislation for defamation, copyright, passing-off, and unfair competition.

Vocal Elements

Figure 27.2 shows a proposed spectrum of misuse of the purely vocal elements of an idiolect (without a verbal element). The proposed order of misuses is as follows:

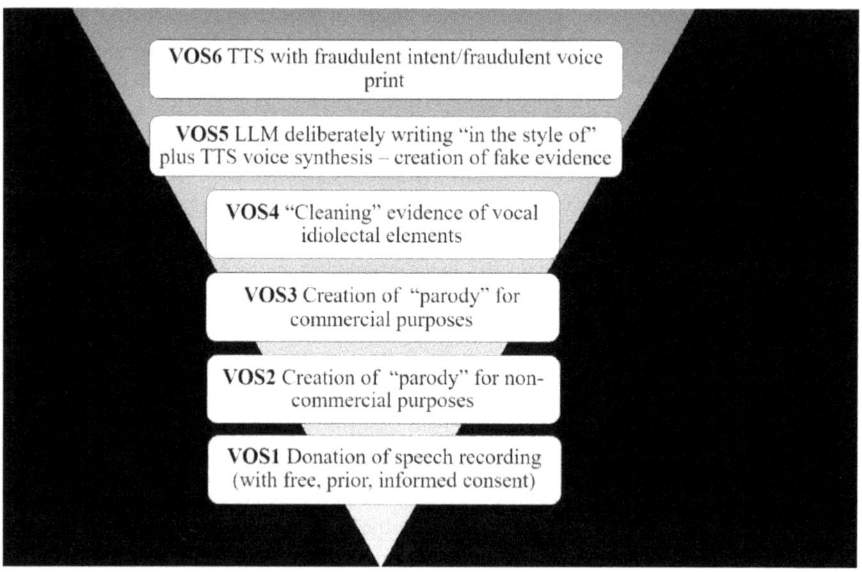

Figure 27.2 Spectrum of Idiolect Misuse (Vocal Element Alone)
Degree of misuse decreases from top to bottom of inverted triangle.

VOS1. As with the VER spectrum above, the base of the inverted triangle pertains to cases where the subject has given their unequivocal prior informed consent to the use of their idiolectal information. In relation to voice-only elements, we are looking firstly at those who "donate" their voice for altruistic, non-commercial, reasons to assist patients to regain a voice through TTS systems. Here we would suggest the scope of prior informed consent is quite narrow in relation to implied permitted use—though for the donor it would clearly be better that the scope is expressly determined within a contract (Scott et al. 2019). We would also include within this category voice "artists" reading words to be used for voice sampling. The scope of prior informed consent in relation to permitted use is more likely here to be determined through express (or implied) contract.

The case of voice artist Susan Bennett (Crockett 2019; Scott et al. 2019) is of interest here. In 2011, Bennett became the female North American English voice of Apple's Siri personal assistant. Bennett had been contracted to TTS recording work with a third-party company in 2005. The third party subsequently sold the recordings to Apple. Though Bennett's voice appeared on millions of iPhones, she received no compensation other than her original (comparatively small) contractual fee. Of course, Bennett did not own the copyright in the recordings. Although her voice was subsequently to become very famous, at the time of recording she was not famous in her own right (outside of the limited voice artist field) and her contract (entirely normally) signed away any further rights that she may have had over the recordings—even though she could not have envisaged their future significance.

VOS2. This is the use of a TTS program to generate a deepfake for pastiche, caricature, or parody purposes for non-consensual but "non-commercial" use. One might imagine this done on a radio program or podcast or via social media. As discussed above, in the absence of defamatory content or a breach of privacy, there may be relatively little, in the United Kingdom or United States at least, that a subject can do about this. In France, however, the doctrine of the "sonorous image" may apply where the subject has some prior element of exposure to the public realm. Clearly, even where such use might cause little harm, it is still non-consensual. Consequently, it sits, we suggest, above consensual uses (see VOS1) in this proposed spectrum of misuse.

VOS3. This is the use of a deepfake pastiche, caricature, or parody for non-consensual and "commercial" use—perhaps, for example, within an advertisement. As discussed above, the law in the United Kingdom, United States, and France is more likely here to protect the commercial interest of the subject where the subject has a distinctive voice and a degree of celebrity (whether through passing-off, rights in publicity, or the sonorous image doctrine, respectively). However, no matter what law applies, on our spectrum there is likely a greater degree of harm to the subject (assuming no defamation) than that seen with a non-commercial pastiche, caricature, or parody. So this activity sits, we suggest, further up the scale of misuse than the non-commercial use mentioned at VOS2.

VOS4. This is the use of voice modification software to "clean" a voice of its vocal idiolectal elements. Whether this type of activity is actually a *use* of idiolectal information is moot. Arguably, if you need to understand the elements of a particular idiolect to effect a clean, then that may, strictly speaking, constitute use; but it is clearly of a different character from the other uses described here. In terms of allocating a position of the spectrum, this decision will clearly rest on the reason for the "cleaning". If I am cleaning my own voice, the question is "why"? If I am intending to somehow mask the origin of the voice such that it is not traceable, the questions are: what do I seek to gain from this disguise? who will suffer harm? and if someone does suffer harm, do I intend to cause that harm? Whilst we have given this activity an arbitrary central location on the spectrum, we acknowledge the exact positioning will likely vary greatly according to whether it is the intention of the person doing the cleaning to do harm and whether harm is actually committed.

If the fake use is clearly satirical then the degree of misuse will likely be less. Where the intent behind the deepfake is fraudulent—for example, where the use is seeking to defraud the subject of their money—this will be at the far end of the spectrum. However, there will be deepfakes that are used for disinformation purposes which may (or may not) cause serious reputational damage. We propose to situate such misuse as sitting between satire and fraud. Such use that is defamatory would sit above that which is not defamatory. However, teasing out disinformation from satire, and determining the degree of reputational damage (if any), will be problematic and would touch upon case law/legislation drawing from defamation, copyright, passing-off, and unfair competition, according to jurisdiction.

VOS5 & VOS6. As with our analysis (above) in relation to the verbal element of the idiolect alone, we have TTS voice synthesis being used to create fake evidence (VOS5) and TTS voice synthesis with a clear fraudulent intent (VOS6). In our opinion, it is clear that a deliberate intent to defraud must sit at the top of the proposed "misuse" spectrum. Where voice synthesis is being used to create deepfake evidence/material, decisions over its position on the misuse spectrum will very much depend on the use to which the deepfake is being put, the *intent* behind the act, and the harm that comes to the subject. The analysis set out above in relation to the verbal element of the idiolect alone applies equally here.

Verbal and Vocal Elements Combined

Figure 27.3 shows a proposed spectrum of misuse of both the verbal and vocal elements of an idiolect used together. We believe that *separately* analyzing verbal and vocal elements on a spectrum of misuse of the *verbal* and then the *vocal* elements of an idiolect assists in identifying the differences which exist between the two. Carefully examining both elements of the idiolect will lead to the most convincing spectra of misuses of idiolectal information (Horvitz 2022; Schick 2020). This is particu-

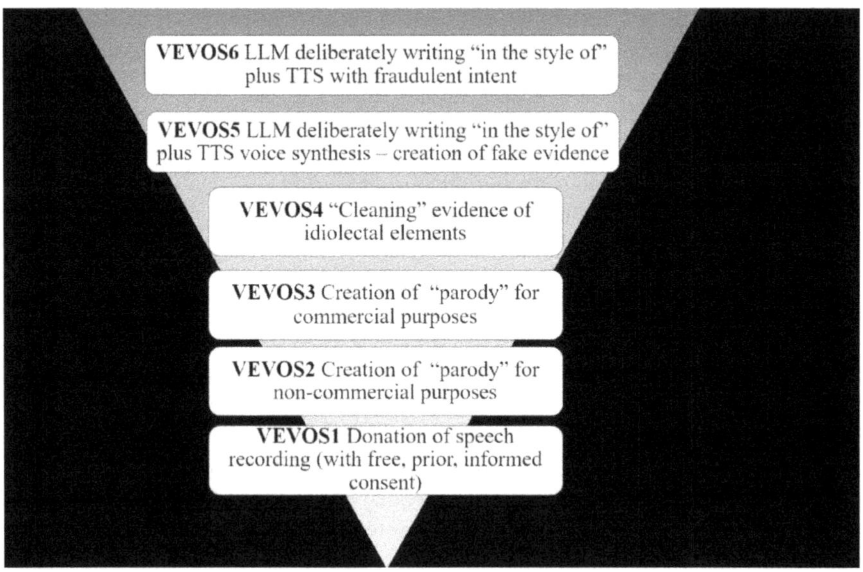

Figure 27.3 Spectrum of Idiolect Misuse (Verbal and Vocal Element)
Degree of misuse decreases from top to bottom of inverted triangle.

larly true in relation to the generation of deepfakes. Potentially, the more convincing a deepfake is the more likely it will serve the purposes or aims of someone seeking to create disinformation or commit fraud. All other things being equal, it certainly seems likely that the reputational damage caused by a deepfake may be easier to demonstrate through TTS than the application of the condition "writing in the style of". Easier still would be a combination of both.

Conclusion

The rapid development, implementation, and combination of LLM and TTS systems will revolutionize both the practice of forensic linguistics and the problems upon which forensic linguistics will be brought to bear. Many of the uses of these systems are yet to be understood, and undoubtedly there will be future uses that we cannot currently imagine. The authors have identified a number of current potential misuses of the verbal and vocal elements of idiolectal information. This chapter has also demonstrated how cases in which idiolectal information has been used can be situated along spectra of relative misuse by examining the fundamental principles of confidentiality/privacy (including data privacy); ownership; and protection against defamation, misrepresentation, or fraud. This approach is intended as a heuristic tool to aid analysis of the use/misuse of idiolectal information. The goal here has been to provide a structure for further analysis of this rapidly developing field. It may assist forensic linguistic experts not only to evaluate where future such uses and misuses fall but also to produce evidence useful for advancing current laws designed to protect subjects from the misuse of their idiolect.

Notes

1 Making modifications of this kind is, for example, positively encouraged by those offering or endorsing "accent reduction" (elocution) services; see, e.g., Watt, Levon, and Ilbury (2023); Watt (2018).
2 See, for example, the UK Copyright Designs and Patent Act 1988 ("CDPA") Section 5A.
3 See, for example, Part II of the UK CDPA which provides that performers have a right to control a sound or film recording of a dramatic/musical performance and control third parties dealing with illicit recordings.
4 See, for example, *Experience Hendrix LLC v. Purple Haze Records Ltd and Ors* [2007] UKSC.
5 See, for example, *Midler v. Ford Motor Co.* (849 F.2d 460 – 9th Cir. 1988).
6 See Margoni (2016).
7 See *Reckitt & Colman Products Ltd. v. Borden Inc.* [1990] 1 WLR 491.
8 See *United Biscuits (UK) Ltd v. Asda Stores Ltd* [1997] RPC 513 (cf. *Moroccanoil Israel Ltd v. Aldi Stores Ltd* [2014] EWHC 1686 (IPEC)).
9 See *Haelan Laboratories, Inc v. Topps Chewing Gum Inc.* (202 F.2d 866 – 2d Cir. 1953; Dogan 2014).
10 See also *Waits v. Frito-Lay, Inc.* 1992. 978 F.2d 1093 - 9th Cir.; Cecchin (1993).
11 See *White v. Samsung Elecs. Am. Inc.* (971 F.2nd 1395 9th Cir. 1992).
12 See the *R v. Bentley* analysis, Coulthard (2006).

Case Law

Baigent and Leigh v. The Random House Group Ltd ([2007] EWCA Civ 247)
Campbell v. Acuff-Rose Music, Inc. 1994. 510 U.S. 569
Campbell v. Mirror Group Newspapers Ltd ([2004] UKHL 22)
Coco v. A N Clark (Engineers) Ltd. ([1969] RPC 41)
Experience Hendrix LLC v. Purple Haze Records Ltd and Ors ([2007] EWCA Civ 501)
Haelan Laboratories, Inc v. Topps Chewing Gum Inc. 1953. 202 F.2d 866 -2d Cir.
Infopaq International A/S v. Danske Dagblades Forening ([2009] ECR I-6569)
Irvine v. Talksport Ltd ([2003] EWCA Civ 423)

Johan Deckmyn and Vrijheidsfonds VZW v. Helena Vandersteen and Others [2014] CJEU C-201/13
Kaye v. Robertson ([1991] FSR 62)
Midler v. Ford Motor Co. 1988. 849 F.2d 460 – 9th Cir.
Moroccanoil Israel Ltd v. Aldi Stores Ltd ([2014] EWHC 1686 (IPEC))
Papillon case (TGI Paris ord., réf., Feb. 27, 1970)
Reckitt & Colman Products Ltd. V Borden Inc. ([1990] 1 WLR 491)
United Biscuits (UK) Ltd v. Asda Stores Ltd ([1997] RPC 513)
Waits v. Frito-Lay, Inc. 1992. 978 F.2d 1093 – 9th Cir.)
White v. Samsung Elecs. Am. Inc. 1992. 971 F.2nd 1395 9th Cir.

References

Augenstein, Isabelle, et al. 2023. *Factuality Challenges in the Era of Large Language Models. arXiv*. https://doi.org/10.48550/arXiv.2310.05189.

Barber, Alex, and Eduardo Garcia Ramirez. 2021. "Idiolects." In *The Stanford Encyclopedia of Philosophy*. Edited by Edward N. Zalta. Retrieved from https://plato.stanford.edu/ENTRIES/idiolects/.

Barlow, Michael. 2010. "Individual Usage: A Corpus-Based Study of Idiolects." Retrieved from http://michaelbarlow.com/barlowLAUD.pdf.

Beaman, Karen V., and Isabelle Buchstaller, eds. 2021. *Language Variation and Language Change Across the Lifespan: Theoretical and Empirical Perspectives from Panel Studies*. London: Routledge.

Bevendorff, Janek, et al. 2022. "Overview of PAN 2022: Authorship Verification, Profiling Irony and Stereotype Spreaders, and Style Change Detection." In *Experimental IR Meets Multilinguality, Multimodality, and Interaction: CLEF 2022. Lecture Notes in Computer Science*, 13390, 382–394. Edited by Alberto Barrón-Cedeño, Giovanni Da San Martino, Mirko Degli Esposti, Fabrizio Sebastiani, Craig Macdonald, Gabriella Pasi, Allan Hanbury, Martin Potthast, Guglielmo Faggioli, and Nicola Ferro. Cham: Springer.

Bloch, Bernard. 1948. "A Set of Postulates for Phonemic Analysis." *Language* 24, no. 1: 3–46.

Bohm, Allison S., Edward J. George, Bennett Cyphers, and Shirley Lu. 2017. "Privacy and Liberty in an Always-On, Always-Listening World." *Columbia Science and Technology Law Review* 19: 1–45.

Boratyn, Daria, et al. 2023. "Machine Learning and Statistical Approaches to Measuring Similarity of Political Parties." *arXiv*. https://doi.org/10.48550/arXiv.2306.03079.

Bowie, David. 2021. "Individual and Group Trajectories Across Adulthood in a Sample of Utah English Speakers." In *Language Variation and Language Change Across the Lifespan*, 101–118. Edited by Karen V. Beaman, and Isabelle Buchstaller. London: Routledge.

Breen, Danielle C. 2021, January 1. "Silent No More: How Deepfakes Will Force Courts to Reconsider Video Admission Standards." *The Free Library*. Retrieved from https://www.thefreelibrary.com/SILENT+NO+MORE%3A+HOW+DEEPFAKES+WILL+FORCE+COURTS+TO+RECONSIDER+VIDEO...-a0652194806.

Brown, Tom B., et al. 2020. "Language Models are Few-Shot Learners." In *Advances in Neural Information Processing Systems (NeurIPS)*, 33, 1877–1901. Edited by Hugo Larochelle, Marc'Aurelio Ranzato, Raia Hadsell, Maria Florina Balcan, and Hsuan-Tien Lin. *arXiv*. https://doi.org/10.48550/arXiv.2005.14165.

Burrows, John F. 1987. *Computation into Criticism: A Study of Jane Austen's Novels and an Experiment in Method*. Oxford: Clarendon Press.

Cafiero, Florian, and Jean-Baptiste Camps. 2023. "Who Could Be Behind QAnon? Authorship Attribution with Supervised Machine-Learning." *arXiv*. https://doi.org/10.48550/arXiv.2303.02078.

Caron, Christophe. 2006, January. "Un Nouveau Droit Voisin Est Né: Le Droit Patrimonial sur l'Image." *Communication Commerce Électronique* 1: 4.

Cecchin, Linda M. 1993. "Waits v. Frito-Lay, Inc. and Tracy-Locke, Inc. 978 F.2d 1093 (9th Cir. 1992)." *DePaul Journal of Art, Technology and Intellectual Property Law* 3, no. 2: 88–90.

Chaski, Carole. 2013. "Best Practices and Admissibility of Forensic Author Identification." *Journal of Law and Policy* 21, no. 2: 333–376.

Chen, Si, et al. 2017, June. "You Can Hear But You Cannot Steal: Defending Against Voice Impersonation Attacks on Smartphones." *Proceedings of the 37th Institute of Electrical and Electronics Engineers (IEEE) International Conference on Distributed Computing Systems*, Atlanta, 183–195. https://doi.org/10.1109/ICDCS.2017.133.

Chomsky, Noam, Ian Roberts, and Jeffrey Watumull. 2023, March 8. "The False Promise of ChatGPT." *New York Times*. Retrieved from https://www.nytimes.com/2023/03/08/opinion/noam-chomsky-chatgpt-ai.html.

Coulthard, Malcolm. 2006. "*...And Then...Language Description and Author Attribution.*" (Sinclair lecture) Birmingham: University of Birmingham ELR (English Language Research). Retrieved from https://user.phil-fak.uni-duesseldorf.de/~bontcheva/WS1213CL/Andthen_Coulthard.pdf.

Coulthard, Malcolm. 2004. "Author Identification, Idiolect, and Linguistic Uniqueness." *Applied Linguistics* 25, no. 4: 431–447.

Cowling, Michael. 2015, February 10. "It's Not Just Your TV Listening in to Your Conversation." *The Conversation*. Retrieved from http://theconversation.com/its-not-just-your-tv-listening-in-to-your-conversation-37409.

Crentsil, Gilbert. 2015. "The Concept of Fraud: A Comparative Analysis." In *Research Handbook on International Financial Crime*, 15–31. Edited by Barry Rider. Cheltenham, UK: Edward Elgar Publishing.

Crockett, Zachary. 2019, January 11. "A Q&A with Susan Bennett, the Original Voice of Siri." *The Hustle*. Retrieved from https://thehustle.co/siri-voice-actor.

Culwin, Fintan, and Mike Child. 2010, June. "Optimising and Automating the Choice of Search Strings when Investigating Possible Plagiarism." *Proceedings of the 4th International Plagiarism Conference*, Newcastle upon Tyne, 1–18. Retrieved from https://goo.gl/RMcDPx.

Cuthbertson, Anthony. 2023, October 9. "Keir Starmer Deepfake Shows Alarming AI Fears are Already Here." *The Independent*. Retrieved from https://www.independent.co.uk/tech/keir-starmer-deepfake-ai-elections-b2426578.html.

de Paiva, Maria da Conceigdo, Maria Eugenia Duarte, and Gregory R. Guy. 2021. "Alignment of Individuals with Community Trends: Subjects from the Portuguese." In *Language Variation and Language Change Across the Lifespan*, 56–76. Edited by Karen V. Beaman and Isabelle Buchstaller. London: Routledge.

de Rancourt-Raymond, Audrey, and Nadia Smaili. 2023. "The Unethical Use of Deepfakes." *Journal of Financial Crime* 30, no. 4: 1066–1077.

Desaire, Heather, Aleesa E. Chua, Madeline Isom, Romana Jarosova, and David Hua. 2023a. "ChatGPT or Academic Scientist? Distinguishing Authorship with over 99% Accuracy Using Off-the-Shelf Machine Learning Tools." *Cell Reports Physical Science* 4, no. 6: 101426.

Desaire, Heather, Aleesa E. Chua, Min-Gyu Kim, and David Hua. 2023b. "Accurately Detecting AI Text when ChatGPT is Told to Write Like a Chemist." *Cell Reports Physical Science* 4: 101672.

Dogan, Stacey L. 2014. "Haelan Laboratories v. Topps Chewing Gum: Publicity as a Legal Right." In *Intellectual Property at the Edge: The Contested Contours of IP*, 17–38. Edited by Rochelle Cooper Dreyfuss and Jane C. Ginsburg. Cambridge: Cambridge University Press.

Feng, Huan, Kassem Fawaz and Kang G. Shin. 2017, October. "Continuous Authentication for Voice Assistants." *Proceedings of the 23rd Annual International Conference on Mobile Computing and Networking*, Snowbird, UT, 343–355. https://doi.org/10.1145/3117811.3117823.

Garrido-Merchán, Eduardo C., José Luis Arroyo-Barrigüete, and Roberto Gozalo-Brizuela. 2023. "Simulating H.P. Lovecraft Horror Literature with the ChatGPT Large Language Model." *arXiv*. https://doi.org/10.48550/arXiv.2305.03429.

George, Alexander. 1990. "Whose Language is it Anyway? Some Notes on Idiolects." *The Philosophical Quarterly* 40, no. 160: 275–298.

Gibson, Johanna. 2022. "All Words and No Performance: A Revolution in Copyright Through Performance in Sound." In *Reforming Intellectual Property*, 92–121. Edited by Gustavo Ghidini and Valeria FalceCheltenham. Edward Elgar Publishing.

Harrington, Jonathan, Sallyanne Palethorpe, and Catherine Watson. 2000. "Monophthongal Vowel Changes in Received Pronunciation: An Acoustic Analysis of the Queen's Christmas Broadcasts." *Journal of the International Phonetic Association* 30, nos. 1/2: 63–78.

Harris, Douglas. 2019. "Deepfakes: False Pornography is Here and the Law Cannot Protect You." *Duke Law and Technology Review* 17: 99–128.

Harrison, Peter S. 2020. "A Good Idea Gone Bad: Can We Still Justify Patent Monopolies?" In *Great Debates in Commercial and Corporate Law*, 62–83. Edited by Andrew Johnston and Lorraine Talbot. London: Macmillan.

Heck, Richard G. 2006. "Idiolects." In *Content and Modality: Themes from the Philosophy of Robert Stalnaker*, 61–92. Edited by Judith J. Thomson and Alex Byrne. Oxford: Oxford University Press.

Higginbotham, James. 2006. "Languages and Idiolects: Their Language and Ours." In *The Oxford Handbook of Philosophy of Language*, 140–150. Edited by Ernest Lepore, and Barry C. Smith. Oxford: Oxford University Press.

Horvitz, Eric. 2022. "On the Horizon: Interactive and Compositional Deepfakes." *Proceedings of the International Conference on Multimodal Interaction (ICMI '22)*, Bengaluru, India, November. New York: Association for Computing Machinery. https://doi.org/10.1145/3536221.3558175.

Ice, Jessica. 2019. "Defamatory Political Deepfakes and the First Amendment." *Case Western Reserve Law Review* 70: 417–455.

Jones, Keenan, et al. 2022. "A Comprehensive Survey of Natural Language Generation Advances from the Perspective of Digital Deception." *arXiv*. https://doi.org/10.48550/arXiv.2208.05757.

Jones, Keenan, Jason R. C. Nurse, and Shujun Li. 2022. "Are You Robert or RoBERTa? Deceiving Online Authorship Attribution Models Using Neural Text Generators." *Proceedings of the International AAAI Conference on Web and Social Media* (ICWSM 2022, 429–440.

Juola, Patrick. 2015. "The Rowling Case: A Proposed Standard Analytic Protocol for Authorship Questions." *Digital Scholarship in the Humanities* 30, no. 1: i100–i113.

Khatsenkova, Sophia. 2023, May 23. "Audio Deepfake Scams: Criminals are Using AI to Sound Like Family and People Are Falling for It." *Euronews.next*. Retrieved from https://www.euronews.com/next/2023/03/25/audio-deepfake-scams-criminals-are-using-ai-to-sound-like-family-and-people-are-falling-fo.

Kontargyris, Xenofon. 2018. *IT Laws in the Era of Cloud-Computing A Comparative Analysis between EU and US Law on the Case Study of Data Protection and Privacy*. Schriften Der Albrecht Mendelssohn Bartholdy Graduate School 6. Baden-Baden, Germany: Nomos Verlag.

Kugler, Matthew B., and Carly Pace. 2021. "Deepfake Privacy: Attitudes and Regulation." *Northwestern University Law Review* 116, no. 3: 611–680.

Lepage, Agathe. 2005, May. "Protection de la Voix à Travers Celle de la Vie Privée." *Communication Commerce Électronique* 5: 48.

Logeais, Elisabeth, and Jean-Baptiste Schroeder. 1998. "The French Right of Image: An Ambiguous Concept Protecting the Human Persona." *Loyola of Los Angeles Entertainment Law Review* 18, no. 3: 511–542.

MacQueen, Hector L. 2005. "'My Tongue is Mine Ain': Copyright, the Spoken Word and Privacy." *Modern Law Review* 68, no. 3: 349–377. https://doi.org/10.1111/j.1468-2230.2005.00542.x.

Mai, Kimberly T., Sergi D. Bray, Toby Davies, and Lewis D. Griffin. 2023. "Warning: Humans Cannot Reliably Detect Speech Deepfakes." *PLoS One* 18, no. 8: e0285333.

Margoni, Thomas. 2016. "The Harmonisation of EU Copyright Law: The Originality Standard." In *Global Governance of Intellectual Property in the 21st Century: Reflecting Policy through Change*, 85–105. Edited by Mark Perry. Cham: Springer.

McLachlan, Hugh. 2016, August 11. "Is Every Human Voice and Fingerprint Really Unique?" *The Conversation*. Retrieved from https://theconversation.com/is-every-human-voice-and-fingerprint-really-unique-63739.

Melzner, Johann, Andrea Bonezzi, and Tom Meyvis. 2023. "Information Disclosure in the Era of Voice Technology." *Journal of Marketing* 87, no. 4: 491–509.

Mostert, Frederick and Sheyna Cruz. 2022. "Image Rights in the Digital Universe." *Journal of Intellectual Property Law and Practice* 17, no. 7: 551–558.

Müller, Nicolas M., Karla Pizzi, and Jennifer Williams. 2022, October. "Human Perception of Audio Deepfakes." Proceedings of the 1st International Workshop on Deepfake Detection for Audio Multimedia (DDAM '*22*), Lisbon, 85–91. *arXiv*. https://doi.org/10.48550/arXiv.2107.09667.

Nagel, David. 2023, March 7. "AI Writing Detection Tool Uses 'Forensic Linguistic' Techniques to Check Authorship." *THE Journal*. Retrieved from https://thejournal.com/articles/2023/03/07/ai-writing-detection-tool-uses-forensic-linguistic-techniques-to-check-authorship.aspx.

Nolan, Francis. 1999, August. "The Devil is in the Detail." Proceedings of the 14th International Congress of Phonetic Sciences (ICPhS'*99*), San Francisco, CA, 1–8.

Ohala, John J. 1989. "Sound Change is Drawn from a Pool of Synchronic Variation." In *Language Change: Contributions to the Study of its Causes*, 173–198. Edited by Leiv E. Breivik, and Ernst H. Jahr. Berlin: de Gruyter Mouton.

Peng, Chun-Yi. 2020. "The Effects of Media Exposure and Language Attitudes on Grammaticality Judgments." *Global Chinese* 6, no. 1: 69–95.

Perot, Emma, and Frederick Mostert. 2020. "Fake It Till You Make It: An Examination of the US and English Approaches to Persona Protection as Applied to Deepfakes on Social Media." *Journal of Intellectual Property Law and Practice* 15, no. 1: 32–39.

Ray, Andrew. 2021. "Disinformation, Deepfakes and Democracies: The Need for Legislative Reform." *The University of New South Wales Law Journal* 44, no. 3: 983–1013.

Reubold, Ulrich, and Jonathan Harrington. 2021. "Accent Reversion in Older Adults: Evidence from the Queen's Christmas Broadcasts." In *Language Variation and Language Change Across the Lifespan*, 119–137. Edited by Karen V. Beaman and Isabelle Buchstalle. London: Routledge.

Rickford, John R. 2021. "Stylistic Variation in Panel Studies of Language Change: Challenge and Opportunity." In *Language Variation and Language Change Across the Lifespan*, 77–98. Edited by Karen V. Beaman and Isabelle Buchstaller. London: Routledge.

Romero, Alberto. 2023. "What You May Have Missed #18." Retrieved from https://thealgorithmicbridge.substack.com/p/what-you-may-have-missed-18.

Romero, Alberto. 2021. "Understanding GPT-3 In 5 Minutes." Retrieved from https://towardsdatascience.com/understanding-gpt-3-in-5-minutes-7fe35c3a1e52.

Roose, Kevin. 2023, February 16. "A Conversation with Bing's Chatbot Left Me Deeply Unsettled." *New York Times*. Retrieved from https://www.nytimes.com/2023/02/16/technology/bing-chatbot-microsoft-chatgpt.html.

Rosati, Eleonora. 2018. "Why Originality in Copyright is Not and Should Not Be a Meaningless Requirement." *Journal of Intellectual Property Law and Practice* 13, no. 8: 597–598.

Sánchez, Damián M., Antonio Moren, and M. Dolores Jiménez López. 2022. "Machine Learning Methods for Automatic Gender Detection." *International Journal on Artificial Intelligence Tools* 31, no. 3: 2241002.

Sankoff, Gillian. 2018. "Language Change Across the Lifespan." *Annual Review of Linguistics* 4: 297–316.

Schick, Nina. 2020. *Deep Fakes and the Infocalypse: What You Urgently Need to Know*. London: Monoray.

Scott, Kristen M., Simone Ashby, David A. Braude, and Matthew P. Aylett. 2019. "Who Owns Your Voice? Ethically-Sourced Voices for Non-Commercial TTS Applications." *Proceedings of the 1st International Conference on Conversational User Interfaces (CUI '19)*, Article 17, 1–3.

Shiffrin, Seana V. 2008. "The Incentives Argument for Intellectual Property Protection." In *Intellectual Property and Theories of Justice*, 94–105. Edited by Axel Gosseries, Alain Marciano and Alain Strowel. New York: Palgrave Macmillan.

Shim, Abigail. 2022. "Forensic Linguistics: Science or Fiction?" *Hastings Law Journal* 74: 207–234.

Shirvanian, Maliheh, Nitesh Saxena, and Dibya Mukhopadhyay. 2018. "Short Voice Imitation Man-in-the-Middle Attacks on Crypto Phones: Defeating Humans and Machines." *Journal of Computer Security* 26, no. 3: 311–333.

Sison, Alejo J.G., Marco T. Daza, Roberto Gozalo-Brizuela and Eduardo C. Garrido-Merchán. 2023. "ChatGPT: More Than a 'Weapon of Mass Deception' - Ethical Challenges and Responses from the Human-Centered Artificial Intelligence (HCAI) Perspective." *arXiv*. https://doi.org/10.48550/arXiv.2304.11215.

Standing, William, and Peter Petré. 2021. "Exploiting Convention: Lifespan Change and Generational Incrementation in the Development of Cleft Constructions." In *Language Variation and Language Change Across the Lifespan*, 141–163. Edited by Karen V. Beaman and Isabelle Buchstaller. London: Routledge.

Tafforeau, Patrick. 2007, May. "L'Être et l'Avoir ou la Patrimonialisation de l'Image des Personnes." *Communication Commerce Électronique* 5: 7.

Tang, Ruixiang, Yu-Neng Chuang, and Xia Hu. 2023. "The Science of Detecting LLM-Generated Texts." *arXiv*. https://doi.org/10.48550/arXiv.2303.07205.

Vaccari, Cristian, and Andrew Chadwick. 2020. "Deepfakes and Disinformation: Exploring the Impact of Synthetic Political Video on Deception, Uncertainty, and Trust in News." *Social Media + Society* 6, no. 1: 1–13.

Watt, Dominic. 2018. "Applied Dialectology: Dialect Coaching, Dialect Reduction, and Forensic Phonetics." In *The Handbook of Dialectology*, 219–232. Edited by Charles Boberg, John Nerbonne and Dominic Watt. Oxford: Wiley Blackwell.

Watt, Dominic, Peter S. Harrison, and Lily Cabot-King. 2019. "Who Owns Your Voice? Linguistic and Legal Perspectives on the Relationship Between Vocal Distinctiveness and the Rights of the Individual Speaker." *International Journal of Speech, Language and the Law* 26, no. 2: 137–180.

Watt, Dominic, Erez Levon, and Christian Ilbury. 2023. "Accent Bias." In *The Routledge Handbook of Linguistic Prescriptivism*, 31–53. Edited by Joan C. Beal, Morana Lukač and Robin Straaijer. London: Routledge.

Yampolskiy, Roman V. 2020. "Unexplainability and Incomprehensibility of AI." *Journal of Artificial Intelligence and Consciousness* 7, no. 2: 277–291.

28

ANALYZING THE LANGUAGE OF NON-HUMAN ANIMALS

Why Should Forensic Linguists Care?

Meike M. de Boer

Introduction

When the police investigate a suspect, they are not typically looking for a cow. And even if there were a criminal cow, she is unlikely to pick up the phone or write a ransom note, leaving linguistic traces. Hence, the focus on human language in forensic linguistics and phonetics is fully explainable and entirely defendable. Yet, in this chapter, I will present three arguments for why forensic linguists and human-animal studies are a perfect match. This chapter then invites readers to rethink how they view the concepts "human", "animal", "language", and "justice".

Postcolonial Linguistics and the Animal Turn

The disproportionate focus on humans in research and everyday life is called "anthropocentrism" (Meijer 2019). This term describes a worldview that centers human experiences, interests, and values above those of all other animals and nature in general—often at the expense of other animals or life forms.[1] For example, humans use and kill other animals for food, clothing, research, profit, forced labor, and entertainment (Proctor, Carder, and Cornish 2013), and destroy the habitats of other animals (Chakrabarty 2012). Because humans have placed themselves at the center of most scientific disciplines, animal studies have long been viewed as a separate field, either completely unrelated to humans or of significance only when human benefits are involved (cf. Proctor, Carder, and Cornish 2013). In many human societies, the distinction between "nature" and "culture" has been held very strictly. As a result, as soon as a topic is related to nature and/or non-human animals, it is perceived as belonging exclusively to biology. However, humans *are*, biologically speaking, animals, and thus part of nature (Linnaeus 1758); and non-human animals *are* part of human society or "culture". We co-exist in this world and human lives are inextricably intertwined with the lives of non-human animals. Hence, the distinction between culture (everything human) and nature (everything non-human) is artificial, incorrect, and unhelpful.

The ability to speak and argue is often seen as exclusively human, which is reflected in one of the most famous adages in linguistics: "language is what makes us human" (Aristotle, quoted in Cornips 2019: 14). Within Cartesian dualism, a dominant philosophy of mind inspired by Descartes, the mind is separated from the body. Within this tradition, speech and language are

 DOI: 10.4324/9781003391074-31

regarded as being directly related to the mind (Schepers-Hughes and Lock 1987). This philosophy of mind is still visible in modern linguistics as the assumption underlying generative linguistics, where language is described as "a system for expressing thought" (Chomsky 2000: 75). In the words of Aristotle, the ability to speak with meaning informed by reason (*logos*) should be distinguished from the mere sound of the voice expressing pain or pleasure (*phoné*) (Meijer 2017).

Aristotle (350 BCE) already described the degraded position of animals in his *Scala Naturae* "Ladder of Being", where he mentioned "a single category of privileged human beings (those who were also male and free)", and described three main categories of beings who were unprivileged: women, slaves, and animals (Martinelli 2010). The attribution of certain characteristics to being uniquely human is called "human exceptionalism" and creates a species hierarchy in which humans are put on top of all other species. It reflects deep asymmetrical power relations that are created and maintained by humans (Meijer 2019). By excluding animal communication from linguistics and categorizing it as merely related to the body, linguistics maintains power asymmetries between humans and other animals and sustains human supremacy and exceptionalism (cf. Cornips 2022). Rethinking the concept of language in itself is thus a step in the direction of justice for all animals. Not only can we as humans not know whether other animals make meaning or can argue rationally, but we must also concede that the definitions we use to decide what counts as language are made by humans and are therefore biased in our favor and against non-human animals.

When we take a more inclusive definition of language, however, other animals' communication can also be classified as language and should therefore also be included in linguistics. Cornips (2022) defines language as "embodied", i.e., not just spoken with the voice but with the entire body (Bucholtz and Hall 2016; Mondada 2016), multimodal (Fricke 2013; Goodwin 2017), and multisensory (Pennycook and Otsuji 2015). Putting these features together, language spoken through the voice alone is not what constitutes language. Language also includes bodily gestures, facial expressions, actions, movements, sensory practices (e.g., tasting, touching, seeing, smelling), and embodied actions such as the position of the body relative to objects and other beings (Cornips 2022). Adopting a more inclusive view of language leads to what has been called an "Animal Turn", a "new nexus of interdisciplinary scholarly interest in the human-animal relationship" (Andersson Cederholm et al. 2014: 5). According to Andersson Cederholm et al. (2014: 5), "current human animal and other animal relations have become so profoundly unequal that they call for inter and multi-disciplinary research". As Cornips (2022, 2019) argues, this interspecies relationship and dismantling of notions of human superiority have implications for the field of linguistics. To some extent, then, this chapter is a call for *any* (linguistic) scholar to consider non-human animals as participants in their next research project. However, there are three reasons why studying speech from other species is especially relevant to forensic linguists. These reasons will be presented in the remainder of this chapter.

Exhibit 1: The End Goal of Forensic Linguistics Is Justice

Ultimately, when working in the field of forensic linguistics, one contributes to justice. With applications in criminal, civil, and asylum proceedings, in counterterrorism, intelligence, and surveillance (Houses of Parliament 2015), the ultimate goal of forensic linguistics is aiding in the process of law and justice. As the purpose of the law is to convict the guilty while at the same time protecting the innocent (Blackstone 1765), fighting injustice may go both ways: aiding the police and prosecution to catch the "bad guys" by providing evidence against guilty suspects and, equally importantly, aiding defense lawyers and judges to prevent the conviction of innocent suspects based on sloppy science or faulty evidence. As Braun and Künzel (1998) put it, lives may literally depend on the work of forensic linguists. This "enormous and immediate impact" on the lives of victims, suspects, and other directly or indirectly involved persons is unlike other fields of applied linguistics (Braun and Künzel 1998: 1350). Hence, forensic linguists are more likely to be justice-driven. Accordingly, one of the most compelling ethical arguments in favor of forensic linguistic professionals to apply their expertise to animals is the criminality endemic to the human treatment of non-human animals. Nowhere is this injustice more apparent than in the animal industry. When we look at the industrial-scale manufacture of animal products, this may be one of the largest yet most invisible examples of injustice of our time (Singer 1975). Animals are routinely kept and killed under the most cruel circumstances for the sole purpose of human consumption and profit (Pennycock 2018; Proctor, Carder, and Cornish 2013). For example, in the Netherlands alone, over 2 million cows, 14 million pigs, and 510 million chickens were killed for consumption in 2023 (CBS 2024).

In the dairy industry, cows are forcibly impregnated to produce breast milk (Cornips 2022). Primarily kept indoors and in chains, female dairy cows are almost always separated from their calves immediately after birth (Boer and Zijlstra 2013). While female calves are retained for an abbreviated life of serial impregnation, male calves—just like male chicks in the egg industry—are routinely separated out and killed as unwanted excess (David and Pardon 2022). This kind of cruelty has been historically justified by the notion that non-human animals are non-sentient. As Descartes (1637, part 5) proclaimed, "animals are like robots: they cannot reason or feel pain". However, it is a well-established fact that non-human animals are sentient beings (Proctor, Carder, and Cornish 2013). The term "sentience" refers to the ability to feel and experience positive and negative emotions, such as joy, pleasure, pain, and fear (Jones 2012). Since animals feel pain, fear, and stress, it is difficult to use the argument of sentience to justify the unethical circumstances they are being kept in.

In addition to incalculable suffering, the animal industry is causing tremendous damage to the Earth, destroying nature, degrading the environment, and depleting vital resources (Pennycook 2018). The meat and dairy industry alone costs a quantity of land, water, and energy that a healthy, balanced planet cannot afford (Malliaroudaki et al. 2022). According to Chakrabarty (2012: 2), humans act as a detrimental "geological force on the planet, changing its climate for millennia to come" due to the amount of energy and fossil fuels necessary for the post-industrial lifestyle. This lasting negative impact causes global inequities and ultimately threatens human civilization as a whole (Chakrabarty 2012). Hence, one could argue that any forensic linguist who is truly committed to promoting justice and bringing an end to injustice would automatically be concerned with applying their skills to address the enormous destructive impact of the animal industry.

One could counterargue, of course, that the boundaries of forensic linguistics are defined by the law; and as long as killing animals for profit or pleasure is not illegal, linguistic investigations aimed at non-human animals may be moral but not forensic. It is true that under most Western systems of law, such as the Dutch jurisprudence, hurting and killing animals is not considered illegal when it serves a "reasonable purpose" (Wet Dieren 2011). However, to this, I would argue that the law is made by humans, and as a result, the behaviors of humans toward other animals are often not criminalized thanks to the widespread bias introduced by human exceptionalism. If one were to see beyond this prejudice and include activities that are not strictly illegal but nonetheless highly unjust, studying the animal industry and its victims quickly becomes a valid area for forensic linguistics. In other words, forensic linguistics may need its own animal turn to remain an inclusive field of study that is truly dedicated to promoting justice as its ultimate goal.

Exhibit 2: Animals May Be Involved in Crimes After All

Virtually any linguistic topic can have a forensic context when it is linked to a criminal case. Since humans and other animals live in close proximity to one another, non-human animals are sometimes involved in police investigations. For example, the African Grey parrot named Bud was a witness to the murder of his human roommate. At first, without any human witnesses present, law enforcement considered the incident to be a case of attempted double homicide, in which a husband killed himself with a firearm after giving his wife a non-fatal shot to the head. However, Bud's imitation of the husband screaming: "Don't shoot! Don't f*cking shoot!" led to a different take. Thanks to the parrot's recounting, the wife became the primary suspect in a potential murder case. Further investigation showed that indeed, the wife had shot her husband and subsequently attempted suicide (BBC 2017). This case raises many important questions about the reliability of non-human animal communication in forensic casework. However, without forensic linguistic investigations, we will be unable to answer these important questions.

In my own casework, I was once asked by a desperate individual to study a chain of events that had occurred in a dispute between neighbors. The question at hand was whether (a) a dog started barking in response to their human neighbor shouting, or (b) the neighbor only started shouting *after* the dog had started barking. Since the recording I was provided had been made through a wall and started in the middle of a heated argument, I ultimately did not take on the case. However, if there had been more research on dog-human communication, dog voices, and the meaning of barks, I might have been able to shed some light on the recording.

Forensic recordings often have all kinds of animal voices in the background, many of them domestic. In the Netherlands, a country with about 18 million human inhabitants (CBS 2025), the most popular pets are cats (2.9 million), songbirds (2.4 million), other birds such as chickens, ducks, geese, and racing pigeons (2.7 million), and dogs (1.7 million) (Voshaar 2020).[2] One in four Dutch households is shared with at least one cat and one in five with one or more dogs (Voshaar 2020). In addition, with humans increasingly taking up living space, our proximity to wild animals such as wild birds, deer, and wolves has also risen. As a result, these animal voices may also be recorded in the background of a forensic phone call.

In other branches of forensic science, the non-human world is already being considered routinely. For instance, forensic entomologists collect and analyze insect evidence in forensic investigations to investigate the minimum time since death, potential causes of death, and information regarding the victim's last meal (Amendt et al. 2011). Likewise, forensic botanists study plant evidence on crime scenes to aid in establishing the time or manner of death or to provide information regarding a primary crime scene (Coyle et al. 2005). Hence, the inclusion of "nature" in forensic science is not so far-fetched.

Similarly, when a non-human animal is recorded in the background of a speech sample, this may give information about where and when the recording was made. Assuming that it is possible to identify the location where the animal's voice was recorded, the next logical question would be whether it can be determined which non-human speaker was the individual who produced the sound in the recording. Although the exact species may be different from case to case, it is relevant to study within- and between-speaker variation in non-human voices to provide potential evidence. Contrary to the widespread stereotype that all individuals within a species of non-human animals sound essentially the same, the research which has been conducted in this area has already produced promising results. For example, in a pilot study conducted with six female individuals of the species *Bos taurus* (cows) and six of the species *Homo sapiens* (humans), de Boer, Cornips, and van Heuven (2024) found that for both species, speaker classification worked similarly well.

While more research on this topic is necessary, this shows that forensic phoneticians might be able to perform forensic speaker comparisons on non-human voices.

Exhibit 3: Forensic Linguists Are Used to Deal with Poor Audio Conditions

Within phonetics, experiments are usually performed under perfect recording conditions (Awan et al. 2023). Under these "gold standards", one voice is typically recorded at a time in a furnished space with carpets and curtains to eliminate or significantly reduce all background noises and echo. A microphone of the highest quality is then used to effectively capture the audio. The microphone tends to be placed as close as possible to speakers' mouths, preferably attached to their heads to keep the distance between the mouth and the microphone consistent even when speakers move their heads. Speakers are often asked to read a fixed text to increase comparability between recordings. By controlling these and other external factors, it is possible to capture speakers' voices as closely as possible and without major interference. Although these steps are very useful when conducting research where one, for example, wants to measure speaker pitch and intonation pattern, voice recordings made in the lab have limited external validity. In real life, speakers talk spontaneously and also quite chaotically—often thinking, repairing, and overlapping their interlocutors (Sacks, Schegloff, and Jefferson 1974). Thus, their speech is often interrupted or overlapped by a variety of competing sounds, including background noises and other speakers.

A major challenge within forensic speech science is to deal with these less-than-perfect recording conditions (Eriksson 2005). For example, speech recordings that are gathered during covert police investigations are often recorded through a hidden microphone or wiretap (Cambier-Langeveld, van Rossum, and Vermeulen 2014). It is not unusual that the recording devices used by law enforcement are not of the highest quality; and the microphones that are used to capture the illicit conversations are often hidden a far distance away from the speaker's mouth. In addition, unlike in a speech laboratory, police recordings often include background noise such as a radio playing music, a roaring engine, and other speakers talking simultaneously. In addition, recordings may be quite short and the phonetic variation which is captured is comparatively limited (Braun and Künzel 1998; Bricker and Pruzansky 1966; Compton 1963). To make matters even more challenging, the speech input may be untypical for the speakers' normal voices due to their efforts to disguise their identity by, for example, pinching their nose or speaking in falsetto (Künzel 2000; Masthoff 1996). Their normal speech patterns may also be altered by emotions like crying (cf. Roberts 2011), screaming (Schreuder 2013), or whispering (Orchard and Yarmey 1995).

An example is the 2013 Florida case of 17-year-old Trayvon Martin, who was shot by the neighborhood watch chief George Zimmerman (Schreuder 2013). In this case, there was considerable discussion over a 911 call made during the incident. On the official audio recording, someone was heard screaming in the distance right before the gun was fired. If the person screaming was Zimmerman, as he claimed, his argument that he had acted in self-defense would become more plausible. However, if the screams could be attributed to Trayvon Martin, the defense's arguments would lose considerable credibility. In the end, the recording was not allowed as evidence, partly because there was no reference material under similar circumstances to perform a speaker comparison and because phonetic measurements may become unreliable when a speaker is screaming (Schreuder 2013). If anyone knows how to deal with gathering and analyzing extreme speech samples for court evidence, it is the forensic phonetician. This training and experience means that professionals within this field are uniquely equipped to handle the challenges that are involved in recording and analyzing non-human animal voice samples.

Recording Animal Voices

As with covert recordings, when recording non-human participants, the person recording the speech has limited control. Firstly, as the human scientist and non-human participant do not share the same language, it is impossible for the researcher to, for example, ask an individual animal to read out a certain text, perform a minute-long monologue, or participate in a human-like conversation. The challenge does not end there, however. When animals speak, they tend to use short or even monosyllabic phrases (van Heuven 2023) that are spontaneously and sparingly uttered. Between the conversational turns of two cows, for example, as much as a minute may elapse (Cornips 2024). This means that the quantity of the speech recorded as well as the phonetic variation, from a human perspective, is especially limited. In this sense, then, recorded meows, coos, and moos can be compared to a suspect merely hm-ing and yes-ing on a short phone call.

Another challenge involves the circumstances under which the recordings are made. Animals are usually not recorded in a phonetics laboratory. While this might be possible with a few species, like domesticated dogs, for other species like cows, pigs, and sheep, taking them out of their comfortable surroundings would be unethical given the stress it would inflict upon them. Due to such ethical but also practical reasons, non-human animals are typically recorded in their home such as a stable or barn. In such environments, acoustic circumstances are poor and often produce an echo. The fact that animal recordings often rely on a single microphone fixed in a single location to record multiple speakers who are moving around and turning their heads is sub-optimal (see, for example, Jahns 2012). However, these circumstances are ultimately comparable to situations where a covert microphone placed in a suspect's house or car is used to record conversations where multiple people have gathered and are talking simultaneously, with their bodies and mouths often away from the recording device.

Thirdly, non-human animals produce a different range of sounds through their different anatomies. This means that the usual phonetic measurements used for humans may be unsuitable. For example, when an animal produces a scream-like sound, spectral measurements such as vowel formants or fundamental frequency may be unreliable (Roberts 2011). However, as mentioned above, forensic speech scientists are well accustomed to dealing with challenging audio conditions, including the collection and analysis of unusual speech samples. All they need to do when gathering speech samples from non-humans is apply the same ingenuity and expertise they routinely utilize when dealing with forensic human speech samples.

Forensic Awareness

Forensic phonetic analysis frequently focuses on the issue of individuality in speech. Whereas in general experimental phonetics, measurements are traditionally averaged across sets of speakers (Politzer-Ahles and Piccinini 2018), the forensic branch centers on idiosyncrasy. This focus is directly applicable when studying a group of speakers from a different species because researchers must be able to differentiate between within- and between-speaker variation. For example, one of the groundbreaking studies in animal linguistics was conducted by Slobodchikoff in 2002. In this investigation of alarm calls made by prairie dogs, Slobodchikoff found that they have different calls for different purposes. For example, the call they gave to warn each other about an approaching human was different from the one they used when a coyote was drawing near. By introducing new threats, Slobodchikoff discovered that the prairie dogs were also able to invent and learn new calls. The ability to create neologisms had never been shown for any animal species besides humans, making this discovery an important step toward rethinking animal communication and cognition. However, as Slobodchikoff did not consider which individual made which utterance, it remains unclear how much of the variation in the alarm calls could have been idiosyncratic rather than semantic. In other words, we still need to determine how many of the differences found were due to between-speaker variation and actually convey the same meaning. This question of idiosyncrasy is the core of forensic linguistics and phonetics.

One of the reasons why this research has yet to be conducted is the misconception that all animals within the same species sound the same and that it is impossible to distinguish one from another. However, as much research has shown, the greater our familiarity with a particular group, be it human or non-human, the more we are able to detect differences that reliably distinguish one individual from another. It is for this reason that, among humans, speaker recognition has shown to be easier in one's own language than in another language (Goggin et al. 1991).

A similar principle is at work when human babies learn a language. As a general rule, they only learn to distinguish those sounds that convey a different meaning and are thus important. Sounds that are phonetically near and do not have to be distinguished are stored in the brain as a single phonetic category (Flege 1995). For example, in German, the voiceless fricatives /s/ (as in s̲ink) and /θ/ (as in t̲h̲ink) do not need to be distinguished because they are not differentiated in the language—German does not use the interdental fricative /θ/. This leads to confusion when a German speaker is learning a second language such as English, and they have to learn how to distinguish between two sounds that sound identical to them. The same difficulty exists when attempting to make cross-species differentiations. A human who grew up with little or no exposure to other animals species may not be immediately able to distinguish the voices of different individuals within another species. However, humans who have lived closely with other species may be readily able to distinguish one individual from another. An excellent example is a farmer who is able to identify the different cows in a herd by their individual moos. Similarly, someone who lives with two or more dogs might learn to distinguish them by the sound of their barks alone. As these anecdotal examples demonstrate, despite the prevalence of cross-species biases, it is theoretically possible to identify individual animals through their utterances. What is needed now is the scientific evidence, and who would be better suited than the forensic phonetician to supply it?

Ethical Considerations

For those who are considering selecting non-human speakers for their next study, there are some ethical considerations to take into account. Using non-human animals merely to satisfy one's own scientific curiosity without any consideration for their well-being is generally unethical. If a legitimate reason can be identified for the study, it is imperative that before the research begins, to safeguard the welfare of the research participants, ethical standards similar to the ones that are in place for human participants must be implemented. Yet, some principles that have become standard practice in research with human participants, such as informed consent and voluntariness (Israel and Hay 2006), are impossible to implement in the same way with non-human research participants. For example, it cannot be verified whether an animal has understood the purpose of the study or to be sure that they understand that they are allowed to leave whenever they wish to terminate their participation. To some extent, studying non-human animals is similar to conducting research with babies or children who are not capable of communicating with the researcher themselves or giving their legal consent. For this reason, research on children requires parental or custodial consent from an adult who is expected to have the minor's best interests in mind. A similar construction could be used with non-human animals (cf. Janssens 2024). The researcher could obtain permission from a human guardian with the provision that that individual can be trusted to make a decision based on the best interests of the non-human animal. This provision is vital as it is questionable whether the human concept of ownership ethically grants a human the moral right to decide over another animal's life. Assigning the participants' rights to their human caregivers, assuming they have one, is therefore not enough to guarantee ethical research.[3]

Along with issues related to informed consent, forensic linguistic researchers interested in working with non-human animals should also aim to design and implement their study in a manner that avoids causing any physical or psychological harm to the participants. Instructive here is the first general ethical principle of the American Psychological Association (APA): beneficence and nonmaleficence (APA 2017). This principle means that not only should human participants be free from stress and pain during the research, but they must also benefit to some extent from their participation, for example, in terms of their overall well-being and welfare or through financial compensation. It is advised that a similar principle is adopted when studying non-human participants: the well-being and comfort of the individual should be prioritized and they should benefit in terms of treats, affection, or pleasant stimulation such as playing—anything that makes them happy. In doing so, the forensic linguist automatically largely follows the other basic APA ethical principles for human research participants, i.e., demonstrating fidelity and responsibility, integrity, justice, and respect for the rights and dignity of others.

After the research has been conducted, the ethical responsibility of the forensic linguist does not end. Researchers must take care to monitor how their results are used once published. For example, de Rooij (2024) studied the language and behavior of chickens to help create a "Happiness Meter" to automatically detect how happy (or unhappy) a chicken is. This particular study was conducted by the Dutch Animal Labor Union [*Vakbond voor Dieren*], a charity promoting rights for non-human animals who are forced to perform labor such as performing tricks in a circus or zoo, working for the police, or being held captive for the consumption of their body or excretions (e.g., eggs and breast milk) (Vakbond voor Dieren 2024). As the Animal Labor Union stresses on its website, research that is designed to improve the welfare of animals caught in exploitative industries carries the risk of being misused to increase productivity and harm animals even more than before. In the case of de Rooij's development of the Happiness Meter, this device could be misused by farmers to increase egg production.

The same ethical caution must be employed in forensic linguistic research involving non-human animals. For example, automatic speaker recognition in combination with speech recognition on cows (cf. Jahns 2012; Jahns, Kowalczyk, and Walter 1997) may tell a farmer which cow needs what. While this sounds like research beneficial for helping increase our understanding of cows' needs, in the wrong hands, the use of speech recognition technology could potentially lead to even less human-cow contact than there already is in modern-day farming, further distancing humans from other animals and increasing the likelihood for more animal abuse. Hence, it is important to consider the potential implications of the study to avoid unintentional and unforeseen harm. Ultimately, studying language in non-human animals may lead to a better understanding and help to build new relations with them (Meijer 2019).

Summary and Conclusion

Historically, linguistic research has focused on human language and communication, marginalizing the roles of other animals. This chapter made a case for the inclusion of non-human animals in the work of forensic speech scientists and forensic linguists. In this chapter, we explored the potential for integrating (forensic) linguistics with human-animal studies, challenging traditional anthropocentric views. A broader, more inclusive definition of language—encompassing not only sounds but also gestures, movements, and sensory practices—opens the door for recognizing animal communication as a legitimate subject of linguistic study. This modern, egalitarian perspective calls for a more equitable consideration of non-human animals in interdisciplinary research, particularly within linguistics.

Moreover, as this chapter also argued, there are three key reasons why forensic linguists are encouraged to apply their expertise to this field of study. The first argument in doing so is in line with the ultimate goal of forensic linguistics: promoting justice and thwarting injustice. By considering animal communication within forensic linguistics, scholars within this field can help to challenge the power asymmetries that harm all life forms. The second argument for integrating animal communication into forensic linguistics is that animals are often (in)directly involved in criminal cases. By conducting research in animal speech, forensic speech scientists may be able to provide speech evidence from non-human animals in the way that they already do for human animals. The third reason forensic linguists should consider animal communication is their expertise in handling poor audio conditions. Their training and familiarity with speech samples of lesser quantity and quality especially equip forensic phoneticians with the skills needed to tackle recordings of non-human speakers.

In view of those arguments, it is concluded that including non-human animals in forensic linguistics can only contribute to justice if ethical standards are implemented and the direct and indirect implications of the research are considered. With this in mind, the field of forensic linguistics has the ability to play a crucial role in addressing injustice against non-human animals and the extreme power asymmetries between humans and other animals. With this chapter, I invite everyone to consider how they view animals and their linguistic practices and to help create an inclusive, postcolonial (forensic) linguistics with space for all individuals.

Notes

1 The dichotomy of human/animal is artificial, as humans are mammals (i.e., animals) themselves. When I say “other animals”, I refer to non-human animals—those often meant when people talk about “animals”.
2 These numbers exclude the 15.9 million fish who are unlikely to be accidentally recorded.
3 Janssens' 2024 proposal of “consent by proxy” offers a framework for how to deal with non-human research participants that is potentially helpful with regard to this issue.

References

Amendt, Jens, et al. 2011. “Forensic Entomology: Applications and Limitations.” *Forensic Science Medical Pathology* 7: 379–392.

American Psychological Association. 2017. “Ethical Principles of Psychologists and Code of Conduct.” Retrieved from https://www.apa.org/ethics/code.

Andersson Cederholm, Erika, Amelie Björck, Kristina Jennbert, and Ann-Sofie Lönngren. 2014. *Exploring the Animal Turn: Human-Animal Relations in Science, Society and Culture*. Lund, Sweden: The Pufendorf Institute of Advanced Studies, Lund University.

Aristotle (350 BCE). *Historia Animalium*. Athens, Greece.

Awan, Shaheen N., et al. 2023. “Smartphone Recordings are Comparable to ‘Gold Standard’ Recordings for Acoustic Measurements of Voice.” *Journal of Voice*. https://doi.org/10.1016/j.jvoice.2023.01.031.

BBC. 2017, August 31. “Pet Parrot Witness Helps Send Husband-Killer to Prison for Life.” Retrieved from https://www.bbc.com/bbcthree/article/2f62b053-9e7a-4e72-8d9d-148bee9688cd.

Boer, Marike, and Jelle Zijlstra. 2013. *Verschillen Tussen Bedrijven in Levensduur van Melkkoeien*. [Differences Between Dairy Farms in Longevity of Dairy Cows]. Wageningen, The Netherlands: Wageningen UR Livestock Research.

Blackstone, William. 1765. *Commentaries on the Laws of England*. Oxford: Clarendon Press.

Braun, Angelika, and Hermann J. Künzel. 1998. “Is Forensic Speaker Identification Unethical – Or Can it Be Unethical Not to Do It?” *Forensic Linguistics* 5, no. 1: 1350–1771.

Bricker, Peter D., and Sandra Pruzansky. 1966. "Effects of Stimulus Content and Duration on Talker Identification." *Journal of the Acoustical Society of America* 40, no. 6: 1441–1450.
Bucholtz, Mary, and Kira Hall. 2016. "Embodied sociolinguistics." In *Sociolinguistics: Theoretical Debates*, 173–197. Edited by Nikolas Coupland. Cambridge: Cambridge University Press.
Cambier-Langeveld, Tina, Maya van Rossum, and Jos Vermeulen. 2014. "Whose Voice is That? Challenges in Forensic Phonetics." In *Above and Beyond the Segments: Experimental Linguistics and Phonetics*, 14–27. Edited by Johanneke Caspers, Yiya Chen, Willemijn Heeren, Jos Pacilly, Niels O. Schiller, and Ellen van Zanten. Amsterdam: John Benjamins Publishing Company.
CBS. 2025. "Bevolkingsteller: Hoeveel Mensen Wonen Nu in Nederland?" [Population Counter: How Many People Live in the Netherlands Right Now?]. Retrieved from https://www.cbs.nl/nl-nl/visualisaties/dashboard-bevolking/bevolkingsteller.
CBS. 2024. "Vleesproductie; Aantal Slachtingen en Geslacht Gewicht per Diersoort." [Meat production; Number of Slaughters and Weight of Slaughters per Species]. Retrieved from https://opendata.cbs.nl/statline/#/CBS/nl/dataset/7123slac/table?fromstatweb.
Chakrabarty, Dipesh. 2012. "Postcolonial Studies and the Challenge Climate Change." *New Literary History* 43, no. 1: 1–18.
Chomsky, Noam. 2000. *On Nature and Language*. New York: Cambridge University Press.
Compton, Arthur J. 1963. "Effects of Filtering and Vocal Duration Upon the Identification of Speakers Aurally." *Journal of the Acoustical Society of America* 35, no. 11: 1748–1752.
Cornips, Leonie. 2024. "The Semiotic Repertoire of Dairy Cows." *Language in Society* 54, no. 4: 1–25.
Cornips, Leonie. 2022. "The Animal Turn in Postcolonial Linguistics: The Interspecies Greeting of the Dairy Cow." *Journal of Postcolonial Linguistics* 6: 209–231.
Cornips, Leonie. 2019. "The Final Frontier: Non-Human Animals on the Linguistic Research Agenda." *Linguistic in the Netherlands* 36, no. 1: 13–19.
Coyle, Heather M., et al. 2005. "Forensic Botany: Using Plant Evidence to Aid in Forensic Death Investigation." *Croatian Medical Journal* 46, no. 4: 606–612.
David, Renaud, and Bart Pardon. 2022. "Preparing Male Dairy Calves for the Veal and Dairy Beef Industry." *Veterinary Clinics: Food Animal Practice* 38, no. 1: 77–92.
de Boer, Meike, Leonie Cornips, and Vincent van Heuven. 2024. "Whose 'Moo' is it? Using Forensic Phonetics on Non-Human Voices." *Roundtable of the Germanic Society for Forensic Linguistics* [Roundtable presentation].
de Rooij, Marjolein. 2024. "Kippengeluk. Een Verkennend Onderzoek naar een Kippengeluksmeter: Indicatoren voor Positief Welzijn." *Vakbond voor Dieren*. Retrieved from https://www.vakbondvoordieren.nl/wp-content/uploads/2024/11/Rapport-Kippengeluk-publiek-DEF.pdf.
Descartes, René. 1637. *Discours de la Méthode: Pour Bien Conduire sa Raison et Chercher la Verité dans les Sciences*. Leiden: De l'Imprimerie de Ian Maire.
Eriksson, Anders. 2005. "Tutorial on Forensic Speech Science." In *Proceedings of the 9th European Conference on Speech Communication and Technology*, 1–14. Edited by Interspeech'2005 - Eurospeech. Lisboa, Portugal.
Flege, James E. 1995. "Second Language Speech Learning: Theory, Findings, and Problems." In *Speech Perception and Linguistic Experience: Issues in Cross-Language Research*, 233–277. Edited by Winifred Strange. Timonium, MD: York Press.
Fricke, Ellen. 2013. "Towards a Unified Grammar of Gesture and Speech: A Multimodal Approach." In *Body – Language – Communication*, 733–754. Edited by Cornelia Müller, Alan Cienki, Ellen Fricke, Silva Ladewig, David McNeill and Sedinha Tessendorf. Berlin: De Gruyter.
Goggin, Judith P., Charles P. Thompson, Gerhard Strube, and Liza R. Simental. 1991. "The Role of Language Familiarity in Voice Identification." *Memory and Cognition* 19, no. 5: 448–458.
Goodwin, Charles. 2017. *Co-Operative Action (Learning in Doing: Social, Cognitive and Computational Perspectives)*. Cambridge: Cambridge University Press.
Houses of Parliament. 2015. "Forensic Language Analysis." *Postnote*, no. 509. London: The Parliamentary Office of Science and Technology. Retrieved from https://post.parliament.uk/research-briefings/post-pn-0509/.
Israel, Mark, and Iain Hay. 2006. *Research Ethics for Social Scientists*. CA: Sage Publications.
Jahns, Gerhard. 2012. "Computational Intelligence to Recognize Animal Vocalization and Diagnose Animal Health Status." In *Computational Intelligence in Intelligent Data Analysis: Studies in Computational*

Intelligence, vol. 445, 239–249. Edited by Christian Moewes and Andreas Nürnberger. Berlin: Springer-Verlag.

Jahns, Gerhard, Wojtek Kowalczyk, and K. Walter. 1997. "Sound Analysis to Recognize Different Animals." *IFAC Proceedings*: 169–173. https://doi.org/10.1016/S1474-6670(17)41265-1.

Janssens, Monique R. E. 2024. "What Would Miffy Do? Applying Informed Consent by Proxy to All Sentient Animals." *Animals: An Open Access Journal from MDPI* 14, no. 18: 2643.

Jones, Robert C. 2012. "Science, Sentience, and Animal Welfare." *Biology & Philosophy* 28: 1–30.

Künzel, Hermann J. 2000. "Effects of Voice Disguise on Speaking Fundamental Frequency." *Forensic Linguistics* 7: 149–179.

Linnaeus, Carl. 1758. *Systema Naturae per Regna Tria Naturae, Secundum Classes, Ordines, Genera, Species, cum Characteribus, Differentiis, Synonymis, Locis*. 10th ed. The Netherlands: Holmiae.

Malliaroudaki, Maria Ioanna, et al. 2022. "Energy Management for a Net Zero Dairy Supply Chain Under Climate Change." *Trends in Food Science & Technology* 126: 153–167.

Martinelli, Dario. 2010. *A Critical Companion to Zoosemiotics. People, Paths, Ideas*. London: Springer.

Masthoff, Herbert. 1996. "A Report on a Voice Disguise Experiment." *International Journal of Speech, Language and the Law* 3, no. 1: 160–167.

Meijer, Eva. 2019. *When Animals Speak: Toward an Interspecies Democracy*. New York: New York University Press.

Meijer, Eva. 2017. "Political Animal Voices." PhD Dissertation, University of Amsterdam.

Mondada, Lorenza. 2016. "Challenges of Multimodality: Language and the Body in Social Interaction." *Journal of Sociolinguistics* 20, no. 3: 336–366.

Orchard, Tara L., and A. Daniel Yarmey. 1995. "The Effects of Whispers, Voice-Sample Duration, and Voice Distinctiveness on Criminal Speaker Identification." *Applied Cognitive Psychology* 9, no. 3: 249–260.

Pennycook, Alastair. 2018. *Posthumanist Applied Linguistics*. London: Routledge.

Pennycook, Alastair, and Emi Otsuji. 2015. "Making Scents of the Landscape." *Linguistic Landscape* 1, no. 3: 191–212.

Politzer-Ahles, Stephen, and Page Piccinini. 2018. "On Visualizing Phonetic Data from Repeated Measures Experiments with Multiple Random Effects." *Journal of Phonetics* 70: 56–69.

Proctor, Helen S., Gemma Carder, and Amelia R. Cornish. 2013. "Searching for Animal Sentience: A Systematic Review of the Scientific Literature." *Animals* 3, no. 3: 882–906.

Roberts, Lisa. 2011. "Acoustic Effects of Authentic and Acted Distress on Fundamental Frequency and Vowel Quality." In *Proceedings of the 17th International Congress of Phonetic Sciences*, 1694–1698. Edited by Wai-Sum Lee and Eric Zee. Hong Kong: City University of Hong Kong.

Sacks, Harvey, Emanuel A. Schegloff, and Gail S. Jefferson. 1974. "A Simplest Systematics for the Organisation of Turn-Taking for Conversation." *Language* 50, no. 4: 696–735.

Scheper-Hughes, Nancy, and Margaret Lock. 1987. "The Mindful Body: A Prolegomenon to Future Work in Medical Anthropology." *Medical Anthropology Quarter* 1, no. 1: 6–41.

Schreuder, Maartje. 2013. "Wie Uitte die Noodkreten, Trayvon Martin of George Zimmerman? [Who Uttered the Screams, Trayvon Martin or George Zimmerman?]" *Expertise en Recht* 6: 185–191.

Singer, Peter. 1975. *Animal Liberation: A New Ethics for Our Treatment of Animals*. New York: HarperCollins.

Slobodchikoff, Con N. 2002. "Cognition and Communication in Prairie Dogs." In *The Cognitive Animal: Empirical and Theoretical Perspectives on Animal Cognition*, 257–264. Edited by Marc Bekoff, Colin Allen, and Gordon M. Burghardt. Cambridge, MA: The MIT Press.

van Heuven, Vincent J. 2023. "Koeienfonetiek. [Cow Phonetics.]" In *Kaas=NL? Melk, Koe, Ras, Kolonie, Taal, Kunst, Mest en Meer* [Cheese=Netherlands? Milk, Cow, Race, Colony, Language, Art, Menure, and More]. Edited by Leonie Cornips, Marieke Hendriksen, and Geertje Mak. Amsterdam: Sterck and De Vreese.

Vakbond voor Dieren. 2024. "Vakbond voor dieren. [Animal Labor Union]." Retrieved from https://www.vakbondvoordieren.nl/.

Voshaar, Gerard. 2020. "Rapportage Huisdierenonderzoek – Nieuwe Cijfers: Meer dan 27 Miljoen Huisdieren in Nederland. [Report Pet Study – New Figures: Over 27 Million Pets in the Netherlands." *Divebo* 5: 28–29.

Wet Dieren. 2011. "Wet van 19 mei 2011, Houdende een Integraal Kader voor Regels over Gehouden Dieren en Daaraan Gerelateerde Onderwerpen (Wet Dieren). [Law of 19 May 2011, Containing an Integrated Framework for Rules on Captivated Animals and Related Topics (Animal Law)." *Staatsblad van het Koninkrijk der Nederlanden* [Official Gazette of the Kingdom of the Netherlands] 2011, no. 345. Retrieved from https://zoek.officielebekendmakingen.nl/stb-2011-345.html.

29

"SO THIS IS HOW YOU CAN MANIPULATE WITNESSES!"

Ethical Considerations in the Teaching and Dissemination of Research in Forensic Linguistics

Karoline Marko

Introduction

Forensic linguistics has many faces. Overall, however, it aims at "improving the delivery of justice through the analysis of language" (Aston Institute for Forensic Linguistics n.d.) for everyone involved in the legal process, be it in the courtroom or in the police interrogation room. Given its real-world applications and consequences, forensic linguistics often requires instructional approaches that are different from those used in other linguistic disciplines (Shuy 2002). The teaching of methods and linguists' contributions to cases can have palpable consequences for the individuals involved in criminal cases or civil disputes and for students who learn about these cases. This potential risk is why ethical considerations should be an integral and iterative part of disseminating forensic linguistic research and methodology, be it inside or outside of the classroom setting.

As is the case in other forensic disciplines, forensic linguists often accumulate and possess procedural and methodological knowledge and skills that should not generally become known to the public, as such methodological knowledge and skills would otherwise be rendered ineffective and useless, and, in some cases, even dangerous. Ding et al. (2022) make this point with regard to the creation of deepfake videos. As they explain, the widespread dissemination of the techniques needed to make such videos "can bring potential harm to individuals or even our community if it falls into the wrong hands with malicious intention" (3429). Similarly, if the "wrong" people learned how they can manipulate weapons so that projectiles cannot be traced back to them, what is the value of the methods used by ballistics' experts? If the "wrong" people learned how to cover up a murder so it could not be detected by medical examiners, what would be the value of the methods used by forensic medical doctors? If the "wrong" people learned how to make their voices completely unrecognizable, what would be the value of the identity detection methods of forensic phoneticians? If the "wrong" people learned how to disguise their writing styles to successfully circumvent authorship analysis and attribution methods, what would be the value of these methods used by forensic linguists?

The possession and disclosure of such sensitive knowledge therefore entail great responsibility, accompanied by many challenging ethical questions. These questions include: Who is the

DOI: 10.4324/9781003391074-32

(in)appropriate audience? Which methods and skills should or shouldn't be presented? What can happen if forensic findings, methods, and skills are disclosed? Who is responsible or liable if the information shared is intentionally or unintentionally misused and people are harmed? This chapter aims at exploring these and other ethical questions involved in the teaching and dissemination of forensic linguistics methods, tools, and skills. To achieve this goal, this chapter presents some of the latest literature on ethics and pedagogy. Importantly, the body of research presented also draws on insights from other related (forensic) disciplines in order to examine how these issues are addressed and handled elsewhere. This literature review is complemented with the insights from two interviews and an anonymous survey of leading international forensic linguists with many years of experience in teaching and researching in forensic linguistics and forensic phonetics.

Background

What inspired the writing of this chapter was an episode recounted by the internationally renowned Australian forensic linguist Diana Eades. Eades has dedicated most of her academic career to researching Aboriginal communication in the legal process and to improving the delivery of justice for these communities. Her work resulted in the writing of a handbook for legal practitioners (Eades 1992). In a 2004 article, she details how lawyers had begun to misuse the communicative strategies outlined in her handbook in a criminal case, the Pinkenba case. This case involved three teenage boys with Aboriginal backgrounds, who were walking near a mall in Brisbane. They were stopped by six police officers and transported to an industrial wasteland outside of town, where they were abandoned. The boys filed a complaint and in the ensuing trial, the lawyer who cross-examined them was found to use the findings of Diana Eades' investigation about Aboriginal communication styles against the boys. In particular, the lawyer made use of what has been termed "gratuitous concurrence", which describes the tendency to answer questions affirmatively (see also Eades 2016, 2006). Eades (2004) explains her experience as follows:

> it appeared that the handbook, which had been written to help lawyers in more effective communication between lawyers and Aboriginal witnesses, was being used in this case [the Pinkenba case] to achieve less effective communication, and even to subvert the witnesses' attempts at communication. And indeed it was very successful in that purpose.
>
> *(501)*

Eades further describes this experience as "disturbing" (501) and recollects that she had had concerns about the misuse of her research prior to this incident but had been conciliated by legal professionals she consulted. As she later recounts:

> when I was writing the handbook, […] I did ask […] is this like giving people tools that they can use against witnesses? And their [the lawyers'] answer was: "don't worry about the legal system, it's adversarial." So they were saying that one of the duties of the adversarial system is that both sides are equipped to fight each other, and either side can use the tool that you have made available to them.
>
> *(Interview with the author, 2023)*

However, as the above-mentioned incident shows, despite the lawyers' reassurances, the adversarial system had not prevented Eades' publication from being put to unintended uses. As Solan (1993) points out, such intentional and/or unintentional misuse can happen in a variety of ways within the courtroom, and this potential is not restricted to forensic linguistics. Bersoff and Glass (1995) describe a court case in which social science research was misused by the court to "augment its opinions" (279). As the researchers boldly state:

> the Court has (1) misused or misapplied data when it believes the data will enhance the persuasiveness of his opinions; (2) ignored or rejected data despite its assertion of empirically testable statements; and (3) disparaged data when the research does not support its views. In some cases, it has done all three.
>
> (293)

It can be argued, of course, that such misuse can be the unintentional result of lawyers and judges not being trained in the proper evaluation and application of research outside their own areas of expertise—an aspect that will be commented on again later. However, there is evidence that some judicial misuse is indeed intentional. Gershman (2003), for instance, discusses a case in which a prosecutor intentionally misrepresented the findings of an expert in order to win the case. To a certain extent, such abuses may come as no surprise given the highly contentious nature of many courtroom settings (Durant and Leung 2016). In a judicial arena, proceedings can become a "'gladiatorial' battle of wits, where the jury gets to decide who fought the better fight" (Brouwer 1981 in Olsson and Luchjenbroers 2014: 242). Indeed, in such high-stakes acrimonious environments, there might actually be more, not less, susceptibility to misusing scientific research. The potential real-life consequences can be horrendous, and can, for example, result in the imprisonment of innocent people.

Given these dangers, the teaching and publication of (manipulative) linguistic techniques in the courtroom should not be taken lightly by forensic linguistic experts. This caution is not limited to the topic of courtroom language. Similarly, ethical questions are encountered with respect to language as evidence. Take, for example, the subject of authorship analysis or authorship attribution, which is a common area of forensic linguistic curricula and publications. Whether automated or manual, this subject area inherently addresses the strategies used to reveal unknown authors, describe writing styles, and identify potential efforts of (identity) manipulation. Hence, the following question arises: can the discussion of authorship analysis methods in the public domain help individuals circumvent authorship attribution or hinder other forms of authorship analysis? According to Brennan, Afroz, and Greenstad (2012), "authors attempting to modify their writing style can reduce the accuracy of these methods [stylometry] from over 95% to the level of random chance" (2). That being the case, a legitimate concern is whether some of the people who are learning about such stylometric techniques are being equipped with knowledge that will enable them to successfully and maliciously distort their writing style for unethical or even illegal purposes, such as engaging in hate speech, making threats and extortion, or committing acts of plagiarism. These kinds of ethical concerns are not limited to forensic linguistics. For that reason, before addressing some of the concrete strategies that forensic linguists have developed to deal with these challenges, the next segment will discuss these concerns as they affect academic teaching and scientific research more generally.

Teacher Ethics, Teaching Ethics, and Ethics in Research

Irrespective of their area of expertise, all teachers carry certain responsibilities for their students (or, in the context of workshops, their attendees). As Ehrich et al. (2011) point out, "the work of teachers [...] is heavily value-laden and, for this reason, susceptible to ethical dilemmas" (173).

In addition to the values of honesty, integrity, and professionalism (Kuther 2003), a central aspect of teaching is the development and building of a relationship with one's students that allows the creation of a safe space and of mutual trust, which, ideally, enables students to act responsibly and in line with moral, ethical, and professional guidelines (Ehrich et al. 2011). It is, however, not a simple undertaking to teach ethical and professional responsibilities. Hobbs (2007) indicates that it is particularly challenging to "find the appropriate blend of conveying sufficient information, presenting sophisticated analytical processes, and inspiring students to take the ideal of professionalism to heart" (1269). Importantly, the teaching of ethical and moral responsibilities not only involves the development of a deep understanding of academic, ethical, and moral issues but also requires high levels of self-reflection.

These skills do not develop independently but must be taught as much as other knowledge and skills must be taught. However, only few curricula (and particularly few linguistics ones) include mandatory ethics classes, not even in the forensic sciences. Therefore, Elkins and Fambegbe (2020) argue that "ethics education is as essential for students of forensic sciences as instrument training, sample handling and quality assurance" (6). Green (2007) further highlights the issue that ethics classes, if they do exist, are often not taken seriously by students as they tend to equate ethics merely with intuition. Therefore, Willis (2023) remarks that "ethics in forensic science deserves much more [conscious] consideration" (5).

Learning how to act ethically is not only essential for students but also for practitioners and academics, as only those professionals who are aware of ethical issues and possess a high level of self-reflection can successfully transfer these skills, knowledge, and awareness to their students/participants. As Boon (2011) points out, "[q]uality teachers are considered to be those individuals whose pedagogy is grounded in values and beliefs that lead to caring, positive teacher-student relationships, embedded in trust and high standards of professional ethics" (76).

Ethical Issues in Other (Forensic) Disciplines

The awareness of potential misuse of knowledge and methods has long been discussed in many scientific disciplines. In the context of information security, for example, Logan and Clarkson (2005) discuss the issue of "ethical hacking" and point out that course curricula often include subjects like hacking tools and methods "without much discussion of the potential for misuse or abuse by students" (158). Stockman (2014) further points out that almost half of all hacking attacks are performed by employees or former employees. The fact that it is often "insiders" who carry out hacking attacks is due in no small measure to the fact that they possess specific inside knowledge about the systems they target. Logan and Clarkson (2005), therefore, ask whether it is possible that "we are training both the 'good guys'—security professionals—and the 'bad guys'—those with knowledge of techniques and tools but without exposure to ethical and legal issues—at the university level?" (160). In the very same vein, there have been intense discussions about whether or not it is beneficial and/or ethically justifiable for students to be taught how to write computer viruses and malware in university courses (Curbelo and Cruz 2013; O'Connor, Sangster, and Dean 2010; Kabay 2004).

For many of the same reasons, in the field of chemistry, Schummer and Borsen (2021) have addressed the ethical issues involved in disseminating the knowledge necessary to construct a variety of weapons: doing so, they argue, might enable students with malicious intentions to do great harm (see, e.g., Contakes and Jashinsky 2021; Schummer 2021; Frank, Forman, and Cole-Hamilton 2018). Debates like these have not gone unnoticed in the forensic disciplines. In the digital forensic sciences, a term that has become widely known is that of "anti-forensics" (Kessler 2007: 1). This term refers to the use of countermeasures to avoid detection in a digital environment. The employment of such methods involves knowing how digital investigative techniques work and how they can be thwarted, disturbed, or, in the worst case, rendered useless. Examples of such "anti-forensic" countermeasures include the hiding or wiping of files and artifacts, trail obfuscation, direct attacks on computer forensic tools, and cryptography (Kessler 2007). Given the extraordinary potential for harm, the public disclosure of such measures through teaching and publishing research is highly contentious. Further, investigative agencies in a variety of countries like the Federal Bureau of Investigation in the United States, the MI5 in the United Kingdom, and the Federal Intelligence Service in Switzerland have issued warnings against the uncontrolled sharing of such sensitive information. Teachers and researchers are cautioned that some foreign students may pass on knowledge they gain during their stay abroad to their home governments (see Dilanian 2022; Swissinfo 2022; Cohen and Marquardt 2019). Similar examples are also found in business contexts (e.g., McEvoy 2023). Even amongst the domestic student population, the potential for misusing the information gained for nefarious purposes cannot be underestimated, with the Unabomber being a prime example thereof (Kaczynski 2016). At the same time, regarding authorship and writing styles in particular, there is a fine line between the misuse of knowledge with malicious intent and the preservation of privacy and anonymity (e.g., Brennan, Afroz, and Greenstadt 2012; see also Buschman 2014; Solove 2003), the latter of which arguably every individual has a right to.

Precautions for Preventing the Misuse of Forensic Linguistic Knowledge and Skills in Teaching and Research

Generally, the precautions a researcher and/or teacher can take in order to prevent the misuse of knowledge and skills fall across a continuum from not taking any precautions at all to trying to control everything (e.g., content) and everyone (e.g., audience). Below, three groups of strategies along this continuum will be outlined.

Let the Chips Fall Where They May

The strategy of "letting the chips fall where they may" is arguably the easiest one to adopt. It does not require much reflection on what is being taught to whom and in which context. Subscribers to this strategy do not concern themselves with any negative effects that might result from their teaching for the audience or for third parties. This, however, can be dangerous or at least negligent. As one participant in the survey for this chapter noted, it is of vital importance to reflect at least on the material used and shared, as well as the potential psychological harm that can come from working with and/or being exposed to forensic data:

> I have also been alert to the possibility that describing certain cases I have worked on, or sharing certain kinds of casework materials, might have a serious psychological impact on the people I am teaching, especially young people of student age who might be suffering from mental illness themselves. And, of course, there is a risk that audience members who have been the victims of crimes such as sexual assault or rape might be triggered even by particular words or images. It is important to take the potential for PTSD episodes seriously under these circumstances.

Thus, if teachers or researchers do not pay attention to the kind of material they share, they risk the psychological harm of their students and thereby violate their ethical responsibility as a teacher to take care of their students' well-being. In addition to being simple, however, there are also some other justifications for adopting this approach. One such underlying rationale was mentioned by a survey participant:

> As we are witnessing today with AI, there is an argument to be made that if scientist A does not do something then scientist B will and therefore it is pointless to worry about consequences. It's a logical argument in my opinion but I still disagree with it because I think we should at least try to do our best to prevent something negative to happen if we have the power to do it.

This statement highlights the need to concern oneself with potential consequences the content and/or style of teaching can have on one's audience. Even if someone else will neglect their responsibilities as a teacher/instructor by failing to acknowledge potential harm stemming from the ignorance of ethical problems, joining in this practice for whatever reason can have a lasting negative impact on students, people in the audience, and, essentially, the discipline of forensic linguistics itself.

A second rationale for adopting this approach is that all research should be available to everyone, regardless of the consequences. However, in the context of disseminating research, experts do face restrictions on what they can publish. Some of these restrictions, as will be discussed later as well, are related to privacy and data protection concerns. Other such restrictions are imposed by third parties involved in the research, and again others are imposed by researchers themselves. Generally, there is a need to share as many details of methods and data as possible—in teaching contexts, we want students to learn the methods, and within the research community, experts want their methods to be tested, developed, and improved. One respondent in the survey stressed this need to balance the amount of detail published:

> I realise that this [i.e., publishing the details of used methods] is not always possible but not publishing a forensic method at all, for example, would also be ethically problematic because research should be open and replicable and other experts should be able to test your claims.

This thought leads to another issue: if researchers do not publish the details of their methods but potentially only transmit them to a certain selected audience, then this will create a monopoly of experts knowledgeable in these methods, which is an undesirable outcome as well. People working in the field should be able to acquire new methods even if they do not belong to a dedicated circle of people.

Select Your Audience, Consider the Context

Moving away from the first approach, all respondents in the survey commented on at least some ethical issues and some considerations that should be taken in both teaching and otherwise disseminating research. Researchers and teachers subscribing to this strategy pay some attention to their audience and select the material they present with care and consideration. This strategy not only takes into account the audiences' potential triggers but also encompasses considerations about the audiences' social and cultural backgrounds. For instance, one respondent pointed out that they had given

> trainings to governmental employees from countries with less-than-optimal human rights records, e.g. ones which make fairly generous use of the death penalty, and/or which are known to persecute political dissidents, LGBTQ+ individuals, members of religious minorities, etc. I have been concerned that information I discussed during these sessions might feedback into the way they conduct forensic casework in those countries. There is certainly an ethical dimension to providing best-practice training to people who might be complicit in abuses of these kinds.

Another respondent commented that they had similar concerns:

> I have had two main concerns related to teaching methods to determine the authorship of texts: first, being able to reveal methods that can help mask the authorship of a text and, secondly, teaching non-linguists (e.g. in police forces) who will not have sufficient knowledge to make linguistic reports with sufficient guarantees and who may dare to make them.

The latter aspect raised by this respondent is of particular interest, as it touches upon a critical issue not yet discussed: educating individuals on forensic linguistics with backgrounds in other disciplines. On the one hand, such interdisciplinary exchanges are vital for the field to thrive, yet it does entail its own ethical challenges. As the respondent pointed out, there might be people in the audience who can potentially use their gained knowledge in their work (e.g., police officers), but who have only received an introduction to the field that does not equate to actual knowledge of linguistics and does not in effect qualify them to apply that knowledge. A similar concern was also raised by another respondent, who stated the following:

> I think the biggest problem is the potential by unqualified people to misrepresent (or deliberately misrepresent) forensic linguistic results, especially when they are complex. [...] The methodology might be misused by people unqualified to use it and therefore misuse it while misrepresenting themselves as experts in this type of analysis (e.g. misrepresentation of qualifications and experience).

This comment emphasizes another important issue: the lack of knowledge in how to apply methods and of how to "read" the results of authorship studies can lead to the unintentional misrepresentation of linguistic findings or the misapplication of methods that ultimately harms the credibility of the discipline. Another research-related comment that appeared concerns the reliance on studies without acknowledging their limitations:

> Results might be misrepresented, excluding the limitations of the study to represent the methodology as more powerful or comprehensive than it is in reality (e.g. using the approach in a context for which it was not developed).

This comment draws attention to two aspects: firstly, it highlights the need for researchers to clearly state the limitations of their studies as well as the need for them to contextualize their method(s). Secondly, it transfers the responsibility to the researcher relying on the findings of the study by stating that they, too, need to assess the study, its limitations, and the context(s) for which certain methods were developed before applying them to their own case. Similarly, a lack of knowledge particularly in respect to the use and interpretation of statistical methods and results was raised by another respondent, who stated the following:

> I occasionally have concerns about the misuse (generally inadvertently) of statistical measures in authorship analysis contexts, but I also have those concerns about academia more generally, and I think those concerns are ethical insomuch as they may be used to misrepresent findings to others.

An interviewee commented on this perception as well:

> evidence is more likely to be misused or misconstrued [if they rely on their intuition about how language works] and so that's what's driven me [...] to try to educate the lawyers about the way language works.

With this, the interviewee points out that in their work with legal professionals in particular, it is crucial for them as a linguist to identify the areas the legal professionals hold misconceptions of in relation to language and educate them about these misconceptions. Reflecting on her long experience in working with lawyers, Diana Eades says in her interview that

> you learn what the gap is between how lawyers and linguists understand language. It's not like I'm trying to hide ways of doing things, and in fact the Pinkenba case was so shockingly political that I realized I had a duty to address those issues head on.

Further, she points out that quite often it's impossible to publish (or teach) about specific cases anyway:

> Getting the permission to publish is quite complicated, especially as the data you work on is confidential. In the cases that I've published on, lawyers had only given permission after the final date for possible appeal.

Inside the classroom, it is also important to consider which aspects of a methodology need to be taught and in what way they need to be taught. One respondent mentioned that

> I jokingly tell my students that through the illustrative casework material I use I will train them to execute the perfect crime using voice disguise, but actually I wonder whether at some point someone might try to act on that information. The risk is small, of course, but it's not zero.

This statement reflects what several respondents have commented on, namely their concern of teaching content that can be used for malicious purposes if it falls into the hands of people with nefarious intentions. In the context of publishing research, similar concerns were raised and one respondent highlighted the responsibility of reviewers in this process:

> I think we can consider the nefarious use of information that we publish, and be sure that we are not making public the techniques used in law enforcement or security without permission. I do think we have some responsibility to avoid the perpetration of crimes using methods that we publish by considering this through the review process.

This respondent particularly commented on the need to acquire permission for the publication of certain methods and/or data, for example by law enforcement officers and/or lawyers, but also by people involved in the research process (e.g., research subjects). The need for permission to publish by involved parties and also the constraints posed by involved parties was mentioned by another respondent working in authorship analysis:

> I would never publish the details of methods or software that I developed with law enforcement agencies as partners. I have no problems in publishing the research that underpins it but not the exact details of how a certain system was built or where it was implemented.

A specific problem with the approach outlined in this section is of a rather philosophical nature. Namely, the question that arises is whether an instructor can actually decide to not teach something to a particular audience that could potentially be misused in detrimental ways, or whether teachers are obligated to teach as comprehensively as possible. The answer to this question also entails considerations as to whether it is ethical to keep any information from students or workshop participants. One interviewee pointed out the following:

> I also think that it would be sort of unethical of us to some extent to try to cheat through this [...] and not that not disclosing everything is cheating, because we can't disclose everything anyway, as we don't have time to do that [...] I think that's probably the best way that we can handle this. Otherwise, we would have to actually not be academics about this, and we would have to say, "this is trade secret, we can't disclose this to any outsiders". And that would be a pity, too.

With this statement, the interviewee clearly emphasizes that keeping information from students (or one's readers) is not ethical, either, and would collide with the academic mindset that involves the open discussion of methods and ideas, which will eventually lead to progress in the field, as was already hinted at above. Importantly, the interviewee argues that, if teachers or researchers did not share their knowledge with others openly, the academic spirit would be lost and turned into one of the trading business, where certain pieces of information are kept from other competitors in the field.

Furthermore, other, more pragmatic problems with this approach exist. In the context of disclosing methodological information in expert reports, Diana Eades mentioned in her interview that not only is there a need to describe the used methods to at least some extent (e.g., to allow replications) but also that knowledge about a topic does not equate to being able to apply that specific knowledge:

> I do mentoring for people who are going to write reports, but I never say to them anything like: "be careful about [what you write], the other side will see your methods", because your method has to be shown to a certain extent. Plus, also the kinds of things that people are giving expert evidence on are relying on some form of analysis that needs training.

However, there are also limits to the responsibilities of the teachers and researchers. As an interviewee argued, for example, it

> can't be my job to also vet people, right? [...] I mean, while I do authorship analysis and other types of forensic analyses, I'm not a detective. I don't even have the skillset to be able to vet people like that.

This statement highlights that even though the audiences' backgrounds might be taken into consideration to a certain extent, it is not the teachers' sole responsibility, and it is not their task to perform background checks. Nevertheless, many respondents did point out the need to pay attention to whom they are teaching and to select the course content with care.

Controlling Context, Restricting Audience and/or Access to Materials, Methods, and Resources

The third strategy to be described here is the strictest one, and researchers/teachers/instructors adhering to it attempt to control as much as possible of the environment or context in which they teach and/or publish, or otherwise disseminate their research. Some of the discussion above implies that a certain control over context and audience is necessary, but the question arises as to not only which precautions can feasibly be taken but also where the responsibility of the researcher/teacher ends, as implied by the interviewee's comment above. Considering that complete control over all aspects of contexts is impossible, the following discussion outlines methods that can be taken by the teacher and/or researcher to exert as much caution as is possible and/or reasonable.

What Does That Mean for Teaching?

The review of the literature as well as the interviews and surveys with the experts have shown that while there are limits to the responsibility of the teacher/researcher, some measures can indeed be taken in the attempt to prevent the future misuse of knowledge and skills. As one specific yet easily implementable measure that teachers can take, the use of trigger warnings was mentioned:

> I have to be careful that audience members are not offended, for example by taboo words or religious slurs. I have had to put together trigger warning texts to warn students that some of the material I use for teaching could upset or offend them.

As one respondent pointed out, however, complete censorship might not be the right option, either:

> I am also concerned that there could be topics which trigger negative psychological reactions in some readers. The use of particular terms (racial, religious, homophobic, or sexist slurs) needs to be handled carefully. I recently published a book chapter in which I reproduced a Twitter thread containing quite a few offensive terms. I don't want to censor them, and nor would I want to include a trigger warning. That seems like overkill to me. But it's conceivable that complaints could follow, and we—or at least our publishers—have to be ready to answer them.

This line of thinking is based on the idea that real-life data is what it is—it can be horrifying, discriminating, and ugly. Therefore, it can be argued that students of forensic linguistics need to be exposed to such data, as working with such data might turn out to be a part of their job description. In this context, the use of trigger warnings and allowing students to leave the room if they feel uncomfortable can still not be stressed enough.

As another strategy, some teachers/instructors require the members of their audience to sign disclaimer forms, as the following comment shows:

> I've asked my students to sign and return a disclaimer form pledging that they won't share case files or otherwise do anything unethical with them. The onus is then on them to behave professionally.

Such a practice, as stressed in the comment, takes the (ethical) responsibility away from the teacher/instructor and puts it onto the students or members of the audience. While there is no guarantee that someone will be prevented from misusing knowledge or data simply because they sign a disclaimer or non-disclosure agreement, it makes them liable if they do so. Unfortunately, however, even if all possible precautions are taken (e.g., anonymizing texts, stressing limitations, caution the use of data),

> in the end, if there are bad actors among the audience/students, none of these precautions will stop them from misusing the material.

But, as mentioned above, the teacher/instructor will have at least done *everything under their control* to prevent potential harm.

Another essential asset in the teaching of forensic linguistics that emerged during an interview is the building of trust and the creation of a safe space in the classroom. Due to the interactive and engaging nature of a teaching environment, the creation of mutual trust is easier in a classroom setting compared to researching and publishing environments. Mutual trust will not only allow the students to trust that the teacher is being honest and reliable; it will also allow the teacher to rely on students' commitment to work ethically and morally responsibly. Connected to mutual trust is the creation of a safe space in which both teachers and students can discuss topics openly, raise potential concerns, and address challenges and (ethical) issues directly. Importantly, also sensitive topics can be addressed and ethical considerations and issues pointed out. Only if students are made aware of ethical issues will they be able to reflect upon them and act accordingly.

Even if teachers create this environment, however, they are limited in what they can do to screen students. This is where the larger institution comes in. Many (though by far not all) universities perform background checks nowadays, thereby limiting the access to classes to a "certified" audience, but access to workshops and classes outside university contexts might not be restricted to the same extent. The preparation of such courses and workshops thus requires high levels of ethical awareness on the instructor's side, and it is at every expert's own discretion to decide whether or not certain topics can be discussed in the respective contexts or not.

What Does That Mean for Disseminating Research?

In the context of the dissemination of research, the matter is less straightforward. Due to the spatial and temporal distance between the researcher and their readers, as well as the lack of direct personal interaction, it remains largely unknown to the researcher who will be the readers of their research papers. Thus, the generation of mutual trust does not work as it does in face-to-face interactions such as teaching; similarly, the creation of a safe space is not only exacerbated but rendered impossible. Here, the researcher has to trust that their readers will adhere to their own professional ethics and standards, and that these standards will disallow at least the intentional misuse and abuse of scientific data, methods, and evidence. A step that researchers can take, however, is to make sure that they publish in journals that value high ethical standards, and that they anonymize their published data to a degree that makes identification of individuals impossible. One complication in this respect has been the endorsement to publish raw datasets. Although it is desirable to replicate findings with the same dataset(s), the publication of raw material creates serious ethical issues. Even if a dataset is published with the best (scientific) intentions and with all possible anonymization measures taken, experience has shown that, for instance, the appearance of a second dataset (which the researcher likely is unaware of) can render all precautions useless and research subjects can be identified. For example, the Netflix Prize dataset containing the movie ratings of half a million Netflix users, even though in anonymized form, was used to reliably de-anonymize the users if they had also posted movie ratings on the Internet Movie Database (Narayanan and Shmatikov 2008; see also Malin and Sweeney 2004). Similarly, a group of researchers (Sweeney, Abu, and Winn 2013) showed that it is possible to de-anonymize individuals based on a medical dataset containing only ZIP codes, birthdates, and sex when compared to a cheaply available voting register. As reported by Lubarsky (2017), such re-identification also works based on merely textual data, such as AOL search queries.

In courses or workshops for audiences other than university students, another precaution can be taken which is related to the potential (unintentional) misuse of half-knowledge that was already mentioned above:

> As precautions, I usually give informative presentations to non-specialized audiences in which I present the discipline in a general way without going into detail about the variables. In the case of teaching in the police forces, I tend to insist a lot that police officers should not make reports if they are not linguists and if they do not have specific training in forensic linguistics, and that they should contact a professional if they need a report in forensic linguistics.

The final point raised in this statement is a central one: although it is important to raise awareness of the discipline of forensic linguistics among members of law enforcement, it needs to be made clear that the aim of workshops with law enforcement officers is not that they themselves should be enabled to conduct linguistic analyses, but that they become sensitized to the issues connected with language and that they develop an awareness for assessing at what point they should consult with a linguist. Another issue is related to this one: the parties hiring forensic linguists need to know what lies within the scope of forensic linguistics and what they can actually demand from a forensic linguist, which is why such interdisciplinary work is highly valuable. Apart from commissioning parties requesting linguists to perform tasks outside of their expertise, there is also a risk that an (inexperienced) linguist will agree to conduct analyses that lie outside their competence, especially when pressured by the commissioning party. In an effort to educate the public as well as professionals on what they can and cannot expect from a forensic linguistic analysis, Sheila Queralt (2019) has published a "Decalogue" for requesting linguistic expertise. Publications such as these are necessary to raise awareness of the discipline but simultaneously also point out its limits.

Some Positive Outlooks

After outlining some strategies that teachers/researchers can use to prevent the unethical or even nefarious use of knowledge and data, this section addresses positive outcomes of previous negative experiences, as well as strategies to encourage ethical behavior.

In the interview for this book chapter, Diana Eades reflected on her experiences after the Pinkenba case. As she explains: "I was very mortified for several days, and then I just thought 'no, that's something of the system, lots of things have gone wrong in that'". Rather than take a defeatist perspective, she used this negative experience as motivation to help make more positive changes. She became even more committed to educating the people involved in the legal system and to conducting detailed research, which resulted in the 2008 publication of the book *Courtroom Talk as Neocolonial Control*. As she later reflected:

> the awful thing in a way is that it was really good that those things were exposed and then that drove me to reach out to training for lawyers, training for judges and magistrates, and to do an in-depth analysis not just of cultural issues but of the way that language was used for manipulation, such as the mechanics of cross-examination and the ideologies behind that.

Despite taking all of these precautions, it is still impossible to completely rule out misuse at every level, however. In one of the interviews, an important point was raised in this respect:

> what I try to comfort myself with when I'm concerned about these things is that even knowing about them [i.e., the methods that we use to reveal any type of criminal intent] does not necessarily make you capable of countering it.

This comment links back to what has been mentioned at the beginning of this chapter: essentially, if someone was interested in finding out about forensic linguistics methods (with and without malicious intent), they would find them on the Internet, in books, and in journals. Importantly, though, even if someone with nefarious intentions finds out about these methods (either by being taught about them or by searching the web), this does not mean that they will possess the skills to apply the methods successfully. And even if people with malicious, unethical intent are introduced to certain forensic linguistic techniques, the chances are still quite low that they will be able to use and apply what they have learned successfully; and if they do, through the help of forensic linguistic experts, the chances are quite high that law enforcement will still be able to detect them.

Discussion and Conclusions: Reconciling All Perspectives?

The preceding discussion indicates the sheer complexity of the topic resulting from a variety of valid standpoints and perspectives. For example, it is valid to argue that research and teaching needs to be open and uncensored in all possible ways. In contrast, it is also valid to vet audiences and to decide not to share all kinds of knowledge in all kinds of settings. The ultimate goal of any strategy, however, should be that no harm results from the teacher's/researcher's behavior. Thus, a certain number of ethical considerations are necessary in every teaching/researching endeavor. This can refer to the use of trigger warnings, the creation of a safe space in which every student can share their concerns, the anonymization of data, the vetting of audiences, and the decision to share only certain pieces of information with certain audiences. The main message that remains is that at least some time should be dedicated to considering which potential harms can result from one's own behavior and to trying to do everything in one's power to prevent that harm.

One of the central questions that emerged during the discussion presented in this chapter is: where does the teacher's and/or researcher's responsibility end? The answer to this question is not straightforward, as the discussion in that chapter has shown. Many aspects need to be considered and the complexity of the matter prevents the establishment of a single true answer. Forensic linguists carry a variety of responsibilities. As teachers, they are obliged to educate their students (and workshop attendees) in a way that will enable the students/attendees to make their own moral and ethical decisions in line with their own professional standards and codes. As researchers, it is important to obtain approval for publications for individual cases, and to consider whether there are any potential harms that can be done to anyone involved. Once the research is published and "out there", what will happen with it evades the linguists' control.

To summarize, it was the aim of this chapter to dive deeper into the ethical issues connected to the teaching and publishing of forensic linguistics skills and methodologies. As pointed out and discussed, it falls not within the realm of the teacher to vet their students, and to a large extent it is outside the realm of the teacher's control what the students will use their acquired knowledge and skills for. Importantly, however, certain measures can and should be taken to reduce the possibility of abuse and misuse of knowledge: raising awareness for ethical issues, creating a save space and trustful environment, and teaching ethical values to enable the students and workshop attendees to reflect upon their own behavior.

Acknowledgements

I would like to thank Tanya Karoli Christensen and Diana Eades who were willing to take time and share their expertise and insights with me in interviews. Further, I want to thank everyone who participated in my survey and the editors of this book for their helpful comments and input.

References

Aston Institute for Forensic Linguistics. n.d. "Aston Institute for Forensic Linguistics." Retrieved from https://www.aston.ac.uk/research/forensic-linguistics.

Bersoff, Donald, and David Glass. 1995. "The Not-so Weisman: The Supreme Court's Continuing Misuse of Social Science Research." *The University of Chicago Law School Roundtable* 2, no. 1: Article 11.

Boon, Helen. 2011. "Raising the Bar: Ethics Education for Quality Teachers." *Australian Journal of Teacher Education* 36, no. 7: 76–93.

Brennan, Michael, Sadia Afroz, and Rachel Greenstadt. 2012. "Adversarial Stylometry: Circumventing Authorship Recognition to Preserve Privacy and Anonymity." *ACM Transactions on Information and System Security* 15, no. 3: 1–22.

Brouwer, G. E. P. 1981. "Inquisitorial and Adversary Procedures: A Comparative Analysis." *Australian Law Journal* 55: 207–223.

Buschman, John. 2014. "The Practical Matter Privacy vs. Anonymity." *Journal of Information Ethics* 23, no. 2: 40–56.

Cohen, Zachary, and Alex Marquardt. 2019, February 1. "US intelligence Warns China Is Using Student Spies to Steal Secrets." *CNN*. Retrieved from https://edition.cnn.com/2019/02/01/politics/us-intelligence-chinese-student-espionage/index.html.

Contakes, Stephen, and Taylor Jashinsky. 2021. "Ethical Responsibilities in Military-related Work: The Case of Napalm." In *Ethics of Chemistry: From Poison Gas to Climate Engineering*, 85–111. Edited by Joachim Schummer and Tom BorsenSingapore: World Scientific Publishing.

Curbelo, Aury, and Alfredo Cruz. 2013. "Faculty Attitudes Toward Teaching Ethical Hacking to Computer and Information Systems Undergraduate Students." Eleventh LACCEI Latin American and Caribbean Conference for Engineering and Technology, 1–8.

Dilanian, Ken. 2022, July 7. "Heads of FBI and MI5 Issue Strong Warning about Threat to the West from China." *NBC News*. Retrieved from https://www.nbcnews.com/politics/national-security/heads-fbi-mi5-issue-strong-warning-threat-west-china-rcna36976.

Ding, Feng, et al. 2022. "Anti-forensics for Face Swapping Videos via Adversarial Training." *IEEE Tranactions in Multimedia* 24: 3429–3441. https://doi.org/10.1109/TMM.2021.3098422.

Durant, Alan, and Janny Leung. 2016. *Language and Law*. London: Routledge.

Eades, Diana. 2016. "Judicial Understandings of Aboriginality and Language Use." *The Judicial Review* 12: 471–490.

Eades, Diana. 2006. "Lexical Struggle in Court: Aboriginal Australians Versus the State." *Journal of Sociolinguistics* 10, no. 2: 153–180.

Eades, Diana. 2004. "Understanding Aboriginal English in the Legal System. A Critical Sociolinguistics Approach." *Applied Linguistics* 25, no. 4: 419–512.

Eades, Diana. 1992. *Aboriginal English and the Law. Communicating with Aboriginal English-speaking Clients. A Handbook for Legal Practitioners.* Brisbane: Continuing Legal Education Department of the Queensland Law Society.

Ehrich, Lisa, Megan Kimber, Jan Millwater, and Neil Cranston. 2011. "Ethical Dilemmas: A Model to Understand Teacher Practice." *Teachers and Teaching: Theory and Practice* 17, no. 2: 173–185.

Elkins, Kelly, and Ibiwunmi Fambegbe. 2020. "Case Studies and Methods for Teaching Professional Ethics for Forensic Science Students." *Journal of Forensic Science Education* 2, no. 1: 1–8.

Frank, Hartmut, Jonathan Forman, and David Cole-Hamilton. 2018. "Chemical Weapons: What is the Purpose? The Hague Ethical Guidelines." *Toxicological and Environmental Chemistry* 1, no. 100: 1–5. https://doi.org/10.1080/02772248.2017.1326633.

Gershman, Bennet. 2003. "Misuse of Scientific Evidence by Prosecutors." *Oklahoma City University Law Review* 28, no. 1: 17–42.

Green, Bruce 2007. *Teaching Lawyers Ethics*. Saint Louis University Law Journal 51, no. 4: 1091-1118.

Hobbs, Steven. 2007. "Hitting the Sweet Spot: Finding the Center in Teaching Professional Responsibility." *St. Louis University Law Journal* 51, no. 4: 1269–1290.

Kabay, M. E. 2004, September 23. "Creating Viruses in a University Course, Part 1." *Networked World.* Retrieved from https://www.networkworld.com/article/2325380/creating-viruses-in-a-university-course-part-1.html.

Kaczynski, David. 2016. *Every Last Tie. The Story of the UNABomber and His Family*. Durham: Duke University Press.

Kessler, Gary. 2007. "Anti-forensics and the Digital Investigator." Australian Digital Forensics Conference, 1–7.

Kuther, Tara. 2003. "A Profile of the Ethical Professor: Student Views." *College Teaching* 51, no. 4: 153–160.

Logan, Patricia, and Allen Clarkson. 2005. "Teaching Students to Hack: Curriculum Issues in Information Security." *SIGSCE'05*: 157–161.

Lubarsky, Boris. 2017. "Re-identification of 'Anonymized Data'." *Georgetown Law Techonology Review*: 202–213. Retrieved from https://www. georgetownlawtechreview. org/re-identification-of-anonymized-data/GLTR-04-2017.

Malin, Bradley, and Latanya Sweeney. 2004. "How (Not) to Protect Genomic Data Privacy in a Distributed Network: Using Trail Re-identification to Evaluate and Design Anonymity Protection Systems." *Journal of Biomedical Informatics* 37, no. 3: 179–192.

McEvoy, Ciran. 2023. *Inland Empire Man Arrested for Allegedly Stealing Sensitive Software from his U.S. Employers to Build a Competing Business in China*. United States Attorney's Office. Retrieved from https://www.justice.gov/usao-cdca/pr/inland-empire-man-arrested-allegedly-stealing-sensitive-software-his-us-employers.

Narayanan, Arvind, and Vitaly Shmatikov. 2008. "Robust De-anonymization of Large Sparse Datasets." *IEEE Computer Society*: 111–125.

O'Connor, T. J., Ben Sangster, and Erik Dean. 2010. "Using Hacking to Teach Computer Science Fundamentals." American Society for Engineering Education, St. Laurence Section, 1–5.

Olsson, John, and June Luchjenbroers. 2014. *Forensic Linguistics*. London: Bloomsbury.

Queralt, Sheila. 2019. *Decalogue for Requesting a Linguistics Expert Report*. Pie de Página.

Schummer, Joachim. 2021. "Ethics of Chemical Weapons Research: Poison Gas in World War One." In *Ethics of Chemistry: From Poison Gas to Climate Engineering,* 55–83. Edited by Joachim Schummer, and Tom Borsen. Singapore: World Scientific Publishing.

Schummer, Joachim, and Tom Borsen, eds. 2021. *Ethics of Chemistry: From Poison Gas to Climate Engineering*. Singapore: World Scientific Publishing.

Shuy, Roger. 2002. "Breaking into Language and Law. The Trials of the Insider-linguist." In *Linguistics, Language, and the Professions: Education, Journalism, Law, Medicine, and Technology*, 67–80. Edited by James Alatis, Heidi Hamilton, and Ai-Hui Tan, Georgetown: Georgetown University Press.

Solan, Larry. 1993. *The Language of Judges*. Chicago and London: The University of Chicago Press.

Solove, Daniel. 2003. "The Virtues of Knowing Less: Justifying Privacy Protections Against Disclosure." *Duke Law Journal* 53, no. 3: 967–1066.

Stockman, Mark. 2014. "Insider Hacking: Applying Situational Crime Prevention to a New White-Collar Crime." *RIIT'14*. http://dx.doi.org/10.1145/2656434.2656436.

Sweeney, Latanya, Akua Abu, and Julia Winn. 2013. "Identifying Participants in the Personal Genome Project by Name (a Re-identification Experiment)." Preprint at *arXiv.* https://arxiv.org/abs/1304.7605.

Swissinfo. 2022, December 26. "Swiss Universities on Guard Against Chinese Espionage." *Swissinfo.ch.* Retrieved from https://www.swissinfo.ch/eng/business/swiss-universities-on-guard-against-chinese-espionage/48161712.

Willis, Sheila. 2023. "The Professionalism of Forensic Science." *WIREs Forensic Science* 5, no. e1478: 1–9.

30

TEACHING SENSITIVE TOPICS IN FORENSIC LINGUISTICS

Ethical Implications for Students and Lecturers

Samuel Larner

Introduction

I worry that I might not be a good lecturer.[1] It is true that I score well in evaluations and I receive positive and encouraging feedback from both students and colleagues on my teaching. However, I am becoming increasingly concerned about the number of students who become upset by the content of my lectures and seminars. I fear that in teaching a subject that I am extremely passionate about, I am causing harm to at least some of my students. This fear arises from two key tensions that emerge in my classes: (1) there is an increasing expectation that my lectures will be engaging, entertaining and authentic, which typically requires the use of real-world, potentially upsetting data; and (2) although I avoid using unnecessarily contentious data, and although I routinely use trigger warnings in an effort to prepare my students for the content, these warnings are seemingly rarely heeded. Aside from the harm that I might be causing my students, I am also concerned about the increasing number of disclosures that my students make to me in light of their experiences in my classroom, and the impact that this has on me: both in terms of my ability to provide appropriate support and in terms of my ability to process and manage the emotional turmoil that it can cause me personally.

The ethical issues inherent in teaching subjects that contain sensitive topics are well documented in the Higher Education (HE) literature. For example, much has been published on this subject when teaching students in medicine, law, psychology, political science, social work, and counseling (e.g., Pinar Alakoc 2019; Hulme and Kitching 2017; Winstone and Kinchin 2017; Kennedy and Scriver 2016; Carello and Butler 2015; Heath 2005). It is my contention that there is nothing especially significant, unique, or special about the subject of forensic linguistics itself that necessitates consideration of a wider, or different, range of ethical issues. However, where forensic linguistics is arguably different as a subject is in two key areas. Firstly, the incidence of mental health conditions amongst linguistics students is generally higher than for other subjects, at least for students in England and Wales. Of people who declared a mental health condition in their university and college applications, the highest proportions were for students applying to Languages (8.8%), and Linguistics, Classics and Related Subjects (7.4%) programs, whereas mental health conditions were lowest amongst Medicine and Dentistry (1.4%) and Engineering (1.4%) (UCAS

DOI: 10.4324/9781003391074-33

2021). It therefore appears that in teaching forensic linguistics, the likelihood of causing harm and/or dealing with student disclosures is potentially higher than for other subjects.

Secondly, the general justification for teaching sensitive topics to students of subjects which have a clearly defined professional route (e.g., medicine, social work, and counseling) is that they are required by the professional standards which govern those subjects. For example, in respect of medical education, the "failure to include sensitive topics in undergraduate curricula results in a serious knowledge gap on the part of the doctor that clearly risks an inadequate response to patients' needs" (Kennedy and Scriver 2016: 193). However, no such professional standards govern linguistics generally, or forensic linguistics specifically. So, recourse to professional standards cannot be used as justification for exposing students to emotionally difficult or sensitive topics. Put another way, Kennedy and Scriver (2016) point out that "[p]riority is always afforded to avoiding harm to the student, whilst at the same time, ensuring that the necessary content is covered" (194), but when we consider forensic linguistics, there is no clear steer on what the *necessary* content actually is. Therefore, arguably, the hurdle we have to jump in order to justify teaching some sensitive topics—notably, those relating to sexual assaults including rape and child abuse—may be higher than for some other subjects. As such, the central questions of concern when considering the ethics of teaching sensitive topics in forensic linguistics must therefore be: (1) is it justifiable to teach such topics; and if so (2) what can be done to minimize the potential for harm to both the student and the lecturer? There are, unfortunately, no clear answers to these questions but I hope to provoke some conversations around some of the predominant issues as I have experienced them. This chapter is structured around two main sections: protecting students from harm and protecting lecturers from harm. Although this chapter explores the potential harms to arise from being a student of forensic linguistics and a lecturer of forensic linguistics, which might indicate distinctly separate issues, as will become apparent, the two are in practice interrelated, and prioritizing the reduction of harm to one potentially causes harm to the other.

Within the field of higher education pedagogy can be found numerous articles, which are often predicated on sharing best practices, reflections, and tips for how to teach sensitive subjects well (e.g., Hill and Zinsmeister 2012; Caswell 2010; Dalton 2010; Lichtenstein 2010; Heath 2005). This chapter does not follow that vein. Indeed, quite to the contrary, I am quite certain that even if I am a good lecturer, I am not doing good enough for my students. My aim in this chapter is therefore to explore the issues that I see as being fundamental to the teaching of forensic linguistics in higher education institutions, with a view to reflecting and introspecting on my own practice, in the hope that it may too resonate with others.

Protecting Students from Harm

Within the classroom will be victims/survivors of crime, as well as potentially perpetrators (Heath 2005: 4). Additionally, students enter higher education with more, and more complex, health conditions than ever before (Campbell et al. 2022), which shapes and influences university student support polices and specifically how lecturers are expected to respond to student needs. Of primary concern to lecturers will be the avoidance of re-traumatization of students; that is, "the triggering or reactivation of trauma-related symptoms originating in earlier traumatic life events" (Carello and Butler 2014: 156). However, it is not just students who have been directly affected by earlier traumatic events who are at risk in our classrooms: so too are students who are, or are trying to be, highly empathic. Such students are susceptible to secondary traumatization (also known as "vicarious traumatization"), whereby hearing about traumatic narratives "can yield trauma-related symptoms in the listener" (156). Indeed, in later work, Carello and Butler (2015) advocate that lecturers "should assume that in virtually every classroom some unknown subset of students will be at heightened risk for re-traumatization or vicarious traumatization as a result of personal trauma histories, mental illness experiences, and current challenges or difficult life transitions" (269). The impact is not just the immediate fallout of causing harm. The effects of a single class may extend well beyond into a student's wider educational experience and may potentially lead to "profound academic disadvantage" (Heath 2005: 4) through, for example, affected students performing poorly in class, missing classes, or dropping out altogether. Consequently, failing to acknowledge students' experiences limits students' access to effective learning, and students may find it to be "profoundly isolating" if it is not acknowledged that some may be having significantly different experiences in the classroom (Heath 2005: 4). Such a context leads Carello and Butler (2015) to argue that lecturers are "obliged" to understand how trauma may impact learning and to be aware of the signs and symptoms of (secondary) traumatization (270). It stands to reason that the risk of causing harm to students should therefore be a key consideration in the way that our forensic linguistics curricula are designed.

Curriculum Design

The pressure on lecturers to make their classes engaging and interesting is a potential source of tension when it comes to considering the ethical issues incumbent in teaching forensic linguistics. One way in which students can be engaged is through teaching sensitive topics (Hulme and Kitching 2017). So, it follows that a module on forensic linguistics which teaches students about, for instance, rape, domestic abuse, and child abuse—indeed any prevalent societal issue—will simultaneously be both potentially more engaging and interesting to students whilst potentially causing a higher incidence of re-traumatization. Consequently, lecturers should consider the rationale for teaching specific topics, and they should ask themselves whether that rationale is sufficient to justify potential re-traumatization. If the answer to that question is yes, then the next question should be whether appropriate safeguards can be put in place to at least minimize the potential harm.

Heath et al. (2017) argue that when teaching sensitive material, the starting point should be to design a learning environment which minimizes harm without removing content which is considered important to the curriculum. Further, the "intensity of exposure to traumatic material should only be to the extent necessary for learning" (Heath et al. 2017: 8). I do not think this is a contentious position to adopt. The issue for forensic linguistics as a subject, however, lies in the determination of what *needs* to be taught. As already established, there are no core curricula or professional standards to be adopted: lecturers typically make their own decisions about what their curriculum should contain. Their decisions will inevitably be governed by what they are comfortable teaching; what they enjoy teaching; what they believe their students will enjoy learning; their own research interests; and what data and resources they can find to support the learning. For example, I teach a session on child sexual abuse because I believe it to be an important societal issue worthy of discussion precisely because it affects so many people. Furthermore, I research in this area (e.g., Larner 2022) and believe it demonstrates the value of applied forensic linguistics in a very real and engaging way. But since child abuse is estimated to affect one in 20 children (NSPCC 2021), I must assume that in a typical class of 80 students, approximately four of them will have personal experience putting them at risk of re-traumatization. This is borne out by the fact that my sessions on rape and child abuse generate the highest number of student disclosures. In lieu of any standardized curriculum, then, lecturers must be acutely aware not only of what they teach and how they teach it, but *why* they teach it. If the actual pedagogic reasons do not outweigh the potential harms, the subject probably should not be taught. But how one decides what is pedagogically valid is no easy question to answer, if indeed an answer is possible at all.

For some lecturers, student ratings can be a determinative factor. On one hand, for lecturers teaching in institutions that espouse rhetoric that learning and teaching should be entertaining and fun, there may be a tendency toward risk-aversion on the basis that upsetting students should be avoided (Hulme and Kitching 2017: 9). This might encourage lecturers to avoid teaching certain sensitive topics in forensic linguistics. However, such an approach can do a disservice to our students who desire intellectual engagement, even if it does create discomfort (Lowe 2015). But of course, "discomfort" is not the same as "trauma". According to Carniel (2018):

> [A] traumatic response to stimuli centres not on the act in question, but on how that act returns the individual [to] that moment when their world was shattered and can manifest itself physically, for example, in the form of nausea or difficulty breathing. This is the process known as 'triggering', wherein something occurring in the present activates—or triggers—the traumatic response. Where an act of offence may, in some cases, be assuaged by an apology or other appropriate form of reparation, a traumatic response may require more complex support mechanisms, such as counselling.
>
> *(144)*

Consequently, even intellectual engagement (whether in the pursuit of higher student ratings or not) cannot be a suitable justification for teaching sensitive topics when the risk of trauma is very real. Where sensitive topics are taught, Heath et al. (2017) argue that including positive content in the curriculum, which foregrounds the resilience of victims rather than focusing only on their difficulties, can transform students' experiences. The decision about whether or not to teach a potentially traumatic subject might more sensibly be thought of as a balancing exercise then: there may be justification for teaching sensitive topics, knowing that some of our students will potentially be re-traumatized, as long as our teaching acknowledges the strength and resilience of survivors of such crimes. However, even this approach potentially creates a new opportunity for harm: student survivors who require therapy, for example, may compare themselves to other survivors who have been highlighted in class specifically for their strength and resilience, and infer that they are somehow "bad survivors".

It should be acknowledged that when considering the ethics of teaching forensic linguistics, students at risk of re-traumatization are not our only concern. As Collins (2013) highlights, students who neither experience re-traumatization nor vicarious traumatization nonetheless need to be considered at potential risk of harm through emotional burnout: "a generic term for the negative outcomes of any highly demanding and stressful situation" (139). Learning about emotionally demanding topics in addition to all of the other pressures may be too much for some students. As Collins (2013) explains:

> Where students are already feeling chronically stressed by the intellectual requirements of university, having to simultaneously deal with emotionally complex and demanding issues in their academic work can be the final straw that pushes them to a point of exhaustion.
>
> *(139)*

This adds another layer of consideration in the design of curricula. Care should be taken to ensure that the study of forensic linguistics does not become a relentless barrage of emotionally demanding topics. Finding the right balance in a subject predominantly concerned with crime is no easy task. However, the onus of responsibility to minimize student harm may not be solely in the hands of lecturers. As adults, university students also carry some of the burden. It is the students who decide whether they want to take a forensic linguistics class.

The Use of Trigger Warnings

Of course, students can only make that decision if they know in advance on what sort of course they are enrolling. Informing students in advance demonstrates an appreciation for students' diversity of experiences, whilst also modeling respect (Petersen Brooke 1999: 26), and this is typically achieved through the provision of trigger warnings: "written or oral notifications of course content meant to provide students advance notice of sensitive material that may produce adverse mental health responses and, therefore, inhibit academic performance" (Cares et al. 2019: 23). The debate surrounding trigger warnings is far too wide to cover in any detail in this chapter,[2] save to acknowledge that "[p]erspectives on trigger warnings are diverse, driven by cultural and institutional contexts, as well as the intersection between personal and professional politics" (Carniel 2018: 144).

In my own experience, it is not unusual for trigger warnings to prompt my students to discuss their personal circumstances with me. In this regard, the trigger warnings work, and it allows me to ensure that students are fully informed about what they will be studying. It also provides a welcome opportunity for me to get to know a student and be able to offer support should the need arise (although see below for a discussion on role strain and the emotional impact of such discussion). Nonetheless, I am repeatedly surprised to find that students who have recognized the significance of the warning and taken the initiative to discuss their concerns with me still continue to take the class, only later to become distressed in class. I do not think a single student who has ever discussed their concerns with me has actually made the decision not to enrol on my modules. This raises the question of why some students knowingly put themselves in a situation which they know will cause them harm. Hulme and Kitching (2017) offer one suggestion from the context of psychology: "because they want to understand their own life experiences, particularly when they are motivated to make a difference" (5). For students, then, putting themselves in a difficult—potentially harmful—situation may be part of their process for healing, or at least coming to terms with their experiences. Hulme and Kitching (2017) advise that lecturers need to be cognizant of the fact that amongst a diverse body of students, some will find exposure to the material helpful, whereas others may find their problems exacerbated. There is clearly some positive potential that can be gained for some students through participating in a class where they are knowingly at risk of trauma, although students should be provided with sufficient information beforehand to enable them to emotionally prepare for classes (Heath et al. 2017: 8).

My personal position is that a student who has survived a traumatic experience should not be prevented from taking my class as a consequence of their trauma: after all, to deny them this experience assumes a paternalistic role that goes beyond my remit as a lecturer and also potentially re-victimizes them, since their victimology both violates their autonomy in making decisions for themselves and prevents them from taking a class of interest. The implication is that the onus should be shared: the student must take responsibility for their own emotional well-being in deciding what weight to put on a trigger warning in relation to their own circumstances. At the same time, the lecturer may need to be prepared to take responsibility for any fallout during, and perhaps after, the class since in provoking emotional responses a responsibility is conferred on lecturers to support our students (Hulme and Kitching 2017).

Respecting Students' Decisions Not to Engage

If the whole purpose of trigger warnings is to make students aware that they will be exposed to sensitive topics, a further strategy to support students who may be uncomfortable studying the material is to offer the option of non-attendance: that is, giving students permission to either not attend the session or to attend but to leave the room if they get upset. In other words, there needs to be some option available to them after the trigger warning. Whilst it is perhaps intuitively agreeable to empower students to self-select which sessions they attend or which assessments they undertake, such an approach does potentially create further problems. One such problem is entirely administrative in nature. The move toward electronic monitoring/registration in each class, certainly in my own institution, means that non-attendance in a session will be logged, and it is no longer the case that simply not turning up to a lecture or seminar will go unnoticed. Indeed, repeated absences are flagged to a personal tutor so that mechanisms can be put in place to support a student's apparent disengagement. For anxious students concerned about their attendance records, the resolution of one problem may exacerbate further problems.

The only real solution is for the student to speak to me so that I am aware of the reason for their non-attendance and can log it (in general terms at least) in the electronic register, so that the absence is "explained". But this seems whole-heartedly at odds with what the approach was trying to achieve: rather than respecting students' decisions not to engage in a class, students may feel that they have to ask for permission not to attend. Some students may reason that it is easier to just sit through a difficult lecture rather than risking their perfect attendance or having to talk about their personal circumstances. Furthermore, once that line of communication has been opened, the lecturer may not know whether this is an instance of a student who does not wish to attend the class and does not wish to talk about their reasons why, or whether the student does not wish to attend class but is desperate to talk to the lecturer about that reason, as a prelude to a disclosure. Not knowing which situation applies inevitably puts the lecturer in a difficult position.

Pedagogic Discomfort

In contrast to respecting students' decisions to withdraw from any problematic classes is the alternative view: discomfort in the classroom can—in some circumstances at least—be positive. Discomfort in the classroom may be "more productive for developing critically engaged and informed graduates who are equipped to tackle the various social and political issues with which a broad humanities education is concerned" (Carniel 2018: 147). That being the case, Carniel (2018) raises the question of whether making students uncomfortable can be both pedagogically and ethically justifiable. However, as discussed above, the distinction between discomfort and re-traumatization must be borne in mind (Carniel 2018). Whilst some topics in forensic linguistics may well create discomfort, which perhaps should be encouraged (for example, the wealth of research exploring the disproportionate outcomes and experiences of non-native speakers of English in English-speaking legal systems), others, such as sexual abuse, are at higher risk of causing re-traumatization. For that reason, it would be a misapplication of the concept of pedagogic discomfort to argue that sensitive topics can ethically be taught to students who have experienced trauma on the basis that discomfort is good.

In summary, how we teach forensic linguistics is instrumental in the impact it has on our students, particularly those who have experienced crime: we either support learning or our learning environments become an obstacle to learning (Carello and Butler 2015). Whilst there are things we can do, both in the design of our curricula and in the preparation of our students for the content, ultimately, they need to make decisions for themselves about whether the subject is an appropriate one for them to study. Only they can make that decision, but we, as lecturers, often find ourselves dealing with the consequences.

Protecting Lecturers from Harm

It might be assumed that since we design our own curricula, we are adequately positioned to protect ourselves. We can make decisions about what we are—and are not—comfortable teaching, and design our curricula accordingly. However, such a position neglects two important considerations. Firstly, often, we do not teach alone and it may be both idealistic and impractical to suggest that courses should be (re)designed at the start of each academic year to take into account the wishes of all members of the teaching team. In some instances, a colleague may be required to teach particular classes and have little say over whether they would actually choose to do so: for example, when a colleague has to cover classes for a lecturer taking a leave of absence. This issue is especially prevalent when a more senior and/or experienced lecturer is working with a more junior, inexperienced colleague since "the degree of familiarity and confidence with content may also contribute to academic vulnerability" (Heath et al. 2017: 6). The position of this chapter has been that many of our students will have experienced sexual abuse, and there is no reason to suggest that the same is not also true of our colleagues. Secondly, although there is no universally accepted core curricula for forensic linguistics, certain topics appear to be integral to the field. For instance, victims of sexual assault are referred to—albeit to differing degrees—in many introductory texts to the field (Coulthard, Johnson, and Wright 2017; Solan and Tiersma 2005; Gibbons 2003); in a handbook (Coulthard, May, and Sousa-Silva 2020); and in edited collections on forensic discourse (e.g., Oxburgh et al. 2016; Solan, Ainsworth, and Shuy 2015; Heffer, Rock, and Conley 2013; Cotterill 2002). To simply avoid teaching such topics may arguably be to gloss over one of the central aims of forensic linguistics: that is, to effect social change through exploring linguistic disadvantage in forensic contexts (Larner 2018). To that extent, some lecturers may find it difficult to avoid teaching topics which they would otherwise avoid. This issue relates only to the content of what is to be taught. There are other wider ethical issues in teaching forensic linguistics that extend beyond the curriculum.

Students' disclosures may be triggering for the lecturer and we are often ill-prepared (Carello and Butler 2014). This can lead to feelings of guilt, sadness, or anger in the lecturer (Branch, Hays-Smith, and Richards 2009 in Hayes-Smith, Richards, and Branch 2010). Heath et al. (2017) suggest that this can impact lecturers in terms of their emotional burden and the additional time it adds to their workload. Furthermore, practical ethical issues may arise such as the lecturer being required to make an assessment about their student's safety, whether and to what extent they should maintain student confidentiality, and whether there are any mandatory reporting obligations that require compliance. As Heath et al. (2017) note, all of these issues bring further concerns about potential legal liability. Consequently, lecturers may find themselves performing multiple roles.

Role Strain

In the context of teaching psychology, Hulme and Kitching (2017) highlight that many lecturers feel it is important to teach sensitive topics because of the benefits to be gained for students, specifically relating to employability and citizenship. However, they also highlight that students may have unrealistic expectations of their lecturers, and that lecturers in turn may feel unable to deal with such expectations appropriately. This mismatch can have an emotional impact on the lecturer. The concept of "role strain" refers to discord between a person's social roles and occurs when "individuals attempt to fulfil multiple, conflicting social roles" (Hayes-Smith, Richards, and Branch 2010: 5–6). Role strain is especially significant in the context of a student making a disclosure because it requires the lecturer to move beyond what they may have considered to be their role. For some lecturers navigating these additional roles may be easier than for others. For example, Petersen Brooke (1999), reflecting on her own experience teaching sensitive topics, found it important "to distinguish clearly between my role as an educator and my very different role of a counsellor" (29). However, for other lecturers such a distinction is less easily maintained, if at all. Indeed, many lecturers find it difficult to establish a clear division between their role as an educator and that of a counselor (Heath et al. 2017).

In their study into role strain, Hayes-Smith, Richards, and Branch (2010) examined the experiences of 26 female academics following student disclosures, on the basis that female academics "may experience a higher level of disclosure given gendered expectations of providing social support" (5). They found that 38% of their sample reported feeling strain in their roles because of the emotional burden placed upon them by student disclosures, and 31% felt strain about how to teach sensitive topics after receiving a disclosure. Additionally, 19% of their sample reported feeling strain since they still had to interact with their students following disclosure, which made it difficult for them to negotiate their role as a lecturer.

In reality, when a student makes a disclosure, the lecturer can often do little more than offer intellectual guidance and the making of referrals where required (Collins 2013). However, in my own experience, I would suggest that providing well-being support is increasingly expected from the institution, if not the student. This situation can be "highly stressful, creating a desire to help students in distress, only to foreground that inadequacy of the lecturer's skills and the limitations of the institutional framework in meeting the students' emotional needs" (p. 139). Mummert, Policastro, and Payne (2014) cite the work of Payne and Gainey (2000) who classify lecturers into three roles: mentors, facilitators, and arbiters:

> As a mentor, professors present course topics to the class in a thoughtful manner and encourage students to contemplate diverse approaches to the topics. The instructor's role as facilitator involves assisting students in the discussion of topics, as well as permitting students to consider key questions and problems related to the issue. Finally, as an arbiter, the instructor is expected to recognize all sides of the issue, while ensuring that course discussion [sic.] remain collegial and civil.
>
> *(Payne and Gainey 2000 in Mummert, Policastro, and Payne 2014: 445)*

Mummert, Policastro, and Payne (2014) argue that these three roles are pivotal to students' reactions both to the course itself and to themselves in relation to those topics. I tentatively suggest that fulfilling these three roles (plus additional roles such as counselor and role model that are likely to be more prevalent now than in the quarter century since Payne and Gainey's work) will—for many—be a further source of role strain. Whilst I recognize these three roles in my own teaching, they are certainly not roles that have been formally taught to me during my academic career, and I suspect that for other lecturers who may have different teaching styles and approaches some of these roles may not come so easily. Heath (2005) argues that lecturers are a "human resource" and "not merely a knowledge bank" and therefore, safety and trust must be built into pedagogical strategies: "[a]cknowledging the centrality of emotion in the classroom and the impact of emotion on teaching and learning is essential to constructing such an approach" (10). These insights serve to highlight that the students' experience—both of the classroom content and any disclosure—is heavily influenced (if not determined) by the skill of the individual lecturer in dealing with sensitive topics. Indeed, Mummert, Policastro, and Payne (2014) are explicit on this point: "how the instructor approaches the material and handles classroom discussion and interaction can have an influential effect on students" (445). This foregrounds the importance of the skill of the individual lecturer and additionally highlights potential role strain for those lecturers who may prefer a more detached role, seeing themselves as a knowledge bank rather than a "human resource", or for those who are unable to engage at this level of emotionality (e.g., because of neurodiversity).

Lack of Training

Role strain may be further exacerbated by the pressure imposed by the institution to handle situations such as student disclosures, without necessarily providing appropriate training on how to respond (Hayes-Smith, Richards, and Branch 2010). In their study (introduced above) Hayes-Smith, Richards, and Branch (2010) found that 42% of their participants reported that they either lacked a clear definition of what the students expected during a disclosure or did not feel "qualified or comfortable negotiating a student disclosure" (12). Similarly, in their investigation of the perceptions of UK psychology lecturers who teach sensitive topics, Hulme and Kitching (2017) found that many lecturers are emotionally affected when a student discloses, but few felt that they had received adequate training to deal with serious issues. Crucially, most also reported that they had not been given adequate sources of support for themselves.

To mitigate some of the effects, Heath et al. (2017) encourage lecturers to engage in peer support, specifically by developing intra-institutional networks so that experiences can be shared, but they also highlight the need for strong mentoring and debriefing opportunities, as well as using self-care strategies. However, recognizing that disclosures create a challenging situation for lecturers, especially those from non-clinical backgrounds who have no experience of patient crises, Kennedy and Scriver (2016) adopt the position that ultimately, it is for the institution itself to take responsibility for its staff:

> [T]here is an over-riding onus on educational institutions to ensure that educators have access to the support they need in order to respond appropriately, to safeguard students, and to ensure the continued wellbeing of staff.
>
> *(194)*

Training and support in how to not only deal practically with disclosures but also manage the emotional impact of receiving disclosures is not always forthcoming.

Heath (2005) highlights that although it is common for lecturers to have concerns about how best to deal with disclosures without having received any appropriate training, people who have experienced sexual assault will not, in the vast majority of cases, disclose to a specialist and may never speak to a counselor. Indeed, students may disclose to the lecturers at least partly because they are *not* trained counselors. Similarly, Hayes-Smith, Richards, and Branch (2010) argue that it is irrelevant whether responding to disclosures is a formally recognized part of the lecturer's professional role because the reality is disclosures are happening. Lecturers occupy a pivotal position because how we react may be instrumental in the student's recovery, especially if the lecturer is the first person the student has told. Given that our subject, classes, and materials may be especially triggering for many students, and accepting the fact that many students may specifically choose to disclose to us, it is arguable that lecturers in forensic linguistics have a heightened need to receive better and more specific training than our colleagues in other branches of linguistics (or disciplines). It is only with this training that we can learn to respond to disclosure—and how we manage our own well-being in response to such disclosures—appropriately.

In summary, the prevalence of mental health conditions amongst our students, coupled with the sensitive nature of the topics we teach, places us in a position of huge responsibility in that students are more likely to disclose personal experiences, which has consequences both for them as the discloser and us as the recipient. As we learn to navigate the increasing demands of the modern academy, we must necessarily engage with conversations about the limits of our roles and how best to be prepared so that we can continue to do our best for our students and for ourselves. Where access to training and support may assist some colleagues, some lecturers may simply be unable to manage these competing demands.

Conclusion

It is unlikely that the answer is simply not to teach topics that might cause distress. Sensitive topics can be both emotionally and physically draining to lecturers who teach these topics, and for students who learn about them (Dalton 2010: 16), but that is insufficient reason to avoid teaching the topics altogether. Indeed, as Dalton (2010) argues, "[t]o leave sensitive topics unexplored (by the teacher) and unknown (to students) leaves us all poorer for the neglect or the avoidance" (16). If one of the goals of forensic linguistics is to create positive change, then as long as sensitive topics such as rape and child abuse are discussed carefully, as in other disciplines, including them in our curricula provides us with "an opportunity to model compassion and integrity and to facilitate conversations which open possibilities for personal and social change around sexual coercion, rather than closing them down" (Heath 2005: 9). Part of that compassion and integrity is to consider how both students and lecturers are cared for before, during, and after such classes.

In writing this chapter, my aim has been to reflect on the wider ethical issues of teaching forensic linguistics rather than producing a practical checklist of "how to teach ethically". The reason is simple. As explained in the introduction, much of the pedagogical literature is reflective in nature, elucidating practical advice based on personal experience, rather than empirical research. That is not to say that such literature is of no value; it is certainly not my intention to create that impression. But what it does mean is that when considering how best to teach our students, we are often limited to sharing ideas, which are presumed to be good, or "best practice", without necessarily considering whether such ideas actually do reduce the harms that can be produced for all participants in the classroom, including the lecturer. One such recommendation is that at the start of each session, the lecturer should openly acknowledge the sensitivity of the topic since doing so:

> can serve to diminish anticipatory anxiety amongst the group by helping individual students to recognise that sensitivity to the issue is not unique to them, but that engaging with the topic is potentially challenging for everyone present.
>
> *(Kennedy and Scriver 2016: 193)*

I do not imagine there is necessarily anything harmful or wrong in adopting this practice, but what concerns me is that implementing such low-level approaches may mask—for some lecturers at least—the bigger ethical issues which I hope to have drawn out in this chapter. It is my contention that as forensic linguistics lecturers, we really need to grapple with the very core of our curriculum and provide strong justification for what we teach and why we teach it, rather than relying on potential "quick fix" solutions which, despite reducing anxiety for some students, may fail to come to terms with the bigger issue: namely, we teach topics which have the strong potential for re-traumatization but we often do not have the level of support, guidance, and training that would be more typical in clinical disciplines.

I began this chapter by questioning whether I was a good lecturer. I would like to end it by re-framing that question: am I a skilled lecturer when it comes to delivering sensitive topics? The real question, then, is not whether we as forensic linguists should teach these topics, but whether *I* personally should teach these topics. My personal decision is that the value of teaching sensitive topics outweighs the potential harm I may cause to my students (and indeed may even help some students). However, it is incumbent on me in making that decision that I deliver my materials in a way that minimizes potential harm to my students as far as possible. That means making the commitment to be fully available to address their concerns, receive their disclosures, and provide appropriate support as required. But honoring that commitment does come at a cost and I have yet to master the reduction of my role strain. Our conversation around the ethics of teaching forensic linguistics, therefore, really begins in one place: with the compromise that each individual lecturer is willing to make in order to protect their students, since I am not confident that there can be equilibrium in balancing both interests.

Notes

1 Throughout this chapter, I have adopted the term "lecturer" as the standard nomenclature in England and Wales for an individual employed by a Higher Education (HE) institution to teach students, although I recognize that the issues raised in this chapter affect all grades of HE professionals engaged in teaching activities. Where individual researchers have used specific terminology reflecting specific grades or education systems, I have standardized throughout by referring simply to "lecturers", unless a distinction is necessary for meaning.

2 See Bentley (2017) for an overview of the debate.

References

Bentley, Michelle. 2017. "Trigger Warnings and the Student Experience." *Politics* 37 no. 4: 470–485.

Branch, Kathryn A, Rebecca Hayes-Smith, and Tara N. Richards. 2009. "'I Just Wing It': Professors' Experiences of Sexual Assault and Intimate Partner Violence by College Students." Unpublished manuscript.

Campbell, Fiona, Lindsay Blank, Anna Cantrell, Susan Baxter, Christoper Blackmore, Jan Dixon, and Elizabeth Goyder. 2022. "Factors that Influence Mental Health of University and College Students in the UK: A Systematic Review." *BMC Public Health* 22, no. 1: 1–22.

Carello, Janice, and Lisa D. Butler. 2015. "Practicing What We Teach: Trauma-informed Educational Practice." *Journal of Teaching and Social Work* 35 no. 3: 262–278.

Carello, Janice, and Lisa D. Butler. 2014. "Potentially Perilous Pedagogies: Teaching Trauma Is Not the Same as Trauma-informed Teaching." *Journal of Trauma & Dissociation* 15 no. 2: 153–168.

Cares, Alison C., Cortney A. Franklin, Bonnie S. Fisher, and Lisa Growette Bostaph. 2019. "'They Were There For People Who Needed Them': Student Attitudes Toward the Use of Trigger Warnings in Victimology Classrooms." *Journal of Criminal Justice Education* 30 no. 1: 22—45.

Carniel, Jessica. 2018. "Insert Image Here: A Reflection on, the Ethics of Imagery in a Critical Pedagogy for the Humanities." *Pedagogy, Culture & Society* 26 no. 1: 141–155.

Caswell, Glenys. 2010. "Teaching Death Studies: Reflections from the classroom." *Enhancing Learning in the Social Sciences* 2, no. 3: 1–11.

Collins, Anthony. 2013. "Teaching Sensitive Topics: Transformative Pedagogy in a Violent Society." *Alternation* 9: 128–149.

Cotterill, Janet, ed. 2002. *Language in the Legal Process*. Basingstoke: Palgrave Macmillan.

Coulthard, Malcolm, Alison Johnson, and David Wright. 2017. *An Introduction to Forensic Linguistics: Language in Evidence*. 2nd ed. Abingdon: Routledge.

Coulthard, Malcolm, Alison May, and Rui Sousa-Silva, eds. 2020. *The Routledge Handbook of Forensic Linguistics*. 2nd ed. London: Routledge.

Dalton, Derek. 2010. "'Crime, Law and Trauma': A Personal Reflection on the Challenges and Rewards of Teaching Sensitive Topics to Criminology Students." *Enhancing Learning in the Social Sciences* 2 no. 3: 1–18.

Gibbons, John. 2003. *Forensic Linguistics: An Introduction to Language in the Justice System*. Oxford: Blackwell Publishing Ltd.

Hayes-Smith, Rebecca, Tara N. Richards, and Kathryn A. Branch. 2010. "'But I'm not a Counsellor': The Nature of Role Strain Experienced by Female Professors when a Student Discloses Sexual Assault and Intimate Partner Violence." *Enhancing Learning in the Social Sciences* 2 no. 3: 1–24.

Heath, Mary. 2005, January. "Encounters with the Volcano: Strategies for Emotional Management in Teaching the Law of Rape." *The Law Teacher*: 1–33.

Heath, Mary, Clemence Due, Wendy Hamood, Amanda Hutchinson, Tania Leiman, Kerry Maxfield, and Jane Warland. 2017. "Teaching Sensitive Material: A Multi-disciplinary Perspective." *Ergo* 4, no. 1: 5–13.

Heffer, Chris, Frances Rock, and John Conley, eds. 2013. *Legal-lay Communication: Textual Travels in the Law*. Oxford: Oxford University Press.

Hill, G. William, and Dorothy D. Zinsmeister. 2012. "Becoming an Ethical Teacher." In *Effective College and University Teaching: Strategies and Tactics for the New Professoriate*, 125–134. Edited by William Buskist and Victor A. Benassi. London: SAGE Publications, Inc. https://doi.org/10.4135/9781452244006.

Hulme, Julie A., and Helen J. Kitching. 2017. "The Nature of Psychology: Reflections on University Teachers' Experiences of Teaching Sensitive Topics." *Psychology Teaching Review* 23, no. 1: 4–14.

Kennedy, Kieran M., and Stacey Scriver. 2016. "Recommendations for Teaching Upon Sensitive Topics in Forensic and Legal Medicine in the Context of Medical Education Pedagogy." *Journal of Forensic and Legal Medicine* 44: 192–195.

Larner, Samuel. 2022. "Facilitating Children's Informal Disclosures of Sexual Abuse: The Role of Online Counsellors at a National Children's Helpline." *Journal of Child Sexual Abuse* 31, no. 3: 276–296.

Larner, Samuel. 2018. "Research Methods in Forensic Linguistics." In *The Palgrave Handbook of Applied Linguistics Research Methodology*, 703–718. Edited by Aek Phakiti, Peter De Costa, Luke Plonsky, and Sue Starfield. London: Palgrave.

Lichtenstein, Bronwen. 2010. "Sensitive Issues in the Classroom: Teaching about HIV in the American Deep South." *Enhancing Learning in the Social Sciences* 2, no. 3: 1–23.

Lowe, Pam. 2015. "Lessening Sensitivity: Student Experiences of Teaching and Learning Sensitive Issues." *Teaching in Higher Education* 20 no. 1: 119–129.

Mummert, Sadie J., Christina Policastro, and Brian K. Payne. 2014. "Revictimization through Education: Does Learning about Victimization Lead to Anxiety and Identification as a Victim Among College Students?" *Journal of Criminal Justice Education* 25, no. 4: 435–451.

NSPCC. 2021. "Statistics Briefing: Child Sexual Abuse." Accessed January 4, 2024. Retrieved from https://learning.nspcc.org.uk/media/1710/statistics-briefing-child-sexual-abuse.pdf.

Oxburgh, Gavin, Trond Myklebust, Tim Grant, and Rebecca Milne, eds. 2016. *Communication in Investigative and Legal Contexts: Integrated Approaches from Forensic Psychology, Linguistics and Law Enforcement*. Chichester: John Wiley and Sons, Ltd.

Payne, Brian, and Randy R. Gainey. 2000. "Developing and Dealing with Controversial Issues in Criminal Justice Courses." *Journal of Criminal Justice Education* 11: 313–325.

Petersen Brooke, Corly. 1999. "Feelings from the Back Row: Negotiating Sensitive Issues in Large Classes." *New Directions in Teaching and Learning* 77 (Spring): 23–33.

Pinar Alakoc, Burcu. 2019. "Terror in the Classroom: Teaching Terrorism without Terrorizing." *Journal of Political Science Education* 15 no. 2: 218–236.

Solan, Larry M., Janet Ainsworth, and Roger W. Shuy, eds. 2015. *Speaking of Language and Law*. Oxford: Oxford University Press.

Solan, L. M. and P. M. Tiersma. 2005. *Speaking of Crime: The language of Criminal Justice*. Chicago: University of Chicago Press.

UCAS. 2021. "Starting the Conversation: UCAS Report on Student Mental Health." Accessed January 4, 2024. Retrieved from https://www.ucas.com/file/513961/download?token=wAaKRniC.

Winstone, Naomi E., and Ian M. Kinchin. 2017. "Teaching Sensitive Issues: Psychological Literacy as an Antidote to Pedagogic Frailty." *Psychology Teaching Review* 23, no. 1: 15–29.

31

WELCOME FORENSIC LINGUISTS!

Ethical Forensic Linguistics Pedagogy and Classroom Practice

Christine Jacknick and M Peregrine Balmat

Introduction

This chapter explores ethical choices we have made in our Forensic Linguistics (FL) course during our decade teaching it, focusing on how our curricular narrative has evolved toward one that is explicitly anti-racist and linguistically liberatory. Our decisions are mindful of our position of privilege, our student population, and their proposed careers. While our context has its own unique features, we argue that our suggestions have broad applicability and that all FL teachers can and should engage in reflexive thinking and writing as a part of ethical pedagogy.

This chapter offers an autoethnographic analysis of course development and teaching, as we place our experiences in conversation with each other and reflect on discussions, emails, artifacts, and curricular shifts. Our class introduces students, many of whom plan to become police officers, to FL in a way that prioritizes careful listening, thinking before acting, and interactional practices that are discourse-informed and intentional. Our classroom practices ethically model a place for care, vulnerability, complexity, and humanity in a profession that often prioritizes quantitative results over the quality of human experiences.

Below, we provide descriptions about our class, our students, our context, and ourselves. With a focus on what we as educators owe to our students, we introduce readers to pedagogy that is linguistically influenced, culturally responsive and sustaining, trauma-informed, and cognizant of discretionary choices made by street-level workers.

The Class

Our course description says we "explore the complex relationship between language and the law" and that "the course critically considers the role of language and its power in the legal process". We examine a wide variety of issues in the criminal legal system where language plays a central role, including sexual assault trials, false confessions, emergency calls, police encounters, and language discrimination in a variety of contexts (e.g., workplace, courtrooms, asylum). To our knowledge, aside from Hofstra University, our college is the only institution actively offering such a course at the undergraduate level in the United States.

DOI: 10.4324/9781003391074-34

Our Students

Borough of Manhattan Community College (BMCC), where we teach, is one of 25 colleges that make up the City University of New York (CUNY), a sprawling behemoth with 7 community colleges and 11 four-year colleges across New York City. According to the Institute for Justice and Opportunity (2020), "forty percent [of CUNY students] come from households with an income of $20,000 or less, and 44 percent are the first in their family to go to college" (9). In 2022, 36% of BMCC students identified as Hispanic, 35% as Black, 15% as Asian, and 14% as white (BMCC Data Dashboard 2024).

While it now also serves as an upper-level research elective for our Linguistics majors, for the lion's share of 12 years, our forensic linguistics course has served primarily Criminal Justice majors, many of whom go on to John Jay College of Criminal Justice and later apply to either the New York Police Department (NYPD) or other law enforcement agencies.

Criminal Justice in the United States

The criminal legal system in the United States is an enormous, complicated network of law enforcement agencies, judicial jurisdictions, and a sprawling prison system. Johnson and Abreu (2019) describe the landscape thusly:

> In the United States there are an estimated 6.6 million people under the supervision of the adult correctional system, of which 2.2 million people are in jail or prison—a 500% increase over the past forty years. National estimates also suggest that there are about 70 million Americans with criminal records (US Department of Education 2016). That racial disparities exist in the US criminal justice system is not new information either (Alexander 2012). Consider that Black and Latinx men are about 6 and 2.7 times more likely to be incarcerated than White men, respectively (Sentencing Project 2015).
>
> *(273)*

As discourse analysts, we understand how linguistic and discursive choices and ideologies in criminal justice contexts have a direct impact on human bodies and lives. We intentionally integrate cases involving language discrimination into our course to illustrate how the legal system systematically produces differential outcomes for speakers of varieties other than the standard, including for Deaf and hard-of-hearing users of American Sign Language. The cases and topics we integrate are both specific to the US criminal justice context (e.g., the prevalence of school shootings) and relevant globally (e.g., treatment of sexual assault victims, cf. Ehrlich 2010). We hope our approach to the course empowers and encourages students to be mindful of language and discourse as they pursue their careers.

Theory, Pedagogy, Praxis

We have three primary concerns in teaching the course: (1) to help students make sense of their own impressions of and experiences with the criminal legal system; (2) to provide them with the tools to bring awareness of language and culture into their careers; and (3) to ethically balance the rigor of the course with a culture of care and vulnerability that creates space for introspection. Our pedagogical perspective, which is reflective of not only the class context and our student body but also our positioning as sociolinguists and as educators, provides a foundation for how we approach these tasks.

Freire (1968/1970) criticizes the "banking model" of education, which views students as receptacles into which "knowledge is deposited". Rather, he argues for more equally distributing power in the classroom, reminding faculty to de-center themselves, share authority, support student leadership and teaching, and create a balanced relationship such that students can hold teachers accountable without penalty. When we make mistakes, we want students to be able to push back and ask for amends.

As sociolinguists, we draw on linguistically and culturally oriented pedagogical practices, including *linguistic pluralism*, *liberatory linguistics*, and *culturally responsive* and *culturally sustaining pedagogy* (Hudley, Mallinson, and Bucholtz 2022; Ladson-Billings 1995, 1992). Table 31.1 presents an overview of the components of our pedagogical practice. These approaches are borne out of legal cases and school board decisions that pushed linguists to generate support for students whose languages are stigmatized (e.g., the Oakland County School Board decision to teach Ebonics as a bridge to Standardized American English, cf. Rickford 1997). Paris and Alim (2017) argue that "asset pedagogy", an element of many of these approaches, repositions the linguistic, literate, and cultural practices of poor and working-class communities as resources and assets for teachers to use in helping their students to access dominant cultural norms expected in schools (4). These approaches center "self-determination", "action and resistance", and "humanization", connecting pedagogy to larger movements for social justice.

Integrating student perspectives challenges negative perceptions they might hold about their own and others' language practices. Their dialects and languages, cultures and histories, are worth studying and contextualizing vis-à-vis established power structures. Linguistic suppression triggers cultural suppression and erasure, and, within the criminal justice system, the privileging of Standardized American English and the often-willful misunderstanding of other language varieties help achieve ends situated in white supremacy, directly impacting the bodies and lives of already marginalized peoples. We therefore advocate for a peopled- (and necessarily bodied-) and culture-centered, intersectional FL.

We maintain that this approach to FL is for *all* students. Students from marginalized communities "have the right to [their] own language", and it is essential they have the opportunity to learn about their language varieties in culturally sustaining ways (Hudley, Mallinson, and Bucholtz 2022: 41). It is equally important for students with privilege to unlearn damaging language ideologies that contribute to racialization (Rosa 2016), housing discrimination (Baugh 2003), racially disproportionate court sentencing (Rickford and King 2016), and physical bodily harm (Vernon and Miller 2005).

While we brought these perspectives to our early teaching, the COVID-19 pandemic introduced us to *trauma-informed pedagogy* (TIP), which helped us mindfully balance how we brought difficult and potentially triggering content to the classroom. New York City was the early COVID epicenter in the United States, and many of our students and their families were "essential workers" of all kinds, required to work while we sheltered in place.[1] To stay mindful of their experiences and sensitive to the nature of our course content (which often includes re-traumatizing audio-video data), we utilized TIP, which extends the framework of "trauma-informed care" (Harris and Fallot 2001) to teaching. The pillars of TIP include "unconditional positive regard for each and every student; restorative approaches to discipline, rooted in relationship; [and] an orientation toward slowing down and considering student motivation, skill and capacity when designing learning experiences" (Venet 2019: 2). Using trauma-informed pedagogy, we acknowledge the complexity of human experience and bring radical empathy to teaching.

Finally, we draw on Balmat's (in press) *street-level pedagogy*, an approach that draws students' attention to the discretionary choices of street-level workers (Lipsky 2010, 1980), including law enforcement officers. Street-level worker discretion often follows patterns, as workers make systematic choices that help manage large caseloads, and this institutional focus forces us to reckon with how workers use discretion in ways that can unwittingly engage implicit biases. Our teaching integrates this perspective across the course, examining how 911 calls are handled, Miranda warnings are delivered, traffic stop encounters proceed, lawyers position others in courtrooms, and linguistic biases and language ideologies inform who is believed or taken seriously. These are all discretionary choices that fall outside of explicit policy. Our pedagogical approach, in which students are supported and validated by liberatory linguistics and culturally sustaining pedagogies, creates an environment where they feel safe considering how their own implicit biases, language ideologies, and discourse ideologies (beliefs about discourse) might get activated through discretionary decision-making on the job.

Table 31.1 Our Pedagogical Approach

Component	*Key Elements*	*Citations*
Linguistic Pluralism (LP)	• exposes students to prestige/standard dialect forms but gives students agency over rhetorical and stylistic choices • rejection of the belief that Standard English is synonymous with academic writing	Matarese and Anson (2010)
Liberatory Linguistics (LL)	• focus on Black languages, varieties, and communicative practices • self-determination and linguistic study led by and for Black people	Hudley, Mallinson, and Bucholtz (2022)
Culturally Responsive/ Relevant Pedagogy (CRP)	• recognition of diversity as a resource for the design and implementation of curriculum, instruction, and policy • focus on valuing students' cultures of origin while socializing them into the culture of schooling	Ladson-Billings (1995, 1992)
Culturally Sustaining Pedagogy (CSP)	• understanding of the inextricable links between culture, language, and identity • moves beyond CRP by not only recognizing students' home language and culture but focusing on sustaining it through schooling	Bucholtz (2017); Paris and Alim (2017)
Trauma-Informed Pedagogy (TIP)	• begins from a place of unconditional positive regard for all students • focus on restorative approaches to discipline • recognition of the (re-)traumatizing effect of schooling	Venet (2023, 2019)
Street-Level Pedagogy	• begins with an understanding of institutionally similar conditions for street-level workers, including pressures from above (administration) and below (clients) and how these impact workers' discretionary decision-making • focus on the impact of worker discretion on outcomes	Balmat (in press)

Who We Are and What We Do

We can both be described in a number of different interconnecting ways that inform our teaching: we are (socio)linguists, discourse analysts, educators, scholars, activists, middle class, white, and Queer. We are also close personal friends; we are in dialogue about teaching, this course, our students, and the world. In this section, we offer key questions for reflexive teaching practice and share our own thinking as we consider them.

Reflective Preparation, Reflective Practice: Peregrine

What Are My Ethical Obligations to My Students?

There are too many to count: preparing them for the analysis we do in the field, honoring diverse scholars, approaches, and cases. Many ethical considerations guided my initial development of the course. One in particular was met was skepticism: I am a staunch advocate for undergraduates reading and conducting primary research. "*Our* students can't…these students aren't at NYU", colleagues would say. I argue that our job as faculty is, in part, to teach students how to read research articles (Shon 2015). My online course platform welcomes students in bold lettering, "WELCOME, FORENSIC LINGUISTS", and in the classroom, I greet students as academic partners who contribute to curriculum, teach sessions, and hold me accountable. Students are treated as scholars and researchers who can and should conduct rigorous analyses of linguistic data.

To What Extent Does FL Require Creating Space for Vulnerability?

Initially I covered as many topics as possible over a 15-week course, and the flow of topics felt ad hoc and lacked continuity. When I look at any curriculum now, I ask, "What's the story this class is telling?", but initially FL had no story. At the time, stories and narrative held little academic value to me. I valued "data", "rigor", and "analysis". Academia and the demands of social science publishing had wrung out my personal style and voice. My shift toward vulnerability in the classroom was borne out of personal and professional need. I knew I was Queer and had fully suppressed and repressed my transness. Reading feminist methodological theory like Behar (1996), which advocates for ethnography that "break[s] your heart" (186), offered a departure from the emotionally anesthetized social science norm. While my graduate work with Walt Wolfram and Ofelia García primed me for social justice-oriented sociolinguistics, Behar's work cracked my professorial heart open.

In my book (Balmat in press), I say, "I can already hear academics opining with an eyeroll, paraphrasing Tom Hanks in *A League of Their Own* and throwing back their heads in unison, 'There's no crying in [social science]' (Marshall 1992 1:09:17)". However, both I and my students reap the rewards of what I see as a necessary vulnerability in the classroom. As an out trans man in transition, my vulnerability in the classroom helps put students at ease about the relationship between my pronouns and my appearance. I want students to also be themselves in the classroom, and so I create space for their own freedom, autonomy, and creative individuality. Language ideologies, discourse ideologies, and inherent biases—activated in the discretionary choices made by criminal justice workers—are generally ugly. Modeling honesty and vulnerability about biases I have personally confronted helps my students confront their own. Lipsky (2010, 1980) states, "Street-level bureaucrats work in situations that often require responses to the human dimensions of situations. They have discretion because the accepted definitions of their tasks call for sensitive observation and judgment" (15). We need spaces for observing and considering our own and each other's humanity to prepare for making human-centered discretionary choices on the job. We need vulnerability and understanding to generate a classroom environment that enables students to look within the self.

Reflexive Teaching, Reflexive Practice: Chris

What Do I Want My Students to Learn? What Kind of Police Officers/ Lawyers/Judges/Jurors/Citizens Do I Hope They Will Become?

The intersection of my scholarship (in autoethnography and social interaction), my identity (as a progressive, Queer, white person), and the identities of my students (mixed political perspectives among students who mostly come from racially "minoritized groups", cf. Rosa 2016) has forced me to grapple with how to be more fully myself in the FL classroom. I struggle with balancing my beliefs about the dysfunction of the legal system and policing in the United States with students' professed desire to pursue law enforcement careers. This struggle fundamentally informs how I conceive of the narrative of the course.

Following a departmental discussion about whether a course on language, race, and ethnicity should be taught from an explicitly anti-racist perspective, I released myself from the expectation to be apolitical. The choices we make *are* political, including any attempt to teach apolitically. In FL, we look at the impact of discursive and linguistic choices; for example, analyses of emergency calls extend to considering how dispatch responses impact communities differently. Recently, in discussing George Zimmerman's 911 call about Trayvon Martin,[2] one of many cases where an emergency call is made to report a person of color engaging in a legal activity, a student questioned why it would be a problem to dispatch police officers "just to be safe". I was surprised but pivoted our discussion to the over-policing of communities of color and the often-deadly results, as well as the discretion exercised by emergency call operators when deciding on appropriate dispatch responses. As a Queer scholar, educator, and activist, I am keenly aware of the more conservative orientation many criminal justice students have, and I work hard early on to establish rapport and goodwill to ensure that students are willing to be vulnerable as we work together to analyze how language affects outcomes in the criminal justice system.

Just as a police officer has the discretion to follow the spirit of the law by accepting "Can I talk to a lawyer"? as an invocation of Miranda rights, or to follow the letter of the law by rejecting it as "equivocal" (Ainsworth 2020), I talk with students about what a "good-faith effort" looks like to explain my ungrading[3] orientation. In so doing, my commitment to trauma-informed pedagogy models for my students how I hope they approach their future careers in law enforcement. In grading participation, teachers often use the heuristic of their overall impression of a student, and discussing how implicit bias can affect these impressions helps students see how their own biases can affect their decisions. Ways of "doing being a good student" are invariably influenced by systems like white supremacy and patriarchy; disinvesting in these systems means we have to see the varied ways our students demonstrate engagement (Jacknick 2021).

When I bring a transcript to class, I encourage students to point out places I need to revise it, not only to position them as research partners but also to model for them what working as a discourse analyst looks like. Rather than evaluating work with rubrics, we practice on data together and then work on transcripts and analysis together in class, just as discourse analysts do in data sessions. I utilize this approach to evaluation not only because it aligns with TIP but also as a form of self-care. Reviewing students' work using a rubric often leaves me feeling like I'm splitting hairs, whereas a focus on whether a student gave a good faith effort not only makes evaluating student work faster and easier, it also establishes a baseline expectation of respect between teacher and students.

Us, on Vulnerability

We align with hooks' (1994) exhortation to "teach in a manner that respects and cares for the souls of our students…to provide the necessary conditions where learning can most deeply and intimately begin" (13). She argues for a radically different positioning that brings more personal vulnerability to our teaching. Going beyond Freire, hooks (2003, 1994) argues that the classroom can become a place for the practice of freedom, joy, and love. In order to get there, we must bring our full selves into our teaching. This is an ongoing project, as personal evolution is constant. Neither of us was out as Queer when we began teaching, and we have been more hesitant about being openly Queer in the FL classroom, in particular. There are many reasons for this hesitancy. Bringing authentic data to the FL classroom requires extensive research and transcription, and scouring the internet for cases demonstrating LGBTQIA+ violence and hate crimes would seriously impact our mental health. This is a small example of self-care for us, but we also see that the absence of such cases in our classes renders violence against our community invisible and "unspeakable" (Nelson 2010).[4] Our awareness of how studying violence against our community impacts us means that we are careful to balance the cases we bring into the classroom, ensuring that no community is oversaturated with examples of violence. Where hooks offers a pedagogical ethos, Behar (1996) offers a feminist revisioning of ethnography that creates space for both our scholar/researcher role and a more personal, vulnerable self, often excluded from academic spaces. We continue to endeavor to make our classrooms spaces where we and our students can bring those vulnerable selves.

While bringing vulnerability to our teaching is newer to us, recognizing the humanity of our students, viewing them as scholars, and providing care for our educational community have always been practices we value. Our focus on humanism and care acknowledges the complexity of human experience and brings radical empathy to our teaching and mentorship. In our early teaching days, both of us would extend care almost without measure to our students, leading to burnout: Peregrine connected students with housing support, and Chris helped students document domestic abuse for immigration proceedings. We would limp to the end of the semester, crashing the moment we submitted grades.

The COVID-19 lockdown in March of 2020 pushed this challenge to the brink. Online teaching, multiple, interlocking crises faced by our students, combined with our own stresses, made it clear that we could not continue caring for our students in a limitless way. Working from home also forced a compression of our identities, making the level of personal withholding we had previously been practicing untenable. We began to take more care with creating boundaries for our time, intentionally carving out ways to advocate for our students in larger, institutional ways, for example, by planning a workshop on housing insecurity for students.

Curriculum Choices as Political Choices

On the heels of COVID, the murders of Breonna Taylor[5] and George Floyd[6]—touchstones for a legacy of structural racism in law enforcement—shifted the criminal justice dialogue and called on white people to educate themselves on how to be better allies. We both reviewed our course over the summer of 2020 to ensure that authors we assigned were diverse and intersectional, and while the topics covered and authors drawn upon remained largely the same, we began to more cohesively tie together the narrative arc of the course with an explicit focus on how interaction and language discrimination lead to differential outcomes in the criminal legal system. As educators, we took up this call to figure out how to bring that allyship and anti-racism into our work. Our pledge fed into the class ethos.

Cases were chosen strategically, each one an inherently political choice. From Peregrine's first FL class, he integrated publicly available data from well-known US cases, like the 911 call from a teacher during the Columbine High School[7] shooting in 1999. One of the first of its kind in public record, this call has become a kind of prototype for school shooting emergency calls. In it, the Columbine High School teacher reports that something has happened, though she doesn't know what. She suggests a window has been broken and notes that she is injured. It is not until the call progresses, however, that she realizes an active shooter situation is unfolding, and our students often struggle to understand her confusion. In the years since the Columbine shooting in 1999, gun violence at schools has become a near constant presence in the US media landscape. Students in American schools today regularly practice "shelter in place" drills; and we as faculty take mandatory online training about what to do in the event of a shooting. Integrating this case allows us to dovetail analysis of FL data with learning about historically important moments.

Choosing topics and examples in FL courses presents a particular set of ethical concerns because they include materials that depict traumatic events. These choices must evolve as similar types of events accumulate and shift cultural attitudes about the legal system and about policing. Many students in our 2025 classes are just out of high school and/or are recent immigrants, often lacking historical and cultural awareness of news-worthy criminal justice incidents and the movements they spawned in the United States in recent decades. After more than ten years teaching the course, we see how these cultural moments function in the classroom in real time, but also how they become historically important, tied to social movements (e.g., March for Our Lives,[8] Black Lives Matter,[9] and #MeToo[10]) and national developments.

We have both brought George Zimmerman's 911 call into class as a way to introduce students to how stance analysis can help to identify race-based ideology in linguistic interactions (cf. van Dijk 1987). Using stance markers, Zimmerman puts Martin in the category of a criminal, suggesting that Martin is "suspicious", "up to no good", and "on drugs". We also look at how the operator suggests that Martin has not done anything requiring police response by saying "let me know if he does anything", as well as how the operator positions the incident as a non-emergency (e.g., through change-of-state tokens and downgrades, saying "oh, he's just walking around the area"). George Zimmerman's eventual trial for the murder of Trayvon Martin also allows us to examine Rachel Jeantel's[11] testimony on the witness stand, to analyze with our students how speakers of non-standard dialects are routinely stigmatized and misunderstood in the legal system (cf. Rickford and King 2016), highlighting another instance of judicial discretion affecting outcomes. The choices by law enforcement and the judge *not* to provide translation for Jeantel's witness interview and testimony, respectively, led her evidence to be discounted by jurors.

The time that has passed since Martin's 2012 murder means that our students now were toddlers then, and it is crucial that we contextualize this case while remembering the humanity of those involved. This year, Chris supplemented this lesson with an introduction to explain how Zimmerman's acquittal of Martin's murder in 2013 sparked the Black Lives Matter movement (Blow 2022). We make it our job to help provide the historical context for these experiences to build on our students' visceral and communal knowledge about how policing affects communities of color.

Another case we originally brought into the classroom as a current issue was the trial of Brock Turner, then a 19-year-old white student at Stanford University, for his 2015 sexual assault of

Chanel Miller,[12] then a 22-year-old Asian American woman. Miller's victim impact statement at Turner's 2016 sentencing hearing provided a first-hand account of the impact of sexual assault trial discourse on victims (cf. Ehrlich 2010). Likewise, we can look at Turner's written statement to see how he presents an image of himself as a naïve young person (e.g., "these things [heavy drinking and promiscuity] were not only accepted but almost encouraged for the freshman to experience"). Students can read in Miller's own words how she experienced the re-victimization so often reported by sexual assault victims during the investigation of their assaults and subsequent trial, as this brief excerpt from her statement shows:

> Instead of taking time to heal, I was taking time to recall the night in excruciating detail, in order to prepare for the attorney's questions that would be invasive, aggressive, and designed to steer me off course, to contradict myself, my sister, phrased in ways to manipulate my answers. Instead of his attorney saying, Did you notice any abrasions? He said, You didn't notice any abrasions, right? This was a game of strategy, as if I could be tricked out of my own worth.

This victim impact statement went viral when it was made public, preceding the #MeToo movement that took social media by storm just a year later. The relatively light sentence Turner initially received (six months of jail time) highlights how judicial discretion, informed by implicit bias, leads to differential outcomes (Gersen 2023).

Without centering on a particular case, we also highlight how the sordid history of oralism sheds light on existing linguistic and cultural resources in the D/deaf community. Our focus on the language and culture in the D/deaf and hard-of-hearing community broadens students' awareness of linguistic discrimination. When they learn how "hearing paternalism" (Reagan 1995: 246) impacts law enforcement (Miller 2004), students consider their discretionary impulses. "If I yell, 'stop', what if the person doesn't hear me? What if I think they're resisting, but they're Deaf or hard-of-hearing?"

In examining data from such cases and contexts, we can illustrate important linguistic and discursive choices that affect outcomes. Our ethical stance requires that we balance that instruction by stressing that these cases of harassment, stigma, assault, and murder involve real people; as forensic linguists, we fortify our humanity by remembering and respecting the humanity of others. Often students are so excited to analyze a case, they can forget that the people in transcripts are real bodies. Moreover, the fact that these individual cases are indicative of larger social issues means that our students are likely to have personal experience with similar incidents. As a result, we make it our job to help provide context for those experiences.

Discussion

Our teaching of FL draws on the extensive repertoire of literature that together calls for being sensitive to and respectful of our students as people: people who may be from marginalized cultures (CRP, CSP) and who may use marginalized and stigmatized language varieties (LP); people with varying levels of trauma (TIP); people who deserve to be treated as intellectuals and as scholars (LL); people who may re-experience trauma in classrooms that address racialized and gender-based violence (TIP); and as people who will enter into professions in the criminal legal system and may struggle to avoid burnout and to maintain their own humanity. Racialized and gender-based violence perpetuated by the criminal legal system is not only codified in policy (ranging from historical examples including the seizure of land from Indigenous peoples to the recent official government erasure of trans and non-binary people through Executive Order 14168) but also enacted through discretionary practices, the in-the-moment decision-making where critical choices are made. For these reasons, to all the approaches discussed above, we add an institutionally aware, human-centered approach based on Peregrine's work on street-level linguistics (Balmat in press).

For the purposes of this chapter, such an approach invites *all* of us in FL (faculty and students alike) to consider that each of the cases and topics we explore highlights choices made by workers, enacted through words and gestures that perform actions. While our students are somewhat unique in that they largely intend to pursue law enforcement careers, FL students planning to become lawyers, judges, analysts, etc. will *also* have to reckon with how their discretionary choices, inevitably influenced by implicit biases, affect outcomes for others. The tenderness, care, dignity, and humanity we bring to the classroom and to the subject matter, we hope, extends outward to the communities being served by the students in our classes. A street-level pedagogy is mindful of how offers of dignity are linguistic choices and actions for which we may be held accountable. Such an approach, then, not only situates the ideals of the discipline within the reality of day-to-day work but also calls on us to be responsible for the choices we make, linguistically and otherwise. With those aims in mind, below is a list of ethical questions we recommend colleagues ask themselves when designing and teaching an FL course:

- What materials and data do we bring to class? How do we create an inclusive syllabus, representing marginalized groups?
- How do we sequence topics in the course, mindful of the formation of classroom community and also of TIP?
- What is my curriculum's politics? What is the curriculum's narrative arc?
- What do we grade? How do we grade it?
- How do we build in support for system-involved/impacted students (Ronda and Rose 2024)?
- How do we maintain "role clarity" (Venet 2019) so we are not attempting to save our students from their own trauma?
- What do we/our students gain when there is space for us all to be vulnerable and human in the classroom?
- How can a street-level perspective help students who intend to pursue criminal justice careers?
- How do we prevent our own and students' burnout?

Conclusion

While we continue to have much work to do as faculty teaching FL, it is not teachers alone who are called to consider their ethical choices and actions. Students should also critically self-reflect. At the end of the semester, we ask students to consider what they've learned and specifically how they plan to maintain beliefs that often are at odds with the institutional values of law enforcement, which expects some team-playing uniformity. Students respond by naming hobbies that give them a sense of individuality; they believe these will inoculate them against institutional, ideological conformity. We spend the end of class grappling with the challenge of maintaining individual beliefs and using what we have learned in class to make informed discretionary decisions. Our hope is that an institutionally aware, human-centered, linguistically and discursively informed approach brings accountability and practical decision-making to theoretical and linguistic discussions around criminal justice and forensic work.

Notes

1 Relatively privileged as we are, we find it difficult to name our own traumas here as well. Though we see that we are engaging in comparative suffering, it felt at the time, and it often was, as though the situation was worse for our students.

2 Trayvon Martin, a 17-year-old African American boy returning from a trip to a convenience store, was followed by George Zimmerman, a 28-year-old Hispanic American man who coordinated the volunteer neighborhood watch in his gated community. Against the advice of the 911 call operator, Zimmerman pursued Martin and approached him, leading to an altercation and his eventual murder of Martin.

3 "Ungrading" is an umbrella term for attempts to decentralize grading by focusing on "creating more effective communication with our students in order to create more learning-focused conditions" (Blum 2020: 15).

4 We have made 911 calls about the Pulse Nightclub shooting available to students as an option for their research papers as a way of including LGBTQIA+ cases without explicitly focusing on them. These 911 calls (including the call from the shooter himself) do not invoke queerness in any way, so it is a superficial inclusion.

5 Breonna Taylor, 26, was an African American emergency room medical technician who was killed March 13, 2020, when plainclothes Louisville Metro Police Department officers forced entry into her apartment. Her boyfriend, Kenneth Walker, had fired a warning shot, believing the police to be intruders. Following her death at the hands of police, and the initial lack of charges against any of the officers involved, the hashtag #sayhername was created by the African American Policy Forum in an effort to bring attention to Black women victims of police brutality.

6 George Floyd, 46, was an African American man from Minneapolis, Minnesota. His murder, captured by a bystander on cell phone video, showed the officer, Derek Chauvin, kneeling on Floyd's back and neck as he lay handcuffed for over nine minutes. His final words, "I can't breathe", became a rallying cry at marches and protests for racial justice and police reform in the United States and worldwide.

7 The massacre at Columbine High School of 12 students and one teacher by two other students on April 20, 1999, is regarded as a watershed moment ushering in the current epidemic of school shootings in the United States. At the time, it was the deadliest shooting at a K-12 school and remained as such for over a decade. Sadly, several school shootings since that time have taken its place as the deadliest in our history.

8 March for Our Lives is a student-led organization started by survivors of the shooting at Marjorie Stoneman Douglas High School in Parkland, Florida, in February 2018.

9 Black Lives Matter was created as a hashtag by three female African American activists following George Zimmerman's acquittal for the murder of Trayvon Martin and has since become a broader social movement.

10 The phrase "Me Too", first coined by activist Tarana Burke, became a widely used hashtag in 2017 as survivors of sexual assault and harassment shared their experiences on social media as part of a movement to both destigmatize these experiences for victims and also to hold perpetrators accountable.

11 Rachel Jeantel, 18, a friend of Trayvon Martin's, was on the phone with him as he was being followed by Zimmerman. She was the star witness at Zimmerman's trial for Martin's murder. Her soft-spoken testimony, delivered in her home dialect of African American Vernacular English (AAVE) with influence from Haitian Kweyol, was largely ignored by jurors in their deliberations despite the fact that she testified far longer than any other witness.

12 Miller was originally known as "Emily Doe" in court documents. She published a memoir, *Know My Name: A Memoir* (2019), about her experiences surrounding the assault and the trial, making her name public for the first time.

References

Ainsworth, Janet. 2020. "Miranda Rights: Curtailing Coercion in Police Interrogation: The Failed Promise of *Miranda v. Arizona.*" *The Routledge Handbook of Forensic Linguistics*, 95–111. Edited by Malcolm Coulthard, Alison May, and Rui Suisa-Silva. New York: Routledge.

Alexander, M. 2012. *The New Jim Crow: Mass Incarceration in the Age of Colorblindness.* New York, NY: The New Press.

Balmat, M Peregrine. In press. *The Subtlety of the Street: The Discourse of Responsibility, the Social Construction of Policy*. Ann Arbor, MI: University of Michigan Press.

Baugh, John. 2003. "Linguistic Profiling." *Black Linguistics: Language, Society, and Politics in Africa and the Americas*, 155–168. Edited by Arnetha Ball, Sinfree Makoni, Geneva Smitherman, and Arthur K. Spears. New York: Routledge.

Behar, Ruth. 1996. *The Vulnerable Observer: Anthropology that Breaks Your Heart*. Boston, MA: Beacon Press.

Blow, Charles. 2022. "How the Killings of Two Black Sons Ignited Social Justice Movements." *CBS News Sunday Morning*. Accessed October 29, 2024. Retrieved from https://www.cbsnews.com/news/charles-m-blow-trayvon-martin-and-the-black-lives-matter-movement/.

Blum, Susan D. 2020. "Why Ungrade? Why Grade?" *Ungrading: Why Rating Students Undermines Learning (and What to Do Instead)*, 1–22. Edited by Susan D. Blum. Morgantown, West Virginia: West Virginia University Press.

BMCC Data Dashboard. 2024. *Enrollment Snapshot*. BMCC Office of Institutional Effectiveness and Analytics. Retrieved from https://public.tableau.com/app/profile/bmcc.oiea/viz/BMCCDataDashboards/Welcome.

Bucholtz, Mary. 2017. "Why Bodies Matter." *Language and Materiality: Ethnographic and Theoretical Explorations*, 260–264. Edited by Jillian R. Cavanaugh and Shalini Shankar. Cambridge, UK: Cambridge University Press.

Ehrlich, Susan. 2010. "Rape Victims: The Discourse of Rape Trials." *The Routledge Handbook of Forensic Linguistics*, 1st ed., 297–312. Edited by Malcolm Coulthard and Alison Johnson. London: Taylor and Francis Group.

Executive Order 14168 of January 20, 2025. "Defending Women from Gender Ideology Extremism and Restoring Biological Truth to the Federal Government." *Code of Federal Regulations*, title 5, section 7301 (2025). Retrieved from https://www.whitehouse.gov/presidential-actions/2025/01/defending-women-from-gender-ideology-extremism-and-restoring-biological-truth-to-the-federal-government/.

Freire, Paolo. 1968/1970. *The Pedagogy of the Oppressed*. New York: Continuum.

Gersen, Jeannie Suk. 2023, March 29. "Revisiting the Brock Turner Case." *The New Yorker*. Retrieved from https://www.newyorker.com/news/our-columnists/revisiting-the-brock-turner-case#:~:text=The%20Santa%20Clara%20County%20Superior,job%2C%20through%20a%20recall%20election.

Harris, Maxine, and Roger D. Fallot. 2001. "Envisioning a Trauma-Informed Service System: A Vital Paradigm Shift." *New Directions for Mental Health Services* 89: 3–22.

hooks, bell. 2003. *Teaching Community: A Pedagogy of Hope* (Vol. 36). London: Psychology Press.

hooks, bell. 1994. *Teaching to Transgress: Education as the Practice of Freedom*. New York: Routledge/Taylor and Francis.

Hudley, Anne Charity, Christine Mallinson, and Mary Bucholtz. 2022. *Talking College: Making Space for Black Language Practices in Higher Education*. New York: Teachers College Press.

The Institute for Justice and Opportunity. 2020. *Mapping the City University of New York: The University's Commitment to Students Impacted by the Criminal Legal System*. New York: The Institute for Justice and Opportunity at the John Jay College of Criminal Justice.

Jacknick, Christine M. 2021. *Multimodal Participation and Engagement: Social Interaction in the Classroom*. Edinburgh, UK: Edinburgh University Press.

Johnson, Royel M., and Joshua Abreu. 2019. "Engaging Justice-Involved Students." *Student Engagement in Higher Education: Theoretical Perspectives and Practical Approaches*, 273–286. Edited by Stephen John Quaye, Shaun R. Harper, and Sumun L. Pendakur. New York: Routledge.

Ladson-Billings, Gloria. 1995. "Toward a Theory of Culturally Relevant Pedagogy." *American Educational Research Journal* 32, no. 3: 465–491.

Ladson-Billings, Gloria. 1992. "Reading Between the Lines and Beyond the Pages: A Culturally Relevant Approach to Literacy Teaching." *Theory into Practice* 31, no. 4: 312–320.

Lipsky, Michael. 2010 [1980]. *Street-Level Bureaucracy, 30th Ann. Ed.: Dilemmas of the Individual in Public Service: Dilemmas of the Individual in Public Service*. New York: Russell Sage Foundation.

Marshall, Penny. 1992. *A League of Their Own*. Directed by Penny Marshall. Los Angeles: Columbia Pictures, film.

Matarese, Maureen, and Chris Anson. 2010. "Teacher Response to AAE Features in the Writing of College Students: A Case Study in the Social Construction of Error." *The Elephant in the Classroom: Race and Writing*, 111–136. Edited by Jane Bowman Smith. Cresskill, NJ: Hampson Press.

Miller, Chanel. 2019. *Know My Name: A Memoir*. New York: Viking Press.
Miller, Katrina R. 2004. "Linguistic Diversity in a Deaf Prison Population: Implications for Due Process." *Journal of Deaf Studies and Deaf Education* 9, no. 1: 112–119.
Nelson, Cynthia D. 2010. "A Gay Immigrant Student's Perspective: Unspeakable Acts in the Language Classroom." *TESOL Quarterly* 44, no. 3: 441–464.
Paris, Django, and Samy H. Alim, eds. 2017. *Cultural Sustaining Pedagogies: Teaching and Learning for Justice in a Changing World.* New York: Teachers College Press.
Reagan, Timothy. 1995. "A Sociocultural Understanding of Deafness: American Sign Language and the Culture of Deaf People." *International Journal of Intercultural Relations* 19, no. 2: 239–251.
Rickford, John R. 1997, December. "Suite for Ebony and Phonics." *Discover*. Retrieved from https://johnrickford.com/portals/45/documents/papers/Rickford-1997a-Suite-for-Ebony-and-Phonics.pdf.
Rickford, John R., and Sharese King. 2016. "Language and Linguistics on Trial: Hearing Rachel Jeantel (and Other Vernacular Speakers) in the Courtroom and Beyond." *Language* 92, no. 4: 948–988.
Ronda, Michelle, and Lisa Hale Rose. 2024. "Project Impact: A Public Community College Campus. Based Support and Reentry Program for Justice-Impacted Students in New York City." *Teachers College Record* 126, no.6/7: 11–23.
Rosa, Jonathan D. 2016. "Standardization, Racialization, Languagelessness: Raciolinguistic Ideologies Across Communicative Contexts." *Journal of Linguistic Anthropology* 26, no. 2: 162–183.
Sentencing. Project. 2015. *Trends in US Corrections.* Washington, DC. Retrieved from http://senten cingproject.org/wp-content/uploads/2016/01/Trends-in-US-Corrections.pdf.
Shon, Phillip C. 2015. *How to Read Journal Articles in the Social Sciences: A Very Practical Guide for Students.* Thousand Oaks, CA: Sage Publications, Inc. Retrieved from https://www.torrossa.com/en/resources/an/5019526.
van Dijk, Theo A. 1987. *Communicating Racism: Ethnic Prejudice in Thought and Talk.* Thousand Oaks, CA: Sage Publications, Inc.
Venet, Alex S. 2023. *Equity-Centered Trauma-Informed Education.* New York: Routledge.
Venet, Alex S. 2019. "Role Clarity and Boundaries for Trauma-Informed Teachers." *Educational Considerations* 44, no. 2: Article 3.
Vernon, McCay, and Katrina R. Miller. 2005. "Obstacles Faced by Deaf People in the Criminal Justice System." *American Annals of the Deaf* 150, no. 3: 283–291.

32

ETHICS AND PUBLISHING IN FORENSIC LINGUISTICS

Policing Credibility

Sarah Frances Gordon and Bernardo Turnbull Plaza

Introduction: The Violence of Linguistic Hierarchies

Linguistic bias in the academic publishing of forensic linguistics (FL) research refers to the systemic privileging of dominant languages, such as English, which serve as gatekeepers of legitimacy. This gatekeeping is not only seen in FL publishing. In academic publishing, this bias is clearly reflected in the overwhelming number of scholarly journal policies that mandate English-language submissions and penalize non-Anglophone rhetorical styles, such as narrative, informal, and non-linear structures. Politzer-Ahles et al. (2016) argue that linguistic bias in academic publishing results in unfair treatment of authors, making it less likely for manuscripts written by people who have English as an Additional Language (EAL) to be published. As this chapter argues, there are strong parallels between linguistic bias in FL academic publishing and the legal proceedings. There is, for example, considerable literature that documents how credibility assessments during asylum hearings or courtroom testimonies also unfairly evaluate non-native English accents, dialects, or code-switching[1] (Dunbar, King, and Vaughn 2024; King and Jacobs 2024; Määttä 2024; Burlando-Salazar 2023; Runcieman 2022; Kurinec and Weaver 2019; Jones et al. 2019; Smith-Khan 2019; Eades 2010). According to Phillipson (1992), this form of linguistic bias can be described as "linguistic imperialism" and involves the perception of English as more important than other languages or dialects, which creates imbalances in how language users are treated and valued. Anzaldúa (2021) goes even further and characterizes this phenomenon as "linguistic terrorism" due to the systemic discrimination that this widespread bias produces. Like most forms of prejudice, this bias operates in our daily lives without our knowing. However, in the case of "linguistic imperialism" (or "terrorism"), this bias also reinforces colonial hierarchies, pressures language users to modify their knowledge to fit dominant frameworks, and forces language users to contort themselves to fit into an Anglophone mindset.

The dominance of English in academic publishing raises several ethical questions: Who benefits from establishing English as the default language for expressing the "truth" in academic research? How do publishers utilize "standard" language to exclude others? This chapter explores linguistic bias in academic publishing as a form of "epistemic violence", a concept constructed by feminist and postcolonial theorist Gayatri C. Spivak as "the remotely orchestrated, far-flung, and heterogeneous project to constitute the colonial subject as Other" (2003: 76). During this chapter, we argue

 DOI: 10.4324/9781003391074-35

that linguistic bias in FL publishing perpetuates colonial hierarchies by regulating knowledge production and policing discursive credibility. We also draw parallels between academic publishing and the legal system, as arbiters of legitimacy. This chapter demonstrates that both employ colonial-era standards to equate "credibility" with adherence to dominant cultural and linguistic norms. While many academic journals maintain English-only or English-first submission policies that functionally discriminate against other language communities, courts also exhibit negatively biased assessments of users of non-dominant languages and dialects that may also limit their rights of access. In both settings, these practices reinforce the idea that linguistic diversity lacks credibility and/or is indicative of intellectual inferiority. To dismantle these discriminatory institutional frameworks in academic publishing, it is argued that meaningful reform will require changing editorial policies to eradicate linguistic bias in FL knowledge-sharing.

Can Language Be Neutral?

Audre Lorde wrote, "The master's tools will never dismantle the master's house" (1984: 112). This is considered one of her most famous quotes, and it tells the story of how upholding dominant language structures will not lead to our liberation. bell hooks (2014) builds on this argument and states that language can serve as both a site of oppression and a tool of liberation. This characterization *feels* uncomfortable for many of us because, as hooks (2014) explains, Western thought focuses on binary oppositions. This perspective makes reconciling opposing ideas within the same concept challenging. However, as hooks (2014) argues, when debating the concept of language, we need to develop a more nuanced and complex understanding of the topic to understand it truly. With this goal in mind, there is an important question that we need to ask ourselves: "Can the languages we use be truly neutral?" If language shapes our identities, as so many scholars suggest (e.g., Anzaldúa 2021; hooks 2014; Foucault 2013; Whorf 2012; Fanon 2008; Freire 2000; Hall 1997; Sapir 1929), then it would be logical to infer that the imposing dominance of English in the scientific community carries considerable consequences. Accordingly, Anzaldúa (2021) asserts that dominant systems of power use (in)formal language requirements and expectations to marginalize and intimidate communities; and in order for those communities to reclaim our/their identities, we/they must subvert language.

Language is not isolated. It is influenced by and reflects social constructs such as race, politics, gender, and the broader cultural context. Thus, language constructs society, and society constructs language. This creates a dynamic relationship that is continuously evolving. Therefore, language choice is not politically neutral; it informs how we perceive others and how they perceive ourselves (Anzaldúa 2021; hooks 2014; Foucault 2013; Whorf 2012; Fanon 2008; Freire 2000; Hall 1997; Sapir 1929). Language use is a generative and regenerative process that helps form identity and perception. We argue that an unethical system of linguistic injustice is produced when one language, like English, is privileged over others. The result is not only exclusion, but also "epistemicide": the systematic erasure of Indigenous and non-Western knowledge systems by colonial and neocolonial regimes (Spivak 2003).

Epistemicide in Academic Publishing: Gatekeeping Knowledge

Elite institutions, journals, and publishers have constructed English as a gatekeeper for knowledge production within the international scientific community. This fact can be clearly seen in today's peer-reviewed literature. As many have observed, most scientific articles for an international readership are published in English, making fluency in the language essential for entering this field (Curry and Lillis 2024; Gordon and Fierros 2022; Olmos-Lopez, Prudencio, and Novelo 2022; Hernández Bonilla 2021; Salatino 2020; Gea-Valor, Rey-Rocha, and Moreno 2014). The empirical data support these research findings. In 2020, for example, over 95% of Web of Science (WOS) indexed articles were published in English (Badillo 2022). Scientific journals typically publish content in a single language, and that language is predominantly English. By comparison, in Scopus, a database for peer-reviewed literature created by Elsevier, only 8.2% of the indexed journals are multilingual. In WOS, this percentage is even lower and equals only 3.4% (Céspedes 2021).

Alduais et al. (2023) conducted a scientometric analysis of 6,490 studies related to forensic linguistics, spanning from 1936 to 2022. This analysis utilized Scopus, WOS, and Lens data. The findings revealed that regarding knowledge production in forensic linguistics, the United Kingdom ranked first, and the United States ranked second in the Scopus and Lens databases (Alduais et al. 2023). At the same time, Spain held the top position in WOS. The top journals in the field of FL include the *Revista de Llengua i Dret/Journal of Language and Law*, the *International Journal of Speech, Language and Law*, and the journal, *Comparative Legilinguistics* (Alduais et al. 2023). The *International Journal of Speech, Language and Law*, arguably the most prominent of all the academic journals in FL, is published by Equinox in England and only accepts English manuscripts. This monolingual policy is symptomatic of the linguistic imperialism seen throughout academia. However, there are alternatives. There are other prominent FL journals that have a more inclusive policy. Based in Spain, the journal *Revista de Llengua i Dret/ Journal of Language and Law*, for example, accepts manuscripts in Galician, Catalan, Dutch, Basque, French, English, Italian, Aragonese, Occitan, Spanish, and German. *Comparative Legilinguistics*, which is based at Adam Mickiewicz University in Poland, also accepts manuscripts in multiple languages (English, French, and German) and guest edited issues of this journal are available in Chinese, Russian, Polish, and Italian. Another emerging journal in the field of FL is *Language and Law / Linguagem e Direito*. Based at the University of Porto in Portugal, this journal publishes manuscripts in Portuguese and English. These multilingual academic FL journals represent an important step in the right direction.

The same pattern can be seen in the academic textbooks. Here again, the vast majority are published in English. John Louth, the Editor-in-Chief of Academic Law at Oxford University Press, reviewed the linguistic distribution of 401 international law books published between April 2014 and March 2015. In his review, he found that 84.79% were published in English, 8.98% in French, 4.74% in German, 1.25% in both French and English, and 0.25% in both German and English (Louth 2015). International law books are overwhelmingly published in English, which suggests a colonial academic mindset, equating linguistic assimilation with intellectual rigor (Gordin 2017). The introduction of a new book series in FL, Global Perspectives in Forensic Linguistics from Africa and Beyond, is also a refreshing addition to the academic literature in FL, and we believe represents an important move in the right direction. Despite these positive examples, there is still much more to be done to combat institutionalized linguistic bias in academic publishing.

Still, all too often, professional privileges are given to research produced by scholars from "elite" universities. Not coincidentally, many of these "high-prestige" institutions are often monolingual, English-speaking, and located in the United States and the United Kingdom (Ammann 2022). Non-Western scholars often struggle to disseminate their research globally—not simply because they lack the required skills in English but also because of the tendency to marginalize non-Western values and knowledge (Noda 2020). Even though most prominent journals have a double-blind peer-review process, in which the identities of the author and reviewers are hidden from each other, the number of scholars who publish in these journals is disproportionately Western and non-Western scholars only represent a small percentage of authorship (Noda 2020).

Noda (2020) explores the bias toward Western scholars in academic publishing within the International Relations (IR) field. Their analysis of seven leading journals in IR (*International Studies Quarterly* [ISQ], *International Studies Review* [ISR], *Journal of Global Security Studies* [JoGSS], *International Political Sociology* [IPS], *International Studies Perspectives* [ISP], *Foreign Policy Analysis* [FPA], *International Interactions* [II]) reveals a significant dominance of authors affiliated with Western institutions, while only a small percentage of contributions come from non-Western scholars. In the seven journals reviewed, contributions from non-Western scholars accounted for an average of only 12% (Noda 2020). There is also a tendency amongst peer reviewers, who are frequently Western scholars themselves, to favor Western epistemology and to discriminate against manuscripts that use non-English sources and favor non-Western knowledge systems (Noda 2020). We have personal experience of this bias ourselves. For example, when we have submitted manuscripts to journals in the past, peer reviewers have suggested we stop referencing Spanish sources and move away from Afro-centric and Latin scholars. We have been urged instead to focus on only English sources and cite more recognized Western scholars in the field, so that the academic community is able to better identify with our study. Are we to believe that Steve Biko and Paulo Freire didn't understand unconscious oppression and liberation discourse, and that it is necessary to consult more Western scholars to fully grasp these concepts? We think not. Ultimately, the publication privileges accorded to researchers from Western institutions help shape the linguistic biases in academic publishing and reinforce pre-existing power imbalances.

The Prestige Economy in Academia and Its Impact on the Global South

Pierre Bourdieu (2018, 1988) argued that prestige is a form of symbolic capital that can signify power and legitimacy within academic fields. The "prestige economy in academia", a construct that originates from the field of anthropology, is associated with elite institutions securing grants, winning awards, and receiving conference invitations, which all help to translate into more opportunities to publish and advance one's career (Blackmore and Kandiko 2011). Historically, these privileges are rooted in 200 years of colonialism and linguistic imperialism. Greek and Latin were widely regarded as the primary languages of science from the late medieval period until the mid-17th century, when they were gradually replaced by English, French, and German (Gordin 2015; Ammon 2001). During the 19th and long into the 20th centuries, British colonization and linguistic imperialism continued to elevate English over other languages in Great Britain, China, India, Africa, Canada, and the United States (Phillipson 2013; Hevia 2003; Mishra 2000; Niranjana 1990). The sharp decline in the international popularity of previous linguistic competitors like German and Russian after the defeat of Nazi Germany and the collapse of the Soviet Union at the end of World War II further facilitated the resulting dominance of English. These developments were also accompanied by a decline in language learning among scholars in the United States and the United Kingdom, as publishing research in English has become the norm (Hernández Bonilla 2021; Gordin 2017). Taken all together, the position of power and prestige enjoyed by English and its users is a product of complex, socio-economic, historical, and political factors reflected in the norms and practices of Western, English-speaking, economically privileged educational centers in the Global North.

By contrast, today, more than 85% of the global population lives in low- and middle-income countries, where fluency in English is lower than in high-income countries of the Global North (González-Dambrauskas et al. 2024). Where academic publishing is concerned, given the prevalence of monolingual English-only policies, non-native English-speaking researchers from these communities regularly face significant disadvantages compared to their native English-speaking academic peers. For example, it has been found that peer reviewers for academic journals often penalize code-switching and rhetorical styles, and metrics such as the impact factor tend to favor Anglophone journals (Olmos-Lopez, Prudencio and Novelo 2022; Ordorika 2018; Lee 2014). Research also suggests that users of English as an Additional Language often experience higher rejection rates from journals due to their limited skills in academic English (González-Dambrauskas et al. 2024; Smith et al. 2023; Amano et al. 2021). The financial burden of professional English-language editing services can also disproportionately affect academics from the Global South, particularly those from resource-poor institutions (González-Dambrauskas et al. 2024; Smith et al. 2023; Amano et al. 2021). This unethical situation worsens existing inequalities that affect academics in the Global South, hindering research production and critical international knowledge-sharing. In this way, linguistic hegemony in academic publishing significantly and unethically threatens the welfare of the FL scientific community by undermining global scientific diversity (Hill 2025).

Importantly, linguistic imperialism is also intricately woven into the practices and policies of institutions where English is not used as an official language. This can be seen, for example, when institutions value research published in English more than publications released in non-English languages, even the official languages of the nations where their institutions are housed. As a result, research published in "other" languages is often marginalized by English-speaking and EAL members of the international scientific community. Ironically, perhaps, many research institutions and universities in Latin America, for example, argue that prioritizing an "international" agenda means promoting English in the national research community. One of the requirements for a doctorate in any field at universities in Mexico, for instance, is the ability to prove proficiency in a foreign language and a heavy preference is given for English, in which a candidate must demonstrate an intermediate level of English proficiency, either through a professional proficiency test such as the Test of English as a Foreign Language (TOEFL) or an exam taken at the university itself (Ibero 2024; UNAM 2022, 2021; López Ramírez 2017). Such requirements align with journal publication policies in Latin America that increasingly encourage English-language submissions from the local scholarly communities (Ferreira 2023; Céspedes 2021; Ordorika 2018; Packer 2016). Bispo (2024) argues that such policies are merely a disguised form of academic colonization in the Global South. Pursuing an "international" agenda should not be made equivalent to an "English-only" research agenda (Ferreira 2023). Rather than instituting policies that erase the presence of other languages and subsequent local identities, true internationalism in scientific publishing should promote multilingualism in academia.

At present, however, the entrenched dynamics of linguistic imperialism remain evident in the unspoken compliance of multilingual researchers from the Global South with "hermeneutical injustice" in academia. "Hermeneutical injustice" occurs when marginalized people struggle to understand their own experiences because they do not have the same access to knowledge or language as others (Fricker 2017; Medina 2017). When people lack the tools or resources to explain what they are going through, their situations may be misperceived or misunderstood by outsiders, which, in turn, can lead to unfair treatment and other disadvantages. EAL users often internalize feelings of linguistic inferiority, viewing their knowledge systems as "second tier" compared to the research published in prominent Anglophone journals, which are frequently and unquestionably praised as epistemologically superior. Phillipson (2013, 1992) argues that such a situation occurs due to gatekeeping editorial practices in academia that are then frequently internalized. These gatekeeping practices not only include English-language submission mandates but also preferences for "global" (often Euro-American) theoretical frameworks, which normalize the dominance of English as the exclusive legitimate channel for knowledge production. The systemic epistemic marginalization in academic publishing also favors journals produced in the Global North over those from the Global South. These biases may compel non-Anglophone scholars to engage in self-censorship, leading them to alter or abandon their local epistemologies to align with imperial standards of "scientific rigor". This compliance is a strategic response within an academic environment that equates English proficiency with credibility.

The internalization of linguistic inferiority among EAL scholars highlights the pervasive epistemic violence of linguistic imperialism within the global academic community. This situation reminds us of Freire's (2000) concept of "internalized oppression", where the colonized mimic the dehumanizing logic of the colonizer and internalize their own oppression. EAL academics may unintentionally marginalize their work because they perceive it (consciously or unconsciously) as less significant than research from Anglophone contexts. This distorted self-perception stems, however, from external systemic hierarchies that associate "scientific legitimacy" with proximity to English-language norms. To challenge these frameworks within FL, we are ethically bound to reject the colonial notion of "prestige" in publishing and prioritize multilingual scholarship. The irony is that researchers in FL have been one of the principal drivers of drawing society's attention to how biases about languages and language users can result in profound social injustice.

Forensic Linguistics and Policing Credibility

According to Fricker (2007), "epistemic injustice" occurs when people are mistreated because of what they know or how they know things. Such epistemic injustice can take two forms: hermeneutical injustice and testimonial injustice. Hermeneutical injustice occurs when dominant linguistic frameworks fail to accommodate non-English perspectives. This can happen, for instance, when Indigenous knowledge systems are excluded from academic discussions and scholarly publications. By contrast, testimonial injustice is slightly different. It can occur, for example, when EAL users face unfair treatment after their credibility is questioned due to perceived linguistic "weaknesses". These so-called "weaknesses" can include limited vocabulary, non-standard grammar, code-switching, or "non-standard English". Biases against these stigmatized forms can help to create power imbalances. Within the criminal justice system, these linguistic biases can result in a witness being considered an unreliable source of information, constituting "testimonial injustice" (Burlando-Salazar 2023; Kurinec and Weaver 2019). Testimonial injustice in the criminal justice system, resulting from biases against marginalized dialects, has been a frequent subject of FL investigation (e.g., Eades, Fraser, and Heydon 2023; McMillan 2021; Moya 2021; Baugh 2018; Eades 2016; Rickford and King 2016; O'Barr 2014; Angermeyer 2013; Hale, Bond, and Sutton 2011; McKeown and Miller 2009; Eades 2008).

Research from the United States and the United Kingdom has demonstrated that transcription and intelligibility errors involving testimony from witnesses who use Creole or African American Vernacular English (AAVE) can undermine their perceived credibility (King and Jacobs 2024; Jones et al. 2019). Brown-Blake and Chambers (2008) investigated a case where a firearm was incorrectly attributed to a Jamaican Creole speaker during a police interview in the United Kingdom. According to the researchers, linguistic biases against the speaker's use of non-standard English negatively influenced the case outcome. The British courtroom's preference for "standard English" testimony over hybrid or vernacular speech reflects a colonial practice linking linguistic conformity with legal credibility (Smith-Khan 2019). The same pattern has been identified in many other legal settings. For example, asylum seekers in Australia have been found to face skepticism when their narrative styles differ from Western conventions (Smith-Khan 2019). Similar findings have been reported in other contexts, such as South Africa (Angu and Gustafsson 2022). Again and again, research has determined that the credibility of witnesses who do not speak "standard English" is routinely and unfairly questioned (King and Jacobs 2024; Burlando-Salazar 2023; Jones et al. 2019; Kurinec and Weaver 2019).

By framing dialects, creoles, or multilingual code-switching as signs of confusion or dishonesty, courts perpetuate discriminatory racial and class hierarchies that associate "credibility" with "standard English". As a result, "standard English" is considered an ideal that courts think we "should" all aspire to (Anzaldúa 2021). As mentioned above, forensic linguists have played a pivotal role in revealing discriminatory language policies inside and outside the criminal justice system and proposing suggestions to correct these societal injustices. We argue that the field of forensic linguistics is ethically bound to do the same to draw attention to and work to correct equally discriminatory publishing policies both within the discipline and more broadly.

In our opinion, the linguistic bias displayed in academic publishing is analogous to the power imbalances FL researchers have exposed in legal contexts, where marginalized individuals are expected to adhere to dominant linguistic standards to be seen as credible. In the same way, non-Western scholars may encounter rejection due to their writing styles or EAL proficiency. By the same token, EAL academics often feel pressured to adjust their language to fit dominant frameworks to ensure that their English-speaking monolingual colleagues understand them and accept their research as credible. Publishers, editorial boards, peer reviewers, and citation practices determine what constitutes legitimate academic scholarship. These gatekeeping practices favor research in English and Eurocentric traditions while marginalizing scholars who speak EAL (Finocchiaro and Perrine 2024). Researchers are often expected to conform to Anglophone norms rather than advocate for widespread systemic change. The biases in academic publishing are far from neutral. Language itself cannot be neutral. Such biases reinforce a prestige economy associating scientific rigor with linguistic and cultural conformity. FL can support linguistic pluralism and dismantle the myth of linguistic neutrality in both contexts by examining the social injustices created by linguistic biases inside both the courtroom and academia.

Achieving true societal justice requires moving away from the colonial fantasy that language and the act of "listening" are neutral. It requires critically examining how linguistic bias influences our reality and taking action to dismantle discriminatory structures. In the same way that forensic linguists are ethically obligated to call attention to language policies in legal systems that unfairly designate certain types of knowledge and language as "valid" while disregarding others, they are ethically bound to resist policies and practices that perpetuate discriminatory global hierarchies within academic publishing and result in the erasure of cultural diversity. This involves challenging gatekeeping practices and amplifying marginalized voices.

Conclusion: Suggested Reforms and Interventions in FL Academic Publishing

Linguistic bias in academia and publishing represents a moral and ethical concern because it creates inequalities and favors specific languages over others, undermining linguistic diversity. If we genuinely value ethics and diversity, we must address this issue. In FL, we need reforms and interventions that promote multilingual publishing, ensure citation equity, and provide training for peer review to combat linguistic bias. Finardi (2022) argues that the invisibility of research production from the Global South is directly related to the languages used in our research evaluation systems at the institutional level. Ultimately, we must change how research is evaluated at this level. The obsession of the Latin American research community with "internationalization" is harmful (Finardi 2022). The focus has shifted away from the quality of a publication to impact factors, metrics, and citation numbers for English-language publications. By making English a prerequisite for international visibility, we have overlooked the importance of local relevance and immediate, meaningful social impact (Finardi 2022). For this reason, we advocate for a balanced approach that includes publishing in English *along with* rather than instead of other languages pertinent to the specific context. For us, as researchers based in Mexico, that means publishing in Spanish, English, and Portuguese.

We also call for academics to engage with sources from various languages and regions, whenever possible (Amano et al. 2021). EAL scholars often heavily cite theorists from the Global North while dismissing vernacular methodological approaches as “primitive”. Makoni (2012) notes that these practices reflect a colonization of knowledge that links language proficiency to academic prestige. We do not want this. To challenge these discriminatory practices and attitudes, we must adopt *conscientização*—critical awareness of linguistic imperialism (Freire 2000). This means actively promoting multilingual, universal epistemologies as acts of resistance against social injustice. For example, this book chapter has endeavored to draw from English, Spanish, and Portuguese sources. We aimed to reference scholars of color, multilingual scholars, and scholars from outside the Global North. We recognize the challenge of balancing local relevance with the desire for international visibility. This can be difficult to achieve; however, we must try.

We also argue that academic journals should adopt more inclusive editorial policies, prioritizing research quality over English proficiency in submissions. The FL community must also draw attention to making people aware of the necessity of separating language skills from research quality (Amano et al. 2021). English writing skills should not be equated with our understanding of research. The impulse to connect the two reflects another instance of linguistic bias. We must take active steps to become aware of our unconscious biases and consider why it has become so common to link English writing skills with intellect. We should also explore whether this tendency is rooted in any colonial mindsets we hold and ask ourselves if our insistence on using English is linked to racist beliefs we may unconsciously hold. Although these discussions are challenging, they are important if our goal is social justice and ethics. Assuming that these are truly our goals, then we argue that our scholarly circles must insist that the journals we patronize take corrective steps. For example, investments should be made to make editing services available for authors who may not be proficient in academic English (Ammann 2022). Efforts should also be made to publish more translations of articles originally written in languages other than English (Ammann 2022; Amano et al. 2021). It is also recommended that EAL scholars be actively solicited to participate in the journal production process as peer reviewers and editors.

Ultimately, the naturalization of linguistic hierarchies in academia has constructed English as a “neutral” and “global” tool for communication. However, this construction obscures the reality that this practice is rooted in colonial regimes and imperialism. hooks (2014) reminds us that we often believe it is natural for the strong to dominate the weak; however, this is not true. Dominance is not accidental. It is meticulously engineered through institutions that align with Anglocentric norms. We must reject the myth of linguistic neutrality and begin to view academia and academic publishing as architects of epistemic violence. We must build coalitions among scholars, activists, legal practitioners, and advocates for social reforms.

Linguistic bias is not just an ethical lapse but a structural feature of our society’s pervasive linguistic bias, which constructs English as an objective measure of intellect and integrity. To decolonize knowledge and the field of forensic linguistics, we need to break the connection between legitimacy and colonial languages. This way, academia and academic publishing will stop acting as barriers and become places that value and promote diverse languages and ways of knowing. The fight for fair knowledge starts when we stop supporting unethical exclusionary practices and begin to take down the barriers to diversity and equality altogether.

Note

1 Linguistic code-switching involves alternating between languages or dialects during a conversation. People who speak more than one language often use this skill to interact in different social, cultural, and institutional settings. For example, a Mexican American speaker might mix Spanish and English in sentences like "I'm going to the store, ¿quieres algo?" which translates to "I am going to the store, do you want anything?" Another form of code-switching (often formally called "code-mixing") involves inserting an English word into a Spanish sentence and then conjugating it as if it were a Spanish word: "Dónde parqueaste el *carro*?", which translates to "Where did you park the car?" This intertwining of linguistic systems reflects a hybrid identity, which Anzaldúa (2021) argues embodies a mestiza consciousness—serving as a form of decolonial resistance against monolingual norms that marginalize Spanglish and other contact languages. This phenomenon is becoming more common in our post-colonial world (Dumitrescu 2014).

References

Alduais, Ahmed, Mohammed Ali Al-Khulaidi, Silvia Allegretta, and Mona Mohammed Abdulkhalek. 2023. "Forensic linguistics: A Scientometric Review." *Cogent Arts & Humanities* 10, no. 1: 2214387.

Amano, Tatsuya, Clarissa Rios Rojas, Yap Boum II, Margarita Calvo, and Biswapriya B. Misra. 2021. "Ten Tips for Overcoming Language Barriers in Science." *Nature Human Behaviour* 5: 1119–1122.

Ammann, Odile. 2022. "Language Bias in International Legal Scholarship: Symptoms, Explanations, Implications and Remedies." *European Journal of International Law* 33, no. 3: 821–850.

Ammon, Ulrich, ed. 2001. *The Dominance of English as a Language of Science*. Berlin: De Gruyter Mouton.

Angermeyer, Philipp Sebastian. 2013. "Multilingual Speakers and Language Choice in the Legal Sphere." *Applied Linguistics Review* 4, no. 1: 105–126.

Angu, Pineteh, and Kristina Gustafsson. 2022. "The Politics of Interpreting during Asylum Interviews. Processes, Challenges and the Implications for the Legitimacy of Claims for Asylum in South Africa." In *Language and the Law: Global Perspectives in Forensic Linguistics from Africa and Beyond*, 63–76. Edited by Monwabisi K. Ralarala, Russell H. Kaschula, Georgina Heydon. Stellenbosch: African Sun Media and the Editors.

Anzaldúa, Gloria. 2021. *Borderlands / La frontera: La nueva mestiza*. Madrid: Capitán Swing Libros.

Badillo, Ángel. 2022, February 17. "El portugués y el español en la ciencia: apuntes para un conocimiento diverso y accesible." Real Instituto Elcano. Accessed December 10, 2025. Retrieved from https://www.realinstitutoelcano.org/monografias/el-portugues-y-el-espanol-en-la-ciencia-apuntes-para-un-conocimiento-diverso-y-accesible/#ver-estudio.

Baugh, John. 2018. *Linguistics in Pursuit of Justice*. Cambridge: Cambridge University Press.

Bispo, Marcelo de Souza. 2024. "Implicações Da Colonização Acadêmica Nos Periódicos Científicos Brasileiros Em Administração." [Implications of Academic Colonization in Brazilian Scientific Journals in Administration]. *Revista de Administração Contemporânea* 28, no. 1: 1–7.

Blackmore, Paul, and Camille B. Kandiko. 2011. "Motivation in Academic Life: A Prestige Economy." *Research in Post-Compulsory Education* 16, no. 4: 399–411.

Bourdieu, Pierre. 2018. "Distinction a Social Critique of the judgment of Taste." In *Inequality*, 287–318. Edited by David Grusky. New York: Routledge.

Bourdieu, Pierre. 1988. *Homo Academicus*. Translated by Peter Collier. Stanford, CA: Stanford University Press.

Brown-Blake, Celia Nadine, and Paul Chambers. 2008. "The Jamaican Creole Speaker in the UK Justice System." *The International Journal of Speech, Language and the Law* 14, no. 2: 269–294.

Burlando-Salazar, Julian A. 2023. "Preventing the Epistemic Harm of Testimonial Injustice in Lay Witness Credibility Assessments." *Boston University Law Review* 103: 1245.

Céspedes, Lucía. 2021. "Latin American Journals and Hegemonic Languages for Academic Publishing in Scopus and Web of Science." *Trabalhos Em Linguística Aplicada* 60, no. 1: 141–54.

Curry, Mary Jane, and Theresa Lillis. 2024. "Multilingualism in Academic Writing for Publication: Putting English in Its Place." *Language Teaching* 57, no. 1: 87–100.

Dumitrescu, Domnita. 2014. "La Alternancia de Lenguas Como Actividad de Imagen En El Discurso Hispanounidense / Code-Switching as Face-Work in the Discourse of the US Hispanics." *Pragmática Sociocultural / Sociocultural Pragmatics* 2, no. 1: 1–34.

Dunbar, Adam, Sharese King, and Charlotte Vaughn. 2024. "Dialect on Trial: An Experimental Examination of Raciolinguistic Ideologies and Character Judgments." *Race and Justice*. https://doi.org/10.1177/21533687241258772.

Eades, Diana. 2016. "Theorising Language in Sociolinguistics and the Law: (How) Can Sociolinguistics Have an Impact on Inequality in the Criminal Justice Process?" In *Sociolinguistics: Theoretical Debates*, 367–389. Edited by Nikolas Coupland. Cambridge: Cambridge University Press.

Eades, Diana. 2010. *Sociolinguistics and the Legal Process*. Bristol: Multilingual Matters.

Eades, Diana. 2008. *Courtroom Talk and Neocolonial Control*. New York: Mouton de Gruyter.

Eades, Diana, Helen Fraser, and Georgina Heydon. 2023. *Forensic Linguistics in Australia: Origins, Progress and Prospects*. Cambridge: Cambridge University Press.

Fanon, Frantz. 2008. *Peau noire, masques blancs, Seuil. Black Skin, White Masks*. Translated by R. Philcox. New York: Grove Press.

Ferreira, Valmeria Brito Almeida Vilela. 2023. "A Construção de Políticas Linguísticas Em Periódicos Científicos".[The Construction of Language Policies in Scientific Periodics]. *Fólio - Revista de Letras* 14, no. 2: 321–325.

Finardi, Kyria Rebeca. 2022. "As Línguas e Rankings No Oscar Da Internacionalização Das Produções Científicas Latino-Americanas." [Languages and Oscar rankings of the internationalization of Latin American scientific productions] *Estudos Linguísticos (São Paulo. 1978)* 51, no. 1: 147–161.

Finocchiaro, Peter, and Timothy Perrine. 2024. "Linguistic Justice in Academic Philosophy: The Rise of English and the Unjust Distribution of Epistemic Goods." *Philosophical Psychology* 37, no. 6: 1483–1512.

Foucault, Michel. 2013. *Archaeology of Knowledge*. London: Routledge.

Freire, Paulo. 2000. *Pedagogy of the Oppressed*. Translated by Myra Bergman Ramos. 30th ed. New York: Continuum.

Fricker, Miranda. 2017. "Evolving Concepts of Epistemic Injustice". In *The Routledge Handbook of Epistemic Injustice*, 53–60. Edited by Ian James Kidd, José Medina, and Gaile, Jr Pohlhaus. New York: Routledge.

Fricker, Miranda. 2007. *Epistemic Injustice*. Oxford: Oxford University Press.

Gea-Valor, Maria-Lluïsa, Jesús Rey-Rocha, and Ana I. Moreno. 2014. "Publishing Research in the International Context: An Analysis of Spanish Scholars' Academic Writing Needs in the Social Sciences." *English for Specific Purposes* 36: 47–59.

González-Dambrauskas, Sebastián, Jorge Ibrain Figueira Salluh, Flávia Ribeiro Machado, and Alexandre Tellechea Rotta. 2024. "Science over Language: A Plea to Consider Language Bias in Scientific Publishing." *Critical Care Science* 36: e20240084en.

Gordin, Michael D. 2017. "Introduction: Hegemonic Languages and Science." *Isis* 108, no. 3: 606–611.

Gordin, Michael D. 2015. *Scientific Babel: How Science Was Done Before and After Global English*. Chicago: University of Chicago Press.

Gordon, Sarah Frances, and Emiliano Gutierrez Fierros. 2022, September. "Combatting Exclusionary Language Practices in Science Publishing: A DEI Concern." *Science Editor*: 84–85.

Hale, Sandra Beatriz, Nigel Bond, and Jeanna Sutton. 2011. "Interpreting Accent in the Courtroom." *Target: International Journal of Translation Studies* 23, no. 1: 48–61.

Hall, Stuart. 1997. "The Local and the Global: Globalization and Ethnicity." In *Dangerous Liaisons: Gender, Nation, and Postcolonial Perspectives*, 173–187. Edited by Anne McClintock, Aamir Mufti, and Ella Shohat. Minneapolis: University of Minnesota Press.

Hernández Bonilla, J. M. 2021, July 30. "How to End the Hegemony of English in Scientific Research." *El País*. Accessed December 10, 2024. Retrieved from https://english.elpais.com/usa/2021-07-30/how-to-end-the-hegemony-of-english-in-scientific-research.html.

Hevia, James L. 2003. *English Lessons: The Pedagogy of Imperialism in Nineteenth-Century China*. Durham: Duke University Press.

Hill, Jane. 2025. *The Everyday Language of White Racism*. Hoboken, NJ: John Wiley & Sons.

hooks, bell. 2014. *Teaching To Transgress*. New York: Routledge.

Ibero. 2024. "Requisito de inglés para posgrado." Accessed April 4, 2025. Retrieved from https://ibero.mx/alumnos-posgrados-requisito-obligatorio-de-ingles.

Jones, Taylor, Jessica Rose Kalbfeld, Ryan Hancock, and Robin Clark. 2019. "Testifying While Black: An Experimental Study of Court Reporter Accuracy in Transcription of African American English." *Language* 95, no. 2: e216–e252.

King, Sharese, and Samantha Jacobs. 2024. "Talk about Testimony: Courtroom Dialogue as Racialized Interactions." *Linguistics Vanguard* 10, no. s3: 265–272.

Kurinec, Courtney A., and Charles A. Weaver. 2019. "Dialect on Trial: Use of African American Vernacular English Influences Juror Appraisals." *Psychology, Crime & Law* 25, no. 8: 803–828.
Lee, Icy. 2014. "Publish or Perish: The Myth and Reality of Academic Publishing." *Language Teaching* 47, no. 2: 250–261.
López Ramírez, Mónica. 2017. "¿En México o en el extranjero?: tipos de toma de decisión de estudiantes mexicanos sobre el lugar para realizar su doctorado." [In Mexico or Abroad? : Types of Decision-making by Mexican Students About Where to Do Their Doctorates] *Sociológica* 32, no. 90: 145–180.
Lorde, Audre. 1984. *Sister Outsider: Essays and Speeches*. New York: Crossing Press.
Louth, John. 2015, August 4. "Guest Post: How Many International Law Books Are Published in a Year?" *OpinioJuris*. Accessed March 10, 2025. Retrieved from https://opiniojuris.org/2015/04/08/guest-post-how-many-international-law-books-are-published-in-a-year/.
Määttä, Simo K. 2024. "Invisible Multilingualism: Language Ecologies, Migration and the Administration of Justice." In *The Routledge Handbook of Translation and Migration*, 13–26. Edited by Brigid Maher, Loredana Polezzi, and Rita Wilson. London: Routledge.
Makoni, Sinfree B. 2012. "Language and Human Rights Discourses in Africa: Lessons from the African Experience." *Journal of Multicultural Discourses* 7, no. 1: 1–20.
McKeown, Cassandra L., and Michael G. Miller. 2009. "Say What: South Dakota's Unsettling Indifference to Linguistic Minorities in the Courtroom." *South Dakota Law Review* 54, no. 1: 33–88.
McMillan, Mason. 2021. "Judges be Trippin: A Legal Analysis of Black English in the Courtroom." *Tulsa Law Review* 57, no. 2: 451–488.
Medina, José. 2017. "Varieties of Hermeneutical Injustice." In *The Routledge Handbook of Epistemic Injustice*, 41–52. Edited by Ian James Kidd, José Medina, and Gaile, Jr Pohlhaus. New York: Routledge.
Mishra, Pramod K. 2000. "English Language, Postcolonial Subjectivity, and Globalization in India." *ARIEL: A Review of International English Literature* 31, no. 1–2: 383–408.
Moya, Sarah. 2021. "Language Barriers and Cultural Incompetency in the Criminal Legal System: The Prejudicial Impacts on LEP Criminal Defendants." *Fordham Urban Law Journal* 49: 401–433.
Niranjana, Tejaswini. 1990. "Translation, Colonialism and Rise of English." *Economic and Political Weekly* 25, no. 15: 773–779.
Noda, Orion. 2020. "Epistemic Hegemony: The Western Straitjacket and Post-colonial Scars in Academic Publishing." *Revista Brasileira de Política Internacional* 63, no. 1: e007. https://doi.org/10.1590/0034-7329202000107.
O'Barr, William M. 2014. *Linguistic Evidence: Language, Power, and Strategy in the Courtroom*. New York: Elsevier.
Olmos-Lopez, Pamela, Fátima Encinas Prudencio, and Ana Novelo. 2022. "Mexican Economics Professors' Publication: Three Case Studies." *English for Specific Purposes* 66: 131–143.
Ordorika, I. 2018. "Las trampas de las publicaciones académicas." [The academic publishing trap]. *Revista Española de Pedagogía* 76, no. 271: 463–480.
Packer, Abel L. 2016, May 10. "The Adoption of English among SciELO Brazil Journals Has Been Increasing." *ScieELO in Perspective*. Accessed March 10, 2025. Retrieved from https://blog.scielo.org/en/2016/05/10/the-adoption-of-english-among-scielo-brazil-journals-has-been-increasing/.
Phillipson, Robert. 2013. *Linguistic Imperialism Continued*. New York: Routledge.
Phillipson, Robert. 1992. *Linguistic Imperialism*. Oxford: Oxford University Press.
Politzer-Ahles, Stephen, Jeffrey J. Holliday, Teresa Girolamo, Maria Spychalska, and Kelly Harper Berkson. 2016. "Is Linguistic Injustice a Myth? A Response to Hyland (2016)." *Journal of Second Language Writing* 34: 3–8.
Rickford, John R., and Sharese King. 2016. "Language and Linguistics on Trial: Hearing Rachel Jeantel (and Other Vernacular Speakers) in the Courtroom and Beyond." *Language* 92, no. 4: 948–988.
Runcieman, Alan James. 2022. "Translanguaging in Court Proceedings: How Interpreter Pedagogy Needs to Address Monolingual Ideologies in Court Interpreting that Delegitimize Litigants' Voices." *International Journal of Interpreter Education* 14, no. 1: 4–16.
Salatino, Maximiliano. 2020. "Open Access in Dispute in Latin America: Toward the Construction of Counter-Hegemonic Structures of Knowledge." In *Digital Activism, Community Media, and Sustainable Communication in Latin America*, 125–148. Edited by Cheryl Martens, Cristina Venegas, Etsa Franklin Salvio, and Sharupi Tapuy. Cham: Palgrave Macmillan.
Sapir, Edward. 1929. "The Status of Linguistics as a Science." *Language* 5, no. 4: 207–214.

Smith, Olivia M., et al. 2023. “Peer Review Perpetuates Barriers for Historically Excluded Groups.” *Nature Ecology & Evolution* 7, no. 4: 512–523.
Smith-Khan, Laura. 2019. “Why Refugee Visa Credibility Assessments Lack Credibility: A Critical Discourse Analysis.” *Griffith Law Review* 28, no. 4: 406–430.
Spivak, Gayatri Chakravorty. 2003. “Can the Subaltern Speak?” In *Imperialism*, 66–111. Edited by Peter H. Cain and Mark Harrison. New York: Routledge.
Universidad Nacional Autónoma de México (UNAM). 2022. “Convocatoria de Ingreso a Doctorado 2022-2.” Accessed April 4, 2025. Retrieved from https://posgrado.unam.mx/filosofia/?p=7884#:~:text=Destacar%20antecedentes%20acad%C3%A9micos%20y%20experiencia%20profesional;%20incluir,RFC%20y%20la%20CURP%20no%20son%20obligatorios.
Universidad Nacional Autónoma de México (UNAM). 2021. “Programa de Maestría y Doctorado en Historia de la Universidad Nacional Autónoma de México.” Accessed April 4, 2025. Retrieved from https://historia.posgrado.unam.mx/idiomas/.
Whorf, Benjamin Lee. 2012. *Language, Thought, and Reality: Selected Writings of Benjamin Lee Whorf.* Cambridge: MIT Press.

33

DIVERSITY IN FORENSIC LINGUISTICS

A Call to Fulfill the Promise

I. M. Nick

Introduction

In courtrooms around the globe, the forensic sciences (FS) have established themselves as a staple part of modern jurisprudence. A direct reflection of this status is the seemingly exponential expansion in the number and variety of scientific disciplines that now have forensic applications. For each and every type of evidence that may be placed before the court, there appears to be at least one scientist who is willing to provide expert testimony. Once primarily restricted to the confines of human anatomy, modern FS encompasses an ever-growing compendium of scientific disciplines From art history, botany, chemistry, and data science to toxicology, virology, xylogy, and zoophysics, the range of scientific fields under the FS umbrella seems to expand with every passing day. Indeed, as Jay Siegel rightly posits (2025), "[a]lmost any science can be a forensic science because almost any science can contribute to solving a crime or evaluating a civil harm" (para 2).

One of the principal drivers of this disciplinary expansion has been the growing recognition of the elucidative assistance scientists can provide during official investigations and subsequent litigation. That development coupled with the general public's seemingly insatiable fascination with all things CSI has led to a veritable boon in FS education. What was once considered an odd to ghastly occupational choice is now widely recognized as an exciting and dynamic career opportunity. Accordingly, the number of FS programs has increased as well. In the United States, for example, since the introduction of a formal system of FS accreditation by the American Academy of Forensic Sciences (AAFS) in 2002, substantial increases have been seen in the number of accredited degree-granting programs. The rates of student enrollment and successful graduation have also risen as well (Jackson 2009). This is welcome news considering the growing demand for FS expertise both inside and outside of the courtroom. Despite these positive developments, FS has continued to suffer from a chronic lack of diversity in the persons who serve as practitioners, consultants, researchers, educators, and students. Importantly, this deficit has not escaped the notice of FS leaders and governmental officials alike.

Already in 2009, an official report issued by the National Science Foundation warned that FS would suffer if concrete steps were not taken to increase the demographic heterogeneity of current and future FS professionals. Nevertheless, nearly two decades later, the problem still exists. Moreover, in view of current developments, there is grave and justified concern that the small

DOI: 10.4324/9781003391074-36

advancements that had been made to increase diversity in FS are not merely at risk of becoming stagnated, but entirely nullified and discredited. The potential ramifications are not to be underestimated. As this chapter will demonstrate, the need for promoting the diversity of persons working within or training to become members of the FS community is not a matter of political cosmetics or the product of a so-called "woke" leftist agenda gone rogue. Rather it is a serious, apolitical, ethical imperative that has direct consequences for the current and future reliability, validity, accuracy, credibility, and propriety of both FS as a whole and each of its individual branches. This chapter will focus specifically on the ethical necessity of diversity as it pertains to the discipline devoted to forensic applications of linguistic evidence: forensic linguistics (FL). More specifically, the chapter will address the reasons why FL has an ethical imperative to actively promote the demographic diversity of the students and professionals who are active within this field. For the purposes of this discussion, this chapter adopts Cinaglia and Rabie-Ahmed's (2025) definition of "ethics". According to them, within applied linguistics, ethics "broadly speaking, can be understood as a set of virtues guiding human behavior to ensure that an individual's actions respect and positively impact both the self and the greater society" (310).

Using this pragmatic and powerful definition, this chapter is divided into three sub-sections. The first presents arguments why diversity in FL is essential to providing the highest level of linguistic analysis possible for the fair and just application of the law for the good of the people. The second sub-section of this chapter offers concrete suggestions for proactively increasing diversity in FL. And, in the third and final segment, the issues raised are contextualized within the ongoing anti-DEI (Diversity, Equity, and Inclusion) movement within the United States and beyond. Although the focus here will be firmly on the first of these, as this chapter will show, the three are inextricably linked.

Why Is Diversity an Ethical Imperative for Forensic Linguistics?

To Counteract Bias

In a word, one of the primary reasons why promoting diversity in FL is such an ethical imperative is "bias". In order to provide maximally reliable, valid, and accurate linguistic analyses for forensic purposes, it is essential that the prerequisite research conducted be minimally tainted by any and all forms of bias. That having been said, it is important to clarify what is precisely meant and not meant by the term "bias". Bias is not used here to refer to the deliberate falsification or manipulation of evidence with the intent to mislead, confuse, defraud, or otherwise deceive. Such forms of professional misconduct are clearly a legitimate cause of ethical concern, but are presumably relatively rare and are not the focus of this chapter. Instead, the term "bias" as used here refers to another type of threat which pervades scientific analyses and may fatally undermine the probative value of FL evidence and expertise. For the current discussion, "bias" refers to "the class of effects through which an individual's pre-existing beliefs, expectations, motives, and situational context influence the collection, perception, and interpretation of evidence" (Kassim, Dror, and Kucklicka 2013: 42). Like so many aspects of the psyche, it remains unclear whether biases are biologically determined, socially conditioned, experientially learned, or some combination thereof (Grawitch 2020; Wilke and Mata 2017). However, whatever their origin(s), it is generally agreed that biases appear to be a universally integral part of cognition. Although "Confirmation Bias" is fairly well recognized in FL and has been thematized (Leeman et al. 2025; Ainsworth 2022; Picornell 2022; Greenlee 2021), there are in fact a great variety of other cognitive "shortcuts" that are routinely used by people to reduce the number of possible paths to a solution and simplify information processing (Anderson 2003; APA Dictionary).[1]

On the one hand, the use of such strategies offers many advantages in our everyday lives. These include decreasing the overall cognitive load, identifying potential patterns and relations, anticipating potential risks or conflicts, and determining what information is important to retain and what may be rejected as "noise" (Pinto 2023; Spellman, Eldridge, and Bieber 2022; Grawitch 2020; Gigerenzer and Gaissmaier 2011; Garcia-Retamero et al. 2007). Clearly, within FS and FL, these advantages are of enormous potential benefit as they may help to increase the overall efficiency and speed of thought-processing. On the other hand, without proper controls, biases may also lead to drawing faulty conclusions on the basis of unforeseen errors and/or unsubstantiated premises. Worse yet, they can also prevent the recognition, reappraisal, and/or reversal of faulty decision-making, causing the person(s) involved to zealously defend previous decisions, even in the face of compelling contrafactual evidence or exposure of significant errors in logic and/or procedure. Ironically, the very same cognitive processes which have evolved to assist in thought-processes may also impede them. The potential repercussions are not to be underestimated. When left unchecked, "[B]ias can impact the actual observation and perception of the data, testing strategies, as well as how the results are then interpreted and how conclusions are reached" (Dror 2020: 7998). While the potential negative impact of bias is great within any field of scientific investigation, within FL—and all the other branches of FS—the repercussion can be nothing less than short of catastrophic. As celebrated linguist John Baugh (2018) explains: "linguistic research has lethal relevance: that is, the linguistic evidence and corresponding expert testimony" provided by the forensic linguist can be "crucial to determining the difference between a murder conviction, and potential death sentence, and exoneration" (26). Although, admittedly, not every case utilizing FL analysis may involve a possible death penalty, for the persons involved, the consequences may still be major and long-lasting.

Acutely aware of the risks and responsibilities involved, it is not uncommon for FL experts to stress the importance of remaining hypervigilant to the effects of bias. However, such admonitions, no matter how well-meaning they may be, betray a fundamental misunderstanding of the nature and operation of bias. As stated above, bias is a natural, automatic, and unconscious part of cognition which cannot be negated through mere concentration or goodwill (Cooper and Meterko 2019; Kassim, Dror, and Kuklicka 2013). Indeed, there is even evidence that heightening people's awareness to the potential of bias may actually increase their propensity to rely on heuristics. In much the same way as telling people not to think of the proverbial Barbie-colored elephant will immediately conjure up images in the minds of most people, the same appears to happen when people are told not to be biased—like a cognitive reflex, where no amount of warning, education, or experience can guard against it. Thus, simply instructing scientists to remain vigilant against possible bias is relatively useless.

The same is true when it comes to another popular, yet ill-informed, piece of advice: namely, that experience and expertise are powerful weapons for combating bias. This misconception may be propagated by those who readily concede that cognitive bias is a legitimate professional concern, but thanks to their extensive training and experience, neither they nor their work are substantively affected by it. The truth is, no amount of education can guard against it. Indeed, "although experts tend to be very confident (sometimes even overconfident), more experienced experts can actually perform worse than novices" (Dror 2020: 7999). Emerging scholars and practitioners in FL may seek to compensate for their lack of experience and expertise by taking care to assiduously follow established protocols and procedures, seek outside assistance from peers and supervisors, ask questions to avoid misunderstandings, consult and check their results multiple times, openly point out areas that they are feeling unsure about, and be receptive to trying new methods or considering alternative interpretations. Though employed to reduce feelings of insecurity, these strategies may also help to decrease errors and cognitive bias (van den Eden, de Poot, and van Koppen

2019; Adelson 1984). By comparison, research has found that, particularly under time pressure, experts may rely too heavily on their past experience, be resistant to trying new methods or considering alternative theories, may skip certain steps which they feel are unnecessary, and draw conclusions prematurely, dismissing the significance of evidence that does not support their initial hypotheses (van den Eden, de Poot, and van Koppen 2019; Almond, et al. 2011; Klein, Moon, and Hoffman 2006; Dreyfus and Dreyfus 1988; Kruglanski and Freund 1983). The point here is not to say that experts perform worse than novices. Thanks to the experience, knowledge, and "expertise-based intuition" (Salas, Rosen, and Diaz Granados 2010) they have been able to amass, they usually outperform novices when it comes to rapidly assessing a situation, devising the most effective solution to a problem or detecting patterns (van den Eden, de Poot, and van Koppen 2019; Salas, Rosen, and Diaz Granados 2010; Andre and Fernand 2008; Ericsson 1998; Lemaire and Siegler 1995; Bédard and Chi 1992; Adelson 1984). Rather it is to say that experts are not immune.

There are certain factors which may make experts especially susceptible to having an illusion of control. For example, as a function of their elevated professional standing and institutional responsibilities, experts may face heightened occupational pressures, such as exceedingly demanding workloads coupled with requests to take on especially challenging high-profile cases that can garner extreme professional and personal scrutiny. Under such high pressure conditions, psychological research has found that there is a heightened tendency for the brain to utilize cognitive biases at its disposal "with the goal of making decisions more quickly, frugally, and/or accurately than more complex methods" (Gigerenzer and Gaissmaier 2011: 454). What's more, thanks to the combination of their professional standing in the field and their unusual discretionary powers to make decisions that can directly affect others' careers, both long term and short term, the potential of biases impacting their scientific actions, opinions, and/or decisions may remain undiscovered or unchallenged out of respect and/or fear (Dror 2020; Kassim, Dror, and Kuckucka 2013).

These external factors, when combined with certain counterproductive internal personality traits (e.g., conflict avoidance, an unwillingness or inability to accept critical feedback, and an inflated sense of self-importance) may serve only to exacerbate the problem (Dror 2020; van den Eeden, de Poot, and van Koppen 2019; Kassim, Dror, and Kuckucka 2013). Still, the misconception that experts are by definition immune to bias is extremely widespread. So much so that psychologists have given this phenomenon its very own name: "the blind spot bias". A study conducted in 2023 brings into sharp relief how prevalent this particular fallacy is in FS. In the investigation, FS professionals across multiple disciplines were asked how much bias was a cause for concern. In total, 70% admitted that bias was a serious concern for FS in general. By comparison, only 52% saw it as being a reason for concern in their own forensic domain; and only 25% responded that bias was an issue in their own research. In other words, although the majority conceded that cognitive bias is "a thing", only a minority saw it as being personally relevant (Adelson 1984). These findings are eerily reminiscent of what survivors and witnesses of a crime exclaim in subsequent news interviews: "We never thought it would happen HERE!" However, the truth is, crimes happen everywhere: no place is immune. And the very same is true of cognitive bias. No one is safe or immune. So, if hypervigilance, education, and experience are not sufficient to counteract the unwanted effects of bias in science, what can FL do?

Luckily, generations of scientific investigation into scientific investigations have revealed concrete strategies to minimize the negative effects of bias in science. These include the institution of systems of oversight implemented by appropriately trained professionals; the independent testing of findings through replication of the same or similar sets of data employing the same or different methodologies; double blind review processes; periodic, unannounced checks of adherence to required protocols; and clear and comprehensive guidelines for best practices and procedures (Rojas Alfaro et al. 2025; Kunkler and Roy 2023; Zapf, Patricia, and Dror 2017; National Research Council 2009). However, if the persons allocated to perform these and other checks and balances suffer from the same set of biases as the original scientists, the effectiveness of these precautions may be substantially undermined—thereby replicating and reinforcing the same errors originally made. This is one of the reasons why the strategy of sending a case to a trusted colleague as an added "check" may not be a sufficient control, especially if that colleague has the same background, training, experience, perspective, as well as a similar to identical demographic background—as is so often the case in people's professional social circles where "like attracts like". As research has determined, this bias toward ingroup favoritism can be unexpectedly counterproductive.

For example, it has been repeatedly found that heterogeneous teams consistently outperform their homogeneous counterparts in terms of the accuracy, reliability, creativity, productivity, profitability, and innovation (Yin et al. 2022; Ding et al. 2021; Jones, Chirino Chace, and Wright 2020; AlShebi, Rahwan and Woon 2018; Smith Doerr, Alegria, and Sacco 2017; Paulus, van der Zee, and Kennworthy 2016; Freeman and Huang 2014; Campbell et al. 2013; Østergaard, Timmermans, and Kristinsson 2011; Herring 2009; Page 2008; Hong and Page 2004). This superior performance appears to be positively related to an elevated propensity to avoid, detect, and correct errors in procedure and logic. As Rosenbaum (2010) explains: "[b]ecause there are many sources of bias when conducting research […] the optimal approach should include multiple perspectives and approaches to conceptualization, development, execution, analysis, interpretation, and application of research data" (146). In short, diversity can act of a natural and effective system of checks and balances.

However, the benefits do not end there. Research has also determined that scientific teams that are composed of a diverse set of members also have significantly higher levels of productivity and reputational status as measured by the number of publications, readers, and citations in high impact references (Smith Doerr, Alegria, and Sacco 2017; Freemann and Huang 2014; Herring 2009). It is important to note that the advantages that come from diverse scientific cooperations may have a shelf-life as familiarity and conformity set in. Like a kind of behavioral regression toward the mean, initial differences may begin to flatten out, as team members fall into established routines over time. Admittedly, it can be extremely comfortable when everyone has their fixed position, follows their familiar routine, and the team itself works together like a well-oiled machine. That level of automaticity, however, is exactly where the danger of personal bias can become the most acute, as assumptions are made and heuristics are used. A regular infusion of new partners, with different perspectives and different ideas can provide a powerful anecdote.

Importantly, the benefits of diversity have been replicated over many decades, by different researchers, across multiple fields, utilizing differing methods, and conducted in many different countries and regions. The findings speaking in favor of promoting diversity in science are consistent, robust, and compelling. For example, in an investigation of 9 million papers and 6 million scientists, Alshebi, Rahwan, and Woon (2018) found that research impact was positively correlated not only with diversity overall, but also with different types of diversity (e.g., ethnicity, gender, academic affiliation, etc.). As a result, the team concluded: "good things happen when people of different backgrounds, cultures, and ethnicities come together to work towards shared goals and the common good" (9). Such findings are of critical importance, as far too often, arguments against promoting diversity are based on the erroneous, often unconscious, notion that welcoming persons from marginalized communities into the FS/FL community entails lowering standards. Aside from the obvious questions of whose standards, established using what norms, this falsehood completely ignores the tremendous contributions made by members of marginalized communities in the past; and it also fails to recognize the tangible advantages these voices bring with an eye toward the future of FL.

However, the promise of diversity can only come to fruition when the systems in which FL students and professionals operate truly value and encourage the expression of different perspectives, opinions, methods, approaches, experiences, and expertise. When the price for participation becomes conformity and unquestioned allegiance to historically imposed norms, then promise of diversity may be effectively nullified or negated. This warning is not only relevant for protecting so-called "visible minorities" but also those whose minority status may not be immediately apparent. There are reasons, for example, why one study determined that 40% of queer scientists elect to keep their sexual orientation a secret from their co-workers (Sawyer 2023). In many cases, it is more than a desire to keep their private life private. Of course, there are heterosexual colleagues who may also choose not to discuss their home lives at work. However, in comparison to peers who are members of the LGBTQ+ community, such disclosures are generally not accompanied by justifiable fears over the possible negative consequences for their livelihoods or even lives. As Sawyer's 2023 investigation revealed, one of the primary reasons why LGBTQ+ scientists decide not to reveal their identity is feeling insecure, unsafe, or fearful of losing their job. The same trepidations may plague scientists whose minority status may also not be readily identifiable (e.g., those with neurodivergence and/or mental health challenges). To truly ensure equal access and opportunity based on talents and skill, it is imperative that factors that inhibit diversity in the workplace be removed.

This includes the dismantling of the very tangible, structural barriers to diversity that can stand in the way of individual and collective progress. It is striking, for example, that despite the fact that Gallaudet University awards amongst the highest number of university degrees in language and linguistics in the United States (Data USA), the number of forensic linguistics studies that address sign languages or the Deaf remains far too low (Carney 2019; Seaborn, Andrews, and Martin 2010; Lucas 2003); and research cooperatives which integrate Deaf linguists are few. Indeed, the same can be said of other forms of disability, and this in spite of the fact that persons with disabilities are often disadvantaged in the criminal justice system and have high rates of victimization (Eades 2008). This means that, despite the skills, talent, and training of this community, they are consistently overlooked by the forensic linguistic community. As Mallinson, Hennert, and Buchholtz (2024) remind: "[b]eing disabled or living with a disability in linguistics, as in the wider world, requires navigating on a daily basis a host of such seemingly small exclusionary practices that together they accumulate into system exclusion" (11). Promoting diversity therefore necessarily also requires ensuring equal access, and not because it is the charitable thing to do, but because it is the right thing to do.

Moreover, as many a member of a marginalized community can attest, access is only the first step along a very long road to justice. It is not, therefore, enough to simply increase the number of minoritized peoples into FL, if the systems, practices, and attitudes which initially served as barriers are either left unchanged, or even reinforced in a kind of reflexive backlash. As Mallinson et al. (2024) sagely remind: "sentiment is not enough to challenge longstanding ideologies, to disrupt structural inequities, and to ensure full inclusion and participation: transformative change requires hard work and skill" (16). What's more, in light of the dramatic increase in the ferocity of attacks against diversity in the sciences and beyond, what is also needed is courage to seek and protect the truth. As Alan Garber, President of Harvard University, valiantly asserts:

> Our motto—Veritas, or truth—guides us as we navigate the challenging path ahead. Seeking truth is a journey without end. It requires us to be open to new information and different perspectives, to subject our beliefs to ongoing scrutiny, and to be ready to change our minds. It

> compels us to take up the difficult work of acknowledging our flaws so that we might realize the full promise of the University, especially when that promise is threatened [...] All of us share a stake in safeguarding that freedom. We proceed now, as always, with the conviction that the fearless and unfettered pursuit of truth liberates humanity.
>
> *(2025, paras 5 and 10).*

Be it in a university, a courtroom, or the seat of government, the wisdom of Garber's statement has relevance for all arenas of truth-seeking; and the bravery demonstrated in issuing the statement, at a time of such great personal and professional peril, is as laudable as it is inspirational. Yes, the price of standing up for diversity, equity, and inclusion is often high, but the price of failing to do so is often much, much higher....

To Meet the Growing Demand for Professional FL Expertise

In the past, it was a common assumption amongst lawyers and judges that seeking the expertise of linguists was unnecessary, if not entirely inappropriate, given their own specialized knowledge of language and the law. Such opinions were frequently based on misconceptions of what the field of linguistics is and what precisely linguists do. As many in the field can confirm, linguists are not infrequently confused with translators, interpreters, grammarians, and lexicographers. However, as understanding about linguistics has increased, so too has the professional appreciation for the sophisticated analyses linguists can provide. Understandably, these changes have been accompanied by a continual demand for their services, both inside and outside of the courtroom. As John Baugh observed already in 2018,

> a large number of police officers, judges, law professors and law students, as well as many social scientists and policy makers, are showing greater interest in the ways in which forensic linguistics might prove beneficial to them in their specific professional contexts.
>
> *(180)*

Nearly a decade later, the appreciation for what FL has to offer has by no means abated.

Today, it is not uncommon to find FL experts working as consultants for individuals and agencies in a wide variety of roles in an increasing number of countries around the world. The rising demand for FL services is a part of the overall trend seen for all of FS. According to data collected by the US Bureau of Labor Statistics, FS "is currently one of the fastest growing industries in the country" (Taylor 2025). Between 2023 and 2033, it is expected that the US employment market for FS will grow at least 14%, while the general growth rate for all other professions in the nation for the exact same time period is projected to reach no more than 4% (Taylor 2025). Despite these positive employment prospects, for many decades now, the supply of trained professionals who can provide these highly specialized scientific services continues to lag behind the demand. In many jurisdictions, the repercussions of this disparity can be seen in the veritable avalanche of backlogged forensic evidence waiting to be processed; the crushing caseloads of FS experts who have the prerequisite expertise; and, sadly also, the retention of consultants who lack the skills to perform the proper analyses (US Department of Justice 2023). In recognition of the risks these circumstances bring with them, there have been repeated and vociferous demands to increase the number of FS professionals across all of its various disciplines.

With that goal in mind, it has been regularly recommended that students from marginalized communities be actively recruited to help meet the ever-growing demand for expert FS services. Such recommendations are not a matter of political expediency as critics of diversity initiatives may charge out of their own ignorance or deep-seeded animosity. To meet its professional mandates, FS quite literally *needs* the talents, skills, intellect, and creativity that the full demographic spectrum of the global community has to offer. This is precisely the point that Smith Doerr, Alegria, and Sacco (2017) make. They write: "the question of why or why not foster diversity in science teams is a false choice. It is a practical reality that the pool of qualified potential scientific workers in the US is increasingly diverse" (140). The failure or refusal to recruit from this expanding pool of talent may therefore constitute a nonsensical act of professional masochism.

In view of that reality, many local, state, national, and international programs have been devised to stimulate greater diversity in the FS workforce. Despite these initiatives and decades of governmental goodwill, only limited gains have been made. In the United States, irrefutable evidence of this ongoing disparity can be seen in the demographic composition of the workforce in the so-called STEM (science, technology, engineering, and mathematics) professions, the basis for many FS disciplines. In Figure 33.1, labor statistics gathered by the US National Science Foundation are presented for 2011 and 2021.

As can be seen in Figure 33.1, while the percentage of persons who identify as being White has dropped by 10% over the space of a decade, only minimal gains have been made across the other demographic categories. Unfortunately, similar statistics are not, to the authors knowledge, kept on the employment or education of persons in FL. However, an examination of data gathered by the US Census Bureau on the demographic distribution of US students who graduate with a doctoral degree in language and linguistics reveals an analogous national trend (see Table 33.1).

Here again, although gains have been made in the percentage of marginalized groups, they have been disappointingly low. It must also be remembered, of course, that the figures given in Table 33.1 are for both languages and linguistics. Thus, it can be assumed that the actual increases for US graduation rates in linguistics are even lower, and the percentage specializing in forensic

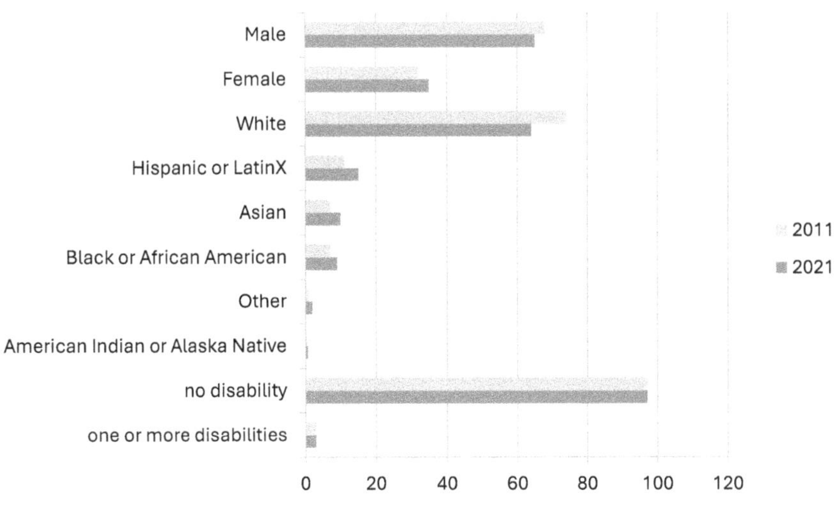

Figure 33.1 Demographic Composition of the US STEM Workforce in 2011 and 2021 by Percentage

Table 33.1 Gender and Ethnoracial Identity of the US University Graduates with a Research Doctorate in Language and Linguistics by Percentage in 2012 and 2022

	Female		*Male*	
	2012	*2022*	*2012*	*2022*
White	27.3	20.8	19.4	17.7
Hispanic/Latinx	5.95	6.33	2.73	3.75
Asian	2.89	2.21	1.29	1.25
Black/African American	0.72	0.77	0.56	0.48
American Indian or Alaska Native	0.16	0	0	0.10
Two or More Races	0.64	0.77	0.24	0.77
Hawaiian or Pacific Islander	0	0	0.08	0

Source: Data USA.

linguistics is lower still. This inference is buttressed by the fact that, according to US Census data, the most commonly reported occupations for persons with a degree in "language and linguistics" are elementary education and middle school education. Other popular jobs for this group of graduates are proofreading, copyediting, and librarianship. When one considers the fact that the overall percentage of language and linguistics graduates entering the US workforce is less than 2%, it is clear that the objective of substantially increasing diversity in FL is far from being met.

These findings should be deeply concerning for anyone interested in securing the longevity and vitality of the discipline. The question to be asked then is *why* has FL—or FS for that matter—made so few inroads in increasing the number and percentage of minorities? As Wagstaff and Laporte already observed in 2018, the continuing "lack of representation is not due to a lack of ability or interest. Instead, it is primarily due to discouragement at every juncture along the educational and career development pathway" (4). Without doubt, the serial micro- and macro-aggressions which many persons from marginalized communities are made to endure may have a cumulative negative effect upon their successful recruitment and retention, particularly when the significance or very existence of these forms of abuse are repeatedly ignored or denied (Mensitieri, Boroş, and Toma 2025; Sawyer 2023; Williams 2021; Galupo, Paz, and Resnick 2016). Research has demonstrated that when left unchecked, such occupational toxicity can "lead to increased burnout, delayed or blunted professional success, or, in the saddest of cases, individuals leaving the profession and taking their unique ideas and perspective with them" (Parikh and Leschied 2022: 1719). In view of such research, advocates for re-humanizing linguistics have argued that the values of reciprocity, rights, regulations, responsibility, social relationships, and respect—the so-called "R-words"—must be brought to the forefront (Holton, Leonard, and Pulsifer 2022; Czaykowska-Higgins 2018). While these calls have focused on improving the ways in which linguists interact with minorities outside of the field, the same applies when interacting with minorities within the field. To be fair, it is important to acknowledge that the prevalence of such indignities is by no means restricted to FL. Across academia, the incidence and severity of offensive, belligerent, intolerant, and injurious attitudes and behaviors continue to increase (Mahmoudi and Keashly 2021; Keashly and Neuman 2013; Hinchey and Rimmel 2000).

Nevertheless, there are issues which are peculiar to FL that may serve as a unique impediment to increasing diversity within the field. The close associations which FL tends to enjoy with the criminal justice system is one factor, in particular, that deserves far more critical attention. It is not unusual, for example, for FL experts to share their expertise with members of the law enforcement (LE) community. Without doubt, such work has been instrumental in improving LE training and procedures in ways that make policing not only more effective but also transparent, fair, and sensitive for the communities they are bound to serve. The same may be said of FL cooperations with other organs of the state. From the courtroom to the halls of governance, FL has been instrumental in helping to bring about substantive and much-needed improvements to legal procedures, policies, and decision-making. This work has not been purely altruistic, however. The resulting professional exchange of information and ideas has been mutually profitable. FL experts who manage to earn the respect and trust of officials may be rewarded handsomely in various ways (Guillaume, Sidebottom, and Tilley 2012; Rojek, Alpert, and Smith 2012). For example, they may be granted access to highly sensitive data; extended invitations to work on subsequent projects; receive requests to serve as independent paid consultants on high-profile cases; and be recommended to other officials and organizations that may yield similar benefits. Overtime, these advantages can translate into more rewards both within their home institutions (e.g., grants, fellowships, raises, renewed contracts, tenure, funds for travel, equipment, publishing, and personnel) and outside of their affiliation (e.g., media inquiries for interviews, requests to serve as a public speaker, invitations to join in policy-making). In view of these advantages, it comes as no surprise that those entering the field may also look for opportunities to form similar professional alliances of their own.

On the surface, there would seem to be nothing nefarious about such networking. However, upon reflection, there may be hidden drawbacks for the field. First and foremost, it cannot be forgotten that the overreliance of information gathered from one type of stakeholder may be biased by the methods, procedures, and perspectives of the original data-collector—even before FS experts begin their own analyses (Buerger, Levin, and Myer 2018). For example, data provided by law enforcement may contain case details which may subsequently trigger cognitive bias in subsequent information users (Zapf and Dror 2017). This contamination can then snowball with one bias leading to and amplifying another (Starr 2022). To guard against this potential, it is imperative to employ procedures which can effectively insulate the specialists responsible for the FL analyses. However, other forms of bias may not be so easily resolved. As Rosenbaum (2010) warns: "there is a risk of crossing the line and engaging in law enforcement (rather than research) functions (e.g., helping to develop lists of suspects or hot spots for targeted enforcement), which raises ethical concerns" (146). Even if these professional lines are not crossed, close collaborations between FL experts and state officials can lead to dependencies and expectations of reciprocity.

It is not unheard of, for example, for forensic experts to feel compelled to "produce" certain results. An equally inappropriate request would be to either downplay the importance of potentially "inconvenient" findings or to refrain from mentioning them at all, unless directly asked. In a UK-based investigation, it was determined that 28.6% of the forensic scientists surveyed admitted that they felt pressure to produce particular results (Geddes 2012). Similar results were obtained in the Bond Solon annual survey of expert witnesses conducted in conjunction with the Law Society of England and Wales. According to the 2024 survey of 557 expert witnesses, 36.96% indicated that in the past year they had come across a "hired gun" and 23.28% reported that they themselves had been placed under undue pressure from instructing solicitors to change their opinion in the last year. Clearly, this is an untenable situation. However, it is important to stress that the danger is not limited to overt pressure to conform. Professional interdependencies and personal affinities may also trigger a host of cognitive biases. As Sir George Jessel sagely warned: "Undoubtedly there is a natural bias to do something serviceable for those who employ you and adequately renumerate you" (Bernstein 2008: 105). Even if adequate precautions are taken to minimize the effects of these biases, the simple existence of collaborations can raise questions of propriety and neutrality. Andrea Roth, professor of law and Chancellor Chair in Criminal Justice at the University of Berkeley California's Department of Law even explicitly advises defense attorneys to consider these potential improprieties during direct and cross-examination of the government's forensic experts. Roth writes (2012): "Counsel should be prepared to cross-examine experts about topics as varied as chain of custody; credentials; limits of expertise; *failure of the government's procedures to correct for cognitive bias; other arguable forms of potential bias, such as affiliation with law enforcement* [...] [emphasis added]" (320). Such questions concerning partiality introduced through collaboration are also critical when it comes to the issue of promoting diversity.

The extremely close and widespread alliances that FL has formed with powerful government stakeholders such as the court and law enforcement may risk alienating individuals and communities who have been historically disempowered by those same entities. Even if great care is taken to maintain appropriate professional distances, the mere appearance of partiality is sufficient cause for concern, particularly during a period of time when so many deeply disturbing incidents involving officials abusing their powers to threaten minority communities have been revealed and so many of the legal protections designed to address discrimination have been reversed. In such a climate, the fact that FL has aligned itself so closely with the state may be decidedly off-putting for marginalized peoples. The point here is not to suggest that FL experts refuse to work on collaborative projects with governmental officials, agencies, or institutions. Instead, it is recommended that more critical attention be given to this practice and its potential impact upon the reputation of the field and the people who do and do not work within it. The same critical questions must be asked with regard to which groups are profiting from the FL research and which groups may be harmed, no matter how indirectly. It is all well and good to strive for neutrality when it comes to conducting research; but it is equally essential that scientists keep their eyes open to see who is using their findings and for what purposes. Precisely this point is made by John Rickford and Sharese King in their seminal 2016 piece, "Language and Linguistics on Trial". As they explain, in view of the millions of people who do not have ready access to linguistics but "who could benefit from the expertise and experience of linguists, we feel that it is not just their languages and dialects that are on trial, but linguistics itself, and that we can and should do more" (981). There are some FL experts who share this same opinion and will occasionally offer their services for free or at a significantly reduced rate. Unlike other professional disciplines like medicine and law, there are no general expectations or requirements to offer pro bono work.

Moreover, frequently, members of minority communities simply lack the financial wherewithal to solicit the services of an FL expert. As a result, the information provided by FL is most commonly out to use by stakeholders with more economic freedom, such as the state. The ubiquity of this reality across the many disciplines of FS has, according to some critics, resulted in a decidedly and somewhat unsavory "pro-prosecution bias" (Whitman and Koppl 2010). The impact on the resulting public image of FL appears not to have been universally positive. As former IAFL(L) President Ronald Butters noted nearly ten years ago in his 2012 "Closing Address", the FL community is often perceived as "'hired guns' who merely pay cynical lip service to ethical considerations" (352). The misconception that FL is only for the powerful is extremely regrettable as the insights and knowledge that FL has to offer can be of enormous benefit in helping to expose societal wrongs (Baugh 2018; Eades 2012, 2010) and protect against miscarriages of justice. The potential is exemplified by "Justice Project", a collaboration between the famed Innocence Project and the Forensic Linguistics Institute at Hofstra University in New York.

Led by FL experts Dr. Robert Leonard and Dr. Juliane Ford, in conjunction with Hofstra Law Professor Eric M. Freedman, this Civil Rights initiative engages faculty and students to re-examine capital cases in which language evidence played a decisive role in the state securing a conviction.

A recent case taken on by the project was that of Melissa Lucio, a Mexican-American mother who was put on trial for suspicion of murdering her two-year-old daughter. During the trial, the jury was presented Ms. Lucio's confession to the police. That confession played an instrumental role in the jury's verdict. In 2008, Ms. Lucio was found guilty and sentenced to death. However, after providing a detailed linguistic analysis of the police interrogation, Hofstra FL experts in the Justice Project were able to uncover evidence consistent with a coerced confession. This evidence was presented to the court and resulted in a stay of execution, just two days before it was scheduled to take place. On October 16, 2024, 16 years after her initial conviction, Judge Arturo Nelson ruled that Ms. Lucio was innocent and that there was such "clear and convincing evidence that no rational juror could convict" her of murdering her child (Death Penalty Information Center 2024, para 1).

This is just one of many cases in which analyses provided by FL experts have proven crucial to protecting civil and human rights. For example, FL has helped to expose linguistic discrimination in employment, housing, and education (Wright 2023; Baumgarten, DuBois, and Gill 2019; Baugh 2018; Skutnabb-Kangas and Phillipson 1998); identify and change long-standing unscientific, unjust, and harmful practices in police interviewing procedures (Haworth et al. 2023; Byrman and Byrman 2018; Ehrlich, Eades, and Ainsworth 2016; Berk-Seligson 2009; Ainsworth 2008; Aldridge 2007; Ainsworth 1993); address problematic practices in the linguistic assessment procedures used for asylum seeker applications (McNamara and Schüpbach 2018; Rosenhouse 2017; Eades 2005; Maryns 2004); improve methods for more effectively and clearly communicating people's legal rights (Eades 2018; Solan 2010, 2001; Shuy 1998); help change discriminatory linguistic practices used in questioning users of minoritized languages in courtroom settings (Eades 2018, 2000; Berk-Seligson 2012; Lucas 2003); and draw attention to and make substantive recommendations for ending inadequacies and inconsistencies in transcribing, recording, interpreting, and translating legal policies and procedures (Olsen and Kermit 2015; Ali and Algane 2013; Berk-Seligson 2012; Hale 2011). Were this aspect of FL brought more to the attention of the general public, it might help attract a greater diversity of talented newcomers who are interested in helping to combat social injustice, which they and their communities of origin have been made to suffer…one generation after another. As Mallinson et al. 2024 observe in *Inclusive Linguistics*, it is not unusual to find that scholars from marginalized communities are "often more interested in exploring applied research, research on education and pedagogy, community-centered collaborations, public-facing work" (435). There are those who may bristle at the idea of FL being used in this manner. Being a social advocate, they might say, should be kept wholly and completely separate from being a forensic scientist. If one were to accept that criticism, however, then one would also by extension have to condemn the use of FL to assist the workings of the state and its representatives. As that appears to be an acceptable, widespread, and valued application, it would seem logical, fair, and ethical to provide the same service for the people the state is sworn to serve. Again, this does not mean manipulating data analysis to serve some political aim, whatever it may be. Instead, this chapter argues that a prerequisite to advancing the call for justice for everyone is providing forensic linguistic services of the highest possible reliability, accuracy, and transparency for all stakeholders and not only a select, privileged few. The more FL succeeds in reaching that goal, the higher its chances of attracting a greater diversity of professionals who will bring with them new questions, new methods, new applications, and new insights that can enrich the discipline and the global community.

To Increase Our Understanding of Languages and Language Communities

According to the latest estimates by Ethnologue, there are approximately 7,000 languages actively being spoken by 3.6 billion people on the planet today. Added to that number are 300 different sign languages used by more than 72 million deaf and hard-of-hearing people around the globe (United Nations 2024); and ca. 4,153 attested writing systems (Ethnologue). Despite this impressive variety of communication, by and large, FL has tended to focus its attention on a small fraction of the world's languages and language users. The findings that have been generated from this select set have then tended to be projected upon the others, for better or worse. While it is reasonable to assume that a certain degree of similarity exists between the languages, it is also logical to expect that are many key differences as well. What precisely these differences are, and if, when, how, and how much they may play a role in FL analyses, is difficult to impossible to determine, given the glaring lack of information we have available about most of the world's systems of communication.

The investigative myopia that has characterized not only FL, but also linguistics in general, has very deep and often disturbing historical roots. For much of the discipline's history, the primary focus has been on spoken and written languages with European roots. When other languages were studied, it was more often than not to point out their perceived relative deficiencies thanks to the presumed intellectual inferiority of the peoples who used them. To describe this widespread form of prejudice and discrimination, in 1988, Tove Skutnabb-Kangas coined the term "linguicism", a concept "analogous…to *racism, sexism,* and *classism*, etc. (and co-articulating with these)"; this term refers to the "ideologies, structures, and practices which are used to legitimate, effectuate, and reproduce an unequal division of power and resources (both material and unmaterial) between groups which are defined on the basis of language" (93). Thanks to generations of sustained social resistance, much of the most vile and virulent vocabulary has been banished from scientific discourse. Many of the underlying tenets of linguicism still, however, remain in force, as evident by the continued privileging of a few languages and language users. Chief among them at present is undoubtedly English.

This means that in order to have their work recognized, non-English-speaking FL researchers are forced to accommodate. While this situation is often excused with the argument that the use of a single language actually makes international communication easier, those who are forced to bear the brunt of this linguistic monopoly rightly ask *easier for whom?* As Ingrid Piller wrote: "as appealing as the notion of a global lingua franc and 'English for everyone' may be, in reality, discourses and practices related to the global spread of English have become a key mechanism to entrench global inequalities" (2016: 334). Although the imposition of "English First" policies can be seen throughout the scientific world, it may pose a particularly onerous obstacle to diversity within FL.

Take for example, publication. Many of the world's leading scientific journals in FL are English-only. That in and of itself is not exceptional. However, the fact that primary readership of these publications is made up of *language* experts, many of whom may also be native English speakers, means that the bar for acceptance may be unusually high. In hopes of overcoming that threshold, it is not unusual for non-English-speaking FL experts to seek out the services of a professional proofreader. However, this option is not financially feasible to all academics. "Academics that choose an alternative route and publish in non-English language journals are at a disadvantage as their work is often cited less and overlooked in the international community" (Gordon and Gutierrez 2022: 84). A similar problem occurs when FL scholars attempt to gain access to potentially important publications for their work. Increasingly, these resources are locked behind expensive paywalls, the fees for which will in all likelihood continue to rise as the publication market continues to try to grapple with shrinking revenues and rising production costs. Conferencing, another traditional avenue to information exchange, is also marked by many of the same hindrances. During international events, English is commonly treated as the de facto lingua franca. Even when other languages are expressly encouraged by the organizers, attendees may still feel compelled to use English, lest they be perceived as being clannish by others. The pressure to conform in these professional spaces may be further exacerbated by departmental or university policies which rate scientific outputs (e.g., publications, conference papers, courses taught) more highly when they are in English, as opposed to any other language.

While for some, this situation may pose no problem at all or only mildly taxing; for others, these circumstances are not simply an issue of inconvenience. They are a manifestation of neo-colonialist attitudes which especially and systematically disadvantage already marginalized peoples. In frustration, those marginalized may create their own professional circles, where their language, cultures, perspectives, histories, and personal identities are better respected; and their scientific research is placed firmly in the center of the discussion. Along with the stimulating and fruitful exchange of ideas and information, these new spaces also offer the emotional supports needed to thrive in what would otherwise be a very isolating and alienating professional world. At the same time, the fact that these needs are not being answered within the larger FL community reveals that there is much more to be done to make FL a space that is truly inclusive and welcoming for all peoples. Again, this not a phenomenon that is restricted to the specific branch of FL but is one that has negatively impacted the entire discipline of Linguistics. Institutional denials of the significance of the scope of this problem in linguistics combined with a perpetual "lack of critical attention on the nature of power in the academy and how it shapes scholarly interaction, whether in publication, in person, or on social media, perpetuates exclusion and marginalization within our field" (Dockum, Rikker, and Green 2024: 96). Given the comparatively limited number of linguists specializing in FL, the negative effects of these developments may be especially acutely felt.

If left unchecked, the progressive fragmentation of FL may further stultify the intra-disciplinary exchange of information, experience, and innovation so that it flows from the center out to the periphery but not in the reverse. According to Nakamura and colleagues (2023), this uni-directionality is ultimately reflective of an underlying neo-colonial paternalism:

> Global North researchers are often regarded as experts in their respective fields, enjoying a reputation beyond their local contexts. Conversely, Global South researchers are often perceived as being confined to their own regions, with their scientific authority seen as deriving from the knowledge and expertise originating in the Global North.
>
> *(1)*

Ironically perhaps, one of the biggest losers in this constellation may be monolingual English speakers who remain woefully ignorant of the truly groundbreaking work that is being conducted in and about other languages of the world. Remaining oblivious to research in and about non-English languages "contributes to incomplete scientific understanding and hinders international collaborations on global challenges" (Gordon and Gutierrez Fierros 2022: 84). The answer to this problem does not lie in expecting all FL researchers to develop multilingual competency but rather to encourage the creation of multilingual scientific cooperatives. This simple strategy can do much to extend the pool of knowledge available to the investigatory practice. Given the high-stakes decision-making that relies on the analyses provided by FL, the discipline can ill-afford to continue policies and practices that hinder our collective understanding of the true diversity of the world's language systems and language users. A philosophical holdover of supremacist theorists like Galton and Gobineau, the maintenance of English First practices risks biasing the questions asked, the data gathered, the methods used, the findings made, the conclusions drawn, and the forensic applications possible for the Anglo-American context and beyond. The resulting holes in understanding are significant as the "scientific investigation of a language, including technical aspects of its grammar, cannot be understood in isolation from its sociopolitical and sociocultural context" (D'Arcy and Bender 2023: 50). Nevertheless, it is not uncommon for scholars from the high-income countries to be invited to consult on cases involving languages they do not speak, about people they do not know, in cultures they have only limited knowledge about. This kind of practice—often referred to as "parachute science" or "helicopter science"—has been increasingly problematized by the international scientific community (Valdez 2024; Nakamura et al. 2023; Genda 2022; Odeny and Bosurgi 2022).

Naturally, it is neither impossible nor unethical for an FL expert to examine linguistic data gathered from languages they do not use, or from communities they do not belong to. Indeed, from the vantage point of an expert outsider, they may be especially sensitive to idiosyncrasies or patterns that an insider may overlook. Still, there are likely to be features of the linguistic data which remain inscrutable to an outsider. These may be the very features where an expert insider's intimate experience with the language and first-hand knowledge of the surrounding social, cultural, and historical context may prove invaluable. Thus, when working on their own, both the insider and the outsider have their respective advantages and disadvantages. However, when they come together, their analyses can overcome these separate weaknesses and capitalize upon their combined strengths (Milch et al. 2009). What is true on the small scale is valid for the entire discipline of FL. By welcoming scholars from around the world, and removing roadblocks to their active participation as equals in the fora of scientific exchange, the wealth of knowledge FL can gain about the world's languages and language communities can be transformative for the discipline, its members, and the public it is ultimately ethically bound to serve.

Conclusion and Suggestions for Increasing DEI in FL

It is not enough to simply point out areas of improvement. It is also necessary to make concrete suggestions to effect positive change. For that reason, this chapter concludes with a number of recommendations to support and promote DEI in FL.

1. Accept that cognitive bias affects everyone, regardless of education, expertise, intention, or intent; and take scientifically established steps to minimize it.
2. Acknowledge injuries in the past and take proactive steps to make sure they are not repeated.
3. Learn from other fields in FS to determine what works and does not work in promoting diversity and removing barriers to inclusivity and equity. This includes the physical barriers to equal access which many differently abled/disabled professionals and students continue to face.
4. Question long-standing traditions, assumptions, and practices that may be the product of historical prejudices, unexamined fallacies, and cognitive bias.
5. Be willing to listen, learn, and take action to bring about positive change.
6. Create professional circles that embrace diversity in all of its various forms. This provides mentorship opportunities, scholarships, and recruitment initiatives to welcome *and* retain both emerging and established members of minoritized communities to the field.
7. Enact ethical codes that include a commitment to DEI and that specify exactly how violations of the DEI principles will be handled.
8. Conduct regular surveys to empirically measure diversity in the FL community and its various activities (e.g., publishing, conferencing, memberships, grants, etc.); and ascertain obstacles to fair treatment and equal access through anonymous qualitative research.
9. Help promote DEI at *all* institutional levels, including those where policy decisions are made.
10. Establish circles of emotional support to weather the backlash that will inevitably come when you work to institute the above recommendations.

If there is anything the recent past has taught, it is how imperative it is that we never forget that civil rights won can also be lost. As disability scientist-activist Morgan Sorenson observed (2023), "relying on laws to handle equity and inclusion is not enough (as we have seen with other diversity issues)" (34). During the writing of this chapter, radical governmental changes have been instituted in the United States reversing years of social progress achieved through generations of sacrifice. At a time when long-established pro-DEI policies and practices are being villainized, we must decide as institutions and individuals what our commitment to justice means. It is the position of this chapter that one of the best courses of action for FL to take during this period of tumult is to remain steadfast in our commitment to provide the highest quality of reliable and accurate analysis, with the lowest level possible of bias. To reach this goal, it is also the contention of this chapter that FL do far more to dramatically increase the current and future diversity of not only the people working within the field but also the languages and language users it examines. It is vitally important that we not lose sight of the fact that there is a fundamental "difference between conducting sound science and remaining 'neutral' in the face of inequity" (Winburn et al. 2021: 3). The promotion of diversity is no way antithetical to our professional role as forensic scientists. Indeed, as this chapter has attempted to show, it is only when we fail to acknowledge and encourage the differences that exist between us that we undermine our capacity to fulfill our mandate to promote social justice as forensic scientists and as human beings.

Note

1 Examples include the following: Actor-Observer Bias, Attentional Bias, Attribution Bias, Authority Bias, Availability Bias, Framing Bias, Hindsight Bias, Omission Bias, Outcome Bias, Plan Continuation Bias, Projection Bias, and Status Quo Bias.

Case Law

Abinger v. Ashton, 17. L.R. Eq. 358, 374 (Ch 1873).

References

Adelson, B. 1984. "When Novices Surpass Experts: The Difficulty of a Task May Increase with Expertise." *Journal of Experimental Psychology: Learning, Memory, and Cognition* 10, no. 3: 483–495.

Ainsworth, Janet. 2022. "Serving Science and Serving Justice: Ethical Issues faced by Forensic Linguists in their Role as Expert Witnesses." In *Language as Evidence*, 35–53. Edited by V. Guillen-Nieto and D. Stein. New York: Palgrave Macmillan.

Ainsworth, Janet. 2008. "'You Have the Right to Remain Silent…But Only If You Ask For It Just So': The Role of Linguistic Ideology in American Police Interrogation Law." *International Journal of Speech, Language, and the Law* 15, no. 1: 1–21.

Ainsworth, Janet. 1993. "In a Different Register: The Pragmatics of Pwerlessness in Police Interrogation." *The Yale Law Journal* 103, no 2 : 259–322.

Aldridge, Michelle. 2007. "The Questioning of Child Witnesses by the Police and in Court: A Linguistic Comparison." *The Language of Sexual Crime*, 63–82. Edited by Janet Cotterill. London: Palgrave Macmillan.

Ali, Siddig Ahmed, and Mohamed Abd Allah Abd Algane. 2013. "The Role of Forensic Translation in Courtroom Contexts." *Arab World English Journal* 4, no. 2 171–180.

Almond, L., et al. 2011. "Heuristics and Biases in Decision-Making." In *Policing Critical Incidents: Leadership and Critical Incident Management*, 151–180. Edited by L. Alison and J. Crego. London: Routledge.

Alshebi, Khalsa, Talal Rahwan, and Wei Lee Woon. 2018. "The Prominence of Ethnic Diversity in Scientific Collaboration." *Nature Communications* 9, no. 1: 5163.

American Psychological Association. n.d. *Heuristic Search*. American Psychological Association Dictionary. Accessed March 1, 2025. Retrieved from https://dictionary.apa.org/heuristic-search.

Anderson, Christopher. 2003. "The Psychology of Doing Nothing: Forms of Decision Avoidance from Reason to Emotion." *Psychological Bulletin* 129: 136–167.

André, Didierjean, and Gobet Fernand. 2008. "Sherlock Holmes: An Expert's View of Expertise." *British Journal of Psychology* 99, no. 1: 109–125.

Baugh, John. 2018. *Linguistics in Pursuit of Justice*. Cambridge: Cambridge University Press.

Baumgarten, Nicole, Inke Du Bois, and Victoria Gill. 2019. "Patterns of Othering Minority Groups in Telephone Gatekeeping Encounters On the UK property Market." *Journal of Language and Discrimination* 3, no. 2: 120–149.

Bédard, J., and M. Chin. 1992. "Expertise." *Current Directions in Psychological Science* 1: 135–139.

Bernstein, David. 2008. "Expert Witnesses, Adversarial Bias, and the (Partial) Failure of the Daubert Revolution." *Iowa Law Review* 93, no. 451: 101–137.

Berk-Seligson, Susan. 2012. "Linguistic Issues in Courtroom Interpretation." In *The Oxford Handbook of Language and Law*, 421–434. Edited by Lawrence Solan and Peter Tiersma. Oxford: Oxford University Press.

Berk-Seligson, Susan. 2009. *Coerced Confessions: The Discourse of Bilingual Police Interrogations*. Berlin: De Gruyter Mouton.

Bond Solon. 2024. "Expert Witness Survey." *The Gazette*. Accessed March 15, 2025. Retrieved from https://www.bondsolon.com/media/jhindyfm/bond-solon-expert-witness-survey-2024.pdf.

Buerger, M., B. Levin, and R. Myer. 2018. *Future Issues in Forensic Science*. Futures Working Group. Society of Police Futurists International. Accessed February 18, 2025. Retrieved from https://sciences.ucf.edu/fwg/wp-content/uploads/sites/24/2018/09/2018.Buerger.PFI_.-Future-issues-in-forensic-science.pdf.

Butters, Ronald. 2012. "Retiring President's Closing Address: Ethics, Best Practices, and Standards." In *Proceedings of the International Association of Forensic Linguistics' Tenth Biennial Conference*, 351–361. Edited by Samuel Tomblin, Nicci MacLeod, Rui Sousa-Silva, and Malcolm Couthard. Birmingham, UK: Centre for Forensic Linguistics.

Byrman, Gunillla, and Ylva Byrman. 2018. "In Evidence: Linguistic Transformations of Events in Police Interview Reports." *Nordic Journal of Linguistics* 41, no. 2:155–181.

Campbell, L., et al. 2013. "Gender Heterogeneous Working Groups Produce Higher Quality Science." *PLoS One* 8, no. 10: e79147.

Carney, Terrence. 2019. "Sign Language and Hate Speech: The Potential Pitfalls of Iconic Signs." In *New Frontiers in Forensic Linguistics: Themes and Perspectives in Language and Law in Africa and Beyond*, 279–300. Edited by Monwabisi Ralarala, Russell Kashula, and Georgina Heydon. Stellenbosch: African Sun Press.

Cignaglia, Carlo, Amr Rabie-Ahmed, and Peter de Costa. 2025a. "Ethical Issues in Applied Linguistics Scholarship: Introduction" in *Ethical Issues in Applied Linguistics Scholarship*, 1–7. Edited by Peter De Costa, Amr Rabie-Ahmed, and Carlo Cinaglia. Amsterdam: John Benjamins Publishing Company.

Cignaglia, Carlo, Amr Rabie-Ahmed, and Peter de Costa. 2025b. "Ethical Dilemmas of Graduate Students Negotiating New Roles and Responsibilities: The Importance of Mindful Engagement, Self-care, and Reflexivity." In *Ethical Issues in Applied Linguistics Scholarship*, 310–327. Edited by Peter De Costa, Amr Rabie-Ahmed, and Carlo Cinaglia. Amsterdam: John Benjamins Publishing Company.

Cooper, Glinda S., and Vanessa Meterko. 2019. "Cognitive Bias Research in Forensic Science: A Systematic Review." *Forensic Science International* 297: 35–46.

Czaykowska-Higgins, Eva. 2018. "Reflections of Ethics: Re-humanizing Linguistics, Building Relationships Across Difference." In *Reflections on Language Documentation 20 Years after Himmelmann 1998*, 110–121. Edited by Bradly McDonnell, Andrea Berez-Kroeker, and Gary Holton. Honolulu: University of Hawaii Press.

D'Arcy, Alexandra, and Emily Bender. 2023. "Ethics in Linguistics." *Annual Review of Linguistics* 9: 49–69.

Data USA. n.d. "Language and Linguistics." Accessed February 26, 2025. Retrieved from https://datausa.io/profile/cip/language-linguistics?redirect=true#:~:text=The%20number%20of%20Language%20%26%20Linguistics,2021%20to%20509%2C182%20in%202022.

Death Penalty Information Center. 2024, November 15. "Trial Judge Declares Melissa Lucio to be 'Actually Innocent', Recommends Texas CCA Overturn Conviction and Death Sentence." Accessed February 15, 20525. Retrieved from https://deathpenaltyinfo.org/trial-judge-declares-melissa-lucio-to-be-actually-innocent-recommends-texas-cca-overturn-conviction-and-death-sentence.

Ding, Jielan et al. 2021. "The Link between Ethnic Diversity and Scientific Impact: The Mediating Effect of Novelty and Audience Diversity." *Scientometrics* 126, no. 9: 7759–7810.

Dockum, Rikker, and Caitlin Green. 2024. "Toward a Big Tent Linguistics: Inclusion and the Myth of the 'Lone Genius'." In *Inclusion in Linguistics*, 95–126. Edited by Anne Charity Hudley, Christine Mallinson, and Mary Bucholtz. New York, NY: Oxford Academic.

Dreyfus, Hubert, and Stuart Dreyfus. 1988. *Mind Over Machine: The Power of Human Intuition and Expertise in the Era of the Computer*. New York: The Free Press.

Dror, I. E. 2020. "Cognitive and Human Factors in Expert Decision Making: Six Fallacies and the Eight Sources of Bias." *Analytical Chemistry* 92, no. 12: 7998–8004.

Eades, Diana. 2018. "Communicating the Right to Silence to Aboriginal Suspects: Lessons from Western Australia v Gibson." Journal of Judicial Administration 28: 4-21.

Eades, Diana. 2012. "The Social Consequences of Language Ideologies in Courtroom Cross-examination." *Language and Society* 41, no. 4: 471–497.

Eades, Diana. 2010. Sociolinguistics and the Legal Process. Bristol: Multilingual Matters.

Eades, Diana. 2000. "'I Don't Think it's an Answer to the Question: Silencing Aboriginal Witnesses in Court." Language in Society 29, no. 2: 161–196.

Eades, Diana. 2005. "Applied Linguistics and Language Analysis in Asylum Seeker Cases." *Applied Linguistics* 26, no. 4: 503–526.

Eades, Diana. 2000. "'I Don't Think it's an Answer to the Question: Silencing Aboriginal Witnesses in Court." *Language in Society* 29, no. 2: 161–196.

Ehrlich, Susan, Diana Eades, and Janet Ainsworth (eds). 2016. Discursive Constructions of Consent in the Legal Process. Oxford: Oxford University Press.

Ericsson, K. 1998. "The Scientific Study of Expert Levels of Performance." *High Ability Studies* 9: 75–100.

Freeman, Richard, and Wei Huang. 2014. "Collaboration: Strength in Diversity." *Nature* 513.

Galupo, Paz Resnick. 2016. "Experiences of LGBT Microaggressions in the Workplace: Implications for Policy." In *Sexual Orientation and Transgender Issues in Organizations: Global Perspectives on LGBT Workforce Diversity*, 271–287. Edited by Thomas Kölen. New York: Springer.

Garber, Alan. 2025 "The Promise of American Higher Education." Retrieved from https://www.harvard.edu/president/news/2025/the-promise-of-american-higher-education/.

Garcia-Retamero, Rocio, et al. 2007. "When One Cure is Not Enough: Combining Fast and Frugal Heuristics with Compound Cure Processing." *The Quarterly Journal of Experimental Psychology* 60: 1197–1215.

Geddes, Linda. 2012, February. "Forensic Failure: Miscarriage of Justice will Occur". *New Scientist.* Accessed March 15, 2025. Retrieved from https://www.newscientist.com/article/mg21328514-600-forensic-failure-miscarriages-of-justice-will-occur/.

Genda, Peter Ami. 2022. "Looking Up and Down: Strong Collaboration is Only the First Step in Tackling Parachute Science." *Conservation Science and Practice* 4, no. 5: e12677.

Gigerenzer, Gerd, and Wolfgang Gaissmaier. 2011. "Heuristic Decision Making." *Annual Review of Psychology* 62: 451–482.

Gigerenzer, Gerd, and Daniel Goldstein.1996. "Reasoning The Fast and Frugal Way: Models of Bounded Rationality." *Psychological Review* 103: 650–669.

Gordon, Sarah Frances, and Emiliano Gutierrez Fierros. 2022, September. "Combating Exclusionary Language Practices in Science Publishing: A DEI Concern." *Science Editor* 45, no. 3: 84–85.

Grawitch, Matt. 2020, November 2. "Heuristics and Biases Related But Not the Same." *Psychology Today*. Accessed March 14, 2025. Retrieved from https://www.psychologytoday.com/us/blog/hovercraft-full-eels/202011/heuristics-and-biases-related-not-the-same.

Greenlee, Mel. 2021. "Capitally Speaking: Language and Bias in Capital Trials." In *The Routledge Handbook of Forensic Linguistics*, 228–244. Edited by Malcolm Coulthard, Alison May, and Rui Sousa-Silva. New York: Routledge.

Guillaume, P., A. Sidebottom, and N. Tilley 2012. "On Police and University Collaborations: A Problem-oriented Policing Case Study." *Police Practice and Research* 13: 389–401.

Hale, Saundra. 2011. "The Need to Raise the Bar: Court Interpreters as Specialised Experts." *Journal of the Judicial Commission of New South Wales* 10, no. 2: 237–258.

Haworth, Kate, et al. 2023. "'For the Record: Applying Linguistics to Improve Evidential Consistteny in Police Investigative Interview Records." *Frontiers in Communication* 8: 1178516.

Herring, C. 2009. "Does Diversity Pay? Race, Gender, and the Business Case for Diversity." *American Sociological Review* 74, no. 2: 208–224.

Hinchey, Patricia, and Isabel Rimmel. 2000. *The Graduate Grind: A Critical Look at Graduate Education.* New York: Palmer.

Holton, Gary, Wesley Leonard, and Peter Pulsifer. 2022. "Indigenous Peoples, Ethics, and Linguistic Data." In *The Open Handbook of Linguistic Data Management*, 49–60. Edited by Andrea Berez-Kroeker, et al. Cambridge, MA: MIT Press.

Hong, Lu, and Scott Page. 2004. "Groups of Diverse Problem Solvers Can Outperform Groups of High-Ability Problem Solvers." *National Academy of Science* 46: 16385–16389.

Jackson, Glen. 2009. "Forensic Science Degree Programs in the United States." *Forensic Science Policy and Management* 1: 2–9.

Jones, Graham., Bernardito Chirino Chace, and Justin Wright. 2020. "Cultural Diversity Drives Innovation: Empowering Teams for Success." *International Journal of Innovation Science* 12, no. 3: 323–343.

Kassim, Saul, Itiel Dror, and Jeff Kuckucka. 2013. "The Forensic Confirmation Bias: Problems, Perspecttives, and Proposed Solutions" *Journal of Applied Research in Memory and Cognition* 2, no.1: 42–52.

Keashly, Loraleigh, and Joel Neuman. 2013. "Bullying in Higher Education: What Current Research, Theorizing, and Practice Tell Us." In *Workplace Bullying in Higher Education*, 1–22. Edited by Jaime Bester. New York: Routledge.

Klein, G., B. Moon, and R. Hoffman. 2006. "Making Sense of Sensemaking I: Alternative Perspectives." *IEEE Intelligent Systems*: 70–73.

Kruglanski, A., and T. Freund. 1983. "The Freezing and Unfreezing of Lay Inferences." *Journal of Experimental and Social Psychology* 19: 448–468.

Kunkler, Kimberly, and Tiffany Roy. 2023. "Reducing the Impact of Cognitive Bias in Decision-Making: Practical Action in Forensic Science Practice." *Forensic Science International: Synergy* 7: 100341.

Leeman, Adrian, Ria Perkins, Grace Sullivan-Buker, and Paul Foulkes. 2025. *An Introduction to Forensic Phonetics and Forensic Linguistics*. New York: Routledge.

Lemaire, P., and R. Siegler. 1995. "Four Aspects of Strategic Change." *Journal of Experimental Psychology: General* 124, no. 1: 83–97.

Lucas, Ceil, ed. 2003. *Language and the Law in Deaf Communities*. Washington, DC: Gallaudet University Press.

Mahmoudi, Morteza, and Loraleigh Keashly. 2021. "Filling the Space: A Framework for Coordinating Global Actions to Diminish Academic Bullying." *Angewandte Chemie* 60: 3338–3344.

Mallinson, Christine, Jon Hennert, Anne Charity, and Mary Buchholtz. 2024. "Inclusion Linguistics: Introduction." In *Inclusion Linguistics*, 1–18. Edited by Anne Hudley, Chrstine Mallinson, and Mary Bucholtz. Oxford: Oxford University Press.

Maryns, Katrijn. 2004. "Identifying the Asylum-Speaker: Reflection on the Pitfalls of Language Analysis in the Determination of National Origin." *International Journal of Speech, Language, and the Law* 11, no. 2: 240–260.

McNamara, Tim, and Doris Schüpbach. 2018. "Fairness and Justice in Language Analysis for the Determination of Origin of Asylum-seekers (LADO)." In *Forensic Linguistics: Asylum-seekers, Refugees, and Immigrants*, 155–174. Edited by I. M. Nick. Wilmington: Vernon Press.

Mensitieri, Delia, Smaranda Boroş, and Claudia Toma. 2025. "Bringing Microaggressions from the Shadows to the Spotlight: Unveiling Silencing Mechanisms and Distinct Patterns in Coping." *Gender, Work and Organization*. Accessed March 14, 2025. Retrieved from https://onlinelibrary.wiley.com/doi/abs/10.1111/gwao.13256.

Milch, K., et al. 2009. "From Individual Preference Construction to Group Decisions: Framing Effects and Group Processes." *Organizational Behavior and Human Decision Processes* 108: 242–255.

Nakamura, Gabriel, et. Al. 2023. "Three Pathways to Better Recognize the Expertise of Global South Researchers." *NPJ Biodiversity* 2, no. 17: https://doi.org/10.1038/s44185-023-00021-7.

National Research Council. 2009. *Strengthening Forensic Science in the United States: A Path forward.* Washington, DC: National Academies Press.

Odeny, Beryne, and Rafaella Bosurgi. 2022. "Time to End Parachute Science." *PLoS One* 19, no. 9: e1004099.

Olsen, Terje, and Patrick Kermit. 2015. "Sign Language, Translation and the Rule of Law Deaf People's Experiences from Encounters with teh Norwegian Crminal Justice System." *Scandinavian Journal of Disability Research* 17, no. S1: 23–41.

Østergaard, C., B.Timmermans, and K. Kristinsson. 2011. "Does a Different View Create Something New? The Effect of Employee Diversity on Innovation." *Research* Policy 40, no. 3: 500–509.

Page, A. 2008. *The Difference: How the Power of Diversity Creates Better Groups, Firms, Schools, and Societies*. Princeton, NJ: Princeton University Press.

Parikh, Ashishkumar, and Jessica Leschied. 2022. "Microaggressions in our Daily Workplace Encounters: A Barrier to Achieving Diversity and Inclusion." *Pediatric Radiology* 52, no. 9: 1719–1723.

Paulus, P., K. I. van der Zee, and J. Kenworthy. 2016. "Cultural Diversity and Team Creativity." In *The Palgrave Handbook of Creativity and Cultural Research*, 57–76. Edited by Vlad Petre Glăveanu. London: Palgrave Macmillan.

Picornell, Isabel. 2022. "Casework in Forensic Linguistics: Looking Outward." In *Methodologies and Challenges in Forensic Linguistic Casework*, 162–169. Edited by Ria Perkins, Isabel Picornell, and Malcolm Coulthard. Hoboken, NJ: Wiley.

Piller, Ingrid, 2016. *Linguistic Diversity and Social Justice: An Introduction to Applied Sociolinguistics*. New York: Oxford University Press.

Pinto, Manuela Fernández. 2023. "Methodological and Cognitive Biases in Science: Issues for Current Research and Ways to Counteract Them." *Perspectives on Science* 31, no 5: 535–554.

Rojas Alfaro, Carolina, et al. 2025. "A Practical Approach to Mitigating Cognitive Bias Effects in Forensic Casework." *Forensic Science International: Synergy* 10: 100569.

Rojek, J., G. Alpert, and H. Smith. 2012. "The Utilization of Research by the Police." *Police Practice and Research* 13: 329–341.

Rosenbaum, D. 2010. "Police Research: Merging the Policy and Action Research Traditions." *Police Practice and Research* 11: 144–149.

Rosenhouse, Judith. 2017. "A Forensic Linguistics Problem: Asylum Seekers' Dialect Indentification Difficulties in Under-documented Adjacent Arabic Dialects." *International Joournal of Legal Discourse* 2, no. 1: 113–123.

Roth, Andrea. 2012. "Ethical Duties of Attorneys and Experts in Cases Involving Forensic Evidence, from the Perspective of the Defense." In *Ethics in Forensic Science*, 313–342. Edited by J. C. Upshaw Downs, and Anjali Ranadive Swienton. Amsterdam: Elsevier.

Salas, E., M. Rosen, and D. Diaz-Granados. 2010. Expertise-based Intuition and Decision Making in Organizations." *Journal of Management* 36: 941–973.

Sawyer, Samantha. 2023. "Inclusion of LGBTQ+ Topics in Forensic Science Education." *Journal of Forensic Science Education* 5, no. 2: 1–5.

Seaborn, Boley, Jean Andrews, and Gabriel Martin. 2010. "Deaf Adults and the Comprehension of Miranda." *Journal of Forensic Psychology Practice* 10, no. 2: 107–132.

Shuy, Roger. 1998. *The Language of Confession, Interrogation, and Deception*. London: SAGE.

Siegel, Jay. 2025. "Forensic Science." *Britannica*. Retrieved from https://www.britannica.com/science/forensic-science.

Skutnabb-Kangas, T., and R. Phillipson. 1998. "Language in Human Rights." *Gazette (Leiden, Netherlands) 60*, no. 1: 27–46.

Skutnabb-Kangas, T., and J. Cummins. 1988. *Minority Education: From Shame to Struggle.* Clevedon, UK: Minority Matters.

Smith Doerr, Laurel, Sharla Alegria, and Timothy Sacco. 2017. "How Diversity Matters in the US Science and Engineering Workforce: A Critical Review Considering Integrating Teams, Fields, and Organizational Contexts." *Engaging Science, Technology, and Society* 3: 139–153.

Solan, Lawrence. 2001. "Causation, Contribution, and Legal Liability: An Empirical Study." *Law and Contemporary Problems* 64, no. 4: 265–298.

Sorenson, Morgan. 2023. "From DEI to DEIA: Why Adding Accessibility is So Important." *Science Editor* 46, no. 1: 33–34.

Spellman, Barbara, Heidi Eldridge, and Paul Bieber. 2022. "Challenges to Reasoning in Forensic Science Decisions." *Forensic Science International: Synergy* 4: 100200.

Starr, Douglas. 2022. "The Bias Hunter" *Science* 376, no. 6594: 686–690.

Taylor, Michelle. 2025, February 26. "Universities Investing in More Degrees to Keep Up with Growing Forensics Industry." *Forensic* Accessed March 15, 2025. Retrieved from https://www.forensicmag.com/3594-All-News/617933-Universities-Investing-in-More-Degrees-to-Keep-Up-with-Growing-Forensics-Industry/#:~:text=Forensic%20science%20is%20currently%20one,Forensics%20in%20the%20Virginias.

United Nations. 2024. "Sign Up for Sign Language Rights!" Accessed February 16, 2025. Retrieved from https://www.un.org/en/observances/sign-languages-day.

U.S. Department of Justice. 2023, December. "Publicly Funded Forensic Crime Laboratories, 2020." *NCJ*: 306473.

U.S. National Science Foundation. 2024. "Annual Performance Report: FY2024." Accessed February 1, 2025. Retrieved from https://nsf-gov-resources.nsf.gov/files/nsf25003.pdf.

Valdez, Jose, et al. 2024. "Strategies for Advancing Inclusive Biodiversity Research through Equitable Practices and Collective Responsibility." *Conservation Biology* 38, no. 6: e14325.

Van den Eden, Claire, Christianne de Poot, and Peter van Koppen. 2019. "The Forensic Confirmation Bias: A Comparison between Experts and Novices." *Journal of Forensic Sciences* 64, no. 1: 120–126.

Wagstaff, R. Iris, and Gerald Laporte. 2018. "The Importance of Diversity and Inclusion in the Forensic Sciences." *National Institute of Justice Journal.* Accessed February 1, 2025. Retrieved from https://nij.gov/journals/279/Pages/importance-ofdiversity-and-inclusion-in- forensic-sciences.aspx.

Whitman, Glen, and Roger Koppl. 2010. "Rational Bias in Forensic Science." *Law, Probability, and Risk* 9, no. 1: 69–90.

Wilke, Andrea, and Rui Mata. 2017. "Cognitive Bias" in *Encyclopedia of Human Behavior*, 531–535. Edited by V. S. Ramachandran. Amsterdam: Elsevier.

Williams, Monnica. 2021. "Racial Microaggressions: Critical Questions, State of the Science, and New Directions." *Perspectives on Psychological Research* 16, no. 5: 880–885.

Winburn, Allysha, et al. 2021. "Responding to the American Academy of Forensic Sciences Vision, Mission, and Values Statements: Comments, Revisions, and Proposed Actions." *Forensic Science International: Synergy* 3: 100197.

Wright, K. E. 2023. "Housing Policy and Linguistic Profiling: An Audit Study of Three American Dialects." *Language* 99, no. 2: e58–e85.

Yin, An, Jessica Juarez, Jesse Goliath, and Isabel Melhado. 2022. "Diversity in Forensic Sciences: Black, Indigenous, and People of Color (BIPOC) Representation in Different Medicolegal Fields in the United States." *Forensic Science International: Synergy* 5: 100280.

Zapf, Patricia, and Itiel Dror. 2017. "Understanding and Mitigating Bias in Forensic Evaluation: Lessons from Forensic Science." *International Journal of Forensic Mental Health* 16, no. 3: 227–238.

34
CONCLUDING THOUGHTS

I. M. Nick

Introduction

As with any book, there are always topics that, in retrospect, one would have liked to have included but was unable to do so for various reasons. The same is true where this reference is concerned. Although we made a concerted effort to encompass a broad selection of ethical issues within Forensic Linguistics, there were certainly topics that we were unable to address. For example, ethical questions that we would have liked to have examined in this Handbook include the following: the development and use of artificial intelligence in the collection, analysis, and interpretation of linguistic data; translating and interpreting minoritized languages in forensic settings; serving as a consultant on controversial, highly politicized cases talking to the media about forensic linguistic cases without inadvertently feeding unhealthy desires for the next true crime sensation. For more information on these subjects, the reader is encouraged to consult other works (Asna 2025; Adkins, Al Bataineh, and Khalaf 2024; Mehta et al. 2024; Sousa-Silva 2024; Wi et al. 2024; Ward 2023; Scott and O'Shea 2022; Longhi 2021; Perkins 2021; Hackman 2020; Watt and Brown 2020; Nick 2018; Howes and Kemp 2017; Grant 2008; Tiersma and Solan 2002). At the same time, it is sincerely hoped that this Handbook will help spark interest in and discussion of these and other important ethical topics which could not be addressed here.

That having been said, there are three ethical issues which we would specifically like to delve into here, in this final chapter. These topics were repeatedly mentioned to us by colleagues and students in the field, as being extremely important. Like the other subjects mentioned above, the three ethical issues to be discussed in this chapter all warrant serious attention. However, unlike the others, these three issues may sometimes be avoided in formal discussions because of unspoken prohibitions against airy dirty laundry in public; out of concern for possible professional fallout; and/or fears over public ridicule or derision by one's peers. Ultimately though, no matter what the reason(s), the continued avoidance of these subjects may end up doing more harm than good. In the hope of encouraging more in-depth, open, honest, and sensitive discussion in the future, this conclusion explores each of these three ethical matters here.

DOI: 10.4324/9781003391074-37

Ethical Issues Surrounding the Use and Supervision of Students in Forensic Linguistic Research

It is not unusual for senior researchers to hire advanced students to work on research projects. For students, being chosen to work closely with a faculty member can open doors to many excellent professional opportunities. Alongside the financial rewards can come outstanding training from experiential learning and individualized mentorship. For the highly motivated, talented, and gifted student, such hands-on work experience can mark the start of a personally and professionally satisfying academic career. However, students are not the only ones who may profit from these work arrangements. Senior researchers who take on student assistance may also benefit from these cooperations. What a senior researcher gives in terms of experience, knowledge, and skills, students may return with enthusiasm, energy, and innovation (McCloskey and Weaver 2020; Johnson 2017; McGinn and Niemczyr 2013; Ragins and Kram 2007; Silva and Tom 2001; Eby et al. 2000). It is precisely this complementary exchange which can make teamwork between emerging and established scholars extraordinarily productive. Research has shown that mentorship between senior and junior researchers is associated with enhanced identification with the scientific discipline (Byars-Winston et al. 2015; Thiry, Laursen, and Hunter 2011; Feldman et al. 2010); greater career satisfaction (Beech et al. 2013; Schapira et al. 1992); increased research productivity (Steiner et al. 2004; Wingard, Garman, and Reznik 2004); as well as enhanced job advancement, including opportunities for hiring and promotion (Eby et al. 2000). For both the mentee and mentor, such relationships have the potential to be transformational cooperations that "inspire mutual growth, learning, and development" (Ragins and Kram 2007: 3).

At the same time, it is essential to remember that such professional alliances are invariably marked by significant power imbalances that can, if left unchecked, easily end in important ethical lines being crossed. The resulting professional misconduct can take many different forms. Some of the most common include supervisors placing unreasonable demands on students' time and energy; giving payment that is incommensurate to the work being provided; failing or refusing to give proper credit to students for the work they have accomplished; neglecting to offer adequate supervision; and expecting students to perform tasks that are far outside of the contractual terms of their work agreements (Johnson 2020, 2017; Devos et al. 2016; McGinn and Niemczyr 2013; Silva and Tom 2001; Eby et al. 2000; Scandura 1998; Goodyear, Crego, and Johnston 1992).

In addition to these unethical work practices, students in financially dependent relationships with their supervisors may also find themselves suffering outright abuse. These assaults may take various forms—from being subjected to commentary designed to chide, belittle, humiliate, or denigrate to being ostracized, gaslit, threatened, coerced, or otherwise physically, emotionally, or sexually attacked (Bloom et al. 2022; Coihen and Beruch 2021; Perry 2018; Devos et al. 2016; Rosenthal, Smidt, and Freyd 2016; Hinchey and Isabel Rimmel 2000; Goodyear, Crego, and Johnston 1992). Though once academia's dirty little secret, thanks in no small measure to international movements encouraging survivors of abuse to organize and demand accountability, research into the phenomenon of bullying in academia has increased dramatically in the past two decades (Mahmoudi and Keashly 2021; Moss and Mahmoudi 2021; Gentry and Whitley 2014; McKay et al. 2008; Hinchey and Rimmel 2000). The results of this work have been sobering, with rates of bullying ranging from 18% to 68% (Keashly and Neuman 2013). In a 2019 international survey of over 6,300 PhD students conducted by the journal *Nature*, it was found that 21% of the respondents had experienced bullying at the institution and the main perpetrators of the abuse were their supervisors. Although statistics for forensic linguistics are currently unavailable, it would be naive to think that incidents in our field do not occur at all. It is therefore imperative that we remain vigilant and take allegations of abuse seriously. As Bowen (2018) reminds us: "Though everyone would like to believe that all forensic scientists are ethical in nature, it is simply not the case" (xiv).

In fairness, these issues are not unique to forensic linguistics or even forensic sciences. There are, however. certain ethical concerns surrounding the use and supervision of students that may be of particular significance for this field. For example, in casework, ethical best practices demand that formal permissions be obtained for students to work with data for forensic purposes. Ideally, that permission should explicitly state not only which person(s) will be allowed to access sensitive data, but also how they are and are not allowed to handle that data (e.g., collection, analysis, storage, report-writing, publication, presentation, etc.). It should not be assumed that permissions given to the senior researcher automatically extend to their students (or colleagues, for that matter). By the same token, it is ethically fraught for senior researchers to tacitly violate the terms of contractual agreements made between themselves and a client, by agreeing to take on a case and then delegating the actual work to students, without the client's knowledge or approval. Though never to be condoned, such ethical misconduct is particularly egregious when the casework in question is being prepared for forensic purposes. If discovered, there may be severe legal repercussions for not only the parties directly involved, but also the institutions affiliated. As Bowen (2018) reminds us, at the end of the day, it is important to keep uppermost in mind that "honesty and integrity are the hallmarks of science and decisions should be made with those core values to maintain the highest degree of professionalism" (17). This is not the only issue that requires particular caution when students are allowed to participate in forensic linguistic research.

Other ethical challenges may arise when student assistants are exposed to confronting material which they are then required to intensively investigate, often for prolonged periods of time. These conditions can result in secondary trauma and/or vicarious trauma (Naufel and Beike 2013; Baird and Kracen 2006). To minimize these potential risks, it is not sufficient for supervisors to ask students if they would like to participate in a forensic linguistic project that may/will contain confronting material. The very fact that a student is new to this sort of subject-matter may mean that they are completely unable to predict how they will react to intensive and/or repeated exposure to graphic data. Furthermore, their excitement over being offered the opportunity to work on a project with authentic data may override any feelings of discomfort they may have before starting their assignment. Moreover, for those who initially agree to take on the position and then later discover that they are indeed having difficulties, their feelings of shame or embarrassment may lead them to try to hide or deny any problems they might be experiencing, in an attempt to avoid disappointing their supervisor and/or jeopardizing their current job or future career prospects (Coihen and Beruch 2021; Naufel and Beike 2013; Rossow and Niemczyk 2013).

To reduce the risk, incidence, and severity of primary and secondary trauma suffered by research assistants, experts recommend supervisors exercise many of the same protective measures they routinely employ to protect their research participants. Accordingly, research assistants should be informed in detail what the investigation will entail; what their specific duties will be; and what risks may be involved. They should also be informed that they are free to withdraw at any point. It is also recommended that supervisors provide adequate training and support before, during, and after exposure to potentially traumatic material. There should also be supportive resources made freely available that all team members are encouraged to use.[1] This encouragement is important to avoid the potential of being stigmatized as the only person who needs help. Supervisors can also set a good example by taking advantage of support services themselves. By modeling positive self-care, supervisors not only help ensure students currently on their team have a positive learning experience. They can also make a positive contribution to establishing self-awareness and self-care as staples of ethically responsible research for the next generation of scholars in forensic linguistics.

Ethical Challenges in Forensic Linguistics Resulting from the Precarious State of Higher Education

Higher Education (HE) is facing an existential crisis. HE institutions around the world are suffering unparalleled financial cuts. In an effort to keep their heads above water, many colleges and universities have been forced to make radical, often painful, slashes to their services and personnel by closing hiring lines, letting treasured members of faculty go, cancelling long-standing programs that served the community, and down-sizing or completely eliminating entire departments. The situation in many countries is both depressing and dire. In the United Kingdom, for example, it has been reported that "Forty per cent of higher-education (HE) establishments are running deficit budgets, more than 50 are cutting jobs this year, and some are facing financial collapse" (Guest 2024, para 2). Speaking specifically for England, the projections are dismal. Adam Habib, Vice Chancellor at SOAS University of London, and Lord Dr. Michael Hastings, Chair of the Board of Trustees at SOAS, project that "72% of English universities will be in deficit by the end of the academic year if they continue as is" (1). The picture is much the same for the European Union. In the 2024 annual report issued by the European Universities Association, it was revealed that one of the most serious threats facing higher education institutions (HEI) in Europe was a critical lack of funding. The statistics are indeed quite concerning. According to the report: "Forty-four per cent of institutions report either continuously low or decreasing funding over the past five years, while 70% of HEIs identify underfunding as one of the top three obstacles for improving learning and teaching" (Gaebel, Zhang, and Stoeber 2024: 10). And across the ocean, in the United States, as Ian Bogost describes in a 2025 article for the *Atlantic*, the situation can best be described as bleak. Bogost writes:

> In the past two weeks, higher ed has been hit by a series of startling and, in some cases, potentially illegal budget cuts. First came a total freeze of federal grants and loans [...] then news that the National Science Foundation, which pays for research in basic, applied, social and behavioral science as well as engineering, could have its funding cut by two-thirds. [...] the National Institutes of Health, which provides tens of billions of dollars in research funding every year, announced an even more momentous change: [...] it would be slashing the amount it pays out in grants for administrative costs.
>
> *(para 1)*

Even at the federal level, institutions that have been foundational to the stability of entire education systems have fallen prey to the chainsaw. At the same time that all of these cuts have taken place, there have not been appreciable drops in the number of student enrollments. In fact, in many countries, just the opposite trend can be seen. The numbers of degree-seeking students have increased as changes in the modern job market have made a university degree a prerequisite for many entry-level positions (Office of Students 2024; Smith 2013). In the wake of this ongoing economic crisis, faculty members who manage to keep their positions often find themselves having to take on an increasing number of work hours in an attempt to compensate for missing colleagues and to safeguard themselves against the next inevitable round of redundancies.

To help make decisions regarding the retention and extension of employment contracts as transparent and fair as possible, many HE administrations have introduced progressively complex work assessment systems that specify the number and type of performance tasks required. In many HE institutions, the general requirement to "publish or perish" has thus been replaced by a complicated algebra of what an academic is required to publish, when, where, how often, and how many. All of these volatile changes and employment insecurities have significantly increased the pressures placed upon today's modern academician. Of course, members of the forensic linguistic community have not been spared. The impact of this climate was felt throughout the preparation of this Handbook. Many scholars who eagerly agreed to contribute a chapter were then forced to withdraw, either having lost their position or receiving news that a book chapter would no longer carry the same weight in their upcoming job review. Across the board, academic scholars are having to re-assess their career plans, as the special stability, security, and intellectual freedoms which HE used to offer under the umbrella of tenure have all but evaporated. The numbers more than support the anxiety which many academics have felt for some time now. For example, using data gathered from the National Center for Education Statistics, the American Association of University Professors (AAUP) determined that "over two-thirds (68 percent) of faculty members in US colleges and universities held contingent appointments in fall 2021" (Colby 2023). In addition, in 2023, the Center for Public Integrity reported, more than 75% of college faculty are not tenured, and the vast majority of persons who are holders of these insecure positions are women and underrepresented minorities (Krupnick 2023). In reaction to these stressors, an increasing number of scholars across the academy are deciding to leave HE in search of a position with less unpredictability and more financial reward (Krupnick 2023).

With respect to forensic linguistics, this trend to leave the HE might be accompanied by a series of troubling ethical issues in the not too distant future. Without the necessary financial security that academia used to offer scholars, it is conceivable that an increasing number of forensic linguists may decide to leave academia and build up a private consultancy instead. Given the hourly rates which scientific experts can command and the growing demand for forensic linguists as consultants, this decision is entirely understandable. However, when one considers the fact that, as mentioned in the introduction to this Handbook, there is no industry-wide accepted definition for what a "forensic linguistic expert" is, this potential trend toward private consultancy could have serious repercussions. If the insecurities and pressures of working in HE do not decrease, and the costs of obtaining a doctoral degree continue to increase, we may see more persons elect to offer their services as a "forensic linguist consultant" without the education, training, and skills traditionally expected. According to Guillén-Nieto and Stein (2022), the increase in practitioners with inadequate training and education is one which has plagued the field for some time now. As the two observe (2022): "It is an open secret in the profession that forensic linguistics has on occasions suffered from presenting a less than favourable and convincing picture of itself to the legal world, and to the judiciary, leading, in part, occasionally to evidence being disregarded" (8).

While the risk to the scientific reputation and credibility of forensic linguistics is without question, even more concerning is the potential negative impact upon the society at large. Moreover, even in situations where the forensic linguist does have the prerequisite academic background to competently provide the expert scientific services they have been contracted to perform, there are still other risks involved. Certainly the pressures that come from having to secure a list of satisfied clients as a sole or primary source of income are not to be underestimated. This risk is also addressed by Guillén-Nieto and Stein (2022) who warn that the more forensic linguistic experts refashion themselves as "service providers" acting on behalf of clients, "the greater the danger is that the 'hired gun' culture will be conducive to seek and produce features that can serve as 'evidence' in the direction of their funders—a situation that calls for an explicit code of professional ethics" (18). To guard against this problem, it is strongly suggested that the general codes of practice used by leading organizations such as the International Association for Forensic and Legal Linguistics and International Association for Forensic Phonetics and Acoustics be expanded to include recommended educational requirements for working as a *competent* forensic linguistic consultant.

It is appreciated that the very idea of oversight might be deemed as unnecessary, inappropriate, or even invasive. However, as Terkourafi (2025) wisely explains, the importance of instituting more explicit and detailed ethical guidelines for linguistics becomes particularly compelling:

> once it is realized that what is at stake is not only an individual researcher's ability to continue carrying out research—whether publicly or privately funded—but also how this research is viewed by society, in terms of its reliability and usefulness, factors which in turn feed back into the public image of academia and the extent to which future generations are attracted to it.
>
> *(1–2)*

It is also understood that these recommendations would not be—at this stage—enforceable. However, just as including recommendations about reporting results is not binding, the simple fact that they have been transparently articulated and are generally accessible offers much-needed guidance. It is understood that drawing up educational recommendations is complex; however, that should not stand in the way of our taking on this important task. After all, who, if not forensic linguists, has the training and the skills necessary to successfully devise transparent organizational policies and procedures, recommendations, and warnings in plain language?

While this project may admittedly take time, the investment may well be worth it. In the meantime, there are other steps that the international forensic linguistic community could take to help safeguard its professional standards. These include the establishment of scholarships and fellowships for members of international forensic linguistic organizations to gain and develop their skills; the creation of mentorship programs to pair a wide diversity of emerging and established scholars to work together on collaborative projects; and the formation of a formal ethics committee made up of seasoned professionals in the forensic linguistic community who can offer ethical guidance and make recommendations. An excellent model for such work is the Committee on Publication Ethics (COPE).

COPE began in 1997, when a small group of editors who were concerned about the ethical issues in publishing decided to form an advisory collective. Since then, the organization now has more than 14,000 members from 97 different countries. Among the many services it provides, COPE[2] offers a variety of ethical supports. These include detailed Codes of Best Practices for editors, journal staff, publishers, and HE institutions; education and training opportunities; a portal where persons facing an ethical dilemma can anonymously submit the details of their problem and receive guidance and support; an ethics discussion forum for members that is carefully and continuously moderated; a host of downloadable multilingual resources freely accessible by the general public; and a searchable databank of ethics cases that describe not only the details of the original dilemma but also offer an in-depth description of possible courses of action and COPE's final official recommendation. What began as an informal meeting of concerned professionals has blossomed into an invaluable positive resource that has helped promote ethical practices in publishing around the world. Something similar could be achieved if the knowledge, experience, and skills of the growing global network of forensic linguists were harnessed to help us face the ethical challenges of today and tomorrow.

Ethical Imperative of Respectful and Considerate Academic Discourse

It has not escaped notice that the tone of public discourse is becoming markedly aggressive and at times revoltingly offensive. In many ways, this trend may well be a reflection, if not a function, of the deep divides that exist today in so many of the world's societies and the truly existential challenges we face in the not too distant future. As a result of this toxic rhetoric, we are in even less of a position to come together to find workable solutions to the very real threats we are facing. As James Hoggan observes in his aptly titled book, *I'm Right and You're an Idiot*, the most critical problem facing our world today may very well be the "pollution in the public square, where a toxic smog of adversarial rhetoric, propaganda, and polarization stifles discussion and debate, creating resistance to change and thwarting our ability to solve our collective problems" (2019: 1).

These toxic changes in discourse have also been clearly seen and felt throughout academia. It is no longer unusual to find that cordial, tolerant, respectful, fact-based dialogue quickly dissolves into personal accusations which can defame as much as they injure. The negative effects of such caustic rhetoric can have a chilling effect at all levels (Lester 2013). Speaking of Forensic Linguistics specifically, some students who witness this verbal volleying may find themselves so repulsed that they elect to pursue another branch of linguistics or completely different discipline altogether. Among those students who decide to stay, this incivility may serve as a model of how to interact with peers, which in time may perpetuate the problem for the next generation of scholars. However, the potential harms are not limited to emerging scholars. The destructive forms of communication that take place in some fora may serve to alienate established colleagues who choose to disengage from larger professional gatherings they perceive as unwelcoming. The breadth and depth of the negative reverberations possible make addressing this issue an ethical priority.

Forensic linguistics has been instrumental in helping to reveal the ways in which language can be weaponized, and it has made many important proposals for thwarting the individual and collective injuries that language can cause (e.g., Gimenez Garcia and Queralt 2024; Requena 2024; Guillén-Nieto 2023; Määttä 2023; Tompkinson 2023; Irimba, Ndambuki, and Mwithi 2021; Hardacker 2020; Christensen 2019; Gales 2011; Infante et al. 2009; Napier and Mardigan 2003; Storey 1995). Given this collective store of knowledge and experience, we may be in a unique position to create safer, more welcoming spaces, where professional discourse is considerate not caustic, and all persons are invited to share their ideas, suggestions, criticisms, recommendations, observations, invitations, and questions. A major step in this direction would be simply to apply many of the already established rules of interaction. For example, many of the universities and organizations that host fora for professional online and face-to-face interactions have regulations about acceptable and inacceptable forms of behavior and communication. In many cases, all that is needed is to consistently enforce pre-existing policies. For the remaining cases that fall outside of the established rules, norms, or guidelines, an appeal can be made to a neutral authority—or better yet a pre-existing ethics committee—to serve as a moderator and make appropriate recommendations for conflict resolution.

This does not mean to say the answer to maintaining civil academic discourse lies in policing communications or stifling spirited, fact-based debates and disagreements. Indeed, one of the hallmarks of academic discourse has been its history of vehement clashes of opinion between scholars who support diametrically opposed theories, interpretations, and methods. Linguistics has certainly seen its fair share of such legendary intellectual disputes between luminaries representing competing positions. However, a clear line must be drawn between heated (and healthy) scholarly debate and personalized derogation. In most cases, this line can be maintained through the consistent, transparent application of basic norms of polite discourse. Granted, there are individual and cultural variations in what is considered polite. It is, however, the very presence and importance of these differences that underscores the need to show others courtesy, respect, and consideration—or, as one senior researcher noted in a recent private conversation, sometimes all we really need is *kindness*.

Conclusion: The Value of Kindness

Now, will kindness alone answer all of the ethical conflicts addressed in this chapter and throughout this Handbook? I imagine, we can probably all agree that the answer to that question is a decisive "No". However, it is equally true that seeking to act through kindness can be a useful tool for facilitating productive, constructive communication; and when sincerely applied, it can help to mitigate unintended injuries when they arise. So, what exactly is meant here by "kindness"? There are seemingly countless theological and philosophical definitions which have been used to operationalize this construct. Some of them are hopelessly kitschy and others are so maddeningly complex as to render them largely useless. For the purposes of this discussion, the following definition from psychology is offered. As Tina Malti, from the University of Toronto, Canada, explains in a 2020 article for the *European Journal of Developmental Psychology*:

> Both Western and Eastern philosophical traditions have conceptualized kindness as *a virtue* [...] and as *a laudable value*. In psychology, kindness has been considered a state of being that *reflects emotions* and inner states as they relate to others and ourselves. Kindness also reflects a *particular sensitivity* for others as well as for the self. In other words, kindness is a *considerate stance towards life*, which *creates meaning and purpose*. It involves a *deep concern* (i.e. *compassion) for both others and the self* and, as such, *reflects an appreciation of the dignity of every human being* [emphasis added].
>
> *(3)*

Within that definition, there are many concepts that have been highlighted in the various chapters of this Handbook: maintaining a clear focus on the meaning and purpose of the task at hand; considering the ways in which one's actions reflect one's personal and professional stance; acknowledging the potential influence of one's emotions and inner state; demonstrating deep concern for safeguarding others and one's self; performing one's tasks to the highest standards possible; and never losing sight of the imperative to respect the dignity of every human being and all life forms. In short, the essence of kindness is to be what we New Yorkers might call being a "mensch": a term which the *Merriam Webster Dictionary* aptly defines as "being a person of integrity and honor". This leads to the driving force behind this Handbook.

Beginning and Ending in Kindness

The idea for this book began rather unspectacularly. It was in some restaurant, at a linguistics conference, somewhere. I don't even remember what time of the year it was. It felt like Spring but there again it could have been Fall or maybe it was a warm day in the Winter…or a cool day in the Summer. The reason why I don't remember any of those details is because I was completely focused on the person who was sitting in front of me. It had been a while since we had seen one another, what with an entire ocean between us and our insane schedules driving us harder than common sense would allow. I remember asking him: "So…when are you gonna slow down… take a break for a while, and just relax?" He burst into laughter and said, "Right after you do, my friend!" And we both laughed together, knowing that neither of us actually wanted to "slow down" because we loved what we did…but still, at the same time, we both also desperately wished that the other one would slow down, just a bit. It was hard seeing the other one struggling. It was scary to think that one day, one of us might not be there.

But those were heavy thoughts and this day, this time, this space was ours, and we were going to enjoy it. Somewhere in between the stories, the laughter, the good food, and drink, we said:

"Wouldn't it be a great idea to put together a book on Forensic Linguistics and Ethics?" And so it began, sitting across from one of the people in the world whom I absolutely adored…still adore… always will adore. Professor Lawrence M. Solan, one of the most brilliant people I ever had the privilege of knowing. But what made Larry so **extraordinary** was not just his incredible intellect and impressively long list of scholarly achievements. After all, there are other geniuses in the world—not many, but there are some, just the same. At the end of the day, what made Larry so utterly irreplaceable was all that **and** he was immeasurably **kind**. Kindness just came naturally to Larry. It was a part of him and everything he did. Everywhere he went, every life he touched, his kindness shone through. It was that Light that made this book possible. Truly, his life is today and will always be a blessing to us all.

Notes

1 The official website for COPE may be found here: https://publicationethics.org/.

2 In their pioneering work, Naufel and Beike (2013) offer a Research Assistant's Bill of Rights. This policy document states that all research assistants have the right to safety, informed consent, refuse participation, withdraw from a research project, counseling, proper training, feedback, debriefing, receive benefits for work performed, choose confidentiality in public acknowledgements. The inclusion of these rights in the Professional Codes of Ethics used by international organizations for forensic linguistics would, in my opinion, be of great benefit for safeguarding both research assistants and their supervisors.

References

Adkins, Jonathan, Ali Al Bataineh, and Majd Khalaf. 2024. "Identifying Persons of Interest in Digital Forensics Using NLP-based AI." *Future Internet* 16, no. 11: 426–445.

Asna, Ntelu, et al. 2025. "The Role of Forensic Linguistics in Investigating and Disclosing Language Crimes on Social Media Platforms." *Theory and Practice in Language Studies* 15, no. 1: 283.

Baird, Katie, and Amanda Kracen. 2006. "Vicarious Traumatization and Secondary Traumatic Stress: A Research Synthesis." *Counseling Psychology Quarterly* 19, no. 2: 181–188.

Beech, Bettina, et al. 2013. "Mentoring Programs for Underrepresented Minority Faculty in Academic Medical Centers: A Systematic Review of the Literature." *Academic Medicine* 88, no. 4: 541–549.

Bloom, Brittnie, Cierra Sorin, Laury Oaks, and Jennifer Wagman. 2022. "Graduate Students Are 'Making a Big Fuss': Responding to Institutional Betrayal Around Campus Sexual Violence and Sexual Harassment." *Journal of School Violence* 22, no. 1: 44–60.

Bogost, Ian. 2025, February 10. "A New Kind of Crisis for American Universities: The Ivory Tower Has Been Breached." *The Atlantic*. Accessed March 25, 2025. Retrieved from https://www.theatlantic.com/health/archive/2025/02/nih-trump-university-crisis/681634/.

Bowen, Robin. 2018. *Ethics and the Practice of Forensic Science*. London: CRC Press.

Byars-Winston, Angela, et al. 2015. "Cultural Diverse Undergraduate Researchers' Academic Outcomes and Perceptions of Their Research Mentoring Relationships." *International Journal of Science Education* 37, no. 15: 2533–2554.

Christensen, Tanya Karoli. 2019. "Indirect Threats as an Illegal Speech Act." In *The Sign of the V. Papers in Honour of Sten Vikner*, 113–130.Edited by Johanna L. Wood, Henrik Jørgensen, and Ken Ramshøj Christensen. Aarhus: Aarhus Universitet.

Coihen, Aaron, and Yehuda Beruch. 2021. "Abuse and Exploitation of Doctoral Students: A Conceptual Model for Traversing a Long and Winding Road to Academia." *Journal of Business* Ethics 180: 505–522.

Colby, Glenn. 2023. *Data Snapshot: Tenure and Contingency in US Higher Education*. American Association of University Professors. Accessed March 25, 2025. Retrieved from https://www.aaup.org/article/data-snapshot-tenure-and-contingency-us-higher-education.

Devos, Christelle, et al. 2016. "Doctoral Students' Experiences Leading to Completion or Attrition: A Matter of Sense, Progress, and Distress." *European Journal of the Psychology of Education* 32, no. 1: 61–77.

Eby, Lilian, et al. 2000. "The Protégé's Perspective Regarding Negative Mentoring Experiences: The Development of a Taxonomy." *Journal of Vocational Behavior* 57: 1–21.

Feldman, Mitchell, et al. 2010. "Does Mentoring Matter. Results from a Survey of Faculty Mentees at a Large Health Sciences University." *Medical Education Online* 15: 5063.

Gaebel, Michael, Thérèse Zhang, and Henriette Stoeber. 2024. *Trends 2024: European Higher Education Institutions in Times of Transition European University Association*. Belgium: European University Association.

Gales, Tammy. 2011. *Ideologies of Violence: A Corpus and Discourse Analytic Approach to Stance in Threatening Communications*. PhD Dissertation. University of California, Davis. ERIC ED520035.

Gentry, R., and Whitley 2014. "Bullying in Graduate School: Its Nature and Effects." *The Qualitative Report* 19, no. 36: 1–18.

Gimenez Garcia, Roser, and Sheila Queralt. 2024. "Grounds for Exemption from Criminal Liability? How Forensic Linguistics Can Contribute to Terrorism Trials." *International Journal of Semiotics and Law* 37: 623–646.

Goodyear, Rod, Clyde Crego, and Michael Johnston. 1992. "Ethical Issues in the Supervision of Student Research: A Study of Critical Incidents." *Professional Psychology: Research and Practice* 23, no. 2: 2023–210.

Grant, Tim. 2008. "Approaching Question in Forensic Authorship Analysis." In *Dimensions of Forensic Linguistics*, 215–229. Edited by John Gibbon and M. Teresa Turell. Amsterdam: John Benjamins.

Guest, Matthew. 2024, September 20. "Education: How Can the Financial Crisis In Higher Education be Tackled?" *Church Times*. Accessed March 20, 2025. Retrieved from https://www.churchtimes.co.uk/articles/2024/20-september/features/features/education-how-can-the-financial-crisis-in-higher-education-be-tackled.

Guillén-Nieto, Victoria. 2023. *Hate Speech: Linguistic Perspectives*. Berlin: De Gruyter Mouton.

Guillén-Nieto Victoria, and Dieter Stein. 2022. "Introduction: Theory and Practice in Forensic Linguistics." In *Language as Evidence: Doing Forensic Linguistics*, 1–33. Edited by Victoria Guillén-Nieto and Dieter Stein. Cham: Palgrave Macmillan.

Hackman, Lucina, 2020. "Communication, Forensic Science, and the Law." *Wiley Interdisciplinary Reviews Forensic Science* 3, no.10: e1396.

Hardacker, Claire. 2020. "Pranksters, Provocateurs, Propagandists: Using Forensic Corpus Linguistics to Identify and Understand Trolling." In *The Routledge Handbook of Forensic Linguistics*, 684–708. Edited by Malcolm Coulthard, Alison May, and Rui Sousa-Silva. New York: Routledge.

Hinchey, Patricia, and Isabel Rimmel. 2000. *The Graduate Grind: A Critical Look at Graduate Education*. New York: Palmer.

Hoggan, James. 2019. *I'm Right and You're an Idiot: The Toxic State of Public Discourse and How to Clean it Up*. British Columbia, Canada: New Society Publishers.

Howes, Loene, and Nenagh Kemp. 2017. "Discord in the Communication of Forensic Science." *Journal of Language and Social Psychology* 36, no. 1: 96–111.

Infante, Dominic, Bruce Riddle, Cary Horvath, and S. Tumlin, 2009. "Verbal Aggressiveness: Messages and Reasons." *Communication Quarterly* 20, no. 2: 116–126.

Irimba, Frederick, Jacinta Ndambuki, and Florence Mwithi. 2021. "A Forensic Interpretation of Hateful Micro-speech Acts and Performative Modality in Facebook and Twitter during 2017 Election-Kenya." *International Journal of Linguistics, Literature, and Translation* 4, no. 3: 74–85.

Johnson, W. 2017. "Mentorship in the Life and Work of the Private Practitioner." In *Handbook of Private Practice: Keys to Success for Mental Health Practitioners*, 222–232. Edited by Steven Walfish, et al. Oxford: Oxford University Press.

Johnson, W. Brad. 2020. "Ethical Consideration for Mentors: Towards a Mentoring Code of Ethics." In *The SAGE Handbook of Mentoring*, 105–118. Edited by David A. Clutterbuck et al. Los Angeles: SAGE.

Keashly, Loraleigh, and Joel Neuman. 2013. "Bullying in Higher Education: What Current Research, Theorizing, and Practice Tell Us." In *Workplace Bullying in Higher Education*, 1–22. Edited by Jaime Bester. New York: Routledge.

Krupnick, Matt. 2023, December 19. "Attacks on Tenure Leave College Professors Eyeing the Exists." *The Center for Public Integrity*. Accessed March 20, 2025. Retrieved from https://publicintegrity.org/education/academic-freedom/attacks-tenure-college-professors-exits/.

Lester, Jaime. 2013. *Workplace Bullying in Higher Education*. New York: Routledge.

Longhi, Julien. 2021. "Using Digital Humanities and Linguistics to Help with Terrorism Investigations." *Forensic Science International* 318: 110564.

Malti, Tina. 2020. “Kindness: A Perspective from Developmental Psychology.” *European Journal of Developmental Psychology* 18, no. 5: 629–657.

Mahmoudi, Morteza, and Loraleigh Keashly. 2021. “Filling the Space: A Framework for Coordinating Global Actions to Diminish Academic Bullying.” *Angewandte Chemie* 60: 3338–3344.

Määttä, Simo K. 2023. “Linguistic and Discursive Properties of Hate Speech and Speech Facilitating the Expression of Hatred: Evidence from Finnish and French Online Discussion Boards.” *Internet Pragmatics* 6, no. 2: 156–172.

McCloskey, Rose, and Kathryn Weaver. 2020. “Working with Research Assistants: Guidelines for Mutually Beneficial Relationships.” *International Journal of Advanced Nursing Education and Research* 5, no. 2: 7–18.

McGinn, Michelle, and Ewelina Niemczyr. 2013. “Research Practice and Research Assistantships: Introducing the Special Issue on Research Assistantships.” *Journal of Research Practice* 9, no. 2 : 1–7.

McKay, Ruth, et al. 2008. “Workplace Bullying in Academia: A Canadian Study”. *Employment Response Rights Journal* 20: 77–100.

Mehta, Vernia, et al., 2024. “Artificial Intelligence in Forensic Linguistics.” In *Artificial Intelligence in Forensic Science*, 79–102. Edited by Kavia Aaini, Swaroop Sonone, Mahipal Singh Sankhla, and Naveen Kumar. Boca Raton: CRC Press.

Moss, Sherry, and Morteza Mahmoudi. 2021. “STEM the Bullying: An Empirical Investigation of Abusive Supervision in Academic Science.” *EClinical Medicine* 40: 101121.

Napier, M., and S. Mardigan. 2003, September 16–19. “Threatening Messages: The Essence of Analyzing Communicated Threats.” *Public Venue Security.*

Nature. 2019, November 13. “PhD Survey.” Accessed March 20, 2025. Retrieved from https://group.springernature.com/gp/group/media/press-releases/archive-2019/nature-phd-survey-puts-spotlight-on-mental-health/17372858.

Naufel, Karen, and Denise Beike. 2013. “The Ethical Treatment of Research Assistants: Are We Forsaking Safety for Science?” *Journal of Research Practice* 9, Article M11. http://jrp.icaap.org/index.php/jrp/article/view/360/318.

Nick, I. 2018. *Forensic Linguistics: Asylum-seekers, Refugees, and Immigrants*. Wilmington, Vermont: Vernon Press.

Office of Students. 2024. “Financial Sustainability of Higher Education Providers in England: November 2024 Update.” Accessed March 25, 2025. Retrieved from https://www.officeforstudents.org.uk/media/nn2fnrkx/financial-sustainability-november-2024.pdf.

Perkins, Ria. 2021. “The Application of Forensic Linguistics in Cybercrime Investigations.” *Policing: A Journal of Policy and Practice* 15, no. 1: 68–78.

Perry, David. 2018, April 16. “How to Hold Predators in Academia Accountable.” *Pacific Standard.* Accessed March 15, 2025. Retrieved from https://psmag.com/education/how-to-hold-predators-in-academia-accountable/.

Ragins, Belle Rose, and Kathy Kram. 2007. *The Handbook of Mentoring at Work.* Los Angeles: SAGE.

Requena, Lucia Sevilla. 2024. “‘She’ll Never be a Man’ A Corpus-based Forensic Linguistic Analysis of Misgendering Discimination on X.” *Languages* 9, no. 9: 291.

Rosenthal, Marina, Alec Smidt, and Jennifer Freyd. 2016. “Still Second Class: Sexual Harasssment of Graduate Students.” *Psychology of Women Quarterly* 40, no. 3: 364–377.

Rossow, Johannes, and Ewelina Niemczyk. 2013. “A Dual Perspective on Risks and Security Within Research Assistanceships.” *Journal of Research Practice* 9, no. 2: M10 1–23.

Scandura, Terri. 1998. “Dysfunctional Mentoring Relationships and Outcomes”. *Journal of Management* 24, no. 3: 449–467.

Schapira, M. et al. 1992.”Mentorship in General Internal Medicine: Investment in Our Future.” *General Internal Medicine* 7, no. 2: 248–251.

Scott, Juliette, and J. A. O’Shea. 2022. “Translation in Libel Cases: Reputations at Stake!” *Comparative Legilinguistics* 50: 123–179.

Silva, Diane, and Alan Tom. 2001. “The Moral Basis of Mentoring.” *Teacher Education Quarterly* 28, no. 2: 39–52.

Sousa-Silva, Rui. 2024. “Fighting Cyber-malice: A Forensic Linguistics Approach to AI-generated Malicious Texts.” *Proceedings of the 1st International Conference on NLP & AI for Cyber Security*, 164–174.

Smith, Marshall. 2013. “ The Global Economic Crisis: Effects on US Higher Education.” *Procedia: Social and Behavioral Sciences* 77: 153–158.

Steiner, John, et al. 2004. "Assessing the Role of Influential Mentors in the Research Development of Primary Care Fellows". *Academic Medicine* 79, no. 9: 865–872.

Storey, Kate. 1995. "The Language of Threats". *International Journal of Speech, Language and the Law* 2, no. 1: 74–80.

Terkourafi, Marina. 2025. "An Ethics for Linguistics? What, Why, and How?" Editorial. *Linguistics* 63, no. 2: 1–31.

Thiry, Heather, Sandra Laursen, and Anne-Barrie Hunter. 2011. "What Experiences Help Students Become Scientists? A Comparative Study of Research and Other Sources of Personal and Professional Gains for STEM Undergraduates." *The Journal of Higher Education* 82, no. 4: 357–388.

Tiersma, Peter, and Lawrence Solan. 2002."The Linguist on the Witness Stand: Forensic Linguistics in American Courts." *Language* 78, no. 2: 221–239.

Tompkinson, James. 2023. *Spoken Threats from Production to Perception*. Cambridge: Cambridge University Press.

Ward, Tony. 2023. "The Forensic Ethics of Scientific Communication." *The Journal of Criminal Law* 87, no. 1: 3–17.

Watt, Dominic, and Georgina Brown. 2020. "Forensic Phonetics and Automatic Speaker Recognition: The Complementarity of Human-and-machine-based Forensic Speaker Comparison." In *The Routledge Handbook of Forensic Linguistics*, 400–415. Edited by Malcolm Coulthard, Alison May, and Rui Sousa-Silva. New York: Routledge.

Wi, Jian, et al. 2024. "Forensic Communication Analysis: Challenges and Opportunities" 2024 IEEE International Conference on Cyber Security and Resilience, 878–883.

Wingard, Deborah, Karen Garman, and Vivian Reznik. 2004. "Facilitating Faculty Success: Outcomes and Cost Benefit of the USCD National Center of Leadership in Academic Medicine." *Academic Medicine* 79, no. 10: S9–S10.

CASE LAW INDEX

Adorno v. Melvin (2017) 213, 222
Al-Hussaini Hussain v. Palmer Communications Inc. 170
Arredondo v. Neven (2014) 177, 186
Associated Press v. Meltwater (2013) 358
Austin v. American Association of Neurological Surgeons (2001) 30, 38

Baigent and Leigh v. The Random House Group Ltd (2007) 384, 393
Baltazar-Monterrosa v. State (2006) 284, 292
Bank of Montreal v. Citak (2001) 247, 251
Bruton v. United States (1968) 194, 203

Calhoun v. U.S. (2013) 184, 186
Campbell v. Acuff-Rose Music, Inc. (1994) 386, 393
Campbell v. Mirror Group Newspapers Ltd (2004) 382–383, 393
Carmen Alfano Family Trust (Trustee of) v. Piersanti (2012) 247, 251
Calhoun v. U.S. (2013) 184, 186
Campbell v. Acuff-Rose Music, Inc. (1994) 386, 393
Campbell v. Mirror Group Newspapers Ltd (2004) 382–383, 393
Coco v. A. N. Clark (Engineers) Ltd. (1969) 382, 393
Commonwealth v. AdonSoto (2016) 284, 292
Commonwealth v. Al-Hussaini (2011) 163, 170
Commonwealth v. Rosario (2017) 287, 292
Commonwealth v. Webster (1850) 218, 222
Consulate Ventures Inc. v. *Amico Contracting & Engineering* (1992) 29, 38
Correa v. Superior Court (2002) 288, 292

Daubert v. Merrell Dow Pharmaceuticals, Inc. (1993) 38, 62, 147, 160, 171, 244, 251
De Angelis v. Siermy (2022) 250, 252
Dean Construction Co. v. M.J. Dixon Construction Ltd. (2011) 247, 252
Detroit Fire & Marine Ins. Co. v. Gagliardi (1934) 27, 38
Dussiaume v. Sandoz Canada Inc. (2023) 252

eBay v. Bidder's Edge, Inc. (2000) 358
English Feedlot, Inc. v. Norden Labs., Inc. (1993) 160, 171
Experience Hendrix LLC v. Purple Haze Records Ltd and Ors (2007) 393

Facebook v. Power Ventures (2016) 358
Faretta v. California (1975) 174, 176–178, 186
Fellowes, McNeil v. Kansa General International Insurance Co. (1998) 247, 252
Frye v. United States (1923) 53, 62, 71, 128

General Electric Co. v. Joiner (1997) 160, 171
General Medical Council & Professional Standards Authority v. Asef Azar (2020) 30, 38
Gideon v. Wainwright (1963) 174, 186
Gould v. Western Coal Corp. (2012) 247, 252
Gray v. State of Maryland (1998) 195, 203

Haelan Laboratories, Inc v. Topps Chewing Gum Inc. (1953) 387, 393
Hernandez v. New York (1991) 285, 292
Hutchingame v. Johnstone (2006) 247, 252

In re (Benavides) Figueroa (2018) 290–292
In re J.F. Guerra (2007) 288, 292
In re Joseph H. (2015) 175, 186
Infopaq International A/S v. Danske Dagblades Forening (2009) 385, 393
Iowa v. Tovar (2004) 177, 186
Irvine v. Talksport Ltd (2003) 386–387, 393

Johan Deckmyn and Vrijheidsfonds VZW v. Helena Vandersteen and Others (2014) 385, 394
Jones v. Kaney (2011) 29, 38
Joseph H. v. California (2016) 175–176, 186

Kannankeril v. Terminix Int'l, Inc. (1997) 38
Kaye v. Robertson (1991) 382, 394
Kirby Lowbed Services Ltd. v. Bank of Nova Scotia (2003) 247, 252
Kumho Tire Co. v. Carmichael (1999) 38, 160, 171, 244, 252

Leal v. State (1989) 283, 292
Liverpool Victoria Insurance Co. Ltd v. Zafar (2019) 30, 36–38

Mak v. Blodgett (1991) 163, 171
Midler v. Ford Motor Co. (1988) 386, 394
Miller-El v. Dretke (2005) 260, 262
Miller v. Minister of Pensions (1947) 210, 222
Miranda v. Arizona (1966) 174, 256, 262
Moffat v. Canada (Minister of Citizenship and Immigration) (2019) 252
Moffitt (Litigation Guardian of) v. TD Canada Trust (2023) 245, 252
Moroccanoil Israel Ltd v. Aldi Stores Ltd (2014) 394

Ocean v. Economical Mutual Insurance Co. (2010) 247, 252

Peña-Rodriguez v. Colorado (2017) 185–186
People v. Arancibia (2013) 285, 292
People v. Brown (1990) 186, 285–286, 292
People v. Coriano 284
People v. Frederickson (2020) 177, 186
People v. Garcia (2022) 282, 292
People v. Garcia Trujillo (1997) 281, 292
People v. Guerra (2006) 287, 289, 292
People v. Harris (2016) 372, 375
People v. Hobart and Konard 289–290
People v. Lee Fat (1880) 282, 292
People v. Polk (1996) 285, 292
People v. Quinonez (2014) 285, 292
People v. Riggs (2008) 177, 186
People v. Robinson (2017; 2019) 183, 186
People v. Torres (1985) 282, 292
People v. Torres (1989) 281, 292
People v. Velasco-Palacios (2015) 287, 292
People v. Waldon (2023) 177, 186
People v. William Rolando 180
People v. Wycoff (2021) 177, 186
People v. Zamora (2012) 282, 292
Person v. Ass'n of Bar of City of New York (1977) 34, 38

Pora v. R (New Zealand) (2015) 34, 39
Postelle v. State (2011) 213, 222

R v. B (2010) 268, 277
R v. Millard (2018) 252
R v. Abbey (1982) 252
R v. Abbey (2009) 244, 252
R v. Bouvette (2023) 252
R v. Chatzidimitriou (2000) 214, 222
R v. Docherty (2010) 247, 252
R v. France (2017) 245, 252
R v. J.-L.J. (2000) 39
R v. Lifchus (1997) 214, 216, 222
R v. M(S) (1991) 241, 252
R v. Mohan (1994) 29, 39, 243–244, 246, 250–252
R v. Momodou & Limani (2005) 36, 39
R v. N.O. (2009) 29, 39
R v. Scott (2018) 247–258, 252
R v. Sekhon (2014) 250, 252
R v. Slade and Ors (2015) 262
R v. Wanhalla (2007) 216, 222
Reckitt & Colman Products Ltd. v. Borden Inc. (1990) 394
Richardson v. Marsh (1987) 181, 186, 195, 203
Royal Trust Corp. of Canada v. Fisherman (2000) 247, 252

Secretary, HUD v. Blackwell (1990) 262
State of Florida v. Zimmerman (2013) 193, 203, 402, 442, 444
State of North Carolina v. Brewington (2000) 195, 203
State of North Carolina v. Ferrell (1980) 195, 203
State v. Gonzalez-Hernandez (2004) 281, 293
State v. Lopez Ramos (2019) 280, 283, 293
State v. Zamora (2022) 183, 186
Stuart v. R (1959) 27, 39

Taylor v. State (2016) 284, 293
The Queen v. Dookheea (2018) 214, 222
Thimmaya v. Lancashire NHS Foundation Trust (2020) 37, 39
Ticketmaster Corp. v. Tickets-com, Inc. (2003) 358
Tyus v. Urban Search Management (1996) 39

United Biscuits (UK) Ltd v. Asda Stores Ltd (1997) 394
United States v. Fields 213, 222
United States v. Garibay (1998) 174, 186
United States v. Granderson (1994) 20
United States v. Hayat (2013) 291, 293
United States v. Hayat (2019) 291, 293
United States v. Long (2002) 282–293
United States v. Onori (1976) 282, 293

United States v. Pena-Espinosa (1995) 293
United States v. Pepe (1974) 206, 222
United States v. Robinson (2024) 363, 375
United States v. Silva-Arzeta (2010) 279, 293

Van Blargan v. Williams Hospitality Corp. (1991) 160, 171
Victor v. Nebraska (1994) 210, 222

Waits v. Frito-Lay, Inc. (1992) 394
Wallace v. State (2020) 284, 293
White Burgess Langille Inman v. Abbott and Haliburton Co. (2015) 29, 39, 245–247, 252
White v. Samsung Electronics America, Inc. (1992) 393*n*11, 394
Woolmington v. DPP 206, 222

Ye v. US (2016) 281, 293

SUBJECT-AUTHOR INDEX

Page numbers in **bold** denote tables, those in *italic* denote figures.

#MeToo movement 444–445
911-call 402, 442, 444; analysis 109, 112; handling 440; operator 109

a priori 35, 45, 50, 55, 60, 127

ableism 323
abuse 8, 11, 273, 304, 306, 330–332, 337, 411–412, 415, 419, 470, 484; allegations of 485; animal 405; child 98–99, 147, 290, 299, 333, 355, 426–427, 434; disability 269; domestic 427, 443; of authority 24; of governmental power 130; of language 107; of power 132, 140; of process 256; outright 484; possibility of 422; prevention of 125; sexual 9, 98–99, 109, 270, 276, 290, 328, 331–332, 334, 338, 353, 427, 430–431; substance 99; survivors of 484; victim 98, 290, 330, 332
academics 3, 32, 36, 61, 79, 85, 105, 154, 214, 322–323, 412, 441, 443, 453–457, 474, 486–487; affiliation 466; background 487; careers 79, 410, 432, 484; codes of ethics 151; colonization 454; commitments 6; community 203, 453, 455; conferences 350; consensus 216; credentials 33; detail 82; disadvantage 427; discourse 489; discussions 455; environment 455; issues 412; meritocracy 255; mindset 417, 452; narcissism 306; opinion 20; peers 453; performance 429; perspectives 145; policies 315; position 29; positionalities 313; prestige 457; programs 3; publishing 1, 450–457, 486; requirements 153; research 347, 450; societies 150; source 1134; spirit 417; standards 151; teaching 411; topics 206; value 441; vulnerability 431; work 203, 215, 428; world 185; year 431, 486
Academy of Royal Medical Colleges (AMRC) 30
accommodation 272; principle of 209
accountability 243, 249, 292, 334, 337, 381, 446, 484; feelings of 23; public statement of 31
accuracy 5, 35–36, 90–95, 98, 101, 113, 117, 207, 209, 211–212, 216–217, 219–221, 259, 272, 281, 283–284, 292, 346, 377, 381, 411, 463, 466, 473
Achieving Best Evidence (ABE) 267–272, 275–279
Acquired Brain Injury (ABI) 227
acts: abusive 332, 337; communicative 118; heinous 243; linguistic 118; non-consensual 330; of cruelty 167; of malpractice 3; of misconduct 22; of plagiarism 411; of recognition 125–126; of resistance 457; of respect 218; of terror 357; of violence 344–345; sadistic 331; speech 107; terrorist 343
Act on Interpretation Services for Persons with Disabilities (2010) 317
activism 44; social 5; personal 442; political 343–361; limitations of 44; prohibitions against 44; recommendations for 488–489
ad hominem attack 50, 127
Adam, R. 322
addressees (participation framework) 207–208
adjacency pairs130–131, 200
adjudication 18, 184, 366; criminal 184–185
admissibility 17, 20, 202, 246; assessment of 245; decision-making 33; legal 53; of evidence 240, 243–244, 246; of expert opinion 244; of expert testimony 20, 29, 37, 147, 160; requirement 246; ruling 160
Adult Witnesses with an Intellectual Disability (AWID) 9, 267–276; vulnerability of 266–271, 275; interviewing 267–272; credibility 268, 272–276; reliability 268–269, 272–273; *see also*

credibility, interviewing, credibility, reliability, vulnerable witness, witness
adversarial system 97, 128, 146, 161, 170, 197, 202, 274, 276, 410
adversary: process 19, 21; system 8, 19, 22, 146–147, 149, 155, 161, 181
adverse credibility indicator 128
advocacy 24, 97, 114, 146–147, 161–162, 167, 169–170; adversarial 97; appearance of 21; ethics of 9, 146, 162; consultancy 1–2, 4–5, 8, 14, 488; ethical limitations of 20–22, 38, 50, 150–152, 250, 488; exhibited 22; guidelines for 4–5, 20–22, 72, 250; hired gun 38, 156, 250, 488; legal 290; position of 162; prosecutor 24; research 364; resistance to 20; self- 317; signs of 22; social 43–45, 48–50; trial 208; zealous 147, 154; *see also* consultancy, ethics, expert witness, forensic linguistics, professional standards
advocates 29, 36, 210, 274, 286, 364, 441, 457, 470; attorney 162; child 306; civil rights 43, 473; human rights 43, 473; Innocence Project 473–474; legal 18–20; personal agency 199; social 43; *see also* advocacy, expert witness, human rights, Innocence Project, social justice
African American Languages (AAL) 259
African American Vernacular English (AAVE) 193, 202, 455
agency 18, 24, 125–126, 199–200, 218, 302, 309, 348–349, 440; government 18, 24, 348–349, 440; immigration 348–349; law enforcement 125–126, 199–200, 218, 302, 309; *see also* asylum, government agencies, immigration, law enforcement, police
agent 132, 199–200, 228, 315, 320, 368, 372; enforcement 83, 90; epistemically reliable 368; epistemically responsible 367; grammatical 195; immigration 312–323; implied 280; individual 368; intentional 232; law enforcement 83, 90, 256; moral 162, 170; of the prosecution 179; recontextualizing 197; *see also* immigration, law enforcement
Ahearn, L. 199
Ainsworth, J. 2, 5, 120, 146, 148, 152, 173, 176, 179, 194, 202
Al-Abdulmohsen, Taleb 343–347, 356–357; Germany 347, 354; lone wolf 344–345, 356–357; mass casualty 343–347, 357; mental health 168; terrorism 343–357; threat communication 345–357; violent extremism 343–357
Albright, T. D. 88, 90
alcoholism 166
Aldridge, M. 9
Alduais, A. 452
Alger, K. 179
Alshebi, K. 466
alternate jury 220
Alternative and Augmentative Communication (AAC) 271–272
ambiguity 126, 133, 200–201; deliberate 33; fundamental 88
American Academy of Forensic Sciences (AAFS) 462
American Association for Applied Linguistics (AAAL) 19
American Bar Association 34; Model Rules 146–147
American College of Surgeons (ACS) 150
American Expert Witness 9
American Legal System 170, 173
American Psychological Association (APA) 405
American Sign Language (ASL) 314, 438
analysis: accurate 17, 356, 477; agency 200; audio 181; auditory-acoustic 56; authorship 32, 409, 411, 416–417; autoethnographic 437; bias 24, 35, 83, 99; cognitive 96; comprehensive 97, 282; conceptual 364; conversation 112, 131; credibility 106, 109; crime scene 10; data 60, 109, 364, **365**, 473; discourse 3, 19, 32, 176; documentation of 154; expert 21, 35, 56, 289, 307; final 3; forensic 17, 33, 44, 56, 61, 249, 307; forensic linguistic 28, 81, 88, 100, 181, 192, 259–260; forensic phonetic 403; forensic stylistic 250; framework 72; independent 59; interpretation 35, 363; language 2, 57–58, 109, 366, 409; linguistic 8–9, 27, 80, 82–84, 124, 149, 151, 207, 219, 225, 228, 290, 305, 363–364, **365**, 366, 384, 389, 420, 463, 473; logical 118, 120; microscopic hair 23–24; mistakes 121; objective 23; of concordances 212; of evidence 44, 48; of physical evidence 44; of speech 43; pragmatic 181; process 151; qualitative 151; reflexive thematic 329; reliability 5, 368; reliable 374; SCAN 109; scientometric 452; sequential 131; social science 304; standards of 29; statistical 126; subjective elements of 19; systematic corpus 373; textual 27, 112; thoroughness of 35; transparency 100, 374; validity 72, 369; *see also* 911-call
analyst 35, 60, 79–80, 82–85, 131, 446; background expertise *365*; crime scene investigation 100; data 83; dedicated 35; discourse 438, 441–442; discretion *365*; DNA 155; experience 257; forensic 55; interpretations 131; personal beliefs 50
Andersson Cederholm, E. 300
Andrews, S. 98
Angelelli, C. V. 282
Anglophone norms 456
animal 183, 398–401, 403–406; abuse 405; bias 399; communication 399, 401, 403, 406, cross-species bias 404; degraded position of 399; human 399, 406; industry 400; linguistic

imperialism 450, 452–455, 457; linguistics 403; non-human 10, 398–406; products 400; recordings 403; relationship 399; species 403–404; speech 406; studies 398, 405; voices 10, 401–403
Animal Labor Union 405
animator (participation framework) 207
anonymization 350–352, 420–421; archival data 351; data protection 346, 382–383, 414; pseudonym 315, 318, terms of use 346; *see also* confidentiality, privacy
anthropocentrism 398
Anya, U. 322
Anzaldúa, G. 450–451
Appeals Court 212; justices 208
Appelbaum, P. 162
appellate court 183, 213, 281–282
area of expertise 3, 32, 45, 57, 68, 150, 411
Aristotle 77–78, 399
Artificial Intelligence (AI) 10, 83, 259, 357, 377–378, 380, 414, 483
Ask, K. 95
Assault and Battery 163, 167–168
Association for the Treatment and Prevention of Sexual Abuse 330
Association of Forensic Science Providers (AFSP) 48, 53
Aston Institute for Forensic Linguistics (AIFL) 409
asylum 318, 437; application 313, 321; authorities 58; cases 50, 57–58, 194; decisions 194, 318, 320; denial 320; government agency 349; hearings 318–321, 323, 450; insane 122; petitions for 483; procedure 58, 313, 315–316, 319–320, 322, 399; process 313, 315–318, 322–323; seekers 57–58, 194, 312–313, 315–320, 455; torture 77, 163–164, 167–168; *see also* advocacy, deaf, immigration, sign language, translation
attribution 89–90, 366, 399; agency 199–200; authorship 249, 411; bias 89; methods 409; of agency 199; of meaning 88; of speech 56
audio 57, 59, 402; analysis 181; conditions 403, 406; evidence 59; files 382; formats 18; materials 59; recordings 8, 43, 90, 96, 131, 139–141, 194, 284–285, 381, 402; sample 57, 59; specialist 181; transcription 57, 60; -video data 439; *see also* forensic speech science, phonetics, phonology
Augmentative and Alternative Communication (AAC) 271–272
Australia 27–28, 30, 33, 36, 160, 194, 213–214, 276, 282–283, 286, 344, 410, 455
authorship: forensic speech science 54; identification 121; phonetics 3, 43, 249; phonology 3; rights 212; *see also* analysis
autoethnography 442
Automated Speech Recognition (ASR) 259
awareness 23, 35, 38, 44, 61, 83, 96, 128, 141, 170, 234, 260, 276, 308, 333, 355, 412, 420, 422, 443, 445, 464; acute 353; critical 457; cultural 169, 444; ethical 86, 419; forensic 403; of language 438; personal 35; professional 22; self- 337–338, 354, 485; training 50

balance 20, 46, 48, 59, 61, 66, 74, 78, 82–83, 114, 152, 195, 219, 272–273, 354, 386, 414, 429, 439, 443, 445, 466; complex 387; delicate 332; ethical 438; evaluation 45; importance of 78; maintaining 50; moral 336; of probabilities 225; professional 22
Balmat, M P. 440
Bang, B. L. 284
Barlow, M. 279
Baugh, J. 2, 261, 464, 468
Beck, G. 165
Behar, R. 441, 443
Behrman, S. 312
beliefs 23, 45, 78, 91, 219, 249, 412, 440, 442, 446, 468; conscious 23; delusional 234; faulty 114; folk 63–64, 72–73, 169; folk-linguistic 44; individual 446; mis- 63, 65, 68–70, 72–73; negative 70; personal 50–51, 257; pre-existing 463; racist 457; sets of 193; unsupported 328
beneficence 405
Benesch, S. 5
Benmaman, V. 280
Berk-Seligson, S. 2–3, 284, 289–290
Berne Convention for the Protection of Literary and Artistic Works (1886) 384
Bersoff, D. 411
best practice 5, 7, 18, 20, 24, 53, 72, 150, 248–249, 251, 268, 271, 415, 426, 434, 466, 485, 488
beyond (a/any/all) reasonable doubt 9, 206–207, 209–219, 221
bias: anchoring 35; anti-client 60; attributional 89; avoiding 84; classism 474; cognitive 23–24, 35, 43, 47, 49–50, 54, 61, 356, 464–465, 471–472, 476; conclusions 35; confirmation 23–24, 50, 109, 179, 192, 463; conscious 100, 240; contextual 109; counter- 58; cross-species 404; discussions of 23; dispositional 89; error 17, 24; ethical 50; ethnic 183; evidence of 23, 183; expert 22, 24, 61; fallacies 35; freedom from 95, 100, 246; gender 183; heuristic 464; hidden effects of 100; impact of 23, 464; impartiality 19, 45, 50, 151; implicit 182, 440, 442, 445–446; influence 23, 58, 100, 184; information 50, 61, 347; investigator 96; language-based 193; LGBTQ+ 415, 443, 467; linguistic 440, 450–453, 455–457; long-standing 285; management 50; methodology 50; minimizing 83–84; mitigating 84; motivation 60–61; norms; perceptual 260; personal 35, 47, 49–50, 65, 366, 466; political

43, 49; possible 88, 100–101, 347; potential 17–18, 24, 35, 81, 84, 98, 100–101, 184, 191, 248, 276, 336, 464, 472; professional 347; psychological 8; racial 183–184; rates 24; research 184; risks of 50; selective attention 89; selective information 35; situational 89; social 43, 49, 183, 185; social desirability 335; source of 43, 50; susceptible to 22; tunnel vision 35, 50; un- 20, 23, 29, 46–48, 88, 120–121, 124, 159, 183, 245, 299; unavoidable 55; unconscious 35, 240, 377, 457; widespread 400, 450; *see also* analysis, gatekeeping, linguistic, priming, stereotype
big data 358n1
biology 18, 398
Black Lives Matter (BLM) 444
Blagden, N. 337
Blandón-Gitlin, I. 8
blind comparisons 23
Bloch, B. 378–379
Blommaert, J. 196
Blondel, D. 27
Bogaard, G. 108–108
Bogost, I. 486
Bolitho, A. 284
bomber 163, 167, 386
Bonjour, L. 367–368
Boon, H. 412
Borough of Manhattan Community College (BMCC) 438
Boston 164, 166; Medical Center 167
Bourdieu, P. 453
Bowen, R. T. 485
Bowman, M. N. 184
bracketing 153
breaches 83, 85; in security 348, 350; of contract 250; of informed consent 346; of privacy 391
Brennan, M. 411
Brewer, W. F. 89
Brinkmann, S. 332–333
British Association of Applied Linguistics (BAAL) 81
Brown-Blake, C. 194, 455
Bruton Rule 194–197, 202
Bryan, V. 354
bubble of disinformation 121–122
Bucholtz, M. 197, **440**
Bull, R. 97, 99
bullying 484
Bundeskriminalamt (BKA) (Federal Criminal Police Office of Germany) 347, 354
burden of proof 206, 208–210, 216
burnout 300–302, 306, 308–309, 428, 443, 445–446, 470
Butters, R. 1–2, 155, 472
by-the-book solutions 170
California Evidence Code 282
California Penal Code 184
California Rural Legal Assistance (CRLA) 281
Cambier-Langeveld, T. 7, 8, 45
Campos, P. 229
Canada 29–30, 36, 239–241, 243, 245–246, 248–251, 453, 490
Canagarajah, S. 320
capability 217; heightened 259; reasoning 162
capital case 177–178, 256, 258, 472; punishment 210; *see also* death penalty, Hofstra University, Innocence Project, Leonard, R., stay of execution
Capuano, A. M. 114
caregivers 114, 241, 405
Carello, J. 426–427
Carniel, J. 428, 430
Cartesian dualism 398
casework 5, 8, 31–32, 35, 48, 54, 68, 73, 76, 79–80, 85, 152, 203, 251, 304, 308, 401, 485; body of 1; expert 50; forensic 5, 45, 49–50, 54, 401, 415; forensic linguistic 27, 79, 81; illustrative 416; investigative 27; legal 28; materials 413; *see also* consultancy, hired gun
categorical imperative 77, 208
cautionary instructions 184
certainty 20, 127, 210, 213, 234–235, 266, 378, 384; absolute 22; assertions of 3; functional 210; high level of 81, 84; linguistic 127; measurable level of 152; moral 210, 212–213; scientific 24; subjective 210–212, 221; unreasonable levels of 37
certification 19, 246; barriers to 11, 267, 274, 457; expertise 150, 468; licensure 3, 121, 149–150, 155; oversight 7, 17, 81, 249
Chakrabarty, D. 400
challenges 9–10, 10, 22, 25, 70, 114, 153–155, 176, 249, 268, 270, 286–287, 291, 322, 328, 338, 402, 411, 419, 427; communicative 268; emotional 336; ethical 6–8, 18, 22, 32, 54, 61, 76, 78, 105, 146, 156, 170, 270, 328, 346, 350, 352, 415, 485–486, 488; existential 488; forensic 119; global 11, 475; legal 5, 315, 347; logistical 270; mental health 467; post-conviction 279, 287, 289; practical 61; pragmatic 105
Charter Standard 4 of the Witness Charter (2013) 270
children 98–99, 108, 175, 228, 241, 243, 281, 299–301, 306, 319, 322, 328–329, 333, 366–367, 379, 404, 427; protection 426, 435, 467; *see also* abuse, credibility, interviewing, reliability, statements, trauma, witness
Christensen, A.-D. 8
Christianson, S.-Å. 92
Chung, C. 111
citation equity 456
City University of New York (CUNY) 438

Civil Procedure Rules (CPR) 29, 30, 81
Clarke, I. 306–307
closure 97, 214, 267; unwarranted discursive 213
Cobian, J. 279
COBUILD 213
co-conspirator 166
codes of practice 5, 19, 47–48, 51, 53–54, 150, 487; *see also* American Psychiatric Association (APA), best practices, guidelines, International Association for Forensic and Legal Linguistics (IAFLL), International Association for Forensic Phonetics and Acoustics (IAFPA), Linguistic Society of America (LSA)
coercion 108, 131–132, 134, 141, 227, 388; dilemma of 133–134; opposing forces of 132; police 135, 141, 256; sexual 434; unethical 107
cognitive: behavioral therapy 330; bias 23, 35, 43, 47, 49–50, 54, 61, 356, 464–465, 471–472, 476; limitations 273, 288–289; processes 90, 94, 257, 464
Collins, A. 428
colonialism 453–454, 457; English first 451, 474–475; English only 286, 451, 453–454; hierarchies 450–451; non-English language 281, 454, 474–475; *see also* Global North, Global South, hermeneutical injustice, linguistic imperialism
Conley, J. M. 8, 145, 149, 154
Committee for Ethnic Diversity in Linguistics (CEDL) 5
Committee on Public Ethics (COPE) 488; *see also* best practices, fairness, guidelines, publishing
communication: aids 267–268, 272–274, 276; of findings 79, 84
Communication of Rights Group (CORG) 72, 176
communicative: adequacy 117; norm 194; success 119; sufficiency 117
compensation 147–148, 285, 391; financial 329, 405; issues 150; *see also* consultancy, hired gun
competency 19, 22, 24, 152, 169; linguistic 475; multilingual 475; professional 242; technical 32
complex 20, 124, 145, 152, **174**, 175, 207, 344, 383, 415, 453, 488, 490; admonitions 178; argumentation 121; array of sensations 88; aspects 338; balance 387; cases 322; challenge 77, 347; communication needs 267; decision 77; definitions 212–213; emotions 336; ethical issues 37; events 89; evidentiary rules 282; health conditions 426; human behaviors 117; intersecting data protection laws 346; issues 24, 279, 428; language 33, 73, 192; legal duties 152; legal landscape 185; legal processes 319; linguistic analysis **365**; medical problems 232; methods 465; nature 289; nexus 173; Oath 73; obstacles 291; questions 98; reactions 269; relationship 11, 437; scientific language 33; situations 338, 380; skills 280; sting operation 66; support mechanisms 428; syntax 213, 215, 272; topics 334; understanding 451; work assessment 486
complexity 4, 6, 94, 185, 289, 332, 344, 381, 421, 437; ethical practice 86; human experience 439, 443; investigations 79; language 33, 114, 177–178; linguistic features 206; linguistics 34, 178; needless 33; proper communication 49; rules 117; sensory input 88
comprehension 33, 72, 174–176, 178, 182, 213, 260; adequate 176; barriers to 176; difficulties 215; improved 218; language 285; level of 99; meaningful 72; skills 98
comprehensibility 20, 32, 207, 209, 212, 215, 220–221, 368, 370; linguistic 206; textual 217
conceptual distortion 121
confession 27, 33, 134–135, 141, 146, 181, 195–196, 473; -based 106, 108; coerced 473; false 106, 108, 176, 437; forced 106, 108, 134; interviews 220; of guilt 134; Reid technique 106–113; -seeking approach 96; unreliable 34; written 108; *see also* interrogation, law enforcement, PEACE, witness
confidence level 209, 211; Daubert 33, 71, 244; Frye 53, 71; statistics 151, 335, 469, 484
confidentiality 5, 19, 71, 148, 150, 152–153, 350, 382–383; agreements 148, 153; concerns 338; participants' 151, 153; principles of 393; promises of 154; student 431
confirmation bias 23–24, 50, 109, 179, 192, 463
conflict of interest 31, 46–48, 71–72, 147, 281, 289; *see also* bias, consultancy, priming
conformity 246, 466–467; cultural 456; ideological 446; linguistic 455; memory 94; social 379
Confucian tradition 78
conscientização 457
consequence 37, 77, 82–83, 85, 120, 146, 168, 175, 195, 197, 220, 255, 261, 269, 274–275, 316, 318, 323, 331, 349, 409, 414, 429, 431, 434, 451, 464; devastating 8, 114, 239; legal 202; linguistic 195, 201–202; negative 82, 84–85, 106, 467; potential 37, 82–83, 411, 414; real-world 124; serious 73; significant 193; unintended 82
consequentialism 76–77, 82, 84–85
consistency 17, 20, 90, 98, 100, 195, 207, 209, 212, 219–221, 254, 273, 351; alleged 290; lack of **365**; message 212; of delivery 217; of examination practices 18; *see also* reliability, validity
conspiracy 165, 180–181; alleged 67; meaning of 181; -minded associates 166; theory 165, 211, 332; thinking 211
constitutional rights 173–175, 178, 195
consultancy 4–5, 9, 32, 248–251; legal 155; private 487; pro bono 80, 84, 472; *see also* Higher Education, hired gun

consulting 125, 141, 202, 334, 351; assignment 2; expert 18; forensic linguist 21, 81, 150, 198; legal 71; linguist 120
Consumer Financial Protection Board (CFPB) 126
consumers 22, 386
contamination 181, 471; cognitive 47; evidence 36, 202; memory 94
context: commercial 386; communication 73; -creating 196; cultural 145, 285, 291, 451; -dependent 196, 314; difficult 232; discourse 258; emotional 90; endorsement 386; English-speaking 90; forensic 45, 61, 94, 149, 401; forensic-relevant 90; forgotten 197; global 7; high-stakes 126, 178; historical 444, 476; information 55, 61; institutional 219; judicial 30; legal 63, 151, 209, 221, 249, 368, 370, 372; linguistic 90, 196–197, 239; military 65; narrative 20; national 322; of patterns 18; real-life 61; religious 213; -renewing 196; -shaped 196; situational 90, 463; social 9, 173, 196; societal 323; trial 214; unspoken 118
contradictions 98, 151, 225; legal 130
Conversation Analysis (CA) 112, 131
conviction 23–24, 27, 99, 132–133, 149, 165, 168, 181–183, 208–210, 215–218, 225, 241, 283, 286–287, 290, 389, 400, 468, 472–473; consequences of 168; criminal 243; erroneous 182, 287; evidence for 289; lack of 109; murder 464; past 23; personal 23, 44; post- 230, 279, 289–291; reversal 183; secure 30; subjective 210; tainted 291; wrongful 140, 210, 287
cooperative communication 118, 121, 123, 126
copyright 383–384, 386, 390–392; infringement 346, 349, 385; law 384–385; ownership 385; protection 384–386; question of 384; subsistence 385; work 385–386
Cornips, L. 399
Coroners and Justice Act (2009) 225–226
corpus linguistics 66–67, 151, 364, 366–367, 372
Corpus of Contemporary American English (COCA) 67, 372
Corpus-Based Legal Interpretation Analysis (CLIA) 363–364, **365**, 367–368, 370, 372–375
Cotterill, J. 232
Coulthard, M. 2, 7, 28
counselling 167, 302; 425–426; 428; wellbeing 433; *see also* mental health, proactive, resilience, self-care, support, trauma
consultancy 4, 251; linguistic 4–5; pressure of 32; private 487; work 155, 249–250
counter report 57–58
countertransference 337
COVID-19 pandemic 211, 283, 329, 439, 443
co-witness influence 94
cows 400–401, 403–405
credibility 1, 22, 31, 33, 45, 49, 110–111, 113, 127–128, 193, 197, 266, 270–272, 276, 283, 290–291, 328, 415, 451, 455–456, 463; analysis 106, 110; assessments 110, 112, 450; considerable 402; discursive 451; hierarchy of 336; indicator 128; lacking 332, 451; legal 455; notions of 275; of forensic linguistics 487; overall 8; perceived 200–201, 455; policing 455; professional 64; Reid technique 106–113; witness 283, 456; *see also* bias, interrogation, PEACE, reliability, validity, vulnerable witness
crime scene 10, 55, 353, 401; investigation (CSI) 462
criminal justice 7, 11, 438, 442, 446; Australia 28; Canada 28, 239; contexts 438; dialogue 443; England and Wales 28, 99, 226, 267; incidents 444; majors 438, Netherlands 134; process 25; reform 149; system 8, 18, 21–22, 28, 43–44, 54, 88, 97–99, 114, 134, 141, 174, 176, 181, 206, 220, 239, 242–243, 275, 279, 284, 333, 344, 439, 442, 455–457, 470; United States 28, 206, 279, 284, 438–439; workers 441
Criminal Procedure Rules (CPR) 29, 32, 46, 47, 53
criminal standard of proof 206–210, 212, 214, 217–221
criminology 7, 11
Critical Incident Stress Management (CISM) 304, 308
cross examination 37, 97, 99, 160–161, 194, 268, 274–275, 421, 472
cross-species 404
Crown Prosecution Service (CPS) 241, 267–268, 271
cultural: attitudes 444; expert 163, 291; issues 163, 169, 421
culturally-responsive 11, 437, 439; pedagogy **440**
Culturally Responsive/Relevant Pedagogy (CRP) **440**, 445
Culturally Sustaining Pedagogy (CSP) **440**, 445
culture 44, 169, 271, 300, 309, 357, 398, 438–439, **440**, 445, 466, 475–476; -centered 439; community 82; differences in 37; diverse 145; dominant 163, 169; hired gun 487; marginalized 445; of care 438; of origin **440**; of schooling **440**; professional 19, 22, 30; scientific 17; sub- 77; wide range of 145; work 22, 71; *see also* deaf, diversity, LGBTQ+, majority, minority, norms, toxicity, workplace
curriculum 427–428, 431, 434, 441, 446; choices 443; design 427; implementation of **440**; narrative arc 446; standardized 427; vitae 250
Cushing-Leubner 322
cybercrime 381

Dalton, D. 434
Dangerous Speech Project 5

darknet 353; *see also* counselling, cybercrime, internet, Interpol, law enforcement, self-care, trauma
dark-web 80
data: cleaning 83, 86; hygiene 348
Daubert: criteria 33, 53; hearing 154; standard 71; test 160, 244; *see also* Frye
Davis, J. 165–166
Deamer, F. 7, 9
de Boer, M. 401
de Rooij, M. 405
deaf: adults 312; applicants 319–320; associations 318; asylum seekers 312–313, 315, 317–318, 320–321; -blind 317; children 322; clients 313, 318; communities 313–314, 445; international gatherings 314; interpreter 313, 315–317, 319, 321; lingua franca 313; linguists 467; people 314, 316–317, 319, 321–322, 474; refugees 10, 312–313, 315–316, 318–319, 321–323; researcher 318; sibling 321; sign language user 316, 319; *see also* culture, discretion, sign language, testimony, witness
death: investigation 241–243; penalty 145, 153, 168, 415; threats 299
decay 91; memory 91–92
deception 108–111, 117, 120, 219–220; detection 106, 108–114; indicators of 111–112; linguistic markers of 112; self- 120; significant 220; strategies for 118; symptom of 112
deceptive argumentation 120
decision-making 35, 45, 100, 182, 219, 230, 300, 471; admissibility 33; authorities 44; discretionary 440; duties 121; expert 35, 44–45; fair 60; faulty 464; group 186; high-stakes 23, 475; informed 267; in-the-moment 446; judicial 351; jury 211, 285; legal 21; medical 267; practical 446; process 20, 199, 318; rational 211; strategic 343; *see also* accuracy, bias, credibility, impartiality
decontextualization 196
defamation 121–125, 382, 387, 390–393
defense 19, 34, 45, 50
dehumanizing language 183
deictic 197, 201
denial 107, 118, 135, 290, 330, 332, 335, 337; innocent 107; institutional 475; of allegations 109
deontic 78, 233, 266–267, 272; deontology 76, 78, 81–83, 85
deportation 168
deposition 125, 148–149, 154, 301
Deoxyribonucleic Acid (DNA) 24, 55; analysts 155; database 54; match 55; sample 55; testing 24, 55; traces 54
Descartes, R. 398, 400
detachment 302–303, 307–308; professional 300; *see also* proximity, support, trauma
Detective (DET) 180–181, 198, 200–201, 287–290, 417
dialogue 97, 110, 310, 441; approach 176; criminal justice 443; fact-based 489; oral 319; respectful 6
dignity 140, 209, 218, 330, 338, 405, 446, 490
dilemmas 76, 131, 134, 138, 140, 170, 270, 275, 300, 338; ethical 6, 8, 10, 54, 76, 78, 131, 146, 155, 271–273, 309, 343, 411, 488; intertwined 134; of coercion and voluntariness 133–134; of silence and truth 134; original 488; preconditions of 131; standard 18
diminished responsibility 9, 225–237
disciplines 3, 11, 18, 22–23, 25, 46, 49, 155, 309, 415, 434, 468, 472; allied 299; clinical 435; comparison-based 23; forensic 2, 18–19, 22–23, 409–410, 412; forensic linguistic 255; forensic science 17–19, 22–23, 343, 469; historically established 156; linguistic 409; multiple 465; professional 472; scientific 398, 412, 462; sub- 18–19, 65; variety of 11; *see also* criminal justice, law, law enforcement, psychology
disclosure 97, 149, 152, 269, 372, 409, 419, 425, 430, 432–435, 467; already-required 148; improper 382; obligations 59; public 413; reciprocal 148; requirement 148; student 426–427, 431–433; *see also* confidentiality, informed consent, privacy
discomfort 6, 428, 430; create 428, 430; feelings of 485; initial 355; pedagogic 430
discourse analysis 3, 19, 32, 176; *see also* Conversational Analysis
discovery 148, 154, 403; breadth of 154; implications 125; legal 67; potential for 351
discretion 155, 207, 212, 243, 245, **365**, 369, 419, 440–442, 446; analyst **365**; choices 437, 440, 440–442, 446; decision-making 440; impulses 445; judicial 444–445; powers 465; practices 446; prosecutorial 168
discrimination 257–258, 262, 268, 344–345, 348, 472; alleged 4; anti- 5; avoiding 270; housing 439; language 4, 437, 444; layers of 323; legally challenge 262; linguistic 255, 445, 473; possible 275; publication 354; systemic 450; *see also* ableism, bias, cross-species, diversity, English First, English Only, equity, evidence, Global, hermeneutical injustice, impartiality, inclusion, linguistic imperialism, non-English language, overt, stereotype
discursive 20, 191; antagonism 21; choices 438, 442, 445; closure 213; credibility 451; informed 446; work 231
disinformation 121–122, 127–128, 358, 381, 390, 392–393; *see also* conspiracy theory, misinformation, pseudoscience
disinformative: argument 128; linguistics 120–121, 128; technique 122, 127

disorders: language 118; mental 99, 228, 329; neurodevelopmental 228; personality 228
dissemination *80*, 409; intentional 8; of conclusion 79; of forensic linguistics 410; of information 85; of opinions 122; of research 419
distortion 91, 111, 179, 216, 281; conceptual 121; crafty 123; deleterious effects of 181; gross 208; intentional 120; memory 93; serious 217; significant 94
diversity 7, 12, 456, 462–463, 466–467, 469, 473, 476–477, 488; barriers to 457; benefits of 466; critics of 469; cultural 456; demographic 90, 463; differently abled 476; disabled 168, 175, 321, 467, 476; Equity, Inclusion (DEI) 463, 468; ethical necessity of 463; expanding 2; fostering 469; increasing 463, 470; inhibiting 467; issues 477; lack of 462; linguistic 451, 456; natural 11; need for 463; neuro- 433; neurological 91; obstacle to 474; of experiences 429; of perspectives 7; promise of 467; promotion of 12, 463, 466–467, 472, 476–477; recognition of **440**; scientific 454; structural barriers of 467; true 475; *see also* ableism, LGBTQ+, sexual orientation
Dodson, M. 195, 198
dogma 111, 213–214, 241, 243
Domanico, A. J. 175–176
drafting committees 207, 212, 214
Dror, I. E. 6, 35, 90, 100
Du, B. 194
due process 173, 256, 292, 372
Dumas, B. 218–219
duty of epistemic care 113, 214
duty to the court 29, 37, 47, 240, 245–249, 251

Eades, D. 3, 72, 194, 196–197, 410, 416–417, 420
Eastman, N. 225
Ebbinghaus, H. 91
Ecker, U. K. 93
ecological validity 219
effective 8, 20, 45–46, 55, 60, 96, 101, 148, 183, 191, 194, 206, 208–209, 212, 216–217, 236, 258, 266, 268, 272–273, 275, 302, 304, 306, 309, 337, 354, 369, 466, 471, 473; case strategy 60; communication 9, 33, 63, 72, 113–114, 118, 274, 319–320, 410; coping strategies 302, 306, 310; cost- 286, 374; cross-examination 37; environment 269; forensic scientist 355; instructions 209; learning 427; participation 267; performance 124; precautionary measures 8; prosecution rates 60; public speaker 353; recruitment strategies 334; rehabilitation 333; remedy 12; response 334; safeguards 358; solution 465; strategy 335; system of checks 466; teaching 11
Eggington, W. 5
Ehrich, L. 411
Ehrlich, S. 196
Elkins, K. 412
Elstein, S. 301–302, 305–307
emergency medical: services (EMS) 304, 306, 308; technician (EMT) 308
emotion 49, 92–93, 182; centrality of 432; impact of 432; inward-directed 334; -laden hearsay 182
emotional 354; attachment 110; baggage 182; balance act 336; burden 431–432; burnout 428; challenges 336; connection 306; cues 334; depletion 301; environment 118; factors 274; harm 82; health 10; impact 429, 431, 433; needs 432; pain 301; paralinguistic features 275; processing 304, 335; proximity 300; pull 64; reactions 92, 337; real-life events 92; reflexivity 334; response 273, 429; risks 300; state 92, 112, 335, 381; strain 338; subject matter 61; support 337, 475–476; turmoil 425; well-being 429; *see also* bias, counselling, credibility, detachment, impartiality, mental health, objectivity, prejudice, reaction, trigger
empathy 10, 97, 106, 337, 356, 439, 443
England 9, 28–29, 32, 37, 46–47, 53, 81, 99, 146, 210–212, 217, 219, 225–226, 266–267, 274–275, 382, 387, 425, 452, 471, 486
England and Wales High Court (EWHC) 51n1, 393n8
English as an Additional Language (EAL) 450, 454–457
English first 451, 474–475
English only 286, 451, 453–454
Ennser-Kananen, J. 10
entextualization 179, 194, 196–197
environmental protection 344
epistemic 364; care 113, 214; demands 372; duties 366; ideal 369; injustice 455; irresponsible discourse 113; knowledge 266; marginalization 455; negligence 113, 214–216, 221; principle 372; responsibility 207, 209, 214, 363–364, 370; rights 267; superiority 323; violence 450, 455, 457
epithets 353; racial 353
equal representation 256
equality 5, 44, 48, 63, 312, 457
equity 255, 268, 270, 318, 476–477; citation 456; issues 315; methodological 254; mutual 261; of the process 185; real 318; *see also* diversity, inclusion
equivocation 123
erroneous 23–24, 73, 90, 94, 120, 181–182, 258, 286–287, 290–291, 466
error 3, 6, 8, 17, 126, 183, 201, 244, 281–282, 290, 310; harmless 282; legal 291; logical 120; perceptual 89; potential for 19; pragmatic 126; rates 17, 22, 24, 33, 93, 244, 259; serious 287; *see also* bias

essentialized characterizations 257
ethical: action 8; application 254, 375; approach 84, 323; arguments 400; awareness 86, 419; barrier 149; behaviour 19, 30–31, 77, 85, 119, 338, 420; best practices 485; biases 50; challenges 6–8, 10, 18, 22, 32, 54, 61, 76, 78, 105, 146, 156, 170, 270, 279, 299, 328, 333, 346, 347, 350, 352, 415, 485–486, 488; choices 120, 216, 437, 446; codes 4, 34, 64, 150; concerns 22, 24–31, 69, 334, 345, 411, 444, 456, 463, 471, 485; conduct 7, 71; conflicts 490; consideration 6, 22, 31, 70, 76, 81, 85–86, 113, 125, 333, 350, 375, 381, 404, 409, 419, 421, 441, 472; conundrum 11; cornerstone 209; decisions 76, 79, 268, 299, 421; demands 146; dilemma 6, 10, 54, 76, 78, 131, 146, 155, 271–273, 309, 343, 411, 488; dimensions 76, 380, 415; duty 128, 146, 149, 155, 191, 202, 279; enterprise 219; experts 161; factors 43; foundations 377; frameworks 29, 76–78; framings 85; ground rule 130; guidance 150, 155, 488; guidelines 4, 78–79, 81, 149–150, 155, 412, 488; hacking 412; imperative 11, 20, 114, 119, 260, 315, 346, 463, 488; implications 9, 79, 332; issues 2, 4, 7, 10, 22, 24, 28, 37, 50, 54–61, 63–64, 76, 79, 83, 85, 105, 153, 170, 212, 219–220, 225, 280, 338, 343, 345–346, 349, 412, 414, 419–420, 422, 425, 427, 431, 434, 483–484, 487–488; judgements 32; mandate 5; misconduct 8, 22, 485; nature 85, 300; necessity 11, 356, 463; need 64; norms 31, 149; obligations 29, 32, 74, 128, 145, 150–153, 211, 216, 441; obstacles 338; opportunities 276; option 303; outcome 76; questions 54–55, 266, 312, 323, 344–345; paths 120–121, 124; pedagogy 437; perspective 148, 207, 221; pitfalls 38; point of view 141; position 146, 148; practice 71, 78, 84–86, 124, 336–337, 488; practitioners 259; prerequisite 10; principles 21, 31, 76–77, 118–119, 206, 209, 220, 337, 358, 364, 405; problems 35, 217, 414; procedures 7, 292; product design 127; questions 1, 3, 5, 8, 82, 147, 152, 162, 169, 254, 351, 378, 409, 411, 446, 450, 483; reasoning 120; recommendation 351; relativism 78; requirements 207, 348, 353; research 79, 405; response 119; responsibility 5, 8, 11, 20–21, 113–114, 125, 153, 304, 333, 347, 350, 352, 366, 405, 412–413, 419; route forward 76; rules 76, 332; scaffold; scientific norm 346; service 141; situation 76; soundness 10; stance 38, 445; standards 3, 24, 38, 71, 110, 149, 251, 281, 332, 338, 374, 404, 406, 419; system 50, 145, 170; template 207; theory 76; thinking 76, 85–86; tightrope 160; topics 64, 483; transgressions 22; values 422; violation 64; *see also* Committee on Public Ethics (COPE), International Association for Forensic and Legal Linguistics (IAFLL), International Association for Forensic Phonetics and Acoustics (IAFPA), Linguistic Society of America (LSA)
ethics: best practice 415, 485; breach of 346; debate 2; morality 105; of advocacy 9; of communication 32, 207; of competence 31; of detaching from stress 302; of experts 162; of independence 34; of litigation 20; of persuasion 207; or preparedness 35; of professional linguistics 146; of research 207, 219; of speech 207; of teaching 426, 428, 435; of truth 36; violation of 5, 64, 125; *see also* deception, Germanic Society for Forensic Linguistics (GSFL)
ethnography 441, 443
European Asylum Support Office (EASO) 320
European Convention on Human Rights (ECHR) 382, 387
European Network of Forensic Science Institutes (ENFSI) 51, 53
European Union's Agency for Law Enforcement Cooperation (Europol) 343–344
Evans, T. 27–28
evidence: -based approach 244–245; compelling 10, 92; contradictory 19; discursive forms of 20; exclusion of 247; expert 9, 28–30, 32–33, 36–38, 43–46, 49, 53–54, 56, 155, 236, 239–241, 243–246, 248, 250–251, 417; fact-based 21; forensic 17, 20, 37, 43–46, 49–51, 53–54, 91, 95, 152, 240, 287, 290, 468; -in-chief 267, 271, 274; infallible 21; language 2–4, 20, 64–65, 68, 72, 93, 106, 114, 191–193, 196, 198, 202, 472; linguistic 18–19, 23, 27–28, 43, 49, 53, 61, 70, 91–92, 94, 96–100, 117, 121–122, 124–125, 127–128, 213, 217, 239, 250, 348, 463–464; of bias 23, 183; pattern 18; presentation of 22; progression of 22; scientific 22, 33, 113, 152, 176, 179, 236, 247, 356, 404; subjective 22
experience: case 61, 148; challenging 274; collective 72; educational 427; embodied 257; human 439, 443; initial 91; intimate 476; lack of 464; learning 485; legal 49; limited 32; linguistics 63, 202, 260, 472; litigation 148; negative 421; past 35, 465; personal 162, 248, 300, 355, 427, 434, 445, 453; positive 6; previous 257; research 10; sharing of 6, 314; significant 81; transformative 11; traumatic 429; unique 152; varied 54; witness 281
Expert Advisers Database (EAD) 81
expert: opinion 1, 4, 21, 30, 33–34, 48, 56, 61, 231, 241, 243–245, 248; role 43–44, 64; -shopping 59; witness 8–9, 18–21, 28–34, 36–38, 47, 66, 71–72, 76–77, 81, 120, 125, 127–128, 145–148, 154, 156, 160–162, 170, 191, 195, 203, 225, 237, 239–240, 242–251, 471
expertise: application of 24, 27, 45, 68; area of 3, 29, 32, 45–46, 57, 150, 411; assessing 31;

complementary 84; demand for 5; guidelines 31, 53, 150; lack of 33, 464; level of 21, 29; limits of 32, 66, 68, 243, 472; linguistic 4, 38, 48–49, 61, 72, 81, 84, 106, 150, 152, 181, 195, 202, 292, 420; non-linguistic 84; objective 44; offering 19; phonetic 61, 84; professed 31; professional 194; providing 7, 202; psychiatric 237; required 30, 38, 56; scientific 3, 66, 244, 287; scope of 32, 243, 247, 249–251; shared 32, 109; skills 240; specialization 249; specialized 2, 150; testifying to 21; value of 21; *see also* consultancy, forensic, hired gun

experts: before the court 160; consultancy 248–249; credibility 31, 33, 45, 64, 127–128; definition 147; forensic linguistic 2, 21, 28–29, 32, 122, 152, 155, 393, 411, 421, 487; impartiality 20, 47, 245–246; licensure 3, 121, 149–150, 155; limitation of 60, 243; reliability 37, 46; scientific 11, 487; *see also* bias, hired gun, lay people, oversight

extremism 354, 357; Al-Abdulmohsen, T. 343–347, 356–357; explosives 356; hate crime 256, 260, 344, 443; hate speech 5, 352, 411; ideological 357; left-wing 347; mass casualty event 6; McVeigh, T. 165–166; National Socialism 348; Nazism 351; political 10, 343–345, 357; racism 443, 474; radical 221, 254, 261, 343, 348, 353, 439, 443, 477, 486; radicalization 357; right-wing 347–348, 354, 356; salad bar 357; terrorism 291, 343–345, 356–357, 450; violent 10, 344, 357

Fairclough, S. 276

fairness 48, 59, 64, 126, 150, 184, 268, 373, 381, 485; importance of 153; indication of 151; legal 280; painstaking 128; perception of 283; principles of 312; *see also* diversity, equity, inclusion, kindness, morality

fallacious argument 121

fallacy 465; of equivocation 123; of inflation of conflict 123; of presenting irrelevant evidence 123; of suppressed evidence 123; scent of 120; *see also* bias, heuristics, prejudice

false confession 176

Faretta: admonitions 177; advisement 178; flaws 176; formulations 178; language 178; procedures 178; waiver 174, 177–178; warnings 176–178

Farrugia, L. 99

fatigue 263; compassion 301–302

Federal Bureau of Investigation (FBI) 23, 67, 165–166, 282, 356–357, 389, 413

Federal Judicial Center 210

Ferguson, A. G. 176

Fetal Alcohol Spectrum Disorder (FASD) 227–228

fiduciary duties 119

Filipović, L. 289

Finardi, K. R. 456

findings: assessment of 46; case 85; certainty of 152; communicating 20, 79, 84, 86; Daubert 147, 160, 244; evidence-based 114; expert 22, 72; forensic 410; Frye 71; initial 149; linguistic 84, 255, 257, 415; objective 152; overconfidence in 35; publishing 350; reporting 84; research 149, 152, 173, 355, 452; scientific 12, 18–19, 91; simplicity of 81; statistics 344, 469, 484, 486; transfer of 338; transparency 100, 150, 372; unhelpful 149; written report 4; *see also* error, hired gun, impartiality, independence

Finegan, E. 2, 146, 151

Finland 313, 315–322; Maahanmuuttovirasto (Migri) 320; Ulkomaalaisasioiden asiankäsittelyjärjestelmä—Electronic Case Management System for Non-Permanent Residents, Finland (UMA) 319

Finnish Immigration Services (MIGRI) 316–317, 319

Finnish Sign Language (FinSL) 313, 316, 318–321

First Language Learned (L1) 65, 280–286, 292

Fishman, C. S. 282, 285

Flesch-Kincaid readability test 177

Flin, R. 99

Floyd, G. 443

Foreign Policy Analysis (FPA) 453

forensic: expertise 51; linguists 2–3, 5, 8, 20–22, 24, 29, 32–33, 38, 49, 51, 76–77, 79–85, 105, 114, 117, 119, 146, 149, 151, 155–156, 181, 191–192, 194–195, 198, 201–202, 249–250, 255–256, 259, 261, 300, 305–306, 346, 400, 405, 410, 420, 464, 487; pathology 239–242, 244, 247, 249. 251; phonetics 32, 249, 410; practitioner 59; psychology 248, 301; science (FS) 17–19, 22–25, 28, 30, 33, 38, 44, 47–48, 51, 61, 100, 124, 401, 412, 462; speaker comparisons 402; speech science 10, 402; stylistics 19, 250

Forensic Linguistics (FL) 88, 343, 351, 352, 357, 437, 441, 450, 452, 463, 468, 470–471, 472,

Forensic Science Activity 47

Forensic Science Regulator 30, 53,

forgetting 91–94

formal register 126

Fraser, H. 282, 286

free: narrative 96, 267; speech 122, 255

Freilich, J. 347

Freire, P. 439, 443, 455

frequency rule 117

Fricker, M. 455

Gabbert, F. 94, 99

gallows humor 306, 308

Garber, A. 467–468

García, O. 321

garden-pathing 121

gatekeeping 37, 243–244, 251, 450–451, 454, 456
Geijsen, K. 99
gender 183, 323, 381
General Medical Council (GMC) 30
generative linguistics 399
German 81, 161, 256, 355–356, 404, 452–453; law enforcement 84; Ministry of the Interior 356; postwar history 356; speaker 404; terrorism 356
Germanic Society for Forensic Linguistics (GSFL) 28, 48–49
Germany 28, 347, 355–357, 453
Gershman, B. 411
Glaas, S. 8
global: academic community 455; challenges 11, 475; community 260, 469, 473; contexts 7; deaf lingua franca 313, 474; hierarchies 456; incidence of violent acts 344; inequalities 474; inequities 400; North 314, 322, 453, 455, 457, 475; population 453; pressures 255; scientific diversity 454; South 322, 453–456, 475; spread of English 474
Global Terrorism Index (2024) 343, 357
Glougie, J. 7, 9
Goffman, E. 208
Golden Rule 9, 207–209, 220–221; argument 208; philosophical 208
golden thread 206–207, 221
Goldstein, N. E. 176
good-faith: argumentation 120; effort 442; respondent 121; service 124–125
Goodman-Delahunty, J. 97
Gordon, S. F. 11
Goudge Report (2008) 239–240, 243–244
Goudge, Hon. S. 239–240
Grant, T. 84, 250
graphic: classroom safety 485; evidence 183; imagery 183, 300–301; presentation to students 485; protection against 485; recommendations for the use of 485; *see also* trauma, trigger warning
gray area 19, 54
Greek antiquity 159
Green, A. 73
Green, B. 412
Green, E. M. 314
Greenlee, M. 73, 258
Grice, P. 118, 123, 201
Grice Speech Act Theory 118, 123; communicative intention 118, 120, 272; cooperative principle 118, 123, 201; maxims 118, 201; perlocutionary success 209; perlocutionary uptake 220
Gries, S. Th. 364
Grimm, P. 160–161
Grisso, T. 176
Gudjonsson, G. H. 34
guidance 6, 28, 30–32, 34, 36, 45–49, 51, 53–54, 56, 60–61, 76, 78–79, 99, 150, 155, 182, 184–185, 192, 225, 245, 274, 366, 432, 435, 488
guidelines 4–5, 18–19, 29, 31, 48, 53, 64, 71–72, 77–79, 81, 98–99, 106, 118, 149–150, 152, 155, 176, 229, 232–233, 267, 271, 274, 281–284, 312, 319, 412, 466, 488–489; best practices 5, 7, 18, 24, 53, 72, 248, 251, 451, 466; enforcement 54, 243, 420; licensure 3, 149–150, 155; teaching, 19, 414, 421; publishing 416, 456, 488; *see also* American Psychological Association (APA), consultancy, International Association for Forensic and Legal Linguistics (IAFLL), International Association for Forensic Phonetics and Acoustics (IAFPA), Linguistic Society of America (LSA)
guiding scaffold 76, 79–81, 85–86
Guillén-Nieto, V. 487
Gulf War (1991) 164

Habick, T. 8
hacking: attacks 412; back 354; ethical 412; tools 412
Haderlie, D. 10
Hagmann, D. 166
harassment 163, 445; sexual 332
Harpster, T. 109
Harrison, P. S. 10
Hashimoto, B. 10, 374
hate 89, 328; crime 256, 260, 344, 443; fuelled 356; group 345; messages 357; speech 5, 346, 352, 411; *see also* extremism, terrorism
Haworth, K. 202
Hayes-Smith, R. 432–433
Heath, M. 427–428, 431–433
Heffer, C. 9, 111, 113, 229–231
helicopter science 476
help-seeking 329
Heritage, J. 196–197
hermeneutical injustice 454–455
Heydon, G. 112, 114
hierarchy 333, 336, 399; of Expert Performance (HEP) 100
Higher Education (HE) 11, 323, 425–426, 486; cutbacks 486; hiring 486; job insecurity 464; tenure 240–241, 471, 487; *see also* consultancy, diversity
Higher Education Institutions (HEI) 426, 486
high-stakes document 118, 120–121, 126–127
Hill, M. D. 97
hired gun 31, 33, 38, 66, 155, 160, 471–472, 487
Hirsch, A. 106
Hobbs, S. 412
Hofstra University 437, 472; *see also* conviction, Innocence Project, Leonard, R.
Hoggan, J. 489

Holocaust 348, 355; *see also* archives, extremism, Germany, right wing, terrorism
Holst, V. F. 90
homeless 163, 166–168
Homicide Act (1957) 225, 226
hooks, b. 443, 451, 457
Horry, R. 95
Hudley, A. C. **440**
Hulme, J. A. 429, 431, 433
human: affairs 215; -animal relationship 399; -animal studies 398, 405; behavior 6, 117, 400, 463; beings 8, 61, 168, 182, 309, 333, 335, 399, 477, 490; benefits 398; caregivers 405; civilization 400; consumption 400; conversations 118, 403; desire 54; exceptionalism 399–400; experiences 398, 437, 439, 443; factors 17–18; interaction 49, 381; interpretation 17, 21; intuition **365**; judgments 54; language 194, 380, 398, 405; lives 398; memory 91, 93–94; mental processes 285; nature 78; observers 260; perspective 403; psychology 248; purposes 18; resource 432–433; rights 5, 44, 48, 194, 256, 260–261, 270, 312, 321, 328, 330, 332, 338, 343, 357, 383, 415, 473; Rights Act (1998) 275; scientists 403; societies 398; speech 381, 403; subjects 218; suffering 10; superiority 399; supervision 381; supremacy 399; trafficking 299; values 322; witnesses 401
humanitarian: grounds 168; principle 275
Hussain, S. 163

ideology 179, 194, 196–197, 257, 369; extremist 357; language 173, 193–194, 258, 261; political 347; race-based 444; referential 196–197
idiolectal information 377–378, 380, 382–383, 388, 390–393
idiosyncrasy 403–404
Ievins, A. 336, 338
illegitimate interpretation 118
image: camera 137–138; change of 2; facial 383; graphic 300–301; photographic 386; public 472, 488; removing 83; right of 387; sonorous, 382–383, 387, 391; vibrant 122
immigration 169; asylum 318; case 168; courts 168, 283; documents 177–178; illegal 183; issues 183; judges 283; officials 73; policy 318; proceedings 283, 443; records 168; services 57, 319–320; status 69
impartial(ity) 5, 18–19, 20, 29, 31, 45–48, 50, 55, 60, 72, 100, 150–153, 208, 245–246, 377; challenge 19; decision 151; focus on 50; goal of 151; lack of 246–248; mindset 47; obligation of 248; officer 275; opinion 151, 246; principles of 8, 46, 54; proceedings 72; quest for 299; researcher 153; reviews 150–152; stance 254; standard of 151, 153; *see also* bias
impeach a verdict 122, 182
imperialism 457; linguistic 450, 452–455, 457; *see also* Anglophone, Global North, Global South, parachute science
implicature 121
implicit bias 182, 440, 442, 445–446
imprisonment 20, 119, 146, 163, 411
inclusion 12, 37, 44, 59, 140, 301, 317–318, 401, 405, 467–468, 477
incriminating 97, 175, 195, 200
independence 29, 31, 34, 47, 51, 67, 125, 215–247, 333
individuals convicted of sexual offending 328, 330, 336
inequality 22, 194; diversity 456; gender 183, 466; repercussions 12; *see also* bias, discrimination, equality, equity, fairness, in forensic science
inference 18, 32, 46, 123, 192, 195, 201, 351, 364, 470; negative 131
inflammatory: evidence or argument 181; speech 353
information: gathering 96–97, 106; management 55, 61; processing 465
informed consent 153–154, 269, 271, 313, 322, 329, 334, 346–347, 381, 388, 391, 404–405; *see also* big data, breaches, challenges, confidentiality, privacy
inner conflict 336–337
Innocence Project 23, 472; *see also* death penalty, Hofstra University, Lenoard, R.
innuendo: false 127; irrelevant 127
inquiry 255, 269, 369; cross-disciplinary 185; field of 3; initial 31; into Pediatric Forensic Pathology 239–243, 245, 247–248, 251; property 258; reliability 244; scientific 119; target of 367; topics of 160
inquisitorial system 160
Institute for Justice and Opportunity 438
Institutional Review Board (IRB) 153
institutional support 300, 308, 310
instructions: abstract 221; cautionary 184; curative 183–184; defined 213; formal 9; handling 82–83; incomprehensible 217; judge's 181; judicial 181, 183–184; jury 182, 184, 207, 209, 213, 215, 218–219; legal 218; letter of 48; limiting 182, 185; misunderstanding 182; pattern 182, 212; proposed 184; proof 206–207, 215, 217–218, 221; specific 89
integrity 12, 19, 35, 47–48, 71, 119, 269, 284, 351, 405, 412, 434, 457, 485, 490; compromise 31; high degree of 7; individuals' 338; professional 5, 19, 22; public 487; scientific 155
intelligence cycle 76, 79, *80*, 85–86; analysis 79, *80*; collection 79, *80*; continuous review 79, *80*; direction 79, *80*; dissemination 79, *80*; processing 79, *80*

intentionality 121, 232
intermediary 267–268, 271, 273–275
internalized oppression 455
International Association for Forensic and Legal Linguistics (IAFLL) 3, 7, 19, 48–49, 54, 71–72, 81, 146, 150, 487; *see also* American Psychological Association (APA), Linguistic Society of America (LSA)
International Association for Forensic Linguistics (IAFL) 1. 5, 28, 49, 152
International Association for Forensic Phonetics (IAFP) 1, 28
International Association for Forensic Phonetics and Acoustics (IAFPA) 3, 5, 7, 19, 48, 54, 150, 487
International Laboratory Accreditation Cooperation (ILAC) 53
International Organization for Standardization (ISO) 48, 53
International Rescue Committee (IRC) 164, 166
International Sign (IS) 313, 315
internet 166–167, 283, 346, 353, 421, 443; articles 166; 346; crimes 300; news channels 346; sources 346; text data 380
Interpol (International Criminal Police Organization) 299–301
interpretation 4, 19, 21, 32, 35, 60, 91, 109, 117, 119, 125–127, 131, 138, 152, 196, 255, 272, 280–283, 286, 288, 319, **365**, 415, 463, 466, 483; analysis 363; casual 126; concept of 289; continual 257; court 280; courtroom 281–282; erroneous 286; errors 290; examiner 22; forensic 19; human 17, 21; ideological 258; illegitimate 118; inept 290; legal 173, 363, 367, 375; legitimate 117–119, 126–127; linguistic 127, 173; literal 280; live 284; of analysis 35; of information 89; of phrases 20; of the data 32; open to 226; oral 280, 283; particular 35–36; personal 89; physics/pattern 18; poor 290–291; prior 100; problems 283–284, 288; professional 289; remote 283; services 317–318; sign language 284, 317; strategies 319; subjective 34, 49; telephonic 282–283; ultimate 192; uncommon 127; use of 194; wishful 118

witness 289
interrogation 106–108, 131–132, 134–136, 138–140, 163, 177, 179–181, 194, 279, 281–284, 287–291; audio-recorded 140; custodial 174; police 97, 107–108, 131, 283, 409, 473; pre- 106; recorded 290; summaries 141; tactics 107; *see also* deception detection, false confession, law enforcement, PEACE, Reid technique, witness
interview suite 269
interviewee 90, 97, 108, 197, 288, 303–304, 306, 317, 319, 329, 333–334, 416–417
interviewing 88, 95, 97–99, 153, 198, 267–268, 271–272, 281, 332–334, 337; approaches 96; cognitive 106; negative 97; police 27, 95, 473; skills 28; strategy 99; suggestive 95
interviews 92, 96, 151, 153, 165, 276, 284, 292, 300, 304, 309, 312–313, 329, 333–335, 410, 418, 421–422, 465, 471; ABE 271; anonymized 299; confessional 220; ethical 333; ethnographic 182, 292; in-depth 329; investigative 90, 97–98, 271–273; juror 153; multiple 95; police 45, 96, 99, 106, 196, 202, 289; post-training 98; pre-training 98; research 337; suspect 96; teller-focused 329; video-recorded 267
intuition(s) 131, 192–193, 237, 238, 363, 366, 368, 412, 416; contextually informed 237; expertise-based 465; human **365**; pre-existing 237; pre-judged 237; speakers' 192; testing 192
investigation 31, 44, 55, 77, 79, 81–82, 84–85, 90, 100, 124, 134, 159, 165, 184, 201, 207, 242, 267, 269, 292, 313–314, 320, 343–348, 353, 356–358, 367, 372, 401, 403, 410, 433, 445, 455, 465–467, 471, 485; avenues of 108; complexity of 79; corpus 127; criminal 44, 90, 94, 165; death 241–243; forensic 178, 181, 346, 401; groundbreaking 93; initial 82; intelligence 79; lack of 24; linguistic 400–401; observational 99; official 165, 462; police 73, 76, 389, 401–402; post-conviction 290–291; practices 76; scientific 345, 464–465, 475; *see also* crime scene, law enforcement
investigative: casework 27; interview 97–98, 267–269, 271–273, 275; practice 85
investigators 22, 80–85, 95–99, 108–109, 271, 283, 299, 343, 347–348
Iran-Iraq War 164
Iraqi 163–165, 168
Irvine, J. T. 193

Jacknick, C. 11
Jacobsson, K. 335
Jacquemet, M. 194
James, W. 88
Jenks, C. 7
Johnson, R. M. 438
Johnstone, B. 98, 257
Judeo-Christian tradition 78
judge: aides 161; appeal 33; evaluation of evidence 236; failure of 33; generalist 37; High Court 36; immigration 283; inquiry 27; instructions 181, 183, 218; modern 21; objectivity 152; practicing 215; questioning 131; trial 71, 177, 182–183, 207, 213–214, 282, 285
judicial authorities 9, 207–209, 214–215, 219–220
Judicial College 207, 210, 218
jury: admonition 184; decision 182, 184, 211, 285; deliberation 218–219; heterogeneous 193;

instructions 181–182, 184, 207, 209, 213–216, 218–219; mock- 184; perception 272; questions 214, 218, 221; reasoning 181; room 181; selection 184; trial system 181; verdict 184, 473
justice: criminal 7–8, 11, 18, 21–22, 28, 43–44, 54, 88, 97–99, 114, 134, 141, 149, 174, 176, 181, 206, 220, 226, 239, 242–243, 267, 275, 279, 284, 333, 344, 438–439, 441–444, 446, 455–456, 467, 470, 472; linguistic 254–256, 260, 312; social 8, 12, 28, 44, 48–51, 55, 63–64, 67, 71, 73–74, 127, 173, 202–203, 306, 312, 321–323, 439, 441, 457, 477; youth 98, 466
justifications 332, 335, 337, 354, 374, 414; authoritative 237

Kant, I. 77, 208
Kaplan, J. P. 20
Karoli Christensen, T. 8, 422
Kassin, S. M. 181
Keane, A. 210
Kebbell, M. 97, 99
Kennedy, K. M. 426, 433
kidnapping 54, 299
kindness 489–491
Klein, D. W. 182
Knowing, Intelligent, and Voluntary test (KIV) 174–175, 178
Komter, M. 8, 96, 181
Komunitas Linguistik Forensik Indonesia (KLFI) 28
Kusters, A. 313–314

Ladson-Billings, G. **440**
Langbein, J. H. 161
language: analysis 57–58, 109, 366; assessment 64, 66–67, 69–70, 73; choice 451; evidence 2–4, 20, 64–65, 68, 72, 93, 106, 114, 191–193, 193, 198, 202, 472; ideology 193–197, 202, 255, 258, 260–261, 439–441; policy 5, 127, 312; *see also* bias, English First, English Only
Large Language Model (LLM) 377–378, 380–381, 387, 389–390, 393,
Larner, S. 11
Larsen, M. 284
Laudan, L. 211
law: Canadian 249; case 9, 29, 36, 47, 155, 194–195, 202, 282, 385; civil 210, 387; common 34, 146, 191, 193–194, 206, 210, 213, 240, 382, 386; contract 374; copyright 384–385; court of 4, 119–120, 124, 128, 178, 226, 275; criminal 130, 133–135, 387; degree 4, 128; English and Welsh 268, 382, 386; enshrined in 79; federal 279; firm 71, 122, 124; international 4; interrogation 194; intricacies of 38; language of the 178; letter of the 105, 236, 442; of evidence 239, 244–245, 248; practice 145; property 274; rule of 206, 209, 212, 344; school 145, 162, 191; text of the 131–132, 136, 140; transgressions of the 105; universal 77; uphold the 128; vagueness of 140
law enforcement 7, 22, 28, 54, 56, 63, 66, 76, 79, 83, 95–96, 105, 108–110, 113–114, 134, 141, 179, 241–243, 268, 271, 279, 289, 300–301, 304, 307, 345, 357–358, 401–402, 416, 420–421, 444–446, 471–472; agencies 96, 99, 106, 347, 417, 438; agent 83, 90; authorities 343, 354; careers 442, 446; concern for 80; contact 69; dubious 8; Dutch 130; framework of 76; German 84; institutions 347; local 66; officers 76, 79, 97, 175, 282, 304, 307–309, 372, 417, 420, 440; officials 96, 354; perspective 85; practices 351; professionals 76, 95; racism in 443
lawyer: -advocate 147; client- 122; courtroom 97; defense 58–59, 96–98, 147, 149, 154, 400; -expert relationship 29, 148; human rights 5; inexperienced 68; instructing 58; license to practice 146; opposition 128, 156; professional 145; retaining 48, 154, 156; trial 145, 208; unfamiliarity of 22; unscrupulous 128
lay-people 49, 192, 214
leading questions 20, 97, 99
Leahy-Harland, S. 97
lecturers 425–427, 429–435
left wing extremism 347, 354; *see also* hate groups, political extremism, right wing extremism, threat, terrorism, violence
legal(ity) 312; anthropology 145; formula 209, 215, 220; interpretation 363, 367, 375
legitimate interpretation 117–119, 126–127
Length of Residence (LOR) 65, 69
Leonard, R. 202, 472
Letson, M. M. 306
LGBTQ+ 415, 443, 467
liberatory linguistics 439–440
Life without the Possibility of Parole (LWOP) 258
Lightbown, P. 65
limitations 7, 17, 19, 21–22, 24, 60, 66, 79, 84, 88, 140, 243, 250, 347, 350, 368, 374, 415, 419, 432; cognitive 273, 288–289; developmental 98; discussion of 84; in-principle 368; mental 288; methodological 3
Limited English Proficient Speaker (LEP) 74, 280–281, 286
limiting instructions 182, 185
Lindegren, S. 10
linguistic: analysis 8–9, 27–28, 80–84, 88, 100, 124, 149, 151, 181, 207, 219, 225, 228, 259–260, 290, 305, 363–364, **365**, 366, 384, 389, 420, 463, 473; anthropology 145, 151; bias 440, 450–453, 455–457; certainty 127; evidence 18–19, 23, 27–28, 43, 49, 53, 61, 70, 91–92, 94, 96–100, 117, 121–122, 124–125, 127–128, 213, 217, 239, 250, 348, 463–464; foreign 285, 321, 454; human rights 256, 260–261, 312; imperialism

450, 452–455, 457; justice 254–256, 260, 312; legitimacy 313; mismatch 255; pluralism 439, **440**, 456; profiling 35, 257–258; rights 323; terrorism 450; truth 127; weapon 261, 344, 358, 489; *see also* research
Linguistic Society of America (LSA) 2, 5, 48, 150, 152
Lipsky, M. 441
literal word 119, 123, 126
Lively, C. J. 98
Loftus, E. F. 93
Logan, P. 412
lone wolf 357; *see also* extremism, McVeigh, T., radicalization, terrorism
Lorde, A. 451
Lubarsky, B. 420
lying 30, 65, 105, 113, 122–124, 127, 133–134, 137–138, 214, 272, 291, 331; *see also* deception

machine learning (ML) 377
MacQueen, H. L. 384
Magdeburg 356
Mainstream US English (MUSE) 254, 259–260
majority 465; California 175; irresponsible 373; jurisdictions 213–214, 383; opinion 372
Makoni, S. B. 457
Mallinson, C. 467, 473
malpractice 3, 121
Malti, T. 490
manipulation 33, 120, 258, 388, 411, 421, 463; *see also* confession, conspiracy, deception, disinformation, misinformation
marginalization 323, 344, 455, 475
Marko, K. 11
Maryns, K. 194
Matarese, M. **440**
Maxims of Paul Grice 118, 123, 127, 201
mayhem 163, 168
McClone, M. S. 123
McNamara, T. 473
McVeigh, T. 165–166; *see also* lone wolf, Oklahoma City bombing, terrorism, threat, violence
meaning: attributions of 88; changed 21, 196–197, 201; context-free 196; determine the 21, 127, 366–367; different 20; discrepancies in 4; expert advice on 51; fixed 110; intended 118, 201, 255; legal 181–182, 212, 215, 217; of conspiracy 181; provision of 21; transfer of 127; unclarifiable 111
measurement 18, 46; phonetic 402–403; spectral 403
Medical Practitioner Tribunal (MPT) 30
medicine 3, 18, 155, 159, 167, 230, 425–426, 472
Meighan, P. J. 322
memory: affect 8, 290; -based evidence 93; beliefs about 91; conformity 94; contamination 94; decay 91–92; degradation 91–92; deterioration 91; distortion 90, 93; episodic 276; error 93, 95; false 95; faulty 89; fallibility of 92; functionality 91; human 91, 93–94; impediments 94; lack of 109; misinformation in 93; phenomena 94; process 91; representation 94; retrieval 93–95; social contagion of 94; storage 91; traces 92, 94
Mental Capacity Act (2005) 268
mental health 82, 164, 248, 303, 305, 443; challenges 467; conditions 228, 233, 425, 434; disorder 98–99; expert 287; issues 166, 169; responses 429; self-care 354
Mental Health Act (1983) 234
mentorship 18, 443, 476, 484, 488
Mertz, E. 191
message consistency 212
meta-agentive discourse 199
methodology 10, 35, 50, 53, 85, 245, 250, 329, 374, 409, 415–416
methods: accepted 185; anonymization 351; assessment 113, 363; attribution 409; authorship analysis 411; cognitive 106; confession-based 106, 108; data analysis 364, **365**; data recovery 345; Daubert 71, 244; debunked 105; deception detection 109; DNA testing 24; evaluation of current 17; experimental 185; flawed 120; forensic 17; Frye 53, 71; identity detection 409; illegitimate 105; inappropriate 3; interpretative 370, *371*; lack of transparency of **365**; language analysis 109; linguistic 105, 119, 207, 260, 410, 421; machine-based 259; misapplication of 415; peer 32–33, 309, 319, 454, 464–467, 483, 489; peer review 122, 244, 248, 451–454, 457; peer support 433; professional 152; primary 351; pseudolinguistic 105–106, 110; pseudoscience 110, 113, 160; qualitative 152; quantitative 151–152, 332; reliable 5, 19, 23, 367, 370; remote 292; replicable 374; representative 261; scientific 17, 114; statistical 415; teaching 415; traditional 357; transparency 36, 141, 251, 366, 368; unreliable 366; validity of 23–24
military attorney 65
Miller, C. 445
mindfulness 10, 337–338
minorities 169, 182, 415, 467, 470, 487; disabled 168, 175, 321, 467, 476; ethnoracial **470**; gender identity 4, 183, 260, 323, 330–332, 381, 451, 466; LGBTQ+ 415, 443, 467; neurodivergent 268, 433, 467; *see also* bias, cultural, deaf, diversity, forensic, linguistic, sign language
Miranda 174–175, 177; court 175; discussion of 178; exchange 289; flaws 174; jurisprudence 173; legal threshold for 176; pre-test of 176; procedures 176; reading 176; rights 175–176, 256, 281, 288, 442; talismanic status of 177; ubiquity 176; understanding of 176; waiver 174, 178; warnings 70, 174–176, 281, 440

mirroring 95, 358n1
miscarriage of justice 184, 239, 242–243, 249, 279
misconduct 22, 24, 36, 183; deliberate 38; ethical 8, 22, 485; professional 22, 59, 463, 484; prosecutorial 287; sexual 108
misinformation 93, 113, 358; effect 93, 179; emotionally charged 93; exposure 93; linguistic 8; resilience of 93; *see also* disinformation
misinformed linguistics 121
misrepresenting 22, 111, 124, 415
misuse of knowledge 412–413, 418, 422
misusing 22, 386, 411, 413, 419
mitigating circumstances 168
Mohan criteria 243–244, 250
molestation 167; *see also* abuse, children, offender, pedophile, sexual abuse, vulnerable witness
moral 77, 232, 364, 400: agent 162, 169; balance 336; battles 336; certainty 210, 212–213; challenges 336; compass 434; concern 456; conflicts 328, 337; considerations 369; decisions 162, 421; deficiency 78; determination 162; dimension 369; discourse 228; duties 366; emotions 328; facts 228; frameworks 229, 231, 234; guidelines 412; ideals 314; identity 336; imperative 352; interests 193; issues 412; judgment 255; laws 77; legitimacy 31; norms 77, 146; obligation 219; orientation 314; philosophy 77; pressure 337; principles 30, 364; purity 259; questions 54, 162; reaction 231; realism 228; realists 228; reasoning 162–163, 170; relativism 78; responsibility 169, 226, 237, 363–364, 370, 412; rewards 156; right 384, 405; saint 78; thinking 78; transgression 78; vacuum 79; value 78; weight 370
Moran, C. 306
Moriarty Harrelson, E. 321
Morley, I. 97
Morrison, G. S. 150
Mouritsen, S. C. 372
multilingual publishing 456
Mummert, S. J. 432–433
Municipality of Residence Act (1994) 317
Myklebust, T. 99
myths 176, 331–332, 338

Nahleen, S. 93
Nakamura, G. 475
narrative authorship 197
narrativization 218
Navarro Beck, S. 8
National Academy of Science (NAS) 17
National Association of Judiciary Interpreters and Translators (NAJIT) 71, 74, 284
National Crime Agency (NCA) 81
National Institute of Standards and Technology (NIST) 18
National Register of Public Service Interpreters (NRPSI) 271
National Research Council of the United States 17–18
National Science Foundation (NSF) 462, 469, 486
National Socialism 348; *see also* extremism, Germany, Holocaust, Nazism, right wing, terrorism, violence
nature: case-by-case 257; coarse-grained 259; collaborative 197; complex 289, confidential 383; contentious 411; contextual 196; descriptive 192; dyadic 228; ethical 22, 85, 300, 474; horrific in 231; human 78; intensive 301; interdisciplinary 7; methodological 61; of alleged activities 65; of language evidence 64; of power 475; of responsibility 225; of scientific findings 19, 364; philosophical 417; sensitive 434; subjective 49; traumatic 307; uncontested 257
Nazism 351; *see also*, extremism, Germany, Holocaust, National Socialism, right wing, terrorism, violence
Nelson, S. 218
neutrality 9, 34, 46, 85, 161, 170, 245–246, 472; linguistic 456–457; scientific 22
New Orleans 356–357
New South Wales Law Reform Commission 213
New York City Police Department (NYPD) 438
Newell, J. M. 300
Nichols, T. 165–166
Nick, I. M. 5, 10, 12
Nine Steps of Interrogation 106–107
Noda, O. 452
nominalizations 215
non-confrontational 96, 337–338
non-English 281, 284, 453–455, 474–475
non-human animals 10, 398–401, 403–406
non-narrative 20, 230, 233
Non-Native Speakers (NNS) 8, 63–73, 72, 430
norms: communicative 194; cultural 4, 439, 451; ethical 31, 149; historical 5; 467; legal 163; linguistic 451; moral 77, 146; of appropriateness 261; of performance 257; of practice 257; professional 7, 9; social 112; wider 77; *see also* bias
novice 464–465

objectivity 5, 8, 19, 21, 31, 35, 46–47, 54–55, 63, 69–72, 74, 114, 150, 152–153, 161, 241, 299; assumed 259; expert's 21, 31, 54–55; heightened 259; impartiality 8, 46–47 72, 150, 152–153, 299; maintaining 72, 152; professional 47, 64, 67; retaining 71; scientific 17; standard of 152; *see also* bias, diversity, illusion of, novice
O'Brien, B. 286, 291
Odinot, G. 92

offender 55, 229, 231, 234, 345, 348, 354, 356–357; actions 232; alleged 269; background 230; crimes 352; description of 230–231; sex 328–330, 333, 335, 337; violent 357; youth 176; *see also* perpetrator
Oklahoma City Bombing 164–167
O'Laughlin, M. 9
omission 17, 152, 226, 282; significant 29, 92
Ontario 239–242, 249
Ontario Pediatric Forensic Pathology Unit (OPFPU) 241–242
open-ended questions 20, 96, 98
operational practice 82
opinion 21, 29, 35, 45–46, 48, 50, 57, 64, 66, 68, 117, 147, 150, 152, 156, 159–162, 169, 242, 244–246, 248, 250, 335, 353, 356, 392, 414, 456, 471–472; a priori 50; academic 20; clashes of 489; differences of 124, 314; evidence 28, 33–34, 51, 146, 245, 249; expressing 45, 154, 242; expert 1, 4, 21, 29–30, 32–33, 48, 56, 61, 150, 231, 240–241, 243–245, 247–248; freedom of 317, 387; impartial 151, 246; lawyers' 149; linguistic 64, 68, 146; majority 372; pathologists' 242; personal 88, 147; professional 68, 242, 333; provider 249; public 366; scientific 46; second 58–60; sufficiently reliable 32; testimony 159; unbiased 47; understated 156
Ordinary Meaning Doctrine 363
Organization of Scientific Area Committee for Forensic Science (OSAC) 17–18
Ortoni, A. 118
overhearers 208
oversight 11, 17, 243, 249, 305, 488; institutional 29; lack of 24; negligent 121; organizations 18; professional 7, 81; regulatory 7, 249; system of 465; technical 81
overstate 20, 22, 140, 156
overt pressure 471
ownership 135, 382–383, 385, 388, 393, 405

Paciocco, D. M. 244
Papillon case (1970) 387, 394
parachute science 476
paradoxical hyperfixation 303
Parfit, D. 208
Paris, D. 439, **440**
participation framework 207
partisan 29, 34; adversaries 36; control 161; evidence 34; experts 37; forensic analysis 33; outcome 151; results 34
partisanship 29, 36
paternalism 445, 475
pathology 240–241; chief 241; discourse 216; experienced 242; forensic 90, 240–242, 244, 247, 249, 251; pediatric 240
pattern: evidence 18; instructions 182, 212
Pavlenko, A. 281, 289
Payne, B. 432
PEACE method 97, 106; *see also* confession, conviction, interrogation, law enforcement, Reid technique
pedagogy 410, 412, 426, 437–440, 442, 446, 473
pediatric: forensic pathology 239, 241–242, 249, 251; pathologist 240; pathology 240, 247
pedophile forum 80, 82, 331–332; *see also* children, forensic psychology, graphic evidence, interviewing, molestation, recidivism, self-care, sex crime, sex offender, trauma, treatment
peer support 433; *see also* psychological support, self-care, trauma
Pennebaker, J. W. 110, 112–113
perception 8, 21, 33, 88–91, 111, 121, 155, 179, 192–193, 196, 229, 255, 258, 283, 331, 335, 416, 450–451, 455, 463–464
Pereira, T. 9
performative 118–119
Perkins, R. 8, 86n3
perlocutionary: success 209; uptake 220
permissions 69, 85, 346, 485
perpetrator 55, 84, 89, 92, 108, 125, 231, 260, 269, 300, 328, 332, 344, 356–357, 426, 484; *see also* offender
persecution 165, 170, 348
person-centered 335
person-first language 330, 338
persuasive engagement 207, 209, 212, 217, 220
Petersen Brooke, Corly 432
Philippine Association for Forensic and Legal Linguistics (PAFLL) 28
Phillips, J. C. 2017. 373–374
Phillipson, R. 450, 454
philosophy 2, 225, 399; dominant 398; moral 77
phonetics 3, 32, 66, 249, 398, 402–404, 410; *see also* forensic, ownership, vocal, voice
phonology 3, 66
Phythian, M. 79
Pi, D. 213
Picornell, I. 7
Piller, I. 474
Pinkenba case 410, 416, 420
Pipe, M-E. 98
Pirelli, G. 301
planning 97, 106, 268, 271–272, 356, 443, 446
Plotnikoff, J. 276
police: administration 175; agencies 60, 347; audio recordings 96; cautions 72, 135; coercion 135, 141, 256; communications 73; custody 98–99, 107, 256; encounters 202, 437; identification 94; interactions 73, 284; interpreters 264, 289–290; interrogations 8, 97, 107–108, 130–132, 134, 136, 138–141, 179, 281–283, 355, 409, 473; interview 27, 45, 68, 73, 92, 95–96, 99, 106, 132, 165, 196–198, 202, 267–273, 275–276,

289–290, 455, 473; investigations 73, 76, 242, 389, 398, 401–402; investigators 85; prioritization 81; questioning 140, 289; reports 90, 92, 131, 134–136, 139–140, 163; staff 78; stations 135; transcribers 181; vehicle stops 89; voice recognition 56, 68; written accounts 96; *see also* Black Lives Matter, confession, crime scene, Federal Bureau of Investigation (FBI), Floyd, G., Interpol, law enforcement, linguistics, Miranda rights, PEACE method, Reid technique, transcription
political extremism 10, 343, 355, 357; *see also* left wing, radicalization, right wing, terrorism, threat, violence
political extremist threat communications (PETCs) 343, 345–358
Politzer-Ahles, S. 450
postcolonial 398, 406, 450
post-traumatic stress disorder (PTSD) 167, 304, 335, 413
power asymmetry 333, 399, 406
Poynter, D. 125
practical wisdom 333–334
pragmatic 6, 19, 122, 127, 181, 343, 345, 463; analysis 181; appropriateness 122; challenges 105; encroachment 369–370; error 126; expectation 127; garden-pathing 121; meanings 200; mechanics 191; principles 118; problems 417; reality 124; rules 118, 127; specialist 181
precautions 7, 10, 23, 55, 64, 100, 309, 350–351, 413, 418–421, 466, 471
precedent 10, 175, 195, 202, 212, 216–217, 244, 363, 372
prejudice 4, 181, 183, 268, 286, 400, 450; against Muslims 167; bases of 73; evocation of 184; historical 476; negative 10; personal 255, 356; racial 184; social 182; unreasonably 385; unwitting 273; widespread 345; without 19; *see also* gender, LGBTQ+, stereotype
Preiss, U. 10
preparedness 35, 317
preservation 18, 170, 202, 413
President's Council of Advisors on Science and Technology (PCAST) 17, 23, 30
pressures 19, 22, 24, 31, 36, 428, **440**, 450, 487; conformity 379; cultural 34; institutional 255; occupational 465
prestige economy 453, 456
presumption of innocence 130, 206, 218, 372
prevention 332, 338
priming 35
principal: charges 99; difference 149; drivers 466, 462; evidence 291; internal motivation 306; interrogator 180; participation framework 207; roles 192; schools of thought 77
principles 44–45, 48, 70, 85, 119, 127, 139, 153, 229, 232, 245, 305, 312, 333, 363–364, 367–368, 370, 372–374, 378, 382, 384, 404, 476; basic 19, 71, 242, 334; communicative 127; democratic 343; ethical 21, 31, 76–77, 139–140, 206, 220, 337–338, 358, 364, 405; foundational 118; general 169; guiding 71; key 45; linguistic 119, 194; moral 30, 364; of equality 5; of human rights 338, 343; of impartiality 8; of preservation 202; of privacy 271, 393; of sound forensic evidence 45; philosophical 382; policing 78; pragmatic 118; scientific 44, 127; theoretical 79; underlying 70
prison 36, 70, 121, 163–164, 183, 231, 287, 291, 329–330, 335–336, 338, 438
privacy 251, 350, 377, 381–382, 414; breach of 391; data 393; expectations of 346; filter 350; individual 78; invasion of 83; preservation of 413; principles of 382, 393; right to 83, 382–383, 387
proactive 91, 347, 354, 476
probability 21, 68, 210–211, 214, 244, 367, 374
professional: acts of malpractice 3; assessment 4; associations 18, 28–30; attitude 61; awareness 22; balance 22; behavior 31, 78; body 30; clientele 5; codes 30–31, 36, 64, 71, 127; competence 19, 242; conduct 31, 146, 148; credibility 64; cultures 19, 22, 30, 71; development 31, 73, 303; distancing 8, 64–65, 68, 70–72, 74, 355, 472; duties 32, 63, 119, 146–147; education 29; ethics 7, 22, 24, 45, 64, 68, 71, 162, 170, 209, 240, 352, 412, 419, 487; excellence 12, 356; expert 15; groups 17, 78; guardrails 3; guidance 54; instincts 68–69; integrity 5, 19, 22; interactions 5; licensure 3, 121, 150; limits 10; malfeasance 121; malpractice 121; misconduct 22, 59, 463, 484; norms 7; objectives 9; objectivity 47, 64, 67; opinions 19, 68, 242, 333; organizations 19, 38, 48, 54, 72, 150; oversight 7; person 35; practice 19, 43, 213, 251; pressures 22, 24; principles 19; privilege 20, 452; qualifications 3, 69–70; regulations 3; relationships 22; services 63, 122; standards 22, 30, 63, 70, 149, 155, 421, 426–427, 488; status 68; work 7, 128, 355
proficiency 18, 24, 65–70, 73, 150, 176, 194, 286, 312–314, 319–320, 322, 454–457
pronouns 110, 197–199, 441; first person 109–110; plural 195; singular 195; use of 113
prosecutorial discretion 168
proximity: close 401; discussion of 308; emotional 300; physical 262; to data 305, 307, 309; to wild animals 401
pseudoscience 110, 113, 160
psychological: assessment 34; bias 8; constitution 191; constructs 261; disengagement 304; drive

335; effect of addiction 248; experiments 113; factors 91; harm 405, 413; impact 85, 413; literature 176, 211; motives 112; overview 88; preferences 112; reactions 418; research 95–96, 182, 184, 465; risks 300; states 112; support 113; symptoms 287; testing 206; theory 112; toll 354; trauma 176; well-being 300
psychology 3, 7, 11, 155, 301, 309, 425, 429, 431, 433, 490; development 490; forensic 248, 301; human 248; neuro- 248; social 108
psychosocial position 208–209
publishing 5, 9, 11–12, 32, 414, 417, 422, 450–451, 453, 455, 457, 471, 476; academic 450–457; benefits of 85; critiques 28; discriminatory 456; ethical issues 488; environments 419; linguistic bias in 456; multilingual 456; research 413, 416; scientific 11, 454; social science 441

qualifications 21, 64–65, 150, 155–156, 169, 246–249, 258, 368; acceptable 31; formal 32; misrepresentation of 415; of experts 20, 32, 148, 154, 240, 245, 250; professional 3, 69–70; questions surrounding 3
qualitative research 77, 153, 332, 476
Queralt, S. 420
questioning 98, 127, 131, 139, 148, 174, 181, 267, 271–272, 279, 288, 473; all night 287; assent to 175; complex 98; confrontational 98; detective 198, 201; guidelines 98; in various settings 88; lawyer's 21, 97; leading 98; line of 138; misunderstandings during 282; narrow 36; of the defendant 177; officer 197, 199–200; one-sided ritual 173; police officers' 132, 140, 283; repeated 289; suggestive 95; swear before 133; witness 256, 289; *see also* interrogation, law enforcement, PEACE method, Reid technique
questions: and answers as action 131; answering 19, 34, 36, 124, 131–133, 135–136, 139, 160, 175, 272, 275, 410; biasing 97, 475; challenging 70; clarification of 99; closed 96; complex 98; difficult 153, 274; ethical 1, 3, 5, 8, 54–55, 147, 152, 169, 300, 312, 323, 344–345, 351, 377, 409–411, 446, 450, 483; first-stage 81; follow-up 70; fundamental 283; leading 20, 97, 99; legal 3, 176, 256, 375; linguistic 4, 64, 68, 155, 173, 363; moral 54; open-ended 20, 96–99; philosophical 225–226; procedural 271; research 191; specific 20; types of 97, 218, 72; verification 18; yes or no 34
Quijano, A. 324

racial: bias 183–184; disparities 438; hatred 355; prejudice 184; profiling 257; stereotypes 184; types 257; unmarkedness 2599
radicalization 357; *see also* extremism, terrorism
Rainof, A. 284
ransom 54–55, 299, 398
rape 290, 329, 427; alleged 65; and child abuse 426–427, 434; and sexual assault 59, 99, 413; cases 60; definition 65, 330; molestation 167; myths 331; perpetrator 328, 331–332, 334; possible 65; sex offender 269, 328–330, 333, 335, 337

trials 59, 97; victims 55, 164, 169, 331
rapport 97, 267, 273, 316, 337, 442
ratified participants 208
reasonable assumption of guilt 130
reasonable doubt 9, 206–207, 209–219, 221
reasoning 63–64, 71–72, 120, 128, 162–163, 170, 182, 193, 230, 245, 250, 284, 352
rebuttal 122–124, 126, 128, 250
recall 89–90, 92, 94–95, 97, 101, 198, 249, 445; ability to 94; accuracy 90, 93; differential 92; false 89; imperfect 92; individual 94; of information 91; personal 88; test 89
reception centers 313, 316–319
recidivism 330, 337
reciprocity 470–471
recollections 89, 93–94, 334
recontextualization 196, 201–202
recontextualizing agent 197
recording hardware 18
recycling of topics 121
redaction(s) 195–202
referent 198–199
referentialist ideology 196–197
reflective journal 337
refugees 312–313, 315–316, 319, 322; application process 320; backgrounds 312; camp 164; crisis 315; deaf 10, 312–313, 315–316, 318, 321–323; Iraqi 164; legitimacy 312; racialized 323; right of 317; *see also* asylum, discrimination, English only, immigration
register 127; electronic 430; formal 126; legal 191; medical 30; voting 420
Registered Intermediary (RI), 271–273
regulations 4, 29, 72, 78, 149, 250, 270, 312, 350, 470, 489; certification 283; enforceable 79; federal 66; governmental 77; institutions' 350; licensure 3, 121, 149–150, 155; national 312; of legal proceedings 20; organization's 48; professional 3; rigid set of 6; violations of 4; *see also* American Psychological Association (APA), experts, forensic, International Association for Forensic and Legal Linguistics (IAFLL), International Association for Forensic Phonetics and Acoustics (IAFPA), malpractice, medicine, oversight, psychology
Reid and Associates 106–108
Reid technique 106–113

relativism 78
relevance 7, 20, 72, 78, 91, 139, 175, 273–274, 457, 464, 468
reliability 17, 19, 22–23, 33, 37, 46, 56, 67, 98, 100–101, 244, 367–372, 401, 466, 473, 488; anonymization 351; court evidence 389; evidence 33, 56; expert scientific opinion 46; forensic evidence 17, 46; future 463; inquiry 244; inter-rater 109, 374; level of 244; linguistic evidence 94; methods 5, 23; rating 69, recording 181; standard of 197; statement 175–176; threshold 244–245, 250–251; *see also* accuracy, analysis, bias, Daubert, error, evidence, Frye, hired gun, impartiality, objectivity, validity
reputation 38, 122–124, 149, 251, 348, 387, 390, 392–393, 466, 472, 475, 487
research: activist 323; archival 348; assistants 485; bias 153, 184; contemporary 91, 256; empirical 174, 182, 220, 299, 302, 309, 330, 337, 347, 434; error 19, 24, 351–352; ethics 79, 146, 207, 219–220, 322–323, 405; ethnographic 153, 283, 316; experimental 281, 283; feminist 333; forensic 10; historical 348; impartial 153; implication 91, 93; independent 248, 333; language 28, 280, 322; limitations in 17; linguistic 28–29, 38, 71, 112, 114, 127, 141, 149, 173, 177, 179, 198, 207, 256, 260, 299, 301, 309, 320, 343–344, 405, 409, 450, 464, 485; multi-disciplinary 399; participants 10, 89, 152–153–154, 313, 317, 319, 322, 337, 346, 404–405; psychological 95–96, 182, 184, 465; qualitative 77, 153, 332, 476; report 17, 38; scientific 94, 176, 364, 411, 475; sensitive 335; standards 151; supervisors 485; terrorist 344; think-tank 77; training 95, 485; trauma 92, 485; *see also* abuse
resettlement 164
resilience 85, 93, 428
responsibility 8–9, 24, 28, 57, 59, 110, 125, 159, 199, 202, 215, 225, 227–228, 269, 275, 306, 320, 322, 347, 352, 366, 368, 370, 405, 409, 415–416, 418, 429, 433–434, 470; centrality of 370; claims 4, 347, 354; concept of 364, 366; conditions of 228, 236; criminal 162, 243; criteria 228; diminished 9, 225–237; epistemic 207, 209, 214, 363–364, 366, 370; ethical 8, 11, 20–21, 113–114, 153, 304, 333, 347, 350, 352, 370, 405, 413, 419; importance of 367; lack of 228; legal 316; moral 169, 226, 237, 363–364; notions of 363–364, 366; of linguists 5; onus of 429; professional 356; removing 228; researcher's 421; sense of 31; social 354; taking 337
re-traumatization 335, 353, 426–428, 430, 435, 439; *see also* counselling, secondary traumatization, self-care
reversal of judgment 182; wrongful conviction 140, 210, 287; *see also* capital punishment, death penalty, Innocence Project, Leonard, R., stay of execution
Rhodes, R. 8, 45, 51n3
Richter, L. L. 285
Rickford, J. R. 181, 185, 193, 202, 472
right to counsel 174, 177
right to silence 130–132, 134, 136–141, **174**, 175
right wing 220, 343, 345; author 165; elements 165; extremism 347, 354, 356
Riner, R. C. 153
risk-, need-, and responsivity model 330
Rogers, R. 175
role: responsibility 215; strain 429, 431–433, 435
Roma, P. 98
Rosenbaum, D. 466, 471
Rosie, J. 8
Roth, A. 472
rule-governed interpretation 117
Rules of Court 246, 282
Rydberg, J. 328

safe space 425, 419, 421
San Diego Foundation 63
Santaniello, L. 286
Sawyer, S. 467
saxophones 161
scalar postmodifier 217
Scandinavian exceptionalism 330
Scanlon, T. M. 77
schema 89–90, 258
Schilling, N. 256–257
Schummer, J. 412
Scientific Area Committees (SAC) 18
Scientific Content Analysis (SCAN) 108–110, 112–113
scope of expertise 247
Scriven, M. 128
secondary traumatization 426–427; *see also* counselling, graphic imagery, re-traumatization, self-care, trauma, trigger
Second or Additional Language Learned (L2) 65
self-: awareness 337–338, 485; care 301, 308, 337–338, 354–355, 433, 442–443, 485; deception 120; disqualifying behavior 127; reflection 337–338, 412; representation 174, 176
sensitive: data 3, 82–83, 191, 305, 345, 348, 350, 352, 471, 485
sentience 400
sex(ual): abuse 98–99, 109, 270, 276, 290, 328, 330–332, 334, 338, 353, 427, 430–431; assault 59, 99, 164, 269, 308, 413, 426, 431, 433, 437–438, 444–445; crime 10; offender 328–330, 333, 335, 337; offender treatment 329–330, 335, 337; offense 60, 287, 329–334, 336–338, orientation 247, 467; violence 328, 331–332, 338; *see also* abuse survivor, abuse victim, pedophile, rape, witness

Shin, H. 322
Shuy, R. W. 2–3, 66, 70, 121, 150–152, 176, 192, 194
Siegel, M. 462
sign language 312–316, 318–322, 467; American 314, 438; communication 284, 316; deaf 316; Finnish 313, 316, 318; importance of 318; interpretation 284, 317; interpreter 315, 317, 319, 321; national 322; proficiency 320; protection of 318; respective 316 users 316
Simons, D. J. 89, 91
sincerity 107, 207, 260, 272
Sivunen, N. 10, 313, 315, 317, 320
skills: college-level 177; communication 65, 98, 268, 318; complex 280; comprehension 98; English writing 457; expertise 240; forensic linguistics 422; interviewing 28; language 67, 281, 287, 289, 457; literacy 313, 317, 320; limited 454; sign language 318; specialized 4, 273
Slobe, T. 193
Slobodchikoff, C. N. 403–404
Smith Doerr, L. 469
Snook, B. 96
Snyder, C. R. 302–303, 309
social: advocate 8, 43–44, 49–50, 473; advocacy 43–45, 48–50; desirability bias 335; justice 8, 12, 28, 44, 48–51, 55, 63–64, 67, 71, 73–74, 127, 173, 202–203, 306, 312, 321–323, 439, 441, 457, 477; relationships 320, 470; responsibility 354; support 357, 432
sociolinguistics 3, 65, 145, 151, 321, 441
Solan, L. M. 33, 127, 150, 179, 192, 194, 216, 410, 491
solution 60, 86, 138, 162–163, 347, 430, 463; acceptable 162; effective 465; equitable 162; legal 130; practical 55; preferred 195; proposed 284; reasonable 168; satisfactory 179
sophistry 124
Sorenson, M. 477
speaker: comparison 53, 55–57, 402; recognition 18, 25, 404–405
special: consideration 168, 170; measures 9, 267–268, 270–271, 273–276
speech 18, 55, 57, 59–60, 98, 107, 137, 179–180, 197, 200, 254, 257, 259, 267, 377, 380–381, 385, 398–399, 402, 455; abilities 273; accented 260; acts 107; analysis of 43; argumentative 191; attributed 56; characteristics 257; community 146; disputed 55; distinctness of 379; domain of 378; ethics of 207, 220; event 123; evidence 61, 406; features 261; forensic 10, 19, 28, 54, 61, 255, 402–403, 405–406; free 122, 255; hate 5, 346, 352, 411; hypothetical 216; impediments 257; incoherent 287; individuality 403; inflammatory 353; injurious 353; input 402; insincere 113; laboratory 402; oral 179; organs 378; patterns 402; perceptions of 261; processing 111; production 378; rate 55; realistic-sounding 381; recognition 405; recorded 146, 402–403; sample 58, 259, 401–403, 406; signal 61, 260; surrounding 181; synthesis 381; uniqueness in 378; *see also* acts, attribution, forensic speech science, Grice, P., ownership, phonetics, transcription, threatening
Spivak, G. C. 450
spoofing 382, 387
scripts 90, 177, 230; waiver 69
standard of proof 206–207, 214, 218–219; criminal 206–210, 212, 214, 217–221
standardization 17–20, 152, 374
standards: accepted 244; accreditation 53; adapting 25; behavior 19, 78; codified 31; colonial-era 451; confidentiality 152; court 71; definitive 152; differences in 19; ethical 3, 24, 31, 38, 71, 110, 149, 251, 281, 338, 374, 404, 406, 419; for review 18; high 119–120, 271, 412; imperial 455; institutionalized 155; lack of 155; legal 44, 71; linguistic 456; methodological 151; of care 244; of forensic methods17; of proof 209, 213; phonetic 48; professional 22, 29, 31, 63, 70, 155, 289, 421, 426–427, 488; relevant 54; reliable 17; research 151; scientific quality 29; selected 71; uniform 24, 283; within linguistics 20; *see also* academics, Daubert, Frye
Stanley, J. 3698
statement: admission 195, 197; concerning 113; direct 66; dishonest 37; edited 197; false 36–37; inadmissible 195; inculpatory 287; interpretation of the 109; involuntary 372; language of the 197, 201; original 196, 198–200; out-of-court 150, 285; public 31; reliability of 175; recorded 175, 179; redacted 197–201; reported 134; signed 287; structure of the 109; untrue 387; verbal 387; victim impact 445; witness 71, 389; written 110, 271, 387, 445
stay of execution 473
stereotype 272, 331, 401
stigma 309, 328, 331, 333–334, 445
stranger danger 331
street-level: bureaucracy 441; linguistics 446; pedagogy 440, 446; perspective 446; workers 437, 440
students 3, 11, 94, 162, 240, 303, 306, 355, 411–412, 416, 419–420, 425–435, 437–446, 462–463, 467–469, 472, 476, 483, 486, 489; abuse 412, 443, 484; as research or teaching assistants 442, 484–485; as research participants 484–485; classroom environment11; difficult subject matter 11, 353, 426–427, 430; disclosure of sensitive data, 417, 419, 425–426, 431–435, 485; foreign 413; graphic images 300–301, 307,

485; informed consent 429–430; negative impact on 414; teaching 11, 409, 411–414, 421–422, 425, 427, 432, 439, 443; trigger warning 418, 425, 429–430, 433; vicarious trauma 426–430; well-being 414, 432
stylistics 19, 250
subjective 35, 210–211, 299, 302; assessment 47, 244; certainty 210–212, 221; confidence 209; conviction 210; decisions 374; elements 19; evidence 22; inherent 7; interpretation 18, 31, 49; mental state 210; nature 49; perceptions 216; time 109
subjectivity 88, 312–313
subpoena 153–154, 177
sufficient appropriate linguistic evidence 121
suicide 4, 90, 299, 335, 401
Sullivan Buker, G. 9–10
supervision 309, 336–337, 381, 438, 484–485
support 1, 6, 17, 24, 45, 50, 60, 64, 71, 83–84, 113, 121, 124–125, 136, 147, 149, 151, 161, 176, 233, 237, 242, 248, 273, 276, 307, 329, 332, 338, 425, 427, 430, 433, 435, 439, 476, 487; collegial 338; contacts 335; coordinated 321; data 452; emotional 337, 476; empirical 109; forensic linguistic 121, 305; institutional 300, 302, 307–310; insufficient 321; interpreting 73; invaluable 236; lack of 24, 336; linguistic 128, 456; mechanisms 428; network 82; peer 433; policies 426; services 334, 485; social 432; structures 323; system 303, 305, 308; well-being 432
sure 9, 66, 169, 211–217, 219–221, 270, 304, 315, 320, 350, 352, 355, 404, 416, 419, 476
Surmon-Böhr, F. 97
suspect 35, 55, 66, 68, 70, 73, 90, 94, 96–97, 100, 107, 130–139, 166, 176, 180–181, 199, 242, 255–257, 281, 287–288, 389, 398, 401, 403, 432
Svartvik, Jan 27–28
Swedish Prison and Probation Service 329
syntactic embedding 215

Tamura, T. 284
tape recordings 284, 286
Taylor, B. 443
Teachers of English to Speakers of Other Languages (TESOL) 73
teaching 5, 9, 11, 19, 28, 307, 355, 409–416, 418–422, 425–428, 431–432, 434–435, 437–446, 486
Terkourafi, M. 488
terrorism 163, 166–167, 291, 343, 345, 356–357, 450; acts 343; attack 165, 343, 357; group 345; organizations 166; plot 67; research 344; suspected 97
Test of English as a Foreign Language (TOEFL) 454
testimonial injustice 455
testimony 1, 3, 19–20, 23, 53, 66, 71, 73, 99, 147, 161, 180–181, 193, 202, 241–242, 246–247, 250, 271, 282, 289–290, 301, 357, 368, 444, 455; accurate 37; competent 150; crucial 193; deposition 148; expert 1, 3, 21–22, 24, 33–34, 44, 48, 54, 56, 106, 147–148, 150–151, 155, 159–160, 246–249, 254, 301, 462, 464; fact-based 9; forensic linguists' 21; incompetent 150; in-court 290; limitations of linguistic 21; oral 33; original 4; psychiatric 162; remote 154, scientific 71; sponsor 147; truthful 98; unscientific 21; witness 96, 159, 161, 231, 247; written 246
The Statutory Code of Practice for Victims of Crime (2020) 270, 272
The Third Terrorist 165–166

The Youth Justice and Criminal Evidence Act (YJCEA) 266–267, 276
therapeutic alliance 337
therapy 167–168, 330, 336, 428
third party use 385
Thompson, M. K. 184
threat 183, 289, 312–313, 318, 328, 344–345, 347, 349, 353, 463; assessment 32; communications 343–345; death 299; escalation 200, 357; evidence 290; extremist 343; letters 346; messages 354; multiple 289; response tactics 304, 308; serious 8; *see also* communication, political extremism, violence
threatening: communication 345; content 346; language 353; messages 347
threshold: for conviction 208–210; reliability 244–245, 250–251
Tiersma, P. M. 184, 192, 216, 219
time-barred debt 125–126
torture 77, 163–164, 167–168
totality of the circumstances test 174, 176
toxic(ity): changes 489; mentality 353; occupational 470; rhetoric 489
training: access to 434; adequate 31, 433, 485; advanced 191; awareness 50; best-practice 415; codified 309; data 380; datasets 259; examiner 24; extensive 464; forensic 50, 61; formal 18, 29, 240; lack of 30, 56, **365**, 433; linguistics 33, 69, 80, 117, 191, 305, 309; needs 18; online 444; outstanding 484; practices 18, 24; procedures 95; programs 107, 162; questions regarding 24; specialist 28, 149, 273; specialized 244; specific 249, 420, 434; variations 22
transcript 136, 148, 179–181, 197, 200–201, 248, 281–282, 284–285, 287, 290–291, 442; analysis of 43–44, 48, 97, 131, 228; audio recording 57, 59–60, 96, 131, 139–140, 194; court 59, 283; Dutch 96, 132–139; of police-suspect interviews 96; police recordings 402; *see also* accuracy,

Conversational Analysis, error, interrogation, law enforcement, omissions
transcription 60, 180, 284, 381, 443, 455; discrepancy 260, 455; independent 60
translation 4, 7, 9, 19, 28, 45–46, 132–133, 135–139, 256, 279–287, 290–291, 444, 457
transparency 36, 45–46, 48, 100, 134–135, 140–141, 249, 251, 284, 292, **365**, 366, 368–372, 374, 381, 473; *see also* analysis
trauma 92, 300–301, 305, 308, 316, 320, 332, 335, 427–428, 430, 437, 445–446; childhood 248; consequence of 429; early life 248; exposure 300; high 301; history of 175, 427; infliction 305; -informed care 439; -informed interviewing 334; ;-Informed Pedagogy (TIP) 11, 439, **440**, 442; management 305; memory distortion 93; past 334–335; practitioners 306; primary 485; prolonged 301; psychological 176; -related symptoms 426; response 300, 336; risk of 428–429; secondary 271, 300–302, 305, 328, 337, 485; situations 306; teaching 11, 409, 414, 425, 435, 443; varying levels of 445; vicarious 82, 300–302, 308–309, 485; victims of 308; workplace 300; *see also* abuse, counselling, graphic imagery, informed consent, trigger warning
traumatic: content 300, 305; data 301; events 92–93, 274, 308, 426, 444; experiences 315, 429; film 93; life events 426; material 427, 485; memories 92; narratives 426; nature 307; populations 301; response 428; subject matter 300, 428
treatment 44, 114; derogatory 384; disengagement from 337; engagement in 335; equitable 184; fair and equal 284, 476; goals 335; hesitation toward 329; of asylum seekers 58; of fraud 387; of sexual assault victims 438; of suspects 130; resistant schizophrenia 236; sex offender 329–330, 335, 337; unequal 291; unfair 450, 454–455
Triano-López, M. 284
trier of fact 44–45, 48, 51, 53, 57, 59–60, 147, 151, 160, 192, 239, 243, 249
trigger: cognitive bias 471; concepts 6; confrontation as a 337; emotions 23; environmental 270; moral reaction 231; potential 414; recognition 258; warnings 352–353, 418, 421, 425, 429–430; word 6
Trinch, S. L. 197
Trump, D. 30, 373
trust 83–84, 106, 220, 245, 363, 367, 412, 432, 471; building of 419; erosion of 332; low levels of 334; mutual 84, 412, 419; public 113, 128; two-way 82; violation of 349
truth 21, 23, 36–37, 46, 60, 97, 109, 112–113, 121–122, 124, 133–134, 138, 140, 151, 159, 161, 166, 215, 218, 272–273, 300, 366–368, 373, 450, 465, 467; discovering 159; factual 124; -finding 134, 154; full 124; good-faith version 122; ground 46, 231, 236; in search of 19, 36; indicators of 106; judicial 229; linguistic 127; objective search for 161; pursuit of 468; seeking 151, 230, 233, 468; -tellers 112; verification of 124; whole 36, 46, 150
Turnball Plaza, B. 11

UK Copyright Designs and Patents Act (1988) (CDPA) 384
UK Police Code of Ethics 78
UK Statistics Authority 77
uncertainty 164, 217, 273; conditions of 210; expressions of 98; objective 211; overcome 21; scientific 20; sense of 200
uncommon interpretation 127
unethical 31–33, 36, 43, 88, 118, 121–122, 124, 126, 147–148, 150, 215, 217, 291, 333, 403–404, 411, 417, 419–420, 476; actions 3; behavior 19, 22, 38, 100; choice 120; circumstances 400; coercion 107; conduct 30; exclusionary practices 457; expert evidence 29, 36; inherently 22; intent 421; leading questions 97; option 120; practices 8; reports 121; situation 454; work practices 484
ungrading 447n3
United Nations' Convention Rights of Persons with Disabilities (CRPD) 317
United States Department of Justice (DOJ) 18
Universal Declaration of Human Rights (UDHR) 383, 387
United Nations General Assembly 63
United Kingdom 19, 28, 30–31, 36, 46, 81, 106, 179, 194, 206, 382, 386, 391, 413, 452–453, 455, 486
United States 73, 173, 179, 256
user profiling 257

validity 17, 2, 46, 69, 113–114, 119, 174, 321, 367–368, 370, 463; assessment tools 113; dubious 367; ecological 219; external 402; of language evidence 72; of methods 24; of the translation 282; of waivers 176; of witness statements 108; scientific 17, 244; *see also* analysis
Valla, L. 27
van Heuven, V. J. 401
van Naerssen, M. 8
Venet, A. S. **440**
verification, 281; adequate 282; of accuracy 281; process for 18; questions regarding 18; user 256
victim 82, 90, 183, 200, 290, 328, 330–332, 337, 350, 352; alleged 136, 269; -blaming 331–332, 336–337; dying 4; -facing 300–301, 308; impact statement 445; perspective 332
Victims' Commissioner 275–276
video recorded evidence in chief 271

violence 233, 300–301, 315–316, 332, 337, 357, 443, 450; acts of 344–345; arbitrary 320; domestic 281; epistemic 450, 455, 457; escalation of 200; gender-based 445; gun 444; high-risk for 329; intimate partner 331; -justifying attitudes 337; of punishment 229; physical 322, 345; power dynamics of 332; sexual 328, 331–332, 338; use of 344
virtue 78, 82, 209, 231, 267, 368, 379, 490
Visonà, M. 306
visual aids 141, 267–268, 270–271, 273, 283, 370
vocal: artist 384; characteristics 380; elements 383, 385, 388, 390, 392; idiolectal elements 391, 393; information 380; samples 385; tract 380; twins 380
voice: cloning 387; comparison 43, 56, 402; data 18; recognition 56
voir dire 66, 160, 248, 250
voluntariness 131–134, 140, 175–176, 289–290, 404
vulnerable 5, 35, 98–99, 267, 275, 352, 446; colleagues 353; communities 315, 322–323; individual 274; members 9, 98; non- 99, 271, 273–274; participants 276; persons 88, 99, 275, 346; populations 300, 306, 312; self 443; suspects 99; to bias 35; to challenge 270; to criticism 141; to misunderstanding 175; waivers of 175; witnesses 9, 267–268, 273–276; youth 185n4, 442

Wagstaff, I. 12, 470
Waldron, J. 369
Wales 9, 28–29, 32, 36–37, 46–47, 53, 72, 81, 99, 210–214, 217, 219, 225–226, 266–267, 274–275, 382, 387, 425, 471
Watt, D. 10, 255, 260
Web of Science (WOS) 452
web crawling 358n1
web-scraping 347, 358n1
Wegner, D. M. 302–303
weight 21, 72, 248, 250, 429, 487; moral 370; of conclusions 24; of evidence 30, 43, 51, 53, 221, 246, 274; of forensic evidence 37; of opinion 243
Whynot, L. A. 314
Wilkinson, S. 227
Willis, G. M. 330
Willis, S. 412
Wing, D. 9, 226, 227
withholding 111, 443; information 180, 209, 220; protection 58
witnesses 29, 92, 98, 133–134, 147, 160, 177, 195, 200, 218, 243, 266, 269, 275, 276, 281–284, 290, 401, 410; Aboriginal 194, 410; accented 285; adult 9; Adult With Intellectual Disability (AWID) 99, 267; alibi 291; assessment 283; attempts at communication 410; bystander 99; child 98; co- 94; coaching 36; competence 266, 268; credibility of 456; crime 92, 94, 465; discussion with others 94; effective participation 267; effects of questioning 97–99; expert 19–20, 28–31, 36–37, 47, 72, 76, 81, 125, 127–128, 146, 148, 159, 161, 170, 191, 237, 239–240, 242–246, 248–251, 471; eye- 89–90, 93–94, 348; faulty claims of 108; interviewing 96, 281; lay 28; memories 94; minimally verbal 275; neurodiverse 268; neuro-typical 272; non-testifying 281; non-vulnerable 273, 275; opposing 22; original testimony of 4; procedural justice for 292; prosecution 147; questioning 289; statements 108, 281; testimony from 96, 159, 231, 455; trial 99, 290; underage 98–99; unreliable 272; vulnerable 9, 268, 273, 275–276; with challenging communication skills 268
Wolf, S. 228
Wolfram, W. 261, 441
workplace 437, 467; toxic 300; trauma 300
World Federation of the Deaf (WFD) 314, 318
World War II 349, 453
Wright, K. E. 9, 259
written reports 33, 96, 141, 154, 349

Xunzi, X. K. 78
Youth Justice and Criminal Evidence Act (YJCEA) (1999) 266–267, 276–277

Zavala, H. 74

For Product Safety Concerns and Information please contact our EU
representative GPSR@taylorandfrancis.com
Taylor & Francis Verlag GmbH, Kaufingerstraße 24, 80331 München, Germany

www.ingramcontent.com/pod-product-compliance
Ingram Content Group UK Ltd.
Pitfield, Milton Keynes, MK11 3LW, UK
UKHW051154130126
466905UK00007B/129

* 9 7 8 1 0 3 2 4 8 8 4 2 4 *